Advanced Concepts and Techniques in the Study of Snow and Ice Resources

An Interdisciplinary Symposium

Organized by the Work Group on Snow and Ice, the Work Group on Remote Sensing, and the Work Group on Nuclear Techniques of the U. S. National Committee for the International Hydrological Decade

Asilomar Conference Grounds
Monterey, California
December 2-6, 1973

Compiled by
Henry S. Santeford
James L. Smith

A United States Contribution to the International Hydrological Decade

NATIONAL ACADEMY OF SCIENCES
Washington, D.C. 1974

Publisher's Note

To expedite the appearance of this manuscript in book form, the text has been reproduced directly from the authors' typescript. Use of this format eliminates the considerable time and expense required for detailed text editing and composition in print - processes that sometimes act to delay or prevent publication of significant manuscripts whose specialized nature or currency of content demand prompt publication.

Readers will find the text edited to a satisfactory level of completeness and comprehensibility, although no special effort has been made to achieve the standard of editorial consistency in minor detail that is typical of typeset books issued under our imprint.

National Academy of Sciences
Printing and Publishing Office

Sponsored by

U.S. National Committee for the International Hydrological Decade,

U.S. Forest Service, Pacific Southwest Forest and Range Experiment Station,

U.S. Army, Cold Regions Research and Engineering Laboratory.

Available from

Printing and Publishing Office
National Academy of Sciences
2101 Constitution Avenue, NW
Washington, D.C. 20418

Library of Congress Catalog Card Number 74-10120
ISBN 0-309-02235-5

Printed in the United States of America

PREFACE

The Interdisciplinary Symposium on Advanced Concepts and Techniques in the Study of Snow and Ice Resources was a forum at which persons involved in the study of snow and ice could discuss measurement techniques with representatives of other technical fields. The Symposium was designed to acquaint people working on one aspect of the snow and ice field with recent needs and developments concerning its other aspects.

Preprints of the papers listed in the program were distributed to all participants. The papers were not actually presented at the meeting but were summarized at the beginning of each session by general reporters, after which the sessions were opened to discussion. Such plenary sessions, consisting of the summary and a discussion period, were held each morning and evening. The afternoons were reserved for informal group discussions, with no structured programs scheduled. These informal discussions took place in the meeting rooms, out on the lawn, at the picnic tables, or even along the beach.

In the hope of making the papers and some of the discussions of the Symposium available to the scientific community as quickly as possible, this proceedings volume was printed directly from camera-ready copies of the papers as submitted by the individual authors. The papers for each session are accompanied by summaries of the associated plenary discussions. Summaries of the large number of informal group discussions were impossible to obtain, but the ideas expressed at many of these informal meetings were later brought up at the plenary sessions and thus are reflected in the summaries presented here.

The Planning Committee hopes that the meeting fulfilled its purpose and served as a catalyst to foster expanded and continuing cooperation among the various people working in the snow and ice field while at the same time acquainting people in other fields of endeavor with some of the problems and complexities of the snow and ice medium.

Symposium Planning Committee

James L. Smith, Chairman
Andrew Assur
Ray E. Isaacson
R. J. P. Lyon
E. Paul McClain
Henry S. Santeford, Staff

ACKNOWLEDGMENT

The U. S. National Committee for the International Hydrological Decade wishes to express its appreciation to the Symposium Planning Committee: James L. Smith, Chairman; Andrew Assur, Ray E. Isaacson, R. J. P. Lyon, and E. Paul McClain, members; and Henry S. Santeford, staff.

Special appreciation is given to James L. Smith for the local arrangements; to Wanda Smith for the family program; to Jim Bergman, Howard Halverson, and Ron Jones for their assistance with the general operations of the meeting; to Margaret Volpe for assistance with administrative matters; to Henry S. Santeford and Sally Santeford for general supervision and organization; and to John O. Ludwigson and Muriel Duggan, editorial consultants, for assistance with the compilation of this volume.

H. Garland Hershey
Chairman, USNC/IHD

After four solid days of contemplating advanced concepts in the snow and ice field, one participant in the Symposium was moved to write the following verse. The various topics and questions mentioned were all points made in the papers submitted to the Symposium, or in the discussions.

WHOOPEE! WHOOPEE! ICE AND SNOW AND THE IHD

Where would a bunch of scientists go
  To further the knowledge of ice and snow
In December of '73
  But Asilomar, by the sea?

The participants, suitably wined and dined,
  Followed the 'Cowpaths of the Mind'*
Over roofs designed to hold that snow.
  And where will Utah mule deer go?

We learned of various measuring toys -
  Gamma rays, black boxes, and white noise.
Where does the alpine snowpack go?
  Could we tell its age by HDO?

It seems it would be very nice
  To know the loading properties of sea ice;
How much ridge can a 'ridge-picker' pick
  If the analyst's really 'on the stick?'

Of course, it appears that a major goal
  Is funding for avalanche control.
The programs that carry the day so far
  Are sensing remotely from Visible to SLAR.

How thick is the ice? How deep is the snow?
  Can a flood peak hydrologist tell when it lets go?
Is the signature of firn unlike that of graupel?
  Do you separate 10-meter ice crystals by scalpel?

Is it a rock glacier, piles of talus, or screes?
  Can you separate snow in the forest from trees?
Should you spend the whole of your summer vacation
  Surveying the rate of your snowfield's ablation?

The answers to all of these questions are few.
  For each problem that's solved, scores come anew.
Thanks, Hank and Jim, for a mighty fine meeting;
  Thus this poor cowpath poem for a last hindsight greeting.

John Hubbard
State University
College, Brockport
Brockport, New York 14420

*The chairman of the Planning Committee opened the symposium with a reading of the poem 'The Calf Path,' by Samuel Walter Foss (Whiffs from Wild Meadows, Lothorp, Lee, and Shepard Co.)

CONTENTS

CONTENTS (Continued)

Session VI
NUCLEAR TECHNIQUES

Session VII
MISCELLANEOUS TECHNIQUES

Keynote Address

Chairman:

James L. Smith, Pacific Southwest Forest and Range Experiment Station

A CHALLENGE IN SNOW AND ICE

Henry S. Santeford
USNC/IHD, National Academy of Sciences

Snow and ice affect every living thing on earth. In the western United States, most of the streamflow originates as snow in the mountainous regions: more than 90 percent of the flow of the Colorado River comes from melted snow. The rivers of the Plains, the Missouri, the Platte, and the Arkansas, also receive more than half of their water contributions from mountain snows, and the situation is similar in the Northeast.

Nor is the influence of snow and ice limited to their importance as a source of water supply. In the heartland of the country, including the Great Lakes region and much of the Mississippi valley, icing conditions often interfere with the operation of waterways, highways, and airports during the winter months. The benefits of each extra, ice-free day in the Great Lakes shipping season are calculated in millions of dollars, and the costs of keeping highways and airports open are measured in similar terms.

Moreover, there has been a great change in peoples' attitudes toward the winter months. What used to be a time of hibernation for many is now a season of rapidly-expanding recreational activities. Skiing - both downhill and cross-country - snowmobiling, ice skating, and other winter activities are dependent on the snow and ice resource.

These examples have centered on only one aspect of snow and ice - the seasonal cover. The field of snow and ice studies also is concerned with permanent snow and ice cover - glaciology, arctic and antarctic studies, and sea ice. For example, understanding the occurrence and nature of sea ice is essential to avoid excessive interference with ocean shipping. Less obvious, but even more significant, is the need to appreciate the very great effect that the permanent snow and ice cover has on global weather patterns. In addition, the water stored in glaciers and in polar snow and ice - nearly 80 percent of the fresh water on the earth - is an obvious potential source to supply man's ever-increasing needs. The nearly 7 million cubic miles of water stored in this permanent snow and ice cover is an overwhelming amount in comparison with the average annual discharge from all the world's rivers - about 7,000 cubic miles. However, use of the permanent snow and ice cover requires consideration of more than just man's needs. Consideration must also be given to the delicate ecosystems of these regions, and to the possibility of inadvertent modification of local, regional, or even global weather patterns.

Increases in use of the polar regions have prompted a widespread increase in multidisciplinary research in these geographic areas. Activities such as the Alaska Pipeline are sure to bring about further development in the Arctic, attended by a host of new problems. The design of large communities in permafrost areas for example, will involve problems not encountered in temperate regions. These problems are not confined to the increased difficulties in construction, but also involve water supply, waste disposal, transportation, and all of the support facilities we have come to accept as part of our normal daily life.

One indication of the increased interest in Arctic research is the response to the Second International Conference on Permafrost held in July (1973) at Yakutsk, U.S.S.R. At this meeting, more than 350 papers on a wide variety of topics associated with the arctic environment were presented. The interest and the amount of research that is going on is so great that, already, plans are being made for a third conference to be held within three years.

Providing answers to the many questions these studies are bringing forth will require not only more but also speedier information gathering, processing, and dissemination. Achieving this will require that those working in the snow and ice field be given access to both advanced concepts and any new technology developed in, but heretofore confined to, other fields of science and technology. Furthering this sort of multidisciplinary interaction is the principal aim of the Interdisciplinary Symposium on Advanced Concepts and Techniques in the Study of Snow and Ice Resources. This meeting will center on technological advances - both actual and conceptual - which may be applicable to one of the areas of interest in the snow and ice field.

All of these areas of interest - both conventional and innovative - can be classified under four main categories - seasonal snow cover, fresh water ice, permanent snow and ice cover, and sea ice. All four categories deal with the same basic resource, yet proceed from different viewpoints. In many instances, communications among - and often within - these categories are not as good as they could, or should, be. This lack of communication has resulted in some duplication of effort, and a general lack of understanding of the entire resource on the part of many of those working in the field. Technological advances adopted by one subdivision, for example, may not be adopted, or even known, to others.

All parts of the snow and ice field are feeling the pressure of increased demands for information concerning the better use of this resource to serve the world's increasing population. The field, however, has not kept abreast of the increased demands. The study of seasonal snow cover is useful as an example of the conditions and problems in snow and ice studies in general.

In the area of seasonal snow cover, recognition of the need for information concerning the nature of the snowpack dates back to the first recorded snow surveys for streamflow forecasting performed in 1834 by the Chief Engineer of the Chenango Canal, Mr. Jarvis. In a report to the Canal Commission of the State of New York, Mr. Jarvis stated that "Snow on the ground which fell in November and December 1834 on the watershed of Madison Brook (9.37 sq. miles) amounted to 87,120,000 ft.^{3} of water." It is not known how he measured the snow depth and water content. Still, his report indicates that a snow survey was made, and as such, it is the first snow survey recorded in the United States, and likely the first anywhere.

In the years that followed, little record can be found of further snow surveys until 1906. At that time, Dr. James E. Church offered to climb Mt. Rose near Lake Tahoe, Nevada, once each month for a year, to obtain temperature readings, if the Weather Bureau would furnish the thermometers. The Bureau agreed, and the Mt. Rose Observatory was founded. During the winter of 1908-1909, Church developed a device that became known as the Mt. Rose Snow Sampler and Scale. With some modifications, this tool is still the most commonly used device for snow measurements. With the development of a method for practical measurement of snow density, depth, and water content came the widespread establishment of snow surveys and their use in flood and streamflow predictions. The early surveys by Church were thus the beginning of a worldwide system from which flood and streamflow forecasts are made for much of the civilized world. In California alone, more than 300 snow courses are measured at monthly intervals from January to June. From these measurements almost all of the streamflow predictions in the state are made.

The technique developed by Church requires on-site visits and is thus both time-consuming and costly. With most of the present systems utilizing this technique, the field data is mailed to a central office where it is combined with similar data from other sites in the region, and from which the results are disseminated to the user agencies by mail.

This method of handling the field data greatly delays distribution of the information to the user and thus decreases the usefulness of the information. For example, when a springtime storm moves into a mountainous area, resulting in a rapid snowmelt and vastly increased streamflow, immediate knowledge of the impending situation would be of great importance in the operation of a downstream reservoir. Clearly, such situations require real-time information; there is rarely time to send a survey crew into the area for ground investigations. Without some sort of a telemetering sensing system, the required operational decision must be based on "judgment," too often without sufficient supporting information.

The problems with the major information systems for seasonal snow cover are not limited to streamflow forecasts - the function for which they were developed. In the past decade or so, the demand for information has greatly expanded, especially in the area of recreation. Here, the needs range from reliable reports on snow conditions for skiing or snowmobiling, to hazard warnings for avalanche danger or thin ice. Almost any skier can testify to the lack of uniformity in reported snow conditions at the different recreational areas. In addition, there is generally no source of information on the snow conditions at other than developed recreational areas, information that is badly needed for such activities as cross-country skiing or snowmobiling. This lack of information is a major factor in the unfortunate safety record of several of these winter sports.

It is clear that most of the existing information systems - the snow surveying system for streamflow forecasts, the snow condition reports for recreational areas, and the avalanche forecast system - often do not meet the needs of today's society. In some instances, new equipment, such as snow pillows and nuclear profiling snow gauges have been used. However, for the most part the equipment, like the methods, consists of simple modifications of the techniques developed at the beginning of the century.

If the present and future demands for information on the snow and ice resource are to be met, new systems must be sought to either replace or augment the existing ones. The development of such systems is, however, not an easy task. Snow and ice form a highly complex system. Not only is it necessary to know how much water is present, but also how much is present in each of the three phases. The general structure of the snowpack, its internal layering, and the ambient conditions above (in the air) and below (in the soil) greatly affect its behavior. Many of these conditions are not now readily measured.

If a device is to measure the necessary parameters, it must operate under some of the most severe weather conditions encountered in field operations. In alpine regions, winds ranging up to 120 mph (50 m/s), temperatures from -50 to +50 F (-45.5 to +10 C), and intense solar radiation are not uncommon and the extremes and rapid fluctuations greatly affect the operation of many field installations.

In areas such as the upper Great Lakes region, soil conditions under the snowpack are also an important factor in determining the overall nature of the deposit. In this region it is common to find the ground temperature well above the freezing point while the air temperature above the snowpack is below zero. The snowpack is then in a constant state of transition, and melt and drainage of water occurs during much of the winter period. Under such conditions temperature and saturation measurements must extend from below the soil surface, up through the snowpack, and into the air in order to measure and predict the behavior of the snowpack. Ideal measuring devices must be capable of providing precise measurements over a wide range of both temperature and saturation conditions, without affecting or being affected by their surroundings.

In the upper Great Lakes region a further difficulty in measurement techniques is encountered during the spring melt. The southern shore of Lake Superior receives large snowfalls - 200 to 300 inches (5.1-7.6 m) per season - yet the lake generally doesn't freeze over and thus provides a constant source of moisture. During the melt period, the atmospheric conditions are often such that large amounts of condensation occur on the melting snowpack. When the proper meteorological conditions occur, the combined runoff from the melting snow and condensate greatly exceeds the seasonal precipitation.

In order to obtain a direct measurement of the moisture exchange between the atmosphere and the snowpack, the measuring device would have to be capable of altering its elevation as the surface of the snow changed, and at the same time offer little, or no, interference with the natural environment of the snow-air interface where the desired measurement is being performed. Ideally, such a device should not interfere with the ambient meteorological conditions - temperature, humidity, pressure, wind, and radiation - nor the properties of the snowpack - albedo, surface texture, internal structure, water content, and drainage network. Since such an ideal device currently does not exist, estimates of the moisture exchange between the snowpack and the atmosphere must be obtained either from indirect measurements and theoretical transfer equations, or from inadequate direct attempts at measurement of the moisture exchange.

These problems associated with the study and management of the seasonal snow cover are representative of those characterizing the entire field of snow and ice. Each component of the field is hampered by reliance on out-dated information systems, coupled with inadequate measurement devices. Advances are being made, yet widespread application of these newer devices is somewhat limited.

A limited application has been made of techniques such as side-scanning radar for mapping sea ice, airborne natural gamma spectrometer surveys for measuring accumulations of snow, and the ERTS multispectral scanner data for a wide variety of snow and ice problems. While many of these techniques are still in various stages of research, they appear to have potential application to many types of snow and ice measurement problems. Other devices, such as telemetry systems - both conventional and satellite-relay - seismic probes, multispectral scanners used in low and medium altitude aircraft, and nuclear sensor systems for obtaining snow profiles have passed the research stage, yet are receiving little or no application in the snow and ice field.

Many reasons have been suggested to explain the apparent lack of application of much of the new technology to snow and ice problems. The large number of technological advances being made makes it impossible for a person in a user field (such as snow and ice management) to understand them all. In many instances, it is difficult even to be aware of those technological advances that have direct application in the user's field. Thus, a user often continues to employ the "tried and true" procedures, or at best, chooses from a limited number of alternatives with which he is acquainted. In either case, the optimal use of present and future technology is not realized.

In some instances, devices have been tried and found unsatisfactory. There is yet, however, the possibility that some of these devices could prove highly beneficial if appropriately modified. However, the user may not know how to modify the device, while those who developed it are not aware that any modifications are necessary.

A further difficulty lies in the fragmentation of the snow and ice field. The technology applicable to the needs of one component may not be applicable to the needs of another. Although the four main components deal with the same basic medium, the problems with which each is concerned differ sufficiently so that communications among the practitioners are not always what they should be. Thus, a technological advance introduced in one component may find little general application in the field simply because of poor intradisciplinary communications.

In view of these constraints, it is obvious that the development of new concepts and techniques for snow and ice study will not be an easy task. One key item that can speed the rate of development is an exchange of ideas among the several disciplines involved. The people working in the snow and ice field must be exposed to the technological advances being made in other fields, while at the same time workers in these fields must be acquainted with the needs and constraints of the snow and ice field. Such an educational process often takes a considerable amount of time, but this may be shortened if some external guidance is provided.

This Symposium hopefully will be a unique and direct approach to encouraging the creative adaptation of advances in other fields to the study and management of snow and ice resources. Our program, which is patterned after the Gordon Conferences, will consider four areas of discussion (with emphasis on area 4): 1) the expanding needs of water resource managers for snow and ice information; 2) the major characteristics of snow and ice that must be measured, or that affect measurement systems; 3) observational systems for snow and ice measurement, data accumulation, and information dissemination; 4) new concepts and techniques from other fields that may be applicable to measurement problems in snow and ice.

Thus, the Symposium should be an educational experience for all participants: the representatives from the snow and ice field will learn about the potential of new technological advances, while the representatives from other fields will learn about the problems and complexities of the snow and ice medium. The specific objectives of the Symposium are to: 1) explore the possible impact of new technology on the study and management of snow and ice resources, especially in the area of measurement, data accumulation, and information dissemination; and 2) explore the application of those techniques that may benefit the study and management of snow and ice resources.

The purpose of the Symposium, is, however, much broader. First, the Symposium is seen as a catalyst for the establishment of a continuing interaction among the various groups participating in the discussions and, thereby, the ultimate advancement of all of the disciplines involved.

For such a meeting to be truly effective, all the participants must be willing to enter into free and open discussion. All have valuable inputs that may not be represented by any other person.

The Symposium, however, is not an end unto itself. The value of the meeting lies not only in the sharing of ideas, but in its potential for development of a nucleus for continued interchanges among those in the various disciplines involved. Without this continuing interchange, this Symposium would be like many others - a good scientific meeting with hydrologists talking to hydrologists, bereft of any cross-fertilization. Although often highly beneficial, such meetings do not generally incorporate new ideas from allied fields. The success of this Symposium rests on the exchange of ideas among various disciplines.

With these thoughts in mind, we ask for your cooperation in making this Symposium a success. We have developed what we hope will be an interesting program. Its success, however, rests with you, and your willingness to actively participate in the discussions.

Session I

Snow - Information Needs and Distinguishing Characteristics

Chairman:

R. B. L. Stoddart, Canadian National Committee for the IHD

General Reporters:

H. S. Santeford, U. S. National Committee for the IHD

D. H. Male, Division of Hydrology, University of Saskatchewan

SNOW LOAD ANALYSIS AND RECREATIONAL USES OF SNOW DATA

Donald W. McAndrew
Soil Conservation Service

ABSTRACT

Architects, engineers, and building officials have long needed a method of estimating the weight of snow that accumulates on a building roof or structure. With this information they can determine the strength requirements necessary to prevent unacceptable strain on the structure or its collapse.

Also, in recent years winter recreation has grown to an extremely large business throughout the western United States. Many ski areas have been built and many more are in the planning stage. The same snow data that have provided a reliable tool for forecasting streamflow, as well as providing information on snow loads, are important in evaluating winter recreation site potential.

SNOW LOAD PROCEDURE

As man moves his homes, playgrounds and related businesses into the higher elevations of the western mountains, many problems of the new environment become apparent. One of these is dealing with the annual snowpack buildup. Questions such as how much snow, date of maximum depth, length of season, etc., need answering for planners, architects, recreationists, and others in the orderly development of the newly developing areas.

The Soil Conservation Service, with its cooperators, has been measuring snow depth and water content throughout the West since the agency's inception; a large majority of snow-course installations have more than 30 years of record. With this background, it seemed logical that the SCS could aid planners and architects by providing the information in a readily understandable form. To this end, the agency has prepared and displayed snow load information either by county or multicounty graphs delineating smaller areas with the same snow load within a county. These two methods can be understood by examining methods used in Nevada and Oregon.

Nevada - Snow-Load Studies

In Washoe County, Nevada, snow-load values have been included in the county building code (Brown, 1970). This is fitting, especially in western states because most building is governed or inspected on a countywide basis. The tables included in the Washoe County Building Code compare the elevation within the county to the snow (roof live) loads in pounds per square foot. "Roof live" snow loads were determined as a percentage of ground snow loads (example: Figure 1 - Excerpt from Washoe County Building Codes, Appendix 1964).

Most localities find that county codes are satisfactory and economically feasible. However, many areas in the West have considerable variance in the annual snowpack at similar elevations within a county size area. If one of these more highly variable areas has considerable construction activity, it is advantageous and justifiable to increase the detail of the county snow-load criteria. This can

Figure 1 - Excerpt from Washoe County Building Codes

ROOF LIVE LOADS - ELEVATIONS AT OR ABOVE 5,000 FT. (1,524 METERS) ABOVE SEA LEVEL

Elevation in feet above sea level - Figures in parenthesis indicate meters	Snow Load in pounds per square foot - Figures in parenthesis indicate killograms per square meter	
	Lake Tahoe Basin	All of Washoe County except Lake Tahoe Basin
5,000 ft. (1,524)	---	40 (195)
5,500 ft. (1,676)	---	80 (391)
6,000 ft. (1,829)	155 (757)	100 (488)
6,500 ft. (1,981)	165 (806)	120 (586)
7,000 ft. (2,134)	175 (854)	140 (683)
7,500 ft. (2,286)	185 (903)	150 (732)
8,000 ft. (2,438)	200 (976)	160 (781)
8,500 ft. (2,591)	225 (1,098)	170 (830)
9,000 ft. (2,743)	250 (1,221)	190 (928)
9,500 ft. (2,896)	275 (1,343)	210 (1,025)
10,000 ft. (3,048)	300 (1,465)	250 (1,221)

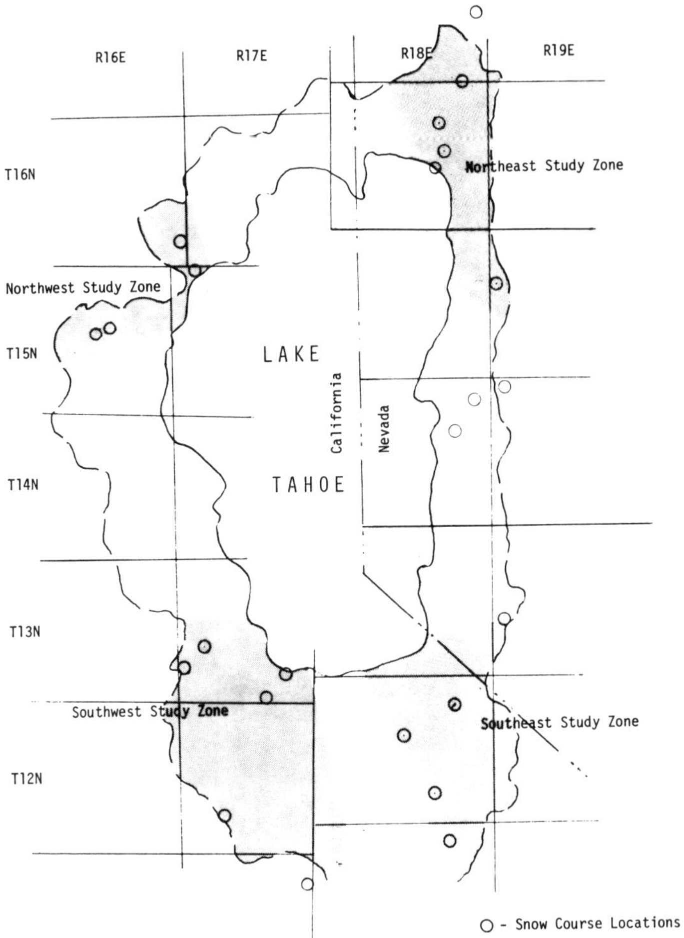

Figure 2 - Study Zones Used in the Analysis of the Lake Tahoe Basin

be handled in several ways. One is by delineating an area within the county and providing snow-load criteria applicable to the area.

The Lake Tahoe area of California and Nevada is a special area that has undergone considerable development during recent years. Many problems caused by the increase in population are very difficult to solve. The problems are complicated legally by having portions of five counties in two states.

Because of pressing ecological problems, the Lake Tahoe Regional Planning Agency was formed. This agency requested the Soil Conservation Service to supply snow-load data to be used in determining building design and other factors such as road and parking lot snow removal. Since considerable development is contemplated, it was necessary to provide snow-load information on a detailed basis.

Lake Tahoe is located at 6,229 feet (1899 meters) elevation in the Sierra Nevada range. The mountain range has peaks in excess of 10,000 feet (3048 meters) around the lake. The majority of the winter's storms are Pacific maritime. The orographic effect of the Sierra Nevada causes the western side of the Tahoe Basin to have a much deeper snowpack than the eastern side, which is in a rain shadow area. The Tahoe Basin also receives heavier snowfall in the north than in the south, primarily due to an orographic effect. Because of these physical factors, a map was prepared delineating the snow zone boundaries as they change markedly from east to west and north to south. For study purposes the watershed was divided into four sections: northeast, northwest, southeast, and southwest (Figure 2).

In each of these areas, snow courses were selected at various elevations. All past records from these snow courses were analyzed, using the Log-Pearson Type 3 Frequency Analysis. The SCS and many other federal agencies have adopted this method of determining the recurrence interval of hydrology phenomena, as recommended by the Water Resources Council, Washington, D.C. The basic formula for this representation of return frequencies is as follows:

$$\text{Log } L = M + KS$$

Where L = computed snow load for a selected occurrence, and $M = \frac{\text{Sum } X}{N}$ (mean of the logs), and

$$S = \sqrt{\frac{\text{Sum } X^2}{N - 1}} \text{ (standard deviation)}$$

Where X = X - M, and K = Pearson Type 3 coordinates which are obtained from standard precomputed mathematical tables using the skewness (g) as a parameter

Wherein $g = \frac{N \text{ Sum } X^3}{(N - 1) \cdot (N - 2) \cdot S^3}$, and N = number of events in the record being used.

Snow depths having a probability of .04 in any one year (4 percent chance) or 1 out of 25 years, were determined from selected snow courses (selected example: Figure 3). These data were converted to pounds per square foot and plotted versus recurrence interval in years (selected example: Figure 4). All snow courses within the four study zones were then plotted, using the Log Pearson Type 3 frequency value at the 4 percent level versus elevation

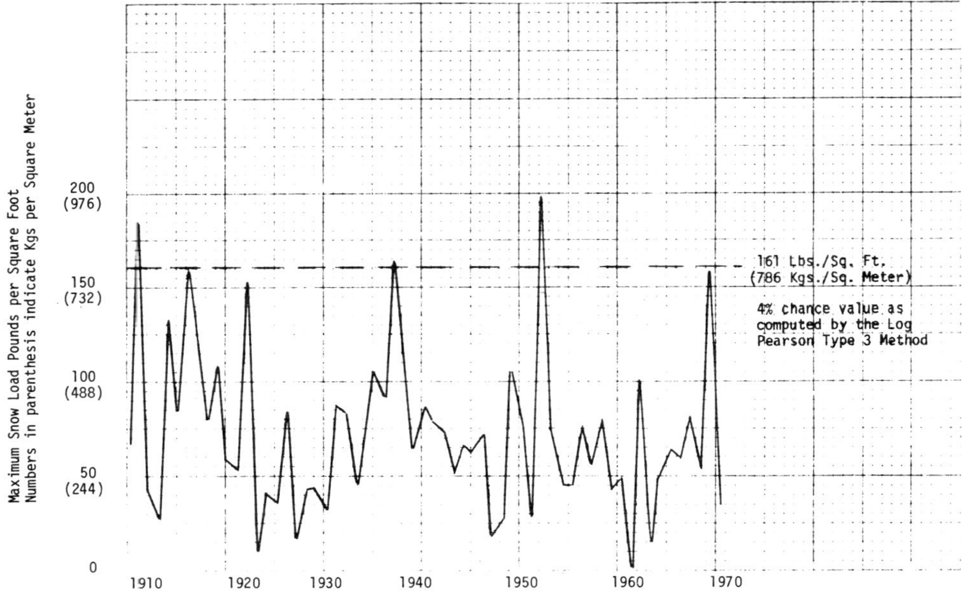

Figure 3 - MAXIMUM YEARLY SNOW VALUES AND 4% CHANCE VALUE AT TAHOE CITY, CALIFORNIA

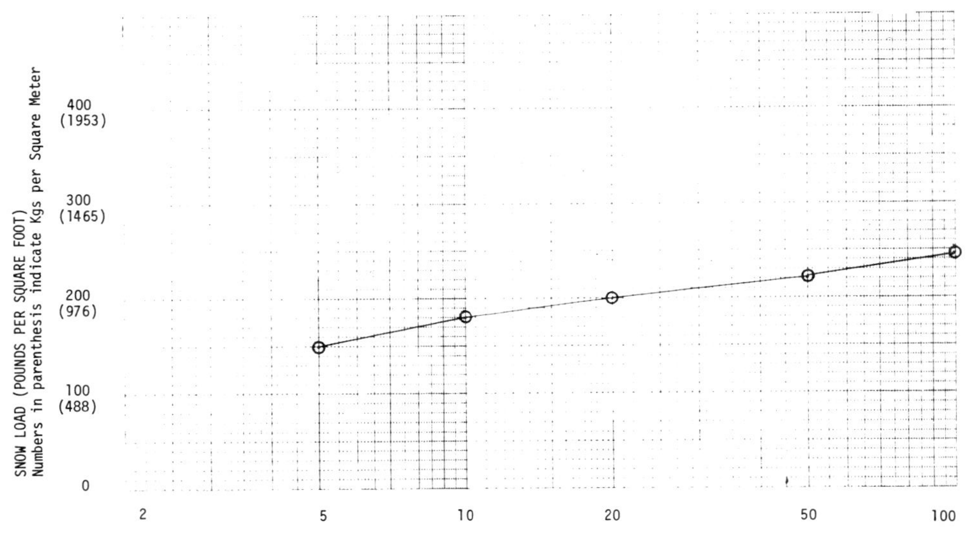

Figure 4 - RECURRENCE INTERVAL (YEARS) Analysis Log Pearson Type 3

Marlette Lake
Snow Course

1.1

(Figure 5). Note the difference in the slopes of the lines for the separate areas. This shows graphically the variation between the areas in the Lake Tahoe Basin. For example, snow load construction criteria for a home built at 7,000 feet (2134 meters) in the northwest portion of the basin would be 400 pounds per square foot (1953 kgs. per square meter), while at the same elevation in the southeast portion it would be slightly less than 150 pounds per square foot (732 kgs. per square meter). This makes separating these areas necessary and illustrates the economic value to the public.

To present this data a map of the area with isolines indicating the various snow load zones was produced (Lake Tahoe Snow Load Zones, 1970). For each 25 pounds per square foot (122 kgs. per square meter) difference, separate isoline delineations were plotted. This map is within limits of accuracy accepted and easily understood by building contractors, county commissioners, etc.

In any deep snow zone area that is having or expects considerable urban development, there is a definite need for this type of snow load data. If basic data are not available from existing snow courses, short-term snow courses throughout the elevation range could be established. After a few years of record, these short-term courses could be correlated to other similar courses in the region with longer periods of record, and be used to determine snow loads with reasonable limits. Example: If a long-term snow course shows 40 percent of its maximum record on the date of sampling, the new snow course located at a similar elevation would show about 40 percent of its maximum load on the same date. This data can then be expanded to form the sectors for developing the snow load criteria. The existing snow courses have been installed primarily for the prediction of streamflow runoff, and they may not be located in places that will yield the best information for analyzing snow loads on buildings and structures. In areas for which building officials feel that snow course data are inadequate, additional short-term snow course sites may be established. The information obtained can then be used to adjust the recommended values.

The snow load procedure used for the majority of the western United States would typically follow the county graph approach. The following description of the "Snow Load Analysis for Oregon" is typical of this method.

Oregon - Snow Load Studies

The Soil Conservation Service and the Structural Engineers Association of Oregon cooperated on a publication, "Snow Load Analysis for Oregon", issued in June 1971. The publication gives the expected snow loads for Oregon and a procedure for using these loads in the design of buildings. The association did the research to find a method of applying estimated snow loads to building design. This research showed that the Uniform Building Code of Canada (NBC, 1960), contained the most rational and useful method of design. The Canadians have done extensive work in comparing ground snow loads to loads found on roofs and structures. They investigated various shapes of roofs and buildings, exposures, etc. and derived a method to apply this information and basic ground snow load data to the design of buildings. Permission was secured from the National Research Council of Canada to recommend their method for use in Oregon and to

Figure 5 - Snow Load Curves for Lake Tahoe Drainage

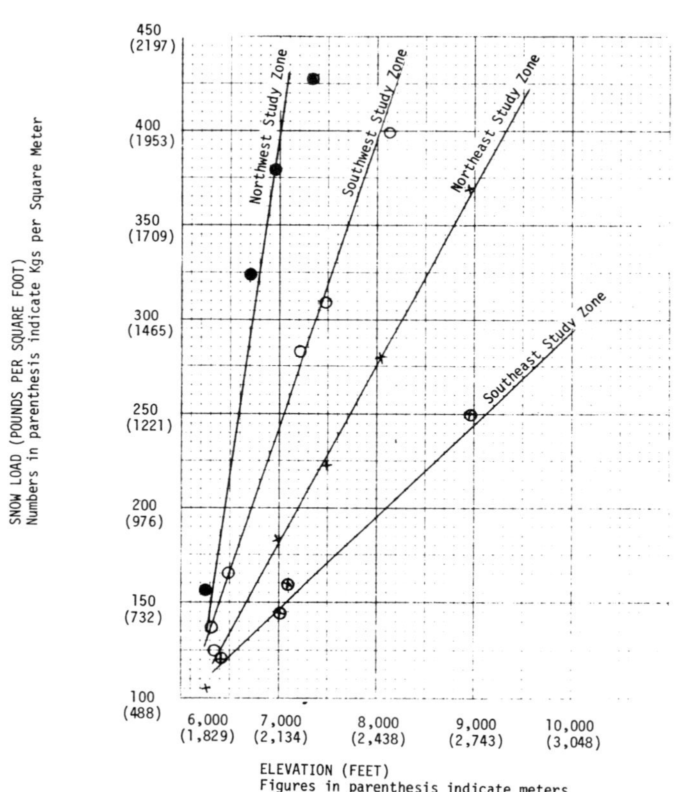

4% CHANCE MAXIMUM SNOW VALUES AS COMPUTED BY THE LOG PEARSON TYPE 3 ANALYSIS reproduce their coefficients and explanations in the "Snow Load Analysis for Oregon" publication.

Previous work has been done by Thom (Thom, H.C.S. 1966) in presenting expected snowpack on the ground in pounds per square foot on a very general basis (a map of the United States). The map he produced was not localized enough to be of value to the Oregon Structural Engineers Association. The 25-year return interval values indicated by Thom range from a minimum of 5 pounds to a maximum of 20 pounds per square foot for the entire state of Oregon. The association was already recommending a minimum of 25 pounds per square foot for the valley areas of the state that receive the least amount of snow, and it needed additional information for the mountainous and desert areas that cover large portions of the state. In order to localize the data as was done by Brown (Brown, 1970) in Nevada,and make it as useful as possible, it was felt that snow measurements for county size areas should be analyzed and presented so the extreme snow accumulation variations that exist in Oregon, as shown in "Summary of Snow Survey Measurements for Oregon" (Frost & George, 1969) would be readily noticeable.

The linear relationship of elevation to snow accumulation has been noted and discussed previously by Schaerer (1970), Garstka (1949), Hannaford (1958), and Seligman (1936). Their work shows this relationship on an individual mountain and watershed basis. It was felt that the relationship would still prevail on a multiwatershed area such as a county.

Maximum recorded weights of snow on the ground as determined from measurements taken by the Soil Conservation Service, Oregon State University, and the State Engineer of Oregon, were taken for various county areas in the state (Figure 6) and plotted against elevation (Figure 7). These maximums were derived from monthly measurements taken during the snow seasons from 1940 to 1969. This interval of years was selected because most snow courses in the state have a period of record at least this long. At a few snow course locations where the period is short, the data were statistically extrapolated, as is often done when insufficient observations are available. All maximum values were then compared to the results of the Log-Pearson Type 3 frequency analysis at the 3 percent level of significance with very good correlation (Figure 8).

In most cases the plotted points (psf vs. elevation) show very small deviations from the curve. However, occasionally there are large deviations due to local orographic effects, exposure, etc., which are noted (Figure 8). When applying the snow load values to structures, the designer needs to adjust the graphical data for local effects, safety, and lifespan considerations.

WINTER RECREATION AREAS

The Soil Conservation Service offices in Montana and Nevada have recently published statewide maps indicating areas within the states having adequate snowfall for ski areas and similar winter recreation sites (Farnes, 1969, and McAndrew & Roda, 1971).

Snowpack criteria for winter recreation developments are dictated somewhat by local needs. The direction and aspect of ski runs, amount of slope, grooming, typical wind drifting, etc., determine an area's snowpack needs. According to ski area operators in

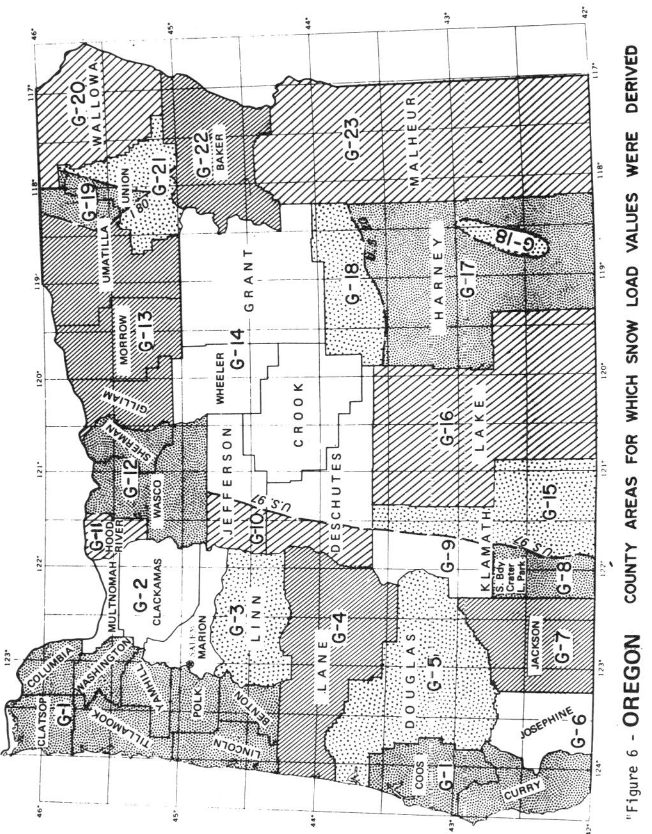

"Figure 6 - **OREGON COUNTY AREAS FOR WHICH SNOW LOAD VALUES WERE DERIVED**

1.1

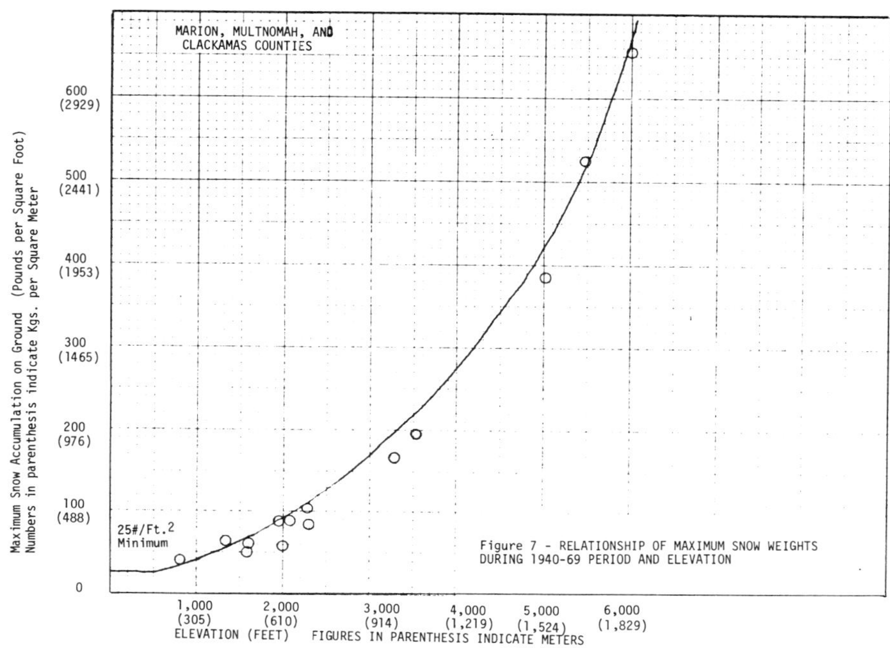

Figure 7 - RELATIONSHIP OF MAXIMUM SNOW WEIGHTS DURING 1940-69 PERIOD AND ELEVATION

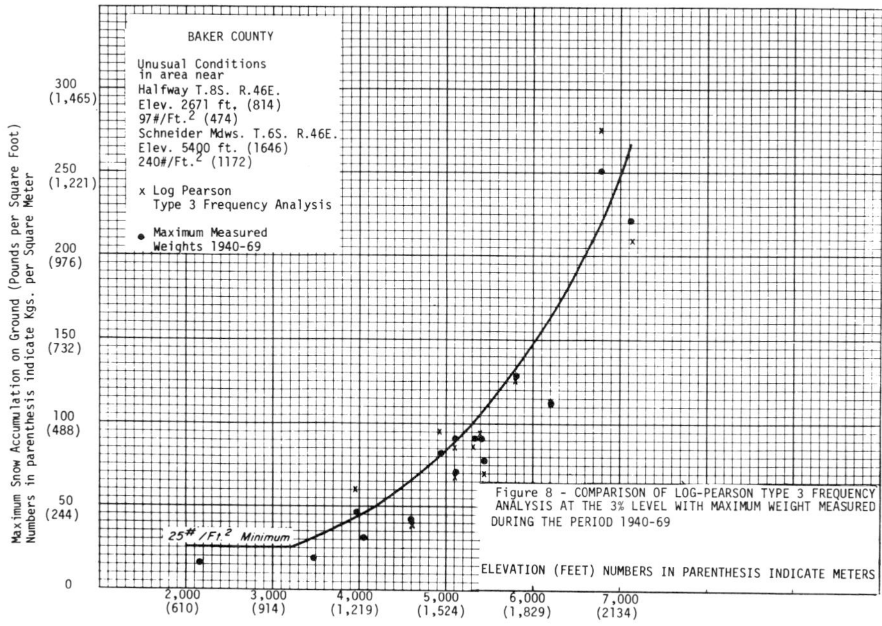

Figure 8 - COMPARISON OF LOG-PEARSON TYPE 3 FREQUENCY ANALYSIS AT THE 3% LEVEL WITH MAXIMUM WEIGHT MEASURED DURING THE PERIOD 1940-69

1.1

Montana, 2 inches of snow-water equivalent, which is the same as 12 to 18 inches of new snow or 4 to 6 inches of packed snow, is the average minimum necessary for skiable snow conditions. It is also important that a ski area be in operation by mid-December at least 18 years out of 20. A closing date is not critical because most skiers do not ski after Easter regardless of snow conditions.

In Nevada many snow courses are measured on January 1, but there are very few December 1 measurements available. Using what were available, average December snowpack increments were estimated. One-half of the December increment for each snow course was subtracted from its respective January 1 value to arrive at an adjusted December 15 reading. Studies of past years' snow pillow data (1965-1972) revealed that snow buildup in the area is fairly uniform throughout December. With the limited data available (7 years), we felt it was reasonable to split the December increment in half. These assumed December 15 snow-water content readings were then analyzed using the Log-Pearson Type 3 analysis.

All areas in the state meeting the criteria of 2 inches of snow-water equivalent in 18 out of 20 years were delineated. A statewide map, indicating adequate snowfall for ski areas and other winter recreation area use in Nevada, was then prepared (McAndrew & Roda, 1971).

When a specific site is being contemplated for a winter recreation area, it is desirable to install a few short-term snow courses for snowpack evaluation of the area. The short-term records collected while the proposed area is in the feasibility study period can be compared directly with long-term data, thereby giving the short-term snow courses more reliability and more meaning. This additional snow course information allows people to develop winter recreation facilities based on data that should improve their chances of having adequate snow.

REFERENCES

Brown, John Webster (1970). An Approach to Snow Load Evaluation, Proceedings of 38th Annual Meeting, Western Snow Conference, pp 52-60.

Farnes, P.E. (1969). Montana Areas with Adequate Snowfall for Winter Recreation - U.S. Department of Agriculture, Soil Conservation Service, Snow Survey and Water Supply Forecast Unit, Bozeman, Montana.

Frost, W.T. and Tommy A. George (1969). Summary of Snow Measurements for Oregon - U.S. Department of Agriculture, Soil Conservation Service, Snow Survey and Water Forecast Unit, Portland, Oregon.

Garstka, W. U. (1949). Interpretations of Snow Surveys. Transactions AGU, Volume 30, No. 3, pp 412-420.

Hannaford, J.F., C.G. Wolfe, and R.W. Miller (1958). Graphical Method for Determination of Area - Elevation Weighting of Snow Course Data. Proceedings of 26th Annual Meeting, Western Snow Conference, pp 73-82.

Lake Tahoe Basin Snow Load Zones (1970). U.S. Department of Agriculture, Soil Conservation Service, Snow Survey and Water Forecast Unit, Reno, Nevada.

McAndrew, D. W. and J. D. Roda (1971). Nevada Areas with Adequate Snowfall for Winter Recreation - U.S. Department of Agriculture, Soil Conservation Service, Snow Survey and Water Supply Forecast Unit, Reno, Nevada.

NBC (1960), National Building Code of Canada, issued by the Associate Committee on the NBC National Research Council, Ottawa, Canada (NRC No. 5800).

Schaerer, P. A. (1970). Variation of Ground Snow Loads in British Columbia. Proceedings of 38th Annual Meeting, Western Snow Conference, pp 44-48.

Seligman, B. A. (1936). Snow Structure and Ski Fields. MacMillan and Company, Limited, pp 456-458.

Thom, H. C. S. (1966). Distribution of Maximum Annual Water Equivalent of Snow on the Ground. Monthly Weather Review, Volume 9, No. 4.

SNOWPACK AND RELATED TRENDS IN MOUNTAINS OF THE SOUTHWESTERN UNITED STATES

Gregory L. Pearson
Hydrologist, Water Supply Forecasting Unit
Soil Conservation Service
U.S. Department of Agriculture
Portland, Oregon

ABSTRACT

Analysis of mountain snowpack in mountains of the southwestern United States indicates a changing climate and major variability within relatively short geographic distances. Examples are presented. Some of the effects of a changing climate are noted, with particular emphasis being given to the effect of the snowline elevation on the deer herd.

The need for more comprehensive data networks in remote areas is noted. Benchmark stations have a potential for pinpointing the progression of climatic changes which may have implications far beyond local areas.

During the early 1960's there was considerable concern among the water users of southern Utah because streamflow had been steadily decreasing for many years. Many felt that this decrease was more a result of increased evapotranspiration losses due to changes in vegetative cover than to a change in climate. The U.S. Forest Service was charged with using land management practices that had caused an increase in vegetation, which in turn had reduced runoff.

Decreased runoff was associated with loss of farm population, as indicated by Census figures for five southwestern counties. In 1930 the population was 43,764. In 1960 it was 34,497 -- a decrease of 9,267 people, or 21 percent. The economic impact was severe.

Because of this an analysis was made of the snowpack on the main headwater areas of the Sevier and Virgin rivers. The area of principal concern was the mountain where Zion National Park and Cedar Breaks National Monument are located. As illustrated below, the snowpack analysis showed that a change in climate was taking place that would account for the decreased runoff. The trend in climate, particularly if continued long enough, could also cause vegetative changes. This report updates the former study.

Figures 1 to 4, inclusive, show how the April 1st snowpack water has varied at four snow courses. These are Harris Flat, elev. 7,700 feet (2,347 meters); Duck Creek, elev. 8,700 feet (2,652 meters); Webster Flat, elev. 9,200 feet (2,804 meters); and Cedar Breaks, elev. 10,390 feet (3,472 meters). Records begin in 1927 at Webster Flat, in 1931 at Harris Flat and in 1935 at Cedar Breaks and Duck Creek. Due to recent wintertime opening of the road at Cedar Breaks, which has affected the snow course, starting in 1970 the values for Cedar Breaks are estimated from the Midway Valley course, elev. 9,800 feet (2,987 meters). The correlation is excellent.

The values plotted represent 5-year moving averages ending in the year shown. Values at all courses for the periods ending in 1916 until 1928 are wholly or partly estimated, determined from snowpack-runoff relationships and using measured runoff values. Points plotted at Webster Flat from 1927 to 1930 use some measured

QUANTITY	ISBN 309-0-	TITLE
1	2235-5	ADV CONCEPTS SNOW & ICE 0774
1	8901-0	POSTAGE

YOUR REF. 5 13 75

PACKING SLIP

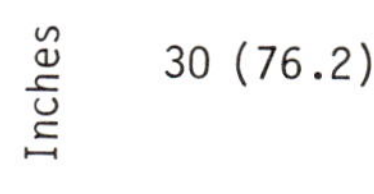

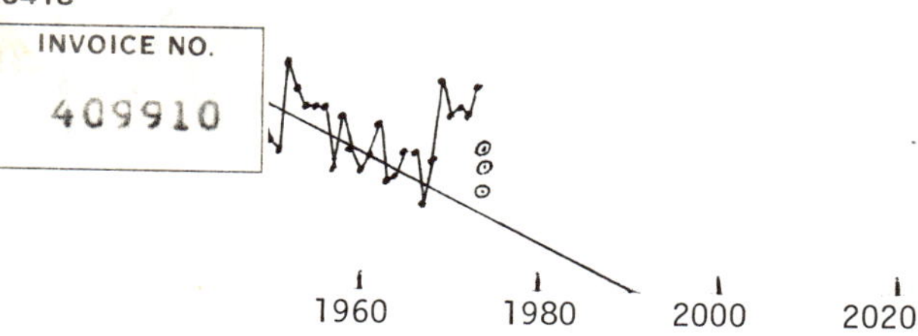

Averages (ending in year shown) of
ter at Harris Flat. Sevier River.
et (2347.0 meters)

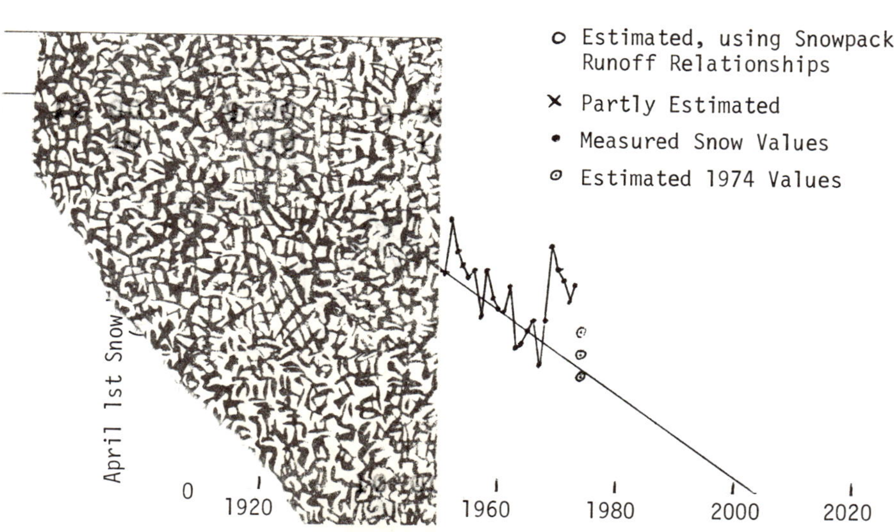

Figure 2 Five Year [illegible] Averages (ending in year shown) of April 1st Snow Water at Duck Creek R.S. Sevier River. Elevation 8700 feet (2651.8 meters)

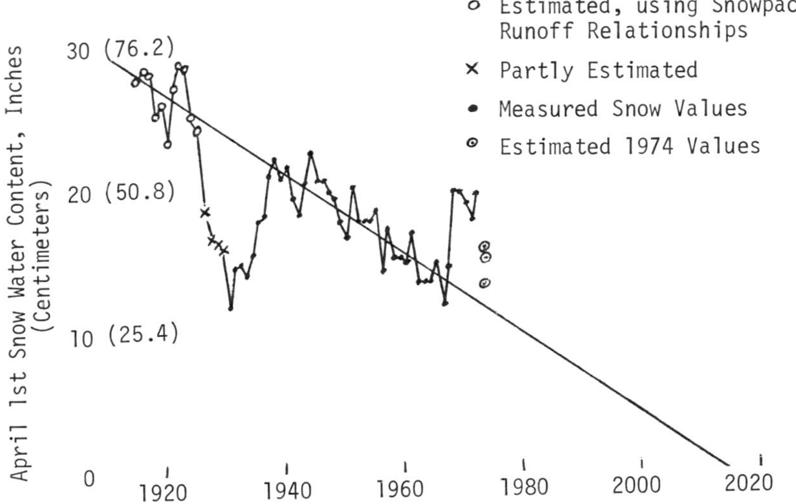

Figure 3 Five-Year Moving Averages (ending in year shown) of April 1st Snow Water at Webster Flat. Virgin River. Elevation 9200 feet (2804 meters)

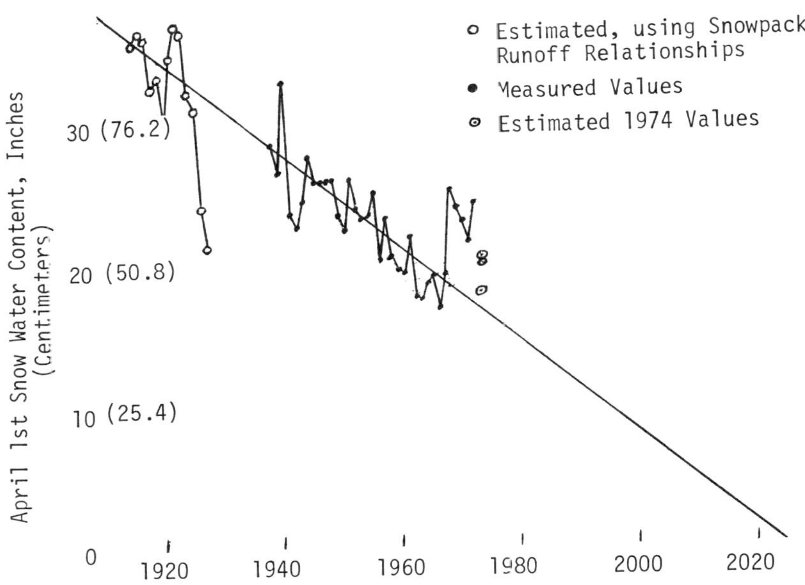

Figure 4 Five-Year Moving Averages (ending in year shown) of April 1st Snow Water at Cedar Breaks. Sevier River. Elevation 10,390 feet (3167 meters)

1.2

snow values and some values estimated from snowpack runoff relationships. Values for Duck Creek from 1929 to 1938 use some measured snow values plus some values estimated from snowpack correlations with Webster Flat and Harris Flat.

All four snow courses indicate a longtime downward trend in which the average water content just prior to 1969 had become about equal to that experienced during the severe drought of the early 1930's. It was near a third to a half of what it had been during the period from 1912 to 1923.

The last 5-year period from 1969 to 1973 indicates a possibility that this trend may soon reverse itself. During no other 5-year period since 1923 has there been a time when the sum of the snow water of the two highest years in the period equalled that of 1969 and 1973.

However, offsetting the above two high years is the fact that, at the two lowest elevation courses (Harris Flat and Duck Creek), during the same 5-year period the sum of the 2 lowest years is lower than for any other 5-year period. At Webster Flat the sum of the 2 lowest has been slightly lower five times, while at Cedar Breaks it has been slightly lower twice.

The next five to ten years should prove very interesting in showing whether or not this time trend reverses itself or continues downward. The figures show the most probable range within which the points to be plotted for 1974 will lie. Three points are plotted for each. The highest point represents the average of the past 4 years and the highest one of the last 11 years, excluding 1969 and 1973. The middle point is the average of the last 4 years, while the lowest point is the average of the last 4 years and the lowest one of the last 11 years.

Unless some more high years are realized very soon, there appears little reason to assume that the downward trend has been broken. If the downward trend continues at about the same rate as it has experienced in the past, it appears that there will be essentially no snow on April 1st at Harris Flat by about the year 1990, at Duck Creek by about 2005, at Webster Flat by 2015, and at Cedar Breaks by about 2025.

Figure 5 shows how the April 1st snow water varies with elevation. Fifteen-year average snow water content values for the period 1953-67, inclusive, are shown for eleven snow courses on this mountain. For any given elevation the snow courses with the most water are those which are in upwind or protected areas. Courses with the least water are in downwind (rain shadow) areas, or areas which are subject to heavier winter snowmelt.

Also shown are the estimated water content bands for three other periods, as follows:

1. Estimated 15-year average for 1912-26, inclusive.
2. Estimated 15-year average for 1993-2007, assuming the downward trend continues at the past rate.
3. Estimated average for a series of years such as 1952 and 1969, assuming the present trend reverses itself and becomes much wetter.

A changing climate such as illustrated here can have a major impact on vegetation native to an area. Under such changing conditions a question may well be asked as to what a climax vegetative condition is in such an area, and when is it reached?

A changing water supply, vegetation, and snow line will affect fish and wildlife, the livestock industry, industrial development, recreation activities such as ski resorts, and human population

1.2

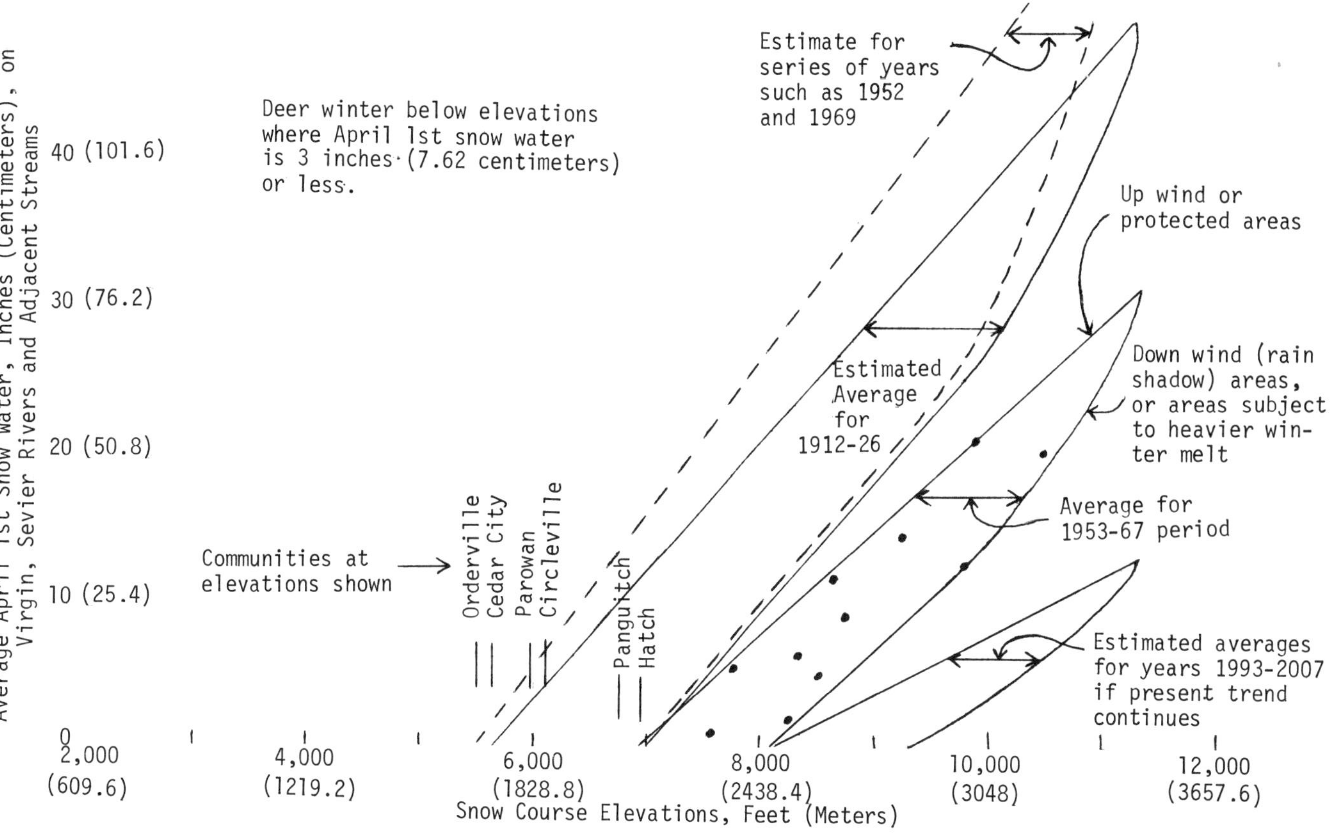

Figure 5 Average April 1st Snow Water versus Elevation

numbers in the area. These changing factors can generate competition between groups for available water supplies and cause controversy between private interests and public land management agencies over land management practices.

One effect of this changing climate is illustrated as it affects mule deer. The snowpack determines the maximum elevation at which deer can winter, thus determining where their winter range will be. Wildlife studies (Gilbert et al, 1970) show that snow depths of 18 inches (0.55 meters) essentially eliminates use of the range by mule deer. Considering snow densities in this area during the winter months, this means that deer will spend most of the winter at elevations where the April 1st snow water is 3 inches (7.62 cm.) or less. Most of the deer will winter in areas where it is 2 inches (5.08 cm.) or less.

Referring again to Figure 5, it can be seen that during the period 1953-67 the winter range of the deer would have been below elevations of about 7,300 to 8,300 feet (2,225 to 2,530 meters). If the present climatic trend continues until about the turn of the century, the deer will be wintering below elevations of about 8,700 to 9,800 feet (2,652 to 4,849 meters). Since this is near the top of the mountain, many of the deer will be spending both winter and summer in the same area.

On the other hand, if the trend reverses itself and returns to the conditions prevailing during 1912-16, or becomes even more severe, the deer will be forced to winter at elevations below about 5,800 to 7,200 feet (1,768 to 2,195 meters). This means they would be forced to winter in the valleys where people are. Available feed would be greatly restricted and the deer herd could be expected to be reduced. The heavier snowpacks would produce a more abundant water supply, allowing for an expanded human population in the area and providing greater competition to the deer herd.

Can the climate change in this area without also changing somewhere else? Storm systems and/or storm types that formerly moved thru this area must be going somewhere else. Figure 6 shows the fluctuations of Great Salt Lake in northern Utah, as reported by the U.S. Geological Survey. Some of the reduction in lake level is undoubtedly due to irrigation developments during the last hundred years, but does this account for all of the loss? Studies such as that of Nelson (1968) indicate that this is principally the result of climatic changes on a much broader scale.

Analysis of records for snow courses on the Beaver River about 50 miles north of the Sevier-Virgin river headwater area shows the same downward trend. In northern Utah, snow courses do not show this trend. However, there is some question as to whether or not low elevation snow cover has changed below the elevation of most of the existing snow courses.

An attempt was made to see if the same climatic trend observed in southern Utah could be found up or downwind from this area. Limited investigation has indicated that this same trend may be going on in the San Joaquin basin of southern California. Figure 7 shows the snow trend for the Big Meadows snow course, elev. 7,600 feet (2,316.5 meters) on the Kings River. Records began in 1930. Records are missing for 1931 and 1967.

Just as in southern Utah, the sum of the water content in 1969 and 1973 is higher than the 2 highest years in any other 5-year period. There have been six times when the 2 lowest have been

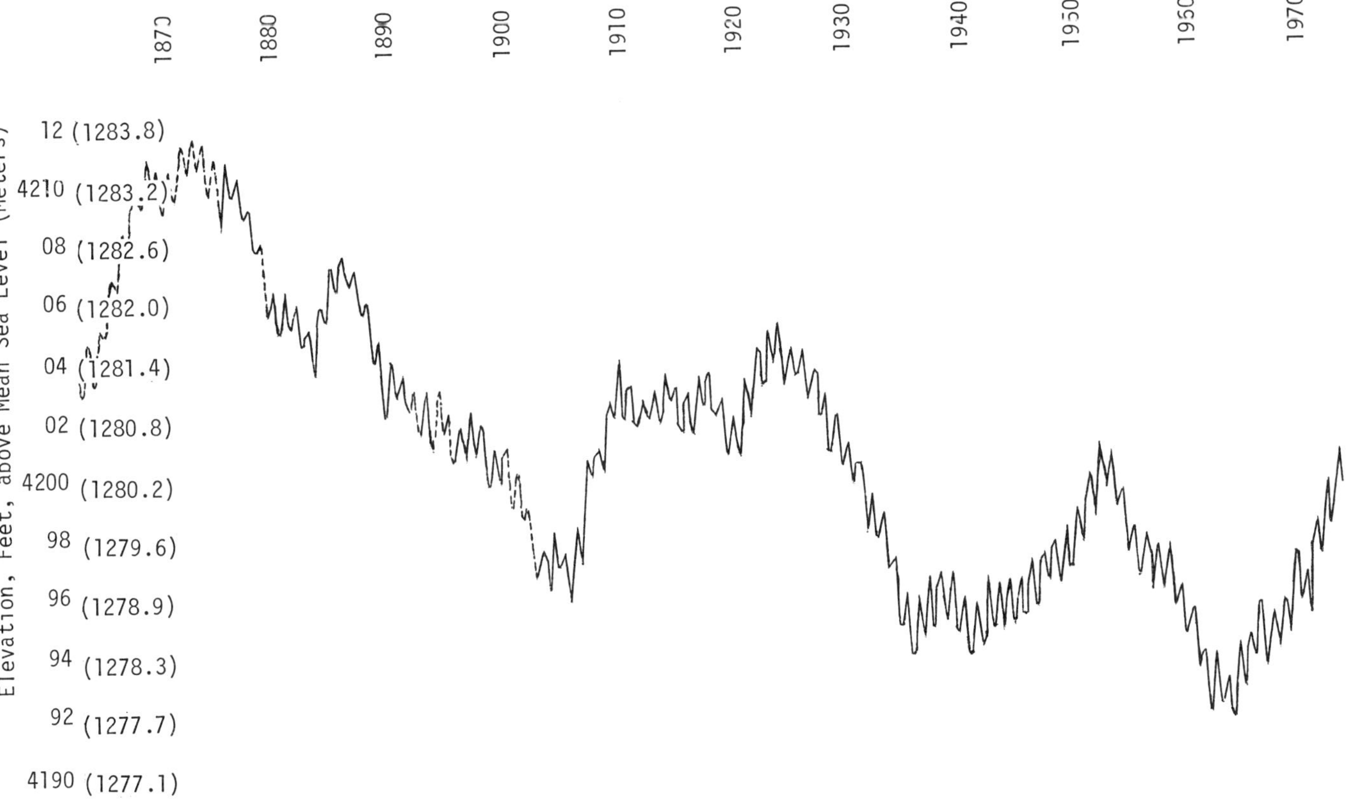

Figure 6 Fluctuations of Great Salt Lake, Utah, as Reported by Geological Survey

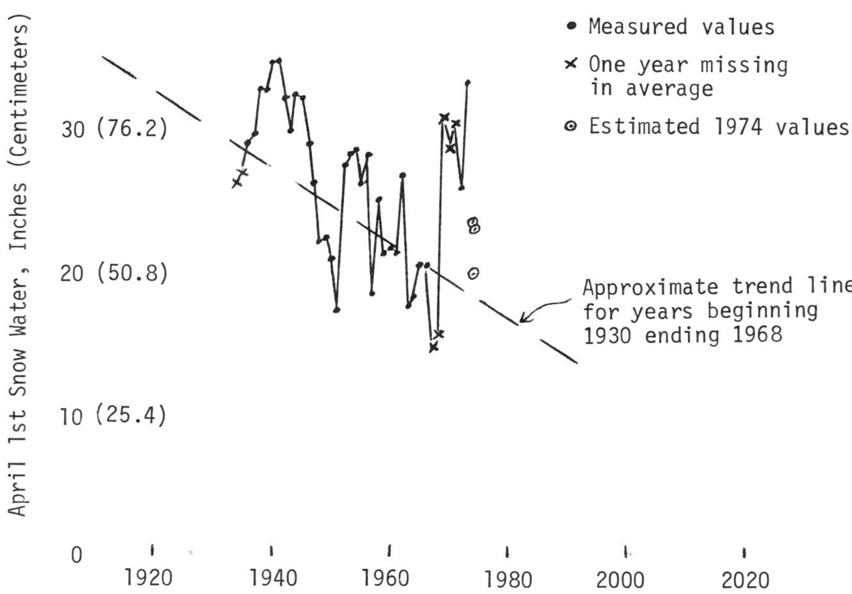

Figure 7 Five Year Moving Average (ending in year shown) of April 1 Snow Water at Big Meadows, Kings River, California Elevation 76 feet (2316.5 meters)

slightly lower. The most probable range for the 1974 value is also plotted. The trend line for the years 1930 to 1968 is very similar to that in southern Utah.

Additional study of such trends is needed to define their extent. A major hindrance to such studies is the lack of adequate basic data networks with longtime, continuous records in the more remote regions. Benchmark stations in such areas are definitely needed.

At present, major benefits come from seasonal water supply forecasts based on relatively short-term snow records. However, when records accumulate from more comprehensive data networks, it may become possible to forecast the progression of climatic changes. Such forecasts would have far greater social, economic and ecological impacts than our existing forecast service.

REFERENCES

Gilbert, Paul F., Olof C. Wallmo, and R. Bruce Gill, 1970. Effect of Snow Depth on Mule Deer in Middle Park, Colorado. Journal of Wildlife Management. Vol. 34, No. 1, pp 15-23.

Nelson, Morlan W., 1968. The Status of the Peregrine Falcon in the Northwest. Published as Chapter 4 in Peregrine Falcon Populations: Their Biology and Decline, edited by Hickey, Joseph J., Univ. of Wisconsin Press, Madison, Milwaukee, and London, 1968.

POLAR REGION SNOW STRATIGRAPHY TECHNIQUES APPLIED TO COMMERCIAL PROBLEMS

William W. Vickers, The MITRE Corp., Bedford, Mass.
Manuel E. Lopez, The MITRE Corp., Bedford, Mass.
James Branch, Sno-Engineering, Inc., Franconia, N. H.

ABSTRACT

Stratigraphic characteristics of snow reflect the weather conditions producing them, for example, a drizzle crust, or melted surface layer occurring on a certain date. Planning a proposed ski resort required a survey of environmental conditions of a remote peak for which no previous historical data existed. A stratigraphic analysis of sections straddling the peak is presented, and the utility of stratigraphic information from the snowpack is illustrated. Experimental work from a second project in Yellowstone National Park, tracing stratigraphy, is covered to illustrate the usefulness of this technique for measuring snow accumulation for large snow courses, particularly where sensitive measurements of cloud seeding are being sought.

SNOWPACK MEASUREMENT

Snowpack measurement seems a simple task. Usually the depth of snow is monitored on a graduated stake. Density samples are taken with a coring device. These measurements are adequate for rough estimates of potential runoff, but the greater demand of users on a limited amount of water, and the precision requirements of experimental programs related to weather modification have pressured us into the development of newer and better measurement techniques. Hopefully these techniques will render increased precision by increasing measurement sensitivity and the number of sampling points. Toward this end, snow pillows and buried radiation sources have entered the field of snowpack measurement; however, field trials of this equipment suggest we have not quite arrived at a simple, reliable, inexpensive way to measure the snowpack. It is particularly difficult, for example, to isolate the minor snowfalls produced by human effort in experimental weather modification. Stake and core techniques, or any similar single point data extractions, provide limited information and precision.

During and after the International Geophysical Year, the stratigraphy of accumulating snow at Little America and Byrd Station, Antarctica was monitored so that one could isolate and identify any particular snowfall in a stratigraphic section later taken from random locations (Vickers, 1966). Physical properties of respective layers (density, grain size, ice petrofabrics of crusts) were measured and re-measured as the layers underwent metamorphic changes due to the thermal conditions of the snowpack. Monitoring reference zones at widely separated stations (700 miles) and comparing them with synoptic meteorological records, gave evidence that the metamorphic changes in the snowpack were widespread, in response to

1.3

the large-scale meteorological systems producing them and that one could trace a given physical identifier, i.e., fused grains formed by snow falling through underlying warm air, for great distances in unmanned areas.

TRACING OF SEEDED SNOWFALL

To demonstrate that the Antarctic techniques could contribute to the measurement of the small practical quantities of precipitation associated with weather modification efforts, several snow pits were excavated in the Yellowstone Park area, and natural snow layers were traced a distance of 30 miles. Artificially produced precipitation from the seeding of Old Faithful Geyser was locally identified in a stratigraphic section, and it was found possible to traverse the fallout area downwind of the geyser, define the limits of the artificial precipitation zone, and establish the sensitivity of stratigraphic measurement on a local scale.

In the tracing of the artificial layers at Yellowstone, it was necessary to work with an unfavorable "signal-to-noise ratio" as the geyser cloud did not contain the precipitable water quantities one might expect in a natural convergent weather system. Tray-type collections of fallout from an early morning seeding showed values of 77.7g/m^2 near the center of the fallout zone; they reduced to 47g/m^2 at a point 330m transversely out from the approximate centerline of the plume, and to about 20g/m^2 at 700m out. This fallout, which had crusted by afternoon and had received modest natural precipitation, was traced as a near-surface layer. Ability to trace the layer faded at 700m from the centerline of the plume, or at a sensitivity of about 20g/m^2 of precipitation.

TRACING OF NATURAL SNOWFALLS

Figure 1 shows three stratigraphic sections at Yellowstone separated by 1.5 and 30 miles. The Black Sands section matches the Ranger Station section quite closely except for slight off-setting due to quantity variation; in the vicinity of 25-35 cm, two fine layers are missing at Black Sands. Greater differences occurred at West Yellowstone, but not enough to interfere with matching the layers. The bottom 20 cm is made up of drift deposits which experienced an irregular distribution of heat (resulting from patches of bare ground and drifted snow) and consequent irregularities in occurrence of fused and sublimated grains. The slight difference in tone between the Ranger Station and West Yellowstone sections is primarily due to lighting. (Natural light was used instead of the superior drilled-hole and artificial-light technique employed in the Antarctic (Anderson, 1960). A Formvar technique for recording stratigraphy as suggested by Schaefer (1942), was given a single trial at Yellowstone; it was found to produce a good stratigraphic record.) Temperature explains the appearance of crusts, clear ice, and the fusion or sublimation of grains. These underwent metamorphosis with changing conditions. There were heavy crusts of fused grains between 25 and 36 cm formed during the period 29-30 December by the 35°F temperature of maritime polar air. Wind often explains laminations

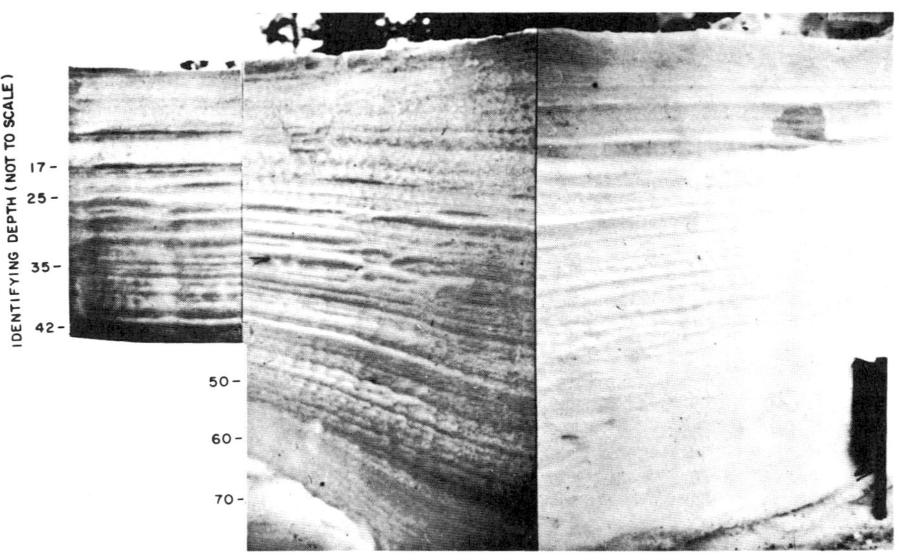

Figure 1. Comparison of Stratigraphic Sections Separated by 1.5 and 30 Miles from Old Faithful Ranger Station.

in the stratigraphy, 36–53 cm, caused by wind sorting of grains. Precipitation quantity at the control station acquaints us with the spacing to be expected between major horizons (found at 10, 17, 25, 36 and 53 cm), when going to unknown areas.

Meteorological events which produce the stratigraphy are an important part of the record as they control stratigraphic anomalies, where much of the information lies. For example, avalanche conditions are produced by sublimation within the pack, and slabbing. Strength of snow is controlled by temperature and grain type. Drifting conditions are influenced by state of the surface, which in turn is determined by wind, grain type and its role in saltation, and temperature profile and its sublimation role within the pack. Most important, the anomalies are the means by which we identify and trace layers over large distances and consequently extract water equivalent measurements for a given snowfall, the total pack, or the time or space variation of same.

Our ability to pinpoint events and measure fine quantities between major stratigraphic horizons will allow us to produce such figures as the latent heat release for a given storm in which we are interested for computer simulation purposes; or an after-the-fact measurement of snowfall produced by seeding — no matter where it may fall. Our ability to identify and trace layers laterally, reveals the degree of spatial variation of respective layers. Measurement variation, as a statistical parameter, indicates to us the sampling density required for a stipulated degree of measurement accuracy (statistical confidence) for a given history of wind and drift, and a given topographic environment.

It is of interest next to examine an additional commercial use to which stratigraphic measurement has been put.

SUITABILITY OF A REMOTE AREA FOR SKIING FACILITIES

Critical to the problem being illustrated is the answer to speculation that the NW facing slope of Mt. Pierce in the Presidential Range of New Hampshire, is subject to severe winds which would erode proposed ski slopes to an untenable condition such as the severe winds occasionally do at nearby Wildcat Mountain and Cannon Mountain (all approximately 4,000 ft elevation). If such speculation is correct, there should be evidence of such snowpack erosion in the snow stratigraphy. Figure 2 is an example of snow erosion at Wildcat as a result of strong NW winds. If buried by subsequent snowfalls, this surface would still retain its scoured characteristics when examined in a stratigraphic cross section through the season's accumulated snowpack. Measurement of snow depth at Wildcat on the date of the photo (March 22), shows, for example, 27" of accumulation on the windward (erosion) side of the upper lift terminal ridge and 108" on the leeward

Figure 2. Snow Erosion — Wildcat Mountain. Sastrugi patterns formed by scouring and deposition of snow by strong winds.

(deposition) side of the ridge, at points 30-40 yards perpendicularly removed from the ridge line, with strong evidence of scouring remaining in the stratigraphic cross-section.

ASSESSMENT OF MT. PIERCE SCOURING

Addressing the speculation regarding the unknown conditions surrounding the summit of Mt. Pierce, the plan of the investigation was to reconstruct the snow stratigraphy in the known inhabited areas adjacent to Mt. Pierce for which critical recorded meteorological information was available so that all snow layers in the cross section could be identified. This was done at the 2,000 ft and 4,000 ft levels to include variation of stratigraphic characteristics with altitude. When the stratigraphic history was completely documented, the investigation switched to the stratigraphy of the unknown areas, the summit region of Mt. Pierce, and a likely upper lift terminal area. Information being sought from the stratigraphy was the magnitude of snow scouring on the upper mountain, and what evidence could be gathered for or against an intolerable wind condition for lift, trails, and the proposed terminal. The basic question was, could trails be developed which could "live with" any negative conditions?

ANALYSIS

Figure 3a shows sections of snow (the brushed walls of excavated pits) at the foot of nearby Cannon Mountain and the foot of Mt. Pierce, respectively. The commonality of meteorological events in both areas is evident. The major snowfalls, rains, drizzles (which form identifiable crusts), etc., were recorded at these locations and the dates labeled in the diagram are a matter of record. These pits of known record are considered reference pits. In addition, the Cannon Summit pit, (elevation 4,000 ft), of Figure 3b, is a documented pit, a reference which is compared to the 4,000 ft summit of Mt. Pierce, the unknown area. Using concurrently recorded data from the manned nearby summit of Mt. Washington (about three miles distant), the succession of meteorological events was corrorborated for Mt. Pierce and exploratory pits were excavated on and near the summit to investigate the meteorological cause and the effect on snow surface, for the summit region.

Since evidence of severe wind erosion was the information being sought, a chain of seven pits was excavated on a NW-SE line across the summit and down to the proposed upper terminal area. Scouring at the immediate summit was obvicus to the most casual observer. The lateral extent of scouring and the specific winds that produced it were to be determined.

Figure 4 shows three pits considered representative of the NW-SE chain of exploratory pits. The small pit to the right of the diagram straddled the summit a few yards off-set from the ground level USGS plaque which was actually scoured bare. The left and center pits were to the

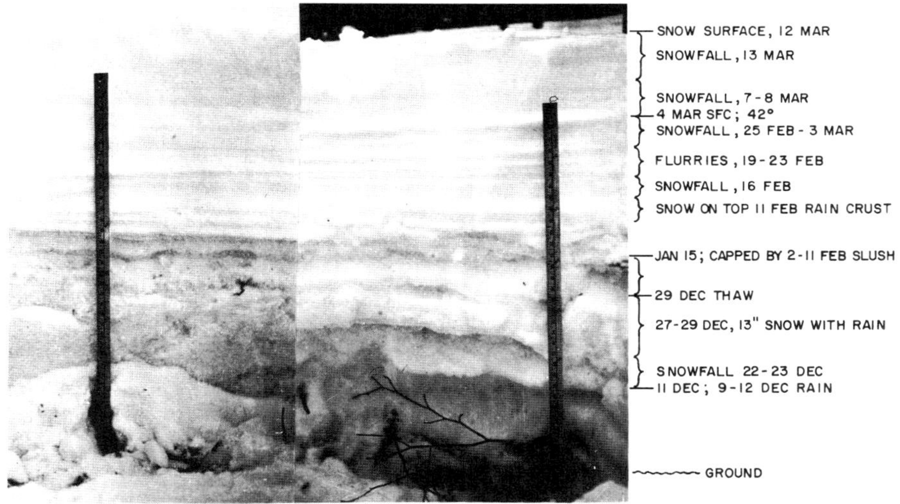

Cannon Mountain (base) Mt. Pierce (base)

a. Note the continuity of individual snowfalls and related meteorological events.

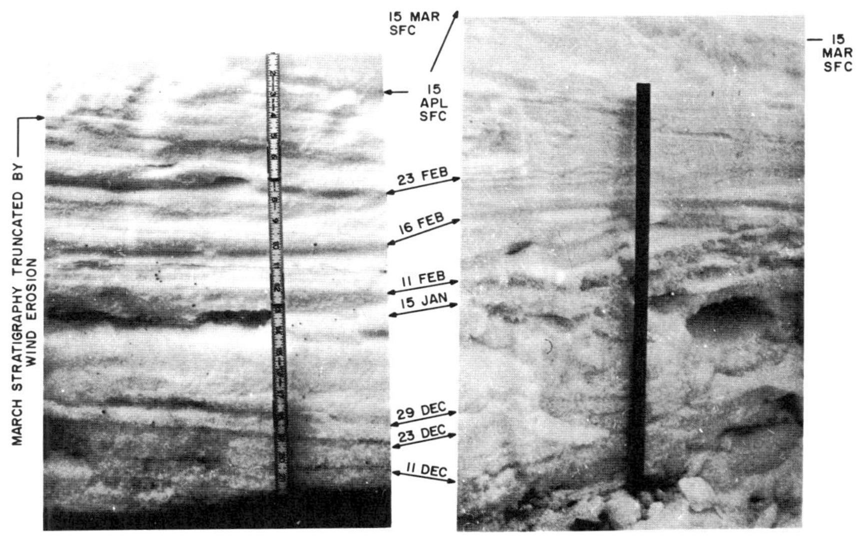

Mt. Pierce (summit) Cannon Mountain (summit)

b. Note the bedding, or laminations, of March snow at the base of Mt. Pierce and Cannon Mountain are seen to be truncated at the top of the lower left photo (summit).

Figure 3a and b. Comparative Stratigraphy

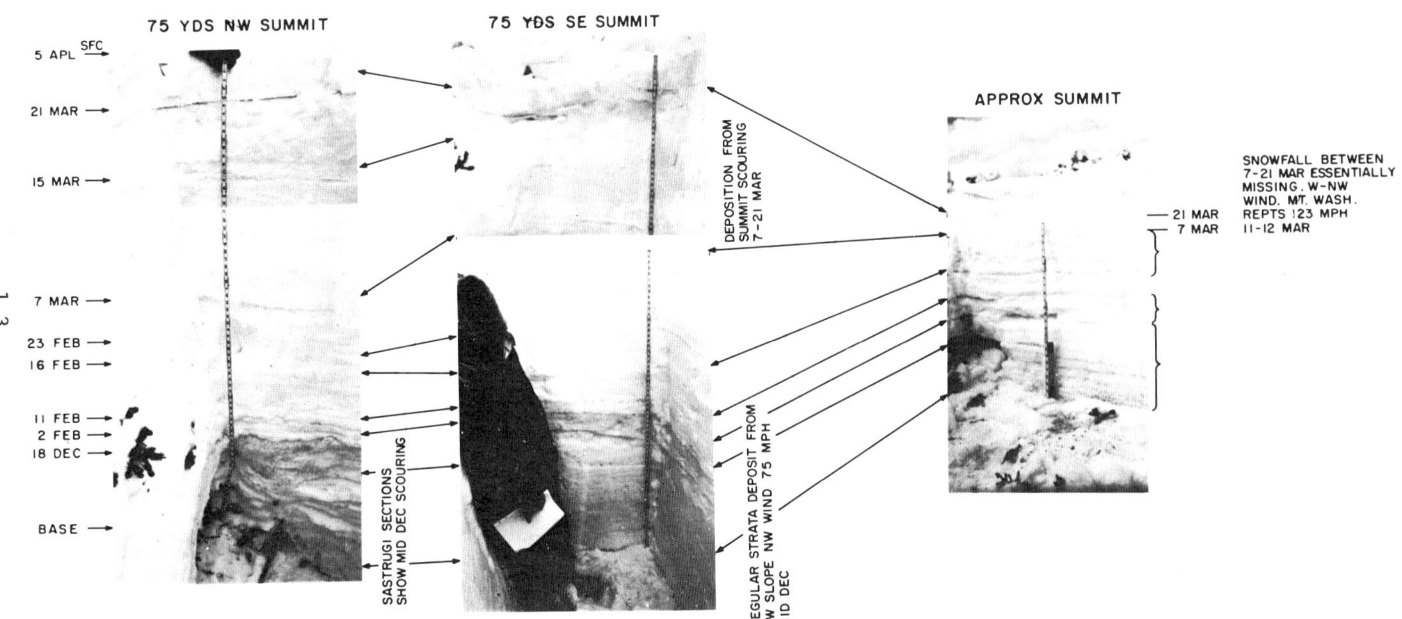

Figure 4. Summit Area Scouring and Deposition. Chain of pits across the summit of Mt. Pierce. Evidence of scouring and filling can be seen in the stratigraphy. Note sastrugi outline in left section of December snow.

northwest and southeast of the summit, respectively. Interpretation of the strata provides the following significant information:

1. Essentially all March accumulation (see notation to right in Figure 4) was blown off the top of the mountain approximately mid-month by west and northwest winds which were recorded as high as 123 mph on top of Mt. Washington, 12-13 March (three miles distant, 2,000 ft higher). Deposition of this eroded snow is evident in the SE pit as noted on the diagram. Northwesterly pits indicated the erosion was confined to the immediate vicinity of the summit.

2. The thaw which occurred in late December was effective at 4,000 ft elevation. A rain produced ice layer consolidated the December base and capped it.

3. High winds (75 mph, Mt. Washington) during the third week of December accompanied heavy snowfall and widespread drifting occurred over much of the general area. Sastrugi patterns (configurations or ridges formed on the surface of a snowfield by action of the wind) are visible in the stratigraphy for this period of time, although this is the only time throughout the winter that such conditions existed. (See left section of Figure 4.)

4. Regular (level, undisturbed) stratigraphy for all but the above mentioned period indicates a generally stable snowpack, apparently well retained by vegetative protection seen in the pit cross-sections (8-10 feet high conifers around the summit area).

5. Snow pit data revealed excellent snowpack, quantitatively as well as qualitatively. Depth of snow near the summit and in the vicinity of the possible terminal area was characteristically 8-9 feet on April 5. This compares with 5 feet on top of nearby Wildcat at the same point in time, and could be interpreted as a 12 percent longer ski season than that expected for nearby active areas.

SUMMARY

Two examples of the useful application of snow pit stratigraphy techniques have been presented. It has been demonstrated in the field that

1. Stratigraphic analysis can provide sufficient measurement sensitivity to aid in the measurement of small quantities of snowfall produced by artificially stimulated precipitation.

2. The stratigraphy contains a bank of information not easily available for remote areas. In the case studied, practical information gained was:

    a) Serious wind scouring was not a problem for the proposed upper lift terminal; it <u>was</u> a problem near the immediate summit area.

    b) Dwarf conifers near the immediate summit appeared essential to snowpack stability (and should not be cut).

    c) The greatest erosion threat was with northwest winds. Cleared trails should be cut across this wind, and glade skiing should be considered for upper areas exposed to these strong northwest winds.

    d) Snow quantity was considerably more than that in active neighboring ski areas. A 12 percent increase in operational days was likely.

REFERENCES

Anderson, V., 1960. A Technique for Photographing Snow-Pit Stratigraphy. J. Geophys. Res. 65 (3), 1080-1082.

Schaefer, V. J., 1942. New Methods for Preparing Surface Replicas for Microscopic Observation. Phys. Rev. 62, 495.

Vickers, W. W., in press. A Study of Ice Accumulation and Tropospheric Circulation in Western Antarctica. Meteorology Volume of IGY Research Series. Published by American Geophysical Union.

NEW CONCEPTS IN SNOW SURVEYING TO MEET EXPANDING NEEDS

Manes Barton
Head, Water Supply Forecasting Unit
Soil Conservation Service
U.S. Department of Agriculture
Portland, Oregon

ABSTRACT

Most of the streamflow of all major western streams originates from melting snow. Amounts of water and timing of delivery of that water is determined from snow surveys normally made at monthly intervals from February 1 to end of snowmelt. Until recently these data have been obtained gravimetrically by men traveling to the site on foot or in oversnow vehicles. Until recently the snow surveys have had a single purpose only--to provide a forecast of the streamflow to come from the measured snowpack.

Increased population and per capita water consumption demand better utilization of available water. This requires more accurate forecasts of amounts and timing of streamflow originating from snowpacks. It also demands "real time" data for use in forecasting for floods, avalanches, etc. Increased emphasis on conservation of natural resources has produced legislation that prohibits the landing of helicopters or the driving of oversnow vehicles on wilderness areas. This increases the difficulty of obtaining "real time" data from some of the heaviest water-producing snowpack areas.

As population pressures increase in the mountains of the West, snow survey data are being increasingly used in design of buildings, etc. Thus, a more sophisticated type of data is now being demanded of snow surveys.

New concepts and tools are rapidly becoming available for increasing accuracy of snow survey forecasts. Currently used methodology for snow surveying cannot be dropped and new, unproved methods substituted. An orderly blending of the new tools and methods with the current system is needed. After the new techniques have proven to be superior, the old can be phased out where appropriate.

This paper suggests ways for orderly transition from the current method of snow surveying to one that uses the most modern of new techniques. The suggested system is flexible so that new types of data acquisition systems can be added as they become available. The system should also provide the new data needed by all interests, e.g., recreation, structural design, flood and avalanche warning, etc. The system suggested in this paper considers new ideas of data needs and acquisition systems, including concepts, equipment, transmission of data, data reduction, and dissemination.

INTRODUCTION

During the early 1900's, Dr. James E. Church initiated snow surveys in the Lake Tahoe region on the Nevada-California border. From this initial work a westwide network of snow courses was established. The data obtained made it possible to forecast, several months in advance, the amount of streamflow that would be available for agricultural as well as other uses. From 1910 until the mid 1950's, principal emphasis was given toward expanding the network,

1.4

improving the manually operated measurement equipment and oversnow travel equipment, and increasing the accuracy of seasonal streamflow forecasts.

The increasing user demand in the 1950's for more frequent and detailed types of forecasts and data resulted in the development of new types of sensors such as isotopic snow gages (Smith et al. 1970) and snow pillows. It should be noted that these sensors could not have been developed had there not been extensive technical breakthroughs in the field of electronics which made it possible to design and operate sensors and their associated solid-state telemetry devices with only modest power requirements.

These new sensors provide more sophisticated data and markedly reduce the lag time between the measurement and use of the data in interpretative analyses and projections. In addition the data can be obtained automatically by remote telemetry techniques on a frequent and timely basis. Except for normal maintenance, these sensors and their associated telemetry equipment can be operated from the base station or central control center without sending personnel into the mountains.

From an environmental standpoint, these sensors can be inconspicuously located in remote wilderness or primitive areas. This removes the monthly impact of man and machines from these areas as is required by manual procedures.

The variety, quality and quantity of data which can be obtained at these automated sites far exceeds that at the typical manual site. A typical automatic site can be instrumented to measure snow water equivalent, precipitation, wind direction and speed, temperature, relative humidity, solar radiation, soil moisture, and many other hydrometeorological variables. In addition, devices such as the isotopic snow gage can be installed to provide expanded data concerning the internal characteristics of the in situ snowpack.

In contrast to the manual measurement techniques, the sensoring instruments used at the automated sites are nondestructive. That is, they do not require removal or destruction of snow or other measured element during the measurement process. However, as will be pointed out later, the shape or support for the sensoring instrument may cause some changes in the "natural" site and data characteristics.

It is the primary purpose of this paper to discuss the orderly transition from the present essentially manual system to one that blends the modern automatic techniques and systems with manual systems.

In the broad perspective, what is required is a "mixed" automatic and manual system that provides a broad variety and quantity of hydrometeorological data to a diversified group of federal, state, and private service agencies which in turn provide these data and their interpretations or projections to their user clientele. This "mixed" system must provide for an orderly transition from the present system. It must minimize obsolescence and/or retrofitting, maximize usage and cost sharing at data sites where there is a large commonalty of interest, and avoid automation when the benefit/costs and extent of data needs indicate that a manual site would be adequate.

It is important to recognize that, although one may generalize regarding the envisioned "mixed" system, in reality such a system consists of a group of subsystems intertied by computer and telecommunication facilities. This is due to the regionalization characteristics and authorities within and between the various service agencies. Although this aspect increases the complexity of planning with appropriate coordination, its impact can be minimized. In fact the need for coordination can result in maximizing the use and cost sharing of data sites as well as joint use of telemetry systems, computers, telecommunication facilities, and software programs relating to system control, reduction, interpretation, storage, and retrieval of data.

Mutual needs and interests of service agencies and their user clientele can generate financial and personnel resources needed for system modernization, and do it more rapidly and efficiently than if the various agencies proceed independently.

Let us now proceed from the broad concepts to more specific details. For this purpose the present and proposed telemetry system of the Soil Conservation Service will be described. Hopefully, this description will provide sufficient information to enable participants at this symposium to analyze and make contributions to the suggestions presented.

SOIL CONSERVATION SERVICE TELEMETRY SYSTEM

For the past several years, the Soil Conservation Service has been active in the design and development of a remote telemetry data acquisition system(s) in the western United States. The purpose of the system is to obtain snow and other hydrometeorological data from the remote and rugged western mountains on a daily or more frequent basis. (Table 1)

The development of telemetered automatic data systems, suitable for wintertime use in the rugged and remote snow-covered mountains of the western United States, is the outgrowth of a series of technological breakthroughs in several scientific fields. The principal and keystone development was the invention and subsequent successful operation of the snow pressure pillow by the Soil Conservation Service in the early 1960's (Beaumont, 1965).

Typically, a snow pressure pillow is fabricated from either neoprene rubber, sheet metal, or stainless steel. The material is welded or soldered into thin circular (6 to 12 feet diameter; 1.83 to 3.66 meters) or rectangular containers (4 by 5 feet; 1.22 by 1.52 meters). They are fitted with two outlets, one for filling and readout and one for releasing entrapped air. The pillows are filled with a methanol (methyl alcohol) and water solution. Dependent on temperature considerations and pillow capacity, which ranges from 5 to 200 gallons (18.9 to 757.1 liters), the solution varies from pure methanol to a 50/50 mixture of methanol and water.

A pillow or a group of interconnected pillows are usually installed flush with the ground surface. The weight of the snow on the pillow configuration causes an increase in the internal pressure of the pillow fluid. This pressure is measured by means of a pressure transducer.

Fluid pressure is fed from the snow pillow outlet by means of rubber, plastic, or metal tubing to the pressure inlet of the transducer. The transducer's electrical output, which is usually converted to voltage, is then injected after appropriate interface modification into a radio for transmission from the remote data site to the base station.

Table 1 Partial Listing of Present and Proposed Western Telemetry Systems*

Name	Location	Sponsors	Purpose
Soil Conservation Service	11 western states	Soil Conservation Service and numerous Cooperators	Hydro-met (see text)
Hydro-met Data Management System	Columbia Basin	Corps of Engineers Bonneville Power Administration Bureau of Reclamation U.S. Geological Survey National Weather Service U.S. Forest Service Environmental Protection Agency and numerous other cooperators including S.C.S.	Hydro-met To include a large data bank and operational control facility
Western Washington	Western Washington	U.S. Geological Survey and numerous other federal, state and private cooperators	Hydro-met To intertie with Columbia Basin Hydro-met system
California	California	California Department of Water Resources and numerous other federal, state and private cooperators	Hydro-met and flood control
Alaska	Fairbanks Anchorage	Corps of Engineers National Weather Service	Flood Warning and Hydro-met Alaska Pipeline
Atmospheric Project	Northern Utah	Utah State University Bureau of Reclamation and cooperators	Weather modification
Colorado River Pilot Project	Durango, CO. area	Bureau of Reclamation and cooperators	Weather modification
Salt River Project	Phoenix, AZ.	Salt River Valley Water Users Association	Hydro-met and supervisory control

* NOTE: This listing is not intended to be all inclusive.

In the past several years, integrated circuits, transistors, solar cells, thermoelectric generators and rechargeable Ni-Cad or Gel/Cell* batteries have revolutionized the use and application of radio transceivers at remote data sites. It is now possible to purchase "off the shelf" self-contained transceivers which weigh less than 20 pounds (9.07 kgs.), cost less than $1,000, and operate for up to 9 months, unattended.

Data are transmitted by such radios via repeater links, as required, to a central base station radio unit. There the data are converted to a digital form either by use of a small process control digital computer and/or predesigned solid state logic boards. The data can then be displayed and/or retransmitted by use of microwave or leased-line teletype circuits.

In a similar fashion--using appropriate sensors--precipitation, temperature, wind speed and direction, relative humidity, and a host of other hydrometeorologic parameters can be remotely sensed and telemetered.

In the design, installation, and operation of an automatic data network system, the quality and quantity of the hydrometeorological data is of primary importance. In our opinion, there are no more severe natural environmental conditions than those encountered in the snow-covered regions of the world. Except for surface use, similar sensing and telemetry equipment in an ocean environment would encounter only four principal problems: namely; pressure/temperature changes, moisture, corrosion, and signal transmission attenuation. If the station were at a fixed location, pressure/temperature would become a static design constant.

Likewise, space applications are primarily affected by temperature and 'g' forces at launch. The vacuum conditions in space are actually most ideal for equipment operation.

Compare this to the principal environmental conditions encountered at a typical data site in the mountainous West:

1. Temperature range . . . . -30° F to 100° F (-34.4^{o} C to 37.8^{o} C)
2. Snowfall depth . . . . . 0 to 25 ft. (0 to 7.62 meters)
3. Wind speed . . . . 0 to over 100 mph (gust)
(0 to over 160.94 kilometers/hour)
4. Relative humidity . . . . 0% - 100%
5. Precipitation forms . . . rime ice, hail and sleet, rain, snow
6. Physiographic factors . . 1000 to more than 12,000 feet elev. mountain barriers
(304.8 to more than 3,657.6 meters)
7. Other . . . . vegetal cover, soils, seasonal climatic changes

To meet these environmental conditions, the equipment must also be compact, reasonably light in weight, transportation shock resistant, modular, and easy to install, service, and maintain. In addition, sensors must be accurate, must not affect the environment or value of the parameter being sensed, and under static conditions must provide a constant reading (i.e. repeatability).

Obviously, cost, equipment availability and a long usable equipment life are equally important factors in such systems.

In the development of the present SCS systems, equipment and operational problems have occurred. This was not unexpected since such systems require extensive interfacing of a wide variety of equipment to operate under severe environmental conditions. At the time the snow pillow sensor was being proven feasible by onsite testing, most of us actively engaged in planning felt that telemetering of the data would be relatively easy. We based this opinion on the idea that telemetry equipment capable of enabling man to communicate from outer space to earth should be able to send data from point A in the mountain to point B 50 to 100 miles (80.47 to 160.94 kilometers) away in the valley.

This opinion did not prove out 100 percent in practice. We soon discovered that telemetry problems equaled, if not exceeded, sensor problems. For example, a high gain repeater omnidirectional whip antenna situated on a windswept ridge will, under certain ice-riming conditions, come to resemble a supersized, inverted snow cone. The result is signal attenuation, temporary loss of signal and, if extreme, total failure and collapse of the antenna and supporting mast. The solution now appears to be to enclose the antenna in conical fiberglass radome such as being used most successfully by British Columbia Hydrology and Power Co. in Canada (Gornall, 1971).

Intermittent or complete radio failure, due to such factors as cold-soldered joints, moisture, corrosion, and excessive standby power drain on the batteries, was common in the prototype systems. This type of problem has largely been overcome by improved design, extensive bench-testing prior to installation, and scheduling installations during the milder summer months.

The effects of rapid and extensive diurnal, ambient air temperature on radios, batteries and interface equipment have been mitigated by insulation in protective housing and/or by burial in the ground.

In the early stages of our program the field interface equipment, which modifies the data from the sensor/transducer for injection into the radio, was a source of telemetry problems and failure. Generally, the interface converts the data into a digital format. This often required a specially designed logic unit. The units were failure prone. The more recent interface logic units are better designed, more universal in application and much more reliable.

These are but a few examples of the telemetry equipment problems we have encountered and, to a large extent, overcome. Another type of problem in telemetering data is the limited number of radio frequencies available for hydro-met data transmission. This, coupled with the limited number of suitable repeater sites, has resulted in a multitude of difficulties. Close coordination between agencies, tone guarding, time and facility sharing, use of microwave facilities, and conversion to the less crowded UHF hydromet frequency band has been utilized to minimize this type of problem. Use of synchronous satellite repeaters or meteor burst telemetry may be the ultimate solution(s).

It would not be correct to leave the impression that the sensoring equipment used has been without fault. Shortly after we had determined butyl rubber snow pillows were good sensors for the measurement of snow water equivalent, we discovered that the fabric lost methanol due to capillary wicking and that under moderate to heavy snow load the welded seams often failed. Neoprene rubber properly laminated and welded has virtually eliminated this problem. Another solution has been the use of carefully fabricated galvanized sheet metal.

Snow pillow size has been determined to be an important consideration. The following minimums are recommended: 40 square feet (3.72 square meters) of pillow surface area for up to 30 inches (76.2 cm.) of snow water equivalent; 60 square feet (5.57 square meters) for 30 to 50 inches (76.2 to 127.0 cm.); 80 square feet (7.43 square meters) for 50 to 75 inches (127.0 to 190.5 cm.); and 120 square feet (11.15 square meters) for more than 75 inches (190.5 cm.).

Pillows should be carefully installed with close attention given to seating them firmly to the soil surface and providing adequate drainage. They should be located at least 15 to 20 feet (4.57 to 6.1 meters) away from any abrupt change in slope, from trees, buildings or other structures. If the snow is likely to be more than 10 feet (3.05 meters) deep, the distance should be increased to 1.5 to 2 times the maximum expected snow depth.

Snow bridged in the orifice of precipitation gages is a well-known phenomenon. Use of 12-inch (30.5 cm.) orifice, tall cylindrical gages charged with a special self-mixing glyco-meth (ethylene glycol and methanol) solution has produced highly reliable results.

The Soil Conservation Service currently operates a network of 1,650 manually read snow courses in eleven western states exclusive of California which has a network of over 200 snow courses operated by their Department of Water Resources. In addition, we have a network of 200 aerial markers, 196 soil moisture stations and 278 precipitation gages. At this time the Soil Conservation Service has operational networks in eight of the eleven western states, acquiring data from over 40 remote sites. Long-range plans call for the installation of over 500 remote data sites in the eleven western states served by our program.

In many instances telemetry installations will be made at existing snow courses having a long, historical record. Others will be new installations at more remote sites which are often inaccessible except in summer using horses or helicopter. Telemetry will obviate or greatly reduce the need for manual readings at those sites to be located at existing snow courses.

The manual network will be reduced from 1,650 snow courses down to 1,000-1,100. Thus, less field personnel will be needed to measure the manual network. It will also be possible to measure the manual network on a less frequent time basis.

Data from the telemetered sites will be read on a flexible schedule with design capability for hourly, semi-daily, daily, or less frequent readings. Base stations will be located in each of our eleven western state office headquarters. These state office base stations, operating under centralized or local digital process computer(s) control, will interrogate a regional network. Typically, this network will consist of the basins in their state and adjoining states that have the best radio telemetry paths via repeater links.

The process control computer(s) will: quantify data, check data for validity, and store data for later retrieval. Data will be exchanged or provided to other users by means of data phone or teletype circuits.

We propose to use larger U.S. Department of Agriculture and/or other federal agency ADP facilities to provide a "central exchange" as well as for processing data, validity checking, and storage facilities beyond the capacity of the smaller mini-computers.

Hydro-met data banks currently being developed by other federal and state agencies will be provided with the data collected by our system either through interconnections to our state base stations or our "central exchange."

Many other federal, state, and private agencies are individually and collectively designing and installing telemetered hydro-met data system networks in the western U.S. A partial listing is given in Table 1. We are working closely with these other groups in coordinated planning, data exchange, equipment requirements, frequency coordination site sharing and in the planning and use of satellites for both remote sensing and as repeaters.

SUMMARY

The need for snow and other hydrometeorological data becomes more pressing each day in the rapidly expanding economy of the West. Utilizing modern data collection equipment, telemetry, and computer techniques, it is possible to meet this need. The resultant "mixed" system will provide the data needed by the users in order for them to make effective and timely management decisions.

*Trade names are used solely to provide specific information. Mention of a trade name does not constitute a guarantee of the product by the U.S. Department of Agriculture nor does it imply an endorsement by the Department over comparable products that are not named.

REFERENCES

Beaumont, R. T., 1965. Mt. Hood Pressure Pillow Snow Gage. Journal of Applied Meteorology, October 1965.

Gornall, J. C., 1971. Advanced Techniques Overcome Severe Environment Encountered by Mobile Communications System. Proceedings of the Western Snow Conference 39th Annual Meeting, April 1971.

Mayo, L. R., 1972. Self-Mixing Antifreeze Solution for Precipitation Gages. Journal of Applied Meteorology, Vol. 77, No. 2, March 1972 pp 400-404.

Smith, J. L., H. G. Halverson, and R. A. Jones, 1970. Development of a Radioactive Isotope Profiling Snow Gage. Contract No. AT(49-11) 2773 Div. of Isotope Development USAEC.

LONG-RANGE GOAL AND INFORMATION NEEDS OF THE COORDINATED SNOW SURVEY PROGRAM IN CALIFORNIA

A. J. Brown
Coordinator, California Cooperative Snow Surveys
California Department of Water Resources

ABSTRACT

The "product" of the snow survey program is the water supply forecast. This forecast provides the water manager with advance information about the probable basin runoff. Good forecasts and wise decisions by the water manager results in many benefits. Poor information or careless action can result in severe damage and high losses.

The proposed long-range goal for the California Cooperative Snow Surveys Program is to develop the capability of "rapid forecasts." Rapid forecasts would be made at any time upon request from a cooperator. An appropriate time might be after a significant storm, or prior to a Board meeting. To achieve the end product of a rapid forecast will require on-site sensors capable of relaying data to a central computer. There is much development work yet to be done to establish reliability and standardization of all equipment components of a sensor system.

Other disciplines can help the snow surveys and forecasting activities in such areas as new ways to measure snow; new instrumentation that is accurate and dependable; newer and easier methods of field installations; new concepts on where snow courses and sensor sites should be located; better use of satellite communications; advice on how to provide capital to convert design ideas into practical operational systems; and suggestions on improving forecasting methods and techniques.

INTRODUCTION

This presentation assumes that pertinent technological advancements will become available and be continually assimilated into the California Cooperative Snow Survey Program, resulting in an evolution from our present status to ever more sophisticated and efficient methods of data collection and water supply forecasting.

This paper illustrates the framework of the present California Snow Survey program through a short historical sketch. Following the historical sketch is a discussion of the purpose of snow surveys and a proposed long-range goal which provide the background to suggest what inputs from other disciplines will be necessary to establish and achieve the long-range goal.

Snow surveying first became a practical method of determining future runoff shortly after the turn of the century, when it was introduced by Dr. James E. Church of the University of Nevada. The first formal snow courses were established in 1910 in the Lake Tahoe Basin. In succeeding years several California public utilities and irrigation districts began to make systematic snow surveys. Recognizing that standardization and coordination of the program by a single agency would enhance benefits to all, the present structure of the cooperative program was requested by participating agencies, and formalized in 1929 by the State Legislature. The function was assigned to the Department of Water Resources with authorization mandated through the State Water Code. Similar snow survey programs were initiated in the other western states during the 1920's and in 1935 the Federal-State Cooperative Snow Surveys were established and assigned to the U. S. Bureau of Agricultural Engineering, now the U. S. Soil Conservation Service.

California's cooperative snow survey and water supply forecasting program, administered by the Department of Water Resources, is based on various cooperative arrangements in order to answer the needs of both individual and collective agencies for the water supply forecasts used in planning water management throughout the State. The nearly 50 cooperating agencies involved provide funds or manpower or a combination to accomplish the field work and data collection necessary to support the forecasting activities. The Department provides the coordination necessary to maximize the benefits from the total effort. This includes providing sampling equipment to all cooperators, providing periodic training to maintain standardization, setting up and maintaining snow measurement courses and aerial snow depth markers, and assisting in the installation of automatic snow sensors. All snow data is received by the Department where it is analyzed and applied in the water supply forecasting schemes for each river basin. The resulting forecasts are published in DWR Bulletin No. 120, "Water Conditions in California", on the first of February, March, April, and May, and distributed to cooperators.

There are numerous other program activities that require the time and effort of both the Department and cooperating agencies, but I will detail just one more of these for the purpose of this presentation. This activity involves the field evaluation of new snow measurement concepts at our "Alpha Instrument Evaluation Site". This facility, located in the American River Basin at an elevation of 7,600 feet, has been used for the last nine years to verify the performance characteristics of various snow sensors and other data collecting devices. During this period there have been 40 automatic snow sensors installed operationally in California's watersheds, with installation techniques and components relying heavily on information gained from studies conducted at the Alpha site. Our experience at the Alpha Evaluation Site has helped trigger some of the long-range concepts proposed in this paper.

PURPOSE AND BENEFITS

The current worldwide shortages of food and energy have highlighted the need to conserve all natural resources. The need to conserve the natural resource of our available water supply is as critical in some areas as is the need to supply food and oxygen. The basic underlying purpose of snow surveys is to provide timely information about the snow water content and potential water supply within the various California river basins. This information, along with other essential data, provides the water and power manager with water supply forecasts that assist him in making meaningful decisions about the conservation of the available water supply.

Proper decisions about how to use the available water supply can result in substantial benefits, or there can be heavy penalties if improper decisions are made or if complete data is not available. For instance, substantial benefits can be realized in the generation of hydroelectric energy as decisions are made concerning the amount of fuel required by steam plants to cover the base energy loads. The cost of energy developed by the State Water Project at the Oroville-Thermalito Complex for instance is just over $2\frac{1}{2}$ mils per kilowatt. Comparable steam generation generally runs over 10 mils per kilowatt, and nuclear generated steam is estimated at about 7 mils per kilowatt. It's easy to see that even small shifts from steam to hydroelectric energy can save many dollars.

The irrigation use of water is another area where accurate decisions, although difficult, can be decisive in terms of money saved or lost. It's not just the farmer though who is interested. Bankers, and other businessmen have a vital interest in the continued health of agricultural productivity. So does the insurance man, the investor, the labor leader, and so forth. Nearly everyone, in fact, benefits from water in some way or another. Yet it is up to the water manager to make the crucial decisions about the regulation and use of water, so it is to him that our water supply forecasts are primarily directed.

How important the agricultural water manager's decisions are, particularly from the economic standpoint, can perhaps be illustrated by the system of allocation of irrigation water to members of the Kings River Water Association, in California's San Joaquin Valley. Many members of the Association, which is comprised of 28 public and private water districts and canal companies, are entitled to Kings River water only when natural streamflow is above 5000 cfs. At all other times irrigation water must be obtained by pumping from groundwater, at a much higher cost. Often, operational decisions must be made several months in advance, concerning such things as pumping schedules, acreages to be irrigated, and even which types of crops may be planted, based on specific crop requirements. Thus, any forehand knowledge of forthcoming water conditions can be of extreme importance.

There are other areas besides agricultural and hydroelectric uses of California's available water supply where benefits are realized. Another very important benefit accrues from flood control. When forecasts indicate high runoff from the snowpack, those responsible for flood control activities can take advance action to minimize peak flows and thereby reduce flood damages. Where flood damage is reduced, the savings due to less damage can easily be measured in the millions of dollars.

I haven't yet mentioned the domestic and industrial use of water. The City of Los Angeles, for instance, is keenly interested in water supply forecasts in order to match the City's use to the water available. In the case of Los Angeles they have several sources from which to choose, including ground water, local streams, the Colorado River, the Owens River, and the California State Water Project. Each source of water has its own cost, so the manager must balance his supplies wisely.

There are also recreational benefits from the water supply forecasts. Benefits accrue from being able to plan ahead for and count on such things as fish releases in the form of minimum streamflows, and from sustained reservoir levels during the summer months to enhance esthetics and such recreational uses of the reservoir as swimming, water skiing, fishing, sailing, and sun-bathing.

Associated with all these uses is a growing concern to maintain adequate water quality and to reduce waste. Workers in this field are increasingly turning to water supply forecasts to predict problem areas and to plan corrective actions. Emphasis on future water programs is from water development to preservation and enhancement of the quality of the existing water supply.

LONG-RANGE GOAL

The current trend in the California Cooperative Snow Surveys Program appears to be leading in the direction of what I'll choose to call "rapid forecasts". The normal monthly forecasts of April-July runoff already are being supplemented by weekly updatings in the Lower San Joaquin Valley stream basins of California. The near future (maybe within 5 to 10 years) may see forecasts updated after each major storm; or it may see the need for new forecasts within several hours after a request is received. The term "rapid forecasts" refers to these sorts of situations.

The end product of the snow surveys program is a water supply forecast. Our long-range goal, then, is to provide a "rapid forecast". But to achieve the goal will require an on-call data system that will provide hydrologic data, such as snow water content, precipitation, and temperature from automated sensors located in the watersheds. This "rapid forecast" system will require four kinds of basic equipment. First, it will require some sort of snow sensor. Existing snow sensors involve devices that are sensitive to pressure, such as the "pillow" or the "tank", or a device which involves a radioactive material, such as the Profiling gage developed by Dr. Jim Smith or the Integrating gage developed by JRB Associates of La Jolla, (see paper 6.9).

The second kind of equipment needed to produce "rapid forecasts" is instrumentation. In the case of pressure sensors, instrumentation is needed to convert changes in pressure into changes in electrical current, such as is accomplished by a pressure transducer. And along with the pressure transducer is needed a source of constant low power that will provide a reliable source of power throughout the snow data gathering season of October through May.

Third, the "rapid forecast" system will require telemetry. The initial installations during the past decade have depended upon the standard line-of-sight, VHF radio repeater. This has the obvious limiting disadvantages of not being able to transmit from desirable sensor sites located down in meadows or valleys that are cut off from a direct line of sight to a mountain top repeater. This disadvantage may soon be overcome, though, through the use of satellite communications, such as the NOAA system, (see paper 5.1).

The fourth kind of equipment is the computer. A central computer could serve as the data bank, receiving information from the various field sensors, and then making this information available to any of the cooperating agencies in the snow survey program. Communications between the central data bank and cooperating agencies in the future most probably would be from computer to computer. Nor should the possibility of COM - computer operated microfilm be overlooked. The central computer would also contain the forecast routines and be capable of providing nearly instant revisions to the previous forecast, as updated data from the sensors is fed into the computer, (see paper 3.7).

All of this equipment, the sensors, telemetry, instrumentation, and the computer, working together to provide "rapid forecasts" would have to meet at least the following two requisites. First, reliability. Sensor sites most likely will be located in remote areas and will be unattended throughout the snow season, except for an occasional visit to take control samples. The sensor equipment, therefore, must be able to withstand extremes in temperature, wind, and wetness, and continue to operate. Computers, satellite access equipment, and associated equipment must be such that uninterruptable and rapid use by cooperators is assured.

Second, standardization. It's advisable to have each of the sensor sites meet some sort of standardization in order to achieve savings to all cooperators in the program. There would be savings in the initial purchase of instrumentation and appurtenances, in the ease of installation, and in easier and faster maintenance and upkeep. In the case of the possible need for portable (seasonal) data sensors in designated Wilderness Areas, standardization would allow cooperators to integrate their efforts in placement and removal of instrumentation. Related training of personnel would also be easier and produce a common pool of talent that would serve everyone's needs to the best advantage.

NEEDS

So far this paper has presented information on how the Snow Survey Program started, what it's purpose is, the benefits to be derived from it, and suggests a long-range goal for the program. At this point I would like to list some specific areas where the various disciplines represented at this symposium might devote their talents. The following examples are some of the more pressing aspects of the snow survey and water supply forecasting activities which need further study and better definition.

We need, for instance, to explore new concepts in ways to measure snow water content. The present standards of physically weighing a snow sample (using scales and snow tubes) taken at points along a preselected course; or measuring the pressure that water exerts upon a fluid filled membrane, or "pillow", are yielding satisfactory results, but there are undoubtedly better ways yet to be developed. Some examples of current investigations involve the use of electromagnetic waves, (see paper 7.6), microwaves, (see paper 7.7), radar waves, (see paper 4.3), infrared photography, (see paper 5.4), and spacecraft imagery, (see paper 5.10). It's not improbable that the principles involved in other basic phenomena such as thermodynamics, heat transfer, stress-strain relationships, and so forth might yield insights into better ways to measure water content of the snow in future years. It's even not impossible to believe that methods might be developed in the future which would detect water content even before the snowpack is deposited!

We need to develop new instrumentation and adapt existing miniaturized instrumentation that would permit compact, transportable, yet dependable data installations. This instrumentation must be able to withstand continuous exposure, be able to complete its responsibilities by measuring accurately and consistently, and be able to do this throughout the winter season using very little power. The transportable feature of instrumentation might grow in importance if pressures from other societal goals become stronger. In California the primary "breeding area" of the runoff producing snowpack is in a north-south belt of the Sierra Nevada above the 5,000 foot elevation. Within this "belt" are the following Parks and Wilderness areas: Warner Mountains, Thousand Lakes, Caribou, Lassen National Park, Desolation Valley, Mokelumne, Emigrant Basin, Hoover, Yosemite National Park, Minarets, John Muir, Sequoia and Kings Canyon National Parks, plus numerous proposals pending.

A glance at the State's map quickly reveals that these Parks and Wilderness areas represent the catchment of approximately 25 percent of the California snow zone, and approaching 50 percent of each winter's snowpack. Compatability with wilderness objectives may well require some data gathering stations and attending instrumentation to be transportable, that is, put together in the Fall and taken down in the Spring. It is not wise, however, to assume that the solution ends there. Recent activities in gathering information on the areal extent and rate of summer precipitation in the upper elevations of the Sierras strongly suggests that summer precipitation can have an economically important effect on sustaining summer flows after the snowmelt process is completed. As this recent work continues it may provide results which will justify continuation of gathering hydrologic data throughout the entire water year, thus negating the transportable feature.

We need to consider the ease with which hydrologic data instrumentation can be installed under field conditions. Easy installation methods are very important. For instance, the standard 12-foot snow sensor "pillow" uses about 200 gallons of fluid - a special mixture of methanol and glycol that has a specific gravity of 1 - whereas the metal membrane snow sensor (SCS "tanks") uses about 20 gallons. This represents a significant saving in weight, making it far easier when packing in the fluid to a proposed site, or having it flown in by helicopter. Connections, and other facets of installation should be as simple as possible, yet still retain reliability. Remote sensor installations should be simple, compact, and of modular construction where possible.

We need to consider the growing use of land. Land which was once remote and unused served to protect the snow course from unnatural disturbances. Now, even the remotest areas are being overrun by people who by their actions are disturbing the snow courses. Experiments have been conducted for the past three years to test and evaluate the effects of snowmobile traffic over a snow course. The results so far are not conclusive, but do tend to indicate there is some effect, even though the degree of effect is not yet certain.

As greater numbers of people invade the "back-country" and the higher elevations of the Sierra, which historically has been the home of snow courses, it has been suggested that the criteria to locate new snow courses be reevaluated. For instance, there is strong argument to locate new installations nearer roads that are kept open during the winter rather than to place an installation more inland and be so inaccessible. Those who favor this approach point out the increasingly higher costs of travel by helicopter, or by oversnow vehicle, plus the additional risks to personnel due to these forms of travel. They also point out that if the new installations were to involve snow sensors then the chances would be minimized that the haphazard movements of winter travelers would create a disturbance to the conditions being monitored by the sensor. As more snow sensors are installed, the forecaster would need to rely on data obtained from fewer snow courses, thus tending to minimize the effect of travel across a snow course. In argument against this approach is the liklihood that a suitable site may not be possible near a road which would satisfy the forecaster's need of getting data representative of the areal and elevation extent of a river basin. I suggest that the need to carefully evaluate placement of installations in remote and inaccessible areas will become more important in the future as complete river basins are automated and as operation and maintenance costs increase.

We need to continue to consider the advantages of satellite communications, and to actively convey our growing needs as water resource managers to those involved in designing satellite communication systems so that our needs to have hydrologic data will be kept in mind as designs progress. Many desirable sensor sites have not been developed because of the line-of-sight requirement of current telemetry networks. Satellite communications would not only remove this restriction but lend itself to using a computer to monitor, record, assemble, and transmit whole series of hydrologic data to a user. The user could get composite printouts of snow sensors, precipitation gages, temperature, wind, and so forth for the whole basin and adjoining basins rather than getting each type of data piecemeal from several sources.

We need to enlist the aid of the economics experts to help suggest ways of providing capital to convert design ideas into practical operational equipment. A long-range goal has been suggested in this paper that if adopted would establish a statewide network of automatic snow sensors with the data transmitted to a central point in the State where rapid forecasts could be made and disseminated to the various users. The establishment of such a network would represent a significant capital investment. Most of the agencies cooperating in the California Snow Survey Program have limited operating budgets, and the governor of the State certainly is trying for economies in state spending! But if the goal were to be obtained it would result in benefits and savings far in excess of the cost. The very crises being experienced by the world now, such as the energy crisis, population crisis, food crisis, and the pollution crisis all have water as a common thread with water playing a vital role in the possible solution of each. The economist could well serve the Nation, the Western States, and California by focusing his particular and essential talents to this financing problem area. No single agency can carry the burden, but the uniqueness of the cooperative nature of the State's snow survey program suggests that collectively a solution can be worked out that would provide the funds needed on a basis that would be fair to all the agencies involved.

And last, we need to concentrate on improving our forecasting methods and techniques. The forecasting technique that has been used for many years by the California Cooperative Snow Survey Program is the "Index Method". Each major river basin whose runoff is to be forecasted has several discrete "points" in the form of snow courses, where several months each winter the water content of the snowpack at that location is measured. After several years a statistical relationship is built up which relates the water content at that point and the other points in the basin at any given time of measurement to the runoff to be expected from that basin. Thus, the collective measurements of water content at several snow courses within and adjacent to a basin provides an "index" to the anticipated April-July runoff.

The actual forecasting procedures take into account other factors, in addition to water content of the snow, such as antecedent precipitation and cumulative runoff from the beginning of the Water Year up to the date of the forecast. Precipitation in future of the forecast is assumed to be equal to the historic normal for those months in the forecast period beyond the date of the forecast.

An approach was first attempted last winter season of substituting long-range meteorological forecasts in place of the assumed "historical normal" for future precipitation on a trial basis for a few river basins. These were the Kings, San Joaquin, Kaweah, Tule, and Kern Basins and was known as "Project Hydrospect". The Project made hydrologic forecasts in December, 1972 for the 1973 April-July runoff from these basins based on long-range meteorological forecasts. The forecasts made in December, 1972 have since proven to be remarkably close to the actual runoff. Project Hydrospect, as a test exercise, has another year to run. Those closely involved with the Project are the first to caution that the apparent first year success may not truly be indicative of the longer term effectiveness of the Project. There is keen interest in Project Hydrospect, however, and many agricultural and other users are watching its results this year very closely.

There is also a growing trend towards the use of "hydrologic models" which take advantage of the computer's capacity to store and manipulate vast amounts of data. The Cooperative Snow Survey Program has two such models operational, one for the Kings River Basin and one for the San Joaquin River Basin. These were developed in 1969 through 1971 jointly by the Department of Water Resources and the consulting firm of Sierra Hydrotech and have been used successfully the past two snowmelt seasons. The Federal-State River Forecast Center in Sacramento, California has recently been developing a general hydrologic model which may be adapted to any major California river basin. The River Forecast models use historic information on precipitation, temperature, and snow observations to synthesize a daily record of snow accumulation and melt suitable for streamflow simulation purposes.

But the challenge remains, even with fairly sophisticated modeling techniques and advanced uses of the computer, to search the forefront of mathematical developments to see if there are any new techniques which could be adapted to forecasting runoff from snowmelt.

The California Snow Surveys Program conducts many activities where imaginative and innovative talents could be applied to achieve increased proficiencies as new and more sophisticated technologies become available. This paper has limited its emphasis to those activities which involve on-site data sensing, data translation and transmission, and the processing of hydrologic data to prepare and disseminate rapid forecasts. All activities of the Snow Surveys program, however, lend themselves to an interdisciplinary approach to problem solving. The disciplines, therefore, are encouraged to direct their talents toward the development and perfection of new equipment, concepts, and procedures which will enhance the runoff forecasting process.

ON PREDICTING WATER RUNOFF FROM A SNOW COVER

S.C. COLBECK
U.S. Army Cold Regions Research and Engineering Laboratory
Hanover, New Hampshire

ABSTRACT

The simultaneous development of remote sensing techniques and the theory of water flow through snow should help improve methods for predicting water runoff from snow-covered watersheds. The flow through snowpacks is considered to occur in two layers with different characteristics and mathematical descriptions. First, vertical seepage through the upper layers of unsaturated snow occurs followed by flow through a saturated layer at the base of the snowpack. Measurements of density and liquid water volume as functions of depth are necessary to provide an initial condition for calculation of flow through the unsaturated layer. In addition, the water flux across the surface must be known either from direct measurements or forecasts of meteorological parameters. Knowledge of the porosity and permeability of the saturated layer must be obtained from direct measurements of the ice content in order to calculate flow through that layer.

For short term forecasts, it may be sufficient to know the thickness of the saturated layer, but for forecasts of more than a few hours, the influx from the overlying snow must also be known. Of the measurement techniques reviewed, only measurements of the dielectric properties by remote sensing can provide the necessary information. Such a system is possible with current technology although much work remains to be done before the scheme discussed here could be operational.

LIST OF SYMBOLS

C_h wave speed or rate of movement of a value of h

C_{S^*} wave speed or rate of movement of a value of S*

h thickness of the saturated layer

k permeability of the pore space

q flow rate of water through the saturated layer

S_+ maximum value of S* at the wave front

S_- minimum value of S* at the wave front

S* effective water saturation or fraction of mobile water

U_w volume flux of water through the unsaturated layer

α constant

θ slope angle

ξ position of the wave front

ϕ porosity

ϕ_e effective porosity or pore fraction available for mobile water

1.6

INTRODUCTION

Much effort has been expended on studying the metamorphism of snowpacks and the routing of water through snowpacks and glaciers. Early in the melt season the presence of large quantities of water cause "ripening," the phenomenon which prepares the snow cover for the ensuing runoff. Water flows rapidly through a mature snowpack in a well established manner following paths which are partly determined during the ripening process. While much descriptive information of this type has been obtained from observations of wet snow, a more fundamental understanding of the physics of water flow through snowpacks, including the mathematical relations necessary for use in simulation models, is needed for the practical purposes of predicting water runoff (Anderson, in press; Viessman, 1970, p. 76).

Many years of work on flow through other porous materials has provided a basis for understanding the movement of water through snow although, since the snow matrix contains all three phases of water, wet snow has many special problems of its own. Flow through homogeneous snow has been described in terms of the gravity drainage of water displacing air from an unsaturated medium; the movement of the water introduced at the surface can be predicted using fairly simple mathematics (Colbeck 1972). This work has been extended to two-dimensional flow and, where two flow regimes occur (see Figure 1), water seeps through an unsaturated layer lying over a saturated layer which in turn overlies the diverting horizon (Colbeck, in press). Although the thermodynamic equilibrium between the three phases of water has a significant role in the evolution of snowpacks, these studies suggest that water runoff from wet snow of known properties can be adequately treated using the standard procedures developed for other materials.

As more sophistication such as inhomogeneous snow, irregular surfaces and infiltration into the underlying materials is added to the modeling of water flow, more accurate descriptions of water runoff can be made. Each advance in the theory provides new insights into the phenomena of flow and increases our understanding of the response of the system to the measurable parameters. However, for the practical purposes of forecasting runoff, it may be more important to emphasize the development of the measurement techniques which are required for the full utilization of the theory rather than emphasizing further development of the theory. It is appropriate at this time to consider what measurements would be required for a forecasting scheme and what capability exists for making those measurements. It is desirable that the development of measuring systems and theory proceed in complimentary stages.

The purpose here is to describe the capability of the existing body of theory and to examine the nature of the input information necessary to fully utilize these abilities. The simplest case of vertical percolation through homogeneous snow is examined first and then two-dimensional flow through snow overlying a boundary is discussed. The information required to forecast the water flow in each of these flow regimes is described. It is shown that existing measurement systems are inadequate to supply that

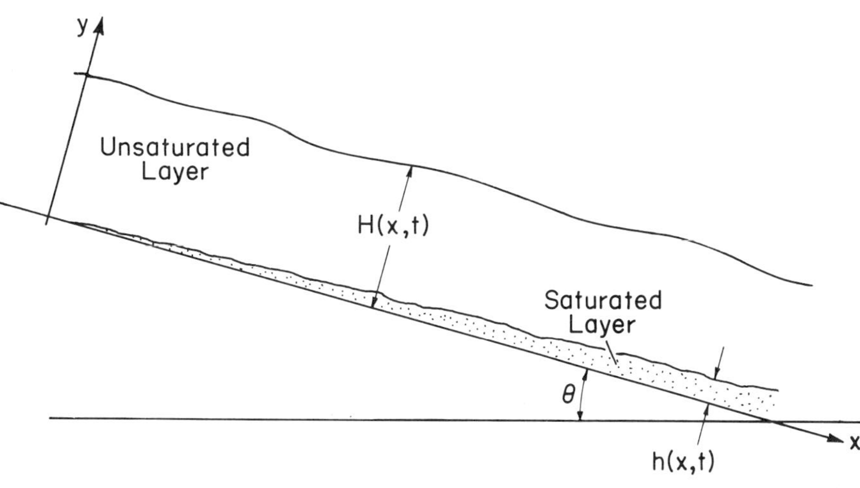

Figure 1. An idealized view of a snowpack showing the two regions of flow - unsaturated flow through the upper layers and saturated flow over the base.

data. Further developments in remote-sensing systems are necessary before adequate information can be provided to accurately predict water runoff from a snow cover.

VERTICAL PERCOLATION THROUGH SNOW

Water flow through snow is inevitably two-dimensional in nature yet the dominant mode of flow is often the initial step — vertical percolation through the unsaturated layer (see Figure 1). The flow through this layer is significant because the wave speed for flow is about two orders of magnitude less than the wave speed for flow through the underlying saturated layer (Colbeck, in press). As a consequence, the travel time through the relatively short path length of the unsaturated layer largely determines the delay in runoff from a snow cover. For a shallow snowpack lying on a gentle slope, the flow along the base dominates the timing of the runoff, whereas for a deep snowpack lying on a steep slope, it is probably sufficient just to calculate the flow through the unsaturated layer.

Flow through the unsaturated zone is difficult to analyze because of the variability of water saturation in time and space. If the degree of saturation can be determined directly, the important variables such as volume flux and wave speed can be calculated from the existing theory of water percolation. The volume flux of water (u_w) varies as the third power of the effective-water saturation (S^*), or (Colbeck and Davidson, in press)

$$u_w = \alpha k S^{*3} . \qquad (1)$$

The volume flux is very sensitive to the volume of water present, i.e. the area of water film available for flow determines the flow rates during the gravity drainage of unsaturated snow and pressure gradients have relatively little influence. This nonlinear relationship between flux and concentration greatly increases the error in the calculation of volume flux resulting from measurement errors of the quantity of liquid water present; further, the difficulties are increased by the fact that the volume of water present during the gravity drainage of snow is normally quite small and therefore difficult to measure.

The effective-water saturation (S^*) also determines the wave speed (C_s^*), or the speed at which a value of S^* moves through the snow.

$$C_s^* = 3\alpha k \phi_e^{-1} S^{*2} . \qquad (2)$$

The distribution of S^* can be predicted by combining Equation (2) with the appropriate boundary and initial conditions. Once the distribution of S^* is known throughout the unsaturated layer, Equation (1) can be used to calculate the distribution of the volume flux. The calculation of wave speed directly from measured values of water flux, while a theoretical possibility, is unlikely in practice since most measurement methods are capable of detecting only the liquid volume and all methods capable of measuring volume flux directly are limited to use as point measurements. This is unfortunate since the error in calculating wave speed from the initial and boundary values expressed in terms of S* is twice as large as the error in the measurement of effective-water saturation itself. Once the wave speed has been used to calculate the distribution of effective-water saturation, the error in calculating the distribution of volume flux is three times as large as the uncertainty in the calculated distribution of effective-water saturation. Clearly if meaningful predictions of water flow are to be made, very accurate measurements of the initial and boundary values of liquid water must be made.

Unfortunately, much more input information is necessary to make predictions of water movement through the unsaturated layer. Each necessary parameter is discussed along with a suggestion for its evaluation. The effective-water saturation must be calculated from a measurement of the actual volume of liquid water present and a determination of the irreducible-water saturation (S_{wi}). The effective-water saturation is a measure of the water available for flow whereas the irreducible-water saturation is a measure of the water held in place as adsorbed or capillary water (Bear and others, 1968, p. 31). It would be very difficult to distinguish between the capillary and free water in a measurement of the liquid water present in snow, hence it is fortunate that the irreducible-water saturation is relatively insensitive to textural variations (Harris and Morrow, 1964). The irreducible-water saturation probably lies within a narrow range of values because the metamorphism associated with the ripening process causes rounding of the individual grains and redistribution of the grain sizes thus producing a snow cover with sphere-like grains with a limited range of diameters. Clearly much experimental work on this topic needs to be done but, for our purposes here, it is sufficient to say that the irreducible-water saturation can be determined with sufficient accuracy from basic knowledge of the behavior of liquid water in snow and knowledge of the porosity of the snowpack.

The effective porosity (ϕ_e), which is a measure of the pore space available for fluid flow, must be known in order to use Equation (2). It seems unlikely that a direct measure of this parameter is possible but it can be calculated from independent measurements of density and liquid-water volume. The permeability (k) of the total pore space of the snow must be known, and again, a direct measurement of this parameter is unlikely. This problem has been handled for other porous materials by obtaining a simple relationship between permeability and the measurable parameters. Shimizu (1970) found that permeability increases as the square of the grain size but decreases exponentially with the density of dry snow and Kuroiwa (1969) related permeability to porosity and grain size. The relationship is somewhat complicated for dry snow since the texture can vary widely but, once wet snow has metamorphosed, a simple relationship between porosity and permeability should give sufficiently accurate results over the narrow range of grain sizes in a mature snowpack. Again experimental observations are necessary to test this hypothesis but, since no direct method of measuring the permeability of wet snow has been found, this may be difficult to accomplish.

The minimum information which must be furnished to calculate all of the necessary parameters is snow density and liquid-water volume. Porosity can be calculated directly from this input data and irreducible-water saturation and permeability can probably be determined to within an adequate limit of accuracy. The other significant parameters and variables can be calculated in turn however, until more experimental evidence is available, the feasibility of using this approach is uncertain. An alternate method of obtaining the necessary information about permeability and porosity is desirable and one of the possible methods has been used successfully in both homogeneous snow (Colbeck and Davidson, in press) and natural snow covers (Colbeck 1972). This method uses a combination of measurements and theory to provide the necessary information about the snow itself. For example, Figure 2 shows idealized profiles of effective-water saturation as functions of depth at the same location; these profiles might represent the distribution of S* several hours apart on a day of fair-weather melting. The rate of propagation of the wave front,

$$d\xi/dt = \alpha k\phi_e^{-1}(S_+^2 + S_+S_- + S_-^2 \qquad (3)$$

where S_+ and S_- are the maximum and minimum values of S* respectively, can be calculated from its movement during the time increment $(t_2 - t_1)$. For depths greater than 1 m, the quantity $(S_+^2 + S_+S_- + S_-^2)$ is approximately constant hence an average value of $(k\phi_e^{-1})$ can be calculated from the information given on Figure 2. Once this quantity is known, Equation (2) can be used to calculate the further movement of this waveform throughout the remainder of the unsaturated layer. However, extrapolation of the calculated value of $(k\phi_e^{-1})$ relies upon the homogeneity of the unsaturated layer while $(k\phi_e^{-1})$ is likely to vary somewhat through most natural snowpacks. When these observations can be made over most of the depth of the unsaturated layer, an average value of $(k\phi_e^{-1})$ can be calculated which is sufficiently accurate for further calculation of the propagation of waveforms of effective-water saturation through that same snowpack. An independent determination of permeability would have to be made in order to use Equation (1) for calculating the volume flux of water from the predicted values of effective-water saturation. This could be most easily accomplished by making independent determinations of the density and liquid-water volume, calculating the porosity and then deducing the value of permeability from the known value of $(k\phi_e^{-1})$.

In a shallow snowpack where wavefronts are not well developed, the propagation of individual values of S* could be used to calculate $(k\phi_e^{-1})$. From Equation (2),

$$k\phi_e^{-1} = S^{*-2}\, C_{S*}/3\alpha \qquad (4)$$

where for any value of S*, C_{S*} is equal to its movement divided by $(t_2 - t_1)$. This can be calculated for a large number of values of S* from the trailing edge of a wave and, if the results are plotted as $\ln C_{S*}$ versus $\ln S^{*2}$, the data should form a straight line whose intercept is $\ln(3\alpha k\phi_e^{-1})$. Once again further knowledge of the dependence of k and ϕ_e on the porosity and grain size of the snow is necessary.

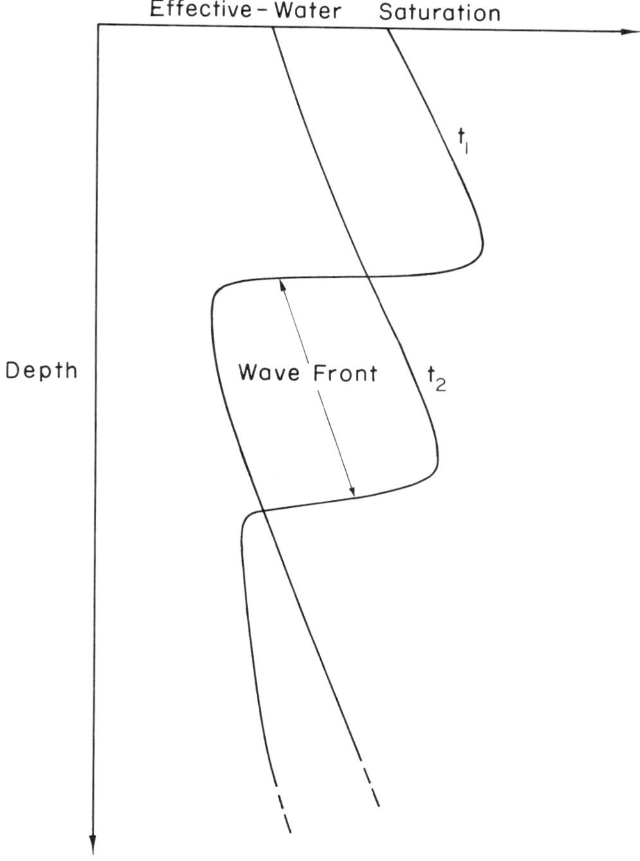

Figure 2. An idealized view of flow through the unsaturated layer. Two sequential profiles of saturation are shown at one location. Note the movement of the wave front during the interval of a few hours.

The minimum amount of data which must be collected in order to make meaningful predictions of water flow through the unsaturated portion of a mature snowpack are the distribution of liquid water and density at some time plus the flux across the surface of the snowpack. These measurements are sufficient to specify the initial conditions, boundary conditions and, with some knowledge of the distribution and grain shape for a ripe snowpack, the required values of irreducible-water saturation, effective porosity and permeability. The distribution of effective-water saturation can then be calculated from Equations (2) and (3) and the variable of real interest, the volume flux of water, can be calculated from Equation (1). While much of this approach to predicting water movement must yet be tested with a natural snowpack, the principles are well known. The difficult part of making an operational system from the scheme outlined here is the development of nondestructive techniques for making the necessary measurements of water volume and density. The current status of these systems is discussed later.

SATURATED FLOW ALONG A BOUNDARY

An idealized view of two-dimensional flow through a snowpack is shown on Figure 1. When infiltration into the underlying soil is negligible, all of the water passing through the unsaturated layer is discharged from the snowpack after moving through the saturated layer. Assuming that flow through the unsaturated layer can be adequately treated by the methods described in the previous section, the prediction of water flow through the saturated layer is relatively straight-forward. The calculation is simplified by the fact that wave speed is approximately constant, or

$$C_h = \alpha k \theta \phi^{-1} \tag{5}$$

The flow rate of water is directly proportional to the height of the saturated layer (Colbeck, in press), or

$$q = \alpha k \theta h \ . \tag{6}$$

Only the thickness of the saturated layer (h) varies in time and space and this can probably be measured with sufficient precision to enable an accurate calculation of the discharge of water from the snowpack.

The determination of permeability and porosity is simplified by the rapid metamorphism of snow when saturated (Wakahama 1968). Grain growth proceeds rapidly up to sizes of 1 to 2 mm therefore the saturated layer should be composed of fairly uniform grains whose properties can be readily deduced. A direct measure of the ice content or "filling factor" of this layer may be required. From this measurement the porosity can be calculated directly and the permeability can be accurately calculated from an assumed grain size. It is unfortunate that a direct measure of the bulk density is inadequate for this purpose. Because of the small difference between ice and water density as well as the uncertainty in the volume of trapped air, only a direct measurement of ice volume will suffice.

Assuming adequate knowledge of the slope (θ), permeability (k), porosity (ϕ) and influx from above, the equations describing water flow through the saturated layer can be solved in a system of coordinates which moves downstream at the wave speed (Colbeck, in press). The solution is given in terms of the flow rates. In addition, if the thickness of the saturated layer can be measured directly, the water flow can be calculated without considering the influx from above. This prediction can only be made for a short period of time about equal to the characteristic travel time through the length of the saturated layer. Unfortunately the movement through the saturated layer does not necessarily control the water runoff but rather the much slower movement through the unsaturated layer often determines the time delay. Emphasis should therefore be placed on development of the measurement systems necessary to make predictions of water flow through the upper layers. By the nature of the required information, however, any system capable of making those measurements could also supply a significant amount of information about the saturated zone.

MEASUREMENT SYSTEMS

Much useful information is provided by making point measurements of snowpacks and using these measurements in schemes for predicting snowmelt runoff. Point measurements are very useful in the scheme proposed here for routing water through the snow cover. However, point measurements are limited by the fact that they may not be representative of aerial averages and the in situ instrumentation may interfere with the processes of ripening and runoff. The inhomogeneous nature of a snowpack requires that measurements of water volume and density should be made over a sufficient area that irregularities in the flow field are averaged. As a first approximation to the size of this area, its characteristic length should be about equal to the thickness of the snowpack.

Remote measurements of wet snow can best satisfy the data collection requirements of the predictive scheme being suggested here and this discussion will be limited to those techniques. Further, for full utilization of the theoretical developments, the complete profile of parameters must be measured throughout the snowpack and not just on the surface. The measurements of surface fluxes alone would be adequate if independent knowledge of the other parameters could be obtained but this seems unlikely. Therefore techniques such as thermal infrared sensing which are limited to detecting the surface wetness of snowpacks and gamma-ray absorption which is limited to detecting the water equivalent of snowpacks are of little value for the present purpose.

Only the microwave techniques (Meier and Edgerton, 1971; Linlor, in press) appear to have the capability of collecting the necessary information. These techniques are promising because, if both the dielectric loss factor and the effective dielectric constant of wet snow can be measured, <u>both</u> the liquid water <u>and</u> ice contents of the snow can be determined. Indirect measurements of these two parameters of the snowpack provide all of the necessary information required to use the scheme described here. Clearly much work remains to be done before a system capable of making these measurements as a function of depth in a snowpack is operational. Progress is being made in this direction, however, and the theory necessary to interpret measurements of the dielectric constants as a function of depth already exists (Linlor, pers. comm.). Recent measurements of the dielectric constants of wet snow and glass beads as a function of liquid water content provide the necessary relationships for use with this approach (Sweeny, unpublished).

DISCUSSION

The prediction of water runoff from snow covered watersheds is of considerable importance. To date the predictive methods have been largely statistical. Recent developments in the routing of water through snowpacks and in the techniques of remote sensing suggest that the forecasting of water flow through snow can be greatly improved. An examination of the existing capabilities of the theory and the measurement systems is given at this time in order to provide direction for the future development of each.

The flow of water through a snowcover generally occurs in two modes - vertical seepage through an unsaturated layer and subsequently, saturated flow along the base of the snow. These two flow regimes are significantly different and require separate analysis. The flow through the unsaturated layer is generally quite slow and for deep snowpacks, the major delay in runoff is caused by flow through these upper layers. The distribution of liquid water as a function of depth must be known at some time to provide an initial condition for the start of a routing calculation. The water flux across the upper surface must be known either from direct determinations of the liquid volume, measurements of the appropriate meteorological parameters or forecasts of future parameters. In the exceptional case of a deep snowpack through which the movement of an existing wave front is being predicted, the surface fluxes may be neglected once the wave front has established itself. The profile of snow density must also be measured in order to calculate such parameters as permeability, porosity and irreducible-water saturation. While these calculations must rely on assumed grain sizes and shape, it is likely that sufficient accuracy can be obtained.

Some uncertainty exists about the distortions of the flow field caused by features like drainage channels and ice layers. Rapid decomposition of ice layers occurs during heavy runoff and it is likely that these features have little effect once heavy runoff begins. The net effect of these structural features is to alter the permeability and porosity of the snowpack. If averaged values of these parameters are taken over a length at least as large as the thickness of the snowpack, the effects of these structural features can probably be adequately treated. An indirect measure of the permeability of a snowpack several meters in depth (Colbeck, 1972) showed that the natural snowpack with ice layers had one-half of the permeability of homogeneous snow of the same porosity. This result supports the assertion made here that the permeability can be accurately calculated from knowledge of the density, water content and the known properties of wet snow.

The flow through the underlying saturated layer is most important where the snowpack is shallow and lies on a gently inclined slope. In such a case the flow occurs quickly through the upper layers and the saturated layer assumes an increased importance. The flow rates through this layer can be computed if the porosity, permeability, slope and thickness of the layer can be determined. Separate measurements of ice and water volumes will be necessary. The wave speed through this layer is very large hence, if predictions of more than a few hours are to be made, the influx into this layer must be predicted from calculations of flow through the upper layer. Once this information is available, the calculation of discharge from the snowpack is straightforward.

Many systems for measuring the properties of wet snow are being used or are undergoing development. Of these, only the determination of the dielectric properties of wet snow from remote sensing has the ability to determine the information necessary for use in the calculation of water flow. From knowledge of the dielectric loss factor and the effective dielectric constant, the volumes of water and ice can be calculated and, in turn, the porosity, permeability and water saturation determined. Further, the variation of these parameters can be calculated throughout the depth of the snowpack. When combined with knowledge of surface fluxes from either direct measurements or forecasts, the flow regime of a snowpack can be completely described.

REFERENCES

Anderson, E., In press. Techniques for predicting snow cover runoff. International Symposia on the Role of Snow and Ice in Hydrology, Banff, 1972.

Bear, J., and others, 1968. Physical principles of water percolation and seepage. J. Bear, D. Zaslavsky, S. Irmay (editor). Paris, UNESCO, Arid Zone Research - XXIX.

Colbeck, S.C., 1972. A theory of water percolation in snow, Journal of Glaciology, vol. 11, no. 63. p. 369-385.

Colbeck, S.C., In press. Water flow through snow overlying an impermeable boundary. Water Resources Research.

Colbeck, S.C. and Davidson, Gail. In press. Water percolation through homogeneous snow, International Symposia on the Role of Snow and Ice in Hydrology, Banff, 1972.

Harris, C.C. and N.R. Morrow, 1964. Pendular moisture in packings of equal spheres. Nature, vol. 203, no. 4946, p. 706-708.

Kuroiwa, D., 1968. Liquid permeability in snow. Union de Geodesie et Geophysique Internationale. Association International d"hydrologie Scientifique. Assemblie geneinle de Berne, 25 Sept - 7 Oct 1967, p. 380-1972.

Meier, M.F. and A.T. Edgerton, 1971. Microwave emission from snow - a progress report. Proc. 7th Intern. Symp. Remote Sensing of Environ., Ann Arbor, Mich, p. 1155-1163.

Shimizu, U., 1970. Air permeability of depositied snow. Low Temperature Science, Series A, no. 22, p. 1-32.

Sweeny, B.D., unpublished. Meausrements of the dielectric properties of wet snow using a microwave technique. Masters Thesis, Dartmouth College, Hanover, 1973.

Viessman, Fr., W., 1970. Discussion 3. Snow Hydrology, Proceedings of a workshop seminar, 1968. Canadian National Committee for the International Hydrological Decade.

Wakahama, G. 1968. The metamorphism of wet snow. Union Geodesique et Geophysique Internationale. Assemblee generale de Bern, 25 Sept - 7 Oct 1967. Commission des Neiges et Glaces, p. 370-379.

THE OCCURRENCE AND MOVEMENT OF LIQUID WATER IN THE SNOWPACK

Edward J. Langham
INRS-EAU, Université du Québec

ABSTRACT

The flow of melt-water in snowpacks containing ice layers has been observed using coloured ice crystals. Experiments were conducted on snowpacks of various slopes, and water-content measurements were made at the same time.

Horizontal or concave ice layers can impound melt-water thus increasing the capacity of the snow to store water. Both horizontal and sloping ice layers divert the flow of melt-water and thus considerably increase the time of passage of the water through the snow.

INTRODUCTION

The melting of seasonal snowpacks may be considered as a release of stored water in which the rate of release is both uncontrolled and uncertain. In an effort to reduce the uncertainty, hydrologic models which are used to calculate river flows include water originating from snowmelt. In such models however there remain problems in obtaining a good estimate of the runoff.

In recent years considerable effort has gone into the study of the melting process, in the belief that the weakness lies in the ability to estimate the quantity of water generated. Unfortunately this has not resulted in much improvement, perhaps because the processes controlling heat transfer to the snow are difficult to model. This may not however be the only reason.

The rate at which snowmelt water appears in rivers is not only determined by the rate of melting but also by the time taken to reach them. The modelling of this process is usually similar to that for rain falling on the ground; that is, little or no attention is given to the role played by the snow pack itself. In fact, delays in the passage of melt-water through the snow have a considerable effect on the relationship between snowmelt and runoff.

The flow of melt-water down through the snow depends on both the meso-structure and the micro-structure of the snowpack. Each of these is highly variable and it is certain that part of the problem with the hydrologic models must be attributed to the consequent variability of this delaying effect.

The relative importance of these structural characteristics depends very much on the type of snow pack. A deep mountain or subartic snowpack may be relatively homogeneous and lacking in meso-structure in which case a "porous medium" model such as that described by Colbeck (1971) may be appropriate.

On the other hand, in areas where the winter temperature rises to near 0°C fairly often, freezing rain, diurnal melting and freezing and other causes can produce a snowpack with many layers which, depending on their properties, may be more important than the micro-structure in controlling the flow of melt-water. This is particularly important in shallow snowpacks where the delay due to capillary flow through porous snow is smaller.

The purpose of the work described here is to study the effect of a layered structure on the distribution and flow of water in the snow-pack.

EXPERIMENTAL TECHNIQUES

The data were obtained using two techniques previously developed for this type of study (Langham, 1971).

During periods of snowmelt the first of these continuously generates a dilute solution of dye from a horizontal line of coloured ice crystals within the snowpack. The molarity of the dyed melt-water is between 10^{-3} and 10^{-5} and thus the effect on the ice/water equilibrium temperature is less than $0.001^{\circ}C$. The additional melting produced by its presence is therefore no more than 6×10^{-6} per unit mass of snow which is a negligeable contribution to the total mass of water in the snowpack.

The second technique allows the water content of the snow to be measured. If the air temperature is above freezing, melting takes place when samples of snow are centrifuged, and measurements do not represent the true water content. This can be accounted for, however, by spinning the samples several times in the centrifuge and plotting a graph of successive measurements of extracted water. This graph is linear and may be extrapolated back to before centrifuging started.

Samples were centrifuged in pairs and water contents expressed as percent of the mass of the sample. Figures 1a and 1b show such plots for air temperature of $0^{\circ}C$ and $3^{\circ}C$ respectively, and figure 1c shows that the slope is a linear function of air temperature as may be expected.

Six sites were used for these experiments. They were all in or near a small wood and were chosen to give a variety of drainage conditions. The snow had a complex stratified structure and similar layers could be recognized at each of the sites. Among these layers the following could be distinguished:

1) Porous snow which was more or less cohesive depending on temperature and wetness.
2) Impermeable layers of ice due to freezing rain. On occasions where part of the rain drained over the sealed surface before freezing lenses formed up to an inch or more thick.
3) Thin ice layers usually close together forming a band up to two inches thick.

Sites were chosen where these layers were horizontal, gently sloping or strongly sloping. About thirty tubes of coloured crystals were placed at these sites and several hundred centrifuge samples were taken.

To obtain the samples, vertical sections were cut through the snow perpendicular to the line sources of coloured melt-water. The free vertical surface has no effect on an unsaturated flow regime. However if the ambient temperature is not $0^{\circ}C$ there is melting or freezing at the exposed vertical surface. To avoid the spurious measurements this causes, it was necessary to cut about six inches off an old surface before taking samples.

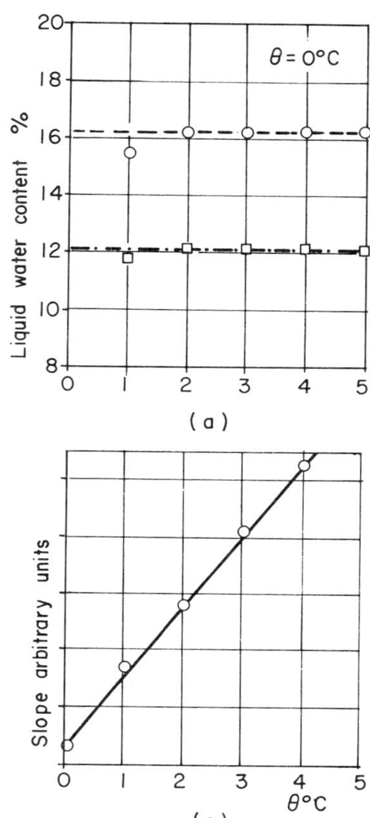

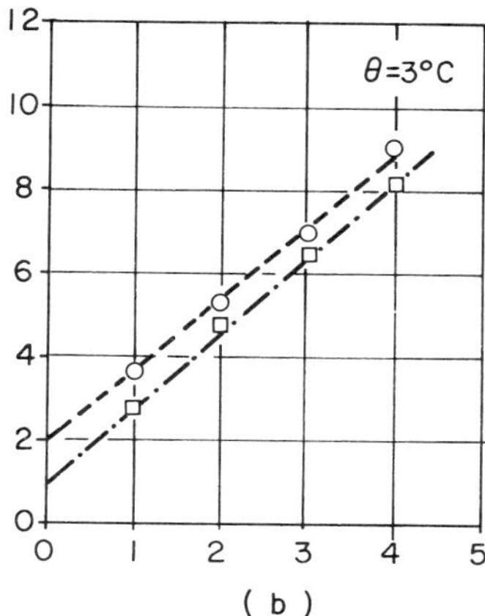

Figure 1. Successive measurements of water extracted by centrifuge and slope of such graphs as a function of air temperature, θ

OBSERVATIONS

The melt-water was observed to pass through the porous snow, subject only to hydrodynamic dispersion on the scale of the crystals, and some meandering on the scale of a few centimetres. On arriving at an impermeable ice layer the water accumulates saturating the snow immediately above it. This is shown dramatically when a new vertical surface is cut through such a layer. At the moment when the cut is made there is white snow beneath the layer and coloured snow above. Since the snow is saturated just above the ice layer the water is not retained by capillary suction and within a few minutes it has spilled down the exposed vertical surface. Figures 2a and 2b show such a sequence of events.

The bands of ice layers seem to have variable permeability and melt-water takes numerous sideways steps in passing through them.

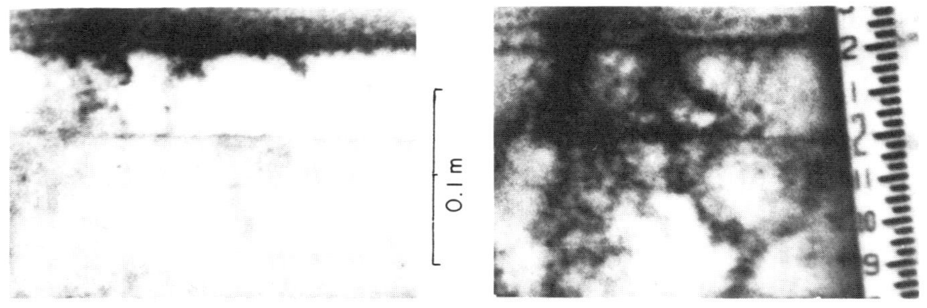

Figures 2 a b
Dyed melt-water spilling from an impermeable ice layer during the minutes following the vertical cut.

When the layers are sloping the flows become strongly directed down the gradient. Figure 3 shows such a flow in the presence of a gentle gradient and figure 4 the same layers in a steeply sloping configuration at another site.

Under static conditions the water content in a homogeneous snowpack depends on the height above the water table and the granular properties of the snow. When continuous melting takes place the flow of melt-water results not only in increased water content but a different vertical distribution since the flow rate is a function of water content. Frequently under spring conditions the melting is intermittent and this produces complicated non-equilibrium flow conditions in the snowpack.

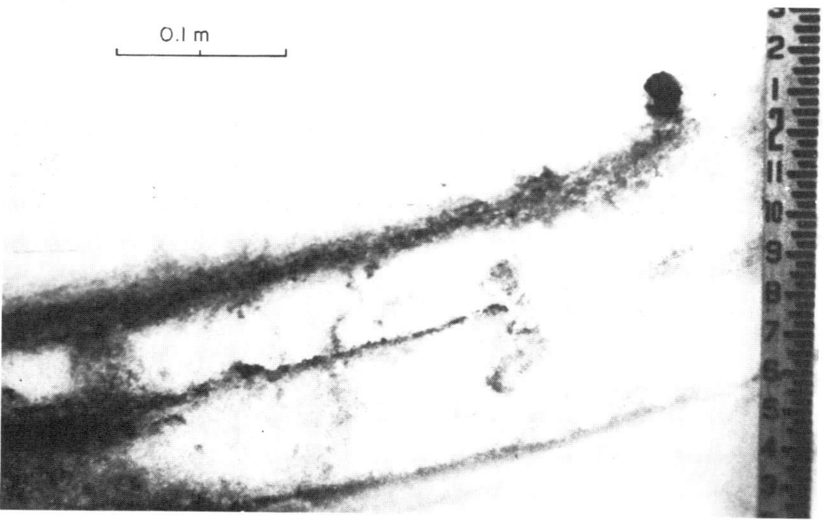

Figure 3. Dyed melt-water flowing down the gradient in gently sloping ice layers.

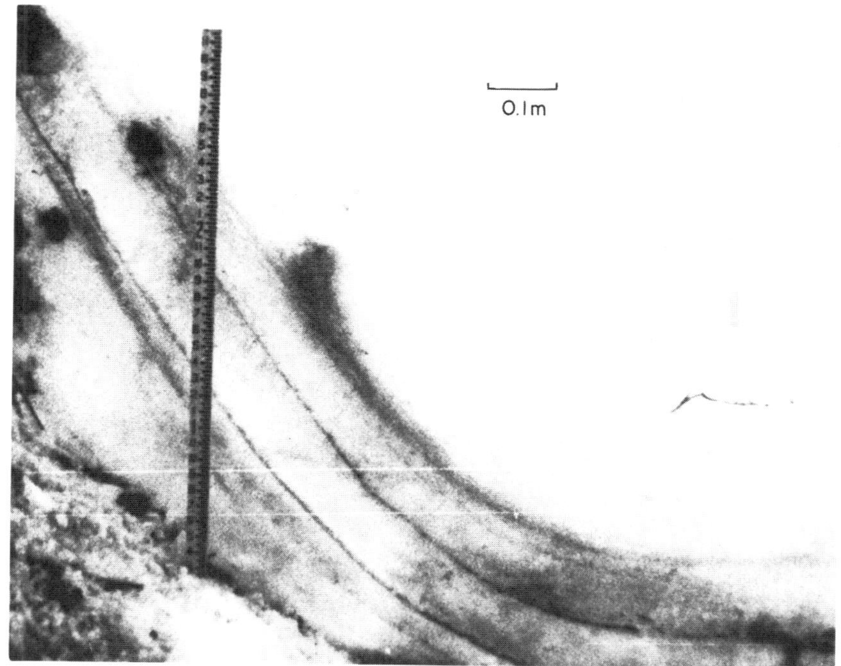

Figure 4. Dyed melt-water flowing down the gradient in steeply sloping ice layers.

Typical flow velocities of melt-water in the snowpack vary between 10^{-5} and 10^{-4}m sec^{-1} (de Quervain, 1972). The fact that the colour streak patterns related to the ice layers persist for several days shows therefore that the ice layers exert a controlling influence over the flow regime in spite of varying meteorological conditions.

It is to be expected that the distribution of water should be related to the structure and the water content measurements were made to determine the nature of this relationship. In general samples were taken along vertical profiles and along ice layers.

Figure 5 shows a vertical profile crossing two ice layers which are near the surface. The measurements were made at about 3 p.m. on the afternoon of 25th March 1973. The air temperature had risen above 0°C at about 8 a.m. and had been steady around 3.5°C for the two hours preceeding sampling.

Samples taken in pairs give similar measurements. Both show a profile in which the water contents above each of the ice layers are considerably above those in the snow beneath them.

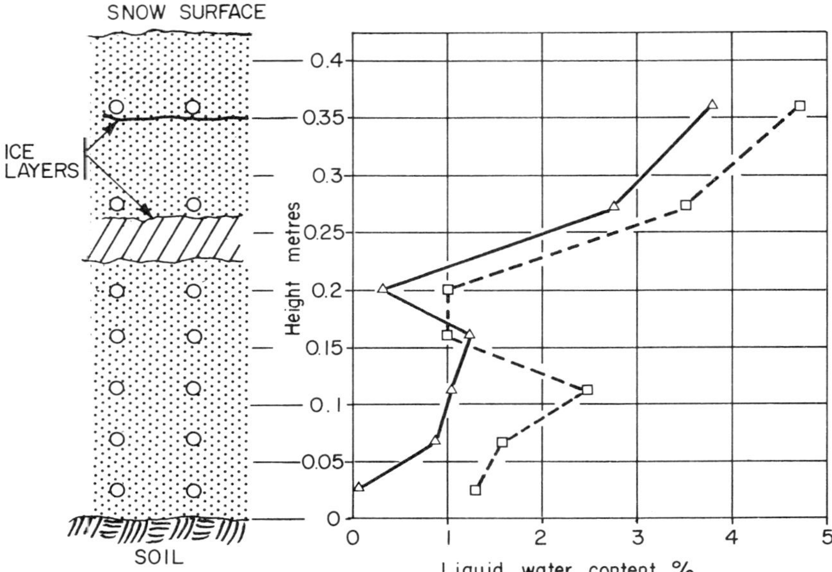

Figure 5. Vertical profile of water content through snowpack containing two ice layers.

It is, at first sight, rather surprising that there should be so much water between the ice layers. However the snow in this region is still only a few inches below the surface and the absorption of radiation is appreciable. (Gray and O'Neil, 1972, have shown that up to 90% of the incident solar radiation may be absorbed in the first 0.1 m of the snowpack.)

Figure 6 shows a horizontal section, again between ice bands, but in this case the lower ice band disappears on the right hand side. The measurements were made at about 3 p.m. on the afternoon of 28th March 1973. On this occasion the air temperature rose above 0°C at about 11 a.m. and had been steady at about 3°C for the two hours preceeding the measurements.

In the region where there is no lower ice layer, the water contents are very low since the water has been able to flow downwards. Over the ice layer the amount of water increases towards the left. The local variations superimposed on this are partly due to irregularities in the snow surface, which affect the penetration of radiation and partly due to minor variations in level of the lower ice layer, which produces drainage over it.

A measurement gives an average water content for a sample and hence the layer of snow sampled, which is about 0.03 m thick, may be saturated at the ice surface though this is not shown in the measurements. The general increase of water content with distance from the edge of the ice layer is due to dynamic retention of melt-water (Langham, 1973).

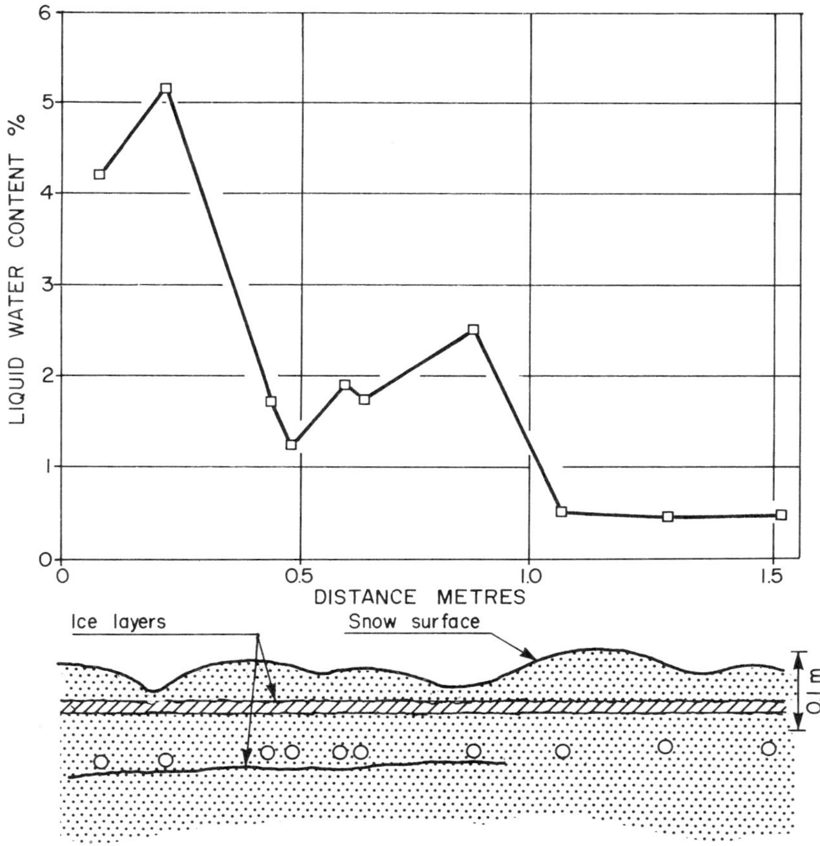

Figure 6. Horizontal profile of water content in snowpack between two ice layers.

Figure 7 shows the effect of steeply sloping ice layers. The samples were taken between 2 p.m. and 3 p.m. on the afternoon of 29th March 1973. The air temperature rose above 0°C at about 9 a.m. and was steady at about 5°C from noon until 3 p.m. This is the same site as that shown in Figure 4 but several days later when only one line of coloured ice crystals remains.

There is a flow streak from the point A which descends until it joins the flow over the ice layer B C just below. The flow follows this layer for a short distance before finding a way through it at point D. It is then lead by a band of thin layers E F between two sample points where the water contents on either side are 1.7% and 1.3%. Further on it leaves this layer and when it reaches the point G the water content is 11.1%, while just above between the layers B C and E F the water content is only 1.8%.

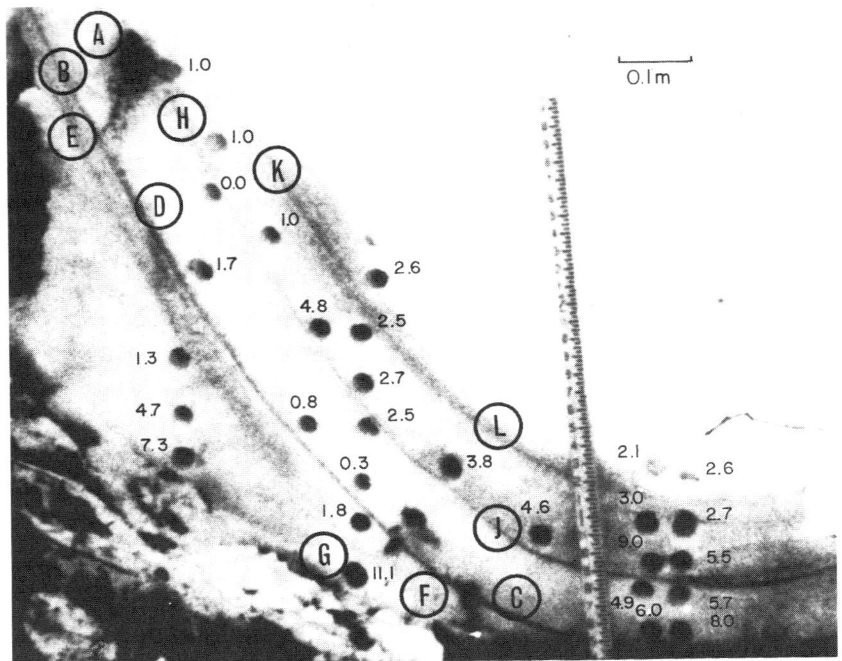

Figure 7. Measurements of water content in relation to stratigraphy and flow pattern for steeply sloping snow pack.

The ice layer H J seems to be relatively impermeable. With the exception of one or two anomalous data, the water contents increase steadily as the water flows over it, whereas below this ice layer the water contents remain very small until the bottom of the slope. The additional water appears to come from the band of ice layers K L since coloured melt-water is seen flowing from it towards the lower right hand side.

CONCLUSION

The data described above are similar to many others obtained during the same period. The results of these observations may be summarized as follows:

1) There is a tendency to detain melt-water above horizontal or concave ice layers. The snow immediately above such a layer may even become saturated. In this way more water can be stored than in homogeneous snow and the hindered passage of the melt-water increases the time response of the snowpack.

2) Where ice layers are sloping the melt-water may flow over them. This results in a longer path of flow and lower flow velocities; both leading to a longer time response than for homogeneous snow.

3) Bands of thin ice layers appear to behave as a layer of reduced permeability which nevertheless allows the passage of water. When such layers are sloping they give rise to flow with a strong component in the direction of the bands. Their effect on the response of

the snowpack is therefore similar to that of impermeable ice layers though less severe.

In a particular snowpack the density of vegetation and drift topography also modify drainage behaviour. In short, there are many parameters which influence the flow of melt-water through stratified snow, not least being the meteorological conditions which continuously modify its structure.

It is clear, however, that the presence of ice layers has a controlling influence on the flow regime which results in the detention of melt-water and a slower time response. In snow packs of this sort, it seems necessary therefore to take this into account when modelling the melt-water runoff.

REFERENCES

Colbeck, S.C., 1971. One-dimensional water flow through snow. Research Report No. 296. Cold Regions Research and Engineering Laboratory. Hanover, New Hampshire.

de Quervain, M.R., 1972. Snow structure, heat and mass flux through snow. International Symposium on Properties and Processes in Snow, UNESCO, (Banff).

Gray, D.M., and O'Neil, A.D.J., 1972. Solar radiation penetration through snow. International Symposium on Properties and Processes in Snow, UNESCO, (Banff).

Langham, E.J., 1971. A new method of using dye to study melt-water movement within a snowpack. Proceedings Volume 2, Symposium No. 8, National Research Council Subcommittee on Hydrology, Ottawa.

Langham, E.J., 1973. Un modèle du manteau nival relié à sa mésostructure. Bulletin of the International Association of Hydrological Sciences. 18:33-43.

HYDROLOGY OF WARM SNOWPACKS AND THEIR EFFECTS UPON WATER DELIVERY...SOME NEW CONCEPTS

James L. Smith
Pacific Southwest Forest and Range Experiment Station
Forest Service, U.S. Department of Agriculture
Berkeley, California

ABSTRACT

Snowpacks can be described as "warm" or "cold" or both on the basis of surface or interior temperatures. The hydrology of "warm" snowpacks differs from that of "cold" snowpacks--largely because of climatic differences. A number of new theories about the hydrology of warm snow are offered. These theories, developed from studies at the U.S. Forest Service's Central Sierra Snow Laboratory, Donner Summit, California, are at variance with older and generally accepted theories. Soil temperature appears to have little effect on snowmelt because melt or rain water from the pack keeps the soil cool and minimizes vapor transfer of heat. Snow temperatures remain near the freezing point in all but the surface few inches in most years. Snowmelt and water movement from the pack occur throughout most winters--regardless of pack density. But density layering and differential maturation of the snowpack will materially affect water movement. Differences in the amount of snow found in contiguous forest and open sites appear to be largely the result of differential aerodynamic placement. Snow interception and subsequent evaporation from tree crowns are probably a source of only minor losses.

Temperature is probably the single most critical factor that affects the condition of a mountain snowpack. Snow hydrologists measure the temperature of the pack, the air above it, and the soil beneath it. On the basis of these readings, a snowpack can be described as "warm" or "cold." In winter, a "cold" snowpack typically ranges from at or near air temperature near the surface to at or near 32°F (0°C) at the soil-snow interface. "Warm" snowpacks--as defined in this paper--are those whose interior temperature below 12 inches (30.48 cm) stays at or near 32°F (0°C) all or most of the snow season.

A cold snowpack is normally quiesent during winter, increasing in density as depth of snow increases. When melt begins in spring, the pack absorbs water until it is saturated and becomes isothermal at 32°F (0°C). Until recently, most snow hydrologists believed that water flowed from the pack only after density reaches at least 0.40 gms/cc^3. The warmth from the unfrozen soil below is transmitted to the lower layers of snow, affecting maturation and melt in these layers. In North America, cold snowpacks are found in the interior mountains of Canada and the United States and along the eastern sea coast.

In North America, warm snowpacks are found in the Sierra Nevada, and Cascades in the United States. They are also common in the northern Rocky Mountains at mid- and lower-elevations (Haupt, 1972). Because of the differences in climate and timber cover between the Rockies and the Sierra Nevada, only part of the theories offered in this paper appear to apply to the northern Rockies.

The Sierra Nevada of California are a "warm" mountain range. Winter temperatures in the timbered snow zone normally range between 0°F (-18.8°C) and 55°F (12.8°C). Snow-covered soils remain unfrozen. The highest plant water use of the entire year occurs during spring, when several feet of snow remain on the ground (Owston, Smith, and Halverson, 1971).

Storms which deposit sizable quantities of snow normally occur at temperatures of 27°F-37°F (-2.8°C - +2.75°C) and precipitation normally falls as wet snow, rain, or both. Snowfalls from a single storm ranges from a trace to over 10 feet (304.8 cm) deep.

Snow maturation is a product of the climate of the area in which the snow falls. In the Sierra Nevada, snow--because it is at the freezing point--is in such a delicate balance that the presence or absence of cover and the aspect play important roles in maturation and melt--even during the early and colder part of winter.

Below the alpine zone of the Sierra Nevada, timber covers 42 percent of the total snowpack area. The management of this resource can have important effects upon the hydrology of the area. Much of the snow research in the Sierra Nevada has been devoted to study of the effect of forest cover upon snow hydrology.

OBSERVATIONS OF FOREST EFFECTS

Observations about the effect of timber upon snow cover in the Sierra Nevada began with Church (1912). Kittredge (1948) brought together the scant amount of data about forest effects upon snow and analyzed it in relation to the broader literature on snow hydrology, particularly work in cold snowpack areas. His theoretical concepts have dominated subsequent research on the effects of forests on the hydrology of snowpacks.

Lacking sufficient data from local studies, Kittredge applied data from other, drier, wind-swept and colder areas. Part of his theories came from studies of snow accumulation under various densities of overhead cover under pine forests in semi-arid Arizona. Other theories came from snow accumulation patterns from wind-break studies in the Dakotas. From these data he tried to explain the differential amounts of snow he found under forest cover and in forest openings in the Sierra Nevada. His basic premise was that snow was aerodynamically differentially placed in forest openings to the lee of a forest wall. From this theory Kittredge developed timber harvest patterns oriented such that the openings resulting from timber cutting would be at right angles to the prevailing wind. Kittredge concluded that the differences between snow on the ground in the open and that found under various densities of overhead cover were related to the interception of snow by the pine and subsequent evaporation-sublimation of the intercepted snow.

When Kittredge applied his theories to analysis of Church's (1912) Sierra Nevada data and to his own data, he found that they explained what had been measured--more snow in open areas than under forests. He assumed this difference was the result of both aerodynamic placement of snow and interception of snow by tree crowns and subsequent evaporation-sublimation.

Most subsequent workers, largely concerned with development of timber harvest patterns, have explained the differential snow amounts they found with Kittredge's theories (for example, Anderson, et al., 1958; Newhall and Smith, 1967).

Anderson's (1963) wall-and-step forest harvest pattern for maximum snow accumulation and delay of melt retained these interceptior and aerodynamic placement theories. However, he added the important theory of snowmelt from "back radiation" from solar warming of the wall of timber to the north of his east-west oriented clear cut strips.

Before the profiling snow gage was developed, workers measured water content by extracting gravimetrically cores of snow (Smith and Willen, 1965). Sampling for most studies has historically been done on April 1, the supposed time of maximum accumulation, at monthly or occasionally at weekly intervals. Since gravimetric sampling is a destructive process, successive samples were obtained from different though nearby locations.

Since development of the isotope profiling snow gage, studies at the U.S. Forest Service's Central Sierra Snow Laboratory have been concerned with the basic variables affecting snow accumulation and maturation. These studies have analyzed snow-gage density profiles recorded daily at the same places over a time span of up to 7 years.

The research at the snow laboratory has shown the magnitude of the changes which occur in warm snowpacks at the same place over a snow season and from year to year, and some of the basic causes. More study will be required, however, before we can accurately predict the numerical snowpack changes which occur in response to all climatic and cover changes.

NEW THEORIES

From our studies have come a number of theories about the hydrology of warm snows that are at variance with older and generally accepted theories. Some of the new theories have been scientifically verified by other investigators, who have been intimately associated with snow measurement and research. These theories have not previously been combined into a comprehensive concept of their interrelated effects upon the hydrology of snowpacks under forest cover and associated open areas. The new theories offered in this paper are these:

1. Soil heat has negligible effect upon melt of the snowpack in west side of the Sierra Nevada.
2. Temperatures in the snowpacks on the Sierra Nevada west side, except in the surface 12 inches (30.48 cm), usually remain at or near the freezing point throughout most of the snow seasons. Infrequent colder snow temperatures, which may occur early in the snow season, can have important hydrologic significance.
3. Snowmelt and water release from warm snows occur throughout most snow seasons. Water release does not depend on the pack reaching a uniform or average density of 0.40-0.50 gms/cc^3.
4. The pack density is more or less heterogenous, dependent

upon aspect, cover, and meteorologic events. Snow maturation occurs in layers. The lower layers frequently mature before the time the upper layers are laid down. Snowpack development related to hydrology varies as a result of changes in cover. Water movement is materially affected by differential density layering in snowpacks. Changes in cover can affect snow hydrology significantly.

5. Water in isothermal snow flows in response to the tension gradient between various sized pores in the different and adjacent density layers.
6. The amount of snow water which reaches the snowpack in the open differs only slightly from that under tree cover. Most of the differential amounts previously measured may be (a) the result of differential melt of the pack, or (b) the result of interception by trees, subsequent melt of this intercepted snow and differential placement of the drip water in a narrow ring around the tree stem where it moved through the snow to the soil and was not measured; or (c) both.

SOIL TEMPERATURE

Snowmelt models applicable to the Sierra Nevada use ground heat as one of the melt producing variables (Amorocho and Espildora, 1966; Anderson and Crawford, 1964; Crawford and Linsley, 1966; U.S. Army Corps of Engineers, 1956).

Soil temperatures have been studied at the Snow Laboratory under a variety of cover and soil conditions since 1966. Since then, temperatures of soil up to 42 inches (106.68 cm) deep have been measured every 30 minutes at the Laboratory. In addition, for a period of 2 years, they were recorded once a month at 30 nearby locations under various combinations of aspect, cover, and soil types.

Statistical analysis of the soil temperature data showed that differences in temperatures between soils derived from granitic and pyroclastic materials and under open and forested sites were statistically and hydrologically insignificant. Thus it appears that cover or its removal will exert little influence over soil temperatures after snowfall begins.

Soil temperature seems to have little effect upon snowmelt except in early fall. At that time, cold air temperatures, or rain or melting snows, or both cool the surface layers of the soil to at or near the freezing point of water. After a permanent snow cover becomes established, the soil is kept cool and the vapor transfer of heat is minimized by the frequent draining of melt or rain water from the pack (figure 1). Heat input from ground water, important in some regions, is not of importance here because of lack of ground water.

Responses of soil temperature to the frequent early and midwinter rain and snowmelt have been similar in all years of the study (since 1966). This frequent melt throughout the snow season prevents the movement of significant quantities of heat from soil to snow. Therefore, we have concluded that ground heat has little or no hydrologic significance in the Sierra Nevada except in the first

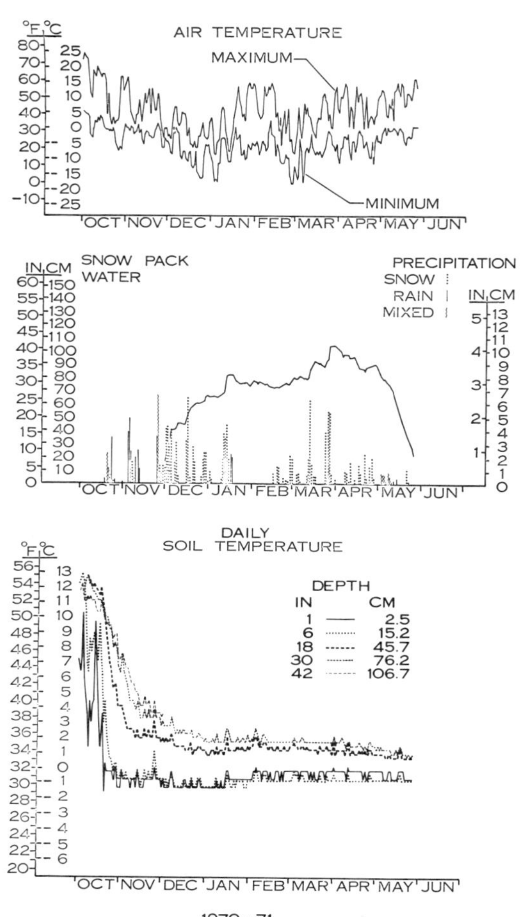

Figure 1. Soil temperature, air temperature, precipitation and snowpack accumulation and melt measured at the U.S. Forest Service's Central Sierra Snow Laboratory, Donner Summit, California for 1970-1971.

snowfalls which occur on warm soil before pack development.

SNOW TEMPERATURE

Snow temperature is a measure of heat deficit, which is sometimes called "cold content" (U.S. Army Corps of Engineers, 1956).

The heat deficit represents the amount of energy that must be supplied to the snow before melt can begin. The deficit may be satisfied by sensible heat transfer from the atmosphere, latent heat transfer, solar radiation, or by heat transfer from the soil.

Most historical snow temperature data have shown gradients of snow temperature within the snowpack (U.S. Army Corps of Engineers, 1956). The older and warmer, usually near 32°F (0°C), snow was found at the soil surface. The snow became progressively colder as the snow-air interface was approached. Such snowpacks became isothermal at 0°C during spring melt.

Internal snow temperatures at the Snow Laboratory have been measured daily each winter since 1965-66 by copper constantan thermocouples with electronic reference compensation. Sensors were placed a fixed distance above the soil surface, or at the surface of each new snowfall.

Results from the 1965-66 through 1971-72 snow seasons showed snow temperature did not vary significantly from 0°C at depths of 12 and more inches (30.48 cm) below the snow-air interface, except for one year (1970-71).

A season-long isothermal snowpack appears to be the typical development at Donner Summit, where the Snow Laboratory is located. Such a state is important hydrologically because the snowpack has little "cold content" to satisfy before melt can occur. In a isothermal snowpack, water yield from snow can occur after a period of high temperatures, rain, or high insolation,--provided the liquid water retention capacity of the snow has been satisfied.

But when the snowpack has a temperature gradient and it occasionally does, the result can be hydrologically important. The 1970-71 snow temperature data from the Snow Laboratory was atypical because temperature gradients did exist until early January. These data are similar to the 1948-49 data reported by the U.S. Army Corps of Engineers (1956).

In January 1971, the snow temperature 67 inches (170.18 cm) from the surface was at 27°F (-2.8°C). Rain water from a storm in January penetrated to a depth of 40 inches (101.6 cm), and froze into a 1 inch (2.54 cm) ice layer. Subsequent rain and melt moved downslope in the snow over this ice layer, depositing large amounts of water into the streams as quick "return flow." In less cold snow, the melt-rain water would have moved vertically downward through the snow to the soil, with less possibility of stream flooding.

MIDWINTER MELT AND SNOW DENSITY

Except during the first major snowfall, melt and water loss from snowpacks we have studied originate at the surface, and move downward through the pack. Sizable quantities of melt water flowed by the end of January in each year between 1966-67 and 1971-72 (table 1). During a 7-year period (1965-1972), melt water drained from the pack in 21 of the 28 months in the December-March period. More than 1 inch (2.54 cm) of water was released in 18 of these 28 months. And more than 3 inches (7.62 cm) were released in 14 of the 28 months.

The effect of so much snowmelt and water movement through the

Table 1.--Years in which melt water drained from snowpacks of the Sierra Nevada of California equaled the amounts shown before April 1, by months, 1966-1972.1/

Month	Melt Water (Inches/Centimeters)					
	<1 (2.54)	1.9 (2.54-5.08)	2-2.9 (5.09-7.62)	3-3.9 (7.63-10.16)	4-4.9 10.17-12.70)	5 (12.71)
Dec.	2	1	--	--	1	--
Jan.	--	1	--	2	--	3
Feb.	1	1	--	1	--	2
Mar.	--	--	1	--	--	5

1/Excludes drainage induced by rainfall.

pack is to "mature" the lower layers of snow early in the season (figure 1). Many snow hydrologists have assumed that density must reach 0.40-0.50 gms/cc^3 before melt water begins draining from the snowpack (Kittredge, 1948). We have found that this is not necessarily true. However, with maturation of the lower layers of warm snows early in the season, snowpack average density frequently reaches 0.40 gms/cc^3 at this time. During the years 1966-1972, the average snowpack density of 0.40 gms/cc^3 was reached as early as January and as late as May. And each intervening month was represented. But the pack does not have to reach 0.40 gms/cc^3 for melt to occur. On April 15, 1968, the pack reached maximum density at three study sites in the open and under forest cover of 0.322, 0.323, and 0.348 gms/cc^3. In January and February, these packs were subjected to a "mist" type rain for parts of three successive weeks. Maximum air temperatures ranged from 34°F-37°F (1.10°C-2.75°C), with minimum temperatures ranging from 17°F-28°F (-8.35°C - -2.2°C). By the end of the rain, density averaged 0.30 gms/cc^3, and the pack had fully matured to a spring-melt condition. Subsequent snowfalls did not materially compress this pack. Sizable releases of water moved from this pack each month, and the entire pack melted without any layer in the pack ever reaching 0.40 gms/cc^3. We have measured layers within the pack that have matured and melted at end of snow season without exceeding 0.23 gms/cc^3 (Smith et al., 1970).

Warm snows increase in density in response to compression and water absorption. Normally as each layer is successively compressed it absorbs all the water it can hold. Additional snowfall further compresses the small pore content of the snow, increasing its water holding capacity. Finally, density increase and structural changes related to maturation changes prevents further compression. Ice lensing, snow bridging, and other formations that produce load-bearing layers protect lower snow layers from further compression.

In warm snows, maturation occurs differentially in time, as melt water moves through the pack all winter long in most years.

We have observed water movement from the pack under a wide range of densities from 0.20 gms/cc^3 to above 0.65 gms/cc^3.

WATER-HOLDING CAPACITIES AND ROUTING

Water-holding capacities of natural snows subjected to natural and artificially applied rain have been studied with the profiling snow gage at the Snow Laboratory. Density increases from water uptake and retention against the pull of gravity for periods up to 2 weeks ranged from 0.03 gm/cc^3 for mature snow, to over 0.20 gm/cc^3 for "dry" snow. Increase was directly related to the original density and wetness of the snow. The more dense and dry the snow, the greater the water retention. The amount may be related to a function of pore space in which the greater the density, the smaller the pore space, and the greater the amount of water held (Smith, et al., 1969). This condition appears reasonable since the water and snow matrix are normally at the same temperature. A snow layer may absorb all the water it can hold against gravitational pull, be further compacted by subsequent snowfalls, and then be able to absorb and retain more water.

Utilizing the data on water-holding capacities, and working with snow density profiles taken before rainstorms, Smith, et al. (1969) were able to successfully predict in advance the amount of water that snowpacks could retain from rain-on-snow storms before water began draining from the packs.

A pack density configuration in open areas of the Sierra Nevada typically consists of a mixture of heavy and light density layers. The lowest layer, normally of high density, consists of snow laid down before Christmas. It is usually well matured and fully wet by January 1. Another major segment of the pack is normally laid down in the heavy storms which occur in late December and January. The surface of this layer frequently becomes a heavy density layer during the 2 to 6 weeks of the January-February drought common to the Sierra Nevada. Subsequent snowfalls will be differentiated by the low density within these layers and the high density surface which forms between storms.

The resultant pack consists of alternating layers of differing density, normally with sharp, narrow lines of demarcation on south-facing slopes and less well defined layering on north-facing slopes (Smith, et al., 1970). Water flow through snow is affected by these layers. Free water will not move from a zone of heavy density to the light density layer beneath it until forced by a hydraulic "head" of water. This condition may be the result of tension differences between the small pores of the heavy density snows and the large pores of the light density snows (Smith, et al., 1969).

In spring, or during heavy rainstorms, drainage patterns that resemble dendritic erosion form on the snow surface as a result of the hindrance to vertical flow. Isotope tracing studies have shown that snowmelt from the elevated humps between the drainage channels flows through the pack downward to the heavy density layer and laterally to the drainage channel, where it moves over the heavy layer downslope in this channel. Such channels are common in open south slopes with their density layering but are almost nonexistent under timber cover and on north slopes. In the south slopes, the rapid delivery of water favor the rise of flashy streams.

Under timber cover and on north slopes density differences are not so pronounced as on south facing open slopes. Water delivery under the north slopes is largely vertical to the soil.

EFFECTS OF COVER UPON SNOWFALL AMOUNT

More snow has normally been found in forest openings than under the forest cover (Anderson, 1963; Anderson, et al., 1958; Church, 1912; Kittredge, 1953; Newhall and Smith, 1967). This difference has generally been ascribed to differential snow placement as a result of aerodynamic dumping of snow or interception of the falling snow by tree crowns and subsequent evaporation-sublimation back into the atmoshpere or both (Rowe and Hendrix, 1951; West and Knoerr, 1959). Our study of these phenomena in the Sierra Nevada lead us to a different explanation for this difference.

Kittredge's original snow placement theory was based upon snow accumulation behind snow fences on the treeless plains of the Dakotas. Our observations of snow placement <u>during</u> snow storms in the Sierra Nevada indicate that the wind sweeping across the forest dips into openings in the timber stands, swirls around the opening, and sweeps out through lanes in the somewhat open stands. The snow dropped out of this swirling wind is uniformly deposited across the openings. In no case have we measured differential snow placement in different parts of a forest opening.

Differential snow amounts found in various quadrants of forest openings can be explained by differential melt caused by emission of sensible heat ("back radiation") from trees on the "sunny" side of the opening. To illustrate this effect, let us consider some data obtained in a study near the Snow Laboratory. In that study, snow water content was measured at eight times during the snow season at Lat. 39°N from transects across east-west and north-south oriented forest openings 132 feet (40.23 M) wide.

The least depth of snow was always found near the wall of timber on the north side of the east-west oriented strips. In the north-south oriented strips snow depth was the same across the opening except for a small equal melt next to both east and west borders of the strip. The decrease of snow from south to north across the east-west cut strip can be explained by solar radiation input to the snow, and to the wall of timber on the north side of the strip and subsequent reradiation of sensible heat to the snow. Height of the forest is about 130 feet (37 M). The snow in a north-south transect was in the sun for only 7 hours on April 1. Since shadows progress from both sides of the clearing, any one point is only exposed for about half this time. On an east-west oriented opening on a moderate north or south slope, the north wall of trees and the north 62 feet (18.90 M) of the snow in the opening were exposed to the sun's heat for the entire 12.64 daylight hours. Direct solar radiation input into the snow amounted to 240 ly/day for the N-S orientation and 786 ly/day for the E-W orientation. Solar radiation impinging on tree boles heat them, and they in turn emit longwave radiation to melt the pack.

Our studies have shown that tree temperatures on June 2 on the north edge of E-W oriented snow-covered strips reached 45.9°C for the south side of the trees but only 29.5°C for the north side and only 32.6°C for the east side. The west side reached 44.8°C in early afternoon for only a short period. Radiation flux to the snowpack from the south side of the exposed trees on the north boundary reached 0.235 ly/min to the snow: four times the average output of the other sides. The effect of this differential in shortwave-longwave radiation helps explain why 2-area inches (5.08 cm) of snow water were left on the E-W strips and 10 inches (25.4 cm) in the N-S strips on the same date late in the season.

On moderate north and south slopes, the greatest melt in a forest opening may be expected to be found in the east and north quadrants of a forest opening. Consequently, the greatest amount of snow is found in the south and west quadrants (Anderson, et al., 1958). Newhall and Smith (1967) also reported faster melt on open southwest aspects than under forest cover, regardless of snow density. However, when we analyzed the snowmelt on true south slopes in 1966-67 and 1972-73, we found more snow on open areas than under forest cover in late season (table 2). We theorized

Table 2.--Snow water under three cover-aspect conditions expressed as a percent of the snow water found in open-adjacent plots.

Date (1973)	Level Site, Lodgepole pine cover	North aspect, red fir cover	South aspect, red fir cover
1-11	94	69	62
2-1	97	70	60
2-7	94	69	61
2-15	93	60	64
2-23	94	71	64
2-27	96	70	64
3-8	96	70	63
3-16	96	70	65
3-23	97	69	63
3-28	94	71	64
4-4	97	69	61
4-12	98	71	64
4-18	114	--	65
4-27	116	76	80
5-4	173	80	68
5-10	220	80	77
5-18	320	88	--
5-21	--	93	--
5-30	--	150	--
5-31	--	--	--
6-1	--	--	--

that on true south slopes, the sun dwells longer on dense (60%) forest cover than on forests on other aspects. The dense forests prevent removal by wind of the heat absorbed through the tree crowns. When snows last until late spring or early summer, the increased heat melts snow faster under forest cover than in open areas too large to be affected by the border trees on the north side.

If there is little or no differential in aerodynamic placement of snow, then orientation of forest openings to influence snow amount does not depend upon prevailing wind direction, but upon the direction from which solar derived heat serves as input to the pack. Accordingly, it may be possible to use forest planting patterns or cutting patterns to influence the time of water delivery from melting snowpacks. And these patterns would not be influenced by prevailing wind direction during storms.

At the Snow Laboratory, we observed that only under the most unusual atmospheric conditions --when dry east winds ("foehn") are blowing-- are appreciable quantities of water lost to evaporation from snow intercepted on tree crowns. Intercepted snow bends the bows of the red fir (_Abies magnifica_ A. Murr.), the dominant species in the forested snowpack zone of the Sierra Nevada. These trees characteristically droop their branches around their stem when loaded with snow. Intercepted snow slides off the branches, or more normally melts and falls off into a circle ranging from about 3-6 feet (0.91-1.83 M) from the stems. The melted snow drips into a small ring, saturating the snow. This narrow, well defined drip ring of matured snow does not retain subsequent drip, but it sinks into the soil. We have measured radioactive isotope tagged water movement through 5 feet of such snow in 15 minutes. It enters the soil beneath the pack where it has not been heretofore measured.

In snow year 1972-73, we collected daily snow-gage profiles under a "highly dense" (80% crown closure) lodgepole pine (_Pinus contorta_ Dougl.) forest and compared water content with that from a nearby gage in the open. Since lodgepole pine boughs are stiff and do not bend, snowfall and snowmelt drip fall vertically to the snow from the point where the snow was captured by the tree crowns. Only insignificant differences were noted in the snow depth, density, or water content of the two packs until spring, when we recorded major differences. That year was ideal for such a study for it was cooler than normal. Therefore, we found less melt differential in early season between forest and adjacent open areas. In fir forests, however, the familiar sight of more snow in the open than under trees was observed until late season, when the melt differential reversed this condition. On both north and south slopes, snow water content under dense (60% cover) fir stands was only 60 to 70 percent of that in the open. In a lodgepole pine forest, the snow water content ranged from 93-97 percent of that in the open (table 2).

In snow year 1972-73, we also recorded evaporation loss. It was slightly below average: 1.09 inch (2.77 cm) of water was lost from the surface of snow in the open compared to a 5-year average of 1.30 inch (3.30 cm). In a warmer year, more melt would have occurred, snow would not have remained on the trees as long, and less evaporation would have occurred.

LAND MANAGEMENT EFFECTS UPON SNOW HYDROLOGY

Cover can have a decisive impact upon snow hydrology. By

changing the forest cover, it is possible to shift the time when snow disappears by 2 to 4 weeks. In general, snow disappears sooner on the south slopes under dense forest cover than under completely open conditions. On all other aspects snow disappears first on open slopes.

Snowpacks under forest cover normally have less profile layering than those in the open. Thus melt or rain-on-snow water could be expected to seep vertically into the soil. Under open conditions, profile layering results in sizable quantities of water being delivered directly to streams without reaching the soil.

Removing fir stands changes the hydrology of the snowpack. Prior to removal of the fir, snow which was intercepted on tree crowns subsequently melted, dropped into the "drip ring" near the tree, and penetrated to the soil. After trees are removed, the usually wet snow uniformly covers the area. Formation of ice lens and differential density layering tends to route more water directly to streams than occurs under forest cover.

Removal of fir on level or northerly exposures can be expected to increase melt rate in late season over that from fully forest covered areas. Snow on forested areas on northerly slopes melts more slowly than in openings and bare areas, extending the delivery time later into the season.

Under lodgepole pine intercepted water is deposited under the point of interception. The density of the pack under the stand has essentially the same density as that in the open, but lacks many of the ice lenses of the open pack.

South facing open packs in early and mid-season have more ice lenses than those under timber cover and on north slopes. Removal of timber cover on south slopes should increase pack depth all winter long, and extend the melt season, but it will increase the likelihood of more rapid delivery of rain-on-snow water to streams. In late melt season, daily snowmelt delivery to streams from open areas will be accelerated over that from forested areas.

Through intensive management of the various parts of a watershed, we should be able to determine when water should be released from snowpacks. And while we have little control over when snowmelt begins, or even the amount of total snowfall, we can move snowmelt timing forward or backward in time by from 2 to 4 weeks, depending upon land-use practices that we adopt. The degree of change in snowmelt time will be related to the total area involved within a watershed which is being managed. The change in one segment of the watershed may be different from that in another segment. With a watershed programmed to fit into downstream reservoir operations, we can realize greater benefits from the entire streamflow. The main benefits are: more water when needed, better end-of-the-season reservoir storage and stream regulation, and a more regulated and greater hydroelectric power potential from the same amount of snow water.

REFERENCES

Amorocho, J., and B. Espildora, 1966. Mathematical simulation of the snow melting process. Water Science and Engineering Papers 3001, University of California, Davis. p. 156.

Anderson, Eric A., and Norman H. Crawford, 1964. The synthesis of continuous snowmelt runoff hydrographs on a digital computer. Technical Report No. 36. Stanford University. p. 101.

Anderson, Henry W., 1963. Managing California snow zone lands for water. U.S. Forest Service Research Paper PSW-6. Pacific Southwest Forest and Range Exp. Stn., Berkeley, California. 28 p.

Anderson, Henry W., Raymond Rice, and Allan J. West, 1958. Snow in forest openings and forest stands. Soc. Amer. Foresters Proc. 1958: 46-50.

Church, J. E., 1912. The conservation of snow: its dependence on forests and mountains. Sci. American Sup. 74:145, 152-155.

Church, J. E., 1913. Snow Surveying: its principles and possibilities. Geog. Rev. 23(4):529-63.

Crawford, Norman H., and Ray K. Linsley, 1966. Digital simulation in hydrology: Stanford watershed model IV. Technical Report No. 39. Stanford University. p. 210.

Haupt, Harold F., 1972. The release of water from forest snowpacks during winter. USDA Forest Service Research Paper INT-114. Intermountain Forest and Range Exp. Stn., Ogden, Utah. 17 p.

Kittredge, Joseph, 1948. Forest influences. McGraw Hill Book Co., N.Y. 394 pp.

Kittredge, Joseph, 1953. Influence of forests on snow in the ponderosa-sugar pine-fir zones of the Sierra Nevada. Hilgardia 22:1-96.

Newhall, George N., and James L. Smith, 1967. Watershed management: effects on basin management. In Development of the total watershed. American Soc. Civil Engineers, Irrig. and Drainage Div., pp. 47-85.

Owston, Peyton W., James L. Smith, and Howard G. Halverson, 1971. Further development of radioisotope techniques for measuring water movement in large trees. U.S. Atomic Energy Com. TID-25463. 38 p.

Rowe, P. B., and T. M. Hendrix, 1951. Interception of rain and snow by second growth ponderosa pine. Amer. Geophys. Union Trans. 32(6):903-908.

Smith, James L., and Donald W. Willen, 1965. Measurement of snowpack profiles with radioactive isotopes. Weatherwise 18:246-251.

Smith, James L., and Howard G. Halverson, 1969. Hydrology of snow profiles obtained with the profiling snow gage. Proc. 37th Meeting, Western Snow Conference. 1969: 41-48.

Smith, James L., Howard G. Halverson, and Ronald A. Jones, 1970. The profiling radioactive snow gage. In Transactions of the isotopic snow gage information meeting. Idaho Nuclear Energy Commission. pp. 17-35.

U.S. Army Corps of Engineers, 1956. Snow Hydrology. North Pacific Div. Corps of Engineers, U.S. Army, Portland, Oregon. p. 437.

West, Allan J., and K. R. Knoerr, 1959. Water losses in the Sierra Nevada. J. Amer. Water Works Assn. 51(4):481-488.

ABLATION CHARACTERISTICS OF AN ALPINE SNOWFIELD IN SUMMER

Louis R. Bartos
Hydrologist, U. S. Forest Service, Provo, Utah

Paul A. Rechard
Water Resources Research Institute, University of Wyoming

ABSTRACT

Ablation characteristics of a summer alpine snow field in the Klondike Lake area, Medicine Bow Mountains, Wyoming, were investigated, and it was found that the mean summer ablation rate was 6.00 $\pm$0.5 cm/day. The total change in volume of water in the snow field was determined using two techniques, mean surface ablation and area-elevation relationship based on topographic survey. Physical characteristics of the subsurface snow, such as snow density, and free water content were also evaluated.

INTRODUCTION

Summer alpine snow fields are a valuable source of water in the Rocky Mountain area, and a knowledge of their characteristics is vital in relation to their possible management.

A snow field which lasts all summer was selected for study in the Snowy Range of the Medicine Bow Mountains, Wyoming. Of interest in the study were techniques for determining or measuring ablation amounts and rates, the variation in the ablation amounts throughout the summer, the relationships between ablation and climatic parameters, and the physical characteristics of the snow.

Techniques for determining ablation included the use of stakes, establishment of a reference horizon, and detailed contouring with resultant area-elevation relationships. The mean summer ablation rate was 6.0 $\pm$0.5 cm/day and the rate varied from a high of 8.9 cm/day early in the season to 2.3 cm/day just prior to the onset of accumulation for the next season. Free water content in the field was found to be from 13 to 17 percent during the middle of the summer, and densities were about 0.65 g/cm^3.

LOCATION

The study area, termed the Klondike Lake snow field, is near the crest of the Medicine Bow Mountains, Wyoming, at latitude 41.4° north, longitude 106.3° west. It is in an alpine cirque basin with a mean elevation 3223 meters (10,900 feet) m.s.l., which is contained in a shallow, sloping, relatively steep-sided, U-shaped drainage way, similar to a couloir. The initial area of the field was approximately 2.31 hectares. The study was undertaken during the summer of 1970.

PROCEDURES

Ablation of the snow surface was measured three times each week using a grid of 21 stakes set in lines approximately perpendicular to the long axis of the field. Several different types of stakes were considered, the prime requirements being that they be durable, light, nonconductive, and inexpensive. Bamboo stakes 2.6 cm in diameter and 2.5 m long were selected and painted white to reduce possible radiative effects of the stake on the snow. These were placed in holes drilled into the snow to a depth of two meters. Drilling was necessary since the highly metamorphosed snow and ice layers made it impossible to drive the stakes to depths beyond 10 cm.

In order to facilitate measurement and reduce errors that might result due to the large (15-20 cm diameter) ablation crater around the stake, a 30 cm diameter aluminum collar was placed on the snow surface at the time of measurement in order to provide a uniform measurement base. A second method for measuring ablation of the snow surface was also tested. Vertical distances from a reference horizon (a 6 m long string line suspended between two deeply implanted stakes) to the snow surface were measured at designated intervals along the horizon.

To determine if stake settlement was causing a variation in ablation amounts, a method developed by LaChapelle (1959) for measuring apparent ablation on stakes of different lengths and depths was used. If the level of the snow surface dropped the same amount for all lengths and depths, it could be assumed that the decrease was due to ablation and not settlement.

To aid in establishing the settlement and physical characteristics of the snowpack, two snow pits were dug, one on the lower part of the field and one on the upper. Density profiles were made using 500 cm3 samples taken from cylinders placed horizontally into the wall of the pit (Benson, 1962). Temperatures within the snowpack profile were found to be relatively isothermal (0°C) throughout the ablation season.

Weather data taken in the area were rather incomplete. Wind run was taken at 1.5 to 2 meters above the snow surface, depending on the rate of ablation between readings. A three-pen thermograph was installed at the anemometer site to measure temperatures at the snow surface and at approximately 15 cm above the snow surface. In addition, a hygrothermograph provided relative humidity data about 10 cm above the snow surface. Radiation and precipitation data were supplied by a climatic station located three miles east of the study site at Little Brooklyn Lake.

In mid-July a topographic map of the snow field was made in order to establish snow surface elevations and contours of the area. The mapping was done by a stadia-topographic survey, using a theodolite and stadia rod. The area was again mapped during the latter part of August in order to compare mass depletion of the snow field with that shown by the ablation stakes.

During the first part of August observation of free water content at various levels within the snow profile was initiated to study water migration through the strongly layered, high-density, snow cover. The calorimetric technique was used to determine the free water content (Bernard and Wilson, 1941).

RESULTS

Ablation Stake Method

The mean ablation rate based on data from stake measurements during the period July 4 to October 3 (92 days) was 6.0 ($\pm$ 0.50) cm/day. However, ablation during this period did not occur at a constant rate. From July 4 to August 10 (37 days), the mean ablation rate was 8.9 cm/day; from August 10 to September 2 (23 days), it was 6.4 cm/day; and from September 2 to freeze-up on October 3 (32 days), it was 2.3 cm/day.

Taking into consideration the changes in density through the ablation season, the average water equivalent ablation rate was 4.6 cm/day. Based on stake data, the total amount of water ablated between July 18 and September 10 (35 days) was 42,000 cubic meters.

Sum of mean stake ablation measurements	= 368.2 cm
Mean firn density	= 0.58
Water equivalent	= 2.14 m
Mean surface area of field	= 3.499 m^2
Volume of water ablated	= 12.877 m^3

A comparison was made of the ablation rate data obtained using the stake and string line methods. For the period July 25 to September 2, the mean ablation according to the string line method was 6.8 ($\pm$ 0.77) cm/day, which was within two percent of that determined from the stakes. From this it was concluded that either method provides a comparative measurement of ablation.

A mean daily ablation - duration curve (Figure 1) was prepared from the data collected during the period July 1 to October 3, 1970. The duration curve is a cumulative frequency curve showing percentage of time a given rate of ablation was exceeded in a given period. The curve is useful for planning purposes and indicates that on alpine snow fields in the Rocky Mountain region, mean daily ablation rates of approximately 6 cm/day can be expected 50 percent of the time.

During the summers of 1955 and 1956 Martinelli (1959) observed an ablation rate in the Front Range of Colorado, of 8.3 cm/day for the months of July and August and the first half of September. This compares with the seasonal mean daily ablation of 6.0 cm/day found on the study area.

Number of Stakes Required

The sample size (number of ablation stakes) required for a $\pm$ 10 percent allowable error in mean snow surface ablation at the 95 percent confidence level was determined for three periods during the season, using the following equation:

$$\text{No. of stakes} = \frac{4\,\sigma}{L^2}$$

L = allowable error 10% of mean ablation
σ = standard deviation

Ablation period	No. of stakes for 95% confidence level
July 4-Aug. 10	2.7
Aug. 10-Sept. 2	18.3
Sept. 2-Oct. 3	81.0

The increase in sample size (number of ablation stakes) required over time was possibly due to an irregular melt pattern caused by increased radiation from the surrounding rock faces and changes in snow surface albedo.

1.9

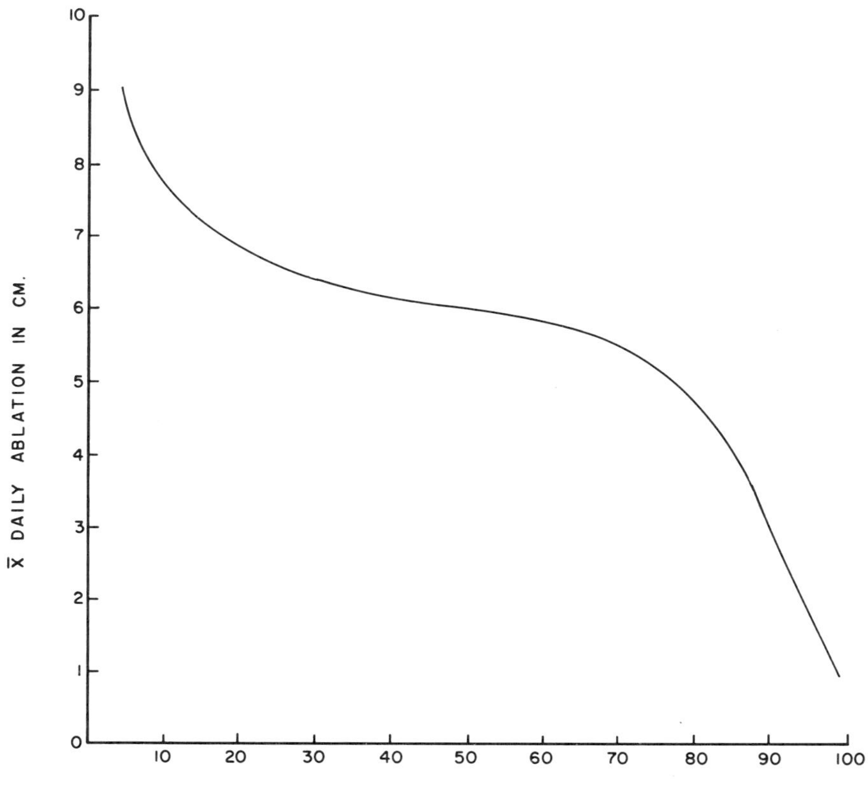

Figure 1. Ablation-Duration Curve, Klondike Lake Snowfield, Summer 1970.

Anomalies

Sizable anomalies occurred in the ablation data several times during the summer for no apparent reason. A possible cause of these anomalies was mentioned by LaChapelle (1959) in relation to this work on the Blue Glacier. He found that the larger rates of ablation during the periods of clear days with strong solar radiation and warm nights accompanied by warm winds, could induce large amounts of sub-surface melting. This melting was not accounted for in the Δ h (change in stake height) or ablation stake readings. A few days after a period of considerable sub-surface melt, the surface snow could collapse causing a large Δ h to occur, apparently within a short period of time.

Settlement

To determine if settlement of the snow were occurring, several small diameter stakes of varying length were driven to different depths in the snow. This technique was described by LaChapelle (1959). Analysis of the data revealed that no appreciable amount of settlement occurred below the 10 cm level.

Mapping Method to Determine Ablation

Both the stake and string line methods of measurement indicate accumulation and ablation at one given point only. This is desirable when obtaining changes on large glaciers or ice sheets, such as those in Greenland or Antarctica. For most temperate glaciers or snow fields, more continuous measurements can be made by using topographic mapping of the snow surface.

Basically, this method requires a contour map of the snow surface prior to the ablation season or other period of interest and another at the end. The latter map must include contours of the ground surface area within the boundaries of the original snow field. The surface area within each elevation band is determined for each time, and an area elevation curve plotted (Figure 2). The shaded area between the two lines on the curve indicates the volume of the snow mass that was ablated during the time interval. The mean snow density times the volume of snow ablated gives the quantity of water that was ablated from the field.

This method was used on the Klondike Lake snow field between July 18 and September 10, 1970 and the following data were obtained:

Total snow volume change of the snow field = 31,860 m^3
Average firn density for the ablation season = 0.58 g/cm^3
Total volume of water ablated = 18,479 m^3

This technique indicated a significantly smaller volume of ablation than that obtained using the stake method. The explanation of the difference is not readily obvious, however work is underway to attempt to discern the reasons therefore.The mapping method should give an accurate determination of mass loss, and yearly comparisons would determine the balance of a particular snow field. Pytte and Ostrem (1965) determined mass balance on the Nigardsbreen in Norway using a method similar to the area-elevation technique; however, they did not calculate mass loss.

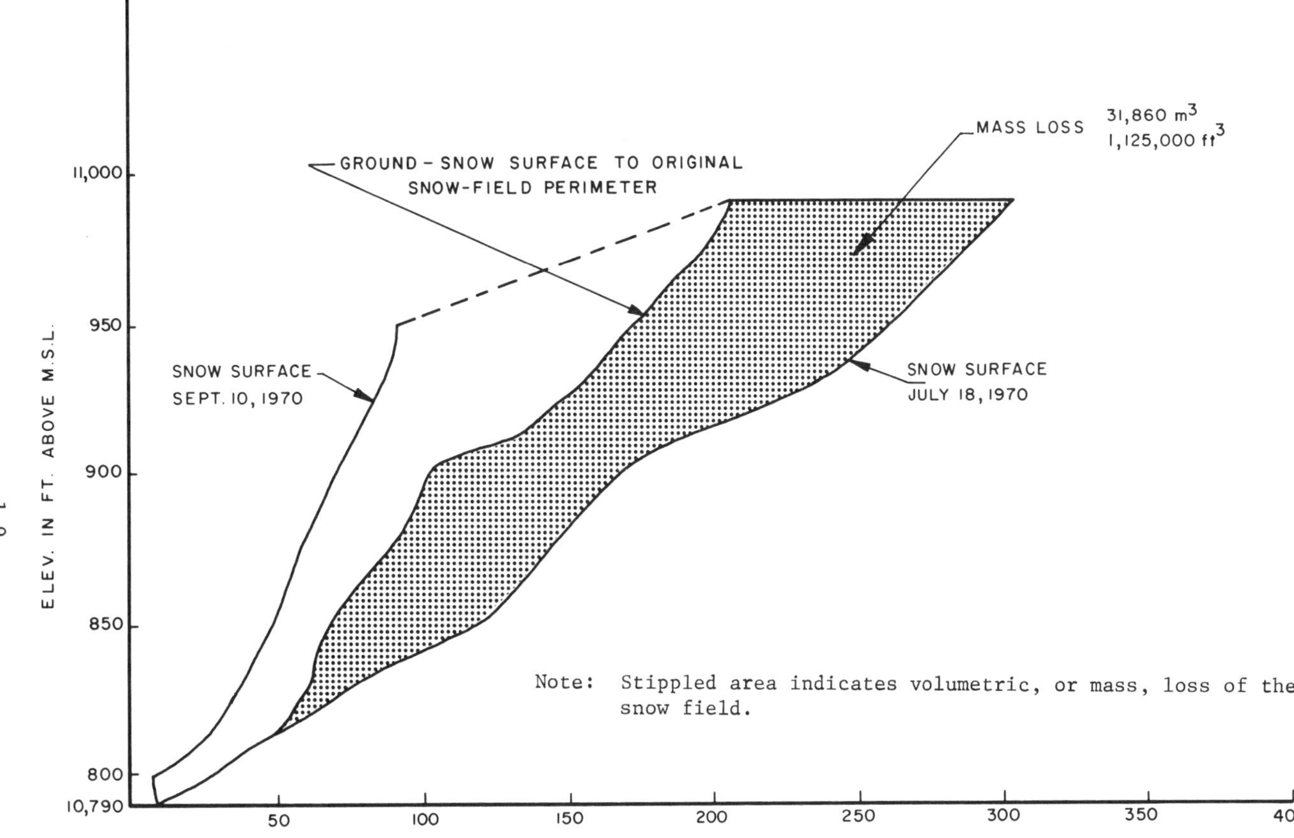

Note: Stippled area indicates volumetric, or mass, loss of the snow field.

Figure 2. Snow Field Area - Elevation, Mass Difference, Summer 1970.

When evaluating the large difference in depleted volumes between the mapping and ablation stake methods, ground surface topography must be taken into consideration. The area-elevation mapping technique takes into account the large areas of shallow snow depths at the upper and lower portions of the snow field, while the average-end-area method used with the ablation stake data considers mean area change and total surface depletion, which in this case more than doubles the depleted snow volume. Thus, mapping of the snow field appears to provide a very usable means for determining the mass loss or change over a given period of time.

Snow Quality and Density

Information on the relative amount of free water in the pack was obtained by means of a calorimeter (in this case a one quart vacuum bottle which was adapted for the purpose). The one-quart size is sufficient for snow samples obtained with the SIPRE 500 cm^3 snow tube. Calorimetric data computations were performed using the method of Bernard and Wilson (1941). The results obtained are shown in the table below.

Date	Location	Depth in cm.	Density	Percent Free Water
Aug. 6	Pit B	0-5	0.58	14
		10	0.58	13
		50	0.63	15
		150	0.61	17
		250	0.61	14
Aug. 25	Pit B	0-5	0.63	13
		40	0.62	15
		70	0.62	14
		90	0.56	4

Since the figures for percentage of free water content are quite high, samples for a particular level were taken three times. The three samples varied less than one percent; therefore, the authors consider these figures quite reliable. The high free water content at the 50 and 150 cm levels was caused by perched water above an ice layer.

There is a striking uniformity in the snow density throughout the vertical profile. This same characteristic was observed by LaChapelle (1954) on the Juneau Ice Field in glacier firn. This uniformity did not occur until August, possibly because of alteration and firn development by the large quantity of free water in the snow pack caused by melt and frequent heavy rains. The densities are comparable to those found on temperate and coastal arctic glaciers.

Snow Pits

Pit A was at an elevation of 3298 meters (10,820 feet) m.s.l. near the toe of the snow field. First excavation of the pit was made July 15, after the ground surface was located by sounding to a depth of 2.64 m. The stratigraphy of this pit was typical of many shallow, spring alpine snow fields with relatively hard bands of old wind-and-sun-crusted surface layers, which during summer melt are zones of free water migration in the pack. These zones did not develop in the shallow young profile as they did in the more mature, late summer firn. Grain size of the firn did not change much throughout the measurement period, which indicated that the ablation rate was more rapid than the metamorphic process. For the most part, grains were irregular ice globules typical of young firn. However, the zone at about 1.60 m was fine-grained and highly compacted.

Pit B was located on the upper part of the field at an elevation of 3325 meters (10,910 feet)m.s.l., and was first excavated on July 29. Since the depth of the firn was beyond the length of the sounding rod, depth was estimated at 4 m. The characteristics of the stratigraphic profiles were more typical of an arctic glacier than of a temperate region snow field. Here again, grain characteristics were relatively uniform throughout the depth of the pack, which is consistent with the uniformity of density.

Some of the more noticeable features of this pit site and profile were well-defined ice layers, lenses, and ice glands. These structures were similar to those observed by Benson (1962) on the Greenland Ice Sheet firn. The glands are pipe-like vertical masses which also extend laterally, and connect ice layers or lenses. In summer snow they are frozen percolation channels feeding or being fed by the ice layers. They develop from percolation water converging to a lower point, forming a first-order channel, and then refreezing. Usually this process takes place in the upper meter, where melt water is initiated. These features are quite prominent in slopes greater than five to ten percent, and usually do not occur on slopes less than this. Ice lenses are zones of perched melt water within the pack, which usually develop on or within an ice layer and then refreeze. They reach a thickness of 3 to 10 cm. Ice bands or layers in snow fields with a slope or grade tend to act as deflecting agents of downward percolating melt water. This was also observed by LaChapelle (1954) and Leighton (1952) on the Juneau Ice Field.

SUMMARY

1. The mean summer ablation rate of the alpine snow field observed in the Rocky Mountain region during 1970 was 6.00 (+0.50) cm/day.
2. The total summer volume of water ablated was computed using two techniques which gave very different results:
    a. The mean surface area method and ablation stakes indicated an ablation of 12,900 cubic meters or 10.5 acre-feet of water.
    b. The area-elevation technique based on topographic surveys indicated 18,500 cubic meters or 14 acre-feet of water were ablated.

    The area-elevation method is considered by the authors to be more reliable, since it accurately determines the volume of water ablated.
3. The free water content of the snow field was about 15 percent. The evident water-holding capacity of the summer snow is much higher than usually experienced in winter snow packs.
4. The summer snow fields maintain a fairly uniform and high density in the vicinity of 60 to 65 g/cm^3.

ACKNOWLEDGEMENTS

Research on which this paper is based was supported in part by the Office of Water Resources Research of the U.S. Department of the Interior under Pl. 88-379, and the University of Wyoming acting through the Wyoming Water Resources Research Institute.

REFERENCES

Ambach, W. and H. Hoinkes, 1963.

Bader, H., et al., 1939. Snow and its Metamorphism. U. S. Army Corps of Engineers SIPRE Translation 14 (1954). 303 p.

Benson, C. S., 1962. Stratigraphic Studies in the Snow and Firn of the Greenland Ice Sheet. U. S. Army Corps of Engineers CRREL Research Report No. 70. 93 p.

Bernard, N., and W. T. Wilson, 1941. A New Technique for the Determination of Heat Necessary to Melt Snow. Trans. Am. Geophysical Union. 22:178-181.

Haefeli, R., H. Bader, and E. Bucher, 1954. Snow Studies on the Juneau Ice Field. J.I.R.P. Report No. 9. 29 p.

LaChapelle, E., 1959. Error in Ablation Measurements from Settlement and Sub-surface Melting. Journal of Glaciology. 3:358-467.

Leighton, B., 1952. Investigation in the Taku Firn, Scientific Observations of the Juneau Ice Field Research Project, Alaska, 1949 Field Season. American Geographic Society J.I.R.P. Report No. 6. 46 p.

Martinelli, M., 1959. Some Hydrologic Aspects of Alpine Snow Fields Under Summer Conditions. Journal of Geophysical Research. 64:451-457.

Pytte, R. and G. Ostrem, 1965. Norges Vassdragsog Elcktrisitetsuesen Meddelelse fru Hydrologisic Audeling. Institute for Special Studies in Hydrology. 14:24-35.

EVAPORATION FROM SNOWDRIFTS UNDER OASIS CONDITIONS

Paul A. Rechard
Charles N. Raffelson
University of Wyoming Water Resources Research Institute

ABSTRACT

Considerable research has shown that it is possible to modify snow accumulation patterns by utilizing artificial wind barriers, such as snow fences. However, economical utilization of the snowdrifts requires that the resultant increase in water supply be obtained at a reasonable cost which assumes an ability to estimate the increase. To determine the quantity of increased water supply, it is necessary to know the amount of water lost from the drift through evaporation.

A study to measure evaporation was conducted on Pole Mountain in southeastern Wyoming, wherein two snowdrifts were formed on top of butyl rubber mats behind two separate snow fences, each 8 feet high (2.4 m) x 100 feet (30 m) long. The water equivalents of the drifts were determined, and the quantity of runoff from the melting of the drifts was measured. Since there could be no infiltration, the difference between the water equivalency of the drift and the runoff was a measure of the evaporation, which during the one-month ablation period, was approximately 20-30 percent of the volume of the drift at the start of ablation.

During the melt season the snowdrifts are "oases" of snow, in that they are surrounded by bare ground. Because of this fact, and the extreme wind packing of the snow, existing models to compute evaporation are not adequate.

Tabler (1973) has shown that a large percentage of the precipitation that falls in the form of snow in windy areas such as southeastern Wyoming, sublimates when it is transported by wind and is therefore lost to beneficial use by man at that site. He has further shown (Tabler, 1973) that snow fences, if properly designed, are a viable watershed management tool, and can be used to decrease the loss by sublimation, and consequently, might increase the water supply available for beneficial purposes.

Before the economic feasibility of building fences for the storage of water in snowdrifts can be established, it is necessary to determine the loss of water from the drifts themselves. Thus, the purpose of this study was to determine the evaporation from two snowdrifts under controlled conditions.

The study site is on Pole Mountain in the Medicine Bow National Forest, 17 miles (27 km) east of Laramie, Wyoming at an elevation of 8,100 feet (2,470 meters). Annual (October - May) snowfall amounts at this site vary from about 6 to 16 inches (160 - 400 mm) water equivalent, with snowpack typically beginning to accumulate in early November, reaching a peak in April and disappearing by June. The area is comprised of grasslands with scattered rock outcrops, and is subject to frequent winds of relatively high speeds and prolonged duration. The terrain slopes gently down to the east (downwind), and there is an effective barrier to blowing snow about 1,100 feet (335 meters) west (upwind) of the fence locations. The soil consists of decomposed granite which is very porous; therefore, overland flow occurs rarely and infiltrates within very short distances.

Two separate snow fences were constructed; each was 100 feet (30 m) long and 8 feet (2.4 m) high with a 6 inch (.2 m) gap between the bottom of the fence and the ground. One of the fences was made of standard vertical=lath, highway-type fencing with an approximate density of 50 percent. The other was a plastic-coated fiber glass fencing material with a density of about 55 percent. Each type of fence perfoermed satisfactorily in forming the drift. A 1/32 in. (.08 cm) butyl rubber mat was laid in the lee of each of the fences with a berm all around to prevent water from running off of or onto the mat. Wind currents did cause the drift to spill over the edge of the mat, which necessitated digging a trench through the drift at the mat boundary. This precluded melt water from following an ice layer and running off the mat, and thus not being recorded as runoff.

The area was selected for the study because the U. S. Forest Service, Rocky Mountain Forest and Range Experiment Station, has an ongoing study of the utilization of snow fences for snow storage on watersheds (Tabler, 1971) in the vicinity. In addition, the persistent winds cause the relocation of snowfall, so that in unforested areas snow accumulation occurs only in topographic depressions, the lee of scattered rocks and trees, and behind snow fences. The remaining area is without snow cover most of the winter, especially during the drift ablation season. Thus, a drift behind a snow fence is isolated from nearby snowpack and is, in essence, an oasis.

When the snow melts, the melt water either contributes to the water supply as runoff in the streams or infiltration to the groundwater, or is lost to the atmosphere through evaporation. In addition, there may also be some loss due to sublimation. For purposes of discussion, evaporation and sublimation will be lumped together as evapo-sublimation.

In order to determine the evapo-sublimation from a snowdrift, it is necessary to measure all the components of the water budget equation:

$$M = R + E + I, \qquad (1)$$

where M = Snow melt ablation of the drift
R = Runoff
E = Evapo-sublimation
I = Infiltration.

Infiltration is almost as difficult to measure directly as evaporation; therefore, in this study infiltration, I, was made equal to zero by the placement of a waterproof butyl rubber mat under the drift. The water budget equation then became

$$M = R + E. \qquad (2)$$

The rubber mat was shaped such that all water on it would drain to one outlet point, where two weirs were placed (one below the other) to measure all runoff, R, from the drifts. Two weirs were used as a precautionary measure in case one of the weirs or its recorder failed. One was a 90° v-notch weir and the other a 45° v-notch weir; both were heated so the water would not freeze. The larger weir (90°) was used in case preliminary calculations of discharge rate were low and the capacity of the 45° weir was exceeded. The weirs were calibrated in the hydraulics laboratory at the University of Wyoming. The heads on the weirs were recorded, and to insure a continuous record of discharge, two recorders were used (one on each weir) on the assumption that both recorders would not fail at the same time.

The change in the water equivalent of the drift, ΔQ, (Q initial minus Q final), is the ablation due to snow melt minus additions to the drift during the time interval, or

$$-\Delta Q = M - P - D, \quad (3)$$

where P = Precipitation on the drift
D = Accumulation on the drift from blowing snow.
Obviously $-\Delta Q$ will equal M for periods when there is no input to the drift. During the spring ablation the blowing of snow is minimal, so during this period $-\Delta Q$ will equal M - P. Substituting equation (2) into equation (3) gives:

$$-\Delta Q = R + E - P - D, \text{ or} \quad (4)$$

$$E = -\Delta Q + P + D - R. \quad (5)$$

The change in water equivalent of the drift, ΔQ, was determined by measuring the volume and density of the drift at weekly intervals. The precipitation, P, was measured by several gages installed at the site. For this study it was not feasible to measure the accumulation to the drift from the capture of blowing snow, D. Therefore M, and thus also the evapo-sublimation were determined only for periods when D was zero.

The measurement of the water equivalent in the drift and the weekly ablation of water equivalent from the drift had to be as precise as possible in order that the evapo-sublimation can be determined with the desired degree of accuracy. To accomplish this precision, the drift volume was determined by surveying, and the drift density was determined using the Federal Mt. Rose snow sampler to take samples at 10-foot (3.1 m) intervals along a transect down the center line of the drift. For purposes of comparison other drifts nearby were intensively sampled at various times during the year. The average densities of all of the drifts did not vary more than 2 percent, which is within the accuracy of the sampler.

Tables 1 and 2 show the computations made to determine evapo-sublimation loss from the two drifts during the melt season from April 26, 1973 through May 30, 1973. April 26 was the date of maximum snow accumulation in the drifts, and the last remnant of Drift No. 1 disappeared on May 30, while Drift No. 2 disappeared one day later.

The first week of the melt season (April 26 - May 3) was preceded by a large storm, and was cold and snowy with strong winds from the north (parallel with the fences). The computed evapo-sublimation rate was high during this period, so cross-sections of the drifts were inspected to determine if some of the unconsolidated snow was eroded during that week. The cross-sections on both drifts indicated that the increase in snow volume between April 12 and 26 occurred at the tail of the drifts, with some consolidation occurring in the main body of the drift. Evidently the tail of the drift was not packed as firmly as the main body and with the high winds out of the North during the week of April 26 to May 3, some of the snow in the drift was relocated. It is probable that the

Table 1

WATER BUDGET OF SNOW DRIFT NO. 1

Pole Mountain, Wyoming - Elevation 8,100 feet

April 26 - May 31, 1973

Date	Drift Vol.	Drift Density	Drift WE	Runoff During Period	Precip. During Period	Drift Surface Area	Apparent Evapo-Sub.	
	ft^3 (m^3)	%	ft^3 (m^3)	ft^3 (m^3)	ft^3 (m^3)	ft^2 (m^2)	ft^3 (m^3)	in. (mm).
	(1)	(2)	(3)	(4)	(5)	(6)	(7)	(8)
Apr.26	67,537 (1,913)	46	31,067 (880)			13,485 (1,253)		
Δ or Σ			-3,927 (111)	129 (4)	776 (22)	13,485 (1,253)	4,574 (130)	4.07 (103)
May 3	58,999 (1,671)	46	27,140 (769)			13,485 (1,253)		
Δ or Σ			-5,304 (150)	4,837 (137)	472 (13)	13,477 (1,252)	939 (127)	0.84 (21)
May 11	38,309 (1,085)	57	21,836 (618)			13,470 (1,251)		
Δ or Σ			-10,068 (285)	5,796 (164)	0 (0)	11,310 (1,051)	4,272 (121)	4.53 (115)
May 17	18,981 (538)	62	11,768 (333)			9,150 (850)		
Δ or Σ			-8,695 (246)	6,895 (195)	146 (4)	6,705 (623)	1,946 (55)	3.48 (88)
May 22	5,392 (153)	57	3,073 (87)			4,260 (396)		
Δ or Σ			-2,049 (58)	2,159 (61)	169 (5)	3,390 (315)	59 (2)	0.21 (5)
May 25	1,626 (46)	63	1,024 (29)			2,520 (234)		
Δ or Σ			-1,024 (29)	869 (125)	371 (10)	1,260 (117)	526 (15)	5.01 (127)
May 30	0	-	0			0		
Total-Apr.26-May 30			-31,067 (880)	20,685 (586)	1,934 (55)		12,316 (350)	18.14 (461)
Total-May 3-May 30			-27,140 (769)	20,556 (582)	1,158 (33)		7,742 (220)	14.07 (358)

Table 2

WATER BUDGET OF SNOW DRIFT NO. 2

Pole Mountain, Wyoming - Elevation 8,100 feet

April 26 - May 31, 1973

Date	Drift Vol.	Drift Density	Drift WE	Runoff During Period	Precip. During Period	Drift Surface Area	Apparent Evapo-Sub.	
	ft^3 (m^3)	%	ft^3 (m^3)	ft^3 (m^3)	ft^3 (m^3)	ft^2 (m^2)	ft^3 (m^3)	in. (mm.)
	(1)	(2)	(3)	(4)	(5)	(6)	(7)	(8)
Apr.26	56,032 (1,587)	46	25,775 (730)			13,370 (1,242)		
Δ or Σ			-3,128 (-89)	602 (17)	791 (22)	13,070 (1,214)	3,317 (94)	3.05 (78)
May 3	49,233 (1,394)	46	22,647 (641)			12,770 (1,186)		
Δ or Σ			-3,015 (-85)	3,268 (93)	482 (14)	11,335 (1,053)	229 (7)	0.24 (6)
May 10	35,704 (1,011)	55	19,632 (556)			9,900 (920)		
Δ or Σ			-7,880 (-223)	6,198 (176)	0 (0)	8,755 (813)	1,682 (48)	2.30 (58)
May 17	19,586 (555)	60	11,752 (333)			7,610 (707)		
Δ or Σ			-7,363 (-209)	5,187 (147)	57 (2)	6,465 (610)	2,233 (63)	4.14 (105)
May 21	7,568 (214)	58	4,389 (124)			5,320 (494)		
Δ or Σ			-1,082 (-31)	908 (126)	92 (3)	5,105 (474)	266 (8)	0.63 (16)
May 22	5,702 (161)	58	3,307 (93)			4,890 (454)		
Δ or Σ			-2,061 (-58)	2,207 (62)	172 (5)	3,845 (357)	26 (1)	0.08 (2)
May 25	2,042 (58)	61	1,246 (35)			2,800 (260)		
Δ or Σ			-1,246 (-35)	1,246 (35)	379 (11)	1,400 (130)	379 (11)	3.25 (83)
May 31	0	-	0			0		
Total-Apr.26-May 31			-25,775 (730)	19,616 (556)	1,973 (56)		8,132 (232)	13.69 (348)
Total-May 3-May 31			-22,647 (641)	19,014 (539)	1,182 (33)		4,815 (138)	10.64 (270)

relocated snow also evapo-sublimated, but because of some uncertainty, the computations of evapo-sublimation were made with and without the data from this weekly period.

The study shows that for Drift No. 1, the evapo-sublimation loss during May was 7,742 ft^3 (220 m^3), or 20 percent of the volume of water equivalent in storage at the beginning of the period. If the figures for the first week (the last week of April) are included the loss was 40 percent. For Drift No. 2 the evapo-sublimation loss for May was 4,815 ft^3 (138 m^3), or 21 percent of the initial drift water equivalent volume. For the entire period April 26 - May 31, the loss was 32 percent of the initial volume.

Although Drift No. 2 had a smaller starting volume than Drift No. 1, by May 17 the volumes were equal because Drift No. 1 had consistently experienced more evapo-sublimation loss than Drift No. 2. Drift No. 2, however, was more isolated from other nearby drifts and therefore, should have been subject to more of an "oasis effect."

Total ablation, M, (evapo-sublimation plus runoff) rather than just evapo-sublimation, per average surface area of the drifts during the weekly periods is shown in Table 3 below:

Table 3

Comparison of Total Ablation per Drift Surface Area

Time Period (1973)	Total Ablation (M = E + R)		Mean Surface Area of Drift (A)		M/A		Ratio M/A Drift 1 / Drift 2
	ft^3	m^3	ft^2	m^2	ft.	m.	
Drift No. 1							
4/26-5/3	4,703	133	13,485	1,253	.35	.11	
5/3-5/11	5,776	164	13,477	1,252	.43	.13	
5/11-5/17	10,068	285	11,310	1,051	.89	.27	
5/17-5/22	8,841	250	6,705	623	1.32	.40	
5/22-5/25	2,218	63	3,390	315	.65	.20	
5/25-5/30	1,395	40	1,260	117	1.11	.34	
Drift No. 2							
4/26-5/3	3,919	111	13,070	1,214	.30	.09	1.17
5/3-5/10	3,497	99	11,335	1,053	.31	.09	1.39
5/10-5/17	7,880	223	8,755	813	.90	.27	.99
5/17-5/22	8,594	243	6,250	581	1.38	.42	.96
5/22-5/25	2,233	63	3,845	357	.58	.18	1.12
5/25-5/31	1,625	46	1,400	130	1.16	.35	.96

From this table it can be seen that the two drifts are ablating quite uniformly. The two time periods of greatest disparity are the first and second. The reason for the variance during these time periods is that Drift No. 1 covered the entire rubber mat (13,485 ft^2, 1,253 m^2) while Drift No. 2 had a maximum surface area of 13,370 ft^2 (1,242 m^2) on a mat of 13,775 ft^2 (1,280 m^2). Through May 11, Drift No. 1 continued to cover most of its mat while the mat under Drift No. 2 was being exposed more and more. The depth of the tail of Drift No. 2 was less than that of Drift No. 1 and the ablation reduced the surface area without producing as much total ablation.

The reason why Drift No. 1 had more evapo-sublimation than Drift No. 2 must be due to some subtle differences in the aspects of the two drifts. These factors will be explored in more detail during the 1973-74 season.

It was hoped that by taking measurements of the drift volume throughout the winter an estimate of winter evapo-sublimation loss could be obtained. There was so much snow during the winter, however, that there was no complete period between measurements when drifting, D, was negligible; therefore no meaningful estimates could be made.

The hydrograph of runoff from one of the drifts is shown in Figure 1. The diurnal fluctuations very closely followed the temperature and radiation sequence, with peak flows occurring daily about 2 or 3 p.m. and minimum flows about 6 or 7 a.m., while the increase in runoff to May 18 corresponded to ablation of about 65 percent of the initial volume, on between 50 and 60 percent of the volumes in the drifts as of May 3. The sharp spikes on the hydrograph are the result of rainfall on the mat.

The evapo-sublimation amounts were converted to inches (or mm) by dividing by the average drift surface area. This is not exactly correct because the major ablation takes place at the edges of the drift. Yet, in order to compare evaporation rates as observed with those computed by previously derived models, it is necessary to have the rate expressed in depth units.

Because of the oasis effect it was hypothesized that existing models of evaporation are not adequate. Models used for testing were those of Light (1941), Kuz'min (1953), Corps of Engineers (1956), and Sabo (1956). Climatological data for application to the models were collected at the site. Incoming and reflected short wave radiation were measured with Kipp and Zonen solarometers, with backup measurement of incoming radiation provided by a Belfort pyrheliograph. Net total radiation was observed with a Fritschen net radiometer. Air temperature and humidity data were obtained from a hygrothermograph placed in a cotton region shelter adjacent to the drifts. Wind speed and direction were measured by an MRI recording gage. Due to malfunction of some of the instruments, some of the daily values had to be estimated from weekly totals; however an attempt was made to apply the models on a daily basis and to compare the computed evaporation with the observed evapo-sublimation on weekly and monthly totals. All of the models significantly underestimated the evaporation as can be seen in the following table.

Table 4

Comparison of Observed and Computed Evapo-Sublimation

May 3 - May 31, 1973

Observed Evapo-Sublimation E				Evaporation as computed by:							
Drift No. 1		Drift No. 2		Light		Kuz'min		Corps		Sabo	
in.	mm.	in.	mm.	in.	mm.	in.	mm.	in.	mm.	in.	mm.
14.07	358	138	270	1	25	0.1	3	2	51	0.3	7

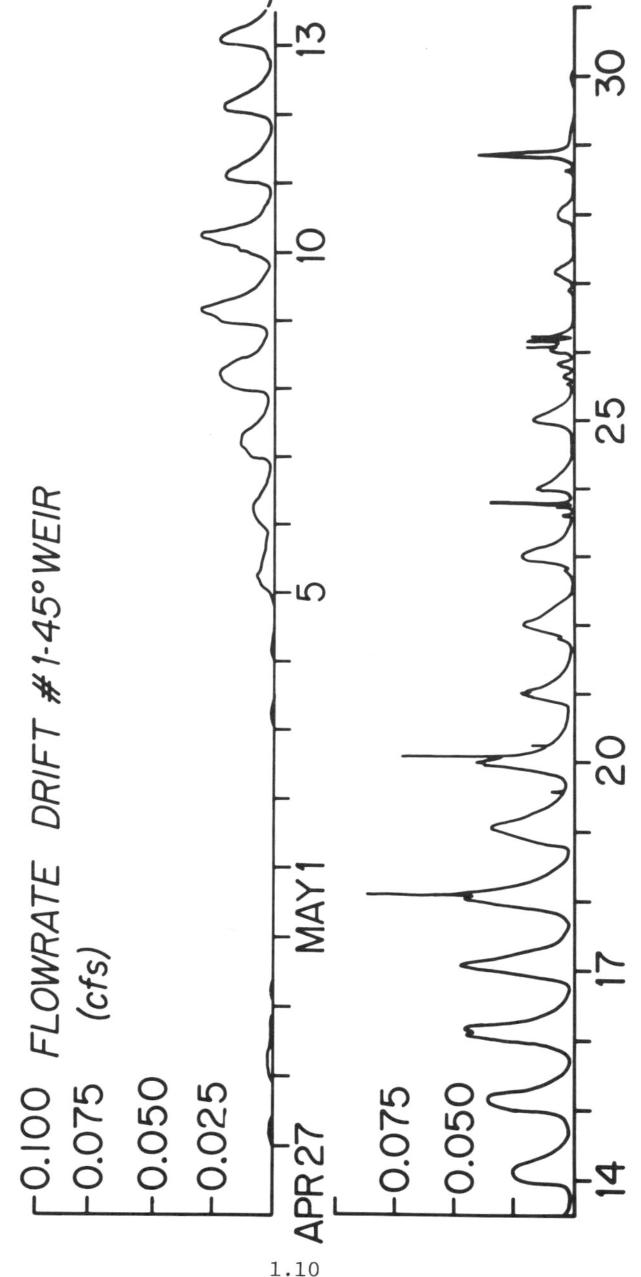

Figure 1. Hydrograph of Runoff from Drift No. 1 as Measured by 45° weir - Pole Mountain Area

CONCLUSIONS

It is concluded from the first year of operation of this study that during the ablation period evapo-sublimation from snowdrifts under oasis conditions approximates 20 to 30 percent of the water equivalent volume of the drift at the start of ablation.

Because the snowdrifts are under oasis conditions and because they are wind packed and therefore more dense than usual snow pack, existing models to compute evaporation from snow are not adequate.

ACKNOWLEDGEMENT

Research on which this paper is based was supported by the U. S. Forest Service, Rocky Mountain Forest and Range Experiment Station, Fort Collins, Colorado.

REFERENCES

Corps of Engineers, U. S. Army, 1956. Snow Hydrology: Summary Report of the Snow Investigations. Portland, Oregon. 437 p.

Kuz'min, P. P., 1953. K metodike issledovaniya i rascheta ispareniya c poverkhnosti snezhnogo pokrova. Trudy Gosudarstvennogo Gidrologicheskogo Instituta No. 41 (95).

Light, P., 1941. Analysis of High Rates of Snowmelting. Trans. American Geophysical Union Reports and Papers, Snow-Survey Conference, Sacramento. 195-205.

Sabo, E. D., 1956. Evaporation from the Snow Cover in the Ergen: District [in translation]. Selected Articles on Snow and Snow Evaporation, 1963. Jerusalem: Israel Program for Scientific Translations. 14-21.

Tabler, R. D., 1971. A Watershed Test of Snow Fences to Increase Snow Accumulation and Water Yield - First Year Results. Proceedings of the 39th Western Snow Conference, Billings, Montana, April 20-22, 1971. 50-55.

Tabler, R. D., 1973. Evaporation Losses of Windblown Snow, and the Potential for Recovery. Paper presented at 41st Western Snow Conference, Grand Junction, Colorado, April 17-19, 1973.

APPLICATION OF THE ENERGY BUDGET FOR PREDICTING SNOWMELT RUNOFF

Don M. Gray
Division of Hydrology, University of Saskatchewan

A.D.J. O'Neill
Atmospheric Environment Service, Department of Environment

ABSTRACT

Energy budget calculations were made during the snowmelt period in 1972 at the Bad Lake Research Watershed in Saskatchewan, Canada. Three separate situations were examined - the energetics of snowmelt for continuous snowcover, the case of an isolated, late-lying snow patch and the melt process on a small treeless basin. The results of the investigations suggest that, on the prairies, net radiation is the dominant energy source for snowmelt at the beginning of melt and that the amount of sensible heat transfer progressively increases in importance as bare ground appears. For a small watershed, it was found that definition of source areas of melt runoff and adjustments to the net radiative flux to account for slope and aspect are essential when attempting to relate energy budget predictions to the measured basin discharge.

INTRODUCTION

On the Canadian Prairies melt from snow is an important water resource - as it replenishes soil moisture needed for germination and growth of crops, serves as a potable fresh water supply for localized domestic use and as its release may result in flood-producing events. The results of the research presented herein are intended to assist in the development of a generalized snowmelt model for Prairie conditions by providing information on the relative magnitudes of the energy budget terms during the melt period.

The energy budget approach is an appropriate framework for the development of a snowmelt model in that the process may be explained on a physical basis and the melt quantity, M, expressed as:

$$M = \frac{Q_M}{L} \qquad \ldots\ 1$$

where Q_M = the amount of heat available for the melt process, and
L = the latent heat of fusion.

Partitioning of Q_M into its components requires examination of the relative importance of radiative, turbulent, conductive and advective heat transfer processes in the melt sequence. The budget for a non-isothermal snowpack, wholly or partly below 0°C may be written as:

$$Q_S = Q_{IS} - Q_{RS} - Q_{GS} + Q_{LD} - Q_{LU} + Q_H + Q_E + Q_V + Q_G - Q_M \qquad \ldots\ 2$$

where Q_S = the increase in internal energy storage within the snow pack,

Q_{IS} = the energy supplied by incoming solar radiation,

Q_{RS} = the energy loss due to the reflection of incoming solar radiation,

Q_{GS} = the energy transferred to the underlying soil surface by penetration of solar radiation through the pack,

Q_{LD} = the energy supplied by longwave radiation to the snow surface,

Q_{LU} = the energy loss by emission of longwave radiation from the snow surface,

Q_H = the gain of energy by turbulent transfer of sensible heat from the air to the snow surface,

Q_E = the gain of energy by turbulent transfer of latent heat (evaporation/condensation or sublimation) from the air to the snow surface,

Q_V = the gain of energy by all vertical advective processes (e.g., rain, the supply of heat in condensate and the mass removal by evaporation/sublimation), and

Q_G = the supply of energy to the snowpack from conduction from the underlying ground and percolation of melt-water and vapour transfer between snow and ground.

When continuous melt is in progress, a snowpack will rapidly reach an isothermal condition at a temperature of 0°C (U.S. Corps of Engineers, 1956) and therefore Q_S is assumed negligible. Similarly, most works (Anderson, 1968; Boyd, et al., 1962; U.S. Corps of Engineers, 1956) have either, implicitly, assumed that Q_{GS} is negligible or can be included in Q_G. Thus, Equation 2 reduces to:

$$Q_M = Q_N + Q_H + Q_E + Q_V + Q_G \qquad \ldots 3$$

in which Q_N = the net all-wave radiation supplied to the snowpack. Equation 3 has been found to be generally satisfactory for synthesizing melt at a point (Wilson, 1941; U.S. Corps of Engineers, 1956; Gold and Williams, 1960; Anderson, 1968; Müller and Keeler, 1969).

DeWalle and Meiman (1971) in their study of the energy budget of an isolated melting snowpack in a forest clearing in Colorado, found that 54 → 64% of the energy supply came from radiation, 41 → 49% from turbulent transfer of sensible heat and that 3 → 5% was used for evaporation. Müller and Keeler (1969) studied melting glaciers in the Canadian Arctic and found that net radiation was followed in importance by sensible heat transfer. A similar conclusion was drawn by Gold and Williams (1960) and Yoshida (1962). Many earlier workers (Orvig, 1954; Hoinkes, 1955; and Adkins, 1958) reported results indicating that net radiation was the main heat source for melt except in cloudy regimes where the turbulent transfer of sensible heat sometimes made the largest contribution.

On the Prairies, the general pattern of snow pack disappearance is that snow first disappears from the fallow fields. This is followed by a gradual shrinkage of the remaining snowcover until only patches or drifts persist in gullies and sheltered areas. Melt from these deep residual drifts frequently is not evident as surface runoff until after the open fields are free of snow. It is important to examine the energy budget of melting snow for these two completely diverse cases; under complete snowcover and for the situation where only isolated residual patches of snow contribute to streamflow. The practical problems involved in such a study necessitate that two separate approaches be taken in the energy budget calculations.

1. When the ground is completely snow-covered the presence of a uniform underlying surface at the bottom of the atmosphere suggests that the use of meteorological turbulence theory may be applied to the calculation of Q_H and Q_E with some validity.
2. When the land is partly snowfree, the application of standard meteorological techniques for calculation of Q_H and Q_E is probably invalid since the underlying surface is patchy and therefore non uniform and heterogeneous - with respect to the heat transfer processes.

THE TOTAL SNOWCOVER ENERGY BUDGET

From 11 March to 15 March, 1972, snowcover was almost complete over the watershed. Energy budget calculations were made during this period and the calculated melt quantities compared with those measured from a snow lysimeter. Details concerning the methods used for evaluating the different components of the energy budget are briefly described below.

1. Incoming, reflected, and net all-wave radiation were recorded continuously on site.
2. Sensible and latent heat components, Q_H and Q_E, were calculated by the Deacon-Swinbank method (1958). The wind and temperature data used for these calculations were measured at 20, 30, 47, 74, 116, 181, 285 and 700 cm and humidity profiles were established from dew point temperatures measured at 30, 47, 74, 116 and 700 cm above the snow surface.
3. Changes in the internal energy of the snowpack were estimated from snow temperature profiles.
4. Estimates of the soil heat flux, Q_G, were made by the temperature profile method described by Slatyer and McIlroy (1961). The results indicated an average heat loss from the snowpack to the ground in the range of 0.02 to 0.6 $MJm^{-2}h^{-1}$ during the afternoons. At night, the calculated flux values were near zero.

Table 1 shows the measured discharge obtained from the snow lysimeter and computed snowmelt for 12, 13 and 14 March, 1972.

Table 1

Comparison of the measured discharge from the snow lysimeter with snowmelt predicted by the energy budget - 12 March to 14 March, 1972. (The units are MJm^{-2} and equivalent water depth in cm^3/cm^2)

Date	Measured Discharge		Deacon-Swinbank	
	MJm^{-2}	cm	MJm^{-2}	cm
12/3/72	0.204	0.06	-0.0285	0
13/3/72	1.419	0.42	1.985	0.59
14/3/72	1.104	0.33	1.325	0.40
	2.727	0.82	3.282	0.99

The association between the energy equivalents of calculated melt and measured discharge on a daily basis provide a measure of support for the Deacon-Swinbank estimates of the terms Q_H and Q_E. Assuming these estimates of Q_H and Q_E to be realistic then the partitioning of total energy supplies to the snow during the period was 93% by net radiation and 7% by sensible heat transfer.

A typical daily trace of measured net radiation and the estimates of Q_H and Q_E is shown in Figure 1. Two features of the data are particularly interesting, namely:

1. The dominance of the net radiation component compared to the other terms.
2. The absence of any diurnal periodicity in the latent and sensible heat terms, Q_E and Q_H.

The dominance of net radiation over Q_E and Q_H suggests that it is the most important single term in the energy budget over a continuous cover of melting snow on the Prairies though examination of several years of data to include a variety of melt events will be necessary to substantiate this finding.

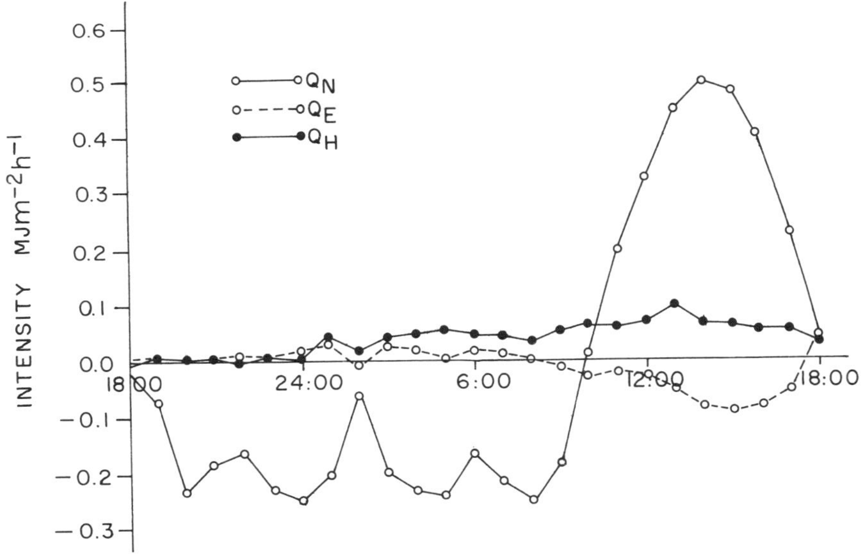

Figure 1. Radiation Components in the Period 18:00/March, 14 to 18:00/ March, 15.

ENERGY BUDGET OF AN ISOLATED SNOW PATCH DURING MELT

As stated previously, frequently on the Prairies a major source of snow melt runoff originates from late-lying snowdrifts. In attempt to clarify the relative significance of the radiative and sensible heat components under such conditions an energy budget study was undertaken on an isolated snowdrift.

The snowpatch was located in a small V-shaped depression on a south-facing slope. The maximum depth of the patch was 125 cm. Its surface area was recorded each day by staking the perimeter and after the snow had disappeared, the area encompassed within each set of stakes was determined by surveying. During the period of study the surface area of the snowpatch decreased from 407.2 m^2 on 8 April to 127.8 m^2 on 13 April, 1972. Other measurements taken included;

1. Incoming and reflected shortwave radiation obtained with Kipp and Zonen pyranometers and net radiation with a Funk-type net pyrradiometer.
2. Air temperatures at different heights above the snow surface.
3. Estimates of evaporation from the snow surface were made using a 3.29 dm x 3.28 dm x 2.19 dm plexiglass pan which was filled with snow and embedded flush with the surface of the pack. The pan was removed twice daily and the evaporative loss determined by weighing.
4. Runoff from the patch was measured directly using a calibrated V-notch weir.

The sensible heat flux, Q_H, was estimated from the Bowen ratio using the measured values of Q_E, air temperatures and dew points. When the air temperature fell below 0°C the temperature of the snow surface was assumed equal to the air temperature. The soil heat flux, Q_G, was neglected.

The magnitudes of the energy budget components are listed in Table 2.

Table 2

Components of daily energy budgets for an isolated snow patch during the melt period, 7 April to 13 April, 1972. (The units are MJm^{-2}).

Date/Time	Energy Budget Components			
	Q_N	Q_E	Q_H	TOTAL
7/1900 - 8/1900 CST	4.51	- 7.88	9.64	6.27
8/1900 - 9/1900	4.12	- 3.56	4.02	4.58
9/1900 - 10/1900	7.07	- 1.80	1.31	6.58
10/1900 - 11/1900	8.78	- 2.15	6.12	12.75
11/1900 - 12/1900	6.22	- 1.93	3.26	7.55
12/1900 - 13/1900	5.09	- 2.76	3.66	5.99
TOTALS	35.79	-20.08	28.01	43.72

The data indicate that except on April 7 - 8 more energy was supplied for melt by net radiation than as sensible heat. It is also evident that the relative amount of the total heat supplied for melt of the snow patch by sensible heat is much larger than for the case of complete snowcover. On the isolated snow patch, 56% of the estimated total energy supply during the period was by net radiation and 44% by sensible heat transfer. In contrast, for the total snowcover situation, the corresponding figures were 93% by net radiation and 7% by sensible heat transfer. These results support the argument that as the Prairie snowpack melts and becomes patchy, significant amounts of heat are advected from these snow-free areas and used to melt the snow on the adjacent snow-covered areas.

Table 3 lists measured discharge and predicted snowmelt.

Table 3

Comparison of the measured daily discharge from an isolated snowpatch at Bad Lake with the melt predicted by the energy budget. (The units are cm^3/cm^2 of surface area of snow).

Date	Measured Discharge (cm)	Predicted Melt (cm)
8/4/72	2.55	1.88
9/4/72	2.00	1.38
10/4/72	1.57	1.98
11/4/72	4.15	3.81
12/4/72	2.96	2.26
13/4/72	1.97	1.80
TOTALS	15.20	13.12

The data presented in Table 3 show close agreement between discharge and predicted melt. Surficially this high degree of association is very encouraging, however, complete acceptance without discussion would be misleading.

1. Infiltration quantities were not added to the measured discharge. This would suggest that the predicted melt quantities were greatly underestimated. From mass balance calculations completed on the patch it was estimated that 43% of the total available water was lost to evaporation and infiltration. Based on the data given in Table 2 it is unlikely that evaporation would account for more than 24% of this total - a potential or maximum value.
2. The underprediction of melt may be attributed to several factors of which the more important are;
(a) No adjustments were made to the radiation measurements to account for slope and aspect. As the pack was south-facing the radiative input would be larger than the value measured by the horizontal pyrradiometer.
(b) The ground heat flux was assumed negligible. For deep packs daily values for this term have been reported in the range, 0.17 → 0.84 MJm^{-2} - equivalent to ~ 0.05 → ~ 0.25 cm per unit area (U.S. Corps of Engineers, 1956; Gold, 1957 and Yoshida, 1962).
(c) No attempt was made to account for radiation penetration along the edges of patch nor energy advected to the patch from adjacent bare areas.

The results presented intuitively suggest that for conditions similar to those under which the study was conducted estimates of discharge which fall within acceptable practical limits can be obtained using a crude, simple energy budget.

ENERGY BUDGET ON A SMALL WATERSHED

It is of practical interest to examine the macro-scale effects of the energy transfer processes as mirrored by the melt and runoff events on a natural watershed. The basin selected for study has an area of 1.91 km^2, is free from tree cover and has a relatively well-defined topographic boundary. The watershed is about 8 km from a main climatological station and the meteorological data observed at the station were taken to be representative of conditions on the basin. Discharge from the watershed was recorded by a V-notch weir. In addition, aerial photographs of the basin and surrounding area were taken daily (weather permitting) during the 1972 snowmelt period.

On 7 March, total snowcover prevailed. By 12 March, a large amount of bare ground had appeared in areas of shallower snow accumulations, particularly on the fallow fields north of the main tributary and on the south-facing slope of the channel. Preferential melting continued on the fallow land and on exposed slopes so that, by 18 March, snowcover was confined to those areas sheltered from the sun and to deeper accumulations in hollows, gullies and the headwaters region.

The runoff hydrographs from the watershed showed strong diurnal cycling in discharge. The first noticeable runoff at the weir occurred on 16 March and the first measurable discharge on 17 March.

Ground surveys of snow conditions were carried out at frequent intervals prior to and during active melt. On 7 March, the first traces of melt were noticed on areas where the vegetal cover protruded through the snow and on the tops of small ridges on fallow fields where the depth of snow was very shallow. Where no exposed ground or vegetation existed, preferential melt occurred on those areas where the snow surface had been coated with dust. On 17 March, when the first discharge was recorded, 90% of the runoff was estimated to have originated from melt on the fallow fields on the north side of the watershed. Excavation of deep drifts in the gullies of the channel on that date revealed that the snow was not ripe and that no discharge occurred directly from these drifts. By 20 March, the stubble fields south of the channel started to contribute to streamflow but the largest input of melt water still originated from the more favourable radiation areas such as the fallow fields. As late as 24 March, only a small contribution could be attributed to the deep packs in the gullies and headwaters region.

APPLICATION

The coupling between snowmelt and runoff is complicated by spatial and temporal variations in the ripening process and by variable storage effects brought about by the freeze-thaw cycle. Detailed analysis of the melt to runoff phase is beyond the scope of this paper but qualitative relationships may be considered. The following work describes the application of a simple budget to a crude topographic model of the watershed.

The watershed was divided into six units each having distinct aspect and topographic features. Hence, the real topographic variations of the basin were represented by six plane surfaces for which representative slope and aspect values were determined. Partitioning of the basin in this manner was an attempt to provide a rational basis for adjusting net radiation to account for these topographic features. In adjusting the net radiation measurements the following procedure was used:

(1) The time over which the energy balance calculations were performed (16 March → 22 March) was divided into different periods according to cloud cover - periods when skies were overcast and periods when the sky cover was less than ten-tenths.

(2) For overcast conditions net radiation was assumed uniform over the basin since the diffuse nature of the solar radiation would not be significantly affected by slope differences.

(3) Under clear sky conditions (cloud cover < 1)

(a) The measured short wave component was divided to direct beam and diffuse sky components uding the experimental data given by List (1966). These calculations were completed on a two-hour time increment.

(b) For each basin unit the direct beam components was adjusted for sun sngle, slope and aspect and the adjusted value added to the diffuse component to provide an estimate of the "global" solar radiation.

(c) The net longwave radiation, Q_{NLW}, over the watershed was assumed constant and equal to: $Q_{NLW} = Q_N - (Q_{IS} - Q_{RS})$.

(d) The net all-wave radiation for a watershed unit was taken as the sum of the global solar component multiplied by the factor (1-A) and the net longwave radiation. The albedo, A, was measured over a snow surface.

Estimates of sensible and latent heat transfer were made by Sverdrup's (1936) method. In these calculations the temperature of the snow surface was taken as 0°C when the air temperature was greater than freezing otherwise the two temperatures were assumed equal.

Because of the lack of data, no attempt was made in the budget calculations to account for internal energy changes in the snow or the amount of ground heat flux.

In determining the total melt volume the change in snowcover with time was accounted for by direct proportioning of the energy flux according to the ratio; area of the watershed unit covered by snow to its total area. The snow-covered areas were defined on the aerial photographs and their area determined by planimetry. For days where no photographs were available the values were obtained from interpolation of the time-disappearance curve constructed.

The energy budget calculations were made on a daily basis (two-hour intervals) from 1700 CST on one day to 1700 CST on the next. The selection of this "chronological day" was based on consideration of the fact 1700 CST would represent an average time when overnight cooling of the snowpack would start. Thus, the calculations include an allowance for the overnight energy loss from the snowpack which should be compensated for on the following day before melt will occur. It is recognized that this procedure represents an over-simplification of the actual freeze-thaw cycle, however, its use was rationalized on the basis that the return of the snowpack to its initial condition (1700 CST) on a given day should mark the time of beginning of melt. In essence, the scheme equates the "cold content" to the overnight heat loss.

RESULTS AND DISCUSSION

The daily estimates of the energy budget components for each unit of the watershed are given in Table 4.

Table 4

Daily energy budget estimates for each of the 6 units of the watershed (The units are MJm^{-2}).

Date-Time	Watershed Units						
	No.	1	2	3	4	5	6
	Slope (°)	4.1	1.8	1.8	20	20	1
	Aspect (°)	101	140	60	9	189	99
	Area (km^2)	0.88	0.32	0.36	0.12	0.12	0.11
16/1700-17/1700 CST		1.48	1.77	1.55	-0.38	3.57	1.67
17/1700-18/1700 CST		2.59	2.88	2.65	0.71	4.56	2.78
18/1700-19/1700 CST		0.40	0.40	0.40	0.40	0.40	0.40
19/1700-20/1700 CST		2.58	2.76	2.65	1.40	3.77	2.72
20/1700-21/1700 CST		5.27	5.67	5.00	2.70	8.11	5.54
21/1700-22/1700 CST		2.96	3.39	3.06	0.31	5.74	3.27

In Table 4 it is of interest to note the large differences in the energy estimates for the north-facing (Unit 4) and south-facing (Unit 5) slopes of the watershed. For the period of calculation, the total heat balance on the two units was 4.94 MJm^{-2} and 26.15 MJm^{-2}, respectively. Also, the daily values of heat input to these two steeply-sloping units differ appreciably from the amounts calculated for the other parts of the watershed which have gradual slopes. This feature clearly demonstrates the necessity of adjusting the net radiation term for slope and aspect in applications of the energy budget to watersheds with high relief.

The predicted melt volume from the total watershed (23,096 m^3) was much greater than the measured volume (2,268 m^3). In comparing these values it should be recognized that several important assumptions were made in the calculations, namely;

(1) That the total heat available produced measurable melt (runoff) from all the snow-covered areas of the watershed.
(2) Infiltration, storage characteristics (mass and energy), ground heat flux and other factors were neglected.

The relative importance of the latter factors (Item 2) as they may affect the magnitude of the difference between calculated and measured values may be intuitively assessed. Of equal or greater significance,however, is the manner of runoff generation from different parts of the watershed. From ground surveys it was observed that the major "contributing" or "source areas" of runoff were the fallow areas of the watershed (Units 2 and 6). A small amount of flow from the stubble fields lying south of the main tributary was also observed. However, no major contributions were observed from deep snowdrifts even at the end of the study period. These visual observations suggest that valid comparisons of predicted melt and measured runoff can only be made after definition of the source areas from which runoff was generated. To exemplify, the calculated melt from the two major contributing areas (Unit 2 and the northern half of Unit 6) was compared with the total measured discharge from the watershed. For the six-day period the predicted melt was 4719 m^3 and the measured discharge 2268 m^3. Although the difference between the two values is still considerable, the direction in magnitude is correct and the results support the use of the energy budget approach for such calculations if information of the infiltration amounts and lag and storage effects are available.

The results from the study are considered more valuable in a qualitatively than quantitatively manner as they reflect the current "State of the Art" leading to development of generalized snowmelt based on energy considerations. In this context the study points out the need for additional study of specific parameters and processes namely;

(1) Identification of major source areas of runoff on a spatial scale,
(2) Evaluation of the effects of lag and storage of snowpacks on the runoff sequence,
(3) Infiltration of melt water to frozen or partially-frozen soils,
(4) Development of methodology and techniques for adjusting radiative terms for slope and aspect and procedures to be followed in partitioning a watershed as a simple terrain model to account for these factors, and
(5) Evaluation of energy budget components; sensible, latent, ground and internal energy changes for different snowcover conditions and stages of melt.

These studies should be directed to gaining an understanding of the dominant physical processes with the view as to how this knowledge may contribute to the development of a simplified model and <u>not</u> as it may lead to increasing the complexity of existing models.

REFERENCES

Adkins, C. J., 1958. The summer climate in the accumulation area of the Salmon Glacier. J. of Glaciology, Vol. 3, No. 23, pp. 195-206.

Anderson, E. A., 1968. The development and testing of snowpack energy balance equations. Water Res. Res., Vol. 4, No. 1, pp. 19-37.

Boyd, D. W., Gold, L. W. and G. P. Williams, 1962. Radiation balance during the snowmelt period at Ottawa, Canada. Res. Paper No. 175, Div. of Building Research, National Res. Council, Ottawa, 10 pp.

Deacon, E. L. and W. C. Swinbank, 1958. Comparison between momentum and water vapour transfer. Climatology and Microclimatology. Proc. Canberra Sym., Unesco Arid Zone Res., No. 11, pp. 38-41.

Dewalle, D. R. and J. R. Meiman, 1971. Energy exchange and late season snowmelt in a small opening in Colorado sub-alpine forest. Water Res. Res., Vol. 7, No. 1, pp. 184-188.

Gold, L. W. and G. P. Williams, 1960. Energy balance during the snowmelt period at an Ottawa site. IASH Gen. Assembly of Helsinki, pp. 288-294.

Hoinkes, H., 1955. Measurements of ablation and heat balance on alpine glaciers. J. of Glaciology, Vol. 2, No. 17, pp. 497-501.

Müller, F. and C. M. Keeler, 1969. Errors in short-term ablation measurements on melting ice surfaces. J. of Glaciology, Vol. 8, No. 52, pp. 91-105.

Orvig, S., 1954. Glacial-meteorological observations on ice caps in Baffin Island. Geografiska Annaler, Arg. XXXVI, Hafte 3-4, pp. 193-318.

Slatyer, R. O. and I. C. McIlroy, 1961. Practical microclimatology. CSIRO, Australia (prepared for UNESCO training course on Arid Zone Microclimatology), 250 pp.

Sverdrup, H. U., 1936. The eddy conductivity of the air over a smooth snow field - results of the Norwegian-Swedish Spitzbergen Expedition in 1934. Geofysiske Publikasjoner, Vol. 11, No. 7, pp. 1-69.

U.S. Army Corps of Engineers, 1956. Snow Hydrology. Summary Report of the Snow Investigations. North Pacific Div., Corps of Engineers, Portland, Oregon, 437 pp.

Yoshida, S., 1962. Hydrometeorological study on snowmelt. J. Met. Res., Vol. 14, No. 12. (In Japanese with English summary).

Date	Drift Vol.	Drift Density	Drift WE	Runoff During Period	Precip. During Period	Drift Surface Area	Apparent Evapo-Sub.	
	ft^3 (m^3)	%	ft^3 (m^3)	ft^3 (m^3)	ft^3 (m^3)	ft^2 (m^2)	ft^3 (m^3)	in. (mm.)
	(1)	(2)	(3)	(4)	(5)	(6)	(7)	(8)
Apr.26	56,032	46	25,775			13,370		
	(1,587)		(730)			(1,242)		
Δ or Σ			-3,128	602	791	13,070	3,317	3.05
			(-89)	(17)	(22)	(1,214)	(94)	(78)
May 3	49,233	46	22,647			12,770		
	(1,394)		(641)			(1,186)		
Δ or Σ			-3,015	3,268	482	11,335	229	0.24
			(-85)	(93)	(14)	(1,053)	(7)	(6)
May 10	35,704	55	19,632			9,900		
	(1,011)		(556)			(920)		
Δ or Σ			-7,880	6,198	0	8,755	1,682	2.30
			(-223)	(176)	(0)	(813)	(48)	(58)
May 17	19,586	60	11,752			7,610		
	(555)		(333)			(707)		
Δ or Σ			-7,363	5,187	57	6,465	2,233	4.14
			(-209)	(147)	(2)	(610)	(63)	(105)
May 21	7,568	58	4,389			5,320		
	(214)		(124)			(494)		
Δ or Σ			-1,082	908	92	5,105	266	0.63
			(-31)	(126)	(3)	(474)	(8)	(16)
May 22	5,702	58	3,307			4,890		
	(161)		(93)			(454)		
Δ or Σ			-2,061	2,207	172	3,845	26	0.08
			(-58)	(62)	(5)	(357)	(1)	(2)
May 25	2,042	61	1,246			2,800		
	(58)		(35)			(260)		
Δ or Σ			-1,246	1,246	379	1,400	379	3.25
			(-35)	(35)	(11)	(130)	(11)	(83)
May 31	0	-	0			0		
Total-Apr.26-May 31			-25,775	19,616	1,973		8,132	13.69
			(730)	(556)	(56)		(232)	(348)
Total-May 3-May 31			-22,647	19,014	1,182		4,815	10.64
			(641)	(539)	(33)		(138)	(270)

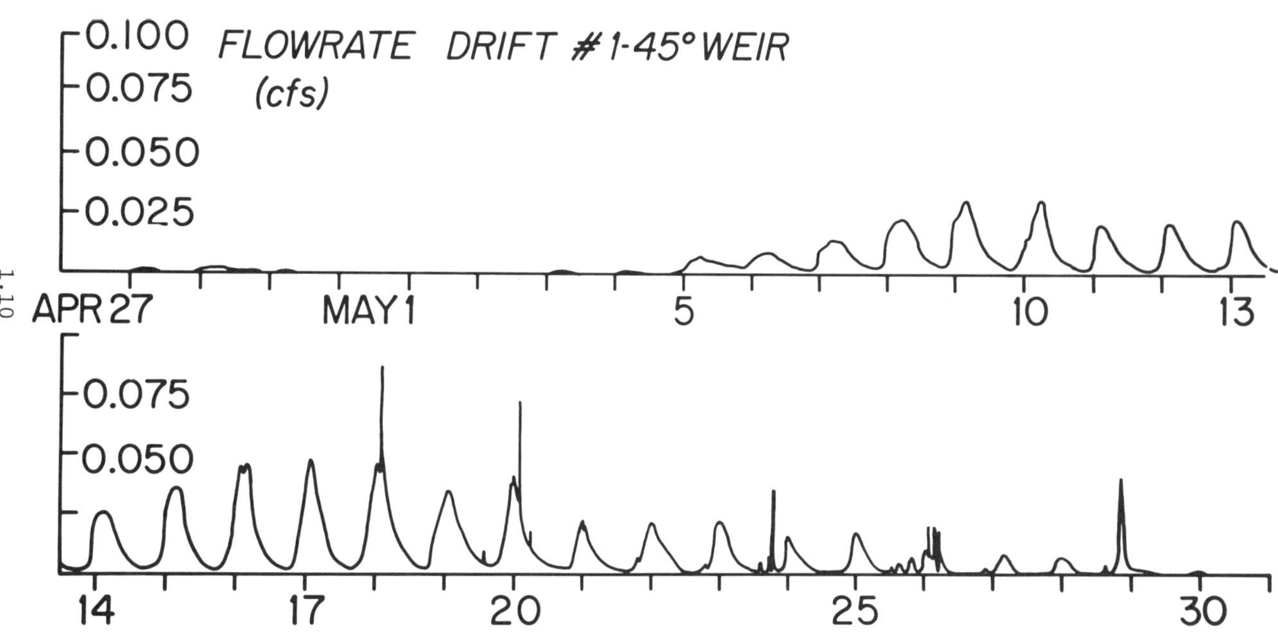

Gosz, J. R., G. E. Likens, F. H. Bormann, and J. S. Eaton. 1969. Leaching losses from leaves of selected tree species. Ecol. Soc. Am. Bull. 50 (2): 72.

Gilbert, Frank A. 1957. Mineral nutrition and the balance of life. Univ. Okla. Press, Norman. p. 52-56.

Johnson, Noye M., Robert C. Reynolds, and Gene E. Likens. 1972. Atmospheric sulfur; its effect of the chemical weathering of New England. Science 177: 514-516.

Johnson, Noye M., Gene E. Likens, F. H. Bormann, D. W. Fisher, and R. S. Pierce. 1969. A working model for the variation in streamwater chemistry at the Hubbard Brook Experimental Forest, New Hampshire. Water Resources Res. 5: 1353-1363.

Likens, Gene E., F. Herbert Bormann, and Noye M. Johnson. 1972. Acid rain. Environment 14 (2): 33-40.

Likens, G. E., F. H. Bormann, R. S. Pierce, and D. W. Fisher. 1971. Nutrient-hydrologic cycle interaction in small forested watershed-ecosystems. In: P. Duvigneaud (ed), Productivity of forest ecosystems. UNESCO. Brussels Symp. Proc. 1969.

Likens, G. E., F. H. Bormann, N. M. Johnson, and R. S. Pierce. 1967. The calcium, magnesium, potassium and sodium budgets for a small forested ecosystem. Ecology 48: 772-785.

Pierce, R. S., J. W. Hornbeck, G. E. Likens, and F. H. Bormann. 1970. Effect of elimination of vegetation on stream quality and quantity. Symp. Results Res. on Represent. and Exp. Basins Proc.: 311-328. Wellington, New Zealand. I.A.S.H.

DEUTERIUM IN ROCKY MOUNTAIN SNOWPACKS

James R. Meiman
Department of Earth Resources, Colorado State University

Irving Friedman
K. Hardcastle
U. S. Geological Survey, Denver, Colorado

ABSTRACT

Eighty-five snowpack cores taken as part of the April 1 snow survey in seven regions of Montana, Wyoming, and Colorado were analyzed for their deuterium concentrations. Mean regional deuterium content ranged from -133 $^{o}/oo$ (relative to Standard Mean Ocean Water) in the San Juan Mountains of southwestern Colorado to -183 $^{o}/oo$ in the Bighorn Mountains of Wyoming. The average coefficient of variation within regions was 4%. Stations in three regions gave varying degrees of differences for the years 1971 and 1972. The results suggest that stable isotope composition of snow can be regionalized for studies to increase understanding of snow hydrologic processes.

INTRODUCTION

Deuterium content in natural water can vary widely with geographic region, season, and topography (Friedman et al. 1964, Friedman and Smith 1970, Moser et al. 1972). Previous work by the authors has shown how variations in deuterium content of snow can be used to help understand processes during metamorphism in a seasonal snowpack (Judy et al. 1970) and in following the path of snowmelt in small forested watersheds (Meiman et al. 1972).

Deuterium content was measured in the spring snowpack near time of maximum accumulation at over 100 snow survey transects in the Rocky Mountains in Montana, Wyoming and Colorado during 1971-1973. These data will be presented in later reports and are not reproduced here. For the purpose of this conference regionalized summaries of the 1971 and 1972 data are presented. The objective of this presentation is to stimulate discussion on the application of such basic chemical characterization of snow to a better understanding of snow hydrologic processes. Two major applications that might be discussed are: (1) Is it possible to regionalize deuterium content of snowpacks for useful extrapola tion of localized studies? and (2) Can deuterium content (or other chemical characteristics) be used to gain a better understanding of the interaction of atmospheric and terrestrial factors influencing snow distribution and snowpack characteristics?

2.3

PROCEDURE

Snow survey courses are located at higher elevations at sites where snow normally is abundant until late spring. This results in a certain degree of "buffering" in the study area; large evaporative losses from shallow snowpacks or shallow snowpacks consisting primarily of snow from a recent localized snowfall could greatly complicate regional analysis of deuterium content.

Samples were collected by snow surveyors during their regular late March (April 1) surveys. Cores obtained with the Federal Snow Sampler were placed in heavy plastic bags and sealed. After returning to headquarters, the cores were allowed to melt in the plastic bags and then after shaking to thoroughly mix the water, a small liquid sample was poured into a bottle, sealed, and mailed to the U. S. Geological Survey Laboratory in Denver, Colorado.

Samples were converted to hydrogen gas and analyzed for deuterium content on a mass spectrometer as described by Friedman (1953) and Friedman and Woodcock (1957). Measurements are reported in δD values where

$$\delta D = \frac{\left(\frac{HD}{H_2}\right)_{\text{sample}} - \left(\frac{HD}{H_2}\right)_{\text{standard}}}{\left(\frac{HD}{H_2}\right)_{\text{standard}}} \times 1000$$

A δD value of -10 indicates that the sample was depleted in deuterium by 10 per mil (1 percent) relative to the standard. The standard used is Standard Mean Ocean Water (SMOW). Laboratory measurements are precise to the nearest per mil. (2 sigma).

Selection of the 85 determinations for this report was based on physiographic region (Figure 1). All 1971 data that could be associated with selected physiographic regions was used. For comparison of 1971 and 1972, all stations within three of the previously selected regions with data available for both years were used. The information on the snow survey locations is from the snow survey index for the states involved (U. S. Soil Conservation Service, 1970, 1971a, 1971b).

RESULTS

The results of the regional comparisons are presented in Table 1 and those for the 1971-1972 comparison are presented in Table 2.

Significant regional differences are evident in Table 1. Comparison of the San Juan region (G) with the Northern Colorado Front Range (Region E), and the difference between Northwest Montana (Region A) and Central Montana (Region B) are examples of such difference in snowpack deuterium content.

Comparison of 1971 and 1972 values from Table 2 reveals a much higher deuterium content (less negative values) in 1972 in the San Juan mountains, a less dramatic increase in deuterium content in 1972 in Northwest Montana and no overall change but high point-by-point fluctuation in Central Montana.

TABLE 1. COMPARISON OF REGIONAL DATA FOR 1971

Region and Elevation Range of Stations (m)	No. of Stations	°/oo D Relative to SMOW Mean	Range	Std.Dev.
A. Northwest Montana, west of 112° and north of Missoula. 1185 - 1950	6	-154	-147 to -165	6
B. Montana 109°-111° west longitude 1585 - 2275	8	-171	-164 to -182	6
C. Bighorn Mountains, Wyoming. 2255 - 2805	5	-183	-179 to -187	3
D. Wind River Range, Wyoming. 2235 - 3050	7	-169	-165 to -177	4
E. Northern Colorado Front Range. 2620 - 3230	15	-169	-164 to -180	4
F. North Central Colorado. 2590 - 3110	8	-161	-154 to -167	4
G. San Juan Mountains southwest Colorado. 2695 - 3505	15	-133	-114 to -154	12

DISCUSSION

The vapor pressure of HDO is less than that of H_2O, therefore water is enriched in D during condensation of water. If we assume a simple model in which an air mass derives its water vapor by evaporation from the oceans ($\delta D = 0$) at 25°C, this air mass will then contain water vapor with a δD of -70 °/oo. If this parcel of saturated air is cooled, either by radiative loss of heat or by orographic lifting it will lose water enriched in deuterium. The δD values of such precipitation are given in Figure 2. Assuming that snow precipitates when the temperature is lowered to 0°C, the first snow that precipitates should have a δD of -64 °/oo. Further

TABLE 2. COMPARISON OF 1971 AND 1972 FOR SELECTED STATIONS

Region and Snow Course	Elevation (m)	o/oo D Relative to SMOW	
		1971	1972
A. Northwest Montana			
Weasel Divide	1660	-147	-143
Brush Creek	1525	-152	-148
Kishenehn	1185	-153	-144
Mineral Creek	1220	-150	-132
Mt. Lockhart	1950	-154	-157
Cotter Mine	1905	-165	-156
		(-154) Mean	(-147)
B. Montana, 109-111°			
Bear Paw Ski	1585	-182	-179
Rock Creek	2000	-166	-183
Spur Park	2440	-171	-174
Bridger Bowl	2210	-165	-161
Shower Falls	2470	-177	-182
Crevice Mtn.	2560	-176	-176
Fisher Creek	2775	-172	-169
Porcupine R. S.	1980	-161	-173
		(-171) Mean	(-174)
C. San Juan Mountains Colorado			
Ironton Park	2990	-150	-123
Red Mt. Pass	3355	-154	-137
Purgatory	3050	-129	-116
Cascade	2695	-114	-112
Wolf Creek Summit	3355	-125	-100
Summitville	3505	-129	-106
Upper San Juan	3110	-117	-102
		(-131) Mean	(-114)

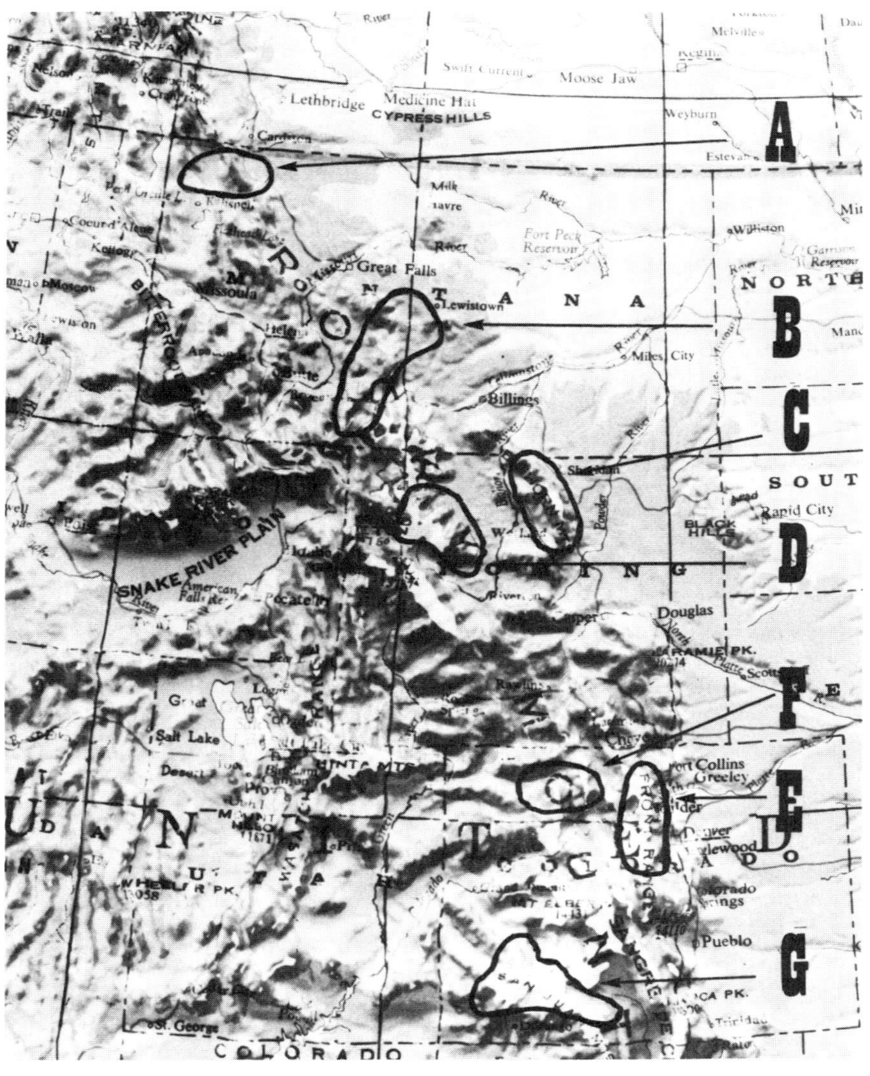

Figure 1. Study area with letters designating regions (see Table I).

2.3

cooling of the air mass will result in snow that is depleted further in deuterium; and snow as "light" as -180 ‰ is not unusual in the higher elevations of the Colorado and Montana Rockies.

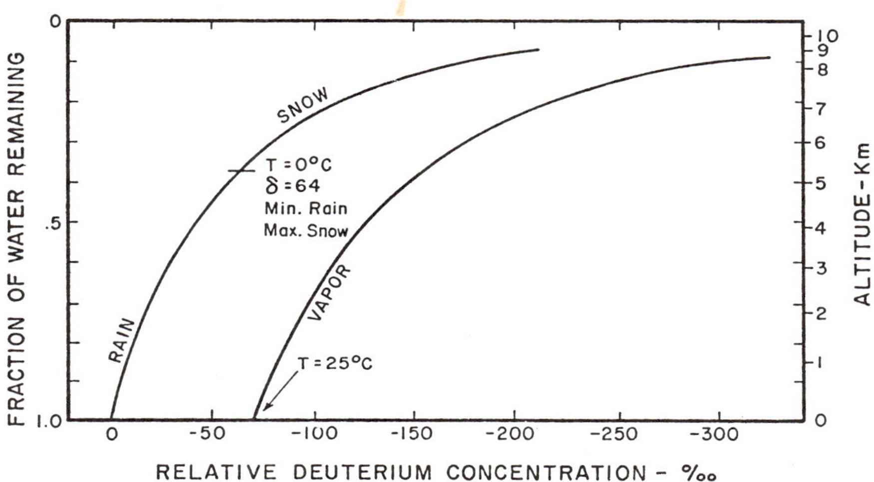

Figure 2. Relative deuterium concentration of precipitation and remaining vapor in atmosphere as function of elevation at which precipitation occurs assuming adiabatic conditions. (from Friedman et al, 1964)

Friedman et al. (1964) have presented data on regional variation in surface waters of North America. They found deuterium concentration to be determined primarily by prevailing temperature; deuterium content decreased with latitude and elevation.

Dansgaard (1961) has identified the factors influencing the isotopic composition of precipitation as:

1. Composition of the precipitating vapor at the beginning of the condensation.
2. The present mixing ratio in the air relative to its initial value.
3. The degree of evaporation from the precipitation until it reaches the ground.
4. The degree to which the re-evaporated material re-enters the precipitating air mass.
5. The temperatures at which the evaporation and condensation processes take place.

In addition to these factors, vapor transfer between snowpack and air can modify further the deuterium content of a snowpack. The variation within the selected regions discussed in this report would be expected to result largely from the interaction of topographic and meteorologic factors. Such variations have been documented by Moser et al. (1972) and Friedman and Smith (1970) among others. It is of interest that, even with these variations within a region, there are still pronounced differences between regions. Furthermore, there are evident year-to-year variations in some of the regions. Thus, there does appear to be potential for combining the deuterium analyses with analysis of air mass characteristics and movements to gain a better understanding of snow precipitation processes.

At an intermediate scale of atmosphere-topography interaction there is the possibility of testing hypotheses on snow precipitation processes. As an example, the companion paper in this session by Rhea and Grant (1973) discusses topographic influences on snowfall in mountainous terrain. It is possible that stable isotope content of individual snowfalls could be used to check Rhea and Grant's assumptions about precipitation efficiency if temperature data were available.

Finally, the data suggests for the area studied that meaningful extrapolations of localized studies to a regional basis may be possible, at least with respect to the snowpack factor. Examples of such studies are given by Arnason et al. (1973), Trabant and Benson (1972), Thorud and Friedman (1970), Dincer et al. (1970), and Meiman et al. (1972). In utilizing stable isotope labels in such studies care must be exerted to distinguish the characteristics of within-region variation.

SUMMARY AND CONCLUSIONS

Analysis of deuterium content of 85 snowpack cores taken for the April 1 snow survey in Montana, Wyoming, and Colorado gave the following results and conclusions:

1.) The mean 1971 deuterium concentration (o/oo D relative to SMOW) for the seven regions sampled ranged from -133 in the San Juan Mountains of southwestern Colorado to -183 in the Bighorn Mountains of Wyoming.
2.) Although within-region variation was evident (coefficients of variation ranging from 2 to 9%) the average coefficient of variation was only 4%.
3.) Comparison of 1971 with 1972 data for the same stations resulted in a mean shift from -131 to -114 in the San Juan Region and no significant difference in central Montana.
4.) These results suggest that the stable isotope composition of snow can be regionalized for studies to further our understanding of snow hydrologic processes, especially as applied to snow precipitation.

REFERENCES

Arnason, B., TH. Buason, J. Martinec, P. Theodorsson. 1972, Movement of water through snowpack traced by deuterium and tritium. Reprint of paper presented at Intl. Symposia on the Role of Snow and Ice in Hydrology. Banf, Alberta, Canada. Canadian Natl. Comm. for the IHD. 16 p.

Dansgaard, W. 1961. The isotopic composition of natural waters. Meddelelser om Gronland. 165(2): 1-20.

Dincer, T., J. Martinec, B. R. Payne, C. K. Yen. 1970. Variation of the tritium and oxygen 18 content in precipitation and snowpack in a representative basin in Czechoslovakia. Isotope Hydrology 1970, Proceedings of the Symposium, Intl. Atomic Energy Agency, Vienna. pp 23-42.

Friedman, Irving. 1953. Deuterium content of natural water and other substances. Geochimica et Cosmochimica ACTA 4(1/2): 89-103.

Friedman, I. and A. H. Woodcock. 1957. Determination of deuterium-hydrogen ratios in Hawaiian waters. Tellus 9(4): 553-556.

Friedman, Irving, A. C. Redfield, B. Schoen, and J. Harris. 1964. The variation of the deuterium content of natural waters in the hydrologic cycle. Reviews of Geophysics. 2(1): 177-223.

Friedman, Irving and George I. Smith. 1970. Deuterium content of snow cores from Sierra Nevada Area. Science 169:467-470.

Judy, Clark and James R. Meiman and Irving Friedman. 1970. Deuterium variations in an annual snowpack. Water Resources Research. 6(1): 125-129.

Meiman, J. R., I. Friedman, and K. Hardcastle. 1972. Deuterium as a tracer in snow hydrology. Preprint of paper presented at Intl. Symposia on the Role of Snow and Ice in Hydrology. Banf, Alberta, Canada. Canadian Natl. Comm. for the IHD, 11 p.

Moser, H., C. Silva, W. Stichler, I. Stowhas. 1972. Measuring the isotope content in precipitation in the Andes. Preprint of paper presented at Intl. Symposia on the Role of Snow and Ice in Hydrology. Banf, Alberta, Canada. Canadian Natl. Comm. for the IHD, 13 p.

Rhea, J. Owen and Lewis O. Grant. 1973. Topographic influences on snowfall patterns in mountainous terrain. Paper presented at Symposium on Advanced Concepts and Techniques in the Study of Snow and Ice Resources. Monterey, Calif. U.S. Natl. Comm. for the IHD.

Simpson, E. S., and D. B. Thorud. 1970. Distinguishing seasonal recharge to groundwater by deuterium analysis in southern Arizona. Intl. Assn. of Scientific Hydrology, Proceedings of the Reading Symposium-World Water Balance p. 112-121.

Trabant, Dennis and Carl Benson. 1972. Field experiments on the development of depth hoar. in Studies in Mineralogy and Precambrian Geology. Doe and Smith (Eds). Geol. Soc. of Amer. 348 p.

U. S. Department of Agriculture Soil Conservation Service. 1970. List and Location of Snow Courses and Soil Moisture Stations. Portland, Oregon. 16 p.

________. 1971a. Water Supply Outlook for Montana (April 1). Bozeman, Mont. 10 p + map.

________. 1971b. Water Supply Outlook for Wyoming (April 1). Casper, Wyo. 12 p + map.

ACKNOWLEDGEMENTS

The authors wish to thank the many snow surveyors in Montana, Wyoming, and Colorado who obtained the snow cores for analysis, and especially the U. S. Soil Conservation Service snow survey supervisors: Jack Washichek-Colorado, George Peak-Wyoming, and Phil Farnes-Montana.

2.3

MEASUREMENT OF ALBEDO OVER POLAR SNOW AND ICE FIELDS USING NIMBUS-3 SATELLITE DATA

Thomas H. Vonder Haar
Department of Atmospheric Science
Colorado State University

ABSTRACT

Effects of variations of snow and ice fields on hemispheric weather and climate patterns are largely accomplished through the albedo of these surfaces. For the first time recent measurements from meteorological satellites have allowed the determination of total spectral albedo (0.3 - 3 μm) over both polar regions at moderately high ground resolution. This paper describes the albedo data-reduction technique, the usefulness of surface measurements to augment the satellite observations, and the patterns and extremes of polar albedo variation measured during 1969-1970.

INTRODUCTION

Although some meteorological satellite measurements were first acquired in 1959, Vonder Haar and Suomi (1969, 1971) presented the first measurements of the planetary albedo from satellites on a global scale and for all seasons. This albedo, an energy ratio, is one component of the net energy exchange between our planet and space. The level of this energy flow and its variation from equator-to-pole in each hemisphere is the forcing function that leads to the circulation of our atmosphere and our oceans. The most current albedo measured from satellites is shown in Figure 1 (Vonder Haar, 1972). It was determined from measurements during 17 seasons during the period 1962-1970; it contains the effects of clouds and air molecules as well as surface reflection and has relatively low ground resolution.

Although Figure 1 shows that the polar regions have average annual albedo values exceeding 55 to 60 percent, the poor ground resolution conveys little detailed information. Thus, the remainder of the albedo data discussed in this paper will be derived from a Medium Resolution Infrared Radiometer experiment, a scanning radiometer (Raschke *et al.* 1973) having ground resolution of about 50 KM and onboard the near-polar orbiting Nimbus-3 satellite. We will confine the discussion to measurements over the polar and sub-polar regions in both the northern and southern hemisphere. Of course, the observations were primarily obtained during local summer. These albedo data will apply to the total shortwave energy spectrum (primarily 0.3-3μm) although the concepts discussed would apply also to spectral albedo data. We will focus on interpreting the satellite data in terms of the surface reflection characteristics and thus will use a term, "minimum" albedo, to do this. The scattering contribution to the planetary albedo from constituents of the polar atmosphere will not be considered and aside from the cloud effect which we attempt to remove, the remaining atmospheric contribution is small (even so, it could be removed with the aid of theory).

Some purposes for using satellite data to investigate the albedo of snow and ice fields and their adjacent regions are:

a) to study the surface radiation budget and its role in air-surface energy exchange, air mass modification and hydrologic processes;

2.4

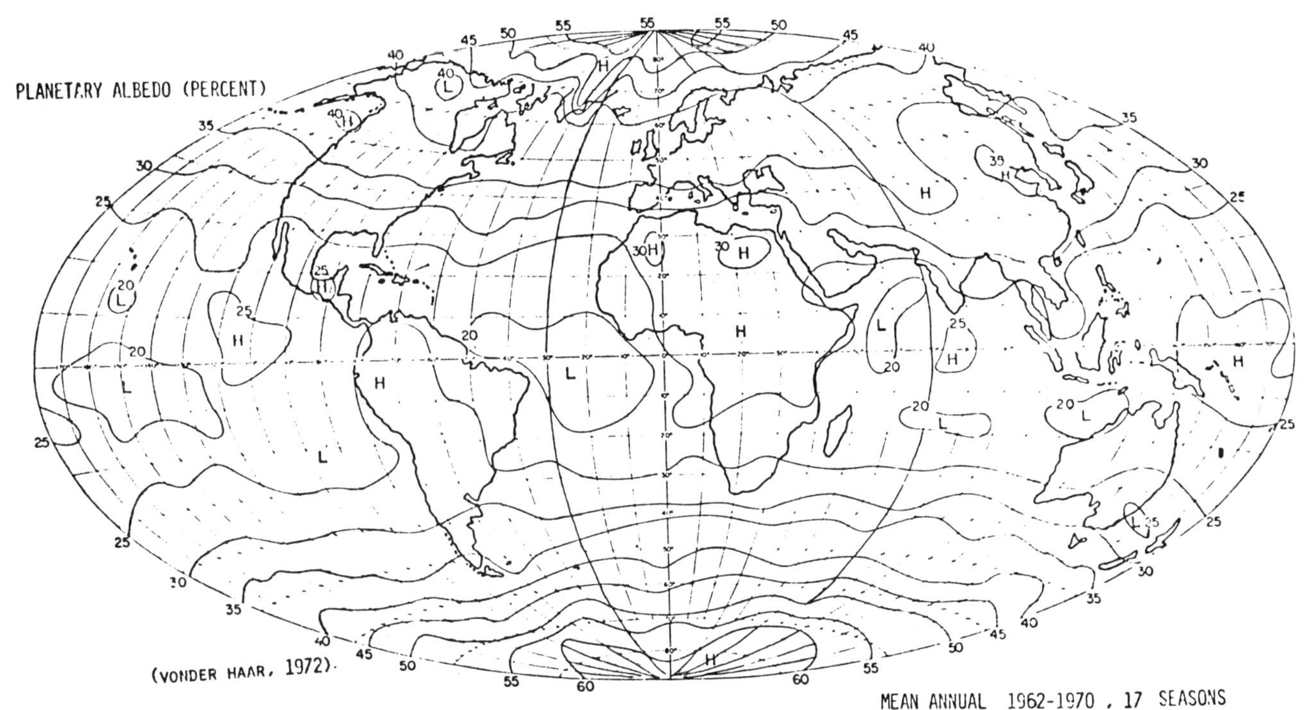

Figure 1. Mean Annual Albedo (17 seasons)

b) as a climate parameter affecting interannual variations of weather over large polar and sub-polar regions; and
c) as a remote sensing indicator of snow and ice conditions and their variation over periods of hours and days.

DETERMINATION OF ALBEDO

One channel of the 5-channel scanning radiometer measures reflected solar radiance (N_r in $w \cdot m^{-2} \cdot sr^{-1}$) over the spectral interval 0.3-3μm. This radiance measurement in (x,y,t) is used to determine the bi-directional reflectance $\rho(\zeta,\theta,\psi)$ from the relation:

$$\rho(\zeta',\theta,\psi) = \frac{N_r}{H_s \cos\zeta'} \ (sr^{-1}) \tag{1}$$

with
- H_s the solar constant adjusted for earth-sun distance
- ζ' zenith angle of the sun at the time of measurement
- θ,ψ zenith angle of satellite and relative sun-satellite azimuth angle, respectively

However the planetary albedo, A (sometimes called the total directional reflectance of a surface) is:

$$A(\zeta') = \int_0^{2\pi} \int_0^{\pi/2} \rho(\zeta',\theta,\psi) \cos\theta \sin\theta d\theta d\psi \tag{2}$$

For snow, assumed to be a nearly isotropic reflecting surface:

$$A(\zeta') = \pi\rho(\zeta',\theta,\psi) \tag{3}$$

and

$$A(\zeta') = \pi \ N_r/H_s \ \cos\zeta' \tag{4}$$

For other surfaces (such as clouds) known to depart greatly from isotropy, a bi-directional reflectance model is needed in order to evaluate Equation 2 for a single radiance measurement. Surface observations (field or laboratory) should also be done for various snow and ice surfaces in order to test the assumption invoked above.

Equation 4 allows determination of the planetary albedo at the time of satellite passage (denoted by ζ') but often it is necessary to determine the average daily albedo.

$$\overline{A} = \frac{\int_{t_n}^{t_s} A(\zeta'(t)) \frac{F(\zeta(t))}{F(\zeta'(t))} H_s \cos (t) \, dt}{\int_{t_n}^{t_s} H_s \cos\zeta(t) \, dt} \tag{5}$$

with $t_s - t_n$ = daylight period for a given day

$F(\zeta(t))$ = $A(\zeta)/A(\zeta=0^o)$ a normalizing factor to account for the variation of albedo of a surface with local time (changing solar illumination angle)

Daily average albedo values will be discussed in this paper. Thus, it was necessary to know the function $F(\zeta(t))$ for land, snow and ice surfaces, as well as for clouds and the open ocean (Figure 2).

THE NEED FOR SUPPORTING MEASUREMENTS AT THE SURFACE

The important values of $F(\zeta(t))$ shown in Figure 2 were obtained from relatively meager theoretical and observational data. For the snow function special measurements were made at 3KM in the Colorado mountains. These data are reported by Korff and Vonder Haar (1972) and by Korff et al. (1973). Their measurements agreed with others taken over Antarctic snowfields, but both sets of data show some variations of snow albedo with local time that need further study, notably a lack of reciprocity between albedo variations in morning and afternoon hours. Thus, just as in the bi-directional reflectance noted above, more field or laboratory measurements of the albedo of snow and ice fields should be encouraged. Interpretation of the satellite data is aided by these "ground-truth" data.

RESULTS

Figures 3 and 4 display the planetary albedo data from Nimbus-3 in terms of a "minimum" albedo. This value is the lowest daily average value encountered during the indicated measurement period. It was chosen for display since the technique allows an objective removal of clouds (which it is assumed would cause an increase over the surface albedo magnitude). Imperfections certainly exist in the method but a scan of the polar maps shows a very tight albedo gradient at the boundary of the ice sheets and continents and thus the method apparently does work to a first approximation.

Note that albedo values over the Antarctic and Greenland icefields often exceed 60% and that the minimum values of circumpolar isolines mark the farthest recession of the snow fields and pack ice. Further discussion of the albedo patterns and use of these data in quantitative energy budget studies will be the topic of future research. In particular, more detailed maps of the seasonal variation of minimum albedo over the northern continents are in preparation.

ACKNOWLEDGEMENTS

Ms. Lyn Koch typed this manuscript and the research was sponsored by the Laboratory for Meteorology and Earth Sciences, NASA, under Grant NGR 06-002-102.

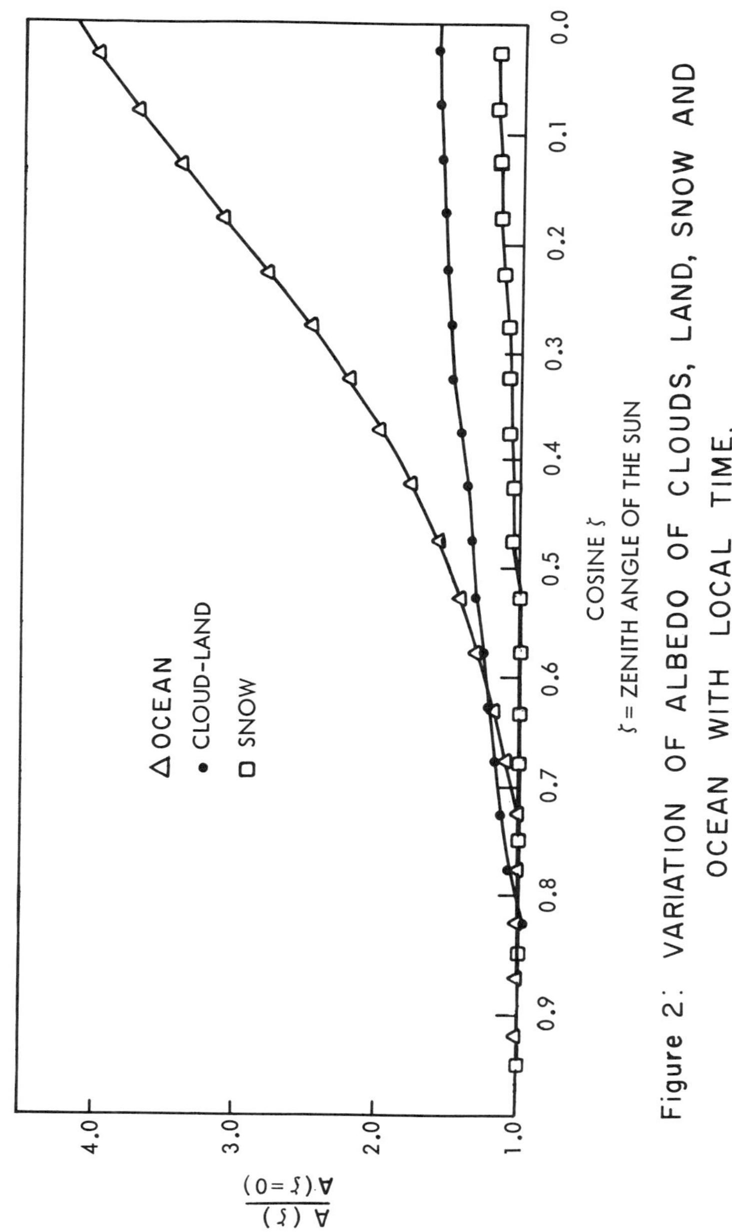

Figure 2: VARIATION OF ALBEDO OF CLOUDS, LAND, SNOW AND OCEAN WITH LOCAL TIME.

2.4

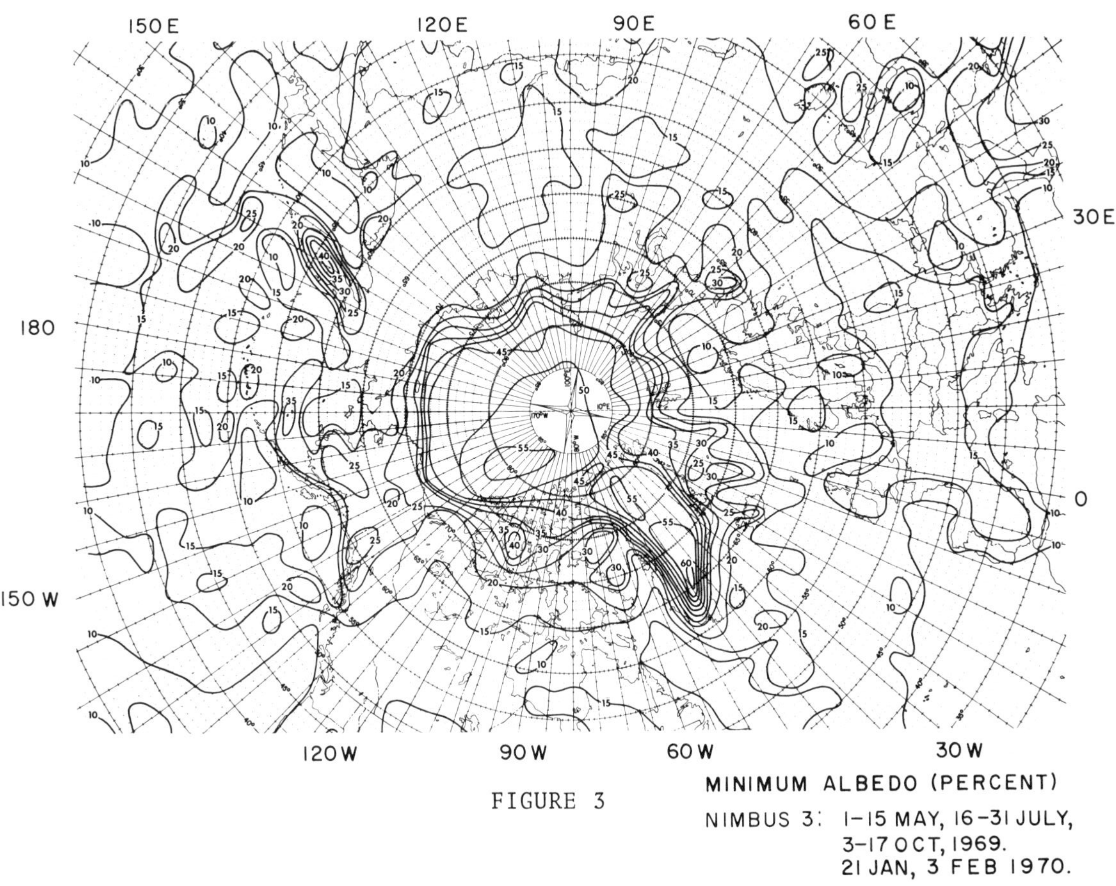

FIGURE 3

MINIMUM ALBEDO (PERCENT)
NIMBUS 3: 1-15 MAY, 16-31 JULY, 3-17 OCT, 1969. 21 JAN, 3 FEB 1970.

60E
90E
120E
150E
30E
180
0
150W
30W
60W
90W
120W
MINIMUM ALBEDO (PERCENT)
NIMBUS 3: 1-15 MAY, 16-31 JULY, 3-17 OCT, 1969. 21 JAN, 3 FEB 1970.

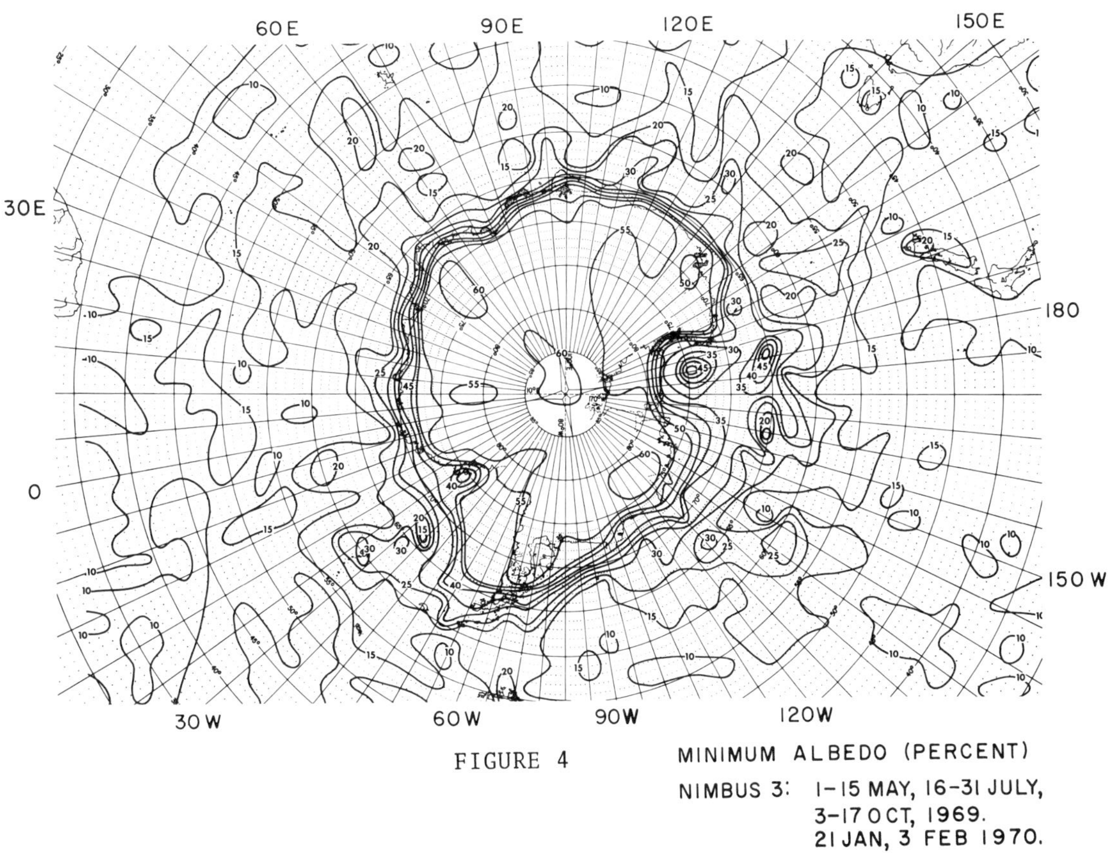

FIGURE 4

REFERENCES

Korff, H. C. and T. H. Vonder Haar, 1972: The albedo of snow in relation to the sun position. Preprint Volume, Conf. Atmos. Rad., Sponsored by Amer. Met. Soc., Aug. 7-9, Fort Collins, Colorado, pp. 236-238.

Korff, H. C., J. J. Gailiun and T. H. Vonder Haar, 1973: Radiation measurements over a snowfield at an elevated site. Manuscript in preparation for publication.

Raschke, E., T. H. Vonder Haar, W. R. Bandeen and M. Pasternak, 1973: The radiation balance of the earth-atmosphere system from Nimbus-III radiation measurements (15 April, 1969 - 3 February, 1970). NASA Tech. Note TN D-7249, April.

Vonder Haar, T. H. and V. Suomi, 1969: Satellite observation of the earth's radiation budget. Science, 163, 667-669.

Vonder Haar, T. H. and V. Suomi, 1971: Measurements of the earth's radiation budget from satellites during a five-year period. Part I: Extended time and space means. J. Atmos. Sci., 28, 3, 305-314.

Vonder Haar, T. H., 1972: Natural variation of the radiation budget of the earth-atmosphere system as measured from satellites. Preprint Volume, Conf. Atmos. Rad., Sponsored by Amer. Met. Soc., Aug. 7-9, Fort Collins, Colorado, pp. 211-220.

ELEVATION AND METEOROLOGICAL CONTROLS ON THE DENSITY OF NEW SNOW

Lewis O. Grant
J. Owen Rhea
Colorado State University

ABSTRACT

Results from a study relating new-fallen mountain snow density over three closely spaced Colorado mountain passes to causal factors (chiefly the vertical temperature profile of the precipitating cloud, windflow, and station elevation) are discussed. A comparison of the snow densities for a subset of the data encompassing approximately half of the events which were affected by weather modification is also presented. Snow data for this study were collected at daily intervals throughout 11 winter seasons from the fall of 1959 through the spring of 1970 over a dense network of stations near Climax, Colorado. Station elevations range from 7,800 ft. msl to 11,300 ft. msl. Snow densities observed ranged from $< .02$ g cm^{-3} to $> .15$ g cm^{-3}.

Results indicate a statistically significant dependency of new snow density on geographic location, upper wind direction and speed, temperature regime, and weather modification. For a given pass, density decreased with increasing elevation. Overall mean density was between .08 and .09 gm/cm^3. A parabolic relationship of density with 700 mb temperature was found. When 700 mb temperatures are adjusted to a reference level 2,500 ft. above station elevation, the temperatures at which minimum densities are observed corresponds well with the dendritic crystal growth range.

INTRODUCTION

The existence of variations in the density of newly fallen snow has long been recognized even though the quick conversion of 1" of water for 10" of snow (density 0.10 gm/cm^3) has had common usage. Variations in density from less than .01 gm/cm^3 to more than 0.30 gm/cm^3 have been reliably documented. Improved information on the actual snow densities for anticipated or actual snow episodes can provide important input for resolution of problems related to: (1) water resource planning and utilization; (2) avalanch forecasting and control; (3) snow removal; (4) ski resort planning and operations; (5) public activities, etc. At the present, however, routine information on and consideration of new snow density is minimal for snow management programs. Several investigators (Bossolasco, 1954; Diamond and Lowry, 1954; Wilson, 1955; LaChapelle,1961; LaChapelle, 1962; Power et al, 1964; Gunn, 1965; Judson, 1965) have reported on variations of density of new snow and some of these have considered certain of the meteorological (primarily temperature) and elevation controls on the snowfall densities for the specific sites that they have considered. Comprehensive definitions of density controls have not yet been clearly established. Snowfall data collected in the central Colorado Rockies in connection with a Colorado State University weather modification experiment provide a unique data base for studying many of the factors that affect the density of new snowfall. The randomized design of the weather modification experiment also makes possible a realistic evaluation of the effects of the cloud seeding on the snowfall density. These snowfall data, used in conjunction with corresponding surface and upper air meteorological information, have been used for this initial consideration of elevation, meteorological, and weather modification controls in the density of new snow.

DATA SOURCES AND STUDY AREA

Snowfall was observed daily at sites over Fremont, Vail and Hoosier passes, all located in the Central Colorado Rockies, during the 11 winter seasons from the winter 1959-60 through the winter 1969-70, (Figure 1). The observation sites were located at approximately 1 mile intervals over these respective mountain passes. Large clearings in the timber were selected for the observing stations. Sites were selected for which the timber skyline elevation angle was in the range of 15° to 40°, at least for the western exposure (prevailing winds westerly), from the observation stations. Snowfall observations were made daily at approximately 9 a.m. Observations were made from specially constructed snowboards. These snowboards had two wedges running the length of the boards in the bottom side to minimize settling in deep snow and a small pole on the upper side to aid in locating the board following snowfall events. Each board was painted white or silver to minimize radiation effects. The temperatures in the study area constantly remain below freezing during the winter months. The data have been screened to eliminate cases during the spring months when melting may have presented a problem. The depth of new snow at each site was determined daily by averaging the depth of new snow measured at several sites on each board. The water content of the new snow was obtained by obtaining the weight of a core of snow taken at a place on the board to correspond to the average snow depth. A relationship of snow density to deposit time on the snowboard is being developed using the large sample of data. This is possible since recording-weighing type precipitation gauges were in general located at every fifth snowboard site.

The emphasis on observations was for days that had been declared suitable as a weather modification experimental day. Randomization was then made to provide for cloud seeding attempts on one-half of the days with the other one-half left unseeded for control purposes. The analyses of snow density consequently are limited to the experimental days for which snowfall actually occurred. The data sample was further limited to those cases for which there was no evidence of drifting snow or melting (notes of such occurrences were made daily for the respective boards). Usable data is available from 64 sites ranging in elevation from 7,770 ft. to 11,300 ft. msl. The data sample varies from a low of 117 to a high of 370 among the various sites. The sites with the greatest number of snowfall cases are those at the higher elevations that experience more snowfall events. Since initial analyses strongly suggested that the snowfall densities for the seeded cases were different from those for the non-seeded cases, initial considerations of elevations and meteorological controls on the density of new snow are primarily based on data from the non-seeded cases only. This reduces the data sample for these analyses by approximately half. This is not a serious reduction since sample sizes for the respective stations still remain in the range from 53 through 178 for the respective stations.

The upper air meteorological parameters were defined from an interpolation of the Grand Junction and Denver, Colorado, radiosonde observations for the time of maximum snowfall for the Climax site. Actual soundings were made at the Climax site for many days and a comparison of actual and interpolated values show good agreement.

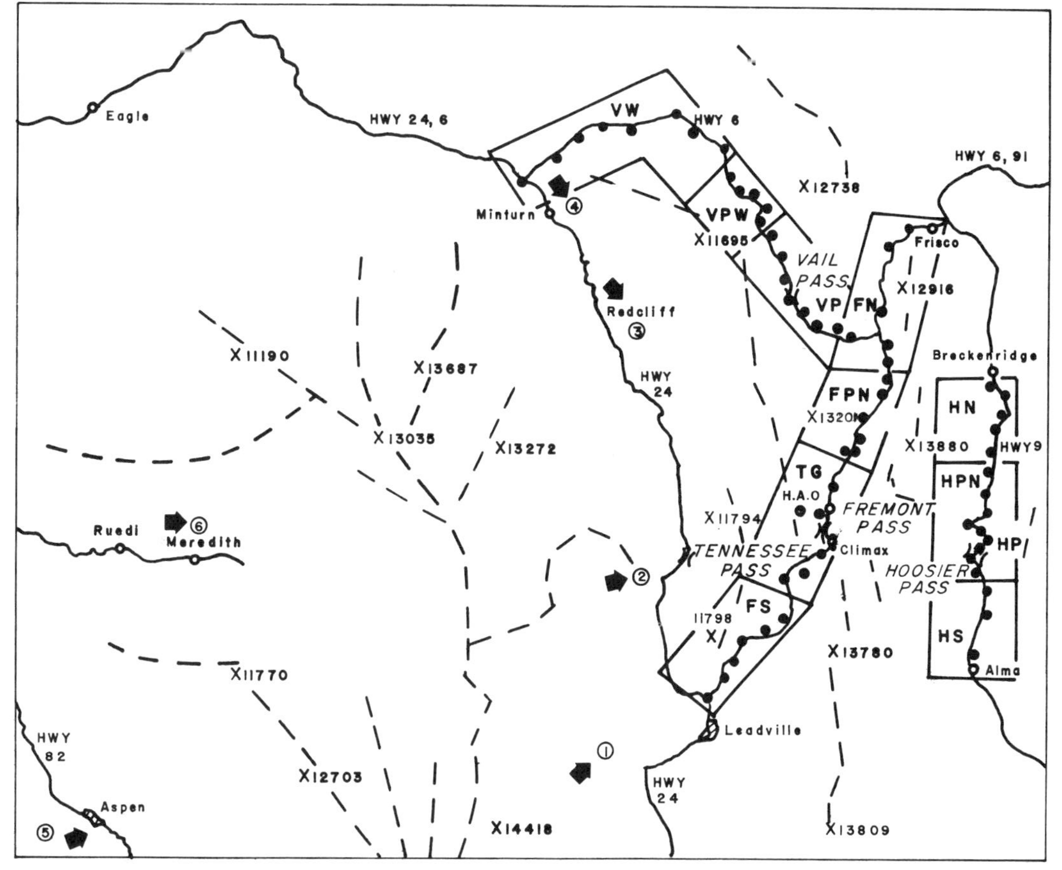

Figure 1. Location of Climax Experiment Area. Snowboard locations are shown as black dots. Seeding generator sites are numbered and circled. Block grouping of snowboards (for certain analyses) are shown. Major ridges are shown as dashed lines.

2.5

PHYSICAL CAUSES FOR VARIATION IN SNOWFALL DENSITY

The degree to which the density of new snow is below that of pure ice is controlled by a number of factors. The primary and controlling factor on new snow density is the amount of air space entrapped in the interstices between the individual snow crystals. The amount of air included within the individual crystals does not, for practical purposes, make a major contribution in the reduction of snowfall density below that for pure ice. The major factors contributing to the size of the interstices between the crystals include: (1) the crystal shape; (2) crystal size; and (3) the forces causing compaction.

Crystal Shape

Clearly, dendritic crystals with spatial extensions from the crystal base provide maximum opportunity for space between crystals as long as the spatial extensions are sufficiently strong to hold the crystals apart. Single crystal snowfalls of this type have, in fact, been observed to produce the lowest density snow (Power at al, 1964). Correspondingly, the more compact crystal forms have a greater opportunity for a tighter fit between crystals thus providing reduced sizes of the interstices. It is somewhat surprising, however, that, while such differences are actively observed, the magnitude of the differences are not as large as one might suppose. This results from the random fitting together of the individual snow crystals as they settle into the new snow cover. Power, et al (1963) have shown, for example, that, while one might expect that needles would pack together to give a relatively high density snow, they, in fact, produce a relative low density snow. Since these crystals form at relatively warm temperature (-3 to - 8°C) they tend to form snowflakes and provide a more open structure. Riming (the freezing of liquid cloud droplets on to individual ice crystals), on the other hand, can reduce the intercrystal space. This in turn should, and frequently does cause higher density snow. Since density is thus affected by both crystal shape and riming, and since both crystal shape and riming are highly dependent on temperature, any process that changes the glaciation temperature of the cloud (such as cloud seeding, crystal settling from higher elevations, etc.) can be expected to change the density of the snowfall.

Crystal Size

Crystal size also should, as with differing sized soil particles which produce "heavy" and "light" soils, produce large differences in snow densities with large crystals producing low density. Individual ice crystals in falling snow can vary in diameter from less than .1 mm to 3 to 4 mm and thus produce substantially different inter-crystal spatial differences. Since crystal growth rate is a function of temperature and humidity (most rapid growth occurs in the dendritic crystal growth regime around - 15°C) and, in general, is inversely related to crystal concentration, crystal size is highly dependent on temperature. Crystal concentrations are, in general, greater at the colder temperatures, since the number of ice forming nuclei activated increases by about a factor of 10 for each 4°C temperature decrease. This increase in crystal concentration, combined with reduced cloud water at the colder temperatures, generally assures reduced crystal sizes. These smaller crystals should allow for better packing and increased snow density. In addition, any cloud process, such as cloud seeding, which produces more ice crystals and corresponding decreases, in crystal size at any temperature should produce a higher density snowfall.

Crystal Compaction

The structural strength of ice is amazingly large. As a demonstration of this, LaChapelle (1962) has reported a 24-hour snowfall of 28" at Steven Pass, Washington, with a density of only .05 gm/cm^3. Clearly little settling had then taken place. Snow densification, however, does start at the time of deposition and appreciable settling and consequent densification can take place within a 24-hour period, as metamorphosis and crystal structural changes take place within a new snow layer. The weight of the snow in itself does not appear to be a controlling factor. LaChapelle (1962) in analyzing the snow density data for Stevens Pass, Washington, for the years 1951-1961 finds, for example, a poor relationship between depth (and consequently weight) of 24-hour snowfall amounts and density. Gunn (1965) reached the same conclusion from studying the snowfall in Montreal, Canada. He found that even when external weights were added that there was no detectable change in the subsidence rate. The temperature and vapor variations that cause crystal structural changes, and strong wind compaction, exert a greater control on densification. Analyses of these effects are in progress in the set of data used in this study.

RESULTS

Several conclusions can now be reached regarding the variations in density of newly fallen snow in the central Colorado Rockies.

Mean Snowfall Density

Mean snowfall density varies considerably for different sites. These variations are clearly related to the locations of the respective stations relative to nearby terrain features as well as to specific elevation and meteorological controls. The mean new snow density for the 64 sites included in this study varied from 0.77 gm/cm^3 to a high of .101 gm/cm^3). The mean density for 58% of the 64 sites was between .08 and .09 gm/cm^3.

Frequency of Various Snowfall Densities

Figure 2 shows the frequency distribution of peak (mode) values of snow densities for the 64 stations. Curve I shows the distribution for stations having a clearly defined single peak in the density while Curve II shows the peak (mode) value for all stations even though most stations exhibit a bimodal distribution with multiple peaks. It can be seen that the peak value (mode) density is almost equally frequent for density values from .07 to .10 gm/cm^3 but with distinct peaks at .07 and .09 gm/cm^3. Individual stations with multiple peaks also exhibit this characteristic.

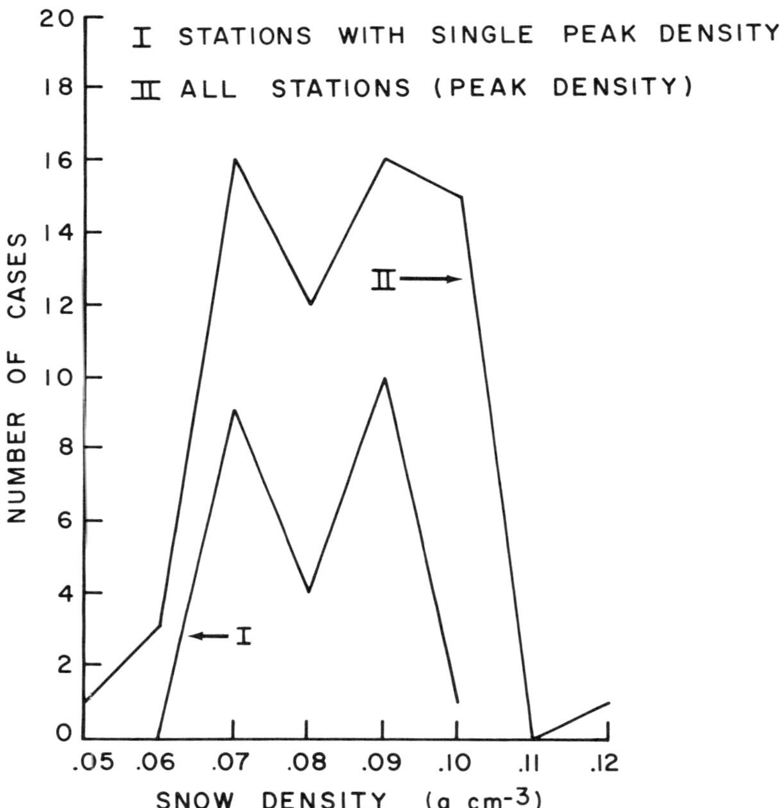

Figure 2. Frequency distribution of Peak (mode) snowfall density for 64 observation stations.

Elevation Controls on Snowfall Density

A substantial variation of density with elevation is evident for the respective mountain passes, although the geographical location of these passes obviously has an equal or even greater influence. Greater densities are observed at the lower elevations. This is probably related to a higher frequency and degree of riming of the ice crystals settling through lower and warmer portions of the cloud systems than for the ice crystals settling to the higher elevations. Evaluations are in progress to more clearly define elevation controls by considering them in combination with other factors.

Geographical Controls on Snowfall Density

The mean snowfall densities are consistently less for corresponding elevations on Vail Pass than on Fremont Pass. Densities at corresponding elevations on Hoosier Pass are still higher. This trend represents an increase in density from West to East. The primary cause for the higher densities on Hoosier Pass are most obvious. The mode values of density ($\approx$ .07 to .09 gm/cm^3) are similar for all three passes. The values are nearly normally distributed for Vail and Fremont Pass and most stations experience few densities beyond the .14 gm/cm^3 range. In contrast, the stations on Hoosier Pass consistently experience a few snowfall events with densities extending to values as high as 0.2 to 0.25 gm/cm^3. These cases are possibly related to deep moist east slope storms that occasionally cover this area. The cases with high densities are now being individually analyzed. The low densities on Vail Pass, particularly on the West side, probably reflect the almost complete absence of an effect from east slope storms. An additional cause for the high snowfall densities at the more eastward pass is likely related to progressively greater ice crystal concentrations and consequently smaller crystal sizes and to the long flat trajectories of small crystals which were initiated high in the cloud. Snowfalls on the western part of Vail Pass are generally from "new" clouds just starting to form on the upwind side of the main Colorado Rocky Mountain massif, and which still have low concentrations of ice crystals and larger crystal sizes. With subsequent passes further east, more residual ice crystals are frequently available in the atmosphere and concentrations are consequently generally higher and particle sizes smaller. As discussed previously this would lead to more dense snow.

Meteorological Controls on Snowfall Density

Meteorological variations clearly cause discernable variations in snowfall density. Density variations related to upper air temperatures, wind direction, and wind velocity have thus far been clearly identified. Testing for the reality of snow density changes thus far have primarily concentrated on testing the significance of differences in the density means for large samples fromdiffering meteorological groupings. Analyses thus far have concentrated on the data for one station, the High Altitude Observatory (HAO) near Fremont Pass. This station is the highest, 11,300 ft. msl, and has a large data base. A "t" test has been used to evaluate the significance of the differences. Table I shows a comparison of the mean snowfall densities for variations in upper level temperature, wind direction and wind velocities. It can be seen that the mean snowfall density is considerably higher for 700 mb temperature of -11°C or colder and for 500 mb temperatures colder than -28°C. There is an extremely low probability that these differences could have occurred by chance and consequently must be considered as real differences. This difference is consistent with physical considerations (although contrary to general impression) since more and smaller crystals which pack closer and produce higher snow densities should be expected at the colder temperatures. It can also be noted that, at least at this location, southwesterly upper level winds (190-250°) and low or high speed winds (in contrast to intermediate wind speeds) also cause higher density snowfall. Investigation of the details of these relationships are in progress using the rest of the network stations.

It is apparent that the relationship between air temperature and snowfall density is very complex. Lowest density snow is observed at HAO with 700 mb temperatures in the -4° to -8°C range. The solid curve in Figure 3 shows a best fit curve for the density of snowfall on not-seeded days at Climax as a function of 700 mb temperature. Snow density increases are observed for both colder and warmer temperatures. Lowest density snow should be observed with unrimed, dendritic-type snow crystals. These are formed at temperatures in the range of -14°C to -17°C. When the temperature at 700 mb (generally slightly below 10,000 ft. msl) is in the -4°C to -8°C temperature range, the -14°C to -17°C temperatures, at which dendritic crystals grow, would be at an elevation of around 14,000 ft. msl. This would be some 2,000 - 3,000 feet above the observing site at the High Altitude Observatory. Since the fall velocity of dendritic crystals is around 1/2 m/sec, the settling of dendritic crystals through this distance would in general take around 20-30 minutes, a time interval which would permit the growth of large dendritic crystals. It is reasonable to assume, that when the fall distance from a dendritic crystal growth regime is less, crystal sizes would decrease and snow density would increase. This would be reflected in a temperature decrease below the -4°C to -8°C temperature range at 700 mb. Such a decrease in temperature would also be associated with a higher percentage of non-dendritic ice crystals and an increased number of crystals with smaller sizes. This would further contribute to an increase in density with colder temperatures. When the dendritic growth zone is at elevations greater than around 2,000 to 3,000 ft. above the station, riming should be enhanced as the crystals fall through the lower, warmer portions of the cloud, and, in addition, a larger percentage of non-dendritic crystals would grow in the lower, warmer parts of the cloud. These factors, again, would contribute to an increase in snow density and would be consistent with the observed increase with 700 mb temperatures greater than about -4°C to -6°C. It thus appears that the general shape of the quadratic curve that best describes the HAO snow density -- 700 mb temperature relationship in Figure 3 is consistent with physical considerations. An "F" test for the significance of having obtained non-zero values for the coefficients for both the X and X^2 terms show that both are highly significant and consequently that the second order term makes a significant

Table I

Comparisons of the differences in mean snowfall densities at the HAO station for variations in upper level temperature and wind conditions.

Temperature	Mean Density	Standard Deviation	Number of Cases	T-value	Prob. value
700 mb temp. 0-10	.07548	.02903	251		
700 mb temp. ≤ -11	.09833	.03402	84	5.995	.0000
500 mb temp. 0-28	.07276	.02866	134		
500 mb temp. ≤ -28	.10063	.02760	44	5.61562	.0000
Wind Direction					
700 mb (190°-250°)	.08714	.02762	76		
700 mb (260°-360°)	.07924	.03342	244	1.86	.0316
500 mb (190°-250°)	.08955	.02945	107		
500 mb (310°-360°)	.07354	.03493	88	3.02	.0014
Wind Speed					
500 mb 12.0 - 17.0 MPS	.07143	.02417	45		
500 mb < 12 MPS	.08248	.03305	55	1.85	.0335
500 mb 12.0 - 17.0 MPS	.07143	.02417	45		
500 mb < 17 MPS	.08214	.03105	85	2.00	.0235

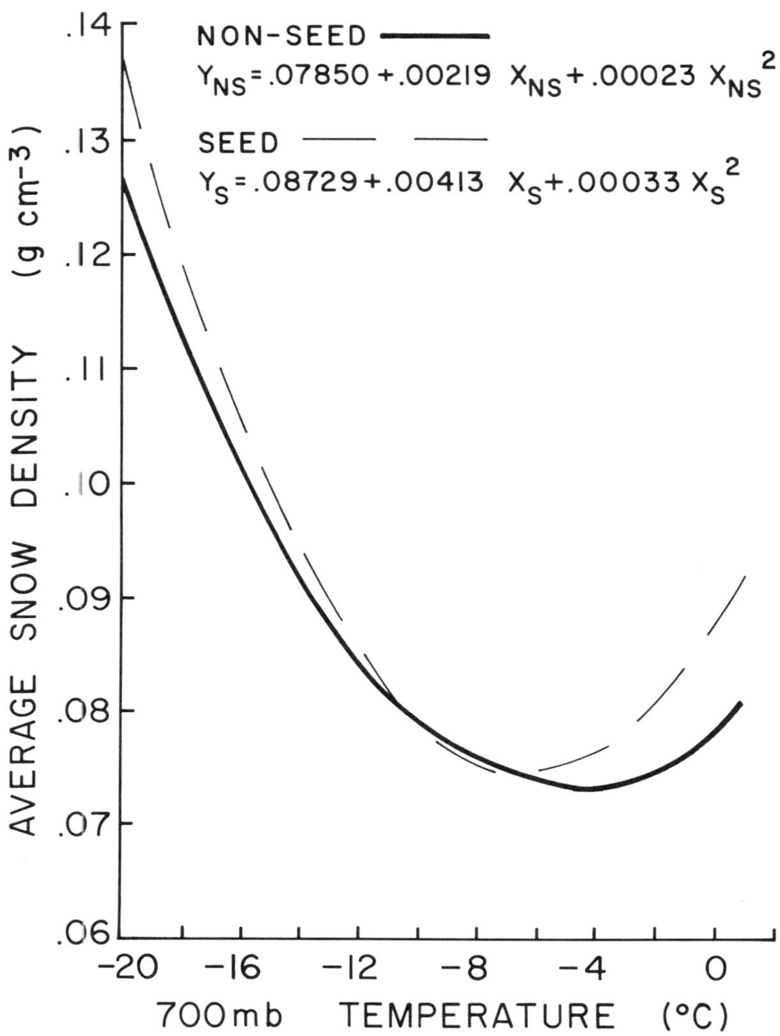

Figure 3. Best fit relationship between 700 mb temperature and snowfal density at the High Altitude Observatory.

contribution to the fit of the data.

Several investigators (LaChapelle, 1961, and Diamond and Lowry, 1954) have obtained results indicating density decreases with temperature at a reference level (e.g. surface or 700 mb). Bossolasco (1954), on the other hand, found a more or less parabolic dependency on surface temperature similar to our Figure 3. When reference temperatures of the present study and those of the above investigators were adjusted to surface temperature, it was found that results such as those of Diamond and Lowry (1954) are reflective of the warm end of our Figure 3 while Bossolasco's data fits ours quite well.

Cloud Seeding Controls on Snowfall Density

The analyses of the weather modification experiments at Climax have been extensively reported and summarized by Grant et al, 1971. It has been found that cold orographic clouds precipitate efficiently naturally under some meteorological conditions and inefficiently under others. There is a high likelihood that precipitation was increased by seeding on randomly selected seeded days for those days when the natural process was inefficient. Physical models have been developed that delineate conditions for which seeding is required to optimize precipitation efficiency. Meteorological conditions favoring cloud seeding effectiveness at Climax include: (1) warmer upper level temperatures (500 mb temperature > -20°C); (2) airflow perpendicular to the mountain, (southwesterly and northwesterly directions); and (3) intermediate wind speeds (700 mb speeds of 12-14 mps). Table II presents the results of a first effort to determine if snowfall densities may have been affected by the cloud seeding efforts. Snowfall densities have been considered for seeded and not-seeded samples for meteorologically defined categories that correspond to (but are not necessarily exactly the same as) ones for which seeding effectiveness would be expected. It can be seen from Table II that snowfall densities on seeded days were consistently greater on seeded than on not-seeded days. In the cases for temperatures and wind direction, the differences in the means are significant at about the 10% level. In the case of wind velocity, which can be very critical in size sorting of ice crystals, the significance is approaching the 1% level.

It can be noted from Figure 3 that the best fit curve for the seeded cases is consistently above the curve for not-seeded cases. This is particularly true at the warmer end of the curve where the alterations from seeding should be the greatest. The curves intersect with a slight shifting of the position near the density minimum. A "t" test has been used to consider the significance of the differences in the means of snow densities determined by these relationships. The mean $\bar{Y}_{ns}$ is .07842 gm/cm^3 and the mean $\bar{Y}_s$ is .08698 gm/cm^3. The T-value for this difference with 333 degrees of freedom is 2.65, giving a probability value of .0043. It is thus probable that there is a real difference in the distribution of snowfall densities between the seeded and not-seeded samples.

Table II

Comparison of snowfall densities for seeded and not-seeded cases within meteorologically defined categories at HAO.

Meteorological Category		Mean Density	Standard Deviation	Number of Cases	T-Value	Prob. value
500 mb temp.	not-seeded	.076	.030	25		
($\geq$ -19°C)	seeded	.088	.037	27	1.25	.109
700 mb wind direction		.077	.035	47		
(310°-360°)		.089	.046	43	1.33	.093
500 mb wind speed		.071	.024	45		
12-17 MPS		.084	.031	41	2.15	.017

ACKNOWLEDGMENT

The assistance provided by Gail Meltesen in making the statistical comparisons and by Sue Robertson in the preparation of the manuscript is gratefully acknowledged. The data utilized in these analyses were collected with support from the Atmospheric Science Section, National Science Foundation, under Grant GA-26401 and previous National Science Foundation Grants.

REFERENCES

Bossolasco, M. (1954): Newly Fallen Snow and Air Temperature. Printed in Nature, August 21, 1954, Vol. 174, No. 4425. pp. 362-363.

Diamond, M. and W. P. Lowry (1954): Correlation of Density of New Snow with 700-Millibar Temperature. Printed in Journal of Meteorology, Vol. 11. Snow, Ice and Permafrost Research Establishment and University of Wisconsin. pp. 512-513.

Grant, L. O., C. F. Chappell and P. W. Mielke, Jr. (1971): The Climax Experiment for Seeding Cold Orographic Clouds. Reprinted from Proceedings of the International Conference on Weather Modification at Canberra, Australia, 1971. pp. 78-84.

Gunn, K. L. S. (1965): Measurements on New-fallen Snow. McGill University Stormy Weather Group Scientific Report MW-44. Prepared for United States Air Force, Contract No. AF 19(628)-249. Bedford, Massachusetts, 27 pp.

Judson, A. (1965): The Weather and Climate of a High Mountain Pass in the Colorado Rockies. U. S. Forest Service Research Paper RM-16. U. S. Department of Agriculture, 29 pp.

LaChapelle, E. (1961): Snow Layer Densification. U. S. Department of Agriculture, Forest Service, Wasatch National Forest, Alta Avalanche Study Center, 8 pp.

LaChapelle, E. (1962): The Density Distribution of New Snow. U. S. Department of Agriculture, Forest Service, Wasatch National Forest, Alta Avalanche Study Center, 13 pp.

Power, B. A., P. W. Summers, and J. D'Avignon (1964): Snow Crystal Forms and Riming Effects as Related to Snowfall Density and General Storm Conditions. Printed in Journal of the Atmospheric Sciences, Vol. 21, May 1964. Weather Engineering Corporation of Canada, Ltd., pp. 300-305.

Wilson, W. T. (1955): The Density of New-fallen Snow. Weekly Weather and Crop Bulletin, Vol. 42, No. 51 (for week ending December 19, 1955), p.7.

TOPOGRAPHIC INFLUENCES ON SNOWFALL PATTERNS IN MOUNTAINOUS TERRAIN[1]

J. Owen Rhea
Lewis O. Grant
Department of Atmospheric Science
Colorado State University
Fort Collins, Colorado

ABSTRACT

Preliminary results from a study which considers total snowfall in mountainous areas to be the resultant of orographic lifting, large scale vertical motion, convection, and upstream barrier interception are discussed. The study utilizes data from (1) a large number of snowcourses and nonrecording precipitation gauges in Colorado and Utah (2) upper air soundings and maps, and (3) local topographic maps. Topographic slope (which determines the magnitude of orographic vertical velocity for a given wind speed) is computed relative to a given precipitation measurement point over scale lengths corresponding to the scales over which snow crystals grow and fall out (10 km to 50 km). Time intervals considered range from 6 hours to long term averages. The end objective of the study is to develop a technique (or techniques) which will (a) improve the short term prediction of mountain snowfall and (b) serve as a diagnostic aid in assessing areal distribution of snow water content on a seasonal cumulative basis (including assessments in ungauged areas).

Preliminary results to date indicate that long term snowcourse water equivalent distribution among points scattered throughout Colorado can be largely explained (r = 0.90) by systematic consideration of (a) the directionally adjusted topographic slope which potential precipitation bearing winds must traverse on their last 20 km of approach to a given station and (b) the number of upstream barriers which the air must pass over.

INTRODUCTION

Specifying the distribution of snowfall in mountainous areas is quite difficult indeed. The physical processes which determine snowfall distribution, including the general importance of terrain as a contributing factor, are known. Quantitatively, though, the overall process is quite complex, with the degree of complexity directly related to the complexity of the terrain.

To obtain usable estimates, hydrologic studies have used the observational fact that precipitation generally increases with

---

[1]This research was sponsored by the U.S. Department of Agriculture Forest Service under Contract No. 31-1371-2027 with Colorado State University.

2.6

elevation to develop local linear regression relationships between precipitation and elevation. A frequent restriction of this approach is that the relationships hold only for a relatively localized area and work particularly well for areas with similar aspect.

Mathematically, orographic precipitation rate is predominantly related to terrain slope and windflow rather than elevation. That is, if the air is saturated the rate at which precipitation is produced is directly proportional to the rate of rise of the air; and the rate of rise of air flowing across upsloping terrain is directly proportional to the product of the windspeed and the magnitude of the terrain slope. The total precipitation process is frequently compounded, however, due to components from (a) large scale vertical motion fields arising from migratory waves (storm centers and upper troughs) and (b) embedded cumulus convection as pointed out by Elliott and Shaffer (1962), Hjermstad (1970) and Chappell (1970).

Additional important effects to consider are (a) the passage of the airstream over major upstream barriers (and partial depletion of the available water for downstream precipitation), (b) simple physical interception of snow crystals by very nearby (1 km - 8 km) upstream barriers, and (c) measurement site location and/or gauge exposure (i.e., this latter is necessary to determine the representativeness of the catch).

Numerous other factors complicate the specification of snowfall distribution in mountainous regions, not the least of which are (1) effects of temperature and wind on snow crystal initiation altitudes and trajectories, respectively, (2) the possible presence of interference patterns between multiple mountain waves and (3) increasingly large errors in snowfall measurement as wind increases, thus precluding accurate knowledge of historical values of precipitation for calibrating models of the process.

Nonetheless, mountain winter precipitation regimes can be studied systematically with the objective of assessing the importance of a given contributing factor(s). It is to this end that this study has been undertaken.

DESIGN OF THE STUDY

General

In mountainous terrain, measured total point precipitation, R_T, can be symbolically expressed in terms of component processes as

$$R_T = (R_d + R_c + R_o)(f(n(z))) - \Delta R_i \pm \Delta R_g \quad (1)$$

Where

R_d = Large-Scale Vertical Motion Precipitation Component

R_c = Convective Precipitation Component

R_o = Orographic (Forced Lifting) Precipitation Component

$f(n(z))$ = Correction for Passage over Major Upstream Barriers

ΔR_i = Correction for Simple Physical Interception of Falling Snow Crystals by Very Nearby (1-8 km) Upstream Barriers

ΔR_g = Correction for Site Location and/or Gauge Exposure

The overall study has been designed to consider each of these components of total snowfall precipitation. Time integration increments for calculated values of each precipitation process component will range from 6 hours to the long term period of record. A large number of snowcourses, precipitation gauges, and all applicable upper air sounding records for Colorado and Utah will be used. Elevation values are being digitized on a 2.5 km horizontal grid increment to allow computation of orographic effects.

This paper describes pertinent results from a few preliminary investigations comparing only R_o, $f(n(z))$, and ΔR_g to total precipitation, R_T. Some information on the magnitude of $^gR_D + R_c$ is obtained, however, since this sum of nonorographic components appears as an intercept value on the R_T axis when plotting R_T against $R_o f(n(z))$.

Orographic Precipitation Formula

Without specifying the exact area of deposition, a generalized precipitation formula can be written as

$$R = \frac{E}{\rho_w g} \int_o^t \int_{P_1}^{P_2} \frac{dq_s}{dz} w dP dt \qquad (2)$$

where

$\frac{dq_s}{dz}$ = the height change of saturation mixing ratio (the ratio of the weight of water vapor to the weight of dry air)

w = the vertical motion, $\frac{dz}{dt}$

$\int_{P_1}^{P_2} dP$ = pressure depth of potentially precipitating saturated air column

$\int_o^t dt$ = duration of the process

ρ_w = density of water ($1 g\ cm^{-3}$)

g = gravitational acceleration

E = precipitation efficiency (proportion of the condensate which precipitates)

For air with horizontal velocity $\frac{dx}{dt}$, moving across terrain having slope $\frac{\partial z}{\partial x}$, the forced orographic vertical motion, w_{os}, (at the surface) is approximately given by

$$w_{os} = (\frac{\partial z}{\partial x}) \frac{dx}{dt} . \qquad (3)$$

This vertical motion changes with height as a function of stability (the more stable the air the more the opposition to vertical motion).

Thus to determine w_o for use in Equation (2), topographic slope computations are necessary. The distance scales over which slope is computed should be commensurate with those for snow crystal growth and fallout processes. Considering the terrain height, ranges of horizontal wind speeds, w_o, altitude of crystal habit temperature growth ranges, and crystal terminal velocities, these scales for orographic snow are found to vary between 10 and 50 km over the study area with mean values of 20-30 km.

Elevation values have been digitized at 2.5 km increments for Colorado and Utah to allow slope computations about selected points. Values of w_{os} computed range from -50 cm sec^{-1} to + 50 cm sec^{-1} with typical magnitudes of ±(5 to 30 cm sec^{-1}). Other investigators who have approached the precipitation distribution problem using topographic slope include (Elliott and Shaffer, 1962; Kusano, et al., 1957; and Spreen, 1947).

Correction for Passage over Upstream Barriers, f(n(z))

One grossly oversimplified formulation for f(n(z)) is

$$f(n(z)) = (1 - E)^{n} \tag{4}$$

where

n = is the number of upstream barriers of height $\geq z$
z = height of precipitation measurement site
E = precipitation efficiency

This exponential relation arises because the condensate, c, forced by the lifting over each barrier of equal height is depleted by proportion E through precipitation, leaving $(1-E)^{1}$ amount of condensate to be depleted by E amount over the second barrier, etc. This assumes among other things, that no resupply by surface evaporation or by upward mixing of sub-cloud higher dew point air occurs and that precipitation efficiency is constant from one barrier to the next.

Correction for Measurement Site Setting and/or Gauge Exposure

Catchment gauge exposure corrections will be very difficult to systematically consider.

Local settings for measurement sites on the other hand, will be placed into the three broad categories (at least for snowcourse sites) of small clearings, larger open areas, and heavy timber since flow separation, evaporation, canopy interception, and redistribution apparently behave so as to create somewhat systematic measurement variations in at least the first two of these categories.

SUMMARY OF PRELIMINARY INVESTIGATIONS AND RESULTS

Northwest Colorado Study

In connection with a weather modification project, special snowcourse and weighing bucket measurements were made along northwest Colorado's Park Range (Figure 1) beginning with the winter of 1964-65 and continuing through the present.

Average ratios of Rabbit Ears Pass (RE) to Buffalo Pass (BP) precipitation for two winters were computed using 4-hourly weighing bucket readings. Using the mean of the Mt. Harris (MH) 4-hourly rawinsonde 10,000 ft. winds and the surface wind atop a 100 ft. tower on Emerald Mt. (EM) 1 1/2 mi (2.5 km) southwest of Steamboat Springs (SS), these precipitation ratios were stratified by wind

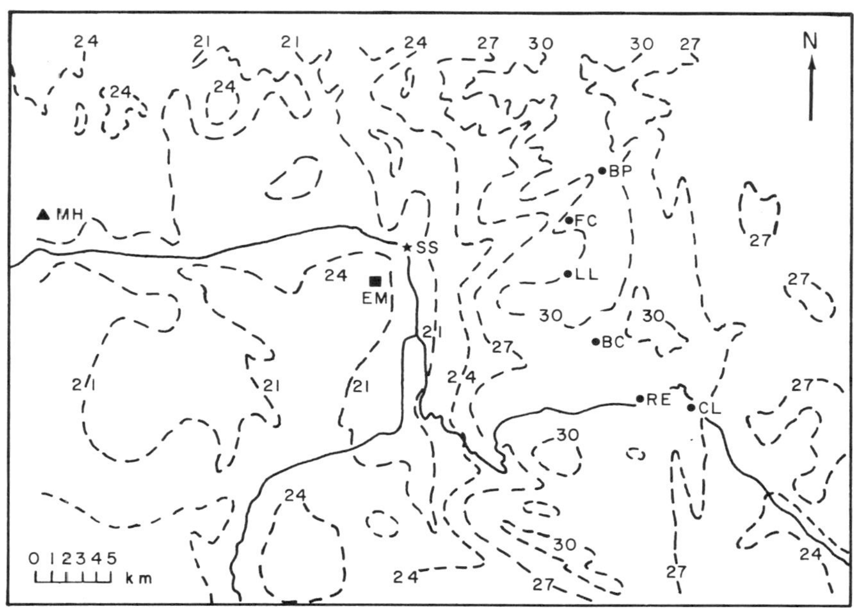

• SNOWCOURSE
■ AEROVANE
▲ UPPER AIR STATION
* PROJECT HEADQUARTERS

Figure 1. Smoothed topography (elev. in m.x10^2) and locations of measurement sites used in Section 3.1.

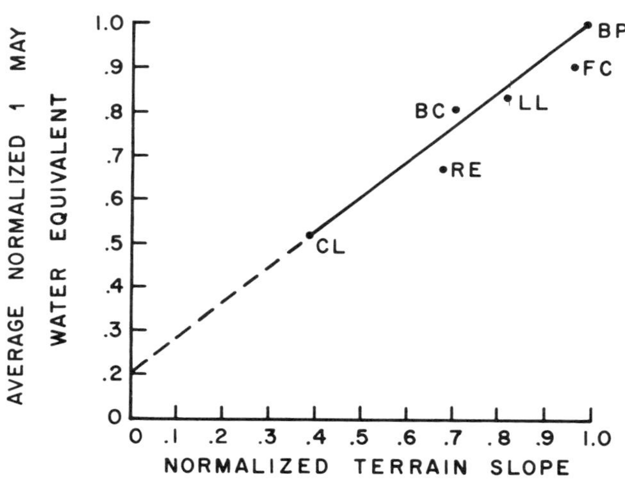

Figure 2. Normalized 1 May water equivalent vs. normalized terrain slope (See text, Section 3.1 for details).

direction (Table 1). Terrain slope values were computed over the first 20 km upwind from each of these stations by each wind direction category. Ratios of terrain slope (RE/BP) were formed and compared to the precipitation ratios (cf. columns 2 and 3 in Table 1). Very close agreement between slope and average precipitation is apparent.

To test this effect further, but strictly on the basis of seasonal snowcourses, terrain slopes were computed over the first 20 km upwind from each of six snowcourse sites on the Park Range (Buffalo Pass, Fish Creek (FC), Long Lake (LL), Base Camp (BC), Rabbit Ears, and Columbine Lodge (CL) in Figure 1). Elevations were noted 20 km upwind at each 15° direction increment over the western semicircle about the station in question. These elevations were subtracted from station elevation. Then these differences were summed. Sums were divided by the sum for Buffalo Pass (the highest and heaviest precipitation station). May 1 snowpack water equivalent (WE) values were also normalized to Buffalo Pass. When scatter plotting normalized WE vs. normalized terrain slope, strong positive correlation was evident (Figure 2).

TABLE 1

Ratios of Rabbit Ears Pass/Buffalo Pass Precipitation and Terrain Slope (computed over the first 20 km upwind of the station) by Wind Direction Class.

10,000 ft. MSL Wind Direction	Precip. Ratio R.E./B.P.	Terrain Slope Ratio R.E./B.P.
210° - 259°	0.65	0.63
260° - 279°	0.63	0.67
280° - 340°	0.84	0.81

Preliminary Snowcourse Study for Western Colorado

Long term average 1 April water equivalent was computed for each of 28 snowcourses scattered throughout western Colorado at elevations $\geq$ 9,000 ft. (Figure 3) and was correlated to actual station elevation (Figure 4).

An "effective" elevation was then computed for each station by averaging the elevation over a 2.5 km radius circle about the station

Topographic slope about each station was computed in 30° direction categories for the distance scale 0 (station location) to 20 km upwind (i.e., slope in a given direction category equals "effective" station elevation minus mean elevation 20 km upwind, with the difference divided by 20 km).

Referring to Equation (2), station slope for a given direction class was multiplied by each of the factors in Table 2 (data taken from an earlier study by Rhea, 1973). These products were then summed over the western semicircle of direction. Each station's sum was then multiplied by (a) 1,000 hrs. for the estimated mean duration of potential snowfall ($\int_0^t dt$) in Colorado's mountainous regions (Rhea, et al., 1969), (b) an average 700 mb wind speed of 9 m sec^{-1} (estimated from the data of Rhea, 1973), (c) a value of E = 0.5, and (d) the appropriate units conversion factors to arrive at computed mean seasonal (Nov. - March) orographic precipitation

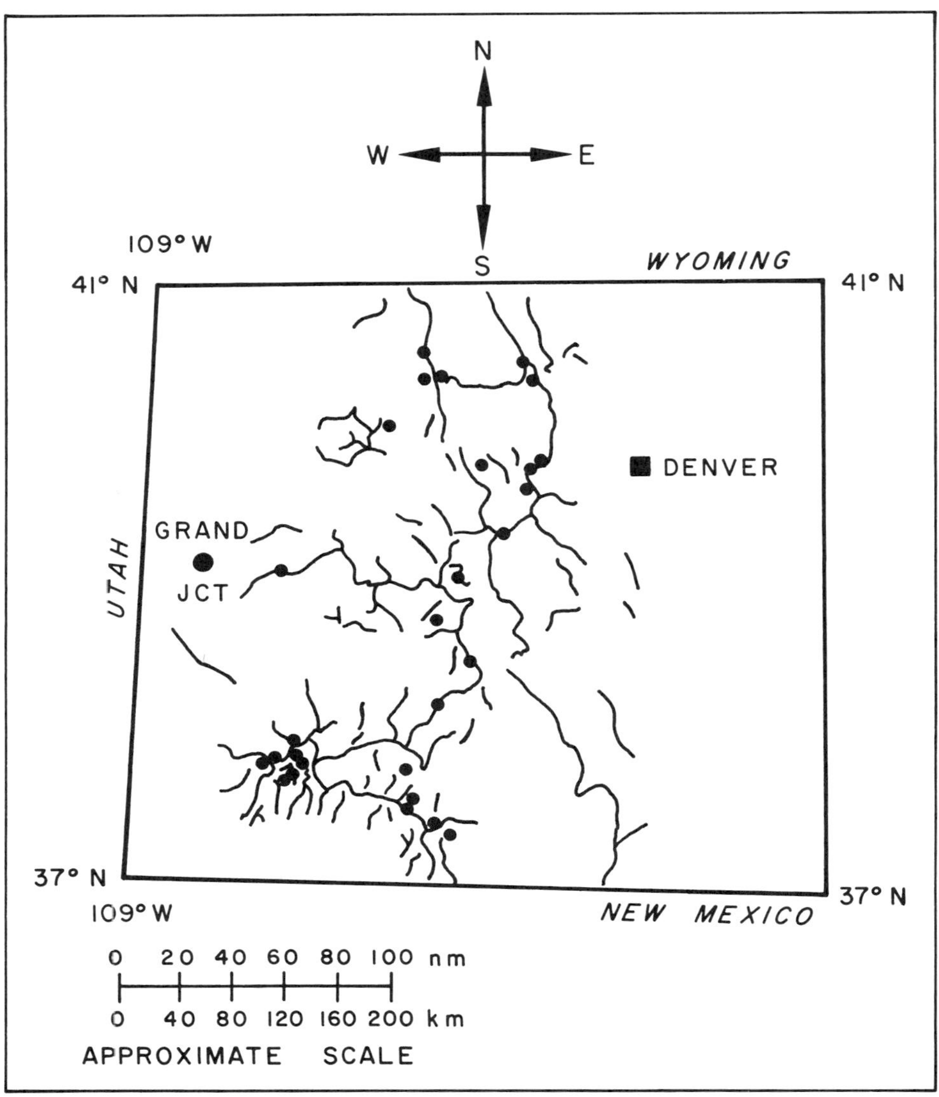

Figure 3. Major ridges over 9000 ft (2700 m) elevation and locations of snowcourses used in Section 3.2.

component for each station. These values were plotted against long term 1 April water equivalent values (Figure 5). The correlation coefficient was 0.82 with a standard error of estimate of 5.6 inches (140 mm).

Correction for passage over upstream barriers was then considered by computing the number of upstream barriers (whose elevations were not less than station "effective" elevation minus 300 meters) with the restrictions that the first barrier must be ≥ 10 km upstream (and at least as high as the "effective" elevation of the station) and each

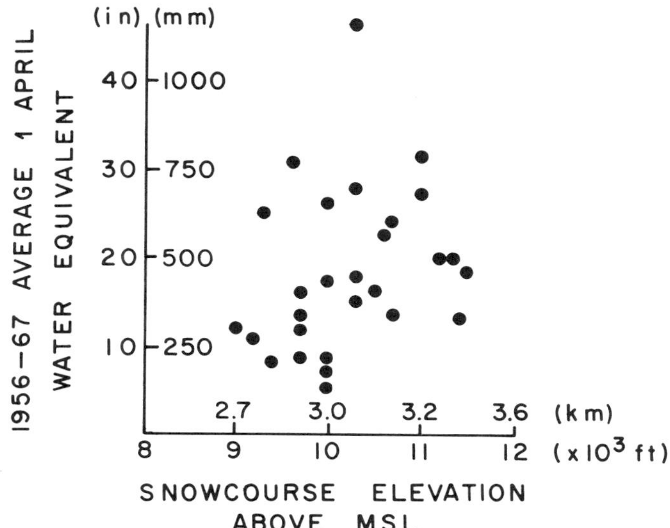

Figure 4. Relation of average 1 April water equivalent to station elevation.

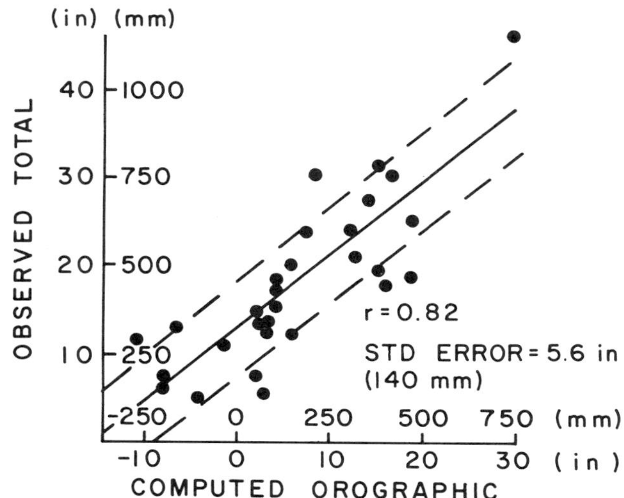

Figure 5. Computed Nov.-Mar. average orographic precipitation (horizontal axis) without correction for passage over upstream barriers versus observed 1956-67 average 1 April total water equivalent (vertical axis).

TABLE 2

	700 MB Direction Class (degrees)					
	175-204	205-234	235-264	265-294	295-324	325-354
Relative Frequency of "Precipitation Days"[1*]	.039	.134	.261	.315	.171	.080
Mean 700 MB $(\partial q_s/\partial z)$ gkg^{-1} km^{-1} [2*]	1.24	1.15	1.13	1.06	1.09	1.0
Estimated Avg. Potential Precipitating Cloud Depth with Cloud Base at 725 mb[3*]	275 mb	275 mb	237 mb	200 mb	162 mb	125 mb

[1*] These relative frequencies are from a 12 season (Nov. - March) precipitation record of 15 nonrecording precipitation stations in mountainous southwest Colorado $\geq$ 0.01" (.254 mm) precipitation at one or more of the stations defining a "precipitation day" each of which was then classed by the 24 hour vector average 700 mb wind direction at Grand Junction, Colorado (Figure 3).

[2*] Assuming 700 MB was saturated and moist adiabatic lapse rate prevailed.

[3*] Based partially on experience and also the reasoning that flow from the southwest quadrant is accompanied by large-scale upward vertical motion and deep clouds while clouds in northwest flow in the mean are mainly orographically forced and thus shallower.

succeeding barrier must be separated from the previous (downstream) one by $\geq$ 30 km (to permit one "snow crystal growth and fallout distance"). Equation (2) was then multiplied by Equation (4) (i.e., the component value of orographic precipitation in each direction category was multiplied by $(1 - 0.5)^n$). The results were again summed over the direction categories of the western semicircle. April 1 water equivalent was then correlated to these computations of orographic precipitation component (Figure 6). Single linear correlation was 0.90, with a standard error of 4.2 in. (106 mm) and line slope 0.95. From the R_T intercept, the apparent nonorographic average WE amounted to 13.8 in. (351 mm).

Departures of WE from the best fit line were then classed into the three site setting categories of Table 3 with the results showing that small clearings catch unrepresentatively large amounts, large clearings or areas of scattered trees are slightly deficient, and heavily timbered snow courses are still more deficient. It should be pointed out, however that these above site setting classifications are strictly approximate, being based only on site descriptions using qualitative and somewhat variable terminology.

CONCLUSIONS

From the preliminary investigations it can be tentatively concluded that over the study area

1. Long term average winter precipitation at a point is strongly positively correlated (r=0.82) to topographic slope computed over the

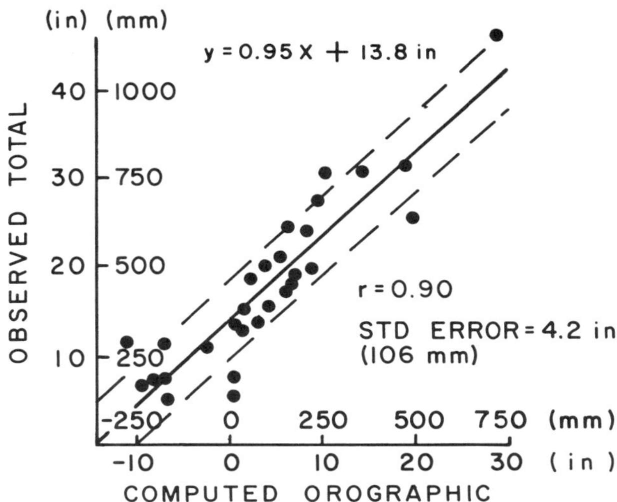

Figure 6. Computed Nov.-Mar. average orographic precipitation (horizontal axis) corrected for passage over n upstream barriers versus observed 1956-67 average 1 April water equivalent (vertical axis).

TABLE 3

SITE DESCRIPTION			
SMALL CLEARING		LARGER OPEN AREAS OR SCATTERED TREES	HEAVY TIMBER
	1.8	−2.3	−3.7
	−0.8	3.8	5.7
	1.5	1.2	−9.5
	3.5	−4.0	
	1.3	−7.6	
	7.4	−1.6	
	1.9	3.4	
	−3.4	−1.8	
	3.2	−3.3	
	1.8	−3.1	
		−8.7	
		−2.6	
		0.1	
		−3.2	
		−1.1	
NO.>0	8	4	1
NO.<0	2	11	2
AVE.	1.8	−2.1	−2.5

2.6

first 20 km upwind of the stations in 30° direction classes over the western semicircle and weighted by (a) direction frequency of "precipitation days" and (b) the estimated cloud depth for each direction.

2. Incorporation of a simple exponential formula $(1 - E)^n$ for the partial depletion of downstream available condensate due to precipitation with efficiency coefficient E upon passage over n successive upstream barriers brought the correlation coefficient between computed and observed long term winter precipitation distribution up to 0.90. Of three values of E used (0.25, 0.50 and 0.75), E = 0.5 best explained the observed snowcourse data.

3. Long term average winter precipitation is not well correlated to station elevation except for points on the same ridge.

4. Snow course local setting appears to make a fairly systematic difference in observed water equivalent value ranging in the mean between +2 in (50 mm) and - 4 in (-100 mm) depending on the local surroundings' effects upon flow separation evaporation canopy interception, and redistribution.

5. Such an approach should be useful in (a) assessing seasonal snowfall distribution in ungauged areas and (b) determining representative measurement network configurations.

REFERENCES

Chappell, C. F. (1970): Modification of cold orographic clouds. Ph.D. Dissertation, Colorado State University, Atmospheric Science Paper No. 173, December, 1970.

Elliott, R.D. and Russell W. Shaffer (1962): The development of quantitative relationships between orographic precipitation and air-mass parameters for use in forecasting and cloud seeding evaluation. Journal of Applied Meteorology. Vol. 1, pp. 218-

Hjermstad, L.M. (1970): The influence of meteorological parameters on the distribution of precipitation across central Colorado mountains. Master's Thesis, Department of Atmospheric Science Colorado State University, Atmospheric Science Paper, No. 163.

Kusano, K., K. Noguchi and M. Sumino (1957): A practical technique of forecasting orographic precipitation. Journal of Meteorological Research. Tokyo, Vol. 9, No. 11, pp. 811-822.

Rhea, J.O., P. Willis and L.G. Davis. Park Range atmospheric Water resources program. Final report to Bureau of Reclamation (Contract No. 14-06-D-5640). E.G. & G., Inc., Boulder, Colorado.

Rhea J.O. 1973: Interpreting Orographic Snowfall Patterns, Colorado State University, Atmospheric Science Paper 192, January 1973, 74 pp.

Spreen, William C. (1947): A determination of the effect of topography upon precipitation. American Geophysical Union. Transaction, Vol. 28, No. 2, pp. 285-290.

CLOUD DROPLET ACCRETION ON SNOW CRYSTALS

Roger F. Reinking
Earth and Planetary Sciences Division
Naval Weapons Center, China Lake, California

ABSTRACT

Accretion on a population of snow crystals is empirically described. The degree of accretion is estimated in terms of crystal growth habit, size and growth time. Improved descriptions of the onset of accretion, the dispersion in the degrees of accretion among crystals, and the formation of graupel are provided. The results are applicable to modeling of precipitation formation as well as densification and metamorphosis of newly fallen snow.

INTRODUCTION

The modes of snow crystal growth in clouds reflect the efficiency of storm systems for producing precipitation and also determine the initial characteristics of snowpack. Snow crystal growth processes are therefore vital links in the physical systems which determine water supply and avalanche potential. Much information with regard to the formation of precipitation and the initial conditions of new snow on the ground can be obtained by examining the nature of cloud droplet accretion on snow crystals. An empirical description of accretion on a population of snow crystals is presented in this paper. The information is derived from a recent project report by Reinking (1973). The information is intended to illustrate certain basic characteristics of the accretion process that have not been adequately described by theory or proven empirically. The analyses include previously unavailable estimates of the degree of accretion on crystals as functions of diffusional crystal growth habit, size and growth time.

BACKGROUND, DATA AND METHODS

Accretion is the process of growth of a precipitation particle by the collision of an ice particle with supercooled liquid droplets which freeze upon contact. The accretional mass growth rate of an individual ice crystal is given by the general equation,

$$\dot{M}_a \equiv \left(\frac{dM}{dt}\right)_a = EAQ\,|V - v| \tag{1}$$

where E is the efficiency of droplet collection and freezing on the crystal (dimensionless), A is the cross-sectional area of the crystal (cm^2), Q is the liquid water content of the supercooled cloud droplets (g/cm^3), and $|V - v|$ is the difference between respective crystal and droplet terminal velocities (cm/sec). The detailed theory of the fundamental parameters of equation (1) has been reviewed, clarified and expanded by Reinking (1973). The theoretical analyses demonstrated that differentiation of ice crystal diffusional growth habits is essential to the description of accretion.

Diffusional snow crystal growth is prerequisite to accretional growth and continues after accretion begins. An ice crystal can grow through the following sequence: (1) initial diffusional growth, (2) onset of riming, (3) growth by accretion plus diffusion, and (4) continued accretion with diffusion to the stage at which the ice particle becomes graupel (see Hindman and Johnson, 1972). Accretion has normally been dealt with empirically by simply categorizing crystals as rimed or unrimed (Ono, 1969; Zikmunda and Vali, 1972). Most precipitation particle growth models have incorporated the assumption that all crystals in a population grow accretionally at the same rate in a given cloud. The E, A, and V parameters, which are very habit dependent, have simply been assumed constant or at least uniform among the habits. Refinements have been gradually forthcoming. Hindman and Johnson (1972) have allowed for basic habit differentiation. Their model does initiate accretion as soon as crystals reach sizes such that $E>0$; i.e., the onset of accretion on all crystals of a particular habit is assumed to occur after a short growth time ($\sim$ 2 min), when the crystals reach sizes of 100-300 μm. This procedure provides a first approximation of the onset.

Empirical measurements of the degree of accretion on individual crystals comprising a large sample can be applied to more accurately describe the accretion process. Effects of the existing nonuniform spatial distribution of cloud water, as opposed to the normally assumed uniform liquid water content, can then be illustrated. The onset and degree of riming for an entire crystal population can be more accurately depicted. A basis for better modeling in terms of individual crystal habits or habit groups can be provided.

The data for this study were collected in conjunction with the field weather modification experiments of Project CENSARE (CENtral SierrA REsearch). The project was designed to study snow formation processes and the corresponding potential for precipitation enhancement by cloud seeding. Some 14,042 snow crystals from 225 individual samples collected during eight synoptically diverse Sierra Nevada snowstorms were analyzed. The sampling site was Tamarack, Ca. (elev. 6918 ft msl), which is located on the windward slope of the Sierra Nevada near the mean elevation of maximum snowfall. A hooded deep freeze was used to ensure crystal data acquisition even when surface temperatures warmed to very near OC. The freezer was positioned in a clearing during snowfall. Crystals settled through collection ports into the freezer where they were replicated on large class slides in plastic (1.5 percent polyvinylformal in ethylene dichloride). Replicas were obtained at 15-45 min intervals during snowfall.

Data reduction were accomplished by projecting the replicas onto a screen with standard overhead equipment. Magnification of the crystal images was set at 100X. The replicated crystals were examined for detail with a microscope prior to projection. Axis size measurements and identification of individual crystals were then made on the screen. A computerized numerical equivalent of the Magono and Lee (1966) crystal classification was used to identify diffusional growth habits. Crystal growth times were computed from diffusional growth rate parameters given by Hindman and Johnson (1972). Estimates of the degree of riming were needed. It is not practical to count droplets on individual crystals when the crystal sample is large, but studies by Hobbs, et al. (1971) and Reinking (1973) showed that it is possible to visually obtain a semi-quantitative estimate of rime coverage on individual crystals on a percent-of-area coverage basis. A person may visually rate amounts from most to least on a scale of four or five and be confident of reasonable reliability in his estimate. The following scale was utilized:

SCALE FOR ESTIMATING DEGREE OF RIMING ON INDIVIDUAL CRYSTALS

Scale (Surface Coverage)	Weight for Computation
Unrimed	0
1-25%	10%
25-50%	37.5%
50-75%	62.5%
75-100%	90%
Obscured (100%)	---

A single percentage representative of each degree-of-rime interval was utilized in computations. The zero rime category, and its separation from the 1-25% category are, of course, absolute. The 25-50% and 50-75% intervals were weighted at the respective medians. The 1-25% interval was weighted slightly below the median, at 10%. The 75-100% interval was weighted slightly above the median, at 90%. The latter two weights do not produce results substantially different from those based on medians, but do serve as indicators of the following factors: (1) Crystals in the light rime category frequently have just encountered the onset of riming and have collected only a few droplets. This is particularly true of small crystals. The weight at 10% indicates this phenomenon. (2) Rime coverage on heavily rimed crystals is particularly uneven. For example, rime near the center of a planar crystal in the 75-100% category may be light, while that near the edges, if spread out evenly may more than cover the whole collecting surface. The weight at 90% indicates that many of the crystals in this heavy rime category have collected enough rime to cover their surfaces, even though the surfaces were not totally obscured. The category designating obscured crystals (100% rime coverage) is predominately represented by graupel which is dealt with separately in the analyses.

DISTRIBUTIONS OF THE DEGREE OF ACCRETION

The field samples of snow crystals are summarized in Table 1. Crystals with similar in-cloud formation temperatures, shapes and terminal velocities comprise the indicated individual habit groups. The sample sizes for all habit groups except S (side planes) and for several individual habits are large enough to permit reliable evaluation of crystal characteristics. Crystal habits that formed throughout the -5C to -30C range were observed. This was expected because Sierra Nevada snowfall normally forms in deep supercooled cloud layers that extend up from the 0C level. However, the concentrations of crystals forming above the -20C level were found by Reinking (1973) to be substantially below those expected from nucleation of natural ice crystal nuclei. An effective maximum cloud top occurs near or below the -20C level in the Sierran storms. The persistence of strong winds aloft, and perhaps a thinning of cloud moisture above the -20C level, prevent fallout of the colder crystal habits on the windward slope of the Sierra.

Three dimensional distributions of the crystal rime coverage have been devised to graphically represent the planar and columnar crystal populations (Figures 1 and 2a, b). This was accomplished by first determining the percent of crystals in a specific size interval as a function of the percentage areal rime coverage. These separate distributions were then plotted as a function of crystal size over the observed ranges. The size intervals of the maximum dimensions

TABLE 1. GROWTH HABITS AND OCCURRENCE OF SNOW CRYSTALS AT TAMARACK

HABIT GROUP		FORMATION TEMPERATURES (C)	SAMPLE SIZE	PERCENT OF TOTAL SAMPLE
N	NEEDLES AND SHEATHS (N1a-d; N2a,b)*	−4 to −8	6804	48.5
C	HOLLOW COLUMNS (C1f)	−8 to −10, −20 to −30	1302	9.3
P1A	HEXAGONAL PLATES (P1a)	−11.5 to −12.5, −17.5 to −18.5	698	5.0
P1B	BRANCHED PLANAR CRYSTALS (P1b-f, P2a-g; P3a-c; P4a,b; CP3a-d)	−12.5 to −17.5	1177	8.4
P1C	FRAGMENTS OF BRANCHED PLANAR (fP1b-f; fP2a-g; fP3a-c; fP4a,b; fCP3a-d; I3.)	−12.5 to −17.5	1484	10.1
P2	RADIATING ASSEMBLAGES OF PLATES AND DENDRITES (P7a,b)	−13.5 to −22.5	263	1.9
CP	CAPPED COLUMNS (CP1a-c)	−8 to −10, −20 to −30 (col) −11.5 to −18.5 (cap)	214	1.5
S	SIDE PLANES (S1, 2, 3)	−20 to −30	39	0.3
R	GRAUPEL (R4a-c)		1040	7.4
I	MISCELLANEOUS ICE PARTICLES (I1)		998	7.1
Total of individual habit groups			14019	99.8
Total sample size			14042	100

*Magono and Lee classification; codes preceded by "f" indicate crystal fragments.

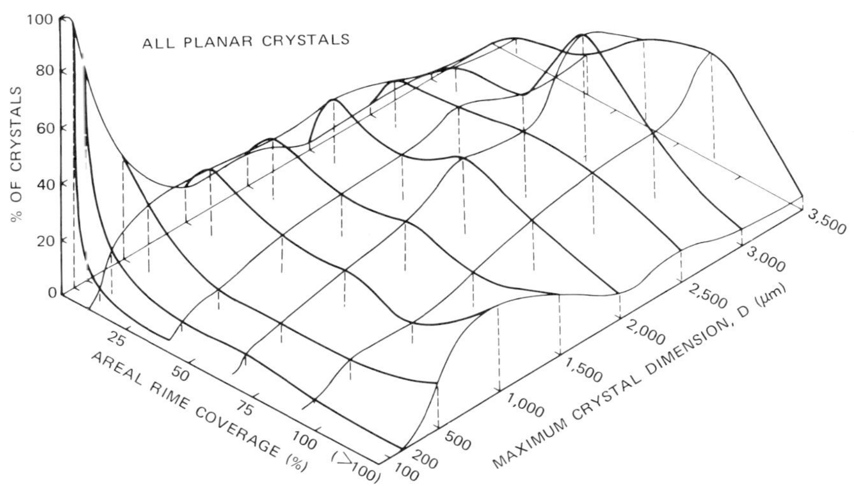

Figure 1. Distribution of the degree of accretion for all planar crystals.

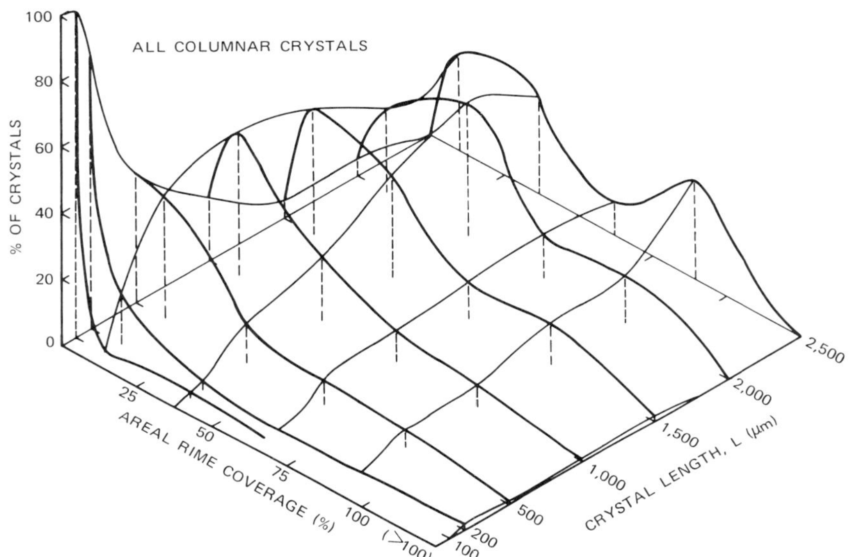

Figure 2a. Distribution of the degree of accretion for all columnar crystals in terms of crystal length.

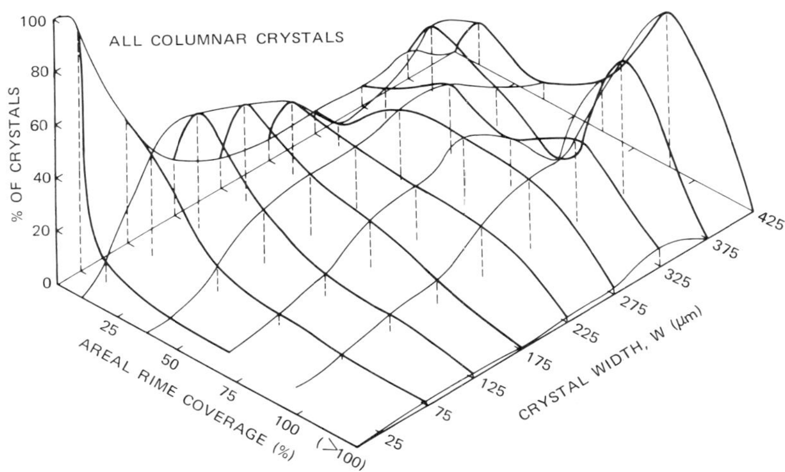

Figure 2b. Distribution of the degree of accretion for all columnar crystals in terms of crystal width.

(a axes) of planar crystals and both the length and width (a and c axes) of columnar crystals are specified on the graphs by their midpoints. Illustration of the degree of riming at and following the first onset has been enhanced by the use of size intervals for the smallest crystals that are narrower than those used at the larger sizes. Each of the three-dimensional distributions defines a surface which shows (a) the spread and maxima in the degrees of riming among the crystals of a given size, as well as (b) the trends of both the dispersion and the maxima with increasing crystal size (or growth time). The 3-D distributions for all planar crystals (predominately habit groups P1A, B, C) and all columnar crystals (habit groups N, C, and CP) are composites which together represent the total data sample. The distribution for all planar crystals includes graupel particles which account for the large percentages of crystals in the heaviest rime category. These all-storm composites do not represent scatter due to differences among individual precipitation periods with crystals having narrow dispersions of rime. The general patterns and magnitudes of dispersion in these accretion distributions are common to the crystals sampled in individual precipitation periods. The composites do of course smooth the data from individual periods, and are therefore more readily interpreted.

For each crystal size interval in the figures, there is a dispersion in the observed degree of accretion about the rime category with the peak percent of crystals. This dispersion is small and centered on the no-rime category for the smallest crystals. The dispersion widens and reaches maximum ranges at intermediate crystal sizes, then narrows again for larger crystals but centers on heavier rime categories. The trend of peak percentages of crystals is to migrate with increasing crystal size from the no-rime to the moderate or heavy rime categories; these characteristic trends represent the continued accumulation of rime throughout all but the initial diffusional growth periods of the crystals.

The lack of accretion on the smallest crystals is predicted by theory. The onset of accretion on some crystals in each sample occurred when the planar crystal diameter and the columnar crystals lengths reached sizes between 100 and 300 μm. The occurrence of the onsets of accretion on crystals of these small sizes is also predicted, and shows that liquid water was present in the Sierra clouds near the formation levels of the crystals and that vapor levels at or above water saturation prevailed to sustain the droplets.

The distributions prove that the sizes and growth times reached by crystals in any given population prior to the onset of accretion span a considerable range. This is not predicted by present theory. The start of accretion on the small crystals, i.e., on those crystals with approximately 2 min growth times, may therefore be defined as the first onset. Further consideration is given to the first onset and the degree of accretion in the next section.

Theoretical and laboratory studies of particle collection on cylinders show that the droplet collecting axis of columnar crystals should be the a axis, or width, since the collection becomes relatively independent of cylinder (crystal) length after initial growth. This is very clearly verified by migration of the peaks in the distributions for the columnar crystals (Figures 2a, b). An examination of this factor for specific crystal habits revealed that the effect of crystal width is more pronounced for hollow columns than for needles and sheaths. This is expected because the needles and sheaths remain much narrower throughout comparable growth periods.

MEAN RIME COVERAGE IN TERMS OF CRYSTAL SIZE

The mean degrees of rime for crystals in specific size intervals were computed for key crystal habits and habit groups. Polynomials were then fit by the least squares method to the sets of means for individual crystal size intervals. The average degree of rime is therefore given in each case as the mean areal rime coverage, in percent. The polynomials are presented by Reinking (1973). Pertinent accretion parameters derived from the equations are presented in Table 2.

TABLE 2. ACCRETION PARAMETERS FROM MEAN AREAL RIME COVERAGE CURVES

	Crystal Dimension	Size of First Onset (μm)	Growth Time at First Onset (min)*	Size at 25% Threshold (μm)	Growth Time at 25% Threshold (min)	Size of Average Crystal (μm)	Growth Time of Average Crystal (min)	Rime Coverage on Average Crystal (%)
ALL PLANAR CRYSTALS	(D)	95	–	1080	–	1034	–	24
ALL COLUMNAR CRYSTALS	(L,W)	120,30	–	900,160	–	689,112	–	20
NEEDLES, N1a	(c,a)	220,24	4.1	1300,205	25	799,77	15	16
SHEATHS, N1c	(c,a)	185,36	3.3	720,110	19	668,97	14	24
HOLLOW COLUMNS, C1f	(c,a)	125,30	3.1	450,155	17	477,149	18	26
HEXAGONAL PLATES, P1a	(a)	150	3.3	None	–	447	14	8
BRANCHED PLANAR CRYSTALS, P1B	(a)	240	1.9	820	6.6	1825	16	42

*Growth times of columnar crystals are calculated from the length dimension.

The crystal sizes at first rime onset (Table 2) are largest for the branched planar crystals (Group P1B). The P1B crystals form in the regime of most rapid diffusional growth (-15 ± 3C). The larger sizes gained prior to the first onset are therefore expected. The growth time to onset is shorter than those for the other crystal types for the same reason. The observed onset size for hexagonal plates, which grow relatively slowly, is correspondingly smaller; the time to first onset is longer. The width-to-length ratios of columnar crystals are respectively greater for needles, sheaths and columns. The widths of these crystals at first onset are remarkably uniform (24, 36, and 30 μm) and reflect the control of the a axis on accretion. Needles must grow to lengths greater than sheaths, and sheaths must grow to longer sizes than columns, to achieve the critical widths required for the first onset. The growth times to first onset of 2-4 min agree with the approximate 2 min estimate from the model of Hindman and Johnson (1972). However, the model allows all crystals to begin accreting droplets at this stage, whereas Figures 1 and 2a, b reveal the delay of onset for many crystals generated in real clouds.

The initial stages of accretion after the onset begin to transform the characteristics of a crystal. Effects on A, V, and E, and therefore on M_a, will become significant as the crystal passes through some intermediate stage of riming. A 25 percent real rime coverage represents a somewhat arbitrary but semi-quantitative reference threshold for this transition stage of accretion. The 25 percent threshold lengths for needles, sheaths and hollow columns are 1300, 720 and 460 μm respectively; corresponding respective growth times are 25, 19, and 17 min. The sequence of growth times is in direct accord with the relative differences in crystal collecting areas, fallspeeds and predicted collection efficiencies (Reinking, 1973). The observed hexagonal plates did not rime to the 25 percent stage. The growth time of about 7 min required for branched planar crystals to reach this stage is short compared to the times required by the columnar types. The effect of the rapid growth of the branched crystals to efficient collecting sizes is again evident.

The average crystal size, equivalent growth time, and corresponding percent areal rime coverage for each of the crystal types are included in Table 2. Growth times to average sizes are very comparable for the simple, regular habits. Riming on these average crystals covers the greatest portions of collecting areas on the branched planar types. Coverage for average columns is greater than that for sheaths which in turn exceeds that for needles. The average hexagonal plates exhibit the least coverage.

The polynomials for mean rime coverage in terms of crystal size were converted to mean rime in terms of crystal cross-sectional area A, by means of equations presented by Reinking (1973), which are based on the work of Auer (1971) and Hindman and Johnson (1972). The resultant curves are presented in Figure 3. Mean rime coverages after crystal growth periods of 10, 30, and 50 min are indicated. A pattern of accretion among the crystal habits is evident for any given crystal area. At $A = 10^{-3}$ cm, e.g., the areal coverage is increasingly larger for the habits taken in the order P1a (or P1A), P1B, N1a, N1c, C1f. The mean rime increases with predicted relative values of the products, EV, for the individual habits; theory and experiment are in agreement (see Reinking, 1973).

The total water accreted on a crystal of a given habit must be considered in terms of the area of the total collecting surface. This surface area for columnar types is approximately equal to πA. The corresponding cross-sectional areas in Figure 3 may be multiplied by π. Planar crystals collect rime predominately on the underside, so the total collecting surface area and the cross-sectional areas are the same The conversion for the columnar crystals would shift the curves toward the P1B curve. The relative positions would still be maintained, except for some overlap of the N1a and P1B curves, at areas less than 6×10^{-3} cm^2. Thus Figure 3 illustrates that the most liquid water for a given M is accreted on the crystals that attain the largest areas. The rapidly growing branched crystals attain the largest areas in the least time. These crystals also gather a given rime coverage in less time than the same coverage is accreted by the columnar types. Assume now that all crystals settle out of the cloud after 50 min of growth. The needles, columns and sheaths then accrete and precipitate very comparable amounts of liquid water. The branched crystals, as compared to the other habits, accrete and precipitate the most liquid water. Hexagonal plates rarely accumulate significant masses of cloud water by the accretion process.

ON THE FORMATION OF GRAUPEL

The numerical model of Hindman and Johnson (1972) correctly allows a differentiation in the final degree of accretion so that some

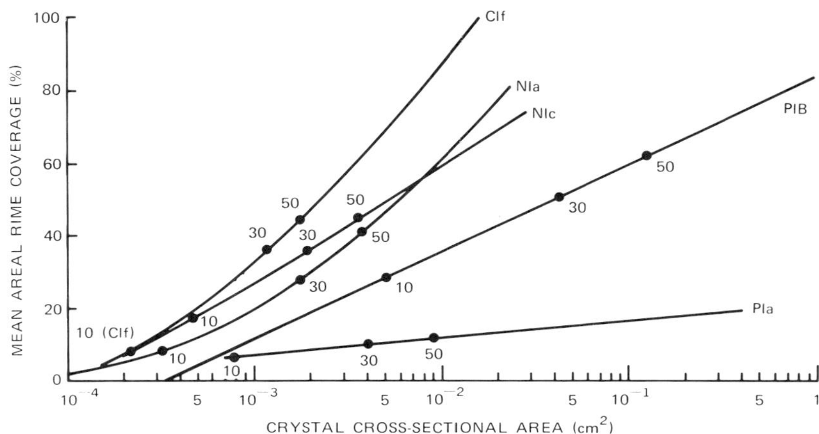

Figure 3. Mean areal rime coverage as a function of crystal cross-sectional areas for specified crystal habits. Points corresponding to 10, 30, and 50-min growth times are indicated.

crystals collect moderate rime while others develop fully to the graupel stage. However, the graupel is modeled to develop only from those crystals with the longest growth time. The following analysis shows that this modeling procedure may be improved.

Tentatively assume that graupel is generated by continued buildup of accretion on core crystals in such a way that an accreted crystal forms the nucleus of a graupel particle. The nature of hexagonal graupel in particular indicates formation on planar crystals. Both hexagonal and lump graupel contributed to the sample. A smaller number of conical graupel particles were also observed. The size distribution of the observed graupel is represented in Figure 1 by the curve at the category of heaviest rime (>100%). Consider the sizes of the graupel relative to the degree of riming of other crystals of the same sizes. The maxima of the 3-D distribution should migrate directly to the peak in the grauple distribution if the graupel is generated by the crystals that grow to the largest sizes by diffusion plus accretion. In Figure 1, the migration of the peak percentage of all planar crystals through increasing degrees of rime does not lead to the peak in the graupel distribution. The graupel peak is at 100 μm but the planar crystal peak at this same size, which migrates out from the accretion onset sizes, is in the lightest rime coverage category. The migration of this latter peak may be traced out to the 75-100 percent rime category at 3500 μm; it bypasses the graupel peak. The graupel distribution may be compared, with the same result, to other 3-D distributions for any of the individual crystal habits or habit groups (Reinking, 1973). Preferential graupel formation on crystals of a specific habit cannot explain the observed graupel size distribution. Graupel that develops with crystals as nuclei must therefore form predominantly on a select few relatively small crystals that rapidly gather heavy rime during early stages of diffusional growth. Such development requires only that relatively large values of $\dot{M}_a$ be realized by small fractions of the total crystal population. Such selective generation of graupel is indeed possible in real clouds, which have nonuniform liquid water contents along any crystal trajectory.

The question remains as to whether graupel does consistently form with crystals as nuclei. There is no doubt that some of the observed graupel particles did grow by this mechanism. Hexagonal graupel was observed; graupel particles with "feather-like" extensions that indicate formation on radiating dendrites were observed. However, the observed hexagonal plates were so lightly rimed that they could not have contributed significantly to the graupel population. Also, the accretion distribution for all planar crystals shows a distinct minimum within the 75-100 percent rime category at the 1000 μm size interval and adjacent size intervals. This minimum separates the main body of the distribution from the graupel distribution. This minimum exists even if all observed crystals are considered by adding the all-columnar distributions to the all-planar group. The minimum is also evident if the graupel distribution is related to crystals of specific habits. Therefore, the data show that the graupel distribution and the crystal distributions are at least partially independent. Knight and Knight (1973) have postulated and provided some empirical evidence for an alternate mechanism of graupel formation, which could produce independent graupel and crystal distributions. Graupel may grow, according to Knight and Knight, from the buildup of rime on localized sites on the underside of planar snow crystals. Cones of rime, attached by their tips to the crystals thereby develop. The cones then break off in the cloud and may gather further rime as they fall as separate particles of graupel. There is no evident reason why this mechanism could not lead to lump graupel as well as the postulated conical graupel. Also, this process does not require graupel size to be as large as nucleus crystals. The smaller of the observed graupel particles can consequently be accounted for, in part, by the occurrence of this process.

CONCLUSIONS

The analyses show the patterns of dispersion in the degrees of accretion among snow crystals that represent a population. The results may be regarded as climatological or as representative of individual precipitation episodes. Representation of the latter is more qualitative. The patterns reveal clear associations with crystal habits, and thus are dependent on the E, A and V parameters. The columnar crystal types are found to accrete and precipitate comparable amounts of water. Branched planar crystals individually precipitate the most water. Water accretion and precipitation by plates is found to be insignificant.

Variations in cloud liquid water content, Q, along crystal trajectories significantly influences the amount of dispersion. The onset of accretion is found to frequently occur well beyond the crystal sizes and growth times when E first reaches values exceeding zero. Graupel appears to form by the Knight mechanism as well as by continued accretion on nucleus crystals.

The observed patterns in the degrees of accretion have not yet been accounted for in models of crystal growth or models of the initial microphysical characteristics and transformation of snowpack.

This research was supported by Agreement No. B-50882 with the California State Department of Water Resources and Contract No. 14-06-D-6592 with the U. S. Bureau of Reclamation.

REFERENCES

Auer, A. H., Jr., 1971. Measurements of the basal face surface area for plate family ice crystals. J. Meteor. Soc. Jap. 49:232-235.

Hindman, E. E., II, and D. B. Johnson, 1972. Numerical simulation of ice particle growth in a cloud of super-cooled water droplets. J. Atmos. Sci. 29:1313-1321.

Hobbs, P. V., et al., 1971. Studies of winter cyclonic storms over the Cascade Mountains (1970-71). Cloud Physics Group Report VI, Atmos. Sci. Dept., Univ. Wash., Seattle. 306 p.

Knight, C. A., and N. C. Knight, 1973. Conical graupel. J. Atmos. Sci. 30:118-124.

Magono, C. and C. W. Lee, 1966. Meteorological classification of natural snow crystals. J. Fac. Sci., Hokkaido Univ. 2:321-335.

Ono, A., 1969. The shape and riming properties of ice crystals in natural clouds. J. Atmos. Sci. 26:138-147.

Reinking, R. F., 1973. Empirical assessment of accretion microphysics. Project CENSARE Final Report, Vol.IV Atmospheric Water Resources Research, Calif. State Univ., Fresno. 309 p.

Zikmunda, J. and G. Vali, 1972. Fall patterns and fall velocities of rimed ice crystals. J. Atmos. Sci. 29:1334-1347.

THE I.H.D. WORLD GLACIER INVENTORY

C.S.L. Ommanney
Glaciology Division, Environment Canada

ABSTRACT

The progress of the world-wide glacier inventory being compiled as an I.H.D. programme is reviewed and aspects requiring new or improved techniques discussed.

INTRODUCTION

The world inventory established as part of the International Hydrological Decade (IHD) programme is, potentially, one of the most comprehensive regional studies of perennial ice and snow. If the required basic data is furnished by all countries with glacierized areas, it will be able to provide detailed information about every permanent snow and ice feature in the world. The inventory establishes the physical characteristics of each glacier at only one point in time. However, by taking the inventory information as a datum and updating it by satellite, airborne and ground surveys, it will be possible to improve on the previous accuracy of record of large scale areal and temporal variations in the perennial snow and ice masses. New values for ice area based on glacier inventories in selected areas have given values of ±25% on the previously accepted figures which do not reflect significant fluctuations in the ice mass but merely the increased accuracy of measurement. The individual identification of each ice mass will facilitate the compilation and use of all historical data on glaciers and the gradual development of a complete information system on them that will maximize the utilization and availability of existing data.

This paper reviews all existing glacier inventory programmes completed or not. Several programmes that were initiated in response to the original International Geophysical Year (IGY) request are included as they partially fulfill the requirements of the IHD inventory. Most of the studies reviewed have been undertaken in direct response to the recommendations laid out in the UNESCO/IASH Guide (1970), but some were started for different reasons. The progress of the IHD inventories was reviewed by Müller and Ommanney (1971) in 1970 but, with the IHD drawing to a close in 1974 and with the availability of new techniques that may enable us to inventory areas presently untouched, it is appropriate to review the substantial progress that has been made in the last three years. A major data gathering system is identified that may provide a useful framework for those wishing to implement or test new techniques and which itself requires the use of new techniques if the original hope for a complete world inventory of perennial snow and ice masses is to be realised.

THE CANADIAN GLACIER INVENTORY PROGRAMME

The Canadian glacier inventory programme has adhered strictly to the basic recommendations outlined in the UNESCO/IASH Guide. This has been possible because the feasibility of many of the recommendations was tested in Canada prior to being incorporated in the Guide and the

Canadian programme did not start until those recommendations had been finalized. The basic data recommended and being collected in Canada for each glacier is - region and basin identification, glacier number, geographical coordinates, Universal Transverse Mercator coordinates, orientation of the accumulation and ablation areas, elevations of the highest, lowest exposed and lowest debris-covered parts of the glacier and of the snow line, date of the snowline estimate, mean elevations of the accumulation and ablation areas, lengths of the various constituent parts (accumulation, ablation and debris-covered areas), mean width of the main stream, surface area of the constituent parts, accumulation area ratio, estimate of mean depth, estimate of volume and glacier classification and description. The classification used is a 10 x 6 matrix that describes in coded form the main characteristic of a glacier (ice cap, valley glacier etc.), the form this takes (compound basin, cirque etc.), the tongue characteristics (expanded foot, piedmont etc.), the nature of the longitudinal profile, the main source of nourishment and an estimate of the activity of the tongue (advancing, retreating etc.). A detailed description of the classification with sources has been given by Ommanney (1969). Experience gained in the inventory of a number of diverse areas has necessitated some additions to the classification. The terms active tributary, stagnant (wasted) and ice-cored moraine have been added to the tongue characteristics and névé-sheathed slope and protalus rampart to the longitudinal profile. These new terms and the procedures used in the Canadian inventory are detailed in a report by Ommanney et al.(1973). Source materials for this work are aerial photographs at a scale of up to 1:25,000 and maps at a scale of up to 1:50,000 except for those few glaciers that have been mapped individually.

The Canadian programme was formally established in 1968 with the setting up of a Glacier Inventory Section. Prior to this it had been decided that there was some information that could be gleaned from aerial photographs and other sources that had not been requested by the Guide but which would add to the general usefulness of the study. Two comment categories were added to the basic Data Sheet. The first consisted of a 5 digit code that described 1) the extent of historical and contemporary information, 2) the nature and amount of photo coverage, 3) special features such as surge characteristics and ogives, 4) moraine types and 5) the number of glacier dammed lakes. The second, a general comment section, provided for some descriptive remarks on each glacier to supplement the coded data. Space was also provided for a glacier name if available.

Following the transfer of glacier detail from the aerial photos to the work map, the latter is placed on a D-Mac Pencil Follower table interfaced with a PDP-8/I computer and the coordinates, lengths, widths, perimeters and elevations digitized. The magnetic tape output is processed on a CDC 6400 computer to provide the basic data input for the Data Sheet. Volumes are calculated using a mean depth based on an area/glacier classification relationship. A computer program has been developed for the analysis of the glacier inventory data (Ommanney et al., 1969). This uses a masking function on the glacier classification and, besides providing basic information such as totals, averages, extrema and histograms of the parameters, can be used to select a specified part of the data file for access or analysis. The selection can be defined by glacier types or by ranges on any particular parameter.

The furthest progressed part of the Canadian glacier inventory is publication of index maps for a Glacier Atlas of Canada, showing the location, region and basin identification & number of each glacier. Maps are now available for all the heavily glacierized areas of the Canadian Arctic (except parts of Ellesmere Island) and for glaciers feeding the Nelson River system. Only data for Axel Heiberg Island has been published so far (Ommanney, 1969) but unpublished data is available for Vancouver Island, the Canadian St. Elias Range (Yukon Territory) and part of the Nelson River basin.

To provide a total information system on Canadian glaciers, all published references to work done on these glaciers is being entered in a computerized bibliography known as ICEREF. Key words identify the contents of an article and glaciers are referenced by their inventory number or their National Topographic Sheet location. Separate lists are being prepared of named glaciers and of those people known to have been in the glacierized areas in the historical past. These lists are used as keys for searching archival sources. Photographic information obtained from such sources will be entered in ICEREF.

Specific data on the perennial snow and ice resources of Canada, to be used in regional water and energy balance calculations, will be the basic output of the inventory. The data file will enable researchers not only to obtain all identified information on a particular glacier but to select a glacier that, according to certain specified parameters, is representative of its type or the region. All glaciers with surge characteristics or glacier dammed lakes liable to jökullhlaup will be identified for possible monitoring in the future.

STATUS OF THE WORLD GLACIER INVENTORY

The Canadian inventory has been discussed in detail because it meets the requirements of the IHD programme, the progress of which is discussed below.

A. AFRICA: A projected inventory of the Ruwenzori Mountains, (Uganda) has been cancelled with the reassignment of the principal investigator (Temple, personal communication).

B. THE AMERICAS:

1. Argentina: Information of the 333 most prominent glaciers in 3 regions of Argentina was provided by Colqui (1962) for the IGY inventory. The data included geographical coordinates, altitude, area and classification. More detailed information, fulfilling most of the IHD requirements, has been given by Feruglio, (1957) for the Cordillera and Bertone (1960) for the Patagonia area.

2. Canada: The Canadian programme has been described above. A partial inventory was carried out for the IGY but does not conform to the IHD recommendations.

3. Chile: Due to the importance of glaciers as a water resource partial inventories have been carried out in the most critical areas (Post, 1970). No overall programme is contemplated at present. However, in 1972 a British expedition visited the Patagonia area to make a feasibility study for an inventory. This will be carried out in 1975/76 and will be based on existing photography but using a revised map. Only features >2 km^2 will be included and snowlines will not be shown (Sessions,1973).

4) Mexico: An inventory of Mexican glaciers was completed for the IGY. Data on maximum and minimum elevations, name and approximate area were provided for 11.5 km^2 of glaciers together with photos and sketches. No estimates of volume were included but glacier classification was discussed in some detail (Lorenzo, 1959).

5) U.S.A.: The first IHD inventory published in the U.S.A. is that for the 756 glaciers of the North Cascades (Post et al., 1971), about half the glacierized area of the coterminous U.S.A. A slightly modified classification has been used and bulk volume estimates are given rather than individual values. Future plans call for a glacier atlas of the Sierra Nevada and an inventory of the Brooks Range in Alaska (Post, personal communication).

C. ANTARCTICA: Although no specific plans exist for an inventory of the Antarctic ice mass much of the present glaciological work will provide improved information on the quantities of ice. While radio-echo sounding techniques are being used to determine depths over a major part of the ice sheet satellite imagery is improving the delimitation of coastal areas.

D. ASIA:

1) Afghanistan: No programme for an inventory of the ice masses of Afghanistan exists at present but there are tentative plans for a U.S. scientist to initiate a programme if funding can be obtained (Müller, personal communication).

2) India: Consideration is being given to starting a glacier inventory of the Sutlej basin that will follow the I.H.D. recommendations (Bahadur, personal communication).

3) Japan: An inventory of perennial snow patches in Central Japan has been completed (Higuchi and Iozawa, 1971). Snow patches were observed from 1968-1970 and tables are given for the changing dimensions from year to year. Snow patches are classified into transverse, longitudinal and circular and photos of many included. The essential IHD data has been collected.

4) Nepal Himalayas: The first partial inventory of the Nepal Himalayas was a pilot study of the Mt. Everest area carried out by F. Müller for the UNESCO/IASH Guide (1970). An extensive programme of glacier research is now being carried out by the Japanese. Photos of over 800 glaciers have been collected from which 80 representative glaciers were selected for further study. Field investigations began in 1971 and data has been collected and analysed for 72 glaciers. The aid of visitors is being solicited to extend the photo coverage and ERTS imagery will be used for the more remote areas (Higuchi, personal communication).

5) Pakistan: Part of the engineering studies carried out for the Tarbela Dam consisted of an inventory of 1,214 glaciers in the upper Indus River basin (Karpov, 1967). The basic requirement of the IHD inventory are satisfied.

E. AUSTRALASIA:

1. New Guinea (West Irian): A glaciological and geophysical programme to study the extent, form and character of the Meren and Carstensz ice fields has been started by a group of Australian scientists. Although no details are available, the second report (Peterson 1973) describes the broad objective as being in accordance with the aids of the IHD inventory.

2) New Zealand: No complete glacier inventory has been carried out to date but 527 glaciers in the Southern Alps have been identified and measured in an attempt to estimate the order of magnitude of water resources of glaciers >0.1 km^2. Volumes have been calculated using an assumed depth/area relationship (Anderton, personal communication).

F) EUROPE INCLUDING THE U.S.S.R.:

1) Austria: Aerial photography of Austria has been completed for glacier mapping at a scale of 1:10,000. Some 5-6 glaciers have been studied intensively to establish a relationship between volume and their physical characteristics (Müller, personal communication). It is understood that Austria intends to compile an inventory of glaciers.

2) Denmark: Work being carried out in Greenland, similar to that in the Antarctic, may provide sufficiently detailed information for inclusion in the world inventory.

3) France: Detailed information on individual French glaciers is being published as "Fiches des Glaciers Français" (Vivian, 1967). More information than that required for the world inventory is provided but volume estimates are not given. Consideration is being given to a complete inventory.

4) Germany: No plans yet exist for a formal inventory of the glaciers of Germany. However, sufficient information is available that it could be done fairly rapidly (Reinwarth, personal communication).

5) Iceland: In an attempt to coordinate the many expeditions leaving the United Kingdom for Iceland the Young Explorer's Trust has introduced an 'Adopt a Glacier Project' which provides the basic framework for an inventory of glaciers and lakes in the Tröllaskagi region. Detailed studies were carried out on 5 glaciers in 1972 and plans called for 15-17 glaciers to be studied in 1973. Primary emphasis in the initial stages is on mapping the glacier under study at 1:5,000. Inventory data sheets are completed for all glaciers in the region being visited (Escritt, personal communication).

6) Italy: The Catasto dei Ghiacciai Italiani consists of 4 volumes of detailed information on every glacier in Italy, with a sketch and photo of each feature (Comitato Glaciologico Italiano, 1959-1960). The data fulfills the requirements of the IHD inventory except for volumes and snowlines.

7) Scandinavia: The Norwegians responded rapidly to the recommendations for a glacier inventory with an Atlas of Glaciers in Southern Norway (Østrem and Zeigler, 1969). Problems with boundary glaciers situated in

one country but flowing to the other led to cooperation between the two countries in the production of the Glacier Atlas of Northern Scandinavia (Østrem et al., 1973). Perennial snowfields have been excluded due to problems of identification. The later publication includes a modified classification which was omitted from the first report. Both publications include volume estimates based on surface area. Some comparisons have been made with values calculated according to the slope angle. As with the Canadian inventory, additional information is given on moraines, lakes and surface features. The Scandinavian inventory is the first national glacier inventory completed in accordance with the IHD recommendations.

8) Switzerland: Switzerland has just started an inventory which will follow the IHD recommendations. Maps are available for most of the country at 1:10,000 which will ensure a fairly high level of accuracy. All Swiss glaciers were photographed within one week during the late summer of 1973 to provide information on the snowline; this will be repeated in 1974 (Müller, personal communication).

9) The U.S.S.R.: The U.S.S.R. glacier inventory programme was developed in response to the IGY recommendations. The data, which is well documented, conforms to the IHD requirements. However, the classification used differs in that 33 morphological glacier types are distinguished which agree only in part with those used by other countries. The procedures for the U.S.S.R. inventory have been published (Academy of Sciences, 1966) as well as almost half of the planned 106 volumes of data. (Vinogradov, personal communication). The intended completion date is 1975.

While it is apparent from the above review that substantial progress is being made towards an inventory of all the world's perennial snow and ice masses, there are still significant gaps. These may be real or due to a lack of information. In considering the distribution of these masses the northern Andes, the southern Pyrenees, the mountains of Africa and the mountainous regions of the Peoples Republic of China are important and interesting areas. Lack of resources, both material and financial, and of trained personnel have restricted programmes in countries where inventories are underway and inhibited the start of other programmes. Hope for the completion of a world-wide inventory lies in the development of techniques that can overcome these constraints. Satellite imagery may be one solution.

REQUIREMENTS FOR NEW TECHNIQUES

In attempting to apply the IHD recommendations, several problems have been encountered and led to questioning of the recommendations by the scientists concerned.

a) Resource Materials: The assumption in the UNESCO/IASH Guide is that all countries have available maps to a scale of 1:250,000, or better, and complete aerial photo coverage of their glacierized areas. For regions where adequate maps and photographs have not been available, techniques may now exist for the provision of these basic resource materials from satellite imagery. Basically, one must be able to identify all drainage basins and individual ice masses of a size >0.1 km^2.

b) Snowlines: The problem of snowline identification is tied to the availability and quality of photographs. In the case of Switzerland it has been possible to arrange for special snowline photography at the end of the melt season, but few countries are in that enviable position. Meier (1973) has discussed the use of ERTS imagery for obtaining an equivalent snowline altitude (ESA) that may well be a more valuable measure of the relative health of glaciers than a number of transient snowlines taken from different years. Techniques for determining the position of the transient snowline from aerial and satellite photography and for assessing climatic equilibrium lines should be investigated for their relevance to the inventory programme.

c) Volumes: Most countries engaged in a glacier inventory programme have established relationships between mean glacier depth and the surface area of the glacier according to its primary classification for assessing volume. The accuracy of such empirical relationships is suspect despite the correlations established between maximum depth and surface area by Lagarec and Cailleux (1972). Paterson (1970) has suggested that the formula $\bar{h} = 11/\alpha$, where $\bar{h}$ is the mean depth of the glacier and α the surface slope, can be taken as a better means of estimating volumes than the empirical relationships. At present insufficient depth data on representative glaciers prevents the determination of volumes accurately. Radio-echo sounding techniques have met with much success in areas of cold ice but most of the inventories discussed are for temperate glaciers for which no adequately tested radio-echo sounding system or other technique of depth assessment has yet been proved, besides the laborious ones of seismic and gravity surveys and drilling. Improved techniques are therefore required for the general assessment of glacier volume.

d) Classification: A morphological classification was adopted for the glacier inventory but suffers two essential drawbacks. First, it has no quantitative basis and so may be subject to varying interpretations of a particular glacier type. Secondly, because the Guide was prepared for a world programme, individual countries are faced with the problem that virtually all their glaciers may fall into one category. There is a need for a system of classification that will convey the essential characteristics of one particular form of glacier (eg. ice cap or valley glacier) but which has some broad quantitative basis. Limnologists use ratios of lengths, widths and perimeters etc. to establish lake classifications and glaciologists might look to a similar technique to describe glaciers more effectively.

APPLICATION OF THE WORLD GLACIER INVENTORY

Several of the applications of the inventory data have already been touched on previously. Energy balance calculations require information on the albedo of various regions of the globe. The glacierized areas represent regions of substantial energy loss the area of which in lower latitudes may be more extensive than previously thought and possibly less extensive in higher latitudes. Glaciers also function as natural reservoirs storing water in cool wet years for release in hot dry years. The inventory will provide much more accurate data on the quantities involved. In Canada, where the technique for measuring the glaciers permits the listing of the amount of ice within a particular elevation belt, the inventory data can be used in glacier melt models to determine the relative contribution of glacier melt to stream flow. Since some of the national programmes identify surge glacier features and jökullhlaup situations, potential disasters can be anticipated and precautionary measures taken. The compilation of glacier data for a far larger region than has previously been attempted and the study of this data may lead to a deeper understanding of the relationship between glacier activity and climatic change. The establishment of a temporary Secretariat for the world glacier inventory, to be located at E.T.H. in Zurich, will facilitate the comparison of data from the various national programmes. This centre will encourage the completion of the inventory and centralize and standardize the data being collected.

ACKNOWLEDGEMENTS

The author extends his sincere thanks to those previously or presently engaged in glacier inventory or related studies who have provided him with the information contained in this paper. Special thanks go to Dr. Fritz Müller, of the E.T.H. Zurich, who as Chairman of the working group on the world inventory has been an invaluable source of information.

REFERENCES

Academy of Sciences of the USSR, 1966. Rukovodsto Po Sostavlyeniyu Kataloga Lednikov SSSR. Gidrometeorologicheskoe Izdatel'stvo, Leningrad. 154 p.

Bertone, M., 1960. Inventario de los glaciares existentes en la vertiente Argentina entre los paralelos 47°30' y 51°S. Institutio Nacional del Hielo Continental Patagonico, Publ. No.3. Buenos Aries. 103 p.

Colqui, B.S. 1962. Argentine Glaciology. In: Antarctic Research, Geophysical Mono. No. 7, Am. Geophys. U., 217-28.

Comitato Glaciologico Italiano, 1959-1962. Catasto dei Ghiacciai Italiani, Anno Geofisico 1957-1958. Vol. I (1959), 171 p.; Vol.II (1961). 324 p.; Vol. III (1961), 389 p.; Vol. IV (1962), 309 p. Consiglio Nazionale delle Ricerche, Torino.

Feruglio, E., 1957. Los glaciares de la cordillera Argentina. In: Geografia de la Republica Argentina 7, 1st pt., Soc. Argentina de Estudios Geograficos GAEA, Buenos Aires.

Higuchi, K. and Iozawa, T., 1971. Atlas of perennial snow patches in Central Japan. Water Research Laboratory, Nagoya University, Nagoya. 81 p.

Karpov, A.V., 1967. Report on the sedimentation problems of the Tarbela reservoir. Unpublished consulting engineering report, Vol. 1, 49 p., Vol. 2, 83 p., Vol. 3. 63 p.

Lagarec, D. and Cailleux, A., 1972. Corrélation entre épaisseur moyenne, épaisseur maximale et surface des glaciers. Zeit. fur Geomorph., N.F., Suppl. Bd. 13: 18-25

Lorenzo, J.L., 1959. Los glaciares de Mexico. Universidad Autonoma de Mexico, Institutio de Geofisica, Mexico D.F. 114 p.

Meier, M.F. 1973. Evaluation of ERTS imagery for mapping and detection of changes of snowcover on land and on glaciers. Paper W-19 in Symposium on significant results obtained from the Earth Resources Technology Satellite -1, Vol. 1, Sect. A, N.A.S.A., Washington, D.C.: 863-875.

Müller, F. and Ommanney, C.S.L., 1971. The contribution of glacier ice to the world water balance: a status report on the world glacier inventory. Symposium on the World Water Balance, Reading, 1970, Vol. 3, I.A.S.H. Publ. No. 94: 6-20.

Ommanney, C.S.L., 1969. The ice masses of Axel Heiberg Island, Canadian Arctic Archipelago: a study in glacier inventory, Axel Heiberg Island Research Reports, Glaciology No. 3, McGill University, Montreal. 105 p.

Ommanney, C.S.L., Goodman, R.H. and Müller, F., 1969. Computer analysis of a glacier inventory of Axel Heiberg Island: Canadian Arctic Archipelago. I.A.S.H. Bulletin, Vol. 14, No. 1:19-28.

Ommanney, C.S.L., Clarkson, J. and Strome, M.M. 1973. Information booklet for the inventory of Canadian glaciers. Glacier Inventory Note 4, Inland Waters Directorate, Department of the Environment, Ottawa. 76p.

Østrem, G. and Ziegler, T., 1969. Atlas over breer i Sør-Norge. Hydrologisk Avedeling Meddel. Nr. 20, Norges Vassdrags- og Elektrisitetsvesen. 207p.

Østrem, G., Haakensen, N. and Melander, O., 1973. Atlas over breer i Nord-Skandinavia. Hydrologisk Avdeling Meddel. Nr. 22, Norges Vassdrags- og Elektrisitetsvesen, and Naturgeografiska Instutitionen Meddel. Nr. 46, Stockholms Universitet. 315p.

Paterson, W.S.B., 1970. The application of ice physics to glacier studies. In: Glaciers, Proceedings of Workshop Seminar, 1970, Vancouver, B.C., Canadian Nat. Comm. for the I.H.D.: 43-46

Peterson, J.A., 1973. Carstensz glaciers expedition (CGE) 1973: preliminary report. Met. Dept., Univ. of Melbourne, April. 21p.

Post, A.S., 1970. Glaciers of the Central Chilean Andes and their importance to the water resources. Unpublished Manuscript in open file, U.S. Geological Survey, Tacoma, Washington. 5 p.

Post, A., Richardson, D., Tangborn, W.V. and Rosselot, F.L., 1971. Inventory of glaciers in the North Cascades, Washington. U.S. Geological Survey Prof. Paper 705-A, 26 p.

Sessions, M.P.M., 1973. Problems of taking a glacier inventory for the North Pentagonian Icefields. Joint Services Expedition to Chilean Patagonia 1972/73, unpublished report. 15 p.

UNESCO/IASH, 1970. Perennial ice and snow masses. Technical papers in Hydrology No. 1, UNESCO No. A.2487. 56 p.

Vivian, R., 1967. Fiches des glaciers Français:le glacier de Sarennes. Revue de Géographie Alpine, Vol. 40, No. 1:247-249.

ICE THICKNESS AND VARIABILITY ON SILVER LAKE, GENESEE COUNTY, MICHIGAN: A RADAR APPROACH

M. Leonard Bryan
Environmental Research Institute of Michigan
and
Department of Physical Geography and Environmental Graphics
University of Michigan - Flint

ABSTRACT

The possibility of obtaining some data concerning lake ice variations using remote sensing instruments led to the need to assess, in a fairly descriptive way, the variations in ice over a small mid-latitude lake. Consequently, during February and March, 1972, Silver Lake was studied for ice thickness, ice type (classified as black and white ice), and the spatial distribution of both over the lake. The survey grid was 100 ft, a grid far in excess of the resolution cell of radar but of a size which would give a reasonable indication of the variations to be expected.

Minor weather changes could present problems concerning the use of radar as a remote sensor for such lake ice studies. This is especially true when the time delay between radar and the ground truth data collection is on the order of a week or more and also where meteorological changes are rapid and fluctuating around 0°C. It has been possible, however, to document and identify some areas on Silver Lake where unsaturated white ice and snow overlie a surface of black ice. Several radar images of the lake are presented to illustrate this point. Continuing studies of remote sensor signatures and the interaction of electromagnetic energy with the earth's interface will increase the utility of these instruments for similar ice and snow studies.

INTRODUCTION

Several papers available in the literature adequately summarize the theory and possibilities of using remote sensing instruments as a data source for both operational and research approaches to ice and snow problems. Meier *et al*. (1966) note the utility of radar for glaciological research. Glacial ice, terminus lakes, old snow, and firn were identified on radar imagery from South Cascade Glacier, Washington state.

Two additional papers concerning imaging radars and their utility for study of fresh-water lake ice and snow are of special interest. Bryan (1972) briefly discusses and summarizes the basic operations of side-looking airborne radar (SLAR), presents imagery dealing with winter ice conditions in eastern Lake Superior, and notes that numerous types of ground truth data relative to radar remote sensing are needed over small experimental plots of ice in order to prepare interpretation methodologies and obtain accurate results. These data include surface roughness, snow density, moisture content, and the surface electrical (dielectric) properties. Two small portable instruments were designed and field tested to ascertain the latter. Bryan and Larson (1973) describe such instruments and discuss both the theory and the field methodology behind their operation. Although data were not available which combine the ground truth and SLAR imagery, the data indicate that observed variations in snow were of a magnitude to cause sufficient alterations in the radar backscatter. This, in turn, should be detectable in the output image of the precision optical processer used in preparation of the radar imagery from the original signal film.

The next logical step in this sequence of work was to determine the variations in the thickness and type of both lake ice and the overburden of snow and slush which existed during a given radar imaging mission. Although the data available are from an earlier period and do not fit exactly into the proper chronological and scientific sequence of events, their availability has shed some light on the questions being raised (Bryan and Larson, 1973) concerning the use of imaging radars for study of fresh-water ice and snow. Specifically, then, this present paper presents several types of data concerning such variations in ice type and thickness together with synthetic aperture SLAR imagery (X-Band system operating at 9.6 GHz (3 cm) in the horizontal - horizontal (HH) mode) for a small inland lake.

STUDY AREA AND METHODOLOGY

The area of study is Silver Lake, Genesee County, Michigan. This lake is located in the east central portion of Michigan approximately 60 mi (97 km) northwest of Detroit and 15 mi (24 km) south of Flint, Michigan (Figure 1). Silver Lake is a small body of water (310 A; 125 HA) which resulted from a dredging operation associated with a cement manufacturing plant formerly located on its northern shore. Only the northern portion of the lake was studied.

Within this section the maximum depth of 65 ft (19.8 m) is located (Figure 2).

DATA PRESENTATION AND DISCUSSION

Data concerning the variability in ice type and thickness and the hydrostatic water level (i.e., level of the water relative to the surface of the total ice) were collected over a surveyed grid at 100 ft (30.5 m) intervals. All data were collected within eight days of the radar flight and although some minor meteorological variations were occurring during this period, there were no changes sufficient to cause significant cracking, ablation, or other changes in the lake ice surface. The majority of the data is presented qualitatively. However, several statistical presentations are possible (Table 1).

It is noted in Table 1 that the percentage of black ice and white ice are, on the average, fairly consistent over the entire lake.[1] Generally, in this bipartite classification scheme, 80 percent of the lake ice is classed as black ice. The thickness data, although of comparable values for both black and white ice with respect to maximum and minimum thickness and for their standard deviations,

---

[1] The terms black ice and white ice are used following the definitions of Andrews (1962), Shaw (1965), and Jones (1970).

Figure 1. Location of study area.

Table 1. Data for Silver Lake based on 990 sample sites

	Maximum	Minimum	Mean	Stan. Dev.
Depth[1]	-65.0(19.8)	-0.5(-15.2)	-21.3(-54.1)	15.7(39.9)
White Ice Thickness	14.0(35.6)	0.0(0.0)	2.5(6.4)	2.1(5.3)
Black Ice Thickness	18.0(46.0)	0.0(0.0)	11.7(29.7)	2.9(7.9)
Total Ice Thickness	21.0(53.3)	2.0(5.1)	14.1(35.8)	2.2(5.6)
Hydrostatic Water Level[2]	-10.0(-25.4)	-0.3(-0.8)	-1.7(-4.3)	0.7(1.8)
Water Level (B/W)[3]	12.0(30.5)	-3.0(-7.6)	0.8(2.0)	2.0(5.1)
Percent White Ice[4]	100.0	0.0	81.6	16.6
Percent Black Ice[4]	100.0	0.0	18.3	16.4

Notes: [1]Data in feet and meters
[2]Water level relative to upper surface of white ice
[3]Water level relative to black ice/white ice interface
[4]Data in percentages
All other data in inches and centimeters.

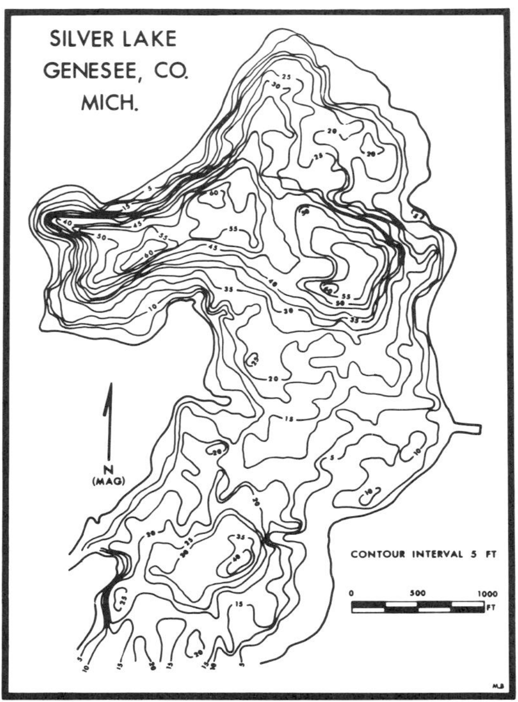

Figure 2. Bathymetric map of Silver Lake, Genesee County, Michigan.

show considerable variation in their means. In addition, the correlation coefficient of black ice versus white ice is -.6; for black ice versus total ice, .73; and for white ice versus total ice, 0.0. These data are interepreted to mean that a considerable period of freezing occurred prior to the initiation of sufficient snowfall and the subsequent melting or otherwise soaking of the snow for the formation of white ice; however, a snowfall in conjunction with or prior to strong winds would result in a similar situation.

Data showing the spatial distribution of ice thickness are presented in Figure 3. This figure and subsequent figures using the block diagram presentation were drawn with a computer program which is available in Tobler (1970). (The program requires that all data be presented in a rectangular grid; consequently, because the lake has an irregular outline, areas of no interest were necessarily included using data values of zero.) Of particular interest in this figure is the distribution of white ice because black ice, having a high density and generally forming a smooth (relative to radar wavelengths, in this case 3 cm) surface on small inland lakes will cause specular reflection of the impinging radar energy. These areas will be black on positive photographs of radar images.

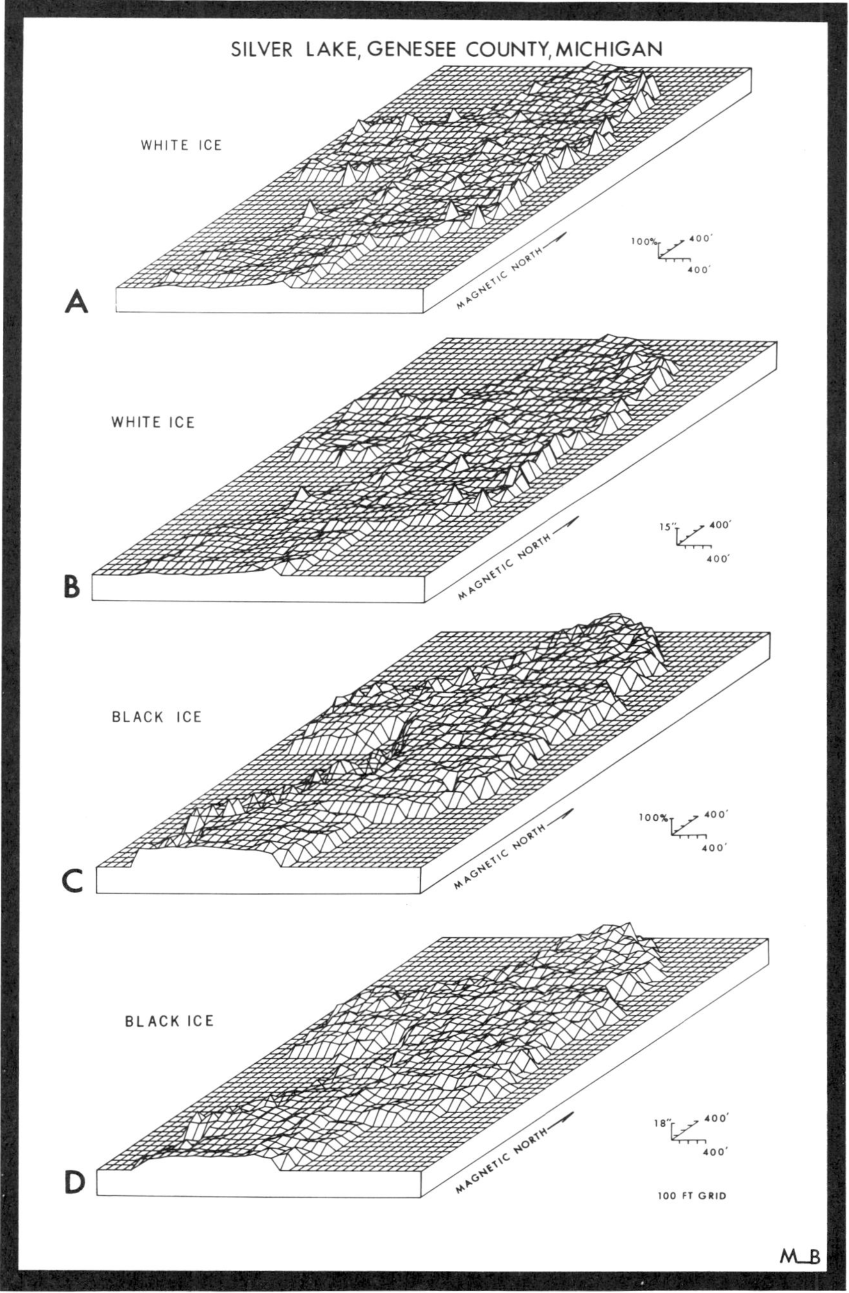

Figure 3. Block diagrams of black and white ice thickness variations.

White ice and snow drifts, on the other hand, are not only less dense but also are subject to the formation of several internal strata as a result of several freezing, snowfall, and soaking cycles. Those strata are thought to be comparable with respect to their electrical properties to layers which form in snow through the normal metamorphism processes. Thus, as discussed by Bryan and Larson (1973), these may well present dielectric boundaries which will cause greater or lesser reflection (i.e., backscatter) of the radar energy to the antenna and will therefore image as various shades of gray on a positive photographic image.

From Figure 3, then, we note the locations of areas in which white ice is of greater thickness, both in terms of the absolute amount and in percentages of total ice. The southern portion of the study area is free of white ice, whereas the central portion contains several peaks of white ice and has generally a thicker cover of white ice. The lake's western embayment, in terms of ice types and amounts, is quite similar to the southern portion of the lake, i.e., black ice dominates except for the small areas around the shorelines where white ice has formed from accumulated snow drifts. The same is true for the northern portion of the lake, where the extreme northeastern shoreline has drifts of snow and subsequent white ice formation. Consequently, these data suggest that the central portion of the lake offers a higher probability of backscattering of radar energy and the subsequent identification of white ice and snow drifts on the radar imagery.

Two additional sets of data are reproduced in Figure 4. These data indicate the level of the water relative to the ice surfaces. Figure 4A shows the hydrostatic water level, that is, the depth of the water surface below the uppermost (white) ice surface. The water level fluctuates considerably over the entire lake, but is, in the central portion of the lake, identified in the earlier discussion, relatively level and high. The range of depth below the white ice/atmosphere interface is from .3 in (.8 cm) to 10.0 in (25.4 cm), as stated in Table 1. The important point here is that in no case is the water at the surface. This is, possibly , an artificial condition because the conduit for water movement from below the ice to the ice surface is a measurement hole which was drilled by the field work team. (Ideally, an instrument as described by Adams and Shaw (1966) should be used for these types of measurements.) Figure 4B graphically presents the level of the water relative to the white ice/black ice interface. Table 1 indicates that the water level varies between 3 in (7.6 cm) below to 12 in (30.5 cm) above that interface. Consequently, in some areas of thin white ice, there is little or no probability of wetness in the basal portions of the strata. Conversely, for areas where the white ice layer is considerably thicker, the basal portions are, in all probability, soaked with water if there is sufficient cracking in the black ice to allow for migration of water to the upper layers. This is a reasonable conclusion, for the loading of the ice with more snow and white ice would cause a greater depression of the entire ice sheet.

As hypothesized earlier, if this low density white ice, or snow drift, is on the lake surface, it should be detected by SLAR. Problems concerning look direction, radar antenna depression, radar frequency, and polarization are important in determining the interaction of the radar energy with the earth's surface. These are, to a great degree, controlled by the investigators and are considered

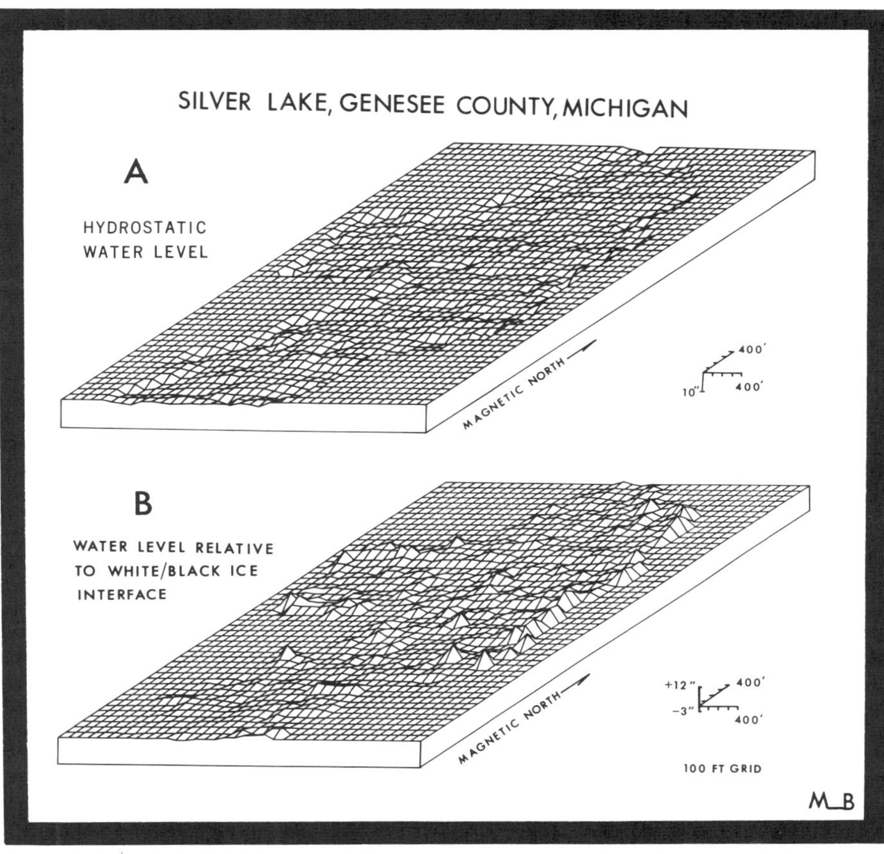

Figure 4. Water levels relative to two ice surfaces.

to be constants for any given set of investigations. These problems have been discussed in different contexts by Lewis (1968) and MacDonald et al. (1969).

Detection of subtle features by SLAR and the appearance of these on a photographic image are not directly comparable. For features of low radar reflectivity, the signal may be below the threshold needed for accurate identification. This may be alleviated by further amplification of the signal and the proper procedure for separating it from system noise and ground clutter. In addition the dynamic range of the radar may be as high as 60 dB, whereas the photographic film on which the final image is placed may have a dynamic range of 10-15 dB. Thus, the range, with respect to the number of identifiable grey levels, would be reduced considerably and the subtleties may be averaged out or otherwise lost. One possible solution is to alter the laser intensities in the precision optical processer. Although high reflectance items may be saturated (i.e., over-exposed), low reflectance features similar to those being discussed may be sufficiently exposed to be visible in the final photographic image. Of course, ideally we would like to study such imagery while it is still in the processor, but that is not always feasible. Figure 5 presents four images of variable laser exposures. The highest exposure is for Figure 5A with the exposure intervals between 5A-5B-5C and 5D being 6 dB. Thus, Figure 5D has an exposure which is 18 dB below that of Figure 5A.

Only in Figures 5A and 5B, the two images with the highest exposures, are some subtle differences in gray tone visible in the central portion of the lake, where it was suggested that white ice would be most probable for detection by SLAR.[2]

THE IMPORTANCE OF WHITE ICE

Fairly extensive work has been conducted concerning the initiation, growth, thickness, variability, and decay of ice in freshwater lakes in both the subarctic and midlatitudinal locations. Much of this has been based upon meteorological parameters and is often related to the concept of degree days. However, the variation of ice over a given lake is not always explained by these arguments. It is readily recognized that each winter's ice cover is unique in many respects. Partly because of less snow accumulation in the lower Michigan area, the direct comparison with earlier subarctic work may not be entirely valid. However, some geographic aspects of the ice thickness and variability (as defined for subarctic lakes) are seen to be valid in the Michigan area.

Earlier work dealing with white ice variability and distribution has been conducted by Jones (1970) and Archer (1966). Archer's primary concern is the nature of the formation of white ice and a discussion of the trends and distribution of snow cover. He also deals with the interpretation of these variations in addition to correlations of various types of ice and snow thicknesses. Comments relative to the prevailing winds and the location of snow drifts, snow thickness and total ice, and the depression of the ice sheet with the subsequent formation of white ice are of particular importance. Jones, although discussing a similar set of relationships, also presents (pp. 103-109) a short discussion entitled "The Hydrological Importance of White Ice." As stated by Jones, "One of the basic theses ... is that the white ice process depletes the amount of snow ... by an amount that can be approximated by assuming that the water equivalent of the original snow cover is equal to about one third that of the resulting white ice. If ... this amount is added to the mean snow water equivalent obtained from a basin survey, this will give a best estimate of accumulated winter snowfall" (Jones, 1970, p. 103). In his study area, a small lake in Labrador-Ungava surrounded by a topographically rough and forested surface, accumulation of drifting and blowing snow may be less on a smooth lake surface than in the adjacent forests. Nonetheless, approximately one fourth of the

---

[2]Not only have some data been lost by placing the image on the photographic film, but the preparation of a third generation data set in the form of plates and their subsequent printing has additionally reduced the radar data. Photographic slides showing these data were available at the symposium.

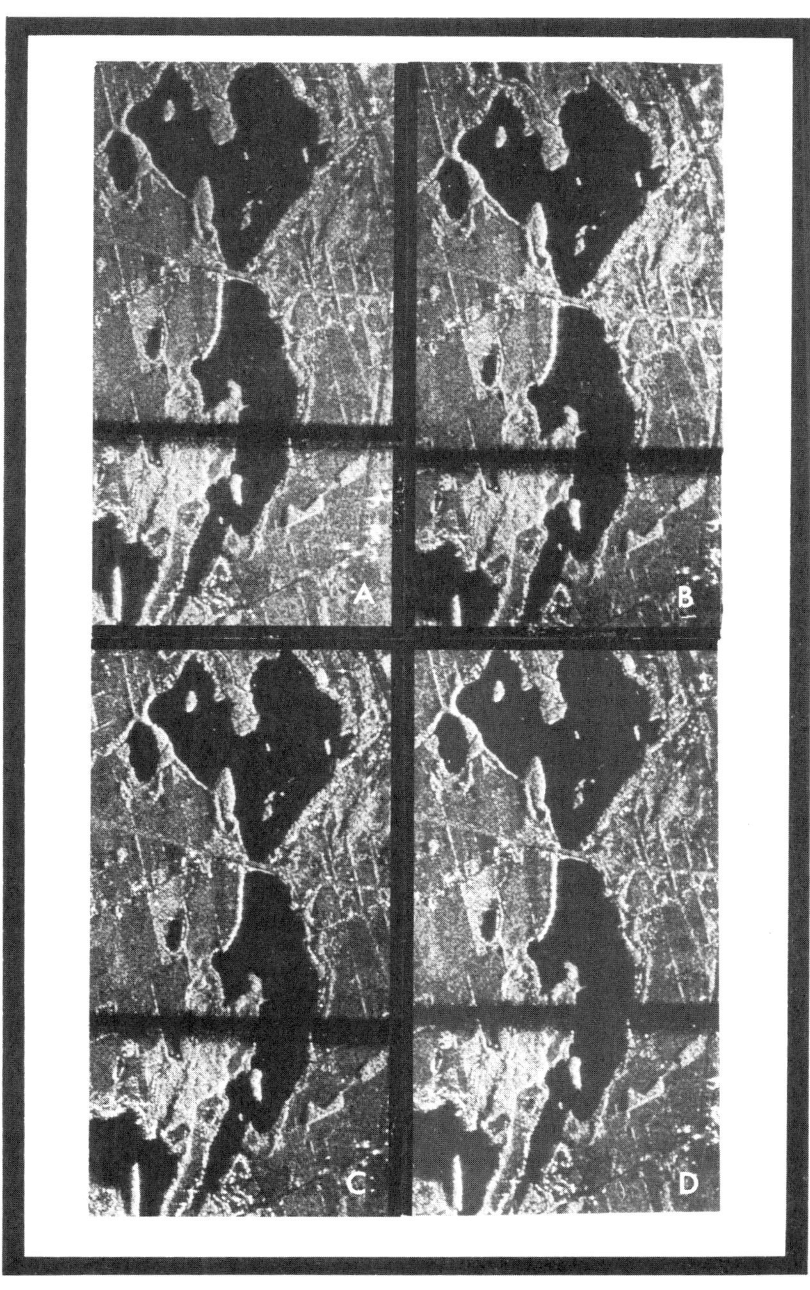

Figure 5. Variable exposures of SLAR (X Band, HH) imagery of Silver Lake, Genesee County, Michigan.

lake ice in the studied watershed consisted of white ice. Of this, Jones determined that approximately 30 percent of the water content was gained from the falling and blowing snow rather than directly from the lake waters. Thus, not only are the snows retained in the basin by being transformed into white ice, but because of the cracking of black ice (by depression or thermally) lake waters are utilized as the final constituent in the snow to white ice transformation. Thus, this water is removed from the water supply needed to maintain stream flow in any given basin. Further calculations by Jones indicate that 17 percent of the water runoff in the study area was caused by ice growth through the depression of the ice layer and the displacement of water out of the lakes. He concludes that "white ice has been shown to be important in considering any assessment of true snowfall in the area, although methods of doing this have not been fully perfected. It is also important in its effects upon stream discharge" (Jones, 1970, p. 109). These have validity and show the need to adequately consider the ice type, distribution and stages of growth when studying the winter hydrology of a drainage basin.

CONCLUSIONS

Both theoretical and empirical research indicate that side-looking airborne radar imagery can be used for the identification of various ice and snow surfaces. The differentiation between snow, black ice and some types of white ice may be feasible using this type of remote sensor. In addition, preliminary research also indicates that white ice may be an important variable when the winter hydrological regime for any given watershed is considered. This may be more acute in areas where drainage is regulated, either for maintain a discharge or for storage for future use. By monitoring white ice accumulation relative to snow falls in a basin, additional knowledge necessary for the more accurate management of water resources may be possible. Much additional research is needed to quantify and verify the suggestions contained in this paper with respect to the types of ice data available through SLAR remote sensing. The ultimate validity of this approach is becoming more apparent and may be increasingly moreso as radar technology improves and a true simultaneous multiple frequency radar of the type discussed by Porcello and Rendleman (1972) becomes available.

ACKNOWLEDGEMENTS

The work reported was sponsored by the National Science Foundation (Washington, D.C., Grant Number GI-30027) through its program, Research Applied to National Needs (RANN). Messers. P. Davis and D. Trune aided in the field work.

REFERENCES

Adams, W. P., and J. B. Shaw, 1966. Improvements in the measurement of lake-ice cover. J. Glaciology. 6:299-301.

Andrews, J. T., 1962. Variability of lake ice growth and quality in the Schefferville region, central Labrador-Ungava. J. Glaciology. 4:337-347.

Archer, D. R., 1966. The ice survey, 1964-1965. In: Adams, W. P. ed, Field research in Labrador-Ungava. Montréal: McGill Sub-Arctic Research Paper No. 21, McGill University, pp. 170-188.

Bryan, M. L., 1972. The utility of imaging radars for the study of lake ice. Paper presented: International Symposium on the Role of Snow and Ice in Hydrology, Canadian IHD Committee, Banff, Alta., Canada, September.

Bryan, M. L., and R. W. Larson, 1973. Application of dielectric constant measurements to radar imagery interpretation. In: Shahrokhi, F., ed, Second Annual Remote Sensing of Earth Resources Conference, University of Tennessee Space Institute, Tullahoma, Tn.

Jones, J. A. A., 1970. The growth and significance of white ice at Knob Lake, Quebec. In: Adams, W. P. ed, Studies of lake cover in Labrador-Ungava. Montréal: McGill Sub-Arctic Research Paper No. 25, McGill University, pp. 6-160.

Lewis, A. J., 1968. Evaluation of multipolarized radar imagery for the detection of selected cultural features. U. S. Dept. Interior, U. S. Geol. Survey. Interagency Report NASA-130. 56pp. (NTIS #N69-28151)

MacDonald, H. C., J. N. Kirk, L. F. Dellwig, and A. J. Lewis, 1969. The influence of radar look direction on the detection of selected geological features. Proceedings, 6th Int. Symp. on Remote Sensing of Environment. Ann Arbor, Mi., U. Mich. Inst. Sci. and Tech. Willow Run Laboratories. October. pp. 637-650.

Meier, M. F., R. H. Alexander, and W. J. Campbell, 1966. Multispectral sensing test at South Cascade Glacier, Washington. Proceedings, 4th Int. Symp. on Remote Sensing of Environment. Ann Arbor, Mi., U. Mich. Inst. Sci. and Tech. Willow Run Laboratories. April. pp. 145-159. (NTIS #AD 638 919)

Porcello, L. J., and R. A. Rendleman, 1972. Multispectral imaging radar. 4th Annual Earth Resources Program Review. NASA 2:35.1-35.18. (NTIS #N72-29327)

Shaw, J. G., 1965. Growth and decay of lake ice in the vicinity of Schefferville (Knob Lake) Quebec. Arctic. 18:123-132.

Tobler, W. R. ed, 1970. Selected computer programs. Ann Arbor, Mi.: Dept. of Geography. vi+162pp.

CHANNELS IN ICE*

Ludvig G. Browman
University of Montana
Missoula, Montana 59801

ABSTRACT

The formation of slush between the ice covering freshwater lake surfaces and the overlying dry snow has usually been attributed to the upward movement of water through sizeable cracks and holes in the ice. However, study of a mid-latitude freshwater mountain lake in western Montana over the past five winters has shown that microscopic or minute channels in the apparently "solid" ice are the primary source of the slush-forming water when the ice is covered by heavy snow. This phenomenon is temperature-dependent, occurring in the ice samples studied only between -0.8°C and 0.0°C (ice temperature 5 cm below the surface).

The microscopic channels--which may enlarge to macroscopic channels--develop originally at the ice grain boundaries. Other channels may form during mild spells from the fusion (largely along crystal boundaries) of air bubbles that are expelled during freezing. The decrease in ice density that occurs as winter progresses can be interpreted to suggest concurrent increase in the degree of channelization and enlargement of existing channels.

INTRODUCTION

The presence of snow-slush on freshwater lake surface ice has usually been attributed to the upwelling of water through fissures, cracks, crevices, and holes in the ice. A commonly held view is that these relatively large openings in the main body of ice are the primary, if not the exclusive, source of the water that soaks the lower levels of the overlying snow.

However, snow-slush may also develop, under certain circumstances, between the upper, dry snow and the ice when the existence of fissures, fractures, or macroscopic openings cannot be demonstrated. The possibility that sheet ice may have minute tubules, or microscopic or larger pores within it would explain this upward movement of water through apparently "solid" sheets of ice, particularly when the weight of the overlying snow is sufficient to produce an adequate hydrostatic pressure.

If channels do exist in snow covered ice, then they would also exist in the ice cover of freshwater lakes with a very light or insignificant snow cover and with no snow-slush. Whether such channels exist continuously throughout the winter period, or come and go in response to temperature changes or other ambient conditions, however, was not known.

---

*This study was supported, in part, by the Office of Water Resources Research (OWRR), U.S. Department of Interior, through the University of Montana Joint Water Resources Research Center (Project A-045-Montana).

Observations, over a period of years, of ice formation and snow-slush on a mid-latitude freshwater mountain lake in western Montana led the author to investigate this concept. The study has shown that minute channels in the ice are responsible for the major portion of slush between the overlying dry snow and the lake ice in years of relatively heavy snow. These channels may, in fact, be the primary source for slush formation on ice-covered freshwater lakes where the ice is covered by heavy snow.

PROCEDURE AND BASIC DATA

Holland Lake, Missoula County, Montana is a mid-latitude lake (Lat. 47° 27'; Long. 113° 36') located west of the Continental Divide at the base of Carmine Peak, Swan Range at an elevation of 1,225 meters. The lake surface is roughly 165 hectares (408 acres) in area with a maximum depth of 48 meters. It is a dimictic, oligotrophic lake surrounded by a complex of Douglas fir-larch forest.

A hollow square raft made of four 25 cm x 25 cm x 2.4 m timbers bolted together and supporting a delay-relay housing unit was anchored over water depths of 14-15 meters approximately 55 m from shore. A vertical, hollow, fiber-glass coated and epoxy covered wooden post, 10 cm in diameter was supported out from the raft approximately 60 cm. Twenty-eight thermistors were positioned at intervals of 6.5 cm and 3.5 cm out from the post. Before freeze-up, 15 of these thermistors sampled air temperatures above the lake surface to a height of about 1 m. Nine thermistors sampled water temperatures down to a depth of 60 cm. One supplemental thermistor was located in water to a depth of 3 m, and another at a depth of 5 m in addition to the 2 controls.

Two controls were used. One was a specially selected resistor suspended 25 cm down into the lake and recording consistently at -1.0° C to determine whether the electrical equipment was affected by extreme temperature variations. The other control was a thermistor positioned at a depth of 12.5 meters and recorded the relatively consistent temperature of the maximum density water layer once that was formed early in winter.

The delay-relay tube unit for selecting 60 second samples of temperature was housed on the raft; the power supply, timer, and recorder were housed in an instrument house on shore connected to the raft by means of an electric cable. Electrical current was available throughout the winter season. Storms and falling trees interrupted the current four times during the winter of 1971-'72, and not at all during 1972-'73.

The air, snow, ice, and water temperatures at the raft were sampled at the 28 levels for 60 second intervals at 0600, 1200, 1800, and 2400 hours daily. Graphic recordings on strip charts were removed weekly for reading and analysis. The recorded line graph was read to within ± 0.1° C. The factory claimed an accuracy for the recorder of ± 0.2° C. The author is confident, based on the behavior of the control resistor and thermistor, that all recordings for the entire system were accurate to within ± 0.5° C even under severe cold conditions, and to within less than ± 0.2° C under ordinary winter conditions.

One or more ice cores, 10 cm in diameter, were taken weekly within 3-4 m of the thermistors, but the ice within the immediate vicinity of the thermistors proper was not disturbed. These ice cores were assumed to represent the ice conditions in the immediate vicinity of the thermistor probes.

Routine measurements were made of snow and slush depths, the thickness of crust ice, and granular ice slush over the solid ice surface, and the thickness of layers of differing kinds of ice within the ice core. Ice cores were examined in the field by means of a binocular dissecting microscope to observe vertical and reticulated channels and to determine the presence of india ink particles.

India ink tests were run on all ice core samples in the field. India ink was finally selected because carbon particles remain discrete and visible within minute channels. Ice core samples were usually obtained mid-day. A section of a rubber motorbike inner tube was slipped over the top 10-12 cm of the core, and an india ink suspension in approximately 0° C melted snow water was poured into the resulting "cup". If no india ink penetrated into the ice core after the elapse of 60 seconds the reading was recorded as negative.

OBSERVATIONS AND RESULTS

Ice can vary from being permeable to an india ink suspension in ice water to being completely impermeable within a very short time. On January 22, 1972 the ice was 42 cm thick with a cover of 13 cm of snow over 5 cm of slush, and had been exposed for the previous 48 hours to an average temperature of about -0.9° C. The ice was permeable to india ink suspension throughout the sample core. The following week (Jan. 29, 1972), with the ice 47 cm thick, 16 cm of snow over 2.5 cm of slush, and the average temperature for the previous 48 hours -25.1° C no india ink penetrated any sample core. All tubules and channels had apparently been frozen shut. The results obtained in the winter of 1971-'72 are considered to be typical of a relatively mild, wet winter, and those of 1972-'73 of a mild, "open", dry winter (Table I).

Table I

Comparison of Ice and Snow Conditions
Winter, 1971-'72 and Winter, 1972-'73

Winter	Precipitation from Nov. 1 to April 1			Min. air temp.	Snow Depth: 40 m back from shore		Snow Depth: Lake surface		Max. ice thickness	Ice density	
	Holland Lake mm	Missoula mm	Kalispell mm	°C	Max. cm	Ave cm	Max. cm	Ave cm	cm	Dec.	Apr.
1971-'72	421	181	148	-34°	117	56	45	20	70	.917	.767
1972-'73	151	58	87	-34°	22	11	9	3	49	.918	.772

There was no correlation either year between the depth of the snow on the lake ice and the penetrability of india ink in ice core samples (Table II). The presence of snow cover probably delayed the influence of ambient air temperature on the patency (i.e. openness) of ice channels.

The bulk of ice formation (1972-'73) was from freezing at the underside of the ice layer and not from freezing slush at the top of the ice as was the case in the winter of 1971-'72. At no time during the winter of 1972-'73 was water ever seen spreading over the ice surface, up or out of cracks, fissures, or holes in the ice. During one thaw (Dec. 21, 1972) water from the lake surface was observed to be actually running down into holes and fractures in the 26 cm thick ice causing, in the resultant surface eddies, floating larch needles to whirl about.

Table II and Figures 3, 5, and 6 indicate that under field conditions when ice temperatures are warmer than -1.0° C india ink suspensions in freshly melted, ice-cold snow water do penetrate into channels of ice cores within less than the 60 second standard. Patency to india ink seems optimal at ice temperatures warmer than -0.8° C. When ice temperatures are below -1.0° C there is no penetration by india ink into the ice core.

When ice temperatures approach 0.0° C the interstices between ice crystals, or the junctions of ice crystals, apparently liquefy and allow vertical movement of water within the ice sheet as determined using microscope observation of the movement of minute suspended particles within ice cores. Thermal conductivity throughout the ice sheet is also increased by the resulting movement of water and thermal convection until the entire ice sheet reaches essentially the same temperature. Temperatures of -1.0° C or less cause the water within the channels to freeze and thus eliminate vertical water movement.

Another source of ice channels, of a reticulated and honeycomb nature, is the air bubbles which are released into and from the ice when water freezes. The fluctuating temperatures during mid-latitude winters permit the movement, fusion, and coalescing of air bubbles into elongated irregular tubules and channels (Figures 4 and 7). Microscopic examination of ice cores reveals continuous minute channels that are formed in this way and frequently along crystal boundaries.

Water temperatures in the spring increase more rapidly below the ice than does the temperature of the overlying ice, resulting in relatively large channels in the bottom ice (Figure 7).

The density of the ice, as measured in the ice core samples, decreases during the course of the winter months (e.g.: Dec. '71 - .917; Jan. '72 - .848; Feb. '72 - .803; Mar. '72 - .767). The decrease in density can be interpreted as further evidence for the increased channelization of ice. By March and April the channels can be easily seen with the unaided eye (Figures 6 and 7).

TABLE II
HOLLAND LAKE, 1972-'73
Temperatures and channels in ice core samples.

Date	Ice Thick- ness	Ave. Temp. 5 cm down for 5 days prior to test date	Ice temp. 5 cm down noon of test date	Ice level above water level	Were ice channels patent to india ink?
	(cm)	(°C)	(°C)	(cm)	
12/ 4/72		3.2			
12/13/72	24.1	-9.8	-7.5	1.2	No
12/20/72	25.4	-1.9	-0.1	1.3	Yes
12/27/72	24.1	0.0	-0.2	1.6	Yes
1/ 1/73	25.4	-2.3	-2.0	2.0	No
1/13/73	31.8	-5.9	-0.5	1.3	No
1/20/73	34.3	-0.8	-1.0	2.3	Yes
1/27/73	35.6	-3.4	-5.0	3.2	No
2/ 3/73	41.9	-3.9	-3.0	3.2	No
2/10/73	44.5	-9.3	-3.5	2.5	No
2/17/73	43.2	-2.5	-0.5	2.3	Yes
2/24/73	44.5	-3.9	-0.8	2.3	Yes
3/ 3/73	44.5	-0.6	-0.0	3.2	Yes
3/10/73	45.7	-1.8	-0.3	3.5	Yes
3/18/73	45.1	-1.0	-2.0	3.5	No
3/24/73	41.3	-0.6	-0.5	2.5	Yes
3/31/73	34.3	-0.0	-0.5	3.0	Yes
4/ 7/73	31.1	0.0	0.0	1.3	Yes
4/14/73	22.9	0.0	-0.5	2.0	Yes

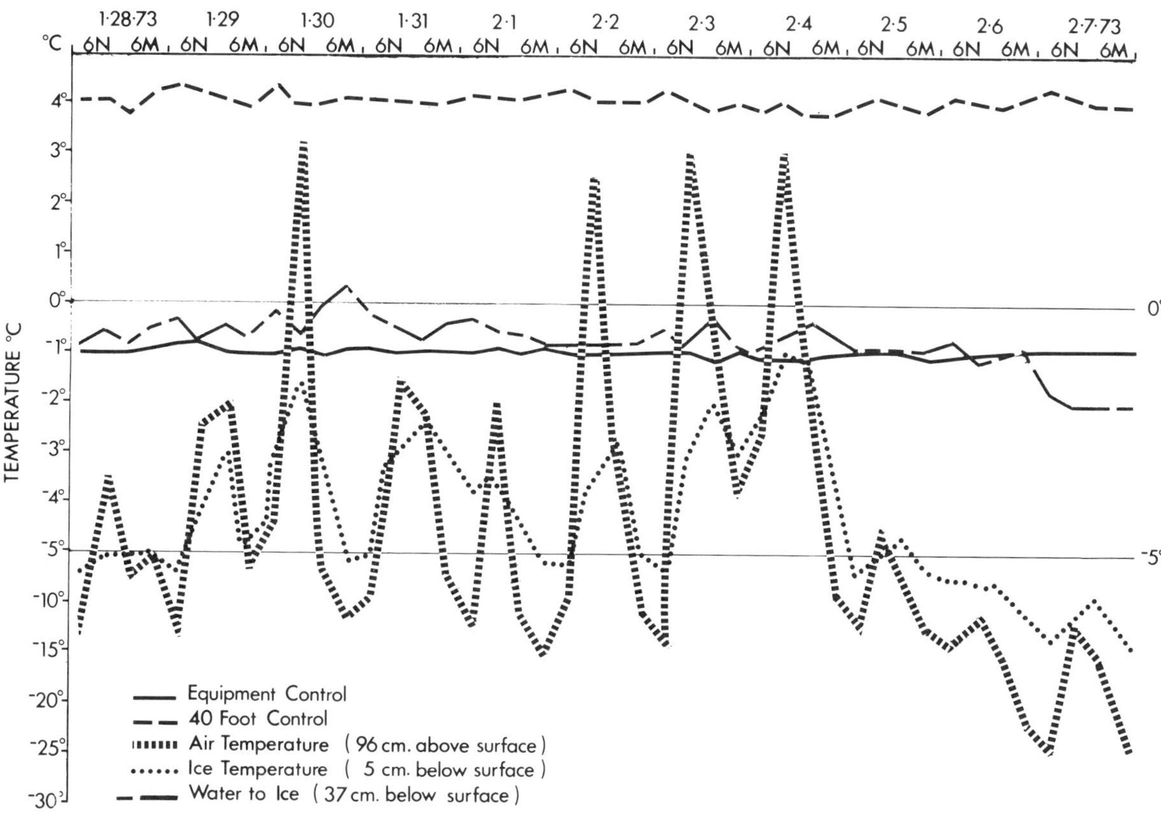

Figure 1 DAILY FLUCTUATION IN AIR AND ICE TEMPERATURES HOLLAND LAKE 1972-1973

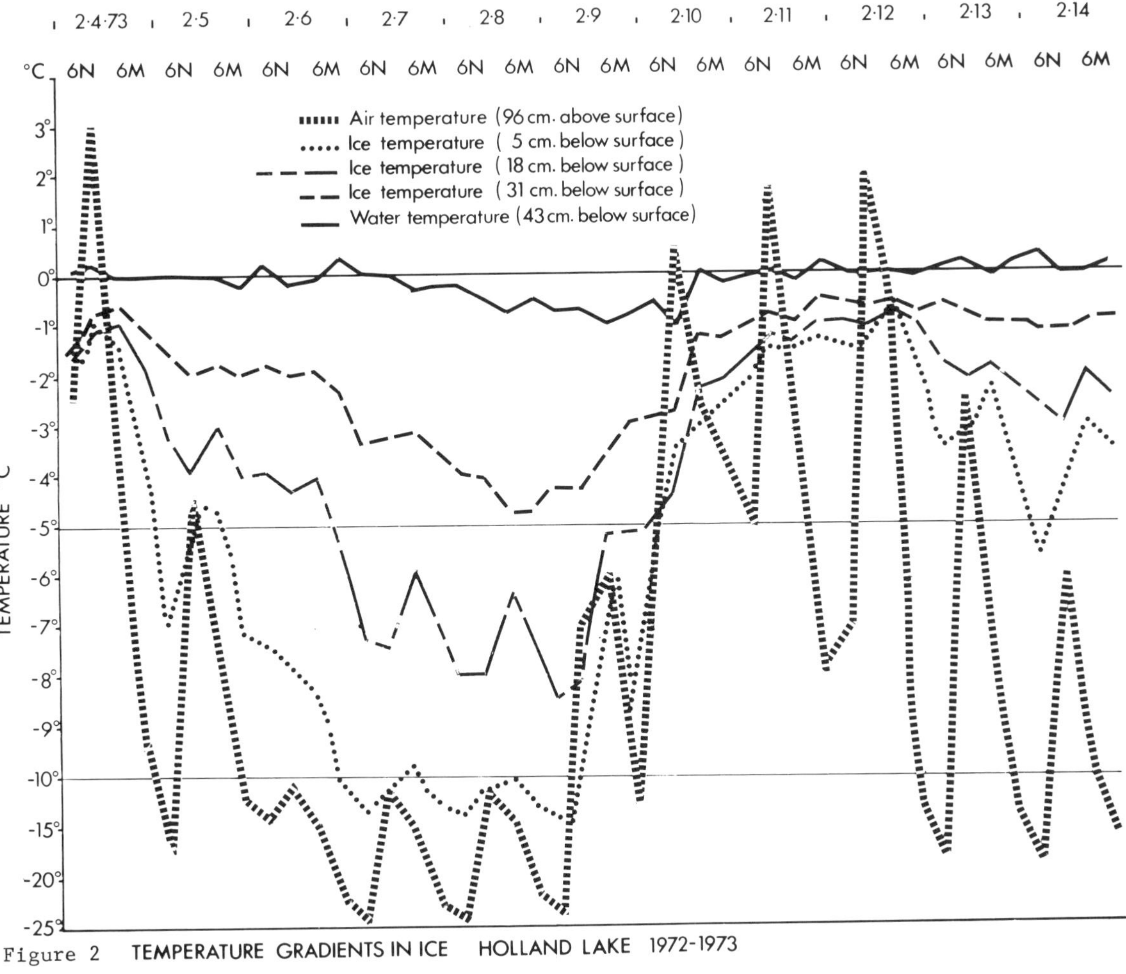

Figure 2 TEMPERATURE GRADIENTS IN ICE HOLLAND LAKE 1972-1973

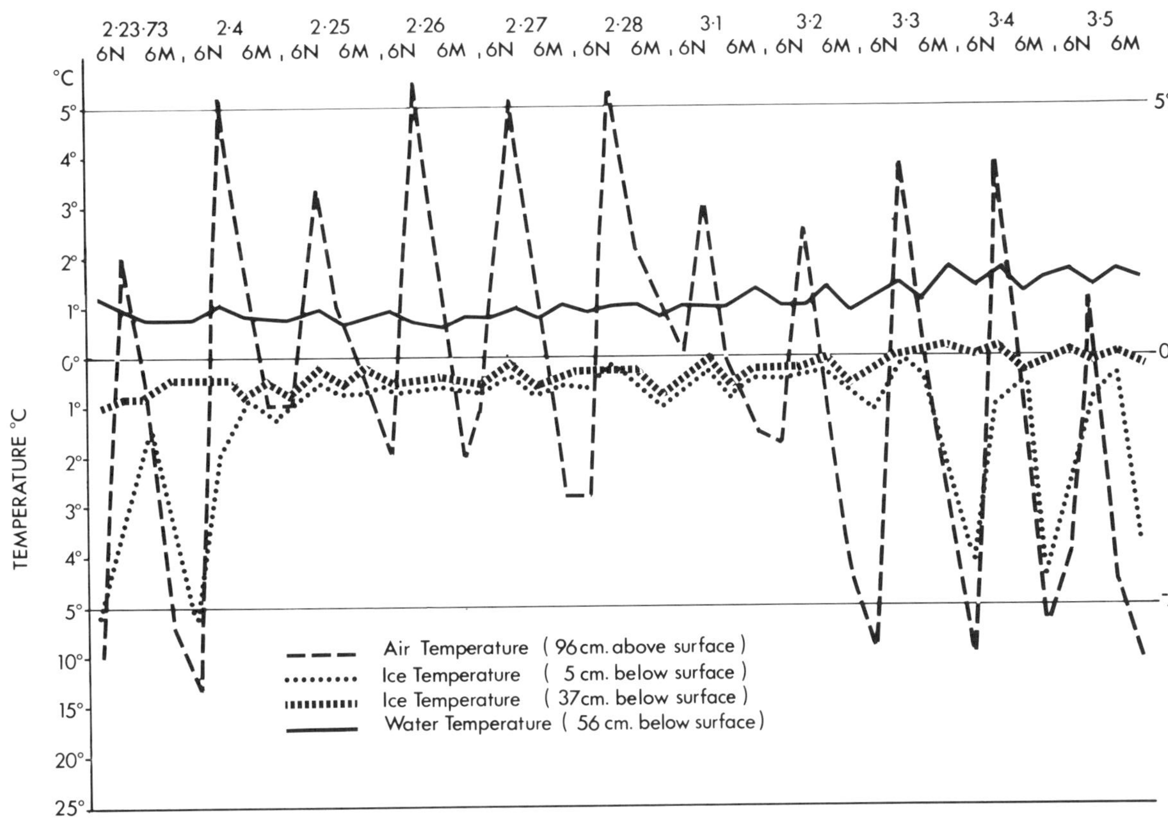

Figure 3 DAILY FLUCTUATION IN AIR AND ICE TEMPERATURES – A WARM SPELL HOLLAND LAKE 1972-1973

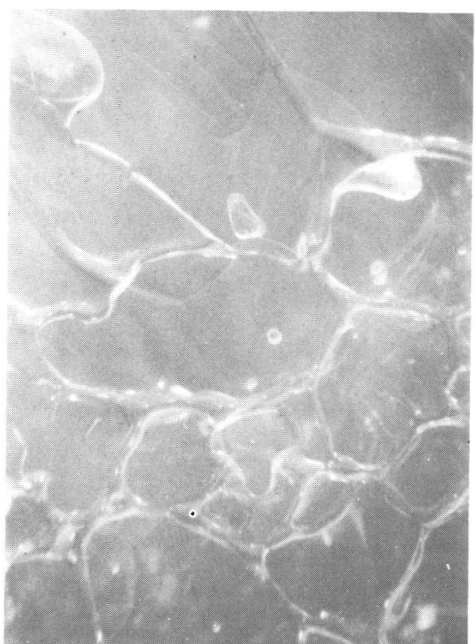

Figure 4. Reticulum of channels and isolated air bubbles. Dec. 8, 1971.

Figure 5. India ink. Network of channels and isolated air bubbles. Dec. 8, 1971.

Figure 6. Meshwork of channels in the upper layer of "white" ice join larger channels in clear ice. Dec. 27, 1972. Temp. -0.2° C.

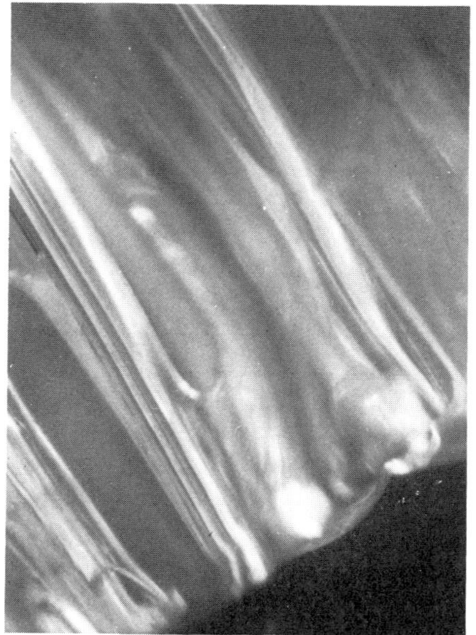

Figure 7. Channels opening at the underside of ice into lake water. Dec. 18, 1971. Temp. -0.3° C.

DISCUSSION

A commonly held view is that most, if not all the water that appears in the lower level of a snow cover on fresh water lakes is the result of the pressure generated by the weight of snow depressing the ice below the hydrostatic water level (Götzinger, 1909, Deruigan, 1967; Williams, 1968), thus forcing the water upward through cracks, fissures, holes, and macroscopic openings in the ice. The "jets" of water that can be seen rising through melt holes after a heavy wet snow (Woodcock, 1965), and the "artesian" flow through holes in Labrador Sea ice with a heavy snow load (Weeks & Lee, 1958) illustrate this concept. Similarly, on Holland Lake with the ice covered by a layer of 2.5 cm of slush, and 3.8 cm of wet snow beneath 16.5 cm of dry snow (March 13, 1971) the water gushed like an artesian well to a height of 17-18 cm through the ice core bore hole, resulting in almost 13 cm of water on the top of the ice in the area cleared of slush and snow.

On the other hand, water can and does pass directly through the sheet ice on fresh water lakes in the absence of fractures and macroscopic holes (Figures 5 and 6). Passage is dependent upon the sheet ice having reached a critical temperature. When the ice temperature rises above -1.0° C microscopic channels gradually develop, through which water rises as a result of hydrostatic pressure, gradually resulting in slush beneath the powder snow. When the ice temperature drops below -1.0° C the channels freeze and no water passes through.

Weeks and Lee (1958) report a similar phenomenon in sea water. When the ice temperature rose to a critical level sea water passed through the ice sheet, probably as a result of hydrostatic pressure forcing the water up through channels created by the fusion of liquid brine cells within the ice. When the ice was "too cold", they reported, it became impermeable to the passage of brine.

Channels, tubules, reticulated thread-like cylinders, etc. develop in at least two ways. Columns of water (channels) form at the intersections of three or more grains of ice (Hobbs & Ketcham, 1969; Kingery & Goodnow, 1963) as its temperature approaches 0.0° C. Pounder (1965) reports that fresh water ice decays first at the boundaries of cyrstals. Microscopic examination of ice at temperatures close to 0.0° C reveals minute suspended particles moving within vertical channels in "black" ice, and within reticulated honeycomb channels in "white" ice (Figure 4).

Another mechanism involved in the formation of channels in fresh water ice is the fusion and coalescing, during moderate temperature conditions, of the air bubbles forced out of the water during the freezing process. Examination of "white" ice reveals the presence of scattered, separated bubbles during the early formation of the ice, and as the ice "matures", the gradual fusion of the stellate and other bubbles into irregularly arranged channels (Figure 4). Shumskii (1964) reports that air bubbles may grow lengthwise, assuming cylindrical or thread-like forms, a phenomenon that the author has also observed. Hayward (1966) reports the formation, under artificial conditions, of ice needles that were hollow with a core of water.

Spring melt further emphasizes the fact of channels in the ice sheet. Ice becomes increasingly porous (Figure 7) as spring advances, forming cylindrical cavities up to 1 cm or more in diameter or vertical "pipiness" (Kratochvil, 1942; Norrman, 1964) particularly from the under side of the ice. Ice density decreases during the spring melt from 0.9 to 0.7, and the ice may have a porosity up to 50% with considerable "melt hole" formation (Poliakova, 1965; Kratochvil, 1942).

The data in Table II indicate that under varying mid-latitude temperature conditions, ice channels may be present at any time during the winter months. They also indicate that channels are not patent if and when ice temperature drops below the critical temperature. Channels develop in ice regardless of the presence or absence of snow cover.

REFERENCES

Deruigin, A. G., 1967. Snow-ice and its significance in computing the thickness of ice cover. Leningrad, Godudorst. gidrologicheskii inst. Trudy. Syp. 148, p. 29-44. (Abstr.)

Götzinger, G., 1909. Studien ueber das Eis der Lunzer Unter-und Obersees. Inter. Rev. Ges. Hydrobiol., Hydrograph, 2:386-396.

Hayward, A. T. J., 1966. Growth of ice tubes. Nature 211 (5045): 172-173.

Hobbs, P. V., and W. M. Ketcham, 1969. The Planar growth of ice from pure melt, in the Physics of Ice. Proc. Int'l Symp. on the Physics of Ice, Munich, Germany, Sept. 3-14, 1968. Ed. Riehl, N., B. Bullemer, & H. Engelhardt, Plenum Press, N.Y.

Kingery, W.D., and W. H. Goodnow, 1963. Brine migration in salt ice, in Ice and Snow, Properties, process, & application. Kingery, W. D., Ed., Proc. Confer. M.I.T., Feb. 12-16, 1962, pp. 237-247. The M.I.T. Press, Cambridge, Mass.

Kratochvil, S., 1942. Ice conditions in a reservoir during the winter period 1941-42. Wasserkraft u. Wasserwirt 37:226-228.

Norrman, J. O., 1964. Lake Vättern. Investigation on shore and bottom morphology. Hydrology. Geografiska Annaler 46(1-2): 53-69. (Abstr.)

Poliakova, K. N., 1965. Characteristics of the melting of the ice cover and the opening of the middle Lena River. Moskva Tsentr. inst. Prognozov. Trudy, 1965. 151:149-170. (Abstr.)

Pounder, E. R., 1965. The Physics of Ice., Pergamon Press, New York, Pp. 1-151.

Shumskii, P.A., 1964. Principles of structural glaciology. (The Petrography of fresh water ice as a method of glaciological investigation, Transl. from Russian by David Kraus). Dover Publications, Inc., New York, Pp. 1-497.

Weeks, W. F., and O. S. Lee, 1958. Observations on the physical properties of sea ice at Hopedale, Labrador. Arctic 11:135-155.

Williams, G.P., 1968. Freeze-up and break-up of fresh water lakes. Proc. Confer. on ice pressures against structures. Laval Univ., Nov. 1966, Pp. 203-215, Div. of Bldg. Res., Tech. Paper #266, N. Res. C. of Canada.

Woodcock, A. J., 1965. Melt patterns in ice over shallow waters. Lim. & Ocean., 10, Suppl. R. 290-297.

ICE AND ICEBREAKERS

N.A. Ehrlich
U.S. Coast Guard Headquarters

J.P. Welsh
U.S. Coast Guard Research & Development Center

ABSTRACT

Contained in this discussion are some of the reasons for U.S. Coast Guard involvement in ice and icebreaking research. Also presented are some examples of the problems encountered in obtaining measurements of the physical properties of ice for engineering use. Included in the discussion is a limited account of Coast Guard applications of remote sensing techniques to ice research.

Ice and Icebreakers

Icebreaking research and development was latent until interest in icebreaking was regenerated by ESSO's MANHATTAN expedition in the late sixties. While this bold expedition was hailed as a great triumph over the environment, it has only to a limited extent advanced the state of the art of icebreaking technology. However, the MANHATTAN experience did manage to bring together results of earlier experiments and knowledge from ice scientists and engineers all over the globe. The ESSO project also gave new life and confidence to modern icebreaker modeling technology which, incidentally, originated with the Soviets during the 1950s but was never heavily relied on or upgraded until 1967 with work at the WARTSILA ice basin in Helsinki, Finland.

The reason for the currently increasing interest is due primarily to favorable economic forecasts for shipping both in Arctic waters and Northern Domestic Waterways (Bates, 1973). On the Great Lakes and Northern Mississippi River, winter navigation has become imperative in order to maintain a competitive economic edge over the southern and western regions of the United States. We are all aware of the impending necessity to develop oil and mineral resources in the northernmost reaches of this hemisphere and we can all be certain that navigation in these areas will sharply increase.

The U.S. Coast Guard traditionally performs two types of icebreaking functions: 1) "Polar" icebreaking primarily for maintaining supply routes to strategic defense locations and supporting scientific expeditions, and 2) "Domestic" icebreaking operations to support commerce along the northern navigation routes in the United States. Until recently almost all of the research and development efforts have been in the Arctic icebreaking technology and sea ice research areas. Recently, however, a greater interest has been evolving in U.S. domestic waterways initiated by Congressional sponsorship of the Great Lakes-St. Lawrence Seaway Season Extension Demonstration Program. As a direct result of the program, the nature and properties of fresh water ice are becoming important for domestic icebreaker design and performance prediction.

The theory developed for icebreaking in Arctic waters may not differ from fresh water icebreaking. An example of an icebreaking resistance equation for continuous-mode ice breaking involves four separately identifiable terms: (From Lewis & Edwards, 1970)

$$Ri = R_1 + R_2 + R_3 + R_4$$

where Ri = total ice resistance
R_1 = resistance due to breaking ice
R_2 = resistance due to forces connected with weight (i.e., submersion of broken ice, turning of broken ice, change of position of icebreaker and dry friction resistance
R_3 = resistance due to passage through broken ice
R_4 = water friction and wavemaking resistance

The semi-empirical equation derived by Lewis and Edwards (1970) is expressed in the following manner:

$$Rim = C_o \sigma h^2 + C_1 \rho i g B h^2 + C_2 \rho i B h v^2$$

where Rim = mean resistance of the ice excluding water frictional and wavemaking forces
σ = flexural ice strength
h = ice thickness
ρi = mass density of ice
B = maximum beam of ship
g = acceleration of gravity
C_o, C_1, C_2 = experimental coefficients (non-dimensional)

The three terms on the right in the above equation account for the resistance forces attributable to breaking the ice (first term), to ice buoyancy (second term) and to momentum exchange between ship and ice (third term). The three constants (C_o, C_1, C_2) account for the coefficients of friction, hull shape and other unknown factors.

The foregoing discussion is presented to demonstrate that the resistance of an icebreaker is not, as commonly thought, due solely to breaking ice but is also influenced by the dynamics of the situation. The presence of h (ice thickness) in all three terms stresses the importance of ice thickness as a factor in this resistance equation The magnitude of the individually identified sources of resistive force is not well known. For example, qualitative observations from full-scale and model icebreaker testing strongly suggest that changes in the coefficient of friction for the ship's hull/ice interaction may affect icebreaking resistance to the same degree as do changes in ice strength. The importance of this qualitative observation is that an alternative engineering approach may be available for reduction of resistance. Specifically, a low friction coating might be applied to the hull, thus reducing the resistance. The low friction hull coating concept will be tested during the 1973-74 winter by the Coast Guard.

Navigation in the Arctic is profoundly influenced by the deformation of the ice cover. For example, the position and orientation of pressure ridges, leads and polynyas are dependent on deformation. Ice in motion, especially the potential pressure that an ice field in motion can exert on a transiting vessel, is a major concern of the ice mariner. Obtaining information which describes the ever changing ice conditions has been the objective of many studies using remote sensing techniques. The Coast Guard has conducted experiments with the AN/DPD-2 Side Looking Airborne Radar (SLAR) to obtain information on the spatial distribution of sea ice features and for iceberg detection.

The major advantage of SLAR for obtaining information is its all-weather capability. This is particularly advantageous for iceberg detection off the Grand Banks of Newfoundland. This area is often zero visibility because of fog. Recent Coast Guard efforts to evaluate

SLAR (the AN/DPD-2) for the International Ice Patrol have resulted in a report (Farmer, 1972) clearly demonstrating the capability of this SLAR to provide imagery from which icebergs and ships can be identified and positioned relative to the aircraft flight path. Earlier efforts by the Coast Guard demonstrated the application of SLAR for mapping sea ice and determining pack ice drift (Johnson and Farmer, 1971). The results of these and other studies provide a clear indication of the ice information value of SLAR for a polar ship routing system.

SLAR can provide spatial and temporal information for a deforming ice cover, but presently available systems cannot provide detailed thickness information. Interpretation techniques can provide categorical thickness. As shown earlier in this discussion, thickness is one of the important parameters to consider for vessel transit in ice-covered waters.

Until recently the only precise thickness measurements that could be obtained were by drilling a hole in the ice and making a direct measurement. Various electromagnetic techniques have been tried, with unsatisfactory results (Adey, 1970). An exception to this may be the development of a technique called "electromagnetic subsurface profiling" (ESP) (Orange, 1973). The only reliable remote thickness technique, excepting possibly ESP, has been the Coast Guard-Sandia developed sea ice penetrometer (SIP) (McIntosh, Young & Welsh, 1972, 1973). The SIP can provide precise thickness measurements for ice up to three meters thick and possibly greater although this has not been unequivocally demonstrated to date. Future study of the SIP is planned with the major objective being the determination of a sampling plan which will give representative ice thickness information for ship routing.

Adequate information on the physical properties of lake ice relevant to the ship-ice interaction for domestic icebreaker engineering is not presently available (Welsh, 1972a).

In response to this need, the Coast Guard ice research program has been concerned with the identification and quantification of the physical properties of ice which affect ship transit. Physical properties, such as the flexural strength, coefficients of static and kinetic friction, density, thermal characteristics of the ice column, and the areal distribution and thickness, have been investigated.

Preliminary results indicate that the frequency distributions of the measurement values obtained for the physical properties mentioned above are not normal distribution functions. The consequence of this observation is that descriptive statistics like the arithmetic mean and standard deviation could be misleading if used as representative numbers for substitution in various equations for engineering purposes.

Figure 1 is an example of a frequency distribution for flexural strength. Figure 2 is a normal probability plot of the same data shown in Figure 1 (Welsh, 1972a). If the data were normally (Gaussian) distributed, all of the points would fall on the straight line--they do not. Similar plots were constructed for data obtained from the literature (Frankenstein, 1959, 1961; Hitch, 1959). This data was also non-normally distributed. As these examples were concerned with lake ice, it may be of interest to know that the same graphic technique was applied to published sea ice flexural strength data (Brown, 1962; Frankenstein & Garner, 1970). The results again

Figure 1. Frequency distribution of flexural strength values.

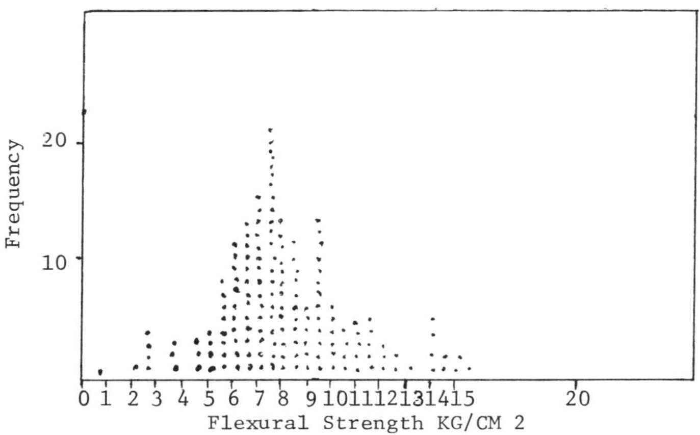

Figure 2. Graphic test of normality of the flexural strength values.

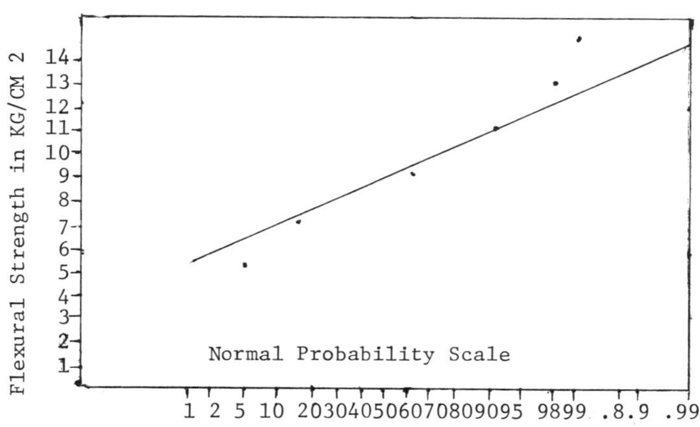

were non-Gaussian distributions. The question arises, is the flexural strength of the floating ice population non-Gaussian in nature or is this a result of measurement technique and/or confounding of variables. An unambiguous answer is not available at this time.

One more example of a non-Gaussian distribution concerns the coefficient of kinetic friction for steel on snow. Figure 3 shows the normal probability plot of kinetic friction data for steel on snow (Welsh & Schooley, 1973). This distribution appears to be skewed, not symmetrical; thus, discriptive statistics like the arithmetic mean and the variance have questionable value for numerical summary of the data. Consider the conclusion that might result if the trend as shown in Figure 4 was accepted without knowledge of the assymmetry of the constituent data. The least squares procedure for obtaining the "best" linear fit is influenced by lack of symmetry. Because the least square line has a positive slope with increasing coefficient of kinetic friction and velocity, it might be concluded that decreasing the sliding velocity would decrease the coefficient of kinetic friction. The magnitude of the assymmetry in this data could have a masking effect, thus the above conclusion may not be realistic.

The above examples serve to illustrate a few of the problems that the naval engineer and naval architect must contend with.

Measurement problems and the ensuing analytical problems are evident in most if not all studies of the physical properties of ice. One area of recent concern for Great Lakes season extension is the problem of navigation in slush ice. Slush ice provides a fine example of measurement problem.

Slush ice is primarily a conglomeration of ice particles of varying sizes and shapes in a water matrix. The sizes (size for the purpose of this discussion is the length of the apparent long axis of an individual particle) range from single ice crystals less that 1 mm up to particles exceeding 4 m. A plot of the size frequency distribution for a sample of slush ice is presented in Figure 5 (Welsh, 1972b). The skewness of this distribution is apparent. One problem here is gaining access to the material to make the measurement. This can easily result in a censored distribution which may or may not be representative of the size population of slush ice. Another problem also concerned with access is the measurement of thickness in a slush ice field. One at least feasible solution has been obtained using an upward-looking sonar technique (Welsh, 1972a). This technique determines the draft of the slush ice passing over the sonar transducer.

Determination of the draft of slush ice using a reliable remote technique would be helpful to winter navigation in a number of harbors and channels in the Great Lakes. For example, based on almost two months of observation in the Muskegon, Michigan, area, it was apparent that inhibition of ship movement by slush ice could have been avoided, the reason being that changing wind direction routinely moved the slush ice in and out of the channel. Thus, a ship en route to Muskegon could be informed of the coverage and draft of slush ice and be advised to proceed based on individual experience with slush ice. The draft, coverage and wind information could be provided to ships en route so that they might either divert or lay off until the channel cleared or it was determined that the slush ice could be transited by the specific vessel.

In conclusion, it appears that there are many measurement problems which challenge the investigator of ice and icebreaking.

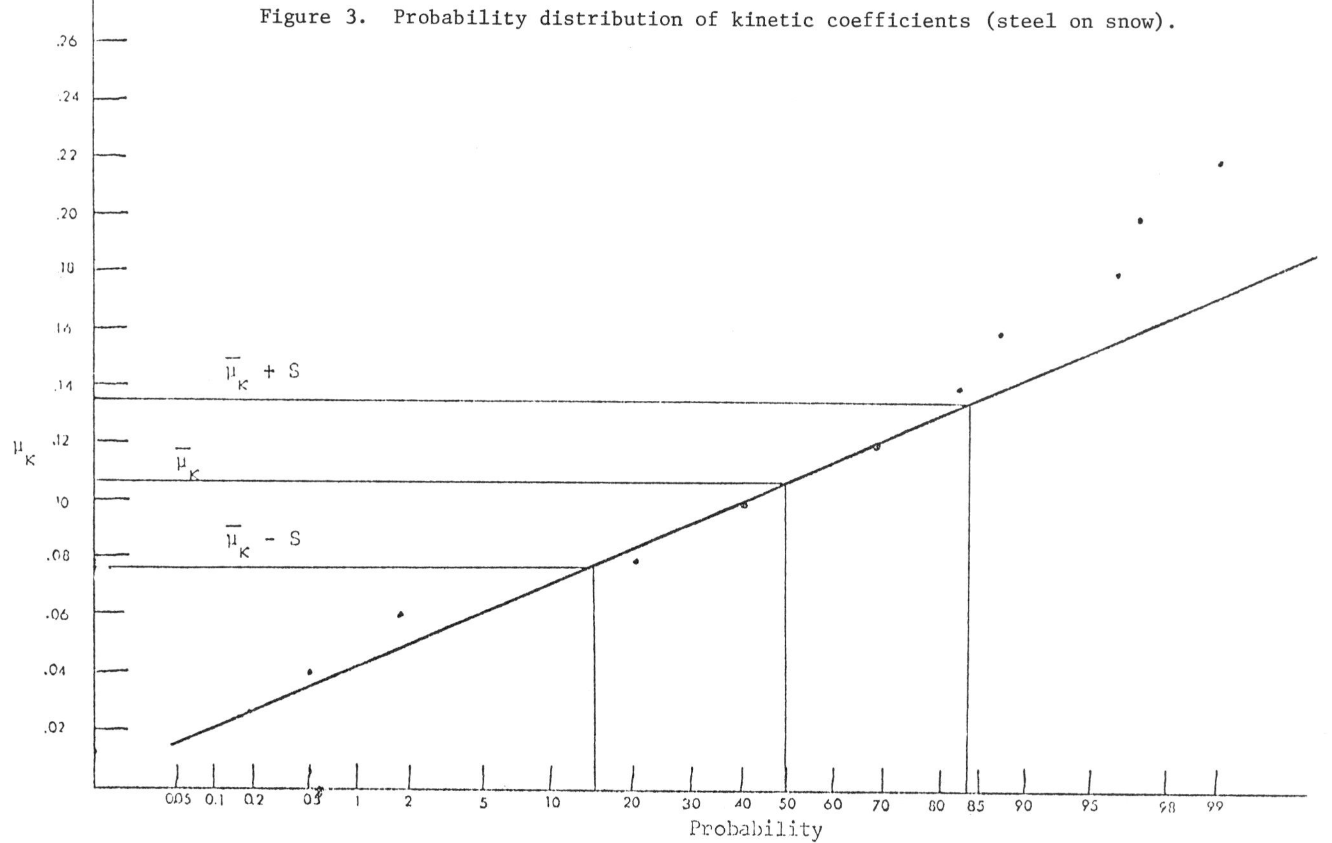

Figure 3. Probability distribution of kinetic coefficients (steel on snow).

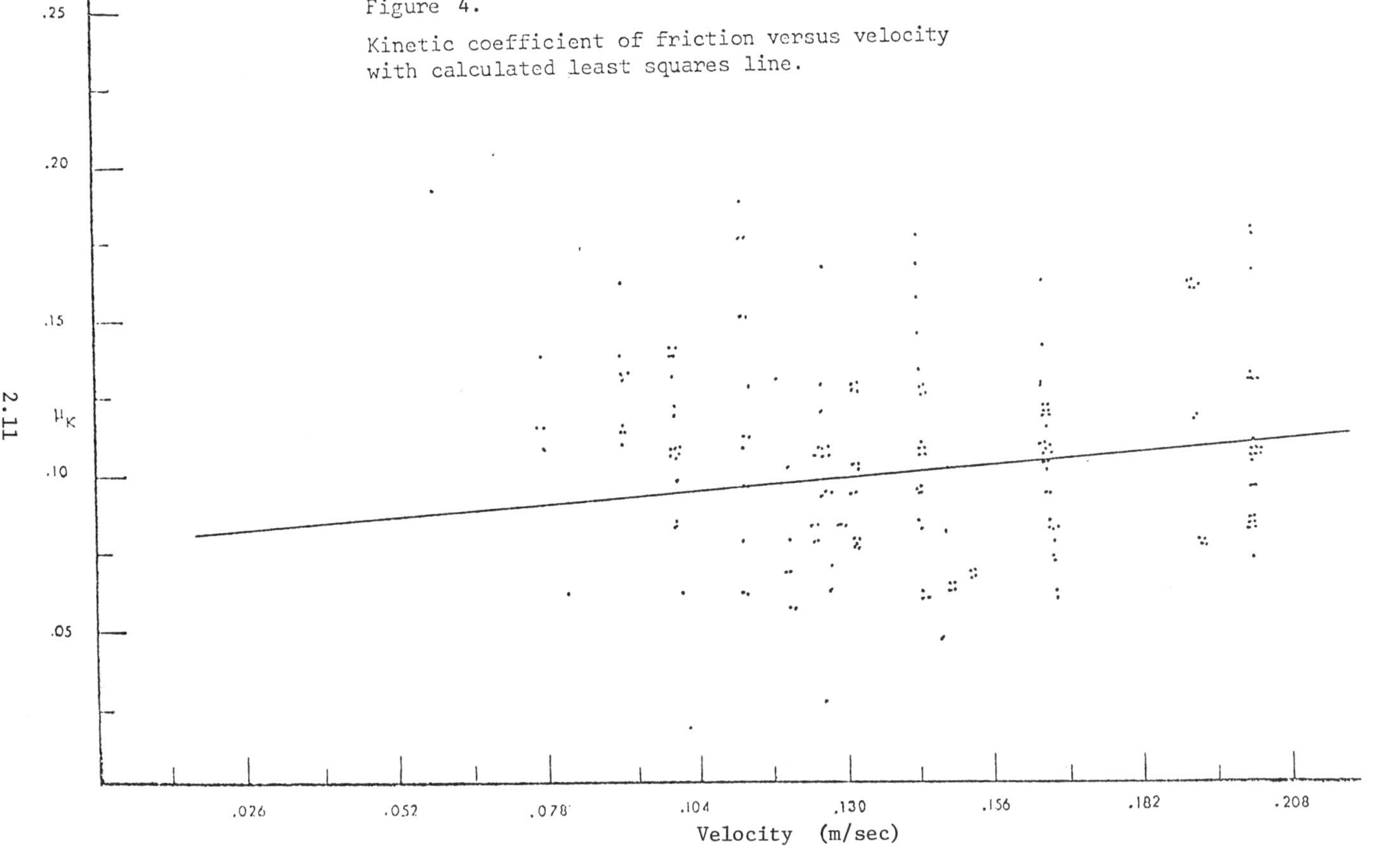

Figure 4.

Kinetic coefficient of friction versus velocity with calculated least squares line.

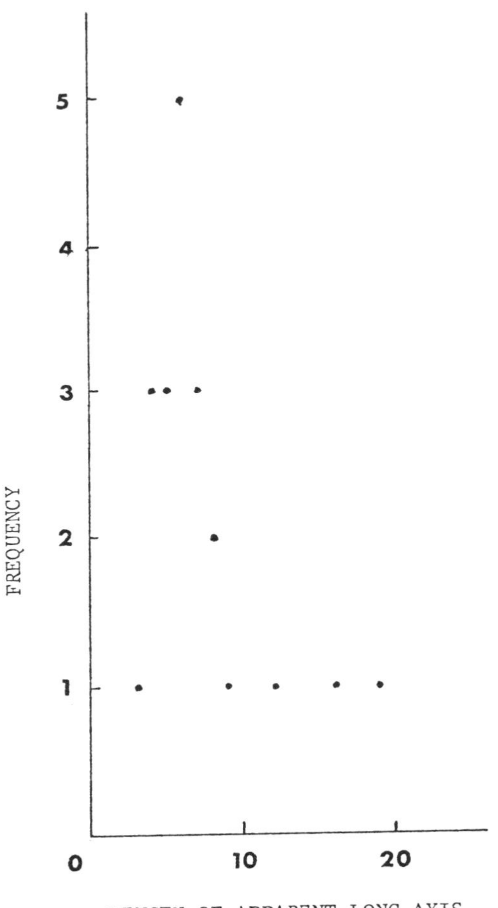

FIGURE 5. Example of size frequency distribution of apparent long axis of slush ice particles.

REFERENCES

Adey, A.W., 1970. A Survey of Sea-Ice Thickness Measuring Techniques, CRC Report No. 1212, Communications Research Centre, Dept. of Communications, Ottawa, Ont., Canada.

Bates, C.C., 1973. Navigation of Ice-Covered Waters. Some New Initiatives by the United States of America, XXIII International Navigation Congress, Ottawa, Ont., Canada.

Brown, J.H., 1963. "Elasticity and Strength of Sea Ice", in Ice and Snow, W.D. Kingery, ed., The MIT Press, Cambridge, Mass.

Frankenstein, G.E., 1959. Strength Data on Lake Ice, SIPRE, T.R. 59.

_____, 1961. "Strength Data on Lake Ice, SIPRE, T.R. 80.

_____, and Garner, R., 1970. Dynamic Young's Modulus and Flexural Strength of Sea Ice, CRREL, T.R. 222.

Farmer, L.D., 1972. Iceberg Classification Using Side Looking Airborne Radar, Final report, Commandant (DAS), U.S. Coast Guard Office of Research and Development, Washington, D.C.

Hitch, R.D., 1959. Flexural Strength of Clear Lake Ice, SIPRE, T.R. 65.

Johnson, J.D., and L.D. Farmer, 1971. Use of Side-Looking Airborne Radar for Sea Ice Identification, Journal of Geophysical Research, Vol. 76, No. 9.

Lewis J.W., and R. Edwards, Jr., 1970. Methods for Predicting Icebreaking and Ice Resistance of Icebreakers, SNAME, Nov. 1970 Annual Meeting.

McIntosh, J.A., C.W. Young, and J.P. Welsh, 1972. Sea Ice Penetrometer Development, Conference on Rapid Penetration of Terrestrial Materials, February 1-3, 1972.

McIntosh, J.A., C.W. Young, and J.P. Welsh, 1973. Development of a Sea-Ice Penetrometer, Journal of Petroleum Technology, March 1973.

Orange, A.S., 1973. "'ESP' saves money in Arctic by finding ice roads for moves," Oilweek, 6 August 1973.

Welsh, J.P., 1972a. Ice Properties and Their Relation to Ship Transit in the Great Lakes, A Progress Report, Interagency Council on Marine Science and Engineering, December 13-15, 1972.

Welsh, J.P., 1972b. Summary Progress Report of Great Lakes Ice Research, Winter 1971-72, Commandant (DAS), U.S. Coast Guard Office of Research and Development, Washington, D.C.

Welsh, J.P., and J.K. Schooley, 1973. A Field Measurement Technique for Determination of the Coefficient of Kinetic Friction for Steel on Snow, Final Report, Commandant (G-DST-2), U.S. Coast Guard Office of Research and Development, Washington, D.C.

CLASSIFICATION OF SEA ICE RIDGING AND SURFACE ROUGHNESS IN THE ARCTIC BASIN

W.D. Hibler III
S.J. Mock
U.S. Army Cold Regions Research and Engineering Laboratory
Hanover, New Hampshire

ABSTRACT

One- and two-parameter classification schemes for sea ice pressure ridging are reviewed. Using these classification schemes the number of ridges above any height may be predicted. 500 km of processed laser profile data flown over the arctic ice pack in November 1970 is used to illustrate the agreement between models and observation. The key parameter is γ_h, defined by μ_h/λ, where μ_h is the number of ridges per kilometer above height h encountered along a straight line path, and λ is the ridge height distribution shape parameter, uniquely determined by $\langle H_h \rangle$, the mean ridge height. Surface roughness spectral characteristics are examined and it is found that ridging intensity correlates well with surface roughness throughout the frequency range. A specific relationship between high frequency roughness (< 13 m) and ridging intensity is shown. Wind form drag values due to pressure ridges are calculated and compared to empirical wind drag values obtained by other researchers for relatively unridged ice.

INTRODUCTION

Sea ice pressure ridge height and spacing distributions can be automatically obtained from airborne laser profiles using digital data reduction techniques. In order to provide coherence and usability to the large number of distributions, Hibler *et al.*(in prep.) discussed a one-parameter classification model for pressure ridges and used approximately 3300 km of laser profile data to validate the models and provide an estimate of the sampling stability of the parameter. We have two objectives in this paper: 1) to review both the one-parameter and earlier two-parameter models (Hibler *et al.*, 1972b), and 2) to examine the relationship between the ridging intensity and the surface roughness spectral characteristics, a relationship having important ramifications for wind stress drag coefficients. For this study we have used a 500-km subset of the 3300 km of laser profile alluded to previously. We use this subset for two reasons. First, it covers a single period of time (November 1970) and is large enough to demonstrate the one-parameter model, including the sampling stability of the single parameter. Secondly, it is a sample of tractable size for analysis of spectral roughness.

Briefly, the data reduction consists of two parts. First, laser data recorded in analog form on magnetic tape are converted to a form acceptable for digital computing by digitizing the analog record and then removing the phase shifts. (See the discussion by Tucker in Mock *et al.*, 1973.) Second, the digitized profile is processed to remove the aircraft altitude variation by using the three-step digital filtering process developed by Hibler (1972). The three-step filtering process provides a leveled profile with a zero fiducial level from which to measure ridge heights. Ridges are digitally identified by declaring a profile peak to be a ridge when the peak is at least 2 ft (0.61 m) above minimum points located both left and right of the peak.

CLASSIFICATION OF RIDGING

There are several ways in which one might wish to classify ridging characteristics. The classification schemes reviewed here are an outgrowth of studies conducted to determine the mobility of vehicles having various height clearance capabilities. Thus the ridging classification was designed to predict the frequency of ridges falling within any given series of height intervals. The one- and two-parameter models both have this capability, but the one-parameter model, in addition to being esthetically more pleasing, has the advantage of being related to the volume of deformed ice, thus having relevance to pack ice dynamics.

Two-parameter ridge classification

The ridging characteristics of a region are determined by the two parameters μ_h and $\langle H_h \rangle$, which allow calculation of the ridge height distribution from the equation

$$P(H)dH = \mu_h \, 2\sqrt{\frac{\lambda}{\pi}} \, \frac{e^{-\lambda H^2} dH}{\operatorname{erfc}(h\sqrt{\lambda})} \qquad (1)$$

where $P(H)dH$ is the number of ridges per kilometer with heights between H and $H + dH$, μ_h is the number of ridges per kilometer above height h, an arbitrary cutoff height, $\langle H_h \rangle$ is the mean ridge height of all ridges above h, and λ is the distribution shape parameter which can be determined from $\langle H_h \rangle$. Equation (1) was derived (Hibler *et al.*, 1972b) by finding the most probable arrangement of a given number of ridges for a given amount of deformed ice, with no restriction placed on the maximum ridge height. We note here that since λ is determinable from $\langle H_h \rangle$, μ_h and λ is an alternative set of the two parameters.

The laser profile data consist of 16 sets, each containing approximately 40 km of track, taken at different geographical locations in the Arctic Basin. We use these 16 sets to formally test Equation (1), setting a cutoff height h of 4 ft (1.22 m) and using the observed mean height $\langle H_h \rangle$ to determine λ. A goodness of fit test was made between the observed and predicted numbers of ridges in 1-ft categories above the cutoff height for each of the 16 sets. The results are summarized in Table I along with values of μ_h and $\langle H_h \rangle$ for each set.

The overall agreement of the theoretical and observed distributions shown in Table I indicates that Equation (1) is a good working distribution model, especially since its derivation has some physical basis.

One-parameter pressure ridge classification

We define a quantity $\gamma_h \equiv \mu_h/\lambda$ which we shall call ridging intensity. This parameter has been found to describe ridging characteristics effectively (Hibler *et al.*, in prep.).

Table I. Summary of ridging parameters and test results for the 16 sample sets. h = 4 ft (1.22 m)

No. of ridges	$\frac{\mu_h}{\lambda}\left(\frac{m^2}{km}\right)$	$\mu_h (km^{-1})$	$\langle H_h \rangle (m)$	*Deg. of freedom*	χ^2	*Test at .05 level*
201	15.4	5.3	1.85	5	8.5	pass
205	13.4	4.7	1.84	4	0.5	pass
301	15.5	7.1	1.73	4	5.9	pass
156	6.9	3.6	1.69	3	8.0	fail
163	8.0	3.8	1.73	3	8.3	fail
148	6.3	3.4	1.68	3	0.9	pass
138	6.5	3.2	1.72	4	5.1	pass
79	2.7	1.8	1.61	3	0.8	pass
64	2.4	1.4	1.65	3	6.0	pass
116	6.4	2.6	1.74	4	5.3	pass
40	2.2	0.9	1.76	3	1.3	pass
67	2.7	1.5	1.66	2	1.4	pass
143	6.6	3.3	1.70	4	1.6	pass
184	9.1	4.3	1.72	4	3.4	pass
304	19.3	6.8	1.84	5	7.6	pass
159	6.9	3.9	1.66	3	2.5	pass
				57	67.1	pass

Let us assume that Equation (1) may be written in the form

$$P(H)dH = F(\gamma_h)e^{-\lambda H^2} dH \tag{2}$$

where the function $F(\gamma_h)$ is estimated by plotting

$$\mu_h 2\sqrt{\frac{\lambda}{\pi}} \frac{1}{\mathrm{erfc}(h\sqrt{\lambda})} \tag{3}$$

versus $\sqrt{\gamma_h}$, and fitting the data with a linear regression line of the form $F(\gamma_h) = m\sqrt{\gamma_h} + b$. Such a plot is shown in Figure 1 for the 16

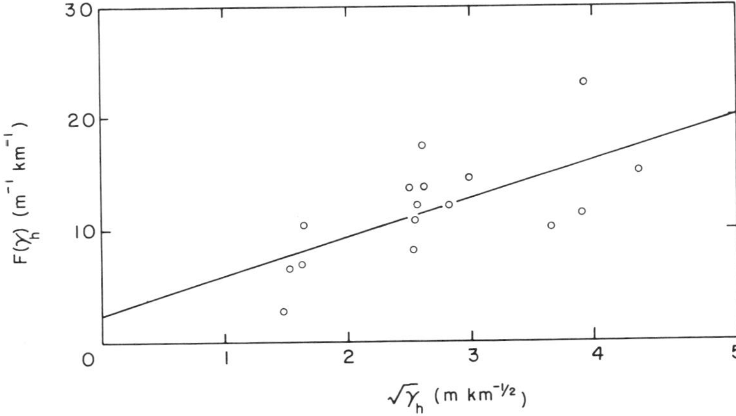

Figure 1. Data and regression line for $F(\gamma_h)$ *versus* $\sqrt{\gamma_h}$; h = *4 ft (1.22 m).*

sample sets. The correlation coefficient is 0.65 (significant at the 5% level) and the regression equation is $F(\gamma_h) = 3.38\sqrt{\gamma_h} + 2.47$. We note in passing that the correlation coefficient between $F(\gamma_h)$ and γ_h is less than 0.65 and that between $F(\gamma_h)$ and λ is very low.

Integrating Equation (2) gives:

$$\gamma_h = \frac{F(\gamma_h)}{2\lambda^{3/2}} \operatorname{erfc}(h\sqrt{\lambda}) \sqrt{\pi} . \qquad (4)$$

Using this expression and the regression line, sets of values for λ, γ_h and $F(\gamma_h)$ can be found, allowing calculation of the number of ridges in height categories from Equation (2). In actual practice sets of unique values of γ_h and $F(\gamma_h)$ were generated for each value of λ, a procedure involving the solution of a quadratic equation for $F(\gamma_h)$ which has one extraneous root.

Figure 2 shows calculated curves of ridge frequency above specified heights as solid curves and observed frequencies from the 16 sample sets as open circles. The figure indicates good agreement between observation and prediction. We shall refer to this figure, specifically the curves, as the ridge picker. A quantitative measure of the effectiveness of the ridge picker is shown by the standard errors of estimate between observation and prediction tabulated in Table II. It should be remembered when examining the ridge picker that there is a certain sampling error associated with each experimental ridge frequency, and consequently the observation would be expected to deviate somewhat randomly from prediction.

Table II. Standard error of estimate for ridge picker.

$$\sigma = \sqrt{\sum_i \frac{(\text{Observed} - \text{Predicted})^2}{N}}$$

Height	σ
> 4 ft (1.22 m)	0.49 ridges km^{-1}
> 6 ft (1.83 m)	0.14 ridges km^{-1}
> 8 ft (2.44 m)	0.06 ridges km^{-1}

Ridging intensity and sampling stability

There are several intuitive reasons why γ_h would be expected to be a very good parameter for describing ridging characteristics. For one, γ_h reflects both the number of ridges per kilometer and the average ridge height. On *a priori* grounds such a parameter would be expected to be useful since it includes different aspects of ridging and consequently would be expected to correlate automatically with both ridge frequency μ_h and $\langle H_h \rangle$. Also the sampling of γ_h should be more stable than either $\langle H_h \rangle$ or μ_h, since fluctuations in either of the latter parameters could occur independently of the other.

Perhaps most important, ridging intensity has a straightforward relation to ice deformation; namely, it scales approximately linearly with the volume of deformed ice per unit area caused by ridges above the cutoff heights. A basic assumption in the ridge height distribution model is that all ridges are geometrically similar; thus, along a line the amount of deformed ice is proportional to $\mu_h \langle H_h^2 \rangle$.

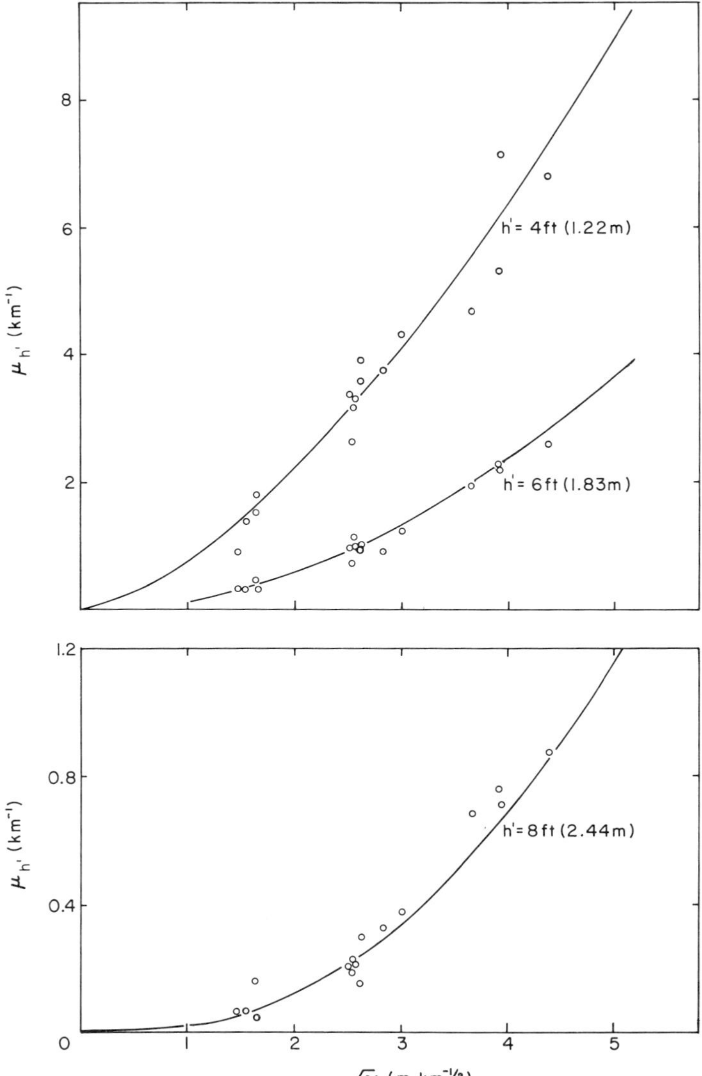

Figure 2. Ridge picker. Predicted (solid lines) and observed values for $\mu_{h'}$*, the number of ridges per kilometer above height* h'*, versus* $\sqrt{\gamma_h}$*;* h = *4 ft (1.22 m).*

Since $\gamma_h \equiv \mu_h/\lambda$, we can plot $\langle H_h^2 \rangle$ vs $1/\lambda$ or, as is done in Figure 3, $\langle H_h^2 \rangle / h^2$ vs $1/\lambda h^2$ (to make dimensionless) to illustrate that γ_h to a first approximation is, in fact, a linear function of the deformation. The solid line in Figure 3 represents results based upon the theoretical height distribution of Equation (1). The open circles are observed data from the 16 data sets where γ_h was evaluated using

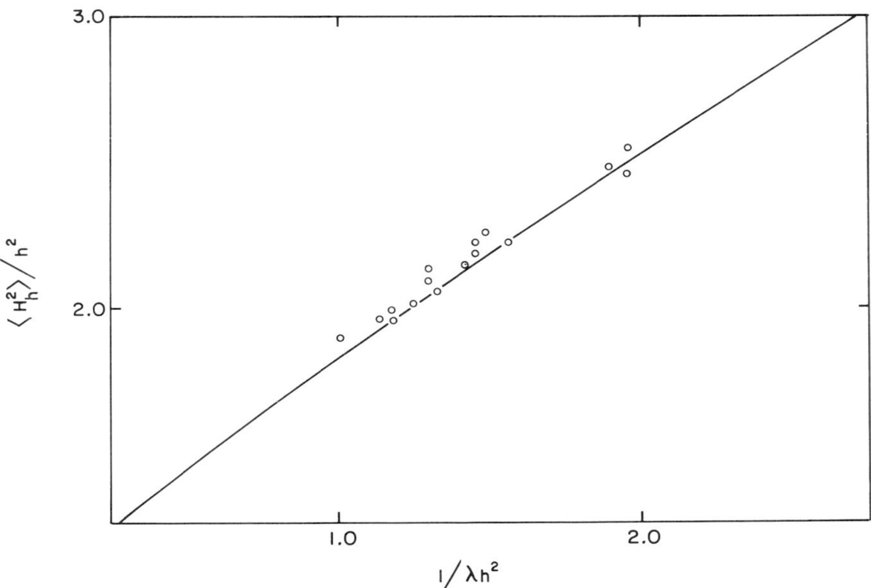

Figure 3. Dimensionless plot of $\langle H_h^2 \rangle$ *versus* $1/\lambda$.

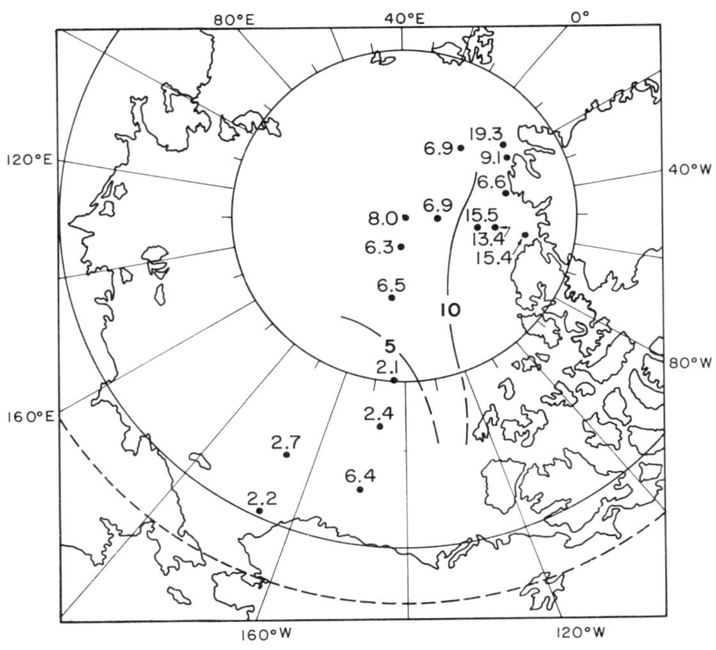

Figure 4. Laser data sample locations and ridging intensities γ_h*(h = 1.22 m) for November 1970.*

the mean ridge height. For randomly oriented ridges, the total length of ridges per unit area is $\pi/2\ \mu_h$ (Mock *et al.*, 1972); thus, γ_h is a linear function of the total volume of deformed ice above h.

Figure 4 shows the ridging intensity of each of the 16 sample sites along with rough contours of ridging intensity. It is not our intent to discuss here the significance of regional variations in this parameter, but rather to demonstrate that they are significant.

We have asserted intuitively that γ_h should be a more stable estimator than the other ridge classification parameters μ_h, $\langle H_h \rangle$ or λ. We would like to demonstrate this point and at the same time show that each of the above parameters does in fact have a regional variation significantly greater than sampling variations within a region.

Each regional segment of laser track (about 40 km) was divided into two parts, and the ridge parameters were evaluated for each segment. An analysis of variance was then made to determine the significance of the regional variation as well as to estimate the relative "strength" of various parameters.

The results are summarized in Table III, arranged in order of decreasing variance ratio. The larger the value of the variance ratio, the greater the between-region variance as compared to the within-region variance. Consequently, larger variance ratios are indicative of more satisfactory parameters for estimating regional characteristics. Table III confirms the stability of γ_h as a ridging characteristic parameter. Note that the residual standard deviation for the ridge frequency μ_h is 0.84 ridges km^{-1}, which is greater than the standard error of estimate for the ridge picker (see Table II).

Table III. Sampling stability estimates for ridging parameters. [F = variance ratio, σ_r^2 = residual variance, h = 4 ft (1.22 m).]

Parameter	*F ratio*	σ_r	*Significance level of F ratio*
γ_h	12.58	2.1 ($m^2\ km^{-1}$)	0.01
μ_h	9.16	0.84 (km^{-1})	0.01
$\langle H_h \rangle$	3.28	0.06 (m)	0.05
$1/\lambda$	3.20	0.04 (m^2)	0.05

SPECTRAL ROUGHNESS CHARACTERISTICS

Classification of spectral roughness

The surface roughness spectrum provides alternative parameters to ridge characteristics for classifying sea ice terrain. The spectral approach has certain advantages and disadvantages. For purposes of examining undulating multiyear ice terrain (Hibler *et al.*, 1972a) and for identifying ice types by their spectral characteristics (Hibler and LeSchack, 1972) spectral densities are useful. For cataloging properties of discrete obstacles and for discussing mobility of vehicles, discrete ridge characteristics are more useful

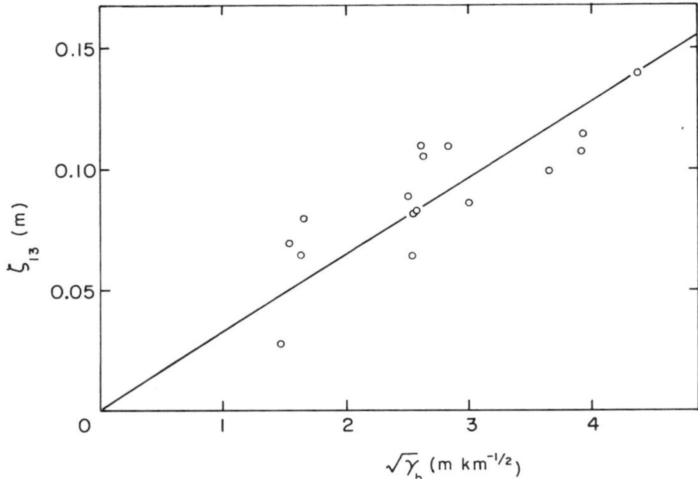

Figure 5. ζ_{13}, the RMS surface roughness for wavelengths less than 13 m, versus $\sqrt{\gamma_h}$; h = 4 ft (1.22 m).

Table IV. Correlation coefficients between spectral and ridge parameters.

	$\sqrt{\gamma_h}$	ζ_{200}	ζ_{13}
$\sqrt{\gamma_h}$	1		
ζ_{200}	0.89	1	
ζ_{13}	0.82	0.92	1

(Hibler and Ackley, in press). It would be particularly useful if any relationship between spectral and ridging characteristics could be demonstrated. We will show that the high frequency spectral components have such a relationship.

High frequency components are also of interest for other reasons. For one, a comparison of wind stress measurements with spectral roughness (Banke and Smith, 1973) indicates that, for terrain without ridges, the wind drag coefficient scales linearly with the RMS roughness with wavelengths shorter than 13 m. For another, the high frequency components are a key indicator of ice type, i.e. either multiyear or first-year (Hibler and LeSchack, 1972).

To investigate and classify such spectral characteristics we calculated, for each of the 16 November sample regions,

$$\zeta_z = \left[\int_{1/2z}^{N_f} P(f)df \right]^{1/2} \qquad (5)$$

for z = 13 and 200 m, with N_f being the Nyquist frequency in cycles per meter and $P(f)$ the power spectrum. The spectral values were calculated from 4000 data points (with spacings of from 1.2 to 1.4 m depending on the region) using 200 lags and a Hamming spectral window. Some results of these calculations are shown in Table IV and in Figure 5.

Figure 5 illustrates the high correlation between ζ_{13} and the square root of the ridging intensity. The least squares line shown in Figure 5 has the form $\zeta_{13} = m\sqrt{\gamma_h}$ with $m = 0.0319$. Although not illustrated, a plot of ζ_{200} versus $\sqrt{\gamma_h}$ shows a similar trend as shown in Figure 5. The implication of these results is that γ_h can be used as an estimator of spectral roughness as well as ridging.

Ramifications of surface roughness: wind stress estimates

In light of the correlation between ζ_{13} and γ_h shown in Figure 5 it is instructive to compare estimated form drag due to ridging (Arya, 1973) with drag coefficients based on surface roughness (Banke and Smith, 1973). Assuming ridges are randomly oriented Arya (1973) obtained a dimensionless drag coefficient C, due to ridges above h', of

$$C = C_D \langle H_{h'} \rangle \frac{\mu_{h'}}{2} \qquad (6)$$

where C_D is the form drag of a single ridge assumed to be in the range 0.2 to 0.9. Using the one-parameter ridge model C may be calculated as a function of $\sqrt{\gamma_h}$. For relatively unridged sea ice Banke and Smith (1973) obtained the regression line

$$C = 0.0012 + 0.0065\zeta_{13} \qquad (7)$$

where ζ_{13} is in meters. Using the relation from Figure 5, $\zeta_{13} \approx 0.032\sqrt{\gamma_h}$, Equation (7) becomes

$$C = 0.0012 + (0.00021)\sqrt{\gamma_h}. \qquad (8)$$

A comparison of the form drag results, Equation (6), with the Banke and Smith (1973) results, Equation (8), is shown in Figure 6. In this figure we have calculated form drag versus $\sqrt{\gamma_h}$, for different values of h', using the one-parameter ridge model and taking $C_D = 0.5$ in Equation (6). In particular curve *a* is the form drag for ridges above 4 ft (1.22 m), curve *b* the form drag for ridges above 2 ft (0.61 m) and curve *c* the form drag for ridges *below* 2 ft (0.61 m) high, where for ridges less than 4 ft (1.22 m) high we have simply extended the distribution model in Equation (2) to smaller ridge heights. The Banke and Smith (1973) curve is dashed. Using typical values of γ_h (Figure 4) the results in Figure 6 generally indicate that form drag is somewhat larger than drag on relatively flat ice.

It is also particularly interesting to note that the slope of the form drag on "ridges" below 2 ft is very similar to the Banke and Smith curve slope. This would suggest that as a rough model for wind stress over sea ice we can consider the sea ice surface to have a flat plate drag (zero intercept of Banke and Smith curve) plus a form drag due to ridging where the ridging model, Equation (1), is assumed to hold down to $h = 0$. Using such a model, wind drag could be estimated from the ridging intensity. Assuming that the ridge model describes small scale roughness is not completely unrealistic in terms of terrain models. Terrain constructed by creating random obstacles down to zero height using Equation (1) has been found to have spectral densities reasonably similar to actual sea ice spectral densities (Aerojet-General, 1972).

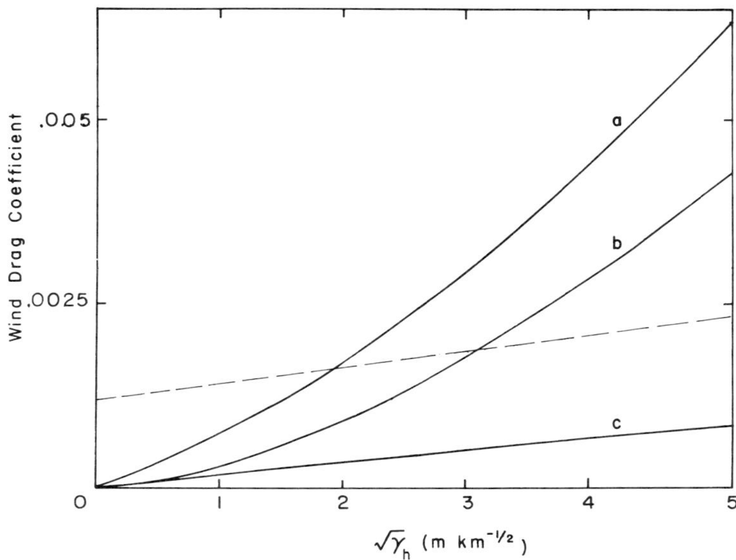

Figure 6. Wind drag coefficient versus $\sqrt{\gamma_h}$; h = 4 ft (1.22 m). Solid curves represent form drag for all ridges (a) above 2 ft (0.61 m), (b) above 4 ft (1.22 m) and (c) below 2 ft (0.61 m). Dashed curve is Banke and Smith (1973) results; see Equation (8).

CONCLUSIONS

In conclusion it appears that the ridging intensity parameter is a very useful parameter for classifying sea ice ridging and surface roughness. The following reasons support such a conclusion.

1. Ridging intensity γ_h may easily be derived from digital processing of laser profile data.
2. γ_h is related to the total ice deformation per unit area and thus in principle may be estimated from drift calculations.
3. The number of ridges per kilometer at any height may be accurately estimated from γ_h.
4. The surface spectral roughness is correlated with $\sqrt{\gamma_h}$.
5. Form drag scales approximately linearly with $\sqrt{\gamma_h}$.
6. The sampling stability of γ_h for a given region is good, that is the variability of γ_h between regions is considerably greater than the sampling variation in γ_h for a given region.

Finally, it is also possible to compare ridging intensities calculated using different cutoff heights. Consequently ridging intensities calculated by different investigators can be compared. It is our hope that as more laser data become available such estimations of γ_h will be carried out.

ACKNOWLEDGMENTS

We would like to thank Floyd Kugzruk for performing much of the numerical computer work for this study and W.B. Tucker of NAVOCEANO for producing processed laser profile data in digital form.

REFERENCES

Aerojet-General Corp., 1972. Arctic Surface Effect Vehicle Program, Quarterly Progress Report No. AGC-T-306.

Arya, S.P.S., 1973. Air friction and form drag on arctic sea ice. AIDJEX Bull. No. 19:43-57.

Banke, E.G. and S.D. Smith, 1973. Wind stress on arctic sea ice. J. Geophys. Res. (in press).

Hibler, W.D. III, 1972. Removal of aircraft altitude variation from laser profiles of the arctic ice pack. J. Geophys. Res. 77(36): 7190-7195.

Hibler, W.D. III and L. LeSchack, 1972. Power spectrum analysis of undersea and surface sea ice profiles. J. Glaciol. II(63):345-356.

Hibler, W.D. III, S.F. Ackley, W.F. Weeks and A. Kovacs, 1972a. Top and bottom roughness of a multi-year ice floe. To be published in Proc. 2nd Symp. Ice and Its Action on Hydraulic Structures, IAAR, Leningrad, 1972.

Hibler, W.D. III, W.F. Weeks and S.J. Mock, 1972b. Statistical aspects of sea ice ridge distributions. J. Geophys. Res. 70(30): 5954-5970.

Hibler, W.D. III and S.F. Ackley, In press. A sea ice terrain model and its application to surface vehicle trafficability. U.S. Army Cold Regions Research and Engineering Laboratory (USA CRREL) Research Report.

Hibler, W.D. III, S.J. Mock and W.B. Tucker, In prep. Classification and variation of sea ice ridging in the Arctic Basin.

Mock, S.J., A. Hartwell and W.D. Hibler III, 1972. Spatial aspects of pressure ridge statistics. J. Geophys. Res. 77(30):5945-5953.

Mock, S.J., V. LaGarde III and W.B. Tucker, 1973. Arctic terrain characteristics data bank. USA CRREL Technical Report (in prep.).

SEA ICE: SCALES, PROBLEMS AND REQUIREMENTS

W. F. Weeks
W. D. Hibler III
S. F. Ackley
U.S. Army Cold Regions Research and Engineering Laboratory
Hanover, New Hampshire 03755

ABSTRACT

Sea ice can be examined on a variety of spatial scales that range over 10 orders of magnitude. Each scale of observation has its characteristic types of problems and these problems have associated requirements for new observational systems and techniques that are necessary for their resolution. The present paper provides a personal view of some of the more important problems and requirements on each scale. The smallest scale, the microscale, is distinguished from other scales by the great importance of changes in the growth conditions on the structure of the resulting ice and the controlling effect of these structural variations on its small scale (<10 m) property variation. The greatest need is for compact instrumentation that is capable of rapidly specifying, by non-destructive methods, the internal state of the sea ice. When observations on the mesoscale (100 m - 50 km) are considered, the micro-structural properties of the ice rapidly become of less importance as the scale length increases, being replaced by effects produced by ensembles of ice features such as floes, leads and pressure ridges. It is within this scale range that the understanding required to bridge the gap between the small scale spatial and temporal behavior of individual ice features and the large scale behavior of the pack as a whole must be sought. Fortunately instrumentation to accomplish most aspects of mesoscale experimentation exists. Its drawbacks are that the equipment is both expensive and relatively untested under Arctic conditions. The mesoscale is also the natural scale for the utilization of remote sensing systems operated from aircraft. However, for the results of such remote sensing flights to be useful for more than a post-mortem requires techniques for rapidly analyzing the data and presenting it in terms of physically justifiable distribution functions. The most important equipment development problem as related to mesoscale studies is the present lack of an instrument that remotely measures ice thickness. On the macroscale (>100 km), conventional remote sensing from aircraft can only be used to obtain random samples of the ice of interest. Most information would therefore have to be provided by satellite based remote sensing systems coupled with arrays of data buoys sited in the ice. Progress in arctic data buoy development is quite encouraging with the main missing element being the inability to obtain adequate oceanographic data. The problems with the satellite based remote sensing data are, as in the mesoscale, primarily related to difficulties in rapid analysis of the images in a format that can be used in current numerical efforts. Other problems are the inability of certain systems to function through a cloud cover and the long holidays in the data acquired by systems such as ERTS.

2.13

INTRODUCTION

A glance at the diverse literature on sea ice shows that there are a wide variety of both scientific and engineering problems of importance related to its occurrence and also that many of these problems have been little studied. The effective scales of these problems range over 10 orders of magnitude, varying from subjects such as the formation of salt crystals within individual brine pockets ($\approx 10^{-4}$ m) to the large-scale circulation of the Arctic and Antarctic ice packs ($\approx 10^{6}$ m). The following paper is an overview of a number of these problem areas that we personally feel are important. In particular we will try to stress not only the problems but the developments, both conceptual and technological, that will be required before these problems can be resolved. We will also, because of space limitations, confine our discussion to problems in which the behavior of the ice itself is an important factor. We hope that this paper will establish an overall framework within which the more detailed results of the other sea ice papers presented at this Symposium can be examined.

MICROSCALE

On the microscale, the more important aspects of sea ice are related to understanding the variations in its internal structure and chemistry as influenced by both initial growth conditions and by subsequent metamorphic changes within the ice. As such, this area of sea ice research is closely related to a variety of investigations in the materials sciences that are concerned with how changes in the conditions during and after solidification affect the properties of the materials that are produced.

In sea ice, the specification of the properties of a given sample requires information on a variety of chemical, thermal and structural parameters that are all strongly influenced by both the initial growth conditions and the state of the ice at the time of interest. For instance, ice properties are strongly affected by the relative volume and chemical composition of the brine and solid salt that are present in the ice; the size, defect structure and orientation of the crystals; and the geometry of the brine pockets, brine drainage tubes and air bubbles.

Similar factors also influence the properties of other, more thoroughly studied materials but there is one principal difference. Most "normal" materials are utilized at temperatures that are well below (less than 50% of) their melting temperature on an absolute scale. In such a temperature range, the characteristics of the material are effectively "frozen" and to a first approximation can be considered to be invariant with time. The same is not true of sea ice which commonly exists at temperatures within a few percent of the melting temperature of pure ice. In fact a liquid phase, brine, exists within the ice even at the very lowest natural temperatures and small changes in temperature cause large changes in the amount of brine and in the resultant physical properties. In addition, a number of processes occur with time that cause rapid changes in ice properties. The foremost of these is brine drainage. Therefore, we will now discuss salt entrapment and brine drainage as an example of this general type of problem.

The initial salinity of sea ice varies quite systematically in a vertical direction. This variation is primarily controlled by variations in the growth conditions. Figure 1 shows a plot of

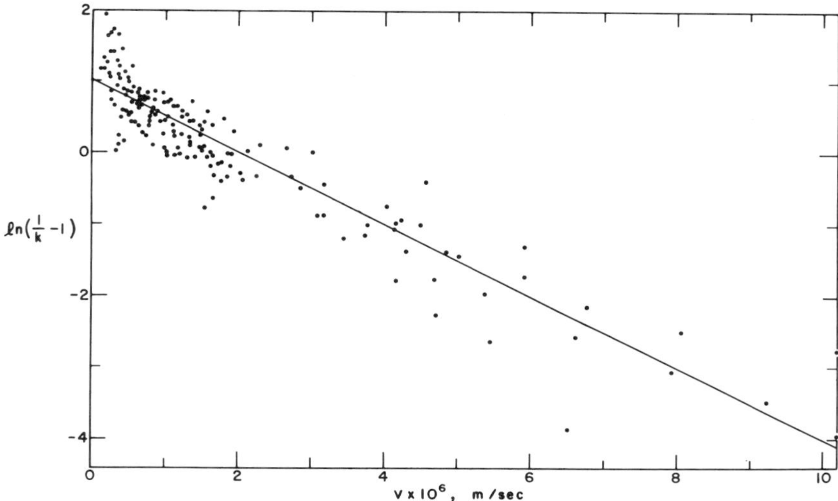

Figure 1. Plot of ℓn (1/k - 1) vs. v for NaCl ice runs where k is the (salinity of the ice/salinity of the water from which that ice formed) and v is the growth velocity (Weeks and Lofgren, 1967). In these tests, k was determined after the test was completed.

ln(1/k - 1) versus v where k is the ratio of the amount of salt trapped in the ice to the amount of salt present in the bulk solution from which the ice formed and v is the growth velocity (Weeks and Lofgren, 1967). The fitted equation is a relation suggested by Burton, Prim and Schlichter that has been widely used in the general crystal growth literature. Although the agreement is quite satisfactory at all but very low values of v, the scatter is quite large. In addition, when the amount of salt entrapped in the ice is calculated using a reasonably complete growth equation, it is found to be appreciably less than the values that are observed (Hasemi and Takano, personal communication). The reason for these problems is clear; both brine entrapment and brine drainage proceed simultaneously. Therefore since Weeks and Lofgren used salinities from ice cores collected at the end of the freezing runs, they effectively made the incorrect assumption that brine drainage during ice growth is zero.

Figure 2 (Cox and Weeks, 1974) shows a plot similar to Figure 1 but in this case k has been determined sequentially during ice growth

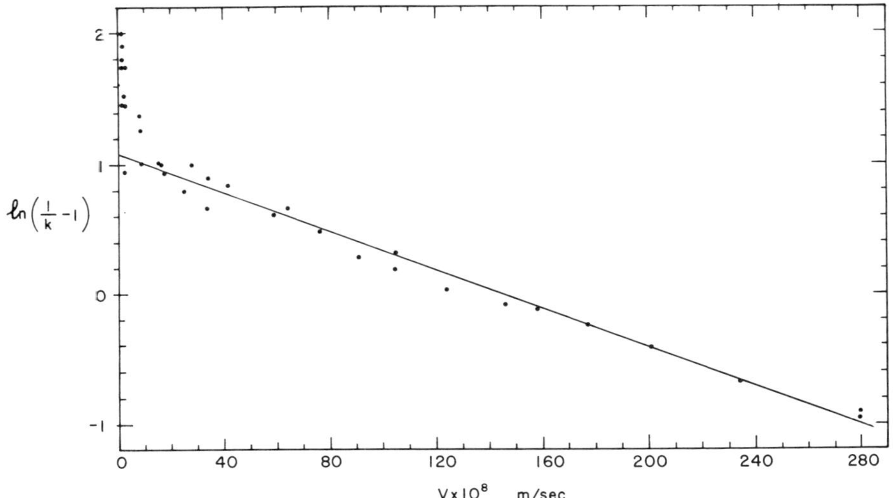

Figure 2. Plot of ℓn (1/k - 1) vs. v for NaCl ice where k is determined in-situ by using Na^{22} as a tracer (Cox and Weeks, 1974).

by using Na^{22} as a tracer. The use of tracers permits one to make replicate chemical analyses at any time at any location in the growing ice specimen without destroying the sample. Note that the scatter has disappeared and more salt is found to be initially entrapped in the ice. Such replicate non-destructive salinity measurements when combined with measurements of the temperature of the ice also allow one to study brine migration at locations well away from the ice-water interface and provide information against which existing brine drainage theories can be tested.

The above problem is representative of a class of problems that occur in the study of sea ice on the microscale in that it requires sequential non-destructive measurements. The resolution requirements for such studies are also interesting. For instance, in studies involving salinity changes, it is known that there are striking and consistent salinity changes in a vertical direction while in the horizontal direction there is a considerable irregular local variation in salinity values (Weeks and Lee, 1962). Because we are primarily interested in the vertical variations, a very high resolution would be quite useful in the vertical direction, while at the same time the horizontal sampling should test a wide enough area to provide some averaging to remove extreme local fluctuations. These two requirements are somewhat mutually exclusive in that the wider the area sampled, the thicker it becomes. This points out the need for highly collimated sensing systems that are ideally capable of variable penetration into the sample. In the Na^{22} brine drainage studies, it has been necessary to use rather involved deconvolution procedures to numerically decrease the angle of the sampling cone. Although such techniques have been successful, they are time consuming and if possible should be avoided. It would also be very interesting to be able to independently tag and follow each major chemical component in sea ice through a major period of brine drainage. Such data would allow one to resolve questions such as the postulated change in the (sulfate/chloride) ratio during brine drainage and the suggested differential entrapment of CO_3^{-2} in the ice.

Another important microscale problem can be illustrated in the same general problem area. When one attempts to analyze observed brine drainage in terms of the internal structure of sea ice, one immediately finds that information is required on the geometry of the brine inclusions, in particular information on their connectivity in the vertical direction. Current methods of examining the structural aspects of sea ice such as the geometry of brine pockets are extremely unsatisfactory. The ice must be cooled to below eutectic temperatures to avoid brine drainage, a thin section prepared, the ice temperature raised to the test temperature, and the brine pockets examined. It is quite possible that this preparation process may even rearrange what is to be examined.

Assuming that there is no change in the geometry of the brine pockets as a result of sample preparation, we then need to tabulate statistical information on their distribution. This is extremely tedious and time consuming even in the horizontal plane where the brine pockets appear as relatively simple shapes (Figure 3). In the vertical plane where the shapes are more complex, it is even difficult to know what to measure to adequately describe the geometry of the system. Also, we undoubtedly need information that will allow us to specify the pertinent distribution functions inasmuch as the average brine pocket will probably not be the most important.

In short, non-destructive methods for determining parameters characterizing both the internal structure of the sea ice as well as its bulk characteristics are needed. Although the problems we have discussed are laboratory problems in their initial stages, it would be hoped that the instrumentation and techniques that develop from such studies can ultimately be packaged so as to have potential field use.

When the observational scale increases to the upper end of the microscale range (1 to 10 m), different problems arise. Nevertheless, most of these problems are related to the problems encountered on smaller scales: the rapid remote non-destructive specification of the internal structure and the bulk properties of the ice. For instance, consider the time-dependent mechanical behavior of a beam of sea ice. To even prepare a beam from a plate of sea ice, one must severely disturb the temperature regime in the beam by placing the sides of the beam in contact with sea water. This causes a gradual change in the temperature distribution across the beam's cross-section, which in turn changes both the brine volume and the brine drainage rate. In long term creep tests, the properties of

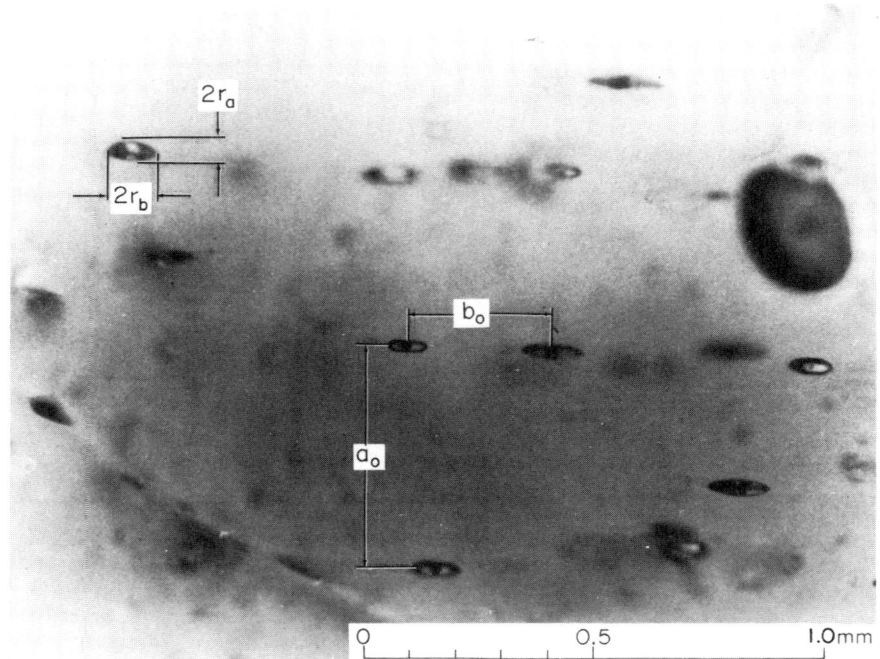

Figure 3. Horizontal thin section of NaCl ice showing brine pockets; temperature - 20°C, salinity ≈ 1%. The parameters r_a, r_b, a_o and b_o are defined in Weeks and Assur (1972).

the beam's cross-section should be monitored continuously. Yet at the present time, the only way we can obtain even one set of measurements is to destroy the sample. Admittedly, one could only test specimens that have been removed from an in-situ condition and have been allowed to reach thermal equilibrium at some set temperature. However, even in these cases there invariably would be a brine volume gradient across the specimen. Also if the ice temperatures are warmer than -10°C, brine drainage during such a test will undoubtedly be extreme.

Another important problem on the 1 to 10 m scale is caused by the variation in the effective failure strength and presumed failure mechanism as the size of the sea ice sample under stress increases. Current thinking (Weeks and Assur, 1972) suggests that these changes are in some way related to the distribution of flaws within the ice, in particular flaws such as large brine drainage tubes or partially rehealed cracks. The probability of such obvious large defects being present in a small specimen is very low. In fact, such a specimen would probably be rejected as being unsuitable for testing. Such flaws, however, must be common in the large blocks of ice that fail during the formation of pressure ridges. Yet, inasmuch as there is currently no easy method for readily identifying and categorizing such flaws, current treatments of this problem will remain as speculation. In addition, when large blocks of sea ice fail, there commonly are pronounced lateral changes in the natural salinity and brine volume profiles within the blocks. This is particularly true in old ice floes that show the development of melt hummocks (Cox and Weeks, 1973). Such melt hummocks usually mirror significant thickness variations which also must be documented.

MESOSCALE

The distinction between microscale and mesoscale is a very important one in that on a scale of more than roughly 100 m, the small scale physical variations in the ice, such as changes in the salinity profile, become less important as the scale length increases. At these larger scales, the observed properties are controlled by sets of interacting elements such as floes, leads and pressure ridges. Therefore, it is on the mesoscale that the understanding must be sought that will allow us to interpret the gross behavior of the ice in terms of mechanistic models that are based on our understanding of the details of the mechanics and thermodynamics of sea ice. It is also in the lower part of the mesoscale that the difficult engineering problems related to ice forces, ice breaking and ice pile-ups are encountered.

There are two primary difficulties in performing a mesoscale study: first to measure the response of the ice to the forces acting upon it, and second to adequately characterize the nature of the ice that is being deformed as well as the products of the deformation. On the small mesoscale where the deformation is primarily produced by motion along one or two leads or ridges, the relative motions can be observed in great detail by using distance measuring equipment such as laser range finders or tellurometers. The laser is clearly the easiest to operate because the remote stations are unmanned. However, lasers are unable to penetrate the thick sea smoke that commonly accompanies the formation of an open lead. Tellurometers are not limited by weather but both the master as well as the slave stations must be manned. Therefore, the collection of a continuous set of data is very exhausting unless a large supply of manpower is available. Clearly more convenient, portable, semi-automatic, all-weather systems for specifying such detailed motions could be developed. Also, after the formation of different ice deformation features, it is highly desirable to be able to determine both the exact geometries of their above-water and below-water portions as well as a variety of indices relating to the state of the deformed ice such as temperature, salinity, and degree of interblock bonding. Although considerable progress has been made in this area (Weeks, Kovacs and Hibler, 1971), current instrumentation is still very crude, bulky, and awkward to use in the field. It is particularly desirable to have a method of examining ridge keels that does not depend upon coring. Also, if a compact side-looking sonar system can be developed for ridge profiling, it should be possible to obtain sequential ridge profiles during actual ridge growth. The comparison of such profiles with the results of current kinematic ridge models (Parmerter and Coon, 1972, 1973), would be extremely informative.

On the upper end of the mesoscale (5 to 50 km), radar systems are available that will, in principle, measure the positions of a large number of remote transponders on an automatic basis. The principal factor limiting the range of such systems is the curvature of the earth. The drawback to these systems is their high cost. Also, the operation of such systems in the high Arctic is far from routine. Nevertheless, the results from such studies would be interesting in that they would specify the detailed, relative short-time scale, motion of a number of points on the ice during different times of the year. Inasmuch as the distribution of ice types would undoubtedly vary within the deformation array (and the transponders could be positioned to accentuate this variation) the different regions of the array would respond differently to identical forcing functions. These varying responses must then be correlated with measurable differences in the gross properties of the ice.

Fortunately the scale of a typical mesoscale study is an optimal scale for utilizing remote sensing since, in principle, the array can easily be covered using conventional aerial-mosaic techniques. Quite adequate characterization of the ice at different times of the year is possible by the use of a combined remote sensing system utilizing photography, infrared scanning, laser profilometry, SLAR, and passive microwave techniques. Also, sufficient detail would be present in the imagery to readily identify marked sites on the ice and the formation of or changes in features related to the deformation of the pack. If a number of such marked sites were accurately positioned by the surface radar system, the aerial mosaics could also be used to resolve a number of current questions regarding the interrelation between observed strain and the size of the deformation array.

It is even possible to obtain surface drag coefficients from the laser measurements of the roughness of the upper ice surface (Figure 5 of Hibler and Mock, 1974). It should also be possible to obtain drag coefficients of the lower-ice surface in a similar manner, inasmuch as current data on the relation between the form of the top and the bottom surfaces of sea ice (Kozo and Diachok, 1973) suggest that the correlation is statistically simple. It is also quite important that any such correlations be experimentally verified by field observations on a variety of ice types. In this case, the missing link is not the equipment which was developed at considerable expense (Francois and Nodland, 1972), but the lack of support for the utilization of the equipment when the development was completed.

The one important missing piece of information that current remote sensing techniques will not provide is data on the ice thickness distribution within the deformation array. This is a critical omission inasmuch as current thinking suggests that the ice thickness distribution is very important in controlling the interaction between the mechanics and the thermodynamics of the pack (Maykut and Thorndike, 1973; Thorndike and Maykut, 1973). Recent work on the development of systems for remotely determining sea ice thickness is, however, encouraging in that radar systems operated from the ice surface are now capable of resolving the ice-water interface. Such "surface" techniques could also be very useful in microscale studies in that they should be capable of resolving and mapping a number of the internal structural features of sea ice such as bubble layers or bands with a large brine volume. Unfortunately there are problems in making a satisfactory air-borne version of such systems. Many of these problems are related to the size of the sea ice area interrogated by the system. We believe that to be highly useful, such systems should sense the thickness of an area no larger (horizontally) than 2 x 2 m and should resolve ice thickness to within 0.1 m. This would permit the resolution of both ridges and the topography of multiyear floes. In one possible technique, UHF microwave radiometry (Adey, 1970), the footprint is tens of meters even when the system is operated from a hovering helicopter. If such a system were to be flown from a height of 300 m, the width of the footprint would be so large as to make the results of little use. Another interesting technique that in principle is capable of "spot" thickness determinations is the measurement of the tilt of VLF waves above the ice surface. Unfortunately, this technique has been plagued by large and apparently erratic variations in the electrical properties of sea ice (conductivity changes over 4 orders of magnitude) that do not appear to correlate simply with either the ice thickness or with other known average properties of the ice. In short, in spite of the large number of possible applications, the current prospects for the rapid development of an adequate remote air-borne sea ice thickness sensing system do not appear promising. As far as instrument development is concerned, it would probably be much easier to design a system that couples with current upward-looking sonar systems and measures ice thickness from below. The problems in this area are unfortunately more related to the difficulty in obtaining a nuclear submarine for scientific work on sea ice, than they are in developing the equipment to make the measurements.

Another aspect of the mesoscale study of sea ice that has received little attention is the utilization of the results of remote sensing to specify parameters that characterize the deformational geometry of the pack at any time. Sea ice, of course, can to a good approximation be considered as a two-dimensional granular media (Coon, 1972). Therefore, it is reasonable to use the results of investigations on other more thoroughly studied granular media such as soils to suggest parameters that are likely to be important. Good first guesses would be the size and shape distributions of the interacting floes, the degree of orientation of the floes, the density of packing (compactness), and the thickness distribution of the thin young ice that forms in the interstices between the floes. With the exception of the last item, all this information can be obtained at present by utilizing remote sensing techniques. There may also be many other parameters that are important. However, the clear identification of the usefulness of such parameters will only come by numerically attempting to relate them to changes in the observed response of the pack. The determination of such indices would, at first glance, appear to be a major job. Fortunately, recent work on the characteristics of ridging suggests that sea ice may be statistically relatively simple and that the distribution functions giving the parameters of interest may, in principle, be rather easy to estimate. For instance, Figure 4 shows two histograms of pressure ridge lengths as measured near the 1971 and 1972

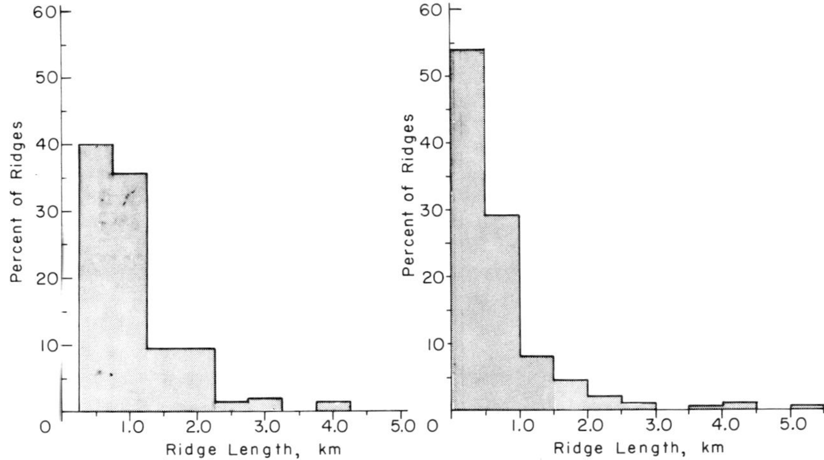

Figure 4. Distributions of the lengths of pressure ridges obtained (left) near the site of the March 1971 AIDJEX camp (74°N, 131°W) and (right) near the site of the March-April 1972 AIDJEX camp (75°N, 145°W) in the Beaufort Sea. The total number of ridges samples were 180 and 307 respectively.

AIDJEX camps in the Beaufort Sea (Hibler and Ackley, 1973). These distributions can be described quite well by a negative exponential distribution. In fact, a negative exponential is physically quite reasonable as the distribution function describing ridge lengths in that it can be derived by assuming that the holes in the ridges occur randomly. Therefore, the spacings between the holes, i.e., the ridge lengths, would be exponentially distributed (Hibler et al., 1972). Using a similar sort of argument, one might also expect that the distribution of floe sizes in the pack would also be a negative exponential. Examination of recent ERTS imagery (Barnes and Bowley, 1973) certainly suggests that this is true much of the time.

We believe that one of the major problems in applying remote sensing to mesoscale sea ice studies is the lack of automatic or semi-automatic techniques for obtaining suitable numerical data from the images. Unless rapid analysis techniques can be developed, remote sensing will never be able to be fully utilized as "near" real-time input into predictive models.

MACROSCALE

When macroscale observations of sea ice are considered, the length scale (hundreds of kilometers) necessitates the use of unmanned surface data collection systems (data buoys). The ability to produce Arctic Data Buoys with suitable positioning capability exists and current field tests of prototype models developed by NOAA for use on the AIDJEX Project are very encouraging (Martin, 1973). Although such buoys appear capable of collecting adequate meteorological information, at the present time only the most rudimentary oceanographic observations can be obtained. Even considering this restriction, at the present the extensive utilization of such buoys appears to be primarily limited by their expense as well as by the expense and difficulty of maintaining an operational array of buoys distributed over a large expanse of sea ice.

To characterize the large areas of ice covered in a macroscale study, it is necessary to utilize remote sensing information collected by satellite systems. The problems are obvious; many sensors such as standard photography and infrared are limited by cloud cover and those that are not, such as passive microwave, possess a very large footprint and are therefore difficult to interpret. In addition, only the more obvious large scale features of the pack such as wide leads can be distinguished on the imagery. Also, there are commonly extensive holidays in the data. For instance, in the ERTS imagery which currently provides the most detailed images, there are 4 days when a given site in the southern Beaufort Sea can be viewed (weather permitting) followed by a 14 day gap before the satellite returns to the same area. Also, no ERTS data is available from latitudes higher than 80°N.

In view of the above comments, one might be tempted to conclude that the prime need in the macroscale characterization of sea ice is for improved satellite remote sensing systems. Such improvements are, of course, highly desirable. We feel, however, that the greatest need in the macroscale area is for a more extensive analysis of the existing data. The papers in Session V of this symposia indicate how profitable such analyses can be. In particular, the paper by Crowder et al (1973) shows that the ERTS imagery can be used to determine divergences, vorticities, velocity profiles across shear zones and information on the scaling laws governing the deformation of the pack. Such information is urgently needed by investigators attempting to develop improved capabilities for numerically modeling the large-scale behavior of sea ice. Also, it is possible to measure a number of indices characterizing the geometry of the pack such as the floe size distribution and the lead pattern from the ERTS imagery. Again, the problem is partially caused by the lack of rapid methods for numerically characterizing the imagery as well as by the fact that the funding for equipment development appears to greatly exceed the funding for utilization when the developmental phase is completed. It is only by thoroughly utilizing current satellite information that the user can develop the necessary experience that is required to properly assist in the development of the next generation of instrumentation.

CONCLUSION

In the above, we have tried to give the reader our personal impression of a few of the diverse problems that confront investigators concerned with the study of sea ice. Striking advances have been made in the past few years in expanding our knowledge of this rather unusual material. With the increasing development of the arctic regions, we can only forsee a rapidly expanding need for research such as we have described here. If the support follows the need, the late 1970's should clearly be a most interesting period for investigators interested in the study of sea ice on all scales.

REFERENCES

Adey, A.W., 1970. A survey of sea-ice-thickness measuring techniques. Communications Research Centre Rept. No. 1214, Ottawa, 26p.

Barnes, J.C., and Bowley, C.J., 1973. Mapping sea ice from the Earth Resources Technology Satellite. Arctic Bulletin 1: 6-13.

Coon, M.D., 1972. Mechanical behavior of compacted arctic ice floes. Offshore Technology Conference Paper OTC 1684, 6p.

Cox, G.F.N., and Weeks, W.F., 1973. Salinity variations in sea ice. AIDJEX Bulletin No. 19: 1-17.

Cox, G.F.N., and Weeks, W.F., 1974. A study of brine drainage in sodium chloride ice. CRREL Research Report (in preparation).

Crowder, W.K., McKim, H.L., Ackley, S.J., Hibler, W.D. and Anderson, D.M., 1973. Mesoscale deformation of sea ice from satellite imagery. This Conference, Paper 5.13.

Francois, R.E., and Nodland, W.E., 1972. Unmanned arctic research submersible (UARS) system development and test report. Applied Physics Laboratory, Univ. Washington, Tech. Rept. No. APL-UW-7219, 88p.

Hibler, W.D., and Ackley, S.F., 1973. A sea ice terrain model and its application to surface vehicle trafficability. J. Terramechanics (in press).

Hibler, W.D., and Mock, S.J., 1973. Classification of sea ice ridging and surface roughness in the Arctic Basin. This Conference, Paper 2.12.

Hibler, W.D., Weeks, W.F. and Mock, S.J., 1972. Statistical aspects of sea-ice ridge distributions. J. Geophys. Res. 77: 5954-5970.

Kozo, T.L., and Diachok, O.I., 1973. Spatial variability of topside and bottomside ice roughness and its relevance to underside acoustic reflection loss. AIDJEX Bulletin No. 19: 113-121.

Martin, P., 1973. Arctic data buoys and AIDJEX. AIDJEX Bulletin No. 22: 1-7 (also see other papers in this Bulletin).

Maykut, G., and Thorndike, A.S., 1973. An approach to coupling the dynamics and thermodynamics of arctic sea ice. AIDJEX Bulletin No. 21: 23-30.

Mock, S.J., Hartwell, A.D. and Hibler, W.D., 1972. Spatial aspects of pressure ridge statistics. J. Geophys. Res. 77: 5945-5953.

Parmerter, R.R., and Coon, M.D., 1972. A model of pressure ridge formation in sea ice. J. Geophys. Res. 77: 6565-6575.

Parmerter, R.R., and Coon, M.D., 1973. Mechanical models of ridging in the arctic sea ice cover. AIDJEX Bulletin No. 19: 59-112.

Thorndike, A.S., and Maykut, G., 1973. On the thickness distribution of sea ice. AIDJEX Bulletin No. 21: 31-48.

Weeks, W.F., and Assur, A., 1972. Fracture of lake and sea ice. In "Fracture, An Advanced Treatise," Vol. 7: Fracture of nonmetals and composites (H. Liebowitz, ed.), p. 879-978, Academic Press, New York.

Weeks, W.F., Kovacs, A. And Hibler, W.D., 1971. Pressure ridge characteristics in the arctic coastal environment. Proceed. First Internat. Conf. on Port and Ocean Engineering under Arctic Conditions, Vol. 1: 152-183, Tech. Univ. Norway, Trondheim.

Weeks, W.F., and Lee, 0.5., 1962. The salinity distribution in young sea ice. Arctic 15: 92-108.

Weeks, W.F., and Lofgren, G., 1967. The effective solute distribution coefficient during the freezing of NaCl solutions. Physics of Snow and Ice, Vol. I, Part 1: 579-597, Institute Low Temperature Science, Hokkaido University, Sapporo.

SUMMARY OF DISCUSSION
(Papers 2.1 to 2.7)

The discussion was opened with questions concerning paper 2.1, by J. W. Hornbeck. Questions were raised concerning the techniques used in getting the SO_4 input and output summation figures; the Cl/SO_4 ratios; and the significance of sea salts as the source of the SO_4. Hornbeck elaborated on the techniques used and noted that the Cl/SO_4 ratios were not the same as sea salt. The high SO_4 concentrations are believed to be the result of pollution - probably from the Pittsburgh area.

During the discussion of paper 2.3, by J. R. Meiman, the near-mystical reliance on "April 1 data" was questioned. Reference was made to the importance of "setting" in determining the reliability of deuterium data gathered on or about April 1. It was noted that for a heavy snowpack such data could be reliable, yet for a thin pack the data were not very good.

Following a few additional comments concerning the "April 1 Syndrome" the discussion moved to an evaluation of the last three papers; number 2.5 by L. O. Grant and O. J. Rhea; number 2.6 by O. J. Rhea and L. O. Grant; and number 2.7 by R. F. Reinking. Reinking showed a number of photographs of snow crystals and related them to the cloud temperature structure, the snow crystal type, and the accretion process. The effects of growth time and surface vapor pressure were also mentioned. Rhea followed this by presenting some data concerning the relationship between the deuterium content of the crystals and the 700mb temperature - the deuterium content increased with increasing 700mb temperature.

This, in turn, led to further discussion concerning the effect of elevation on snowfall patterns (papers 2.5 and 2.6 by Grant and Rhea). In response to the question - "How can you correlate elevation with water content in the U.S. while in Canada you cannot?" - Rhea responded that the correlation exists for a single mountain (ridge) but not for a series of mountains (or ridges). A method for explaining the lack of correlation between elevation and water content for the situation where a series of mountains (ridges) is encountered is presented in Paper 2.6. Here, the product of (horizontal) wind speed and slope is used as a control parameter. Given the relationship, (horizontal) wind speed x slope = vertical wind speed, a questioner asked, is the observed effect the result of reduced uplift of the air mass as it moves from ridge to ridge, or actually a decrease in moisture from slope to slope? Rhea responded that under the imposed condition (as outlined in the paper) the ridges were sufficiently far apart that there would not be a significant reduction in the uplift from ridge to ridge. The influence of a resupply of moisture into the air mass when the precipitation is associated with a major cyclone was also discussed. However, no consensus was reached as to the possible importance of this factor.

SUMMARY OF DISCUSSION
(Papers 2.8 to 2.13)

(Paper 2.10 by L. G. Browman) H. R. Krouse suggested that a stable isotope technique might provide an alternative approach, since the isotopic ratios for water passed up through the ice would probably differ from the ratios for water passing down through the snow cover to the ice surface. W. F. Weeks commented that the observations made in this study are contrary to the behavior predicted by the phase rule, but there was no further discussion of a phase rule problem. M. A. Billello asked if the cracks or holes might be of some other origin, e.g., thermal, thus indicating that the water might come from a source at some distance from the measurement site. This problem was anticipated by the author. His solution was to clear the snow from a large area of ice and examine this area for cracks or holes. None were found during the field measurement program. A. Assur commented that the type of work done by Browman should be encouraged because the knowledge obtained is basic and therefore useful for solution of related problems.

(Paper 2.11 by N. A. Ehrlich and J. P. Welsh) Some elaboration was requested of the technique of measuring flexural strength of lake ice and of some of the probable sources of the dispersion in the data. The technique of preparing an in-situ cantilever beam was described and a few of the probable sources of dispersion of the flexural strength data were presented. The sources discussed were the inability to measure temperature in the upper surface of the ice beam and the inability to measure pre-stressing of the sample beam.

(Paper 2.12 by W. D. Hibler and S. J. Mock) A description of the appearance of hummock fields in the laser profile data was requested by R. M. Koerner. Hibler replied that hummock fields were not considered separately in this analysis. Koerner then suggested that perhaps a "hummock picker" similar to the "ridge picker" presented in the paper should be developed, since hummock fields may present a mobility problem. The discussion that followed was concerned with the definition of a mobility problem, and this was then specifically related to the type of vehicle involved (e.g., dog sled or large air cushion vehicle).

(Paper 2.13 by W. F. Weeks, W. D. Hibler, and S. F. Ackley) E. P. McClain suggested that the imagery obtained from polar orbiting satellites might be useful in resolving some of the macroscale sea ice problems. Weeks pointed out that the very large quantity of detailed information contained on the imagery requires computer-assisted analytical techniques, and that the development of techniques for numerical analysis of satellite imagery is in the beginning stage. He concluded that it will be very interesting to observe future developments in this. Of particular interest, Weeks observed, will be the use of the results of such analyses as either input to or checks on the numerical models of pack ice behavior presently in use and in the development stages.

Session III

Information Systems

Chairman:

E. L. Peck, Research Branch, National Weather Service, NOAA

General Reporters:

Eric Dyck, Division of Hydrology, University of Saskatchewan

A. J. Brown, Coordinator, California Cooperative Snow Surveys, Calif. Dept. of Water Resources

THE ARCTIC AND SUBARCTIC SEASONAL SNOWPACK: RESEARCH AND MANAGEMENT APPROACHES IN ALASKA

Charles W. Slaughter
Research Hydrologist, Alaskan Division, USACRREL
Fairbanks, Alaska

Arthur G. Crook
Snow Survey Supervisor, Soil Conservation Service, USDA
Anchorage, Alaska

ABSTRACT

The seasonal snowpack of Arctic and Subarctic regions differs from that of temperate regions in several ways, including persistence (six to ten months of continuous snow cover), snowpack density (0.05 to 0.35 g/cm^3 in "taiga" regions, and up to 0.55 g/cm^3 in "tundra" regions), and intensive depth hoar development. Information needs are many. Basic data concerning amounts, timing, and areal disposition of snowfall are exceedingly sparse, yet indispensable for evaluation of the snow resource. Information on snowpack metamorphism and melt processes is lacking; need is seen for research into: ablation processes and meltwater yield over frozen soils (permafrost); snowpack metamorphism under steep vapor pressure gradients, minimal solar radiation input, and limited reworking by wind; the role of vapor transfer at the snow/air interface; and seasonal snowpack influence on local environments in relation to animal habitat, plant succession, and permafrost regimen. Study of snowpack management opportunities should include regulation of snow accumulation, redistribution, and ablation for such ends as water supply augmentation, streamflow regulation, drift control, site rehabilitiation enhancement, and improvement of snow-based recreation. Obtaining snow data in a harsh winter environment is a basic concern; conventional snow survey techniques should be supplemented by such advances as real-time telemetry, utilization of satellite-based imagery, and radioisotope snow gages.

> "Onset and disappearance of seasonal snow cover mark a drastic change of the physical characteristics of the earth-atmosphere interface and of the way in which the two media interact. A layer of snow lies on sea ice, glaciers, and frozen ground much of the year. An understanding of mass and energy balances for any of these ice features is dependent on knowledge of the timing and the thermal and radiative properties of the snow layer. Such knowledge is of general importance in studies of atmospheric circulation and the world water balance and is of specific importance to utilization of the Arctic by man."
>
> (Committee on Polar Research, 1970)

INTRODUCTION

Alaska yields about 40% of the total United States freshwater runoff, about 9.84×10^{11} m^3 water per year. A large portion of this water is contributed by Arctic and Subarctic regions - the subarctic Yukon and Kuskokwim river systems alone provide some 2.48×10^{11} m^3 runoff per year. While there are 73,000 km^2 of permanent snow and ice in Alaska, this is concentrated in regions of maritime climatic influence; the Arctic and Subarctic are almost devoid of major glaciers (though the glaciers of the Pacific Mountains system obviously do occupy many headwater regions of tributaries to both the Yukon and Kuskokwim Rivers). Seasonal snow is thus of overriding importance in the annual runoff.

Snow commonly covers the landscape for six to ten months of the year in these northern regions. Unlike much of the western North American snowpack zone, this snowpack is a factor encountered daily in the lives of all the populace - in the valleys as well as in the uplands - for at least 50% of the year. It is thus proper to understand the environmental, as well as hydrologic, role of seasonal snow in cold regions.

SNOW TYPES

Two primary northern snow types (and regions) have been distinguished by Pruitt (1970): "taiga snow" and "tundra snow". The taiga is a broad ecotone, or a distinct environmental system, transitional between boreal forest and treeless Arctic tundra. Pruitt considers the snowpack of interior Alaska, between the Brooks Range and the Alaska Range, to be "typical" taiga snow: densities at initial deposition are low, 0.05 to 0.10 g/cm^3 (Benson, 1967); movement and reworking by wind is not common, save on ridges and high plateaus; during winter months solar radiation influence on the snowpack is minimal at this high latitude; snowpack metamorphosis is primarily through vapor transfer in response to strong vapor pressure gradients. Relatively warm soil temperatures at the base of the snowpack (rarely below -5°C) and cold ambient air temperatures (-40°C is not uncommon) produce steep vapor pressure gradients, with upward water vapor flux within the snowpack of up to 0.07 g/cm^3/day (Trabant, et al, 1970; Benson, 1967). The result is extensive development of "depth hoar", large loosely-bonded crystals with snowpack density commonly 0.20 to 0.28 g/cm^3.

"Tundra" snow, on the other hand, is almost invariably redistributed and modified by wind action. Particles are abraded and rounded, exposed surfaces are swept free of snow, and drifts accumulate in depressions and behind flow obstructions. Stream valleys may be completely filled with dense (0.35 to 0.40 g/cm^3) hard snow while adjacent slopes are snowfree (Benson, 1967; Pruitt, 1966). Drifts may have mean pack density of greater than 0.50 g/cm^3 even in mid-winter (Slaughter, et al, 1973).

Hare (1971) has confirmed this distinction between taiga and tundra snowpacks, based in part upon climatological considerations. He showed that density of taiga snowpacks in Canada reached 0.25 to 0.30 g/cm^3 in late winter, while tundra snow consistently has a density greater than 0.35 g/cm^3. McKay and Findlay (1971) note the same distinction.

Within the taiga of Alaska, typically "tundra" snowpacks may be expected at higher elevations where a treeless alpine environment is found and wind action is greatly increased relative to the lowlands (Benson, 1967).

We would expect to find gradations within these two major snow types. In Alaska we can identify a marked gradation within the taiga zone, in addition to high-elevation alpine tundra. Moving west toward the Bering Sea, we can expect less snow, more mixed rain/snow events, increased wind, and higher snow densities.

We thus conclude that Subarctic and Arctic seasonal snowpacks may be considered rather unique as compared with temperate snowpacks, including those of the western United States snow zones. Distinguishing characteristics include (1) an extended seasonal distribution of snowfall and duration of the seasonal snowpack, (2) intense depth hoar development and vertical vapor migration within the snowpack in response to very steep temperature and vapor pressure gradients during winter, particularly in regions of continental climate, (3) low total snowpack density in sheltered taiga settings, 0.15 to 0.25 g/cm^3 being common in interior Alaska, (4) extremely high snowpack density even during mid-winter in Arctic regions, as a result of intensive and prolonged reworking and packing by wind action, (5) relative shortness of the ablation season in the interior Subarctic - from first onset of melt to snowpack disappearance may take only ten to fifteen days.

Another distinguishing feature of these northern snow regions might be a relatively high proportion of total land area above treeline. This is certainly true of western and northern Alaska, beyond the regional treeline; it is also true of the uplands of central Alaska. For example, nearly 20% of the Chena River basin of central Alaska is above treeline.

The foregoing attributes of Arctic and Subarctic snowpacks do not, of course, mean that the snow responds in a unique manner to the physical laws governing pack metamorphism and ablation. Rather, they are consequences of differing climates and depositional environments. Nevertheless, we would suggest that information needs for both research and management are notably different from those for other locales.

INFORMATION NEEDS

Information needs are classified into "research" and "management" sections, with full awareness of the many overlaps between the two categories.

Snowpack Research

Knowledge of areal and seasonal distribution of snowfall - how much it snows, where and when - is lamentably lacking for most of the Arctic and Subarctic. This is probably the primary and most basic snow information need in the North. Much effort has gone into improving this situation in Alaska, primarily through expansion of the snow survey network. In 1963 there were ten snow courses in Alaska; in 1973 there are 130. Still, this measurement network does not extend to the northern and western parts of the state, and higher elevations are inadequately represented. Data on snow and ice thickness for a number of remote sites, virtually all at lowland locations, are also available (Bilello and Bates, 1972) and provide a starting point for snow cover evaluation.

Continuing detailed studies of snowpack metamorphism during winter are needed. Benson (1967) and Pruitt (1970) have convincingly pointed out some direct and possible indirect consequences of pack maturation under low solar radiation loads, steep temperature gradients, and extended time periods. Extension of such studies to environments other than lowland settings, documentation of environmental effects on pack metamorphism in differing settings, and examination of physical and biological consequences of these factors are all needed.

Hydrology of the northern regions is inseparable from snow problems. Specific aspects of snow hydrology which need early examination, in addition to the simple question of how much snow where and when, include understanding the snowpack "ripening" process under Arctic and Subarctic conditions; understanding snowmelt in terms of energy partitioning is a high-priority item only beginning to receive attention from hydrologists in Alaska. Disposition of meltwater within and beneath the snowpack is a related question. Colbeck (1972, 1973) is furthering our understanding of snow meltwater behavior; how is this affected by conditions of intense depth hoar, and cold snow and soil (often permafrost) directly beneath the melting snow surface?

In areas of continuous and deep organic litter and moss ground cover, a common situation in the Subarctic (Dingman, 1971), disposition of meltwater following release from the pack is a related problem.

The role of vapor transfer in snow hydrology has not to our knowledge been directly examined in the Arctic or Subarctic. Jeffrey (1968) points to conflicting evidence on the role of latent heat (and water vapor) flux from snow in mountainous terrain; the long period of snow cover in northern regions would at the very least increase exposure time of snow and ice for possible vaporization. With a large amount of wind movement in alpine and tundra settings, the potential for evaporative loss during lateral transport may similarly be high (Schmidt, 1972; Tabler and Schmidt, 1972).

In addition to examination of energy budget parameters in snow hydrology, cognizance of the effects of the snow cover itself on the energy budget of the landscape is required. Imposition of a high-albedo reflective covering on virtually the entire surface of a watershed or a region, and maintenance of that high short-wave reflectivity for six months or more during periods of already low total availability of solar radiation input, must have significant effects on local and regional environments. Energy loss through direct reflection of half or more of the incoming direct-beam solar radiation would certainly affect potential productivity of plant communities, as well as thermal stability of any setting. Further, during periods of melt the energy utilized in phase change from solid to liquid, and in vaporization, is energy that is not available for other purposes (photosynthesis, warming of soil, warming plant tissues) in that local environment.

Relationships between snow cover and stream and valley icings ("naleds", "aufeis") pose other problems for hydrologic analysis. Such icings occur during winter, but are dominantly ice or ice/snow composites. Formed from interflow or groundwater sources, they may comprise significant storage components in the hydrologic cycle of upland watersheds (Kane and Slaughter, 1972).

The role of the winter snowpack in ameliorating or otherwise modifying the near surface and subnivean environment is almost completely open to investigation. Pruitt (1970) and Benson (1967) have suggested that insulation provided by snow makes overwintering possible for some organisms, while dessication through vertical vapor migration of the surface plant/litter and upper soil layers may be a significant environmental stress.

In more southerly permafrost-underlain landscapes, a shallow winter snow cover may be essential to existence of permafrost (Brown, 1970); deep snow may sufficiently retard long-wave radiation cooling during winter so that permafrost either does not form, or warms and degrades. Along with better understanding of this area, knowledge of the effects of snow cover changes under use on subnivean ecologic relationships should be developed. An obvious case in point is the possible consequences of snow packing under traffic by snowmobiles. Where snowmobile traffic is concentrated and occurs throughout the winter, it seems that pronounced effects on underlying vegetation and fauna (as well as on the subsequent melt behavior of the snowpack itself) must result. The whole subject of "snow ecology" has received but scant attention to date, even though snow is a ubiquitous and dominating presence in northern regions.

Snowpack Management

Little attention has been paid to active management of high-latitude seasonal snowpacks. Some possibilities have been suggested by the Committee on Polar Research (1970) and Slaughter (1972), among others, but active snow management programs remain scarce or non-existent in the Subarctic and Arctic.

Common watershed management objectives of snowpack manipulation include modifying accumulation and ablation patterns to effect changes in quantity, quality, and timing of streamflow. Techniques of vegetation manipulation toward these ends developed in more temperate regions may well be adapted to the Arctic and Subarctic. One factor which will affect such adaptation is the simple lack of forest cover in much of the North. In the absence of forests to manipulate, emphasis will be on modifying snow accumulation through man-made drift control structures such as snow fences or temporary berms constructed of snow. In the forested taiga allowance must be made for slow growth rates and lower species diversity in planning vegetation manipulation for snow management. Modifying ablation patterns through surface additives has the same potential in the Arctic and Subarctic as in more temperate zones, with due allowance for differences in solar radiation input during winter and spring months.

In explaining the hydrologic functioning of many high-latitude watersheds, the seasonal snowpack may prove the most significant independent variable. On catchments of glacially-affected streams, the snowpack may strongly affect ablation of the ice field. Cursory examination of selected hydrographs from Alaskan basins indicates that heavy snowfall winters are often followed by reduced glacial melt. On other watersheds free from glaciers, the winter snowpack may explain up to 75% of annual stream discharge variation. A better understanding of the entire snowmelt hydrology area is fundamental to providing better short- and long-term predictions of water yield and flood flows.

An effective operational program to induce snow and ice melt at key locations on rivers is needed in Alaska. Most spring flooding damage is caused by ice jams. Since spring is a low-precipitation period in much of the state, classic rain-on-snow floods are rare. The US Army Corps of Engineers and the Alaska State Disaster Office have cooperated in an operational "ice-dusting" program for several years. Pre-selected critical areas in river channels are dusted with sand and coal dust to accelerate melt and reduce the potential for channel blockage and resultant ice jam flooding. Evaluation of such a program is difficult and research upon which to base management decisions is generally lacking.

Snowdrift control techniques offer great promise in Alaska. Transportation routes, particularly on the rapidly developing North Slope of the Brooks Range, are subject to snowdrift blockage. Snow fences to accumulate and control wind-transported snow, such as those described by Tabler (1972, 1973), may be used successfully to keep these roadways open to travel. The same induced-drifting techniques may be used for snowpack management on both alpine and nordic recreational skiing areas. Northern towns and villages commonly have inadequate supplies of potable water; snowpack accumulation by fencing or other entrapments and catchment of the meltwater is an alternative with untapped potential (Slaughter, et al, 1973).

Construction sites, burned forests, and other disturbed areas where rapid revegetation is necessary offer yet another opportunity for snowpack management techniques. More favorable soil moisture and temperature conditions for seed germination and survival may be created through snowpack regulation.

NEW APPROACHES IN DATA ACQUISITION

Heavy emphasis should be placed on the need for developing or improving methods of acquiring basic information commonly available in more temperate snow regions. Severe winter climates with -40°C common from November through March, high winds in the treeless alpine and arctic, extremely restrictive transportation and communication facilities, and high operating costs all contribute to a very real need for new and improved approaches to data collection and research efforts.

Real-time telemetry of data on snowpack depth, water equivalent, and related environmental data has been viewed as a partial solution to these problems. Pre-programmed or on-call telemetry from unattended sensors in the field could significantly reduce the manpower, time, and transportation costs now imposed in field data collection.

The first operational real-time telemetry network in Alaska was installed in the Tanana River basin of central Alaska, following the August 1967 flood at Fairbanks. That radio-linked network provides real-time river stage and precipitation readings from six field locations (Slaughter, Freeman, and Audsley, 1973). Since the network operates only during the open-water season, snowpack sensors have not been incorporated into the system. A separate radio telemetry system is currently being installed in the Caribou-Poker Creeks Research Watershed, north of Fairbanks, Alaska, to transmit microclimatology data (temperature, dewpoint, wind, short- and long-wave radiation) from a field research site to a Fairbanks base. That system is designed to allow future incorporation of snowpack sensors such as liquid-filled pressure pillows.

Two statewide data telemetry systems are also being considered, which would telemeter snowpack and related parameters from a significant portion of the state. First, the Soil Conservation Service has plans for a network of forty data sites for which real-time information would be transmitted via VHF radio to base station facilities in Fairbanks, Juneau, and Anchorage (Crook, 1971). Funding to initiate this and similar plans developed for other western states will be available this year (1973-74). Secondly, recent congressional activity regarding the proposed Trans-Alaska oil pipeline has stimulated interest in another telemetry system. The pipeline project, more costly than any previous single private engineering feat, will be subjected to sufficient hydro-meteorological and related hazards that a telemetry system for warning and monitoring of hydrological events along the pipeline will be necessary. The National Weather Service and Soil Conservation Service are developing a joint plan for this proposed system. Alternative methods for transmission include VHF radio via ground or satellite repeaters, and micro-wave systems.

The ability to telemeter data increases the need for reliable sensors. Requirements for operation over a wide range of field conditions, often without electrical power and far-removed from transportation corridors, imply rugged, relatively simple, sensor/recorder systems. In our experience, there is much room for improvement in this realm.

The isotopic snow profiling gage offers an expensive but useful method of gathering research data useful in both snow physics and hydrology. This sophisticated device currently can be used in a relatively few locations, due to the lack of line power and adequate sites in a large portion of the state. The first such gage in Alaska was installed during the fall of 1972 at the Mt. Alyeska ski area near Anchorage. This high-avalanche-hazard area provides an ideal setting for use of the radio-isotope gage in avalanche research.

Utilization of satellite-based imagery in analysis of regional snow distribution will be treated in other papers in this Symposium. It is sufficient here to suggest that satellite-based imagery is expected to be of great value in monitoring snow input and ablation in the great expanses of the North which are not served by conventional data acquisition systems. Among limitations of this approach which should be recognized are the limited amounts of daylight during winter months, with consequent reduction of the possible time for obtaining imagery, and the low sun angles encountered during that same period. The potential advantages in terms of extensive areal coverage and repetitive imagery, all at low cost (following initial satellite system installation), argue for increasing use of satellite imagery in conjunction with ground-based data networks.

The use of instrumented research catchments in snow studies is new only in the sense that little use has been made of the "research watershed" approach in high latitudes. Research watersheds can include either "representative" or "experimental" watersheds, or both; in the former attention is devoted to process study, observation of natural change, and developing predictive techniques; while "experimental" watersheds are utilized primarily for study of the hydrologic and environmental effects of planned cultural changes to the landscape (Toebes and Ouryvaev, 1970). Snow studies would be an indispensable component of research in representative and experimental watersheds in northern latitudes. Observation of snow accumulation and ablation under natural conditions is included in activities in the Caribou-Poker Creeks Research Watershed of central Alaska, the only such research watershed in Alaska's subarctic. Such instrumented catchments are logical sites for study of snowpacks on a meso-scale, as well as for many of the detailed process studies suggested elsewhere in this paper. The call of the Committee on Polar Research (1970) for "...establishing basic snow research stations in a few carefully selected but widely different climates..." could well be met through the use of appropriately selected and operated representative and experimental watersheds.

CONCLUSION

As the environment, the snowpack, and appropriate snow management techniques are better defined and understood, improvement should be realized in man's rational occupation and use of the North. Some potential benefits have been noted or implied in preceding sections of this paper. Our emphasis has been on information needs. Meeting these needs and applying results to cold regions problems should be the goal of those concerned with seasonal snow in the North.

REFERENCES

Benson, C. S., 1967. A Reconnaissance Snow Survey of Interior Alaska. Scientific Report UAGR-190, Geophysical Institute, University of Alaska, College, Alaska. 71 p.

Bilello, M. A., and R. E. Bates, 1972. Ice Thickness Observations, North American Arctic and Subarctic, 1967-68, 1968-69. Special Report 43, Part VI, US Army Cold Regions Research and Engineering Laboratory, Hanover, New Hampshire. 97 p.

Brown, R. J. E., 1970. Permafrost as an Ecological Factor in the Subarctic. Pp 128-140 IN Ecology of the Subarctic Regions: Proceedings of the Helsinki Symposium. UNESCO, Paris, France.

Colbeck, S. C., 1972. A Theory of Water Percolation in Snow. Journal of Glaciology 11(63): 369-385.

Colbeck, S. C., 1973. Effects of Stratigraphic Layers on Water Flow Through Snow. Research Report 311, US Army Cold Regions Research and Engineering Laboratory, Hanover, New Hampshire. 13 p.

Committee on Polar Research, 1970. Polar Research - A Survey Committee on Polar Research, National Research Council. National Academy of Sciences, Washington, DC. 204 p.

Crook, A. G., 1971. Telemetry of Mountain Snowpack and Related Data - Proposed System for Alaska. Soil Conservation Service, USDA, Anchorage, Alaska. 10 p.

Dingman, S. L., 1971. Hydrology of the Glenn Creek Watershed, Tanana River Basin, Central Alaska. Research Report 297, US Army Cold Regions Research and Engineering Laboratory, Hanover, New Hampshire. 112 p.

Hare, F. K., 1971. Snow-cover Problems near the Arctic Treeline of North America. Report Kevo Subarctic Research Station 6: 31-40.

3.1

Jeffrey, W. W., 1968. Snow Hydrology in the Forest Environment. Pp 1-19 IN Snow Hydrology: Proceedings of the Workshop Seminar. Canadian National Committee, The International Hydrological Decade, Ottawa, Canada.

Kane, D. L., and C. W. Slaughter, 1972. Seasonal Regime and Hydrological Significance of Stream Icings in Central Alaska. Paper presented at International Symposia on the Role of Snow and Ice in Hydrology, Banff, Alberta, September 1972. (In press)

McKay, G. A., and B. Findlay, 1971. Variation of Snow Resources with Topography and Vegetation in Canada. Pp 17-26 IN Proceedings of 39th Annual Western Snow Conference.

Pruitt, W. O., 1966. Ecology of Terrestrial Mammals. Pp 519-564 IN Environment of the Cape Thompson Region, Alaska. US Atomic Energy Commission, Oak Ridge, Tennessee.

Pruitt, W. O., 1970. Some Ecological Aspects of Snow. Pp 83-89 IN Ecology of the Subarctic Regions: Proceedings of the Helsinki Symposium. UNESCO, Paris, France.

Schmidt, R. A., 1972. Sublimation of Wind-transported Snow - A Model. Research Paper RM-90, Rocky Mountain Forest and Range Experimental Station, US Forest Service, Fort Collins, Colorado. 24 p.

Slaughter, C. W., 1972. Snowpack Management Potential in Alaska. Pp 175-190 IN Watersheds in Transition. Proceedings Series No. 14, American Water Resources Association, Urbana, Illinois.

Slaughter, C. W., T. G. Freeman, and G. L. Audsley, 1973. Cooperation in Water Resources Programs: Alaska's Example. Paper presented at Ninth American Water Resources Conference, Seattle, Washington, 24 October 1973. (In press)

Slaughter, C. W., P. V. Sellmann, M. Mellor, J. Brown, and L. Brown, 1973. Snow Accumulation for Freshwater Supply Augmentation, Barrow, Alaska. Paper presented at 23rd Alaska Science Conference, August 1973. Alaska Division, American Association for the Advancement of Science, College, Alaska. (In press)

Tabler, R. D., 1972. Evaluation of the First-year Performance of the Interstate-80 Snow Fence System. Final Report on Contract 16-247-CA. Rocky Mountain Forest and Range Experiment Station, US Forest Service, Fort Collins, Colorado. 84 p.

Tabler, R. D., 1973. Evaporation Losses of Wind-blown Snow, and Potential for Increasing Snow Storage with Snow Fences. IN Proceedings of 41st Western Snow Conference. (In press)

Tabler, R. D., and R. A. Schmidt, 1972. Weather Conditions that Determine Snow Transport Distances at a Site in Wyoming. Paper presented at International Symposia on the Role of Snow and Ice in Hydrology, Banff, Alberta, September 1972. (In press)

Toebes, C., and V. Ouryvaev (Editors), 1970. Representative and Experimental Basins. An International Guide for Research and Practice. Studies and Reports in Hydrology 4, UNESCO. 348 p.

Trabant, D., C. S. Benson, and G. Weller, 1970. Physical-thermal Processes in the Seasonal Snow Cover of Northern Alaska. Pp 328-332 IN Proceedings of the Twentieth Alaska Science Conference, Alaska Division, American Association for the Advancement of Science, College, Alaska.

SURFACE MEASUREMENTS OF SNOW AND ICE FOR CORRELATION WITH DATA COLLECTED BY REMOTE SYSTEMS

Michael A. Bilello
U.S. Army Cold Regions Research and Engineering Laboratory
Hanover, New Hampshire

ABSTRACT

Reconnaissance by aircraft or satellite is one way to determine the areal extent of the earth's snow and ice cover. However, determining the depth and physical properties of the snow cover and the thickness of ice on lakes and rivers and along coastlines by this remote method is in an early stage of development. Data collected from such remote systems could be correlated with actual surface conditions by using an existing network of over 100 snow and ice observing stations located in North America above 45°N latitude. This network, established by the U.S. Army Cold Regions Research and Engineering Laboratory in cooperation with other U.S. and Canadian Government agencies, provides weekly snow and ice measurements during the winter. This paper identifies the stations in the network, reviews the types of measurements made and equipment used, and summarizes the results of studies derived from the collected data. The 10 to 20 years of record, as well as the data currently being received, provide an extensive and reliable source of information for comparing or verifying observations obtained by other methods.

INTRODUCTION

Photographs taken from aircraft and received from orbiting satellites have confirmed the effectiveness of these remote systems in surveying the cover of snow and ice on the earth's surface. From initial examinations of satellite pictures, for example, Fritz (1963) found that "snow fields can be distinguished from clouds in satellite pictures when the photographs are taken on successive days because cloud patterns change from day to day, but the show fields generally remain unaltered over a period of a few days." After analyzing Tiros II photographs Wark *et al.* (1962) reported that "many details of an ice cover can be distinguished, such as type of ice, amount, and presence of leads and cracks." The authors further state that although some clouds of a certain type and amount resemble ice, there is usually enough difference in appearance to allow identification by an experienced interpreter.

Since 1960 snow and ice reconnaissance from satellites has continued to improve as the sensors became earth-oriented and the vehicles were placed in quasi-polar orbits. A recent investigation of snow extent using photographs taken from unmanned space flight was conducted by Barnes and Bowley (1970). Another, which described sea ice conditions using satellite vidicon data, was conducted by McLain (1973). The results of these and other similar studies show that satellite imagery provides reliable measurements of the extent and in some instances the conditions of the snow and ice cover on earth.

As the sensors on board the satellites continue to improve, measurements such as the depth and density of the snow cover and the thickness of ice on lakes, rivers and oceans will eventually become a reality. Meanwhile it would be beneficial to be able to verify the data collected from the present remote systems with concurrent ground measurements. An extensive observation network which could provide information on snow cover properties and surface ice conditions and ice thicknesses throughout northern North America is the subject of this paper.

NETWORKS FOR OBSERVING SNOW AND ICE

Because of the lack of reliable and continuous data on the formation, growth and decay of sea, lake and river ice, and the regional variations in the physical properties of the snow cover, several networks for observing these winter surface conditions were established throughout North America starting in 1946. The number of stations gradually increased so that the networks now extend from the west coast of Alaska to the east coast of Canada, and from just below 83°N latitude south to the Great Lakes (Figures 1, 2).

The data obtained at these stations are valuable to commercial fishermen, loggers, aircraft pilots, government agencies, and the military. For example, the ice information is used to determine the opening and closing dates on navigable water, and whether or not men and vehicles can safely travel on the ice. It is also useful in determining when aircraft can land on lake and river ice, and for planning the activities of icebreakers during the winter. Information on the physical characteristics of the snow cover is used to determine the insulating effects of snow on soils and pavements, to calculate the severity of snow loads on roofs and other structures, and to estimate water content and runoff rates to be expected in the spring. The data have also proved useful for long-range analyses of climatic and environmental regimes in the Arctic and Subarctic.

The first five stations in the primary network were established in northern Canada between 1946 and 1950 as a joint program of the Canadian Department of Transport's Meteorological Division and the U.S. Weather Bureau (now National Weather Service). During the period 1955-65, in cooperation with these two agencies and the U.S. Air Force Air Weather Service, the U.S. Army Cold Regions Research and Engineering Laboratory (CRREL) expanded the network in Canada and in states across the northern conterminous United States. Starting in 1961 CRREL, in association with the U.S. Department of Commerce-Soil Conservation Service and the U.S. Army Alaska Eskimo Scouts, also established additional snow and ice observing networks in Alaska. The measurements at all the sites were, and still are, made by government employees or military personnel, and by local residents such as schoolteachers, clergy, Eskimos, homesteaders and lodge-keepers.

Although minor changes in the networks, such as site replacement and/or relocation, have taken place over the period of 10 to 20 years of observations, the total number of stations in the program has steadily increased. Since 1952, therefore, snow cover data have been compiled for 19 locations in Canada (10 of which were administered by the National Research Council of Canada), 30 in Alaska (16 of which were administered by the Soil Conservation Service), and 7 across the northern part of the U.S. mainland.

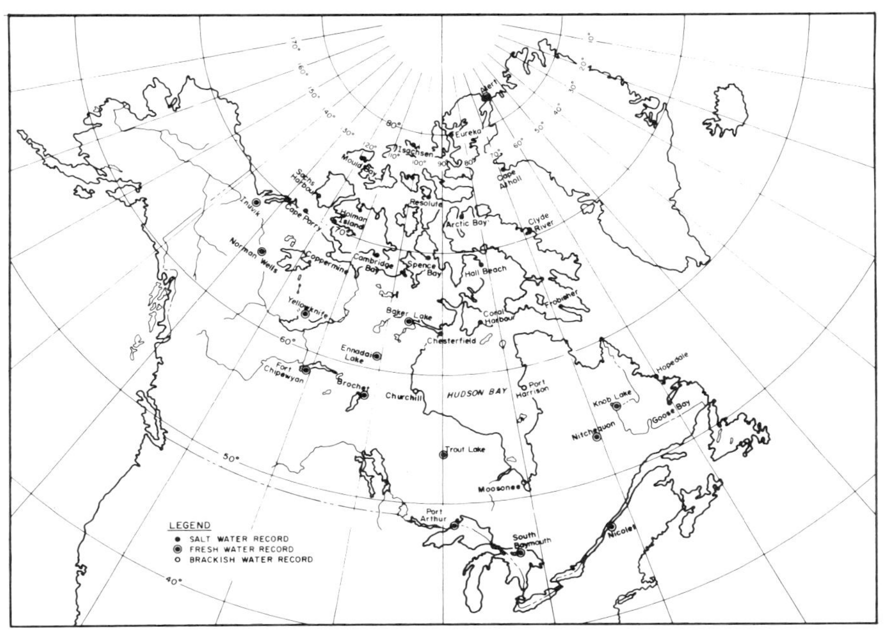

Figure 1. Canadian stations.

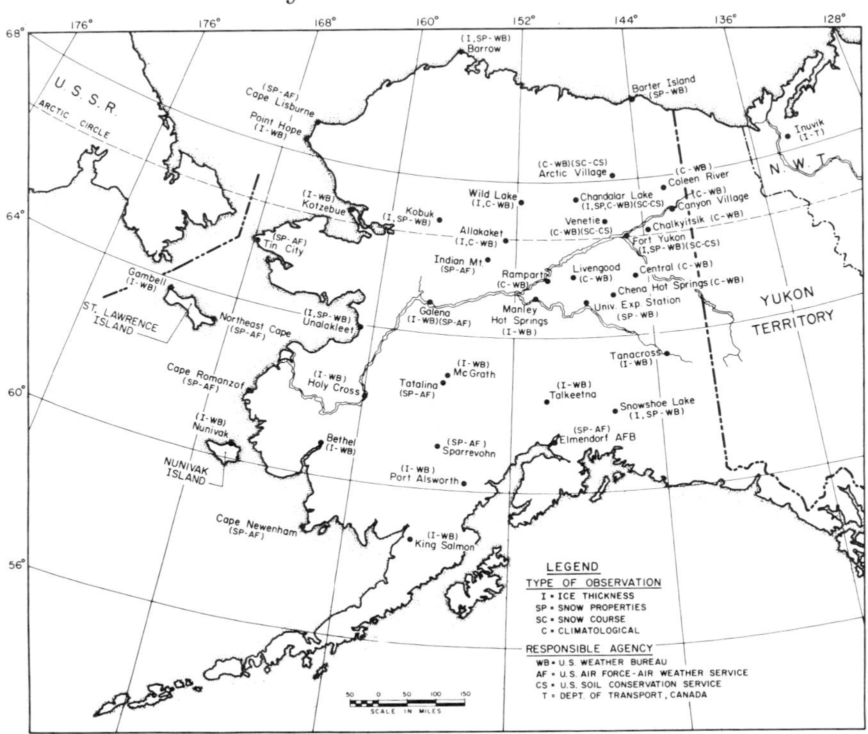

Figure 2. Alaskan stations.

In the ice observing program, there were, for example, 64 stations providing ice conditions and ice thickness data during the winter of 1969-70, 41 in Canada and 23 in Alaska. About one-third of the sites in Canada and about two-thirds of those in Alaska are inland freshwater sites. The remaining sites are near salt and brackish water bodies such as harbors, bays and river deltas, where the ice thickness measurements are made on land-fast sea ice.

SNOW OBSERVATIONS

The snow observations are made in accordance with the standard procedures described in USA CRREL Instruction Manual No. 1 (USA CRREL, 1962). Once each week, a snow profile is exposed at the sample site

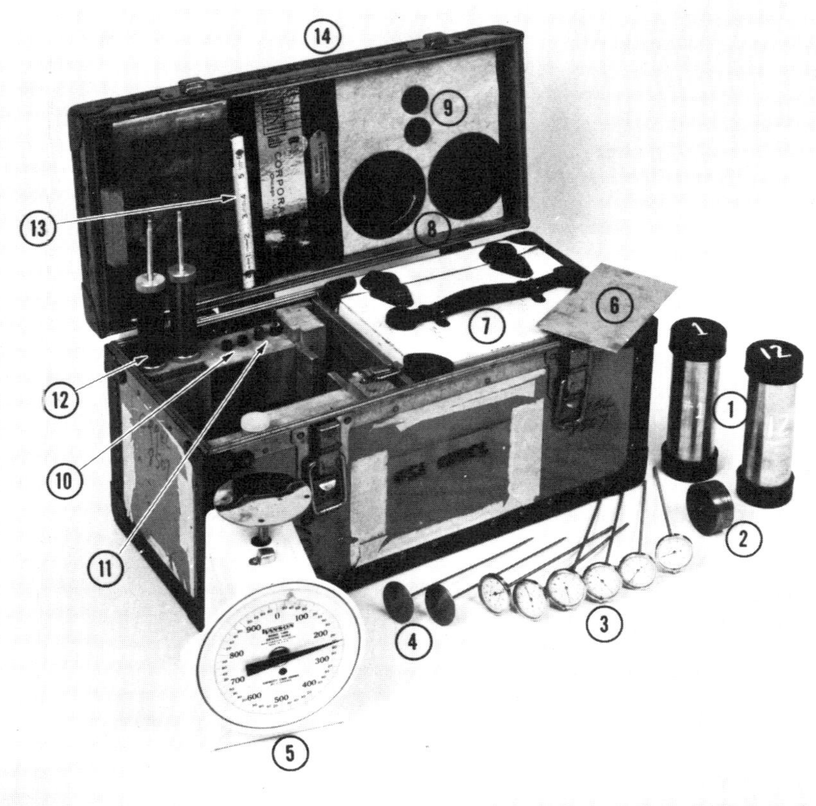

1. SNOW TUBES (12)
2. MANDREL
3. DIAL THERMOMETERS (6)
4. PROBE (2)
5. SCALE
6. SNOW CUTTER
7. SNOW TUBE CASE
8. SNOW HARDNESS DISK X1 (2)
9. SNOW HARDNESS DISK X10 (2)
10. SNOW HARDNESS DISK X100 (2)
11. SNOW HARDNESS DISK X1000 (2)
12. GAGE HARDNESS, RED AND BLACK
13. RULE (6 FT)
14. FIBER KIT

Figure 3. Snow observation equipment.

by making a vertical cut through the snow cover to the ground. The snow profile usually consists of several identifiable layers which may be the product of storm precipitation, metamorphic processes due to temperature changes, or intensive drifting. Textural differences and crusts at the boundaries help to identify and delineate the different layers. The weekly snow measurements include the thickness of each layer, the crystal classification, the temperature, the density and the hardness index.

The National Research Council of Canada provided the original snow-measuring equipment and CRREL later designed its own kit (Figure 3). The CRREL kit differs only slightly from the Canadian kit and contains the following equipment:

1. A set of 500-cm^3 snow sampling tubes with rubber caps to measure density in g/cm^3.
2. A scale to weigh the snow samples.
3. Two gages with disks to measure hardness indexes of 1 to 100,000 g/cm^2.
4. A set of dial thermometers with long stems to measure the snow temperature.
5. A snow cutter, a mandrel and snow temperature probes for use in making the observations.
6. A rule to measure the thickness of each layer.

ICE OBSERVATIONS

Initially, ice-thickness measurements were made through holes chiseled in the ice. When the ice cover was too thick for chiseling, dynamite was used to form a cavity in the top of it. To facilitate the ice thickness observation and to improve the accuracy of measurement, CRREL developed a hand-operated ice auger (Figure 4) and, in

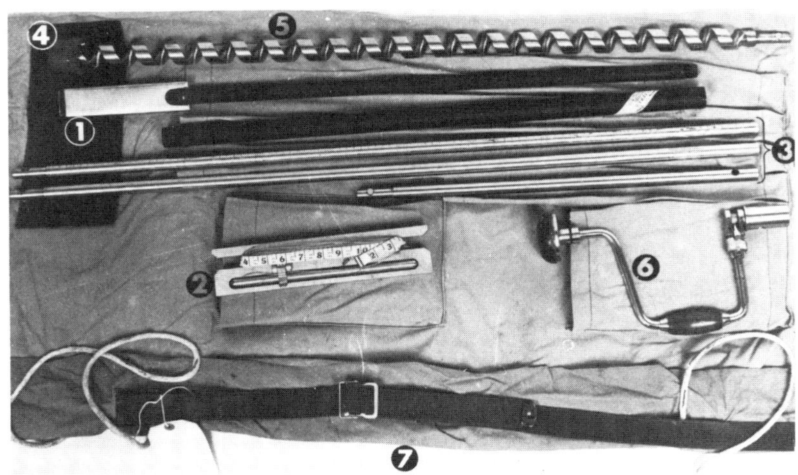

1. ICE CHISEL
2. MEASURING TAPE, WITH ROD
3. EXTENSION RODS
4. AUGER PROTECTIVE CAP
5. AUGER
6. BRACE
7. CANVAS COVER

Figure 4. CRREL ice-thickness measuring kit.

April 1956, distributed one to each station in the network. With this auger, a 3.8-cm- (1.5-in.) diameter hole can be bored by hand through 2 m of ice in about 15 minutes. When the freshwater ice or sea ice (older than 1 year) is thicker than the length of the auger (1 m) the auger may stick. Frequent removal of the ice cuttings during the drilling or the pouring of some nonfreezing liquid into the hole generally solves this problem.

To measure ice thickness, a measuring tape with a short, pivotable rod attached to its end is lowered into the augered hole. When the rod is below the ice, it swings to a horizontal position. The tape is then lifted until the rod contacts the bottom of the ice. The thickness of the ice is then recorded. A separate wire attached to the end of the rod is used to return it to the vertical position so that the rod and tape can be withdrawn from the hole.

The ice thickness observations are made once each week in bodies of water close to the station. The snow depth on the ice as well as ice surface conditions, e.g. occurrence of cracks and leads and dates of first ice, freeze-over and breakup, are also recorded.

In a separate observation program in Alaska personnel of the U.S. Department of the Interior, Geological Survey, Water Resources Division periodically measure flow beneath the river ice in Alaska during the winter. The thickness of the ice for this program is measured less frequently during the year but the observations are made every few meters across the river. These measurements have been included as supplementary data in some of the biennial ice thickness reports published by CRREL (e.g. Bilello and Bates, 1966, 1971).

During April, May and June 1970 the *SS Manhattan* navigated the sea-ice packs of Davis Strait, Baffin Bay and other bodies of water off northeastern Canada. Included in the work schedule on this second voyage of the vessel into the Arctic were almost daily measurements of the thickness of the sea ice. The observations were made by boring through undisturbed ice floes or by visually estimating the thickness of upended, exposed ice blocks. The data collected during this voyage are reported in the most recent issue of CRREL's series of ice thickness reports (Bilello and Bates, 1972).

RESULTS OF DATA RECEIVED FROM THE NETWORKS

A preliminary analysis of snow density, temperature, and hardness data obtained from the networks showed the snow cover in the Canadian Archipelago to be colder, denser and harder than that of interior Alaska (USA SIPRE, 1957). The average snow cover temperature from November through March at the Alaskan stations was 4° to $9^{o}C$ higher than the average air temperature, whereas at the Canadian Archipelago stations the snow was $4.5^{o}C$ warmer than the air. The range in snow cover density for interior Alaskan stations is approximately 0.13 to 0.20 g/cm^3 in November and 0.23 to 0.27 g/cm^3 in March. The densities along the north coast of Alaska and the northern islands of Canada range from approximately 0.30 to 0.36 g/cm^3 in November, and from 0.33 to 0.39 g/cm^3 in March. Investigation of snow cover hardness revealed regional variation similar to that for density, i.e. lower hardness index values in sheltered regions and higher values at exposed sites.

In a subsequent study, in which additional snow data received from the networks were used, a relationship between regional variation in snow cover density and observed temperatures and wind speeds

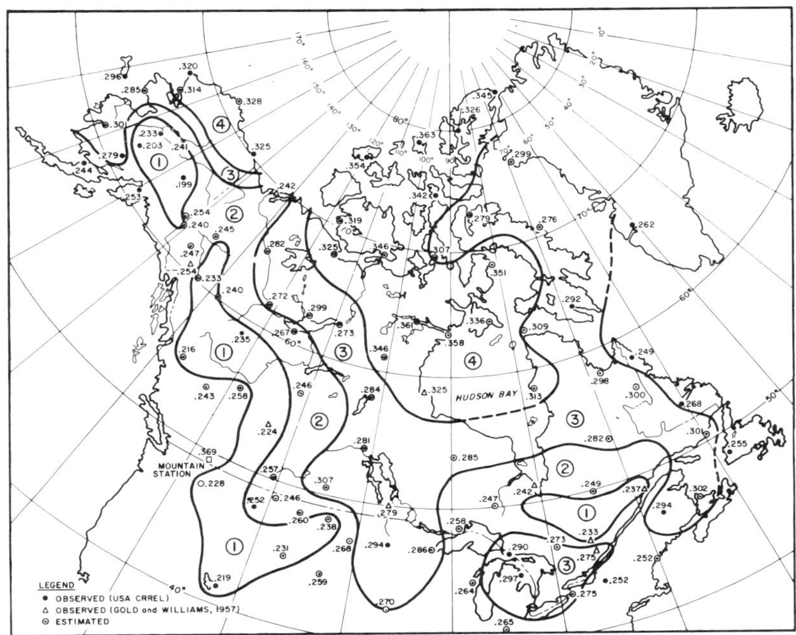

Figure 5. Average seasonal snow-cover density. Categories 1-4 are separated at densities of 0.24, 0.27 and 0.31 g/cm³.

was obtained (Bilello, 1966). The relationship, derived by a multiple regression analysis, resulted in the following equation:

$$\rho = 0.152 - 0.0031T + 0.019W$$

where ρ = average seasonal snow cover density (g/cm³)
T = average seasonal air temperature (°C)
W = average seasonal wind speed (m/sec).

The term *seasonal* refers to November through March inclusive, unless the average air temperature for the month is above freezing. The correlation coefficient and the standard error of estimate are 0.84 and 0.025 g/cm³, respectively. A test of the relationship, using densities for 10 locations not included in the initial study, showed good agreement between observed and estimated densities. Consequently, 61 stations with 10 years of climatological data were used to obtain estimated snow cover densities throughout North America. This information, as well as the observed densities from the network stations, was used to develop an average seasonal snow cover map (Figure 5). The continent was divided into four snow density zones separated at density values of 0.24, 0.27 and 0.31 g/cm³. The regions delineated in Figure 5 are described in general terms in Table I.

Naturally, the local topography and vegetation create differences in the density from point to point within a region defined by each category. Deviations from the average value for each region can also be expected from month to month and year to year.

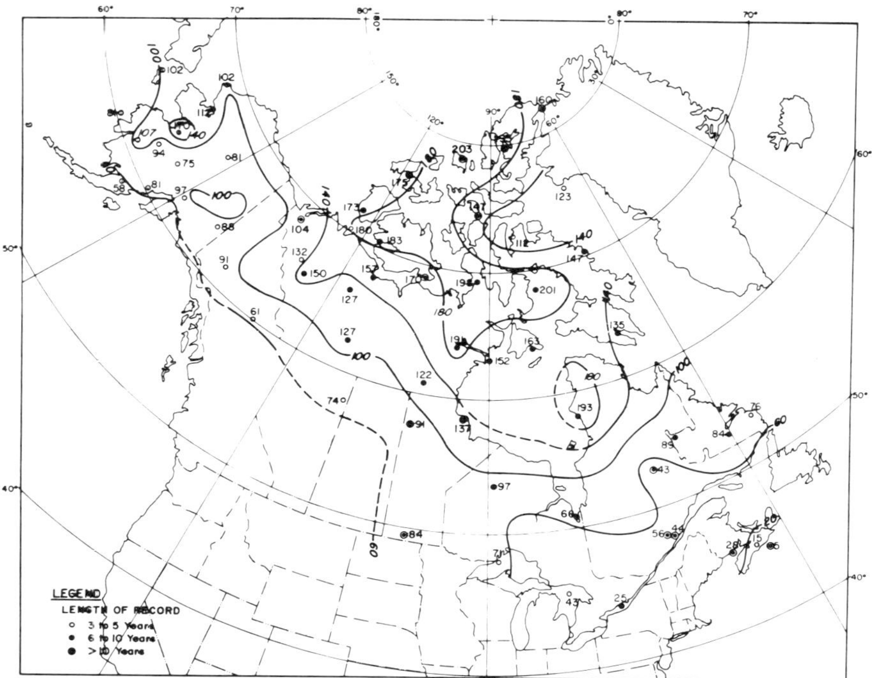

Figure 6. Least ice thickness (cm) observed at the time of maximum thickness for the years of record.

Table I. Snow cover density regions of North America.

Region	*General location*	*Winter wind conditions*	*Average seasonal snow cover density* (g/cm^3)
1	Inland	Light	< 0.24
2	Variable	Moderate	≥ 0.24 to < 0.27
3	Inland and coastal	Moderate to strong	≥ 0.27 to < 0.31
4	Arctic and Subarctic	Strong and frigid	≥ 0.31

The data on ice thickness and ice conditions on lakes and rivers and along the coast of North America as received from the networks are given in USA CRREL Special Report 43, Parts I-VI (Bilello, 1961, 1964; Bilello and Bates, 1966, 1969, 1971, 1972). To indicate the magnitude and distribution of ice growth, isoline maps of maximum ice thicknesses observed during each year are also presented in these reports, starting with the winter of 1956-57.

The collected ice thickness data have also made possible an analysis of the least ice thickness observed at the time of maximum thickness (Figure 6) and the greatest ice thickness observed at the time of maximum thickness (Figure 7). Some of the information used in Figures 6 and 7 was obtained from the Department of Transport,

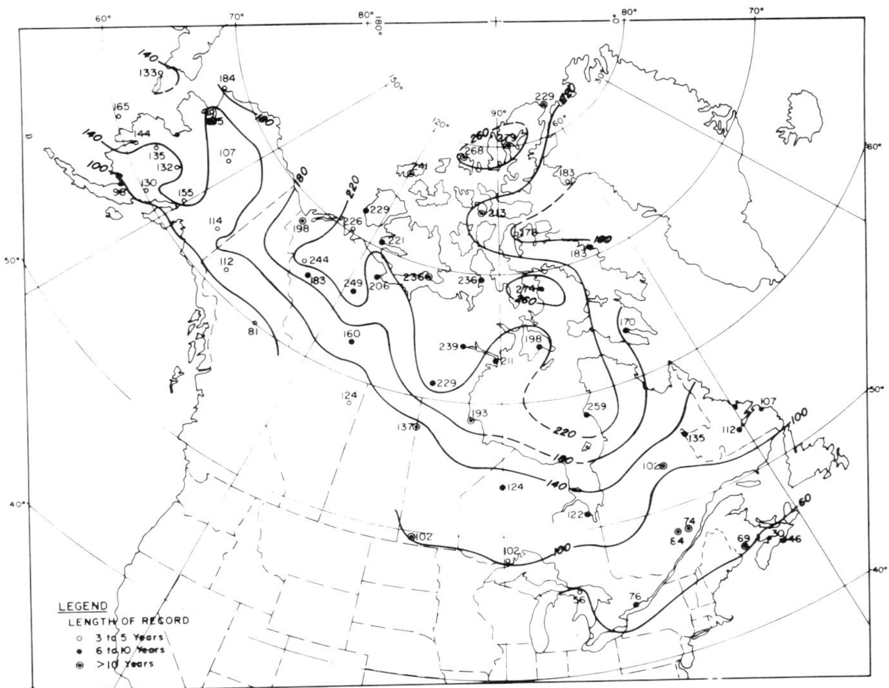

Figure 7. Greatest ice thickness (cm) observed at the time of maximum thickness for the years of record.

Canada (1959). A recent review of six years of ice data which had accumulated since the maps were drawn showed so few changes in the plotted values it was concluded that the maps would not have to be redrawn. Therefore, in Canada the least ice thickness observed at the time of maximum thickness ranges from less than 20 cm in the vicinity of Nova Scotia to nearly 200 cm in the north-central region (Figure 6). In Alaska, excluding the Aleutian chain and the southeast panhandle regions, the value ranged from near 60 cm in the southern part of the state to 170 cm on the west coast. From Figure 7, we find that the greatest ice thickness ranges from 30 cm near Nova Scotia to over 260 cm in parts of the Canadian Arctic.

In some instances, extreme variability in ice thickness, caused by ice rafting, is observed. Ice rafting in oceans is caused by wind, waves, tides and currents. On lakes and rivers thawing conditions, stream flow and changes in water level also cause the ice to break up and raft. For example, exceptionally high ice thickness values which indicate major rafting on some rivers were observed in Alaska during November and December 1956 (Bilello and Bates, 1966).

DISCUSSION

The main purpose of this report is to describe a network of stations in North America where observations on the properties of snow and the thickness of ice are being made. Results of studies using the data collected from these stations show that despite areal and yearly variations, definite patterns in the magnitudes of the values emerge when they are plotted on small scale maps.

Aerial infrared and passive microwave systems are being studied for distinguishing clouds from snow-covered terrain and for penetrating clouds to obtain snow and ice data. Nimbus II High Resolution Infrared Radiometer pictures, taken at midnight, show the Gulf Stream in the Atlantic Ocean by differentiating surface water temperatures (American Meteorological Society, 1966). A method for determining the temperatures of rivers by means of narrow-angle radiation-measuring devices installed in satellites has also been investigated (Dmitriev and Evnevich, 1966). Rough estimates of snow depth up to about six inches can be estimated in non-forested areas from satellite photographs (Barnes and Bowley, 1969).

Other methods currently in use for measuring snow density in remote areas are snow pillows (Beaumont, 1968), and nuclear devices such as that described by Smith *et al.* (1969). Such instrumentation could be equipped with telemetering devices to periodically send information to satellites, which in turn would relay it to a central station on command. The same procedures could be used to obtain ice thickness measurements in isolated regions. These approaches require considerable planning in the installation of equipment, maintenance and calibration of the sensors, longevity of the power pack and consistency in operation of the telemetry.

Development of advanced equipment to measure snow depth and ice thickness from aircraft or from space is currently under contract by private companies and several government agencies, such as NASA and NOAA. It is certain, therefore, that such information will be available from such sources in the future (Carter, 1967). When one of the systems proves promising, visual on-site observations as described in this paper will be useful to verify its reliability.

REFERENCES

American Meteorological Society, 1966. Cover photograph. Bulletin. 47(9).

Barnes, J.C., and C.J. Bowley, 1969. Operational snow mapping from satellite photography. Proc., Eastern Snow Conf., Portland, Maine, February 1969.

Barnes, J.C., and C.J. Bowley, 1970. The use of environmental satellite data for mapping annual snow-extent decrease in the western U.S. Final Rpt. for Dept. of Commerce by Allied Res. Assoc., Inc., Concord, Mass.

Beaumont, R., 1968. The present status of snow pillow instrumentation. Proc., Eastern Snow Conf., Boston, Mass., February 1968.

Bilello, M.A., 1961, 1964. Ice thickness observations, North American arctic and subarctic, 1958-1962. U.S. Army Cold Regions Research and Engineering Laboratory (USA CRREL) Special Report 43, Parts I (1961) and II (1964).

Bilello, M.A., 1966. Relationships between climate and regional variations in snow-cover density in North America. Intnt'l. Conf. on Low Temp. Sci., Hokkaido, Japan, 15 pp. Also USA CRREL Research Report 267, 1969.

Bilello, M.A. and R.E. Bates, 1966-1972. Ice thickness observations, North American arctic and subarctic, 1962-1970. USA CRREL Special Report 43, Parts III (1966), IV (1969), V (1971) and VI (1972).

Carter, L.J., 1967. Post-Apollo: NASA's plans get boost from LBJ and PSAC. Science. 155(3766):1084-1087.

Department of Transport, Canada, 1959. Maximum winter ice thicknesses in rivers and lakes in Canada. Canadian Meteorological Branch, CIR-3195, Ice 4.

Dmitriev, A.A. and T.V. Evnevich, 1966. A modeling approach to the problem of determining the temperature of rivers from satellites. Atmospheric and Oceanic Phys., 2(8):897-899.

Fritz, S., 1963. Snow surveys from satellite pictures. In Rocket and Satellite Meteorology, pp. 419-421. Amsterdam: North Holland Publishing Co.

Gold, L.W. and G.P. Williams, 1957. Some results of the snow survey of Canada. Div. of Bldg. Res., Nat'l. Res. Council, Canada, Res. Paper No. 38, 15 pp.

McLain, E.P., 1973. Quantitative use of satellite vidicon data for delimiting sea ice conditions. Arctic. 26(1).

Smith, J.L., H. Halverson and R. Jones, 1969. The profiling snow gage and its utility in water supply forecasting. Amer. Nuclear Soc. Trans. 12(2):500.

USA CRREL, 1962. Instructions for making and recording snow observations. Instruction Manual 1, 10 pp.

USA SIPRE, 1957. A survey of arctic snow-cover properties as related to climatic conditions. U.S. Army Snow, Ice and Permafrost Research Establishment (USA SIPRE) Research Report 39, 9 pp.

Wark, D.Q., R.W. Popham, W.A. Dotson and K.S. Colaw, 1962. Ice observations by the TIROS II satellite and by aircraft. Arctic. 15(1):9-26.

A PRELIMINARY WATER BALANCE EVALUATION OF AN INTENSIVE SNOW SURVEY IN A MOUNTAINOUS WATERSHED

D. Storr
Atmospheric Environment Service, Calgary, Alberta.

D. L. Golding
Canadian Forestry Service, Edmonton, Alberta.

ABSTRACT

Using water equivalent values from an intensive snow survey and subsequent precipitation as input, and measured streamflow, calculated evaporation and calculated change in storage as output, a water balance from the time of the survey to the end of the snowmelt period has been attempted for each of the five years 1969-73 in Marmot Creek Experimental Watershed. The net errors were less than 10% of the input values, but a bias in one or more of the components is suspected.

INTRODUCTION

The title of this paper is somewhat misleading in that a water balance evaluates the accuracy of all its components, but this is the first time that snow survey data have been incorporated in the input side of the balance at Marmot Creek. A simple balance was used:

Snowpack + Subsequent Precipitation = Run-off + Evapotranspiration ± Change in Storage ± Net Error (1)

A water balance for an appropriate area and time interval is not only the most reliable means of evaluating the accuracy of the components, but also frequently points out areas of need for further study. Annual balances, using measured precipitation and streamflow, and calculated evapotranspiration and change in storage have been proved possible and valuable at the Marmot Creek Experimental Watershed (Storr, 1973a). This study reports the results of water balances for the snowmelt period for five years, 1969-73. The results are not only valuable as part of the calibration of the basin in its natural state, but will be very useful in explaining the effects of the planned logging operations.

SITE DESCRIPTION

Marmot Creek is a small ($9.4 km^2$) experimental watershed at latitude 50°57' N, longitude 115°10' W, on the west side of the Kananaskis Valley, just east of the continental divide in Alberta. In this rugged terrain, elevations range from 1585 to 2805 metres, with an average slope of 39% causing numerous access problems. The general aspect is easterly.

The basin is relatively water-tight (Stevenson, 1967) with little or no sub-surface inflow or outflow. As reported by Jeffrey (1965) from the work of Green and Jones (1961), the relatively impermeable bedrock is exposed in the headwaters area, but almost all the remainder of the basin is covered by a deep layer of coarse and permeable surficial materials which allows rainfall to infiltrate rapidly. There is very little surface flow into the channels and the effects of rainfall on streamflow are smoothed out by its transition through sub-surface storage (Davis, 1964). The soil mantle is very shallow and confined mainly to the area below treeline.

The lower reaches of the basin are covered with a dense stand of lodgepole pine, then mature spruce up to 30m tall extend to treeline at 2135 to 2285m. In the alpine zone, shrubs and grasses give way to talus slopes and bare rock. A complete inventory of the vegetation has been given by Kirby and Ogilvie (1969).

The instrumentation network is shown in Figure 1.

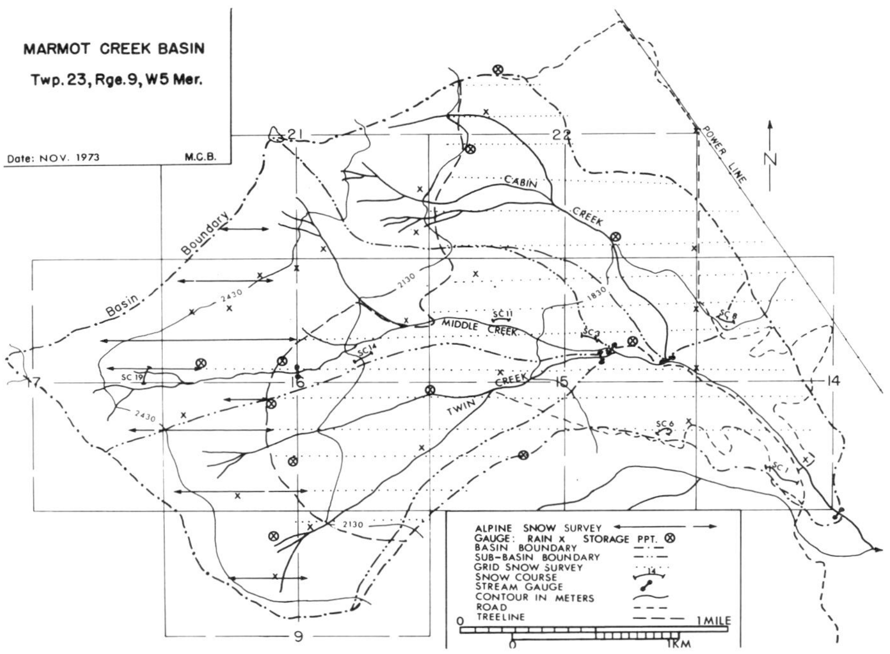

Figure 1. Instrumentation and snow-sampling points on Marmot Creek basin.

COMPONENT ANALYSIS

The period covered by this balance extends from the date of the survey in March to July 31 when in most years all the snowpack has melted. For a valid balance, each of the components should be measured or calculated independently, with none of them estimated by remaindering. The following sections describing the method and problems in computing each component in the Marmot Creek snow-melt period balance are therefore necessary for the evaluation of the accuracy of the results.

1. Snowpack Measurement

Various snow-sampling schemes have been used on Marmot basin since its designation as an experimental basin in 1962. The first measurements were of snow depth read by telescope from graduated stakes established in the alpine area. A test of the photogrammetric technique was carried out in 1964. Neither method provides a measure of snow density, a serious drawback especially in alpine areas where density variations are extreme due to differential exposure to wind and radiation. In 1962, two 10-point snow courses were established, the number increasing to 20 by 1967. Based on an analysis of the correlation between snow courses, the number was reduced to eight in 1969. Although providing a good index of snow accumulation on the basin, the snow courses give little insight into patterns of accumulation across the basin.

One of the sub-basins is scheduled for logging in clear-cut blocks in 1974 to determine the effect on streamflow regime and sediment production. The changes in streamflow regime will be, to a large degree, the result of changed patterns of snow accumulation and rates of snowmelt. To provide information on pre-treatment accumulation patterns, an intensive snow-sampling program has been conducted since 1969. The sampling is carried out in March on that part of the basin below timberline. Snow depth and snowpack water equivalent (s.w.e.) are measured at 1500 points on a 20.1 m (east-west) x 201.2 m (north-south) grid between 1,700 m and 2,300 m elevation (Figure 1). For 250 points (i.e., every fifth point, east-west) topographic information, including elevation (to nearest 30 m), was available. The grid-sampling provides complete coverage of the snowpack in the forested part of the basin but no data on alpine snowpack.

Snow accumulation is a function of many physiographic and meteorologic factors including elevation, slope, aspect, vegetative cover, snowfall amount, wind speed and direction, temperature, and exposure to radiation. To extrapolate from the known patterns of snowpack in the forest to the unknown patterns of the alpine zone proved very difficult, because little is known relating the above factors to snowpack. Because there is a direct relationship between several of the meteorologic factors and elevation, it was decided to study only the relationship between snowpack and elevation, use this to estimate the alpine snowpack, and accept the inherent errors from ignoring the other factors.

Two sets of data were available, the first based on the 250 grid-sampling points within the forest for 1969-73, and the second based on five 10-point snow courses located from 1,900 m to 2,200 m elevation on a central ridge having similar slope, aspect, and forest cover. Snow-course data were available for 1965-1968. In 1969 all but one of these courses were discontinued. The mean elevation of each snow course was ascribed to each of the 10 points in that course. The relation of s.w.e. to elevation was linear in both cases with no improvement using a higher degree polynomial. For the grid data, correlations were low, with coefficients from 0.25 in 1969 to 0.68 in 1972. This relationship varies from year to year (Golding, 1969) mainly because it is a function of total snowpack. The rate of increase in s.w.e. with increasing elevation (i.e., the slope, b, of the regression) increases with greater snowpacks (Figure 2).

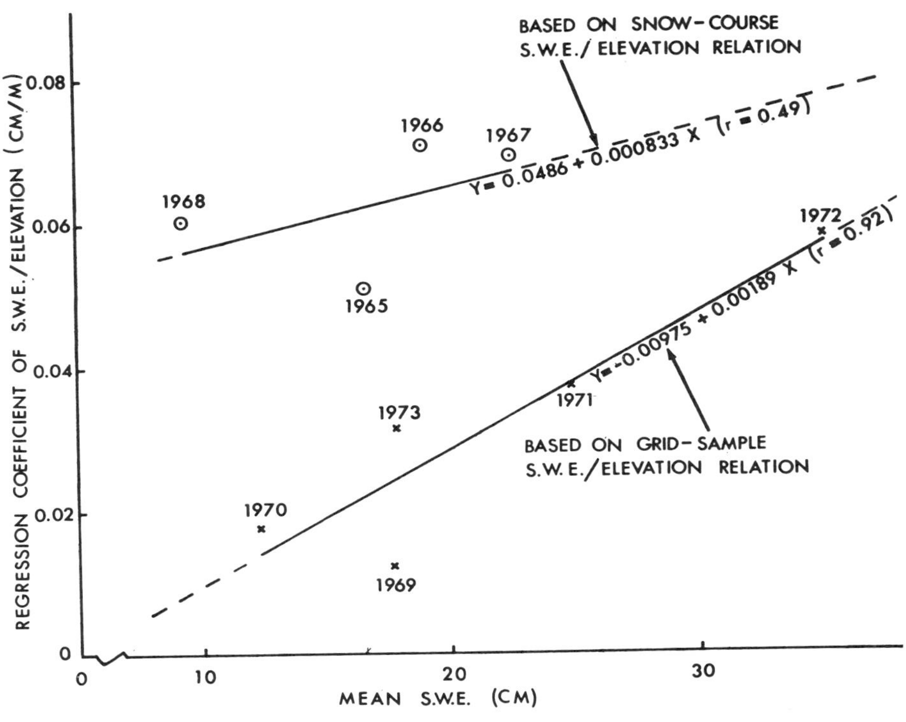

Figure 2. Relation of regression coefficient of s.w.e./elevation to mean snowpack in forest zone.

For the five snow courses, the correlation of s.w.e. with elevation was much higher than for the grid data. For the March snowpack, correlation coefficients were greater than 0.98 for each of the four years, 1965-1968, for which data are available. The relationship was also calculated for monthly measurements and showed that both intercept and slope of the regression increased during the season (Golding, 1968), because (1) individual storms exhibit the same general tendency of greater snowfall with greater elevation and (2) as spring progresses snow continues to accumulate at higher elevations while melting occurs at lower levels. This implies that a season-dependent (rather than time-dependent) relationship should be developed for slope and intercept. The snow course data could then be adjusted to the grid sampling date taking account of climatic variations in the intervening period of approximately 10 days. This refinement may be worth further study.

Comparing the two s.w.e. vs elevation relationships, the one based on the grid sampling represents a wider elevational range which extends slightly higher than that derived from the snow course data. The latter has fewer points than the former, but a higher mean elevation and a higher and more consistent correlation (Figure 3) with elevation. For both sets of data, the increase in s.w.e./m

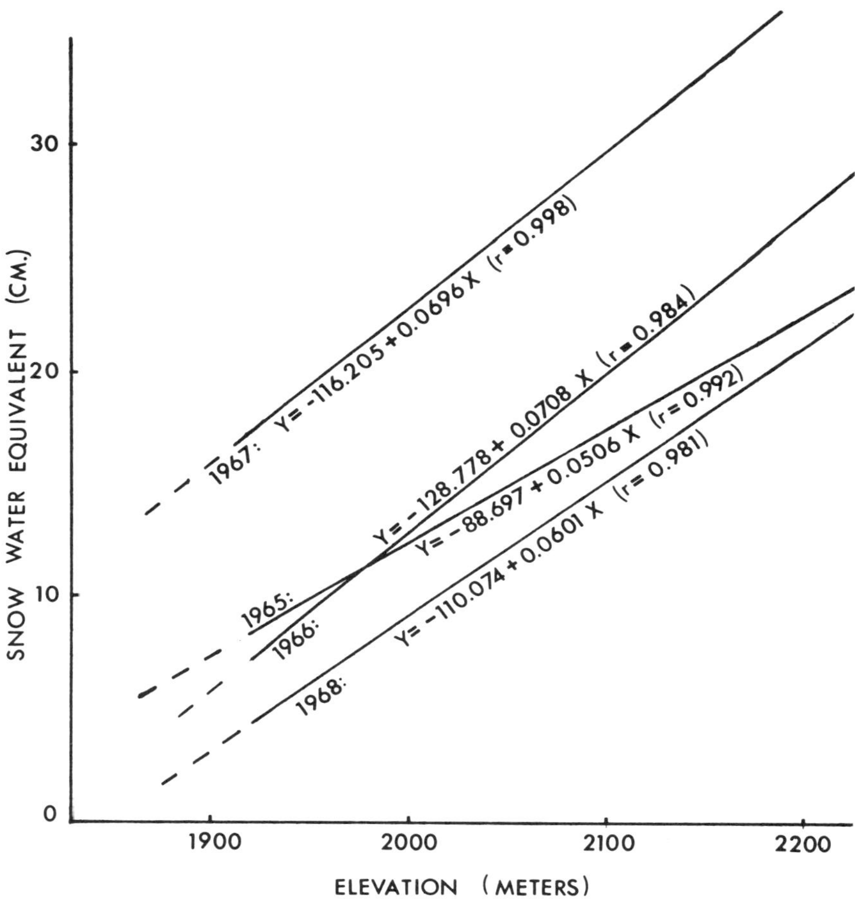

Figure 3. Change in snowpack with elevation, from snow-course data.

increase in elevation/cm of mean s.w.e. was applied to the mean s.w.e. of the forested zone and an estimate of the alpine and basin snowpacks obtained from the topographic map and the proportionate areas of the forested and alpine zones (65 and 35% respectively). It is noted (Table 2) that the basin average s.w.e. derived from the snow course relationship with elevation (Snowpack A) is 17% greater than that derived from the grid sampling data (Snowpack B).

As mentioned earlier, both methods estimate the alpine pack solely on the s.w.e. vs elevation relationship. It is suspected that the relative importance of the other controlling factors increases greatly in the alpine zone.

A partial sampling of the alpine area was carried out in 1973, on a 40 x 400 m grid where access was possible (Figure 1). The alpine measurements averaged 196 mm s.w.e. compared to 277 mm and 404 mm estimated with grid and snow-course data respectively. The measurements confirmed a high variability in s.w.e. in the alpine zone. On one 600-meter line of 16 points, values ranged from 38 mm to 724 mm with mean and standard deviation of 272 mm and 196 mm respectively. From visual observations it is suspected that some very high accumulation areas were missed by the sampling, resulting in a low estimate of alpine pack. A more complete sampling program is planned for the alpine zone in 1974.

2. Subsequent Precipitation

The measurement of subsequent precipitation may be divided into the measurement of subsequent snowfall and the measurement of subsequent rainfall. Two networks of precipitation gauges (Figure 1) are used at Marmot Creek (Ferguson and Storr, 1969).

Before 1971, one network of 10 Sacramento storage gauges and one recording gauge measured seasonal and annual precipitation; in 1971, one of the upper storage gauges was replaced by another recording gauge. These gauges are mounted on stands with the orifice about 2.5 m above ground and unlikely to be overtopped by the snowpack. They are equipped with Alter shields and most are located in small clearings in the forest to obtain additional protection from the wind. Fortunately, wind speeds during precipitation in the forested zone are low (Storr, 1973b) so any undercatch should be small.

Another network of 33 Atmospheric Environment Service (AES) standard raingauges measures rainfall in the snowfree season. This network is the densest in Canada outside of urban areas and one of the densest in North America. The AES rain gauge is quite small, with a capacity of 115 mm and an orifice of 64.5 cm^2 standing 30 cm above ground. It therefore causes little obstruction to the wind, and even in the relatively windy alpine zone the undercatch is thought to be less than 5%. It is, however, of very limited use for measuring snowfall. In nine years of use at Marmot Creek, its capacity has been sufficient for weekly readings on all except two occasions. Unfortunately these occurred in 1969 and 1970, so affect the data for this study. For these events, the precipitation was estimated (with a consequent decrease in accuracy) from the storage gauge network and data from a few experimental gauges being tested in the basin.

Unfortunately, it has not been possible to read all the storage gauges at the time of the snow surveys, so it has been necessary in this study to estimate a basin average of precipitation in the period after the survey and before mid-June when the rain gauge network is activated. From 1969-70, this was done by correlating basin averages (using Thiessen polygons) with data from the lowest recording gauge at Con 5;

$$\text{Basin snowfall (mm)} = 5.8 + 1.669 \text{ Con 5}; \quad r = 0.96 \qquad (2)$$

$$\text{Basin rainfall (mm)} = 0.5 + 1.17 \text{ Con 5}; \quad r = 0.96 \qquad (3)$$

changes (ΔW) have been related (Storr, 1973a) to changes in streamflow and mean streamflow. If there is no direct surface flow into a stream, the amount of flow (N) is a direct function of the amount of water in the sub-surface reservoir, so changes in storage are a direct function of changes in streamflow. It was also found that because the sub-surface reservoir is not a rectilinear box, they are also a function of streamflow itself. The relationship between ΔW, ΔN, and $\overline{N}$ (the mean of the daily discharges at the beginning and end of the period) at Marmot Creek is given by:

$\Delta W = 2154.4\ \Delta N e^{-8.11\overline{N}}$, $r^2 = .939$, where N and $\overline{N}$ are in m^3/s and ΔW is in mm.

It was developed from 68 recession periods over nine years when firm data on streamflow and rainfall were available, snowmelt was finished, and evaporation could be calculated from net radiation with the least chance of error. ΔW was then obtained as a residual for each period.

The relationship is not valid for periods beginning or ending with direct surface flow into the channels, but this is extremely unlikely for the periods considered in this study. The annual water balances for Marmot Creek which were mentioned earlier are shown in Table 1 to illustrate the accuracy possible on this time scale.

TABLE 1

ANNUAL WATER BALANCES AT MARMOT CREEK (MILLIMETERS)

Water Year	Precipitation	Streamflow	Evapotrans.	ΔW	Net Error
15/10/63-30/09/64	1039	488	438	101	12
30/09/64-02/10/65	1027	510	498	29	-10
02/10/65-03/10/66	834	495	436	-117	20
03/10/66-03/10/67	871	442	487	- 15	-43
03/10/67-10/09/68	855	353	407	86	9
10/09/68-09/09/69	857	523	389	- 54	- 1
09/09/69-08/09/70	736	333	418	- 5	-10
08/09/70-07/09/71	909	445	453	3	8
07/09/71-06/09/72	972	487	435	36	14

The net errors are quite small, average less than 2% of the precipitation input and are almost equally divided between positive and negative. This suggests that either there is no obvious bias in any of the components, or that there are cancelling biases in two or more.

RESULTS AND DISCUSSION

The water balances for the five snowmelt seasons 1969-73 are presented in Table 2. At time of writing, the 1973 data are incomplete so discussion is limited to the first four years.

Although the net errors are smaller than we have any right to expect from such a complex balance in such a non-homogeneous area, and although it is dangerous to generalize from such a small sample, it is of interest to speculate on the possible sources of the errors. The consistent sign of the error (using either estimate of the snowpack) suggests a bias in one or more of the components.

The omission from the water balance of condensation on the snowpack after the survey and on vegetation after snowmelt (as noted by Fritschen and Doraiswamy, 1973) is a contributing factor to the

For the 1971-72 seasons, a multiple correlation with the two recording gauges was used:

$$\text{Basin precipitation(mm)} = -41.9+0.633 \text{ Twin 1} + 0.52 \text{ Con 5}; \quad r = .992 \qquad (4)$$

Whether the precipitation occurs as rain or snow makes a marked difference in the accuracy of the correlation estimates on two counts. First, the basin average for rainfall events is based on the 33 gauge network whereas the snowfall relationship is based on the 11 gauge network, and errors in one or more gauges will have a much larger effect. Secondly, the rainfall relationship is based on 87 weekly readings whereas the snowfall relationship is based on only 10 seasonal readings. Although the correlation coefficients for the two relationships are equal, the standard error of estimate for snowfall (59.7 mm) is almost 30 times as great as that for rainfall (2 mm). These two differences compound the well-known inaccuracies in gauge measurement of snowfall.

Surface temperature is only an approximate method of determining whether the precipitation falls as rain or snow (U.S. Corps of Engineers, 1956) and is made even less reliable by areal and temporal variations in temperature and by imprecise timing of both temperature and precipitation. This reduces confidence in the correlation method of estimating precipitation at Marmot Creek, but it is the only method which has been devised to date, so has been used in spite of its limitations. The problem is further compounded for events of mixed rain and snow, or of rain at Con 5 and snow at higher elevations for which no relationship can be developed.

3. Streamflow

Streamflow is the most accurately measured component of the water balance at Marmot Creek. The main stem stream gauge is a concrete, 120-degrees, sharp-crested, V-notch weir of 4.53 m^3/sec capacity, sealed to the impermeable Rocky Mountain formation which crosses the lower end of the basin. It has been rated both volumetrically and by current meter, and conforms closely to the theoretical rating. A heat lamp keeps it ice-free in winter.

The total discharge in the period from the date of the snow survey to July 31 each year was divided by the basin area to provide the basin average run-off.

4. Evapotranspiration

Evapotranspiration estimates were obtained using an energy budget technique (Storr, 1973c). In brief, this involves adjusting the measured point net radiation for the average slope, aspect and vegetative cover to obtain basin averages of the available energy each month (Storr, 1972); budgeting appropriate amounts for photosynthesis (Denmead, 1964), the flux into or out of the soil and forest (Baumgartner, 1956; Rauner, 1958), and the energy used for the melting of the snowpack; and separating the remaining fluxes of latent and sensible heat by the use of the Bowen ratio.

This technique while apparently very complex, is relatively straightforward, and produces results which fit the annual water balance very well.

5. Storage Changes

Because there are no lakes or glaciers in the basin, all the storage changes occur below the surface. Sub-surface storage error, but no estimates of the magnitude of this component have been made.

TABLE 2

SNOWMELT SEASON WATER BALANCES AT MARMOT CREEK (MILLIMETERS)

Year	Snowpack A	Snowpack B	Later precip.	Total Input A	Total Input B	Stream-flow	Evapo-trans.	Storage Change	Total Output	Net Error A	Net Error B
1969	254	211	562	816	773	416	249	144	809	7	-36
1970	179	148	495	674	643	270	277	118	665	9	-22
1971	361	299	477	838	776	382	294	147	823	15	-47
1972	508	419	369	877	788	410	278	167	855	22	-67
1973	259	214	478	737	692	322	282	130	734	3	-42

Simple correlations between the net errors and the components used give the following coefficients:

	Error using Snowpack A	Snowpack B
Snowpack	.943	.996
Subsequent precipitation	-.957	-.781
Storage change	.778	.968
Evapotranspiration	.563	.270
Streamflow	.369	.719

The snowpack data appear to bear the major responsibility for the net errors, with subsequent precipitation and storage change next in line. These three are given further consideration.

Most errors in snow surveying result in an over-estimate of the snowpack, supporting estimate A which appears about 4% too high. Estimate B appears to be about 15% too low. The very dense network of sampling points and the low variability in the survey in the forested zone leave very little room for sampling error, forcing attention on the estimates of the alpine snowpack. Dr. W.T. Dickinson, University of Guelph (personal communication), using an estimate of the alpine snowpack similar to "A" in a computerized and stratified run-off model of the basin, reports that the observed run-off would require a greater alpine snowpack. However, both estimates "A" and "B" are based on extending the snowpack-elevation relationship found in the forested area into the alpine zone. No measurements are available to justify either approach. Many other investigators have found a relationship between snowfall and snowpack and elevation, but it appears from the 1973 alpine measurements that the scatter in the relationship at Marmot Creek increases several fold above treeline. Because of the very great variability of the alpine pack, and the difficulty in obtaining an unbiased sample, it is questionable whether an intensive grid survey of the alpine snowpack will produce results commensurate with the costs. Some system of remote sensing to obtain an areal average in the alpine zone appears much more promising.

The possible sources of error in the precipitation data have already been discussed. To summarize them, the data on subsequent precipitation is comprised of two parts; rainfall measurements from the 33 gauge network where there is minimal chance of error, and early season precipitation estimates by correlation techniques which are inherently less reliable, and are made more so by the problem of differentiating rain and snow. The errors in the precipitation data could be of the same order of magnitude as the snowpack errors.

It is difficult to analyze the reason for the high correlation between error and change in storage. Ice formation in the stream channel in March may cause an over-estimate of ΔW by reducing streamflow at that time, thus creating an artificially large ΔN and low $\overline{N}$. Unfortunately, no measurements are taken of ice formation so we do not know if this factor is significant or not.

REFERENCES

Baumgartner, A., 1956. Untersuchungen uber den Warme und Wasserhaushalt eines jungen Waldes. Ber. Deut. Wetterdienstes 28(5).

Davis, D.A., 1964. Collection and compilation of Marmot Creek streamflow data in Marmot Creek Basin. Proc. Hydrol. Symp. No. 4, Nat. Res. Counc. of Can., Ottawa, p. 105-116.

Denmead, O.T., 1964. Evaporation sources and apparent diffusivities in a forest canopy. J. Appl. Meteorol. 3(4):383-389.

Ferguson, H.L., and D. Storr, 1969. Some current studies of local precipitation variability over Western Canada. Proc. Amer. Water Resources Assoc. Symp. on Water Balance in N. Amer., Banff, Canada.

Fritschen, L.J., and P. Doraiswamy, 1973. Dew: an addition to the hydrologic balance of Douglas fir. Water Resources Res. 9 (4):891-894.

Golding, D.L., 1968. Snow measurement on Marmot Creek experimental watershed. Can. Dept. Forest. & Rural Develop. Inform. Rep. A-X-18, Calgary, Alta. 16 pp.

Golding, D.L., 1969. Snow relationships on Marmot Creek experimental watershed. Can. Dept. Fish. & Forest. Bi-Monthly Res. Notes 25(2):12-13.

Green, R., and R.F. Jones, 1961. Geology and groundwater study of Marmot Creek basin. Alta. Res. Counc.,Edmonton. Typed report.

Jeffrey, W.W., 1965. Experimental watersheds in the Rocky Mountains, Alberta, Canada. IASH Publ. 66, Symp. of Budapest.

Kirby, C.L., and R.T. Ogilvie, 1969. The forests of Marmot Creek watershed research basin. Can. Forest. Serv. Publ. 1259, Can. Dept. Fish. & Forest., Ottawa.

Rauner, Yu. L., 1958. Some results of heat budget measurements in a deciduous forest. Izu. Acad. Sci., USSR, Georg. Ser. 5:79-86 (Engl. Transl.).

Stevenson, D.R., 1967. Geological and groundwater investigation in Marmot Creek experimental basin of Southwestern Alberta, Canada. M. Sc. Thesis, Univ. of Alta., Edmonton. 106 pp.

Storr, D., 1972. Estimating effective net radiation for a mountainous watershed. Bound.-Layer Meteorol. 3(1):3-14.

Storr, D., 1973a. A hydrometeorological method for computing changes in subsurface water storage at Marmot Creek (in review).

Storr, D., 1973b. Some aspects of wind-snow relationships at Marmot Creek, Alberta, for the design of snow-trap forest clearings. Can. J. Forest Res. (in press).

Storr, D., 1973c. Preliminary monthly and annual estimates of evapotranspiration at Marmot Creek, Alberta, by the energy-budget method (in review).

U.S. Corps of Engineers, 1956. Snow hydrology. U.S. Army, Portland, Oregon.

A DATA COLLECTION AND REDUCTION SYSTEM FOR SNOW ACCUMULATION STUDIES

G.J. Young
Glaciology Division, Inland Waters Directorate
Department of the Environment
Ottawa, Ontario

ABSTRACT

The paper describes a method to provide maps of snow accumulation from irregularly spaced point sampling locations. In generating the distribution maps, use is made of associations between snow depth and terrain characteristics. A grid square technique is employed to describe terrain characteristics within the area (altitudes at grid line intersections are used to generate measures of surface slope, azimuth and local relief); these characteristics are then used to stratify the area and select field sampling locations. Relationships are obtained between terrain characteristics and snow depth values at sampling locations and are applied to each grid point in turn to produce an areal summary of snowpack depth. The grid square technique is particularly powerful for computer operations making summation and mapping procedures fast and simple. While the method has been developed and tested on I.H.D. glacier basins in Western Canada, it is sufficiently flexible that it can be applied with little modification to any area. A standardized data bank structure and the high speed of data reduction make the method suitable for real time melt prediction studies or as an independent check on other snow accumulation models.

INTRODUCTION

Knowledge of the spatial distribution of the snowpack within a basin immediately prior to spring melt is essential for assessing the amount and timing of spring and summer river discharges. While the necessity for such knowledge has long been recognized no satisfactory standard solution has yet been devised for fast and accurate production of snow maps covering extreme cases. However, in some areas of relatively simple terrain, where patterns of accumulation remain reasonably constant from one year to another, satisfactory solutions have been obtained. Understanding the reasons for particular spatial distributions of the snowpack has proved elusive because the aerodynamic processes controlling snowpack accumulation are so complex.

The purpose of this paper is to outline a system which is currently being used to describe the snowpack distribution within five glacier basins in Western Canada. Given only a small number of depth and density samples obtained manually from the glacier surface the method makes use of relationships between snowpack characteristics and glacier surface geometry to produce maps and tables of predicted snowpack distributions. The considerable advantages in speed and flexibility of data handling of the new method over the standard method are discussed later.

3.5

While the eventual aim of accumulation studies should be a complete understanding of the aerodynamics of blown snow and hence the way in which the total glacier snowpack is built up, the aim of the method described here is to produce reasonably accurate snow distribution maps quickly and efficiently. The ability to describe accurately snowpack distribution will be useful in investigations designed to elucidate the physical processes involved. Further, the principles of the method described here may be applied to nonglacier basins and thus provide reliable input that is potentially very useful to melt simulation models.

BACKGROUND TO THE PRESENT STUDIES OF GLACIERS

Figure 1 shows the geographical locations of the five International Hydrological Decade (I.H.D.) glacier basins in Western Canada studied by the Glaciology Division. The basins represent a west-east transect from the maritime conditions of the western side of the Coast Range to the continental conditions on the eastern side of the Rocky Mountains. There is considerable variety within the group of five glaciers as shown in Table 1. The variety includes: (a) elevation above sea level; (b) mean surface slope; (c) size, although no really large glacier is included; and (d) regime. Compared with the eastern glaciers the west coast glaciers are more dynamic, with larger winter accumulations and larger summer melts. Any satisfactory standardized method has to be able to take these considerable variations into account.

The total hydrology and energy balances of each of these glaciers has been studied according to I.H.D. recommendations and standards since 1965. The studies are scheduled to continue in the present manner until the end of the decade and similar studies will probably continue thereafter. The snow accumulation studies have, therefore, constituted only a part, albeit an important part, of the overall studies. Details of the mass balance studies on these glaciers have been given by Østrem, (1966), Stanley, (1970, 1971), Young, (1970, 1971, 1973), and Mokievsky-Zubok, [1971, 1973(a), 1973(b)].

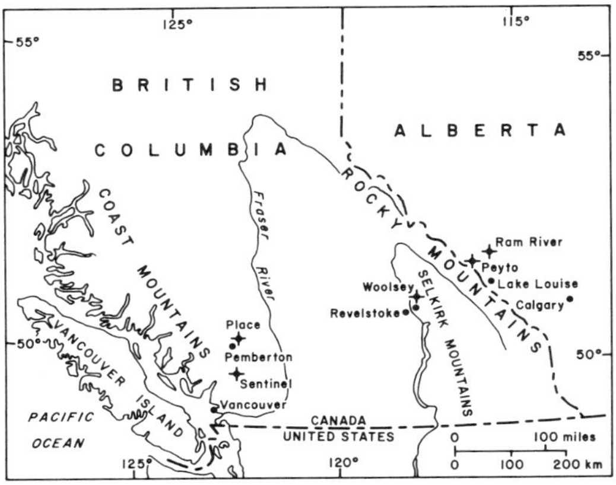

Figure 1. Map of glacier locations.

Table 1. Comparison of glaciers.

	Sentinel	Place	Woolsey	Peyto	Ram
Area of glacier (km^2)	2.03	4.01	3.92	13.40	1.80
Highest elevation (meters a.s.l.)	2100	2543	2640	3185	2990
Lowest elevation (meters a.s.l.)	1548	1828	1930	2125	2565
Mean slope of surface (degrees)	15.7	10.5	17.5	12.9	13.5
Mean azimuth of surface (degrees)	333	5	22	33	10
Mean net winter accumulation (m^3 x 10^6 w.e.)	6.84**	7.90**	9.85*	19.71***	1.67*
Mean specific accumulation (meters of w.e.)	3.37**	1.97**	2.52*	1.47***	0.93*
Mean net summer loss (m^3 x 10^6 w.e.)	6.92**	10.09**	11.97*	28.11***	2.61*
Mean specific net loss (meters of w.e.)	3.41**	2.52**	3.05*	2.30***	1.45*

* Mean 1966-70 ** Mean 1966-71 *** Mean 1966-72

THE STANDARD METHOD OF COMPILING ACCUMULATION MAPS

Tangborn et. al. (1971) have reviewed the various ways in which mass balance calculations can be made. The standard procedure which has been adopted for the Canadian I.H.D. glaciers is one of direct measurement and has been described in detail in the field manual by Østrem and Stanley, (1969). As illustrated in Figure 2, it consists of establishing, on the glacier surface, a network of fixed markers at which depth soundings of the snowpack are made. Soundings are also made on profiles between and around markers to give a high density and fairly uniform coverage over the glacier. A large number of snow depth measurements have been considered necessary to account for the considerable variability in depth values. Bulk density of the pack is measured at only a few points on the glacier for density remains fairly uniform over an entire glacier especially before the onset of melt. Density is usually measured by laboriously digging pits and taking samples from the exposed pit walls. The procedure may become unnecessary in the future when portable radioisotope snow gauges are more readily available. Density measurements are assumed to be valid for large areas surrounding the pit and the density values are applied to the snow-depth values to obtain water equivalent values at all sounding locations. Isolines of net winter accumulation are then interpolated by eye between sounding locations and, from the map so produced, total quantities of water added are calculated for the whole glacier and for altitude zones on the glacier.

This standard system is subject to three main problems which vary in their relative importance from one glacier to another: (1) the difficulties in making depth and density measurements at specific locations; (2) the sampling problem of how many measurements to make and exactly where to make them; (3) the difficulties associated with interpolation from the measuring sites to unvisited parts of the glaciers.

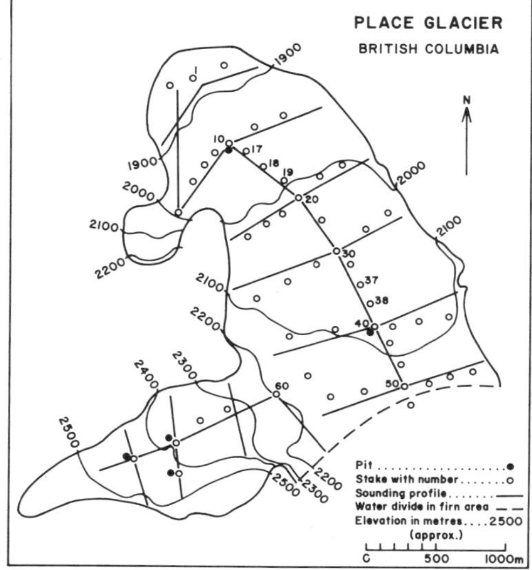

Figure 2. An example of a standard sampling network.

RESULTS OBTAINED FROM PAST STUDIES

Apart from the important results showing when the glaciers have experienced net gains or losses in mass over the years and how they have contributed to river discharge downstream of the glacier snouts, the studies have shown important relationships between accumulation patterns and terrain geometry. One of the most important results has been the confirmation of the often expressed view (Meiman, 1968) that while absolute amounts of accumulation on any one glacier may vary widely from year to year, the overall patterns at the end of winter are fairly constant.

A detailed study conducted by the author on Peyto Glacier has, in general, confirmed this conclusion, but modified it in detail. The study on Peyto Glacier has shown that while the strong association between snow depth and altitude can always be closely approximated by a regression line, the slope and position of the line are not constant from year to year and are closely associated with the local configuration of the glacier geometry. The snow depth at any one point can be determined by an equation of the form:

$$sd = a + bZ + cR + dS \qquad (1)$$

where sd is the snow depth, a is the intercept value, b, c, and d are regression coefficients, Z is the altitude of the point, R is the local relief and S is the surface slope angle. For any one glacier the coefficients c and d tend to vary little while the values of a and b vary considerably. The conclusions can be summarized as:

(1) total accumulations vary from year to year;

(2) there is always a strong relationship between snow depth and altitude but this is not the same relationship from one year to another;

(3) local patterns are fairly constant from year to year and are associated with the configuration of the glacier surface.

Furthermore, it can be demonstrated (Figure 4) that the total accumulation for an entire glacier calculated by the standard method can be closely approximated by an equation based on altitude alone. Equation (1), above, can be applied to each glacier for each year. Mean values of the coefficients c and d can then be calculated for each glacier. These mean coefficients describe persistence in the local patterns of snow distribution for the glacier. These important results provide the framework around which the new method, outlined below, is built.

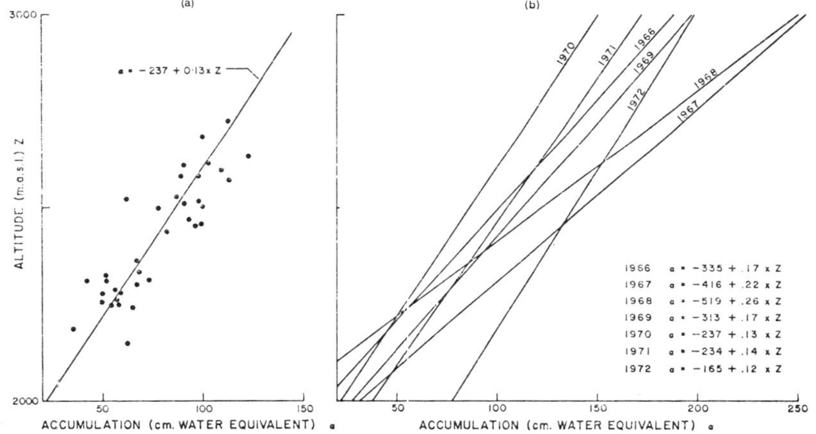

Figure 3. Peyto Glacier. Altitude - accumulation relationships
(a) 1970 Net winter balance
(b) Net winter balances 1966-72.

THE NEW METHOD OF COLLECTING AND REDUCING ACCUMULATION DATA

The method outlined here, like the standard method, utilizes depth and density measurements made on the glacier at a network of locations. However, the assumptions that there will always be a strong linear relationship between altitude and the amount of snow, that patterns will be constant from year to year, and that patterns will be dependent on surface shape of the glacier are used to rationalize the sampling network and to interpolate between sampling locations when producing the accumulation map. The outline of the method is:

1. The shape of the surface is described on a regular array of points on the glacier surface.

2. From the many shape measures, that are thus generated, a factor analysis procedure selects a small number of measures which together adequately characterize the whole assemblage of variables.
3. A small number of locations are selected from the many locations on the glacier surface for which shape measures have been generated so that a wide range in altitude is covered and that any particular sampling location is representative of the altitude zone within which it lies.
4. Snow depth and density are sampled at this small set of locations.
5. The relationship of density to altitude is established by means of a least squares linear regression model. The equation is applied to the depth soundings made at various altitudes to obtain water equivalents at sounding locations.
6. The relationship of water equivalent (w.e.) to altitude is established by a least squares linear regression.
7. The equation is applied to all points on the regular array of locations over the whole glacier to obtain a simple map of w.e.
8. The w.e. values so produced are weighted in accordance with local surface shape measures to simulate local patterns of variability.
9. The results are mapped and tabulated.

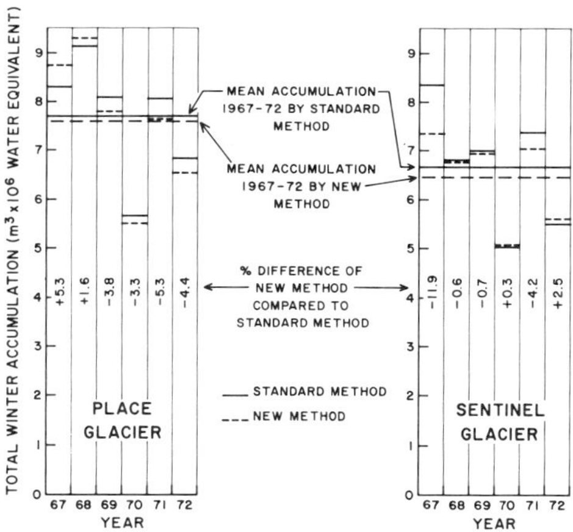

Figure 4. A comparison between accumulation totals calculated by the standard method and the new method using altitude as the only predictor.

THE DESCRIPTION OF GLACIER SURFACE SHAPE

A program has been written by the author, (Young, 1973) to describe the shape of the surface at a regular array of points within a basin. A square grid (Universal Transverse Mercator in this case) is laid over the basin, with a grid interval of 100 m and altitudes at grid intersections are read. These altitudes are used in the generation of surface shape measures. A large number of surface shape parameters are generated for each point on the grid including surface slope angle, slope azimuth, and relief, each measured for facets of different sizes centered on the point in question. As these parameters are highly intercorrelated, a factor analysis is used to reduce the number of dimensions in the system of variables. As a result, it is found that slope angle and relief measured on a local scale (i.e. for facets of the surface within 100 m of the grid points) and altitude are highly correlated with the three most important factors in the factor matrix. It was, therefore, decided to work with these three variables in the subsequent analyses. Slope angle and local relief are calculated for each grid intersection by fitting a regression plane to the altitude of the point itself plus the altitudes of the four nearest neighbour points (Figure 5). The angle of maximum slope of the plane is given to the grid point as its slope angle and the extent to which the grid point is above or below the plane is taken as the measure of local relief.

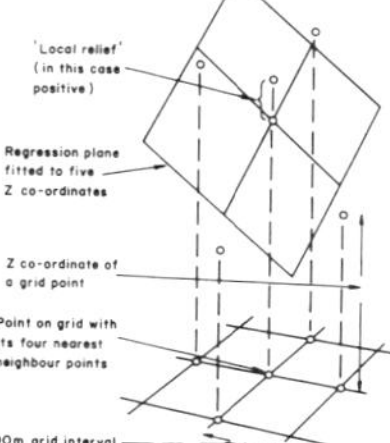

Figure 5. Diagram showing geometrical relations used to derive "local relief".

SELECTION OF SAMPLING LOCATIONS

Using the procedure described above, an altitude, a slope angle, and a local relief index are given to each point on the regular grid. The question is then asked whether a given point is representative or not in the terms of slope angle and local relief for the altitude zone in which it lies.

To this end the following procedure is applied to each point on the grid: within a 40m altitude interval, centered on the altitude of the point in question, the mean and standard deviations of slope angle and local relief are calculated. The point in question is then characterized as being near to the mean (i.e. typical of the altitude zone) or far from it in terms of slope angle and local relief. "Typical" is defined as being within $\pm 0.319\ \sigma$ of the mean; this figure is chosen because, if the frequency distributions for slope angle and local relief are normal then about 25% of points should be classed as typical. Such a figure provides a convenient number of sampling locations. Maps of areas whose slope angles and local reliefs are near average for any given altitude are then composed and, superimposing these maps, a composite map is produced of areas which, for their altitude zones, are close to the mean for these parameters. Figure 6 shows a typical map of such locations. From such a map a much smaller sample of points is subjectively made (maximizing range in altitude and avoiding areas on the glacier known to have other peculiarities, such as medial moraines). Accumulation measurements are made in the field at this small subsample of 2 to 10 points. The locations of the selected sampling sites are determined by field survey with an accuracy of 25 m.

Samples are taken in places which are typical in terms of two surface shape parameters (local relief and slope angle) which are known to be strongly associated with snow depth. This stratified sampling approach allows snow amount to be associated with altitude with less fear that the samples may be biased than is the case using the standard method.

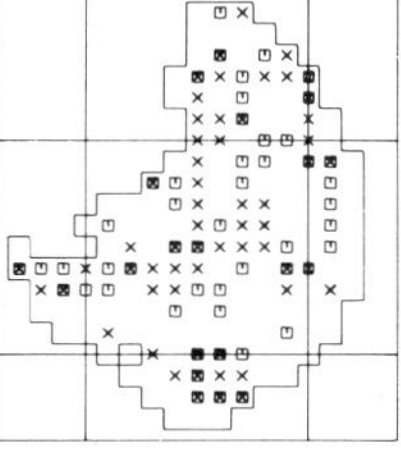

▢ BEST LOCATIONS IN TERMS OF LOCAL RELIEF
× BEST LOCATIONS IN TERMS OF SLOPE ANGLE
⊠ BEST COMBINED LOCATIONS

Figure 6. Ram River Glacier. Stratified sampling design.

CALCULATION OF ACCUMULATION QUANTITIES FROM SAMPLE DATA

After depth and density measurements have been taken at the small number of sampling locations water equivalent values are calculated. If, as is usually the case, more than one density measurement has been made, a linear change of density with altitude is assumed. Care is taken that such an assumption would not produce an equation which would yield obviously incorrect values at very high or very low altitudes. In such a case it is usually better to apply a mean density value to the whole glacier. The appropriate density for the altitude is applied to each depth sample to yield the w.e. value.

A linear regression model is then used to associate w.e. with altitude. It has the form:

$$\text{w.e.} = a + bZ \tag{2}$$

where w.e. is water equivalent, a and b are constants and Z is altitude. This equation is then applied to all points on the square grid to yield w.e. estimates for the entire glacier. As stated above, it has been confirmed on the five glaciers under study that a calculation of total w.e. amount for the glacier in this way usually yields results within ±5% of the value calculated by the standard method.

The aim of the next part of the analysis is to redistribute the snow according to the mean coefficients c and d in (1) above and in accordance with the variability of local relief and slope angle values. The total amount of water is kept constant by altering the intercept value in equation (2). In other words, the total amount of accumulation for the whole glacier is calculated on the basis of altitude alone; redistribution of that amount is then effected in accordance with knowledge of persistence in local patterns.

Maps and tabulations are made of the final w.e. accumulation pattern.

THE CAPABILITIES OF THE COMPUTER PROGRAM

The success or failure of a project which deals with data collected over many years from several glaciers depends on the way in which the data are stored and the way in which the reduction program is structured. These aspects are not discussed in detail for they are of secondary importance to the method used. The importance of the technical aspects, however, should not be neglected, as most of the problems have been due to organization rather than due to the formulation of the basic concepts which are, in fact, very simple. An organizational flow chart is presented in Figure 7. It shows that nonvariable information is stored in a tape data bank, the information, i.e. the snow depth and density measurements (any number of each presented in any order) with their relevant spatial coordinates are, at present, read in from punch cards. Header cards indicate the glacier to which the data pertain and also various options, e.g., the size of the map and the class intervals for map shading. Any number of analyses, either from the same or different glaciers, can be run in one job.

The data bank is so structured that it is very easy to add more glaciers or nonglacier research basins to it without necessitating a restructuring of the computer program. It is envisaged that the ease and speed with which data can be reduced using this system will make possible new avenues of research, for example, better melt simulation modelling using the more reliable accumulation input maps; and will certainly have a considerable effect on the utilization of manpower.

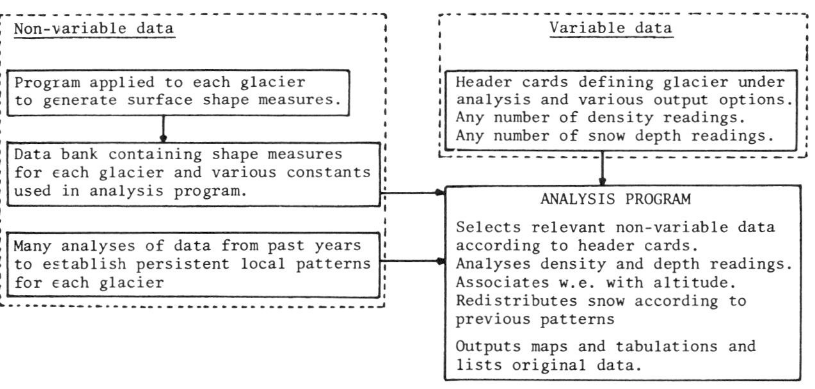

Figure 7. Organizational flow chart.

ACKNOWLEDGEMENTS

The author wishes to acknowledge the advice and encouragement given by Dr. A.D. Stanley throughout the preparation and implementation of this project. Dr. G. Holdsworth and Mr. D. Sellars gave useful critical advice when the manuscript was being drafted.

REFERENCES

Meiman, J.R., 1968. Snow accumulation related to elevation, aspect and forest canopy. Proc. of workshop seminar on snow hydrology, Canadian National Committee for I.H.D., pp. 35-47.

Mokievsky-Zubok, O., 1971. Half decade study of mass balance at Sentinel Glacier, B.C., Canada. I.A.S.H. General Assembly of Moscow.

Mokievsky-Zubok, 1973(a). Study of Sentinel Glacier, British Columbia, Canada within the International Hydrological Decade (I.H.D.) program; procedures and techniques. Inland Waters Directorate, Technical Bulletin No. 77.

Mokievsky-Zubok, 1973(b). Methods used to determine the mass balance on Sentinel Glacier, British Columbia, Canada, Inland Waters Directorate, Scientific Series No. 30.

Østrem, G., 1966. Mass balance studies in glaciers in Western Canada, 1965. Geogr. Bull., Vol. 8, No. 1, pp. 81-107.

Østrem, G. and A.D. Stanley, 1969. Glacier mass balance measurements. Canadian Department of Energy, Mines and Resources and Norwegian Water and Electricity Board, 129 p.

Stanley, A.D., 1970. Combined balance studies at selected glacier basins in Canada. Proc. workshop seminar on glaciers, Canadian National Committee for I.H.D., pp. 5-9.

Stanley, A.D., 1971. Mass and water balance studies at selected glacier basins in Western Canada. I.A.S.H., General Assembly of Moscow, 1971.

Tangborn, W.V., R.M. Krimmel and M.F. Meier, 1971. A comparison of glacier mass balance by glaciologic, hydrologic and mapping methods, South Cascade Glacier, Washington. I.A.S.H., General Assembly of Moscow, 1971.

Young, G.J., 1970. Mass balance measurements related to surface geometry on Peyto Glacier, Alberta. Proc. of workshop seminar on glaciers. Canadian National Committee for I.H.D., pp. 11-20.

Young, G.J., 1971. Accumulation and ablation patterns as functions of the surface geometry of a glacier. I.A.S.H. General Assembly of Moscow, 1971.

Young, G.J., 1973. A computer program using the grid square technique to describe terrain characteristics within a drainage basin. Inland Waters Directorate, Technical Bulletin No. 76.

ESTIMATING TRUE BASIN SNOWCOVER

H. Steppuhn
G. E. Dyck
Division of Hydrology, University of Saskatchewan,
Saskatoon, Saskatchewan, Canada

ABSTRACT

Hydrologic studies often require accurate estimates of the mean depth, density and water-equivalent of the snowcover on a basin. Measurements in Canada indicate consistent similarities in the areal variations of snowcovers within areal units having similar landscape features. Forests, pastures, cultivated fields, ponds, etc., within the same climatic region tend to accumulate snow according to recurring patterns unique to specific terrain and land use.

Snow samples collected within similar areal units usually exhibit similar frequency distributions. These results, combined with the observation that snowcover variance for density is consistently lower than for depth, form the base for a ready, but statistically valid, method of estimating true snow water-equivalents. The method is operationally orientated and features flexibility depending on data requirement, funding and desired level of confidence.

VARIABLES AND VARIABILITY

Hydrologic studies often require knowledge of mean depth, density, and water equivalent of the snowcover on a basin. Since snowcover is an areal phenomenon and basin areas are rather extensive, all measures of absolute, or true, snowcover are estimates derived from areally-distributed snow surveys. Valid use of these estimates require concurrent measures of associated confidences. The degree of confidence depends directly on the number of observations per sample and inversely on the inherent variability of each snowcover variable.

Areal variability occurs over three geographic fields; (I) regional, (II) local, and (III) micro.

(I) Regional variability reflects climatic differences in precipitation, radiation, air temperature, etc. Regions with similar climate measure from 1 to 1,000,000 km^2 depending on latitude, elevation, orographic influences, and proxcimity to large lakes and seas. Burns (1973) outlined climatic regions for Northwestern Canada (Figure 1). A 1972 snow survey (Steppuhn, et al.), covering the Mackenzie Valley portion of Northwest Canada, showed a definite correspondence between climatic regions and large-scale snowcover patterns.

(II) Snowcover variability over local, within-region fields, stem from numerous influences and are often observed and reported. Anderson (1967) and others, working in forested mountains, have related snow accumulation to local terrain variables of elevation, slope, and aspect and to forest variables of canopy density, tree species, tree heights, and forest openings. MacKay (1963) and Kuz'min (1960) emphasized the interaction of terrain and wind in establishing snowcover patterns over prairie. Snowcover variances caused by locally-induced differences in melt rates are common. · Co-operative snow investigators (1956) listed influences causing local differences in melt rates, including hillslope aspect and forest cover. West (1961) added cold air drainage as a factor of snowmelt, and Zotikov and Moiseeva (1972) considered differential melt caused by soil on snow. Natural events, such as avalanches,

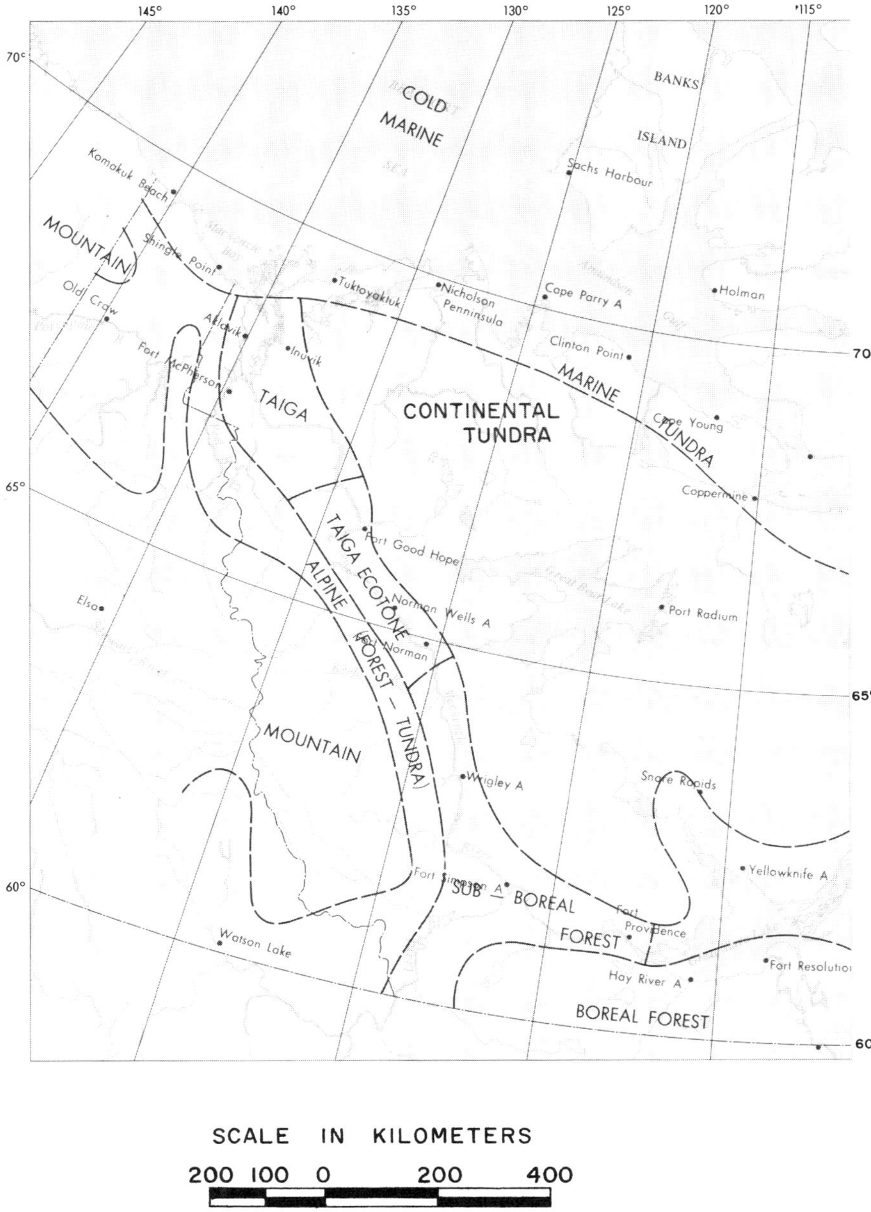

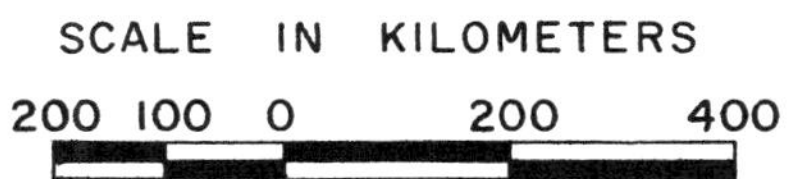

Figure 1. Climatic Regions of Northwest Canada (Burns, 1973)

also alter local snowcover patterns. Lakshman (1973), for one, found distinct relations between snowcover and agricultural land use, while Meiman (1968) concluded from his literature review that forest management practices usually, but not always, affect snow accumulation.

3.6

(III) Micro scale variability in snowcover refer to differences which occur within a few square meters. Causes of these differences include: surface roughness of vegetation and soil; internal alterations, such as depth hoar development; processes related to transport, such as dune deposition.

SAMPLING

Statistical sampling is predicted on two major requirements; samples (1) must be randomly selected and (2) must include a complete, accurate representation of population variability. When an areally-distributed population is sampled, the random requirement is satisfied (1) by systematically sampling the population which is assumed randomly distributed over the area, or (2) by completely randomizing the selection of those areal units to be sampled, or (3) by some combination of (1) and (2).

Applied to basin snow surveys, neither random selection nor complete coverage is easily assured. If we accept research results linking snow accumulation to terrain, vegetation, and land use, the assumption of an areally random snowcover is invalid; landscape distributions are not random. Randomizing the selection of sample units is also difficult because the practice demands free access to all basin subunits. Any restriction in basin coverage would limit representativeness of the survey.

Clearly, we must adopt some form of stratified sampling, where snowcover variability is distributed somewhat uniformly throughout each stratum. The unitized sample design reported in this paper combines stratified sampling with subsampling (Steel and Torrie, 1960). This tri-level design consists of subunits, j, within areal units, i, within strata, h. The area of each basin is partitioned into K number of strata, where each stratum, h, contains M_h number of areal units and each areal unit, i, contains N_i number of subunits. Approximate minimum sizes for strata, areal units, and subunits comprising a 100-km^2 basin approach 100 ha, 1 ha, and 1 m^2, respectively. Snowcover observations are taken in n number of subunits (points), m number of areal units (snow courses), and k number of strata (landscape classes). Usually, $m < M$, $n < N$, and $k = K$.

Designed primarily to sample local and micro snowcover variability, this sampling scheme functions most effectively when strata-subdivision is confined to areas within a single climatic region. Maps, aerial photographs, satellite imagery, and previous surveys are used to define and delineate areal units which exhibit similar landscapes and are scattered throughout the basin. Similar units form distinct landscape classes or sampling strata which reflect basin terrain and winter land use. Landscape classes from three snow surveys are recorded in Table 1.

Under this unitized sample design, randomization is partially derived from each of four sources: (I) Micro variability of snowcover is of such a scale that its areal distribution within each areal unit is strongly random. For example, systematic sampling of snow depth in Canadian forests frequently exhibits a random component of variance about the snow course mean, which relates to the peripheral fall of intercepted snow around coniferous trees. (II) Local variability also tends to be somewhat randomly distributed. Snow depth data sampled from 100 different areal units were tested for normality by the third-moment test for statistical symmetry. A symmetrical

Table 1. Landscape Classes for Three Snow Surveys in Canada

I. Bier Basin near Watson Lake, Yukon

CLASS	AREA	
	Hectares	% of Area
(a) Upland Forest of Pine & Spruce	1121	67.3
(b) Organic Forest of Spruce	472	28.3
(c) Large Brushy Clearing	73	4.4
Total	1666	

II. Bad Lake Basin near Rosetown, Saskatchewan

CLASS	AREA	
	Hectares	% of Area
(a) Upland Crop Stubble	117	3.2
(b) Plains Crop Stubble	160	4.4
(c) Lowland Crop Stubble	60	1.6
(d) Upland Pasture	1400	38.5
(e) Plains Pasture	1200	33.0
(f) Lowland Pasture	700	19.2
Total Surveyed	3637	

III. Battle Basin near Ponoka, Alberta

CLASS	
(a) Stubble Plain	(h) Fallow Undulating Plain
(b) Stubble Topland	(i) Herbaceous, Broad Lowland
(c) Stubble Undulating Plain	(j) Brushy Broad Lowland
(d) Hayfield Plain	(k) Brushland Plains
(e) Hayfield Undulating Plain	(l) Woodland Undulating Plain
(f) Fallow Plain	(m) Pasture Plain
(g) Fallow Topland	(n) Recent Clearing Plain

frequency distribution was indicated in over 40% of those units located in a prairie environment and in over 60% of those units within a forest. (III) Sample areal units within each landscape class are selected randomly irrespective of location except in cases where access is difficult and costly. (IV) The directional orientation of each snow course and the spacing of sample points along the course may be chosen randomly. In practice, these choices must also insure adequate sampling coverage of the areal unit.

Incorporation of stratification into the sampling scheme tends to reduce sample variance. Table 2 provides information on the comparative gain in variance reduction resulting from establishment of (1) areal units and (2) landscape strata. Snow data, obtained from Bier Basin, Yukon and expressed as ratios of variance among groups to variance within groups, indicated a ratio gain for the landscape grouping which was four times greater than for areal unit groupings. A similar comparison in Table 3 for snow depth data from Battle Basin, Alberta again showed significant gains in variance reduction for landscape stratification. Snowcover depth and density statistics from Bad Lake, Saskatchewan also support stratification (Table 4). Comparative variation, as expressed by the coefficient-of-variation (standard deviation/mean), averaged considerably lower for individual class-values than when combined.

Table 2. Summary of Variance-Ratios by Landscape Class for Snowcover Depth, 1973, Bier Basin, Yukon

CLASS	Variance-Ratio*	
	(Among Areal Units / Within Areal Units)	(Among Classes / Within Classes)
Upland Forest	9.19	
Organic Spruce	1.58	
Large Clearing	4.43	
Classes Combined		44.05

*Ratio of variance between means of the groups to variance between observations within the groups

Table 3. Summary of Variance-Ratios by Landscape Classes for Snowcover Depth, 1973, Battle Basin, Alberta

CLASS	Variance-Ratio*	
	(Among Areal Units / Within Areal Units)	(Among Classes / Within Classes)
Stubble Plain	3.42	
Hayfield Plain	9.01	
Brushland Plain	6.26	
Classes Combined		16.62

*Ratio of variance between means of the group to variance between observations within the groups

Table 4. Summary of Snowcover Statistics by Landscape Class, 1972, Bad Lake, Saskatchewan

CLASS	No. of Snow Courses	Depth (cm)		Density (g/cc)	
		Mean	Coeff. of Variation	Mean	Coeff. of Variation
h	m_h	$\bar{d}_h$	Cv_h	$\bar{\rho}_h$	Cv_h
Upland Stubble	2	34	.291	.14	.163
Lowland Stubble	1	53	.282	.16	.215
Plain Stubble	3	21	.320	.22	.154
Upland Pasture	9	14	.703	.16	.370
Lowland Pasture	9	62	.634	.27	.339
Plain Pasture	7	27	.608	.19	.363
Classes Combined		34	.935	.20	.440

The lowering of sample variance by stratification has merit; statisticians (Cochran, 1953) recommend the practice wherever possible. Reducing snow sample variance increases confidence about the sample mean, reduces the number of snow courses required, increases the probability that snowcover is similarly distributed over each areal unit comprising each class, and diminishes the importance of complete class-wide dispersion of snow courses. Consequently, data obtained from accessible snow courses may be extrapolated with greater validity to remote reaches of the basin.

Areal frequency distributions of snowcover observations over areal units and classes often are not symmetrical. Figure 2 illustrates the non-symmetry associated with snow depths measured in the upland-pasture class at Bad Lake, Saskatchewan. Under the unitized sample design, normal statistics still apply to non-symmetrical depth data. The central limit theorem states that the means of all samples taken from the same population will tend to follow a normal distribution irrespective of the population distribution.

Snow surveys are rarely designed to separately sample depth and density. Separate sampling is advantageous, if the areal variability of the two variables differ appreciably. If density variance were less and relatively normal, the advantage would compound, considering the comparative difficulty associated with density measurement. The similarity between water equivalent and depth frequency distributions, shown in Figure 2, suggests a dominance of depth over density in influencing resultant water equivalent. The Figure also indicates what repeated symmetry tests have confirmed, namely, that class-wide density distributions are usually symmetrical, if not normal.

Results from studies of prairie snowcovers give additional support for separate sampling of depth and density. Lakshman (1973) related water equivalent to snow depth for 1970, 1971 and 1972 data from four basins in Saskatchewan. He consistently found linear correlations of 0.85 or greater, inferring a high covariance between depth and density or a relatively low density variance. Coefficients-of-variation from the 1972 survey at Bad Lake, Saskatchewan, compared in Table 4, averaged twice as great for depth as for density.

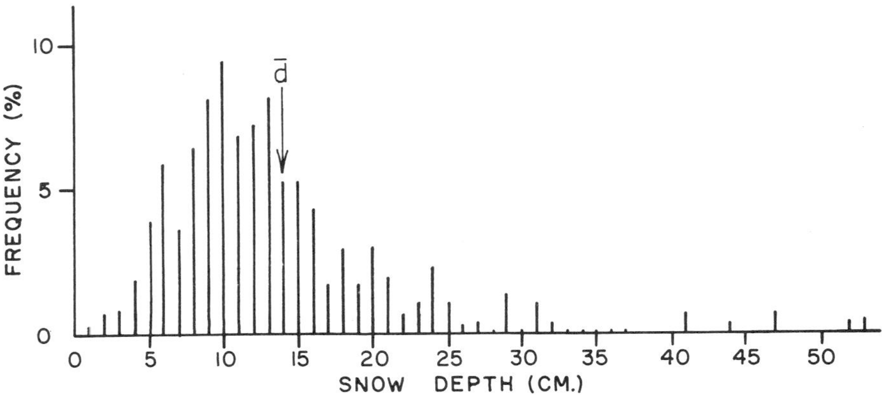

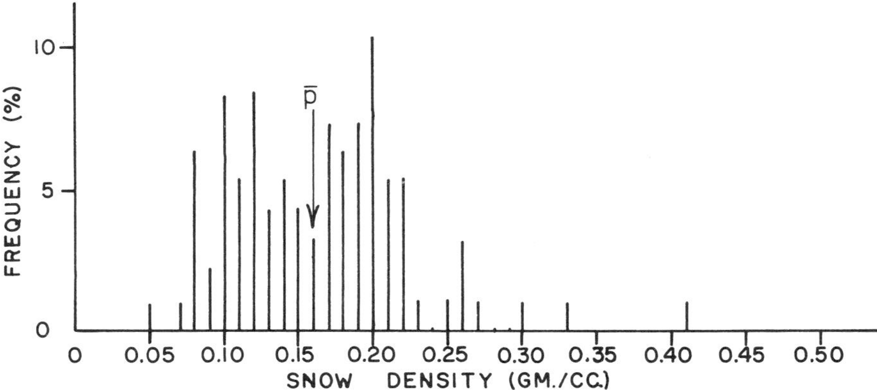

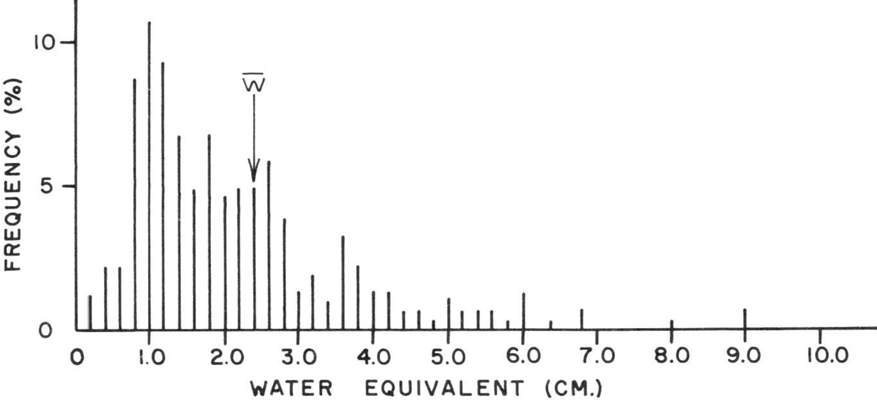

Figure 2. Frequency Distributions of 312 Snowcover Measurements over Upland-Pasture, Bad Lake Basin, Saskatchewan

Statistics associated with the unitized sample design exist for each unit. An observation is denoted by x_{hij}, where h refers to the landscape class, i, to the snow course, and j to the observation point. Equations for computation of sample means for snow courses, $\bar{x}_i$, landscape classes, $\bar{x}_h$, and the basin, $\bar{x}_b$, and of corresponding error variances, $s^2_{\bar{x}i}$, $s^2_{\bar{x}h}$ and $s^2_{\bar{x}b}$ are

$$\bar{x}_i = \frac{1}{n}\sum_j x_{hij} \quad , \quad s^2_{\bar{x}i} = \frac{\sum_j (x_{hij} - \bar{x}_i)^2}{n(n-1)}$$

$$\bar{x}_h = \frac{1}{m}\sum_i \bar{x}_i \quad , \quad s^2_{\bar{x}h} = \sum_i \frac{\frac{1}{n_i}(\bar{x}_i - \bar{x}_h)^2}{m-1}$$

$$\bar{x}_b = \sum_h \frac{A_h \bar{x}_h}{A_b} \quad , \quad s^2_{\bar{x}b} = \sum_h \left(\frac{A_h}{A_b}\right)^2 s^2_{\bar{x}h}$$

where A_h and A_b are areas of class and basin, respectively. Equations for computation of confidence intervals, CI, at levels, i, h, or b are of the form, $CI = 2|t_\alpha s_{\bar{x}}|$, where t_α is the t-value at the selected α-level of confidence and $s_{\bar{x}}$ is the sample error of the mean.

METHODOLOGY

Unitized sampling has been used to estimate the maximum winter snowcover over basins in four climatic regions; prairie, parkland, forest and tundra. The methodology for this sampling can be illustrated by reviewing its application to Bier Basin near Watson Lake, Yukon, in Northwest Canada (Figure 1).

Snow Survey

An aerial reconnaissance preceded the snow survey. This reconnaissance supported the supposition that three major groups formed the basin's landscape. Distributed as shown in Figure 3 and tabulated by area in Table 1, these landscape classes were (1) upland forest of pine and spruce, (2) organic spruce wetland, and (3) large brushy clearing.

Figure 3 locates the twenty snow courses selected for survey. Allocations according to landscape class were ten, five and five for upland forest, organic spruce, and large clearing, respectively. Although random selection of course location was partially successful, access influenced some site choices.

Snowcover depths and densities were observed at points approximately ten paces apart along randomly oriented courses. A scaled-rod was used to read 36 or more depths per course. Vertically-integrated snowcover densities were obtained by weighing snow cores of known length taken with 4.6 cm or 8.0 cm diameter sampling tubes. Although the number of density observations per course were not consistent, the total number per landscape class was ample to yield a statistically-valid class mean. Rational for this practice hinges on the assumption that class-wide, snowcover densities are symmetrically and near-normally distributed about the mean of the class.

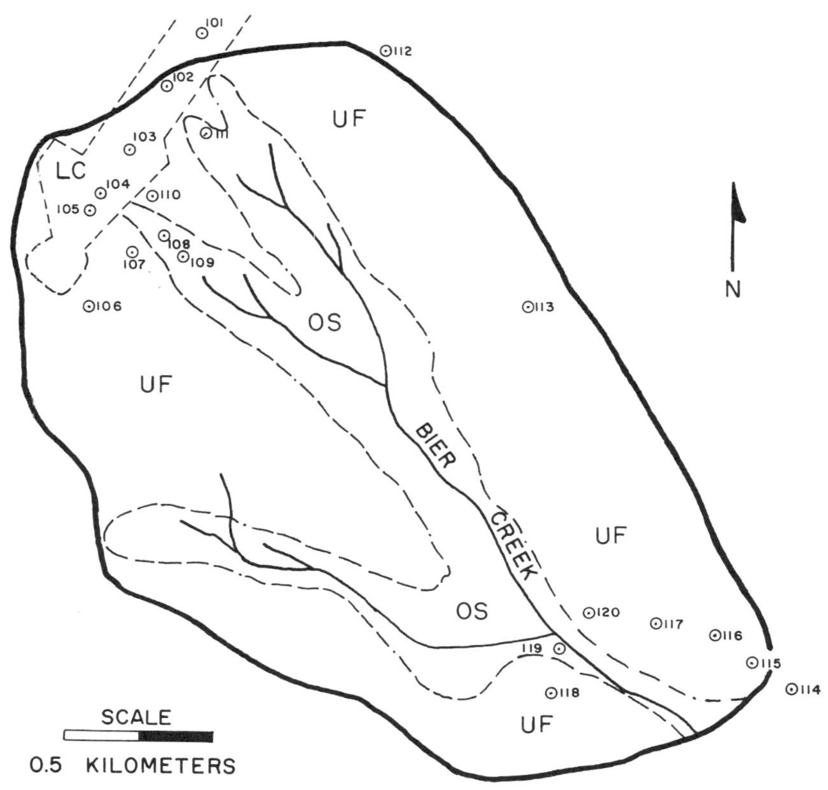

Figure 3. Bier Basin near Watson Lake, Yukon, Canada

Course and Class Statistics

Snowcover depth and density data from the Bier Basin survey were initially summarized and statistically analyzed according to the manner in which each variable was sampled. Depth data were summed by snow course and density data by landscape class. Statistical means and variances for depth are presented by course in Table 5, while density statistics, including error variances and confidence intervals, are listed by class in Table 6. Table 6 also includes class-wide means, error variances and confidence intervals for snowcover depth derived from the course means recorded in Table 5.

Table 5. Snowcover Depth Statistics by Snow Course for 1973 Survey of Bier Basin near Watson Lake, Yukon

Depth in cm.			Depth in cm.		
No. of Obs. n	Mean $\bar{d}_i$	Variance s_i^2	No. of Obs. n	Mean $\bar{d}_i$	Variance s_i^2
Upland Forest			Organic Spruce		
36	60.5	29.16	36	69.9	42.20
59	61.7	62.26	37	70.9	77.10
37	58.9	72.00	36	70.6	54.00
36	57.4	74.20	38	71.6	71.42
48	66.0	56.00	37	67.6	47.49
36	66.0	61.55	Large Clearing		
36	67.6	58.78			
36	65.5	41.23	36	74.4	34.13
37	60.7	24.90	36	76.2	34.26
38	59.2	45.42	40	79.3	36.58
			36	76.5	22.13
			36	77.5	14.90

Table 6. Snowcover Statistics by Landscape Class, 1973, Bier Basin, Yukon

Variable	Statistic*	Landscape Class, h		
		Upland Forest	Organic Spruce	Large Clearing
Depth (cm)	$\bar{d}_h$	62.36	70.10	76.76
	CI/2	5.77	2.62	3.05
	$s^2_{\bar{d}h}$	12.25	2.51	3.44
	m_h	10	5	5
Density (g/cm^3)	$\bar{\rho}_h$	.206	.197	.172
	CI/2	.008	.011	.021

Continued ...

Table 6. Snowcover Statistics by Landscape Class, 1973, Bier Basin, Yukon (Continued)

Variable	Statistic*	Landscape Class, h		
		Upland Forest	Organic Spruce	Large Clearing
Density (g/cm^3)	$s^2_{\bar{\rho}h}$	.0000223	.0000482	.000156
	$z_h N$	41	21	18

* Snowcover Statistics Legend:

$\bar{d}_h$ Class mean depth

CI Confidence interval at $\alpha = 0.1$

$s^2_{\bar{d}h}$ Error variance of mean depth

m_h Number of snow courses

$s^2_{\bar{\rho}h}$ Standard error of mean density

z_h Number of density observations

Class Verification

The next step in analysis centered on verification of the chosen landscape classification. Within each class, depth samples from each snow course should exhibit some uniformity in variance about the course mean. Bartlet's test for homogenity of variances (Steel and Torrie, 1960) was applied to the Bier Basin depth data. Test results are given in Table 7. Depth variances for the organic spruce and large clearing classes showed within-class homogenity at α - probabilities of 0.05 or more. Sample variances within the upland forest were about as uniform ($\alpha = 0.01$) as when this class was combined with organic spruce. All other class combinations proved non-homogeneous.

Table 7. Summary of Homogenity Tests for Depth Variance

Class Group	Highest α-Level of Significance Indicating Homogenity: Homogeneous ⟷ Non-homogeneous							
	0.2	0.1	0.05	0.025	0.01	0.005	0.001	< 0.001
Large Clearing LC			X					
Upland Forest UF					X			
Organic Spruce OS	X							
LC + UF								X
LC + OS								X
UF + OS					X			
LC + UF + OS								X

Any combination of classes whose depth variances are homogeneous should be tested for equality of means. That is, do these class means diverge sufficiently to justify the retention of more than one unified class? Applicable statistical tests include the t-test, the Wilcoxan rank sum test, and the F-test. Applied to the upland and organic combination, the F-test significantly inferred the existence of two divergent means each representing a unique class. Had the results indicated an equality of means, formation of a single class from the two would have been justified.

Future Snow Surveys

Snow surveys provide information to suggest statistically-sufficient sample sizes for future surveys. These suggested sizes include the number of depth observations per snow course, n_i, the number of snow courses per class, m_h, and the number of densities per class, z_h. Procedures necessary to determine suggested sizes are usually based on temporal extrapolation of variances.

Fixed n_i-values for all snow courses may not insure adequate sampling coverage over each areal unit. The optimum n_i is that number yielding a sample which defines the depth-frequency of areal unit i with a confidence of about 5% for its mean and 10% for its standard deviation. Means, $\bar{d}$, and standard deviations, s, computed for a series of snow depth samples are tabulated in Table 8; each succeeding sample contains fewer observations but covers the same snow course. The 5 and 10 percent deviation levels for $\bar{d}$ and s suggest an n_i-value between 26 - 32.

Table 8. Statistics from Repeated Samples of Snow Depth Taken Over an Areal Unit (B1), Upland-Pasture Class, Bad Lake, Saskatchewan, 1972

No. of Obs. n	Mean Depth $\bar{d}$	Deviation %	Standard Deviation s	Deviation %
64	13.85	0	8.817	0
57	13.97	1	8.925	1
54	14.10	2	8.990	2
51	13.31	4	7.838	11
48	14.31	3	9.461	7
46	13.56	2	7.956	10
43	14.37	4	9.285	5
38	13.33	4	8.093	8
32	13.26	4	8.052	9
26	14.61	5	9.897	12
22	14.61	5	10.241	16
18	14.59	5	10.940	24
16	12.47	10	6.581	25
13	16.02	16	12.144	38
11	12.00	13	6.757	23
8	16.12	16	14.371	63

The suggested number of snow courses, m_h, depends on the within-class variance, $s^2_{(wt)hi}$, which is equal to the remainder of total class variance, s^2_{hi}, minus the variance between snow course means, s^2_h. Values for m_h may be found by the relation

$$m_h \geq 2(t_\alpha + t_\beta)^2 (s^2_{(wt)hi})(CI)^{-2},$$

where t_α and t_β are t-values for α and β at ∞ degrees of freedom, and CI is a chosen confidence interval based on a specified α-probability. Confidence intervals relate to error tolerance, E, by CI = $2|E|$. Calculated m_h-values for Bier Basin at $\alpha = 0.1$, $\beta = 0.8$ and E = $\pm 0.1\bar{d}_h$ cm were upland forest 5, organic spruce 5, and large clearing 2.

The basis for determining z_h number of future density observations is the class-wide density variance, s^2_h. The computational equation is $z_h \geq t^2_\alpha s^2_h E^{-2}$. Suggested z_h-values for upland forest, organic spruce and large clearing classes in Bier Basin for an E of ± 0.02 g/cc were 6, 7 and 19, respectively.

Water-Equivalent Statistics

Verification tests of selected landscape classes in Bier Basin were followed by determination of class-wide water equivalents, $\bar{w}$. If variance between class mean depth, $\bar{d}_h$, and density, $\bar{\rho}_h$, is negligible, $\bar{w}_h = (\bar{\rho}_h)(\bar{d}_h)$. Concurrently, if $s^2_{\bar{w}h}$, $s^2_{\bar{\rho}h}$, and $s^2_{\bar{d}h}$ designate error variance of mean water equivalent, mean density and mean depth, respectively, $s^2_{\bar{w}h} = (s^2_{\bar{\rho}h})(s^2_{\bar{d}h}) + (\bar{d}^2_h)(s^2_{\bar{\rho}h}) + (\bar{\rho}^2_h)(s^2_{\bar{d}h})$. In practice, this equation reduces to

$$s^2_{\bar{w}h} \approx (\bar{d}^2_h)(s^2_{\bar{\rho}h}) + (\bar{\rho}^2_h)(s^2_{\bar{d}h}), \text{ because}$$

$\bar{d}^2_h >> s^2_{\bar{d}h}$ and $1.0 > \bar{\rho}^2_h >> s^2_{\bar{\rho}h}$. If covariance between $\bar{d}_h$ and $\bar{\rho}_h$ **is** significant,

$$\bar{w}_h = (\bar{d}_h)(\bar{\rho}_h) + (r_{d\rho})_h (s_d)_h (s_\rho)_h \quad \text{and}$$

$$s^2_{\bar{w}} \approx (\bar{d}^2_h)(s^2_{\bar{\rho}h}) + (\bar{\rho}^2_h)(s^2_{\bar{d}h}) - 2 (\bar{d}_h)(\bar{\rho}_h)(r_{d\rho})_h (s_d)_h (s_\rho)_h,$$

where $r_{d\rho}$ is the correlation coefficient between within-class, common-point observations of snow density and depth, while s_ρ and s_d are the respective standard deviations of the sample observations. Class confidence intervals, $CI_{\bar{w}h}$, at selected α-probabilities are determined in the usual manner, $CI_{\bar{w}h} = 2|(t_\alpha)(s_{\bar{w}h})|$.

Table 9 records Bier Basin water equivalent statistics by landscape class. Although class mean water equivalents averaged within one cm of each other, a review of Table 6 will reveal significant class differences in depth and density; these data suggest that equal masses of snow formed snowcovers of differing compositions depending on landscape.

Basin landscape may also significantly influence snowcover water equivalent. Class mean water equivalents from the 1972 Bad Lake, Saskatchewan survey were:

Upland Stubble	5.1	±	0.9 cm	at	$\alpha = 0.1$
Lowland Stubble	9.3	±	2.5 cm	at	$\alpha = 0.1$
Plain Stubble	4.5	±	0.6 cm	at	$\alpha = 0.1$

Upland Pasture	2.4	±	0.2 cm	at	$\alpha = 0.1$
Lowland Pasture	18.6	±	1.6 cm	at	$\alpha = 0.1$
Lowland Pasture	6.1	±	0.5 cm	at	$\alpha = 0.1$

Table 9. Water Equivalent Statistics by Landscape Class, 1973, Bier Basin, Yukon

Variable	Statistic*	Landscape Class, h		
		Upland Forest	Organic Spruce	Large Clearing
Water Equivalent (cm)	$\overline{w}_h$	12.85	13.82	13.21
	CI/2	1.285	0.973	1.667
	$s^2_{\overline{w}h}$	0.606	0.348	1.021

* Snowcover Statistics Legend:

$\overline{w}_h$ Class mean water equivalent

CI Confidence interval at $\alpha = 0.1$

$s^2_{\overline{w}h}$ Standard error of mean water equivalent

Basin Statistics

A weighted summation procedure for Bier Basin resulted in statistics describing true basin snowcover. Class mean values for depth, density and water equivalent were weighted by area ratios, R_h, derived as the quotient of class area, A_h, over basin area, A_b, i.e., $R_h = A_h/A_b$. Class error variances for each variable were also weighed, but by R_h^2. Weighted values were summed for each class, yielding the following estimates of the true snowcover over Bier Basin:

Basin mean depth in cm = 65.3 ± 4.1 at $\alpha = 0.1$
Basin mean density in gm/cc = 0.202 ± 0.020 at $\alpha = 0.1$
Basin mean water equivalent in cm = 13.1 ± 0.9 at $\alpha = 0.1$

CRITIQUE

The unitized sampling method features flexibility. For example, only preferred landscape classes need be surveyed; classes such as lakes and ponds may be ignored. Also, to lower costs, classes may be combined, or the assumption of normality may be extended to snow depths. The method offers a relatively rapid snow survey by keeping the number of density measurements to a minimum.

Possible disadvantages of unitized sampling include: (1) the method is predicated on the existence of snowcover differences between basin landscapes. If for specific storms, or specific times during winter accumulation, basin snowcover were areally uniform, landscape stratification would be superfluous. Experimentation with the unitized method have been limited to maximum winter snowcovers. (2) As developed, unitized sampling applies only to ground surveys; adaption to remote surveying has not been attempted. (3) The method has never been used in climates where significant winter precipitation includes rainfall. (4) Landscape classes which affect snowcover are unique to each climate region and often to each basin. Some degree of discretion must accompany basin classification. (5) The method has not been applied to basins with large ranges in elevation.

Estimates of absolute or true snowcover depth, density and water equivalent find ready use in management and research. Hydraulic models often require absolute values for precipitation inputs. Basin-wide winter energy budgets require estimates of snowcover depth and density for heat flow determination. The areal extrapolation of data from precipitation gauges or indexing snow courses depends on calibration by true snowcover values. True snowcover estimates also provide guidance in selecting suitable locations for these gauges and indexing courses. Finally, this method can fulfill a need for a valid, reliable and ready means of determining "ground truth" in evaluating new techniques and instruments in snow surveying.

REFERENCES

Anderson, H. W., 1965. Snow accumulation and melt in relation to terrain in wet and dry years. Proceedings 33rd. Western Snow Conference, Colorado Springs, U.S.A. pp. 73-82.

Burns, B. M., 1973. The climate of the Mackenzie Valley-Beaufort Sea, Volume I. Climatological Studies No. 24. Atmospheric Environment Service, Canada Dept. of Environment.

Cochran, W. G., 1953. Sampling techniques. New York: John Wiley & Sons, Inc.

Kuz'min, P. P., 1960. Snowcover and snow reserves. Gidrometeorologicheskow Izdatelsko, Leningrad. Translation National Science Found. 1963. Washington, D.C. pp. 99-105.

Lakshman, G., 1973. Drainage basin study. Progress Report No. 8. Saskatchewan Research Council. Saskatoon, Canada.

McKay, G. A., 1963. Relationships between snow survey and climatological measurements. International Assoc. Scientific Hydrol. IUGG Assembly, Pub. No. 63. Berkeley, U.S.A. pp. 214-227.

Meiman, J. R., 1968. Snow accumulation related to elevation, aspect and forest canopy. Snow Hydrol., Proceedings Workshop Seminar, Canadian Nat. Comm./I.H.D.

Steel, R. G. and J. H. Torrie, 1960. Principles and procedures of Statistics. Toronto: McGraw-Hill Book Co.

Steppuhn, H., Beattie, C., and D. Gray, 1972. Designing snowpack survey techniques for the Mackenzie Valley. Progress Report for Climatological Division, Atmospheric Environment Service, 4905 Dufferin Street, Downsview, Ontario, Canada.

U.S. Army Corps of Engineers, 1956. Snow Hydrology. Summary Report of Snow Investigations. North Pacific Division, Portland, U.S.A.

West, A. J., 1961. Cold air drainage in forest openings. Pacific Southwest Forest and Range Experiment Station, Research Note 180.

Zotikov, I. A. and G. P. Moiseeva, 1972. A theoretical study of ice surface dusting influence on melting intensity. Symposia on the Role of Snow and Ice in Hydrology, Sept. 1972. Banff, Canada.

A NEW CALIFORNIA DEPARTMENT
OF WATER RESOURCES TELEMETRY SYSTEM

George W. Barnes, Jr.
California State Department of Water Resources

ABSTRACT

The California Department of Water Resources recently installed a new telemetry system to collect hydrologic and meteorologic data automatically. The system is computer controlled and uses the State's microwave system and mountain top VHF radio repeaters for communications. The remote data stations are modular in construction and may accept both analog and digital sensors. It is envisioned the system will be expanded and will also tie into other computer systems so as to organize and process telemetered data more effectively in the future.

In recent years there has been an increasing effort to manage our existing resources more efficiently. Emphasis has grown in the use of computer modeling techniques for activities such as snowmelt forecasting, river forecasting, reservoir operations, water quality simulation, and many others. Many of these activities create the need for hydrologic, meteorologic, and water quality data on a real-time basis. With the advent of computers, the need has grown to not only collect data, but to collect it reliably in a manner that will minimize the data handling to reduce errors and meet critical time requirements.

This past summer the Department of Water Resources installed a new telemetry system which was engineered and manufactured to specifications by the Astro-Met Plant of Thiokol Chemical Corporation. The purpose of the system is to obtain operational hydrologic and meteorologic data for flood forecasting and snowmelt forecasting, and for use in operation of the State Water Project.

Initially, seventeen remote data stations were installed. However, an additional twenty-four stations have been delivered and will be installed this winter and spring. Thus, there will be a total of forty-one stations installed by next summer. In addition, it is anticipated the system will be expanded to about sixty remote data stations in the next few years.

The telemetry system will provide precipitation, streamflow, temperature, snow water equivalent, and water conductivity data from various areas of the Sierras and the Central Valley of California. Figure 1 shows the general location of the initial seventeen remote data stations and the communications paths used. The equipment at each remote data station will accept either a parallel digital output or a zero to five volt analog output from the various sensors. The stations will transmit the data over VHF (very high frequency) radio and state microwave UHF (ultra high frequency) communication channels to the Central Interrogation Station in Sacramento.

The Central Interrogation Station consists principally of a control consol, a Data General Corporation 1220 Computer, and two Teletype Corporation ASR 33 terminals. The Teletypewriters provide data printout and computer control. One Teletype is located in the Federal-State River Forecast Center and the other Teletype

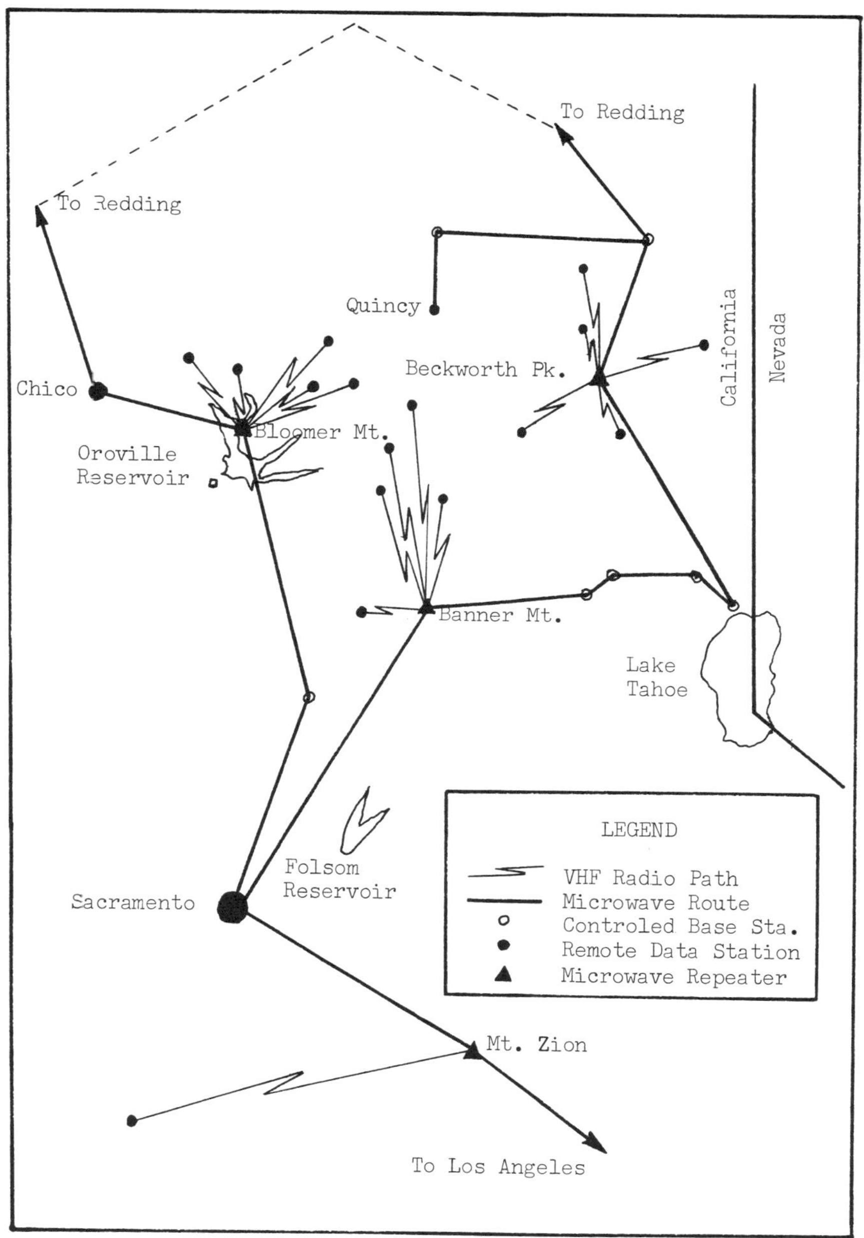

Figure 1. Telemetry System Network For The Initial Seventeen Remote Data Stations.

is adjacent to the control console and computer which are in the State Water Project Control Center. Figure 2 shows an overall view of the Central Interrogation Station.

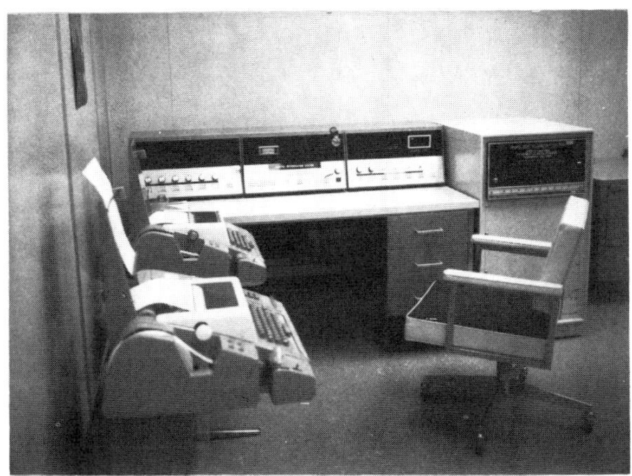

Figure 2. Central Interrogation Station.

In addition, three Receive Only Teletypewriters are to be connected to the system to provide data printout at the Department's Central District and two levee maintenance centers. A block diagram showing the present configuration of the Central Interrogation Station is shown in Figure 3.

The Central Interrogation Station is connected to the state microwave multiplex system. Data communications is therefore from the control console via the microwave system to several mountaintop VHF base stations which each control one or more remote data stations by VHF radio. The microwave system provides the capability of a closed loop communications route so that many of the remote data stations may be accessed by more than one communications path.

Under normal operation the Nova 1220 Computer controls the functions of the telemetry system automatically. The operational program is written so as to allow for the automatic interrogation of up to six groups of remote data stations at any desired preselected time interval from 1 minute to 24 hours. Selected time intervals are easily resettable for each group by entry on the local Teletypewriter at the control console. The software was initially written so that a total of up to 45 remote data stations can be interrogated. Sufficient unused core memory was also provided so that this number of data stations can be increased easily at a later date.

In addition to the automatic interrogation capability, the computer allows for the manual interrogation of each group of stations and the manual interrogation of any individual remote data station. Either type of manual interrogation may be initiated easily through either of the two ASR 33 Teletypewriters. Modification of the interrogation sequence, addition of remote data stations in the group interrogation sequence, and changing of the printout formats are all easily accomplished through the Teletypewriter located at the control console without rewriting the basic control program.

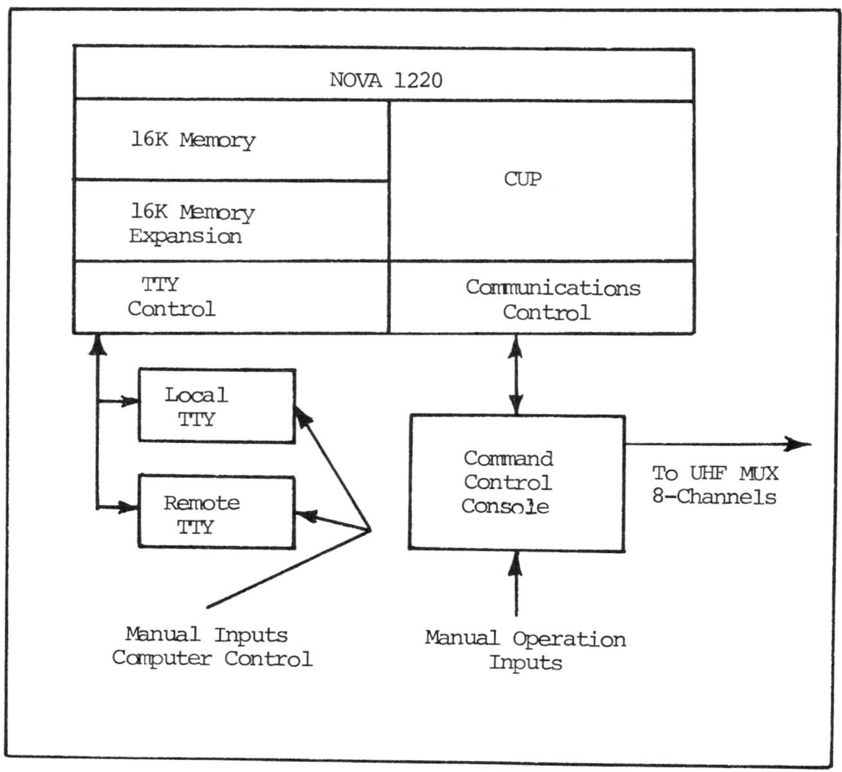

Figure 3. Block Diagram of Central Interrogation Station.

A sample printout for a group interrogation is shown in Figure 4. When an automatic group interrogation is initiated, each station in the group is polled in a predetermined sequence via its primary communications path. The response from each remote data station is logged only at the Teletypewriters at which the printout for that group of stations is desired. Each data sensor value from each remote data station may be adjusted by the computer prior to printout by a programmed linear equation Y = MX+B, where "Y" equals the printed value, and "X" is the actual data sensor reading, and "M" and "B" are predetermined constants. In the event that a station does not respond properly or is missing, the station is interrogated a second time via its secondary communications route. If the data is received correctly on the second interrogation attempt, the figure "2" will be printed in the "R" (route) column. When a remote data station interrogation is successful, the data for each sensor is printed in successive rows with each data type printed in the same columnar position. The date and time of each interrogation sequence is logged at the top of each printout page. In addition, the sensor data and time of each group interrogation are stored in the computer memory. On any successive interrogation the time elapsed since the previous group interrogation is logged at the top of the page. Also, the change in each data value from the previous group interrogation is printed below the current data reading as shown in Figure 4. When an individual station is manually interrogated, the delta values are printed, but the current readings do not replace those in computer memory. Thus, the delta values printed always correspond to the elapsed time since the last group interrogation. Following a group interrogation, the printout advances to a new eleven inch page.

If a data station does not respond after two interrogation attempts, the reading "No Response" is printed following the station three letter mnemonic. If one or more of the sensors from a remote data station reports missing or invalid data, then slashes "/////", will be printed for the change or delta value for that sensor. When a sensor reports data where the change or delta exceeds questionable preprogrammed limits, a question mark, "?", will be printed following the data.

If there is a failure of the primary power supply at the remote data station, an "A" will be printed in the "A" (alarm) column of the printout for that station.

The central station including the computer has backup power supplied by a diesel generator. However, in the event of a power interrupt the computer will initially request an operator to reset the time and date on the control console clock which in turn provides the timing for the computer. If an operator does not reset the clock, a question mark, "?", is printed after the time and date of each succeeding printout until the clock is reset or the time is reentered in the computer program. In the event that the power failure occurs during an interrogation, the remaining data will be lost for that interrogation sequence. The next scheduled automatic group interrogation will commence at the proper preselected time although the actual clock time may be in error until reset.

The control console provides the interface between the State Microwave System and the Nova 1220 computer. The console contains provisions for connecting to and controlling eight separate microwave channels although the system is initially only connected to three channels. In addition, the control console contains the real-time solid state clock which provides the timing for the Nova 1220 computer program.

Normally, the control console is used for system maintenance which requires voice communications to technicians in the field. The two-way voice communications between the control console and each remote data station and each mountaintop base station are selected by a thumbwheel dial on the console.

The second major function enabled by the control console is that it provides a manual interrogation capability to be used in the event that the Nova 1220 computer fails and is temporarily out of service.

The State's microwave system provides the primary communications network to access the remote data stations. The operational computer program provides the information during a station interrogation to select the correct microwave channel, address the particular mountaintop base station repeater on that channel, and select the correct address for the remote data station from that mountaintop. The message format for transmission is in the American Standard Code for Information Interchange (ASCII). The transmission technique used is Frequency Shift Keying (FSK) modulation. The message character output from a remote data station is 100 ASCII characters in which

STATE OPERATIONS HYDROLOGIC DATA
FEATHER RIVER

DATE	TIME	HOURS SINCE LAST GROUP INTERROGATION
Ø2/Ø1/74	8:45	Ø:15

STA	AR	STAGE PRECP	SNOW	TEMP	WTR/C
BRS		NO RESPONSE			
CAM	A	8.21? -.Ø1?		68.2 .4	
SBY	2	3.68 .ØØ		84.3 .3	
BKL		///// /////	///// /////		
QCY		14.03 .15			
BEK		13.61? .Ø1?			
MER		NO RESPONSE			
PRD		NO RESPONSE			
FRD		78.2 .Ø			
DAV		17.6 .Ø			

Figure 4. Printout Sample for an Automatic Group Interrogation

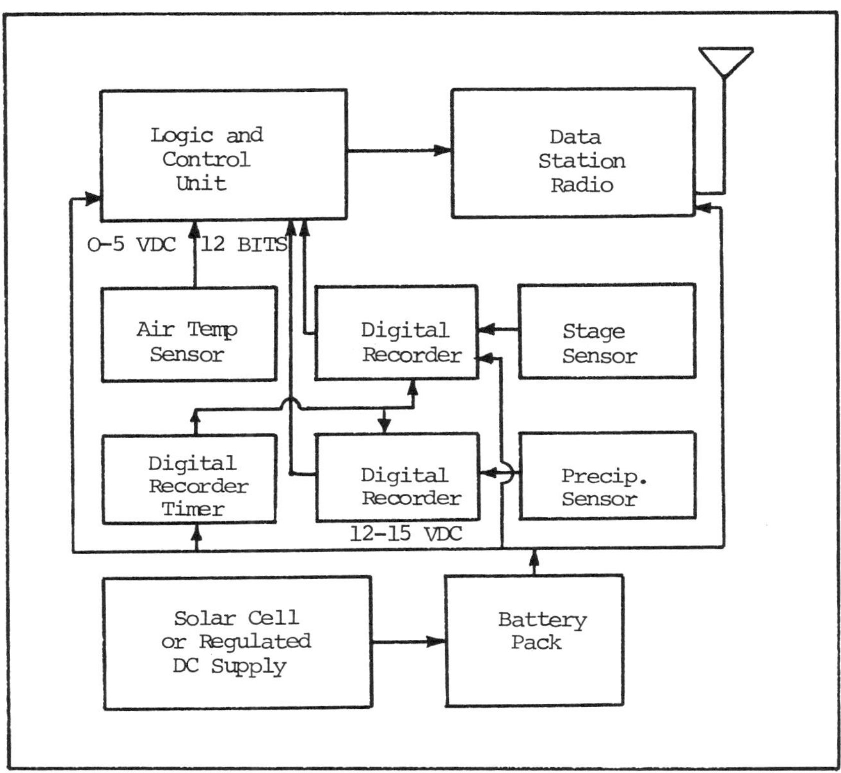

Figure 5. Block Diagram of a Remote Data Station.

parity is provided character by character within the data string. Under normal operation, the transmission rate is 110 baud from the central station to each data station and 220 baud from the data station back.

The remote data station will respond only to a proper address code sent from the central interrogation station. The station then scans all of its sensors and transmits the data back to the central interrogation station. Both analog and digital output sensors may be utilized at a remote data station. The number of sensors that may be used at a remote data station varies depending on the number of analog versus digital sensors desired and the current capacity of the power distribution system. From a practical standpoint, up to eight analog and four digital sensors may be easily accomodated at each remote station, although these numbers may be increased. The analog sensors must provide an output that ranges from 0 to 5 volts. Digital sensors must provide either a parallel contact closure or voltage levels. Figure 5 shows a simplified block diagram of a remote data station, while Figure 6 shows a picture of the remote data station telemetry equipment.

Figure 6. Remote Data Station Telemetry Equipment.

The primary power source used at the remote data stations is either AC commercial power or solar cells. The primary power supply is used to float a 12 DC volt gelled-electrolyte lead-acide type battery. The battery has a nominal capacity of 20 ampier-hours in the event the primary power fails. Thus, data may be obtained from the station for about fourteen days in the event of a power failure. The solar panels are designed to provide up to one amphere of charging current depending on the atmospheric conditions and state of charge of the battery. The types of data initially required for the system are precipitation, snow water equivalent, river and reservoir stage, air temperature, and water conductivity. Some of the precipitation sensors are for areas in the Sierras which experience a mixture of rain and snow. Therefore, these sensors were designed to measure total precipitation in a snow zone where it was anticipated the sensors would not have to be visited during the winter. The typical design of the precipitation gage and shelter cabin is shown in Figure 7. The gage itself consists of twelve inch diameter aluminum pipe beveled at the top edge, and it includes a wind shield. The height of the gage varies depending on the maximum snow depth for the area and the design of the equipment shelter cabin used. At some sites the gage was 24 feet high. A bulkhead is provided near the bottom of the gage and plumbing is run to a twelve inch standpipe in the shelter cabin. A Stevens Model 7000 Digital Recorder that records in inches and hundredths of inches is used to measure the change of height of the liquid in the standpipe. The Stevens Recorders provide the parallel digital input to the telemetry system. The gages are charged with a mixture of ethylene glycol and methanol to a specific gravity of ninety-seven hundredths. It is anticipated that water

Figure 7. Precipitation Gage and Shelter Cabin.

entering the gage will thus be slightly heavier than the charge and begin to sink. While sinking, the water will mix with the antifreeze solution. Nitrogen is slowly bubbled into the precipitation gage to also assist in preventing freezing of the liquid.

The snow pillows which provide snow water equivalent data are being interrogated at two locations, Gold Lake and French Hill. These gages are circular twelve-foot diameter pillows filled with approximately 200 gallons of ethylene glycol and methanol mixed to a specific gravity of one. Plumbing from the pillow connects to a standpipe in the shelter cabin in which the change in liquid level is again measured by a Stevens Model 7000 Recorder which provides the contact closures for telemetering.

River and reservoir stages are also telemetered using a Stevens Model 7000 Recorder which records in feet and hundredths of feet. Either three or four significant digits may be telemetered. The ambient air temperature sensors consist of a Hy-Cal Engineering platinum resistance probe (Model RTS-4147-D) and bridge amplifier (Model SC-2867-A) to provide a zero to five volt output for the temperature range of zero to 100°F. The probe is placed inside a small vented enclosure which is itself shaded and placed high on the outside wall of the shelter cabin.

The water conductivity sensors used to index water quality provide a shaft position which is then connected to a Stevens Model 7000 Recroder for telemetry.

Presently, the telemetry system is operating independently of other telemetry and computer systems. The Nova 1220 computer will provide the flexibility necessary to meet many of the Department of Water Resources' future telemetry needs. First, the Nova computer has an interrupt priority system so that it may be connected to the output terminals of other telemetry systems and thus more operational data can be automatically collected, edited, and organized. The Nova Computer has a direct memory access channel. This will permit a direct tie to a larger computer system being planned for the Federal-State River Forecast Center. Thus the data may be collected and quickly stored in large data files on the RFC computer system and used in computer models for flood forecasting, snowmelt forecasting, and other purposes. The present Federal-State RFC computer system does not have an interrupt capability but larger more flexible systems are being evaluated.

It is also anticipated the telemetry system will be connected to the California State Water Project's control system computer, the Univac 418. This will provide automatic transfer of data for use in the operation of the Project's reservoirs and pumping facilities.

The Nova 1220 operating system also has the capability of running user programs that, for example, could distribute to on-line terminals hydrologic reports, river forecasts, and other useful information. The sixteen level priority interrupt capability enables a user program to be interrupted during a telemetry interrogation and then continue after the interrogation is completed. Finally, the Nova computer capacity of sixteen thousand words (16K) can easily be expanded to a memory capacity of thirty-two thousand words (32K), and the addition of perifial storage such as magnetic tape or disk drives is also possible. Terminals, such as line printers and CRT (cathode ray tube) devices for data entry and information output may also be connected if desired.

The Department of Water Resources new telemetry system incorporates flexibility for future growth in the design of the central interrogation station, the communications, and the remote data stations. It is envisioned that most operational data for snowmelt and river forecasting will eventually be collected and organized in an efficient manner by the mini computer controlled system and the data then transferred automatically to larger computer systems for processing during operational use. This will reduce the present time consuming processes and errors associated with collecting and organizing the data by hand. The new telemetry system will, therefore, assist in providing for improved river forecasts, snowmelt forecasts, and other operational data uses in the future.

A REAL-TIME DATA NETWORK FOR AVALANCHE FORECASTING IN THE CASCADE MOUNTAINS OF WASHINGTON STATE

Edward R. LaChapelle
Geophysics Program, University of Washington

Terence Fox
Department of Atmospheric Sciences, University of Washington

ABSTRACT

A cooperative network has been established for collecting a daily flow of data on snow, weather and avalanche conditions in the Cascade Mountains together with a full spectrum of synoptic weather information. This network furnished the basis for operational avalanche forecasting for mountain highways and a feedback of regional avalanche data and mountain weather forecasts to the participating observers. From the research standpoint, two years of data from this network have served to identify snowfall distribution patterns over the Cascades, the relation of these patterns to weather systems, and the patterns of weather which lead to major avalanche cycles. Intense frontal systems with open warm sectors which produce rain following large quantities of snowfall produce the clearest such pattern. Cold snowfalls produce conditional instability amenable to artificial avalanche release, but few natural releases. The progression of avalanche occurrence through the Cascade Mountains can be correlated with the progression of frontal systems.

The Cascade Mountains of Western Washington are an active region for snow avalanches. Systematic acquisition of avalanche occurrence records in recent years has shown that a higher incidence of avalanches occurs in the Cascades than in any other mountain range in the Western United States (Judson, 1972). Highways and railways traversing the mountain passes of the Cascades have a long history of avalanche problems, including the Wellington disaster, the most serious recorded avalanche accident in United States history in which 96 persons perished (El Hult, 1960).

Over the past two decades there have been limited and local programs of avalanche forecasting and control in the Cascades, both on the part of the Washington State Department of Highways at the mountain passes and by the U. S. Forest Service at several of the major ski areas. A new dimension to the highway avalanche problem has now been added with the opening in September, 1972 of State Highway SR-20, the North Cascade Highway. This highway traverses a wilderness area where historical records of weather conditions and avalanche occurrence are non-existent and snow data are limited to a few hydrologic snow courses. Over a distance of 108 kilometers, the highway is beset by some 70 major avalanche paths.

The completion of the North Cascade Highway, the possibility that it might later be kept open throughout the winter, and the introduction of additional avalanche problems at Snoqualmie Pass through relocation of I-90, have all prompted the Department of Highways to embark on the development in cooperation with the University of Washington of a comprehensive avalanche forecasting

3.8

and control program. This development includes the preparation of avalanche atlases for the major Cascade Passes, a program of basic research in snow mechanics related to avalanche defense construction, and the establishment of a data network to support hazard forecasting. This present paper describes the data network and its product.

The preparation of regional avalanche evaluations and forecasts depends on the acquisition of information about mountain weather conditions, state of the snow cover, and current avalanche activity, usually on a daily basis. Such information can be provided by an avalanche data network operating on a real-time basis. Key requirements are observation sites selected to represent as near as possible avalanche release zone conditions, adequate instrumentation, reliable observers in residence at or near the sites, and efficient means of communicating the observations to forecasters at a central station. Input of current synoptic weather information is also essential. Two-way communication allowing the forecaster to feed back data and inquiries to the observers is highly desirable. Data networks like this have operated for a number of years in Europe, the pioneering Swiss avalanche warning system established in 1945 being the best known (Swiss Federal Institute for Snow and Avalanche Research, 1958). Similar systems operate today in the Austrian provinces of Vorarlberg, Tyrol and Salzburg. Although avalanche forecasting at individual ski areas and mountain passes is well-established in North America, no regional network intended to generate daily avalanche forecasts has previously operated on this continent. The U. S. Forest Service does collect snow, weather and avalanche data on a monthly basis throughout the Western United States for purposes of statistical analysis and research (Judson, 1970).

Within the past two decades individual avalanche forecasting and control programs have been established in the Cascades by the U. S. Forest Service at the winter sports areas of Mt. Baker (part-time), Stevens Pass, Snoqualmie Pass, Crystal Mountain and Mission Ridge. The aim of the regional data network, which embraces these sites as well as others, has been two-fold. First, an identification has been sought of the regional distribution and variations of snowfall and avalanche activity throughout the Cascades and the relation of these to different weather patterns in the Pacific Northwest. Second, the relationship of snowfall and avalanche activity along the hitherto unknown areas of the North Cascades Highway to more familiar patterns elsewhere in the Cascades has been investigated. Such knowledge is expected to lead to a sound basis for operational avalanche forecasting on Washington State Highways, including the North Cascades Highway.

The Cascades Avalanche Network has operated for the past two winters of 1971/72 and 1972/73. It will operate again during the winter of 1973/74. After that time the data-collecting function may be absorbed into the forecasting operation of the Department of Highways.

The five primary observation sites have been located at Paradise in Mt. Rainier National Park, Crystal Mountain Ski Area, Stevens Pass, Mt. Baker Ski Area, and Washington Pass (North Cascade Highway). All of these stations have provided daily reports throughout each winter to the central office located on the University of Washington campus where up-to-minute data from the National Weather Service are available via teletype and facsimile equipment. Data were obtained part-time from the Alpental Ski Area near Snoqualmie Pass, and the winter records from this site and from Mission Ridge

Ski Area were later obtained from the Forest Service Avalanche Data Network for further end-of-winter analysis. In the winter of 1971/72, Washington Pass was the site of recording instruments but was visited only occasionally, the daily reports coming from Early Winters Creek farther to the east. In 1972/73 a full-time observer was stationed at Washington Pass to provide daily reports from this site. Site locations are shown in Figure 1. Further site details and the observations made at each site are summarized in Table 1.

The Data Network was established through cooperative effort of several agencies. The Paradise observations were included in the regular part of the snow safety programs at Crystal Mountain, Alpental and Stevens Pass. The Washington Pass observations were executed by the University of Washington, which also provided a full-time observer at Mt. Baker to supplement Forest Service activities normally conducted only on weekends. The Forest Service furnished radio communication links from Washington Pass and Mt. Baker. In

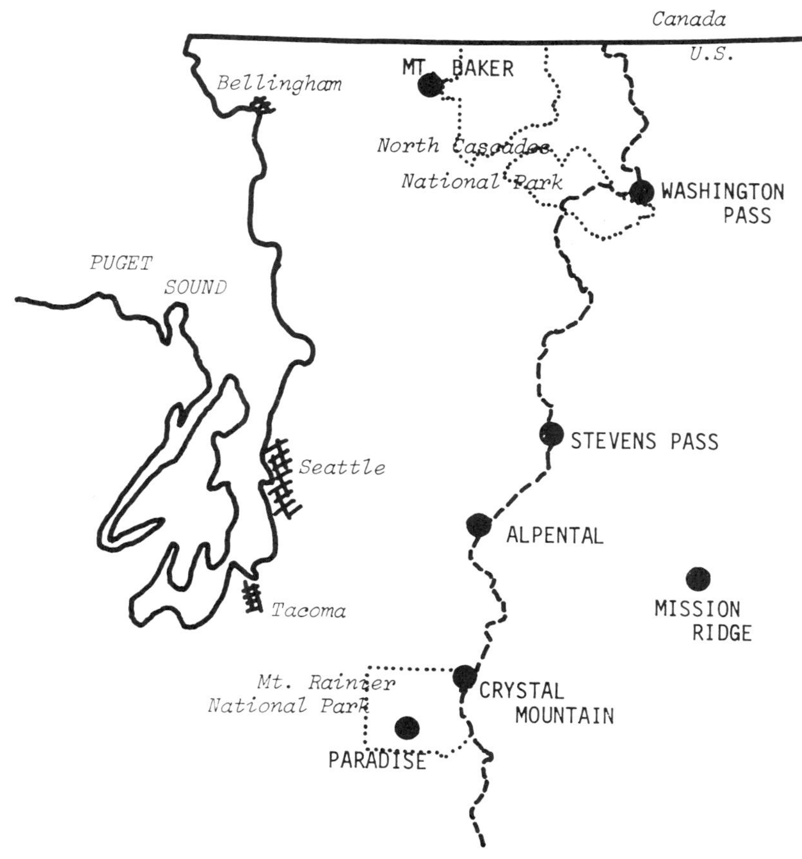

Figure 1. Location of observation sites in the Cascade Mountains of Western Washington State.

STATION		PRECIPITATION					TEMPERATURE			6-HOUR WIND AVERAGES		PRESENT WEATHER				LENGTH OF RECORD	
	elevation (feet)	total snow depth	daily new snow	liquid water in new snow	water from rain	precip intensity by 6 hr. periods	maximum-minimum	temp at time of observation	20 cm snow depth temperature	direction	speed	wind direction	wind speed	sky cover	weather	from	to
Paradise	5550	X	X	X	X	X	X	X				X	X	(X)	(X)	11/ 1/71	4/30/72
Crystal Mountain	4400	X	X	X	X	X	X	X	X	X^a	X^a	X	X	X	X	12/ 6/71	4/30/72
Stevens Pass	4050	X	X	X	X	X	X	X	X^b	X^c	X^c			(X)	(X)	11/ 1/71	4/29/72
Early Winters	2200	X	X		X	X	X	X	X^b	(X)	(X)	X	X	X	X	12/ 1/71	3/31/72
Mount Baker	4200	X	X	X	X	X	X	X	X	X^d	X^d	X	X	X	X	12/ 6/71	5/31/72
Alpental	3200	(X)	(X)	(X)	(X)	(X)	(X)	(X)	(X)							12/ 1/71	3/31/72
Mission Ridge	4600	X	X	X	X	(X)	X	X	X^e	X^f	X^f					12/11/71	3/31/72

X - indicates that the observation was made regularly

(X)- indicates that the observation was made intermittently

a - The anemometer at Crystal Mountain was on a north-south ridge at an elevation of 6900 feet

b - Snow temperature record at Stevens Pass and at Early Winters began 12/19/71

c - From 11/1/71 to 1/30/72, wind speed only was recorded at "Big Chief", elevation 4850 feet. From 1/31/72 to 4/30/72, both speed and direction were recorded, but in a new location, at the top of Chair #5, elevation 4875 feet.

d - The anemometer in the Mount Baker ski area is on Panorama Dome, at an elevation of 5000 feet.

e - Snow temperature record at Mission Ridge began 1/1/72

f - The anemometer at Mission Ridge is at an elevation of 6750 feet.

Table 1. Index of observations from Cascade observing sites.

1972/73 the Department of Highways supplemented the Forest Service observations at Stevens Pass. The University furnished office, telephone and the weather data facilities and one of us (Fox) coordinated the observations and handled the daily data flow.

The practical daily routine of the Data Network involved a series of morning telephone calls from the central office to collect the early-morning snow and weather observation from the Network stations. Reports of avalanche activity were also transmitted at this time. (Mt. Baker and Washington Pass depended on the intermediate radio link.) In 1972/73, a second data transmission from Washington Pass was obtained late in the afternoon. The most recent weather maps were then examined to relate the general weather conditions to the local ones observed at the mountain stations. During the second winter, special mountain weather and avalanche forecasts were returned to the field stations, with emphasis for Stevens Pass where a duty forecaster for the Department of Highways was stationed.

The Cascade Avalanche Data Network initially was established as a research tool to provide better understanding of weather and avalanche patterns in the Cascade Mountains, with particular reference to the North Cascade Highway. It immediately provided useful operational feedback to the field stations and has served as a pilot model for coordination of National Weather Service data flow with that from the mountain field stations. Even in the absence of formal avalanche forecasts, the distribution to Forest Service snow rangers and highway foreman of current regional avalanche activity reports, which are not normally available to the individual sites, has provided a useful additional tool in the management of snow hazards. We now summarize our significant findings from two year's operation of the Network.

The winter of 1971/72 was one of record snowfalls and extreme avalanche activity in the Washington Cascades. The basic sequence of events which has been known to be associated with avalanching in the Washington Cascades, that of substantial snowfall followed by a warming to temperatures near or above freezing, was a dominant feature of each of the five major periods of natural activity. In most cases, the initiation of avalanche activity coincided with the time of the warming, indicating that the rapid rise in temperature may have been the key avalanche trigger. In particular cases, a geographic progression of the initiation of avalanche activity appeared to be associated with a similar progression of the initial warming.

Although the initiation and in most cases a substantial portion of the avalanching was apparently related to the initial temperature rise in these situations, in every case of serious activity the avalanches continued to occur well after the arrival of the warm air, and well after the initiation of the activity itself. In each of the cases studied, the initial warming was followed by a period of continuing precipitation at temperatures which remained relatively warm, and avalanching continued to occur throughout this period. In some cases this later avalanching appeared to come as rather definite time groupings of activity. These secondary releases may have been associated with some secondary atmospheric warming that was small in magnitude but possibly large in geographic scale.

The evidence from all of the cases indicated that, as long as significant precipitation continued at warm temperatures (<u>warm</u> meaning that temperatures at the elevations of slide release zones were near freezing), the possibility of avalanche release was high. The chance of rain falling on large amounts of newly deposited snow was always present in these situations. This, together with the additional loading and possible lubricating effects of the rain on the snow pack at lower elevations apparently led to the continuing slide activity.

This meteorological sequence, the large snowfall ahead of a rapid rise in temperature, followed by an extended period of continuing precipitation at warm temperatures, was produced by a variety of synoptic weather situations in the cases studied. The most common situation was that of an intense, maturing warm front - cold front system, with a definite open warm sector present as it passed through the Cascades. In each of the cycles of January 11, February 16, and March 5, the new snow ahead of the warm front fell on substantial amounts of snow which had been deposited in the previous two or three days by preceding systems. Fairly intense orographic precipitation continued in the Cascades while the warm sector was present, as relatively strong westerly winds responded to the intense pressure gradients of these storms.

In contrast to the previous year, snowfall in the Cascades during the 1972/73 winter was well below normal. A recurring split upper level flow pattern tended to divert much of the intense storm activity to the north and to the south of Washington State. There was also significantly less avalanche activity, and there were no cycles which approached in severity any of the five serious cycles in the previous winter.

The most severe and widespread cycle of this second winter occurred late on the night of December 25 and on the morning of December 26. The avalanches were associated with rain from the warm sector of a frontal system which passed quickly through the state at that time. This was the most well-defined open warm front - cold front system of the winter.

During January 12 through 14, 1973, another period of natural avalanche activity occurred when a series of occluded fronts brought precipitation and gradually warming temperatures throughout the period. Rain occurred in many areas on the 13th and 14th.

Two separate invasions of cold, dry arctic air produced a different kind of avalanche condition. Strong pressure gradients associated with an arctic high pressure area northeast of the region produced, at Washington Pass and at Stevens Pass, prolonged periods of strong easterly winds which formed wind slab conditions at these times. The release of avalanches in these situations was in some cases independent of any frontal activity. In other cases, new snow or warming, as the arctic high pressure broke down, apparently weakened these slab layers to produce avalanche releases.

Most of the avalanches in these cycles, and in the entire winter in general, were relatively small, due to the absence of intense frontal activity and the resulting lack of large new snowfall amounts compared with those of the previous winter.

In summary, significant snowfall followed by a rapid rise in temperature and a period of intensive warm precipitation repeatedly led to the release of widespread natural avalanches in the winters of 1971/72 and 1972/73. These conditions resulted from the passage of intensive frontal systems, the distinctive patterns being frontal systems with open warm sectors.

Snow deposited during cooling trends led to noticeably less natural avalanching. Such snow was often susceptible to artificial release or to later natural release under the influence of warm storms. The value of avalanche control procedures in such a situation is clear.

It is obvious that an operational avalanche forecasting and control program in the Washington Cascades must include continuous and direct information on the developing large-scale weather situation. Approximate arrival times of frontal systems, and the duration and intensity of their effects, are key data for avalanche forecasting and control. Direct and rapid communication between different areas in the Cascades through a central data collection and forecast office can be of great use in recognizing and controlling avalanche hazard.

REFERENCES

El Hult, Ruby, 1960. Northwest disaster: Avalanche and fire. Binford and Mort, Portland. 228 p.

Judson, A., 1970. A pilot study of weather, snow and avalanche reporting for Western United States. Ice Engineering and Avalanche Forecasting and Control. National Research Council of Canada, Associate Committee on Geotechnical Research. Technical Memorandum No. 98, Ottawa. 163 p.

Judson, A., 1972. Personal communication.

Swiss Federal Institute for Snow and Avalanche Research, 1958. Der Lawinendienst. Interner Bericht No. 156, Weissfluhjoch. 26 p.

ACKNOWLEDGEMENT

Operations of the Cascade Avalanche Data Network and associated research have been supported by the National Science Foundation, Grant GU-2655.

AVALANCHE HAZARD EVALUATION AND PREDICTION IN THE SAN JUAN MOUNTAINS OF SOUTHWESTERN COLORADO

Richard L. Armstrong
Institute of Arctic and Alpine Research
University of Colorado

ABSTRACT

The Institute of Arctic and Alpine Research, University of Colorado, has begun a study of the causes of snow avalanches in the San Juan Mountains of southwestern Colorado in order to develop a forecast model to predict avalanche occurrence. The project is also evaluating the possibilities of increased avalanche activity which could result from winter cloud seeding activities of the Bureau of Reclamation's Upper Colorado River Basin Pilot Project. A general discussion of the various aspects of avalanche research is presented. The specific research methodology of the San Juan Avalanche Project is described with particular attention given to studies regarding the physical and stratigraphic properties of the snowpack. Problems regarding instrumentation associated with avalanche research are also discussed.

INTRODUCTION

Research relating to the nature of snow avalanches in the San Juan Mountains of southwestern Colorado is being conducted by the Institute of Arctic and Alpine Research (INSTAAR), University of Colorado, under contract to the Bureau of Reclamation, U.S. Department of the Interior. During the second winter of research (1972-1973), the San Juan Avalanche Project continued studies directed towards the development of a methodology which seeks to determine the causes of avalanches in the San Juan Mountains and to begin to develop a forecast model for the prediction of avalanche occurrence. In addition, the project is designed to evaluate the possibilities of increased avalanche activity which could result from winter cloud seeding activities of the Bureau of Reclamation's Upper Colorado River Basin Pilot Project. At present, seeding by the pilot project is not directed at the area of this study. It is hoped that a practical capability to forecast avalanches will become an important part of operational control of seeding activities, and that when a high avalanche hazard is forecast, seeding of the area will be suspended until the snow slides have been controlled or the snowpack has stabilized.

The study area is defined by the avalanche hazard along U.S. Highway 550 between Coal Bank Hill and Ouray (58 km) and Colorado Highway 110 between Silverton and Gladstone (14 km). Approximately one hundred avalanches cross these two highways with varying frequency.

Basically, the project has undertaken the study of the problem of avalanche initiation by analysis of the complex relationship which exists among terrain, climate and snow stratigraphy. When the project was initiated, only a limited amount of climatological data was available for the study area. It was recognized that an avalanche prediction model would rely heavily upon data gathered from highly accurate, reliable instruments installed on carefully selected sites. Therefore, a network of fixed instrumentation is utilized to measure meteorological parameters, determine certain physical properties within the snowpack, and detect avalanche events.

The primary snow study site is located at Red Mountain Pass (3400 m) and includes instrumentation providing such basic information as air temperature, temperatures within the snowpack, wind speed and direction, precipitation rate and amount, snow settlement rate, and net all-wave radiation at the snow surface. In addition, an isotopic profiling snow gauge provides snow density and water equivalent values throughout the snowpack at 1.0 cm intervals. An experimental network of seismic and infrasonic instrumentation was maintained during the first two winters research in order to determine the value of such equipment as avalanche event recorders. In addition, trip wires have been installed in the paths of frequently occurring avalanches in order to acquire accurate event times.

The current primary objective of the San Juan Avalanche Project is to determine the dominant, and hopefully predictable, processes which contribute most directly to the initiation of avalanches within the study area, to establish empirical relationships among these processes, and to expose these relationships to a detailed statistical analysis in order to provide the desired forecast model.

Such avalanche studies are a necessary part of any responsible program to augment snowfall for it is assumed that an increase in winter precipitation may well result in an increase in the number or severity of avalanches in a region.

However, due to the relationship which exists between any given period of avalanche activity and a complex and often prolonged series of meteorological events, there appears no practical way to determine the effect of any single cloud-seeding event on the subsequent avalanche patterns. The results of seeding or not seeding a storm in November could not be predicted in terms of how the particular stratigraphic layer within the snowpack composed of that precipitation event would react to an avalanche cycle two or three months later. However, public safety within the study area could be enhanced by providing accurate stability evaluations based on the meteorological events as they actually occur, whether or not they were influenced by cloud seeding.

If a winter seeding program is to be continued in the San Juan Mountains, or in any mountain area, it then becomes necessary to learn how to evaluate the avalanche hazard at a given time, to learn how to predict the effects of a given storm on the avalanche hazard as evaluated (to forecast the avalanche potential), and to learn how to control avalanches so as to present the least hazard to life and property at an economically justifiable cost.

AVALANCHE RESEARCH - A GENERAL OVERVIEW

Concern about the avalanche problem is by no means new, but in recent years there has been an intensification of efforts to find effective ways of controlling the hazard. Several useful techniques have been developed, and are operational while other methods are under study. Fundamentally, there are two different ways to prevent or control avalanches: (1) modification of the terrain and (2) modification of the snow, which includes deliberate release of the slides when and where they will do little or no harm.

Within the first category, the intent is to either anchor the snow on the slope in order to prevent the initiation of an avalanche or to divert the flow of the moving avalanche away from threatened areas. The retention or diversion structures required for this method are expensive and require continued maintenance, but are reasonably permanent and offer maximum protection. Modification of the terrain is usually chosen when the problem is to protect a large area or fixed installations. This method is used to a much greater extent in Europe than in North America.

The second category includes one of the oldest and simplest strategies, that of artificially triggering small avalanches before the potential for a large one has developed. Properly applied, artificial release can bring down piecemeal the snow load deposited along a path where avalanches are likely to occur. In addition, the deliberate release of snow at chosen times makes it possible to give adequate warning in order that the danger zone be cleared. The artificial release of avalanches is usually achieved by detonating explosives in the zone of potential avalanche release. Generally, this is accomplished either by hand-placing explosives in the snow or by transporting the explosive to the slope by some form of artillery. Modification of the actual snow cover in order to induce stabilization has also been undertaken by mechanical compaction as well as chemical means. Techniques involving the modification of the snow are most commonly applied in situations where smaller areas require protection such as highways and ski slopes.

A further method of avoiding avalanche hazard involves zoning ordinances which prevent all development related to human activity within endangered areas. While the task of defining which areas are in fact hazardous is by no means a simple one, the major difficulty with such a plan involves the legal complications of enacting zoning ordinances.

Perhaps the most elementary defense against avalanches is to attempt to forecast their occurrence so that people and transient traffic can be warned to stay clear. Although physical studies and experience have provided numerous clues, avalanche forecasting is often still considered an art rather than a science. The variables that determine when conditions are ripe for the triggering of an avalanche are numerous and complex. It is difficult to measure the mechanical properties of snow samples with any precision, whether in the laboratory or in the field, because the characteristics of the sample change rapidly while it is being handled.

Perhaps the greatest single problem regarding the advancement of avalanche research results from the anisotropic nature of snow as a material. The inhomogeniety of a natural snowpack has thus far prevented any systematic, detailed analytical treatment of snow mechanics and thermodynamics. Such basic properties as the tensile strength, elasticity, and viscosity of snow are highly dependent on temperature and therefore experiments done in the laboratory regarding such properties are valid for only one temperature condition. Problems also arise in attempting to relate strength values obtained from relatively small laboratory samples to the much larger areas and volumes associated with avalanche release zones within a natural snowpack. Consequently, there is currently no well established body of scientific knowledge to explain the causes of avalanches in general, or no universally accepted set of failure criteria for snow as a material.

Research relating to snow avalanche as a human hazard is taking place at various facilities at the present time. The Swiss Federal Institute for Snow and Avalanche Research has been an international leader in the investigations of avalanche problems for over thirty years. In Japan, research activities are maintained by the Japanese Disaster Prevention Institute, the Japanese National Railways, and the Institute of Low Temperature Science, Hokkaido University. Within the United States, the U.S. Forest Service has been extensively involved in avalanche research for over ten years, while several universities including the University of Colorado, University of Washington, and Montana State University have more recently become involved in this type of work. In Canada, the National Research Council is concerned with avalanche research and a thorough investigation of the problem is underway in the U.S.S.R.

At facilities such as those mentioned above, research related to snow avalanches is currently being conducted in the following general areas:

1) Basic physical and mechanical properties of snow
2) The avalanche release mechanism (failure criteria in natural snow)
3) Avalanche occurrence forecasting
4) Mountain weather forecasting
5) Methods for the artificial release of avalanches
6) Avalanche defense structures
7) Reforestation of avalanche slopes
8) The destructive force of avalanches
9) Structural requirements for buildings exposed to avalanches
10) Avalanche zone mapping and the implementation of avalanche zoning ordinances.

THE SAN JUAN AVALANCHE PROJECT 1971-1973

The INSTAAR project is unique in the sense that it is the only field research project of this type to be conducted in the United States which does not possess some type of direct avalanche control responsibility. Most previous studies have been associated with ski areas and operated on the principle that whenever an uncertainty existed regarding the stability of the snow within the area, control measures were undertaken to initiate artificial release of avalanches by explosives. While several of the avalanche paths being observed by this project are subject to artillery control by the Colorado Department of Highways, we are nevertheless offered the unique opportunity to study numerous slide paths which are allowed to release naturally. In the study of this natural type of avalanche event, it is hoped that our research will contribute to the basic body of knowledge regarding the release mechanism of avalanches. This would seem most appropriate at this time due to the increasing need to develop methodology for comprehensive or "back-country" forecasting as opposed to localized forecasting within the relatively confined limits of a ski area. In order to provide for the general safety of increasing numbers of people becoming involved in back-country winter recreational activities such as ski touring, helicopter skiing, snowmobiling and winter mountaineering, a better understanding of the phenomena leading to the initiation of natural avalanches becomes a primary concern.

As previously noted, much of the research which has been done in the United States regarding avalanche formation has been associated with the operational requirements of a ski area. Avalanche occurrence forecasting schemes within ski areas derive a twofold advantage from the fact that only a relatively small area is being considered. Field personnel charged with avalanche hazard responsibilities often have the opportunity to become intimately familiar with known avalanche release zones within the ski area. Specific control measures may be undertaken in regard to a certain avalanche path based on knowledge resulting from frequent investigations of the location in question. Secondly, such frequent investigations are direct in nature; that is, one is able to actually be present in the potential release area and by means of cursory stratigraphic investigations, or simply by ski-checking, valuable information regarding snow cover stability is obtained. Such an opportunity is not available when the hazard is of the magnitude associated with a highway, such as the San Juan Avalanche Project's responsibility concerning Highway 550. The number of individual avalanche paths, as well as their wide distribution, prevent such direct investigations from occurring except on rare occasions. In addition, responsibility for hazard control within a ski area often exists only during those hours when skiers are actually on the mountain, while a highway situation continues twenty-four hours a day.

Basically, personnel concerned with a highway situation must, in the interest of economy and logistical efficiency, develop a capability which will allow them to forecast a condition of instability within the snowpack of an avalanche release zone without the opportunity to actually investigate the situation at that specific location. In order to accomplish this, data acquired at more accessible snow study sites is extrapolated to release zone situations.

During the first winter's research (1971-1972), the San Juan Avalanche Project accomplished the initial objective of establishing a research procedure capable of adequately observing and recording the various phenomena associated with avalanche initiation. The next step was to attempt to determine the relative contribution of each factor and to isolate those processes which contribute most directly to avalanche formation.

During the second winter, considerable emphasis was given to the study of snow stratigraphy. Detailed investigations into the physical properties of the snow within the study area were prompted by the fact that the San Juan Mountains exhibit climatic extremes not found in more northerly latitudes where most practical and scientific knowledge of snow avalanche formation has been accumulated. The combination of high altitude, low latitude and predominantly continental climate produces a radiation snow climate.

Generally, this condition is the result of two factors. First, the extreme nocturnal radiation cooling occurring on all exposures produces snowpack temperature gradients of a magnitude sufficient to cause significant recrystallization or temperature-gradient metamorphism. The second factor is the substantial amount of solar energy available to south- and west-facing slopes. This daytime condition causes melt just below the surface and subsequent freeze-thaw crusts. These two situations continue to influence the snow cover throughout the winter and the resulting stratigraphy is highly complex. Therefore, during the second winter's research, considerable emphasis was directed towards a better understanding of the snow stratigraphy within the study area through the acquisition of abundant snow pit data.

These snow pits are of three types. One type is located at standard level snow study sites. A second type is located on a test slope or avalanche release zone which possesses an elevation and a slope angle and orientation comparable to actual slide paths in the immediate area, but which in itself is relatively free of hazard to the observer. The third type of snowpit is associated with the actual fracture line of the avalanche, whether naturally or artificially released. This type of investigation provides data which is the closest approximation to the idealized research objective, that of determining the actual physical and mechanical properties of the snowpack at the zone of failure. Fracture line profiles are obtained as soon as possible after the avalanche has released.

These stratigraphic investigations provide data regarding snow strength (rammsonde values), density, crystal type and size, and temperature gradient within the snowpack. These studies are further augmented by frequent measurements throughout the study area of snow strength and temperature gradients without employing the relatively time consuming pit method. Field personnel obtain such stratigraphic information with the aid of a ram penetrometer and a portable temperature probe.

The following general information is drawn from the stratigraphic studies of the past two winter seasons. The annual snow accumulation in the release zones, generally reaching depths of 1.5 to 3.0 m, is not sufficient to suppress strong temperature gradients with the prevailing radiation regime, and constructive, or temperature-gradient (TG) metamorphism predominates in the snow cover. During the winter of 1972-1973, in 21 release zone pits dug through the snow cover prior to 1 March, 71% of the snow layers (total layer thickness in centimeters) exhibited crystallographic evidence of either partial or advanced TG metamorphism. These same pits showed 18% new or partly-metamorphosed new snow and only 11% snow which was primarily produced by destructive, or equi-temperature metamorphism. The percentage of snow reported as TG-derived drops to 55% for 18 pits in March and April, as would be expected when spring melt and equi-temperature metamorphism have destroyed some of the crystallographic evidence.

Depth hoar formation is most prevalent at lower elevations within the research area (below 3200-3300 m) where snow accumulation is shallower and nocturnal radiation cooling leads to persistent temperature inversions in the valleys. On northerly slopes and in the forests a very weak depth hoar structure envelopes most of the snow cover. In the higher avalanche release zones, mature depth hoar is less extensive, although some is found in the lower parts of many pit profiles. During winters of light snow accumulation, such as 1971-1972, developed depth hoar layers appear more frequently in the release zones.

On the whole, snow cover in the release zones is mechanically weak throughout the winter. The deeper accumulation sites do ameliorate temperature gradients enough to allow some stabilizing gain in strength during the winter, but even at these sites certain weak layers tend to persist into spring. Of the 39 release zone profiles produced during the 1972-1973 winter, the mean ram resistance of 21 profiles prior to 1 March is only 6.1 kg. Ten of these 21 profiles have mean ram resistances of less than 5 kg, 7 lie between 5 and 10 kg, and only 4 lie between 10 and 20 kg. For the period March-April, the mean ram resistance rises to 13.8 kg. Throughout the winter, excluding new or partly-metamorphosed new snow, significant (over 5 cm) layers with ram resistances under 5 kg can be found in all but 4 pits, the latter exceptions occurring late in the spring. In 22 pits, a major part of the profile shows a ram resistance of less than 2 kg, with half of these less than 1 kg. The pervasive influence of TG metamorphism apparently maintains low snow strengths even in those layers where its crystallographic effects cannot readily be distinguished.

Weak bonding between layers occurs frequently. This often appears as thin layers of low cohesion snow between crusts. Intense surface radiation cooling and consequent recrystallization by strong TG metamorphism close to the snow surface appears to be the most common cause of such layers. Occasional falls of very soft, low-density snow, sometimes only a centimeter or two thick, also contribute to this layer formation. Crust disintegration through TG metamorphism also adds to poor layer bonding. In some instances, surface hoar layers have been identified as the lubricating layer in slab avalanches. All of these factors stem from strong radiation cooling at the snow surface except the low density new snow, and even this tends to be preserved by such cooling.

The moderate snowfalls and radiation snow climate of the San Juan Mountains combine to produce a snow cover which in most avalanche release zones and for most of each winter is conditionally unstable. By conditionally unstable it is meant that at any given time while the snow cover is at sub-freezing temperatures, it is only marginally unstable in respect to spontaneous slab avalanche release through internal causes, but it remains throughout each winter highly susceptible to either load-induced or thaw-induced avalanche release. The addition of new load to the snow cover by precipitation or wind-transport, or the introduction of meltwater to subsurface snow layers, normally by thaw but also possible by rain, are in themselves a necessary but not sufficient condition for avalanche release. The conditionally unstable nature of the snow cover resulting from predominantly low mechanical strength and poor layer bonding is a sufficient condition to generate wet snow avalanches following introduction of meltwater, but it is neither necessary nor sufficient to generate load-induced avalanches, for load-induced avalanches may fall through instabilities in the new snow alone, while a snow cover existing prior to the load event remains stable and does not necessarily participate in the avalanche formation (direct-action avalanches). But a conditionally unstable snow cover such as that so common to the San Juan Mountains can set the stage for load-induced avalanches which would not otherwise fall if the load had been deposited on stable snow.

An avalanche resulting from the failure of some weaker zone within the pre-existing snowpack can often be composed of numerous layers representing several previous storms. This type of avalanche, often referred to as a climax avalanche, is directly related to some instability within the old snowpack rather than being the result of poor bonding between the new snow and the old snow surface as is the case with direct-action avalanches. These direct-action avalanches occur during or immediately after a storm and incorporate only the new snow of that single storm. The climax type of avalanche appears to be a frequent occurrence within the San Juan study area. Of the 24 mid-winter fracture line investigations undertaken during the past two winters, 14 involved climax type avalanches.

Similar studies regarding snow stratigraphy will continue during the 1973-1974 winter. In particular, an effort will be made to learn more about the processes associated with temperature gradients. A limited body of field and laboratory evidence suggests that the vapor transfer and recrystallization process associated with depth hoar formation is initiated when temperature gradients exceed 0.1°C per centimeter but few data are available relating the magnitude of this gradient to snow density and crystal type. One phase of next winter's research will attempt to relate the rate of recrystallization and the ultimate size of the ensuing crystal to density and crystal type of the original snow. Fixed thermocouple arrays will be used to determine the temperature gradients in the natural snow cover at selected sites representing various elevations and slope orientations. The crystal structure of the surface snow layers will be investigated daily at those sites and regular snow pit profiles will be obtained at approximately ten day intervals. In addition to this field work, laboratory experiments will be performed in the cold laboratory located at Red Mountain Pass. Snow samples will be exposed to a varied spectrum of artificially imposed temperature gradients. The density, mechanical characteristics, and crystal types of these samples will be observed periodically. Crystal type evolution will be recorded with photomicrographic apparatus. The basic purpose of the above study is to observe the role of recrystallization in causing poor layer bonding which can lead to avalanche initiation. Prof. E.R. LaChapelle, University of Washington, consultant to the INSTAAR Project, will determine the methodology and direct the experimental procedure of this work. In addition, Prof. LaChapelle is to be credited for the primary contribution regarding the interpretation and analysis of snow structure data obtained during the 1972-1973 winter season.

INSTRUMENTATION AND AVALANCHE RESEARCH

Field projects engaged in avalanche research require instrumentation with the following characteristics. Durability of equipment is of primary concern since this type of research is located in a mountainous environment during the winter season. Those periods of weather during which instrumentation is most likely to fail, high winds, low temperatures, heavy amounts and rapid rates of both precipitation and rime ice deposition, are periods of critical interest to this field of research. Instrumentation developed for conventional climatological studies in less severe environments where an isolated loss of data during severe weather may be acceptable, can fail to produce data during those periods when the respective information would be most valuable. Due to the fact that study sites are frequently required at locations which are remote and without easy access, instrumentation that is light in weight, compact and independent of 110 VAC power sources is often essential.

To be most compatible with the needs of avalanche research, data should be available on a real time basis at uninterrupted or frequent intervals. A data acquisition system which provided, for example, an instantaneous value for wind speed at intervals of one hour would be of little or no use to an avalanche forecast scheme. Due to the turbulent nature of storm winds an instantaneous sample becomes statistically invalid. The stability of a snowpack can deteriorate at a rapid rate under conditions of intense loading by wind transport. Data regarding wind speed must be available continuously in order to adequately update an avalanche hazard forecast. Similarly, instrumentation designed to measure parameters associated with the physical and mechanical properties of the snowpack can only be put to optimum use when data is made available at whatever frequency personnel determine necessary. The subtle nature of the physical changes taking place within the snowpack cannot be properly analyzed by an instrumentation network possessing a predetermined data retrieval frequency.

While a more simplified approach to instrument design and manufacture possesses obvious advantages when reliability is considered, another positive feature of this consideration becomes apparent regarding field research in remote areas. Field equipment which is the product of an unnecessarily complex design suffers not only due to the increased potential for failure in one of its numerous components, but also, once such a failure has taken place, significant data loss can be incurred due to lengthy repair time. This situation is often caused by the highly complex nature of the instrument which precludes the possibility of field personnel implementing the necessary repairs, and further time is lost in either transporting the equipment to the necessary laboratory situation, or transporting the qualified technician to the field site.

Another problem encountered in the instrumentation of avalanche research, one which relates more to the nature of the phenomenon being investigated rather than the specific instrumentation, should be noted. The object of the study, the snow avalanche, occurs in an area which does not lend itself to ease of instrumentation. It has become common practice to locate primary instrument sites in level areas, below tree line, protected from the wind and easily accessible to the observer. An actual avalanche starting zone is certain to be in sharp contrast with this type of location.

A common dilemma in avalanche research results. While it is highly desirable, theoretically, to locate instrumentation within an avalanche starting zone, the practical limitations are obvious. In addition to the need to avoid the destructive force of the avalanche itself, sloping surfaces offer additional difficulties. Local snow structure is influenced by solar radiation and wind patterns, according to slope angle and orientation. Mechanical processes involved in creep and glide of the snowpack are related to the type and shape of the ground surface beneath the snow as well as to the slope angle and orientation. The difficulties involved in determining a representative slope become apparent and one returns to the concept of a standard snow study site for instrument location.

It then becomes necessary to extrapolate data obtained at a level, sheltered site to what may actually be happening on adjacent slopes, both above and below timberline. Such empirical relationships can be determined between study plot and release zone, although many years of observation may well be an essential requirement.

In one location, the San Juan Avalanche Project has departed from this standard approach by installing a secondary instrumentation site within 50.0 m of the starting sone of a series of avalanches which frequently affect the highway. The opportunity to measure precipitation, air temperature and wind speed and direction adjacent to these avalanche paths has provided much useful information. However, it still remains that the ideal would be to receive data from the actual starting zone. For example, in this particular instance the avalanches run in individual gullies, which when exposed to storm winds approaching at right angles can be expected to experience a pattern of new snow loading quite unrelated to the more level and unobstructed instrument site. In order to properly measure the amount and rate of loading, and thus to actually monitor the changing stress pattern, measurements must be made within that zone where, perhaps in direct response to this increasing stress, the failure will occur.

Optimum location of instrumentation in an alpine zone is a complex problem. Precipitation patterns are constantly being altered according to turbulent and orographic effects introduced by the mountainous terrain. Temperature gradients within the snowpack and incoming solar radiation values at the surface as well as related snow structure vary greatly according to slope angle and aspect. Under such circumstances, any single site could well be considered anamolous in terms of being representative of a larger study area.

THE COMPUTER AS AN AID IN AVALANCHE HAZARD FORECASTING

Len Miller
Dian Miller
Avalanche Control Consultants
Olympia, Washington

ABSTRACT

Beginning with twenty one years of daily weather and avalanche observations, we are attempting to develop mathematical models for avalanche hazard forecasting. Observational systems, data accumulation, and computer technology are combined to forecast and plan snow and ice avalanche control for winter recreation and transportation.

Preliminary results indicate that rapid, accurate hazard forecasts can be made with the use of a selective regression program. We are currently running forecasting programs for twenty-four-hour periods and developing programs for continuous storm periods. These programs will be further tested and expanded during the coming winter.

Most of the work for this paper was done while on an Individual Student Contract with THE EVERGREEN STATE COLLEGE in Avalanche Forecasting by Computer.

Man has been the primary factor in avalanche hazard forecasting while data collected by scientific instruments have been secondary tools. Forecasting seems to involve about twenty percent fact and eighty percent educated intuition on the part of the forecaster. A lot of talent has been directed toward experimental and theoretical investigations into the mechanics of snow. At the same time, significant advancements have been made in computer programming and telemetry. We felt that all of this knowledge must somehow be combined and computer technology applied to active forecasting.

The increased winter usage of mountain highways and recreational areas calls for more reliance on instruments and quantitative techniques. The complex mechanical and physical properties of the snow cover; the meteorological influences of wind, temperature, and temperature changes; slope and exposure of the terrain; and other factors all combine to complicate a quantitative approach. It is further complicated by the fact that data collected on these variables qualitatively changes from site to site, year to year, and observer to observer.

Avalanche observers and forecasters vary considerably in educational backgrounds. Some have advanced degrees and specialized training, but many have only a high school education. Thus, an effective system, while dealing with complex factors, must be easy to use and to understand. If a forecaster, utilizing a computer terminal, could type out the current observations and get back an evaluation, it would make forecasting faster and easier, not to mention the reduction in strain to nervous systems. While we cannot rely entirely on machines for our decision-making, we can considerably assist the decision-making process.

Dr. Robert Barringer, then with The Evergreen State College, felt it was possible to take a selective regression program he had written a number of years previously, and apply it to the problem. It had potential for the development of a system for forecasting of avalanches by computer. With Dr. Barringer's program, the technical assistance of Dale Baird at The Evergreen State College, and the Evergreen Hewlett Packard 2000B/C computer, our project began to take shape.

For the initial work we chose to use the data collected on Stevens Pass in Washington for the Forest Service West Wide Data Collection Network at Fort Collins, Colorado. The data was programmed with SNOW and SNOWUP. SNOW places the data in a file in a form useable by the computer. SNOWUP edits and updates the file. The first program developed to actually work with the data we call NORM. This program makes it possible to study the relationship between a single variable and the actual occurrences of avalanches. The following three figures are samples of runs of NORM. NORM builds a matrix from the data file. We can select any variable and see the relation between any variable and numbers of avalanches. We also select maximum and minimum values for that variable and the increment or step. In these figures the data is for winter 1971-72.

wind velocities	avalanches
+1.0000****	8
+3.0000****	0
+5.0000****	36
+7.0000****	23
+9.0000****	51
+11.0000****	87
+13.0000****	288
+15.0000****	83
+17.0000****	41
+19.0000****	57
+21.0000****	34
+23.0000****	19
+25.0000****	0
+27.0000****	193
+29.0000****	80
+31.0000****	0
+33.0000****	9
+35.0000****	121
+37.0000****	0
+39.0000****	23

new snow depth	avalanches
+0.0000****	226
+1.7500****	169
+3.5000****	151
+5.2500****	44
+7.0000****	126
+8.7500****	72
+10.5000****	146
+12.2500****	68
+14.0000****	99
+15.7500****	33
+17.5000****	99
+19.2500****	0
+21.0000****	0
+22.7500****	0
+24.5000****	0
+26.2500****	0
+28.0000****	0
+29.7500****	8
+31.5000****	0

minimum temperatures	avalanches
-6.0000****	6
-4.0000****	0
-2.0000****	2
+0.0000****	0
+2.0000****	0
+4.0000****	14
+6.0000****	0
+8.0000****	0
+10.0000****	30
+12.0000****	7
+14.0000****	54
+16.0000****	64
+18.0000****	231
+20.0000****	154
+22.0000****	326
+24.0000****	99
+26.0000****	72
+28.0000****	57
+30.0000****	66
+32.0000****	59

Figure 1. Sample runs of NORM for the variables of wind velocity, new snow depth, and minimum temperature. Wind velocity in this sample is from one mile per hour (.45m/s) to 39 mph (17.4 m/s) with an increment of 2 mph (.9 m/s). For example, for wind velocities from 11 to 13 mph there were 288 avalanches. In this example, new snow depth varies from 0 to 31.5 inches (0-80 cm) with an increment of 1.75 inches (4.4 cm). Minimum temperatures in this sample run vary from -6° F (-21° C) to 32° F (0° C) with an increment of 2° F. Notice the concentration of avalanches for minimum temperatures in the vicinity of 16-22° F (-9 to -6° C).

wind direction	avalanches
+0.0000****	103
+1.0000****	0
+2.0000****	0
+3.0000****	342
+4.0000****	8
+5.0000****	0
+6.0000****	487
+7.0000****	301
+8.0000****	0

crystal type	avalanches
+0.0000****	971
+1.0000****	424
+2.0000****	33
+3.0000****	609

date	avalanches
71-11- 1***	0
71-11-21***	39
71-12-10***	86
71-12-30***	104
72- 1-19***	148
72- 2- 8***	475
72- 2-28***	224
72- 3-17***	57
72- 4- 6***	61
72- 4-26***	47

Figure 2. Sample runs of NORM for the variables wind direction, crystal type, and date. Wind direction in this sample considers 8 directions and 0(no wind). 1 is N, 2 is NE, 3 is E, 4 is SE, and so on. Avalanches seem to be concentrated at E, SW, and W, representatives of the prevailing storm winds in this area. We are using the Forest Service standard notation for crystal type by number. The high degree of avalanching for type 1 seems unusual, and bears investigation. Looking at numbers of avalanches by date from 11/1/71 to 4/26/72 with a 20 day increment, the worst of the slide season seems to be from mid-December to the end of February with the high point from mid-January to mid-February.

water content - density	avalanches
+0.0000****	284
+0.0600****	572
+0.1200****	272
+0.1800****	84
+0.2400****	0
+0.3000****	0
+0.3600****	0
+0.4200****	0
+0.4800****	0
+0.5400****	0
+0.6000****	0
+0.6600****	0
+0.7200****	2
+0.7800****	0
+0.8400****	0
+0.9000****	0
+0.9600****	27
+1.0200****	0

total snow depth	avalanches
+16.0000****	1
+27.0000****	17
+38.0000****	21
+49.0000****	0
+60.0000****	13
+71.0000****	43
+82.0000****	59
+93.0000****	38
+104.0000****	22
+115.0000****	104
+126.0000****	218
+137.0000****	363
+148.0000****	225
+159.0000****	117
+170.0000****	0

maximum temperatures	avalanches
+2.0000****	6
+5.0000****	0
+8.0000****	2
+11.0000****	0
+14.0000****	4
+17.0000****	15
+20.0000****	13
+23.0000****	68
+26.0000****	70
+29.0000****	241
+32.0000****	524
+35.0000****	206
+38.0000****	53
+41.0000****	0
+44.0000****	23
+47.0000****	0
+50.0000****	16

Figure 3. Sample runs of NORM for the variables of water content, total snow depth, and maximum temperatures. Water content in this sample run varies from a density of 0 to a density of 1 with a step of 0.06. In this example, snow depth varies from 16 inches(.41 m) to 170 in.(4.32 m) with an increment of 11 in. (28 cm). This data has been used to help highway department determination of man-power requirements. Maximum daily temperatures in this sample run vary from 2° F(-17° C) to 50° F(+10° C) with a step of 3 degrees F(1.1° C). Note the concentration of avalanches in the range of 26 to 35° F. (-3 to +2° C) maximum daily temperatures.

Using NORM, different maximum and minimum values and steps may be chosen for any variable to investigate more closely within some range of that variable. It is limited, however, to comparing a single variable with numbers of avalanches. SNOWOP was built to look at combinations of variables. It can determine those ranges of any combination of variables that might result in the release of avalanches. The comparison is on a day to day basis. If variable one has six days that significant numbers of avalanches occur, and a second variable may have 45 days with avalanches occurring at a particular value of that variable, and the day count of the combination of the two is 3 days, we would conclude that the combination of the two had little likelihood of influencing avalanche occurrence. Both NORM and SNOWOP work from matrices which they build. It takes only a few seconds to build a matrix from the ten variables initially recorded for 179 days of observations for winter 1971-72.

Next comes the actual multiple regression analysis of the data using Dr. Barringer's program as the base. Dr. Barringer is somewhat unconventional in his use of concepts normally associated with equations of multiple regression. For example, the F-test is the conventional test from analysis of variance used in a slightly unconventional way: it is the ratio of a)the reduction in the sum-of-squares due to the inclusion of the variable under consideration, and b) the average variance associated with the residual degrees of freedom. Thus, if we have 100 data points at a stage where there are five independent variables already in the equation and we are considering inclusion of a sixth, and if the current sum-of-squares is 25 and the reduction due to the acceptance of the new variable is 4, then the F-test for this variable will be 4/1 divided by 21/94, which is 18. The test gives an idea of the worth of any given random variable. However, we list our variables in order of their F-test, biasing the choice of variables. Since the variables are not chosen at random, we must demand an F higher than the 3.96 found in F-tables using the 5% criterion for 10° of freedom. The program asks for an F-value, and we usually choose 5, considering it to be 'very good' in terms of significance of the variable. An F-test of 1 doesn't tell us much about the significance of the variable, but if the value were, for example, 10, then we could be fairly certain that the variable was the right one.

In our selective regression program, either the programmer or the computer can choose the variables for evaluation. If it is felt that the computer can make a better evaluation of the data, we can ask the computer to pick the best variables and make a forecast. With twenty-one years of data we made forecasting runs by picking a variable or variables ourselves as well as making runs with variables chosen by the computer. Again we will use a run from the 1971-72 winter as a sample. Figure 4 is an example of variables chosen by the computer for the 179 observations of the 1971-72 winter.

The selective regression program lets us arrange the variables into any sequence we choose for the purpose of checking the significance of any combination of variables. If we feel that there are only two variables influencing avalanche activity on a particular day, we can ask the computer to build an equation using these two variables and plot the results. We will again use an example from the data of 1971-72 winter; Figure 5 is a reduction of a sample computer forecasting run using only crystal type #3 as the independent variable. The computer was asked to plot the results. For each day we can compare the forecast with a function of the actual number of avalanches that actually occurred on that day.

After the variables are chosen, an equation of the form

$$y = a_o + a_1x_1 + a_2x_2 + \quad . \; . \; . \quad + a_nx_n$$

is developed. The y is the dependent variable, numbers of avalanches. The x_i are the independent variables. The a_i are the coefficients to be determined. The multiple regression solution gives the best value of these coefficients for a particular sample of observations. The solution also gives a measure of the reliability of these coefficients.

F TEST VALUE?5

	VARIABLE	EQU CØR	SØS REDUC	F TEST	CØEF
5	NEW DEPTH	0	40.6155	28.8782	.100475
15	WIND VEL-2	0	33.14	22.8761	1.50837E-03
8	WIND VEL	0	27.9777	18.9315	4.54234E-02
12	NEW DEPTH2	0	22.0182	14.5671	3.74847E-03
4	TØT DEPTH	0	20.7294	13.6486	7.14378E-03
11	TØT DEPTH2	0	18.6091	12.1567	3.99586E-05
21	EAST	0	16.6374	10.7901	1.10028
18	ANY 1	0	14.0989	9.05952	.686306
6	WATER	0	10.1287	6.41592	.300691

Figure 4. A sample of computer selection of variables from among a winter's observations. Observations given the computer as independent variables were date, total snow depth, maximum temperature, minimum, new snowfall, water equivalent of new snowfall, wind direction, wind velocity, and crystal type. The variables chosen by the computer for inclusion in the equation, in order of importance, are new snow depth, wind velocity squared, wind velocity, new snow depth squared, total snow depth, total snow depth squared, east wind, snow crystal type #1, and water content of new snow. The coefficients of these variables are given in the last column.

A number of intermediate regression equations are obtained as the variables are added one at a time. The variables are added, as we indicated earlier, according to a biased system rather than a random one. The apparently most important variables are added first, and those variables with no significant correlation to the number of avalanches are not added at all. In some cases a variable felt to be significant may be added at some point only to be deemed insignificant after several other variables are added to the selective regression equation. This insignificant variable is removed before more variables are added. Only significant variables are included in the final equation.

Correlation appears in the program when listing candidates and when listing coefficients for the equation. In both cases it is the correlation of the variable with the other factors already in the equation. Statistics usually assumes that the mean has been subtracted from each variable; we do not necessarily. Statistics usually discusses variability from the mean; we can use variability from zero.

The original calculation time on the computer for the selective regression equations to be developed and a forecast printed was on the order of twenty minutes. This was too expensive in terms of computer time. Two new programs were developed to create files with the data compiled into a cross-product matrix and a data matrix. This put the data into a form that was more readily accessible by the selective regression program. The selective regression program is now working the data into the equation of significant variables and their coefficients in much less than a minute. We can now type out the morning observations at a computer terminal, and in less than three minutes the terminal will type out a forecast.

At any time we may wish to consider only certain variables or even a single variable. The selective regression program will develop an equation accordingly. Also, at any stage, an evaluation in graphic form can be requested. Figure 5 is an example of an equation being developed for a single variable, and daily forecasts made from this equation. The computer has plotted this daily forecast and compared it graphically with a function of the actual avalanche occurrence on each day.

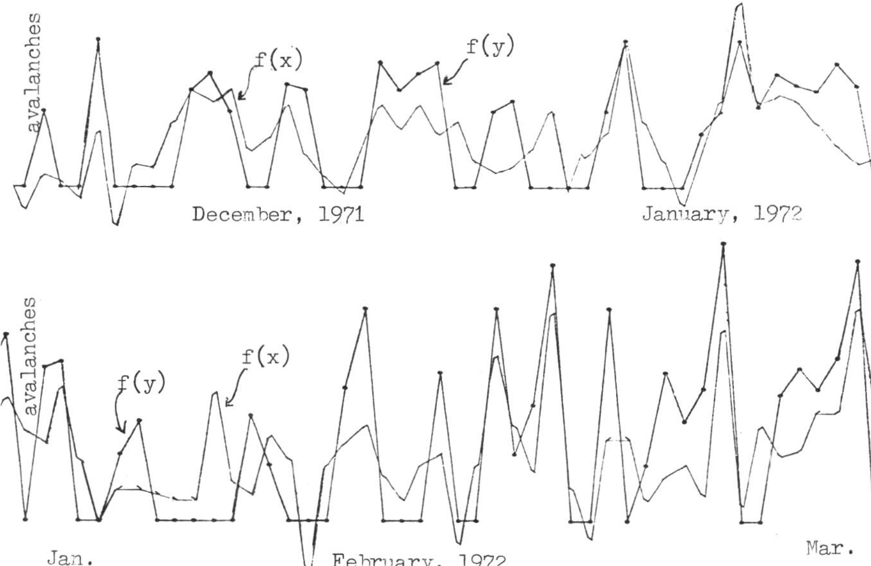

Figure 5. A reduction of a sample graphic comparison of a forecast with the actual occurrence of avalanches. The daily forecasts are printed with a slash, the direction of the slash indicating increasing or decreasing hazard. The dots are a function of the actual avalanche occurrence. In this figure, the dots and slashes are joined by lines although, in fact, this is not a continuous forecast. The dependent variable y is the avalanches observed on that day. The independent variable x is the snow crystal type #3. In this case the programmer developed an equation and made a forecast on the basis of a single variable.

There are some problems in all of the selective regression runs using the twenty-one years of data from Stevens Pass prior to the fall of 1972. The numbers of avalanches observed are those noted in casual observations from the ski area and an entire length of mountain highway. There was no control group of avalanches and no systematic means of observation of slides. Since data was recorded on a daily basis, the forecasts are necessarily in terms of an estimate of the numbers of avalanches expected during the next 24 hours. There are severe limitations on this kind of forecast in terms of its use for control work. It is, however, valuable as a 24-hour forecaster for use in a regional warning system for forecasting avalanche hazard. In this sense it can be operational now. Snow profiles, slope, exposure, and other potential variables were not a part of the twenty-one years of data. This winter we will be using these and other variables in our selective regression equations in an attempt to develop precise, continuous forecasts. Preliminary results along these lines with some very limited data from the North-Cascades sub-tropical winter of 1972-73 were encouraging.

Development and research will continue. This winter will see the addition of new variables, previously unavailable in our data. In addition, more frequent, precise data, collected expressly for inclusion in a forecast, will make possible an extension to continuous up-to-the-minute forecasting by computer.

ICE INFORMATION COLLECTION, DISSEMINATION AND USES ON THE GREAT LAKES

Frank H. Quinn
Lake Survey Center, NOAA

ABSTRACT

The waters of the Great Lakes are affected by ice conditions for approximately four months each year. These conditions influence navigation and power production in the lakes and connecting channels as well as contribute to the natural regulation of the Great Lakes system. The emphasis on ice information was greatly increased in 1971 with the advent of the inter-agency Great Lakes-St. Lawrence Seaway Navigation Season Extension Program. The ice information collected as part of this and other Great Lakes ice programs include ice type and thickness measurements, aerial and satellite imagery, time lapse photography, and ice movement measurements. The overall purpose is to provide reconnaissance information and detailed scientific data on the formation, type, movement and decay of ice in the Great Lakes. The information that is necessary for operational requirements such as ice forecasts and vessel routings is transmitted to an Ice Navigation Center which dissemates the data to the various users. The ice information that is used for scientific studies such as ice forecast developement, geochemical and petrographic analysis, and winter navigation problems is archieved at the Lake Survey Center in Detroit. Further ice research as well as additional development and more efficient use of remote sensing are necessary to obtain optimal use of the water resources of the Great Lakes.

INTRODUCTION

The purpose of this paper is to present a broad over-view of the current ice information activities on the Great Lakes. While the activities described will be primarily those undertaken or sponsored by U.S. government agencies, there are additional programs by Canadian agencies and university groups. The collection of ice information on the lakes probably began with the Indians inhabiting the region and continued on a piecemeal basis, based on individual needs, until 1963. At that time the Lake Survey Center established a comprehensive program for Great Lakes ice research. Under this program the collection of a suitable data base for both scientific and planning purposes was begun. The archiving of pertinent data from all sources was initiated and a program began to collect data relating to ice formation, accumulation, composition, decay, distribution and structural and crystalographic features. Out of this program has come an atlas on the Great Lakes ice (Rondy,1968) and a fresh water ice bibliography (Lake Survey, 1971). In addition a continuing series of reports on the Great Lakes ice cover are published on a yearly basis. The latest in this series covers the winter of 1971-1972 (Assel, 1972).

The emphasis on ice information was greatly increased in 1971 with the advent of the inter-agency Great Lakes - St. Lawrence Seaway Navigation Season Extension Program. This is a three year program authorized by Congress in 1970 to demonstrate the practicability of extending the navigation season on the Great Lakes. The Corps of Engineers is the lead agency of the program. The ice information portion of the program is addressed by the Ice Information Workgroup with the Lake Survey as lead agency. Other participating agencies in the workgroup are the Coast Guard, the National Weather Service, the Corps of Engineers, and the St. Lawrence Seaway Developement Corporation. This work-group is funded to augment and expand the ongoing ice information activities of the various agencies. Many of the activities presented in this paper are sponsored by the program, reports from which are made on a yearly basis.

ICE DISTRIBUTION CHARACTERISTICS

The ice formation, growth and decay characteristics of the five Great Lakes vary widely due to their different geographical locations, hydrometeorologic conditions and physical characteristics. The Great Lakes are situated between the United States and Canada with a climate varying between continental and semimarine. The wide range in latitude between the northern and southern lakes, 41°20' to 48°50' north latitude, leads to a wide range in winter temperatures. For example, the winter period ranges from about 82 days on the southern shore of Lake Erie to about 150 days on Lake Superior (Rondy, 1968). Also, for Lakes Michigan and Huron the wide range in latitude provides for the situation where ice can be forming in the northern ends of the lakes while decaying in the southern extremes.

The relative extent and time distribution of ice is also governed by the physical characteristics of the lakes themselves. The most important characteristic relative to ice formation on the Great Lakes is the ratio of lake volume to surface area, or average depth. These depths are shown for each of the lakes in "Table 1." The lakes with the smaller average depths deplete their stored heat and freeze faster than the deeper lakes. This is of greater importance than the latitudinal effect on heat loss rates as illustrated by Lakes St. Clair and Erie. These lakes, although the most southerly in the system, are the first to form substantial ice covers. The maximum ice covers which can be expected to form during a normal winter are given as a percentage of the total lake area in "Table 2" along with the probable dates (Rondy, 1969).

"Table 1. Selected Lake Characteristics."

Lake	Volume km^3	Surface Area km^2	Average Depth M
Superior	12,230	82,100	149
Michigan	4,920	57,750	85
Huron	3,537	59,500	599
St. Clair	4	1,113	4
Erie	483	25,657	19
Ontario	1,637	19,000	86

Thus it can be seen that each lake has its own characteristic ice formation, growth, and decay, depending primarily upon its physical characteristics.

"Table 2. Maximum ice covers, normal winter."

Lake	Maximum ice cover	
	Date	Percentage
Superior	March 25-April 5	60
Michigan	March 15-25	40
Huron	March 20-30	60
St. Clair	January 20-30	99
Erie	February 20-28	95
Ontario	March 10-20	15

SURFACE OBSERVATIONS AND MEASUREMENTS

The most numerous surface measurements are for ice thickness, which are taken throughout the Great Lakes and their connecting channels. Lake ice measurements are taken around the periphery of the lakes at 35 locations on a weekly basis by the Lake Survey and at about 43 locations on a daily basis by the Coast Guard. The measurements quantify the lake ice, snow ice, slush or water layer and snow cover thicknesses. The Corps of Engineers takes weekly measurements of the total ice thickness near the navigation channels in both the St. Marys and St. Clair Rivers. In addition ice samples are taken from each lake at continuing stations for crystalographic analysis.

Ice movement and distribution are monitored in critical areas by time lapse photography. The Corps have operated time lapse cameras at two locations on the St. Marys River and at one location on the St. Clair River. The cameras are super 8mm automatics equipped with time lapse mechanisms. Electric timers were used to shut the cameras off at night to conserve film. The film was run at a speed of approximately one frame per minute. The primary subject matter were ferry crossings and an ice bridge. The film provided valuable chronological sequences of the ice conditions. The main problem with the cameras is that they are a daytime system only.

An important part of the data collection program has been the monitoring of the environmental effects of the ice on shore structures and river banks. This effort, conducted by the Corps on the St. Marys and St. Clair Rivers, was designed to document the effects of ice movement on shore structures such as docks and piles under both natural conditions and those resulting from ice breaking or commercial navigation under ice conditions. A number of structures were selected for monitoring by both visual and photographic means to determine the type of damage, if any, the ice conditions causing the damage and the date the damage occurred. The impact of ice on shore erosion was analysed by conducting surveys on selected ranges throughout the winter season. Ice movement measurements were made using ice targets constructed of plywood and through the use of various dyes. The targets were placed in the St. Marys River using an airboat and were monitored by visual shore observations and aerial photography. They were identified by large numbers which could easily be observed during aerial reconnaissance. the targets are generally found to be effective but problems occur when major ice movements or breakup occurs during nighttime or storm conditions when visual observations are impossible. The dye marking programs are conducted using both yellow fluorescent and red Rhodamine B dyes. This type of monitoring was found to be successful for short time periods when little or no snowfall occurred and the dye strips could be identified from aerial reconnaissance.

In addition to monitoring the ice, a water temperature recording network was established for scientific studies related to ice freeze-up. Thermographs were installed in four bays and harbors on the lakes as well as at two locations on both the St. Marys and St. Clair Rivers. The thermographs are run year round to monitor both the heating and cooling cycles of the lakes. Series of transects were also made along the vessel routes across Lake Superior prior to freeze-up. Along these transects temperature profiles were taken from U.S. Steel freighters using a bathythermograph system with expendable probes.

REMOTE SENSING

The remote sensing program consists primarily of various forms of aerial imagery. The greatest portion of this consists of visual observations of the primary ice formation areas, generally taken on a weekly basis by the Coast Guard. Aerial overflights are also taken over each of the lakes during the period of maximum ice cover. The output from these flights consists of ice charts depicting ice location, concentration and type. A second form of aerial sensing which is proving adaptable to the lakes is side looking aerial radar, SLAR. SLAR overflights have been successfully demonstrated over Lake Erie for the past few seasons by NASA. Their principal advantage is that one flight over the axis of the lake yeilds a complete synoptic detailed overview of the ice cover independent of weather conditions. The radar's ability to see through cloud cover is a particular advantage during the winter season on the lakes. Developement is now ongoing to make the SLAR imagery available on a near real time basis for operational purposes.

The ice conditions in the connecting channels and some of the harbors are also documented through the use of aerial photography. Both low level oblique photography from small aircraft and high level vertical photography taken from Air Force photo-recon aircraft are used to provide photographs for mosaics to permanently record the ice conditions. In addition extensive use is made of 35 mm slides for documentation purposes.

Two supplementary forms of aerial sensing used to determine the water temperature regime of the system prior to freeze-up are infrared photography and aerial radiation thermometer profiles, ART. These are usually flown in the connecting channels and critical lake areas. Mosaics have been prepared from the infrared photos for the St. Marys River. This data generally serves as a critical input into various freeze-up forecasts.

The application of satellite imagery to depict the Great Lakes ice cover is still in its infancy. Two of the major problems associated with the satellites are resolution and frequency of coverage. The operational weather satellites, in general, do not have sufficient resolution to accurately depict the ice cover while the ERTS satellite is in orbit to pass over the lakes on only a eighteen day interval. Both types of satellites have problems seeing through the heavy cloud cover which blankets much of the Great Lakes during the winter season. Currently visible and IR data are being collected from the NOAA-2 satellite and from the DAPP Air Force satellite program during the winter seasons. These data are being analysed for both scientific and real time operational applications.

DATA DISSEMINATION

The ice information data can be broadly separated into two categories: the data required for operational purposes and that necessary for scientific and engineering research and studies. To provide for prompt dissemination of ice information and associated products the Coast Guard has established the Ice Navigation Center on Lake Erie in Cleveland, Ohio. This center which is in operation from December 15 until the end of the ice season serves as the focal point for operational information on the lakes. All real and near real time data are transmitted to the center by facsimile, teletype or telecopier for distribution to various users. The center is staffed by Coast Guard personnel and by an ice forcaster from the National Weather Service. They are responsible for keeping abreast of all vessel movements, for scheduling ice flights and other Coast Guard ice activities and for the dissemination of ice information to marine users. The primary output of the center is a daily ice summary accompanied by an ice forecast and outlook. This product is disseminated to various Coast Guard stations and to vessel agents. In addition, daily composite ice charts are transmitted on a daily basis to the National Weather Service Forecast Office in Detroit, Ice Forecasting Central in Ottawa, and to Coast Guard Radio Station Boston. This station rebroadcasts the ice analysis charts by radio facsimile to Coast Guard and commercial ships operating on the lakes. The center also provides on request, routine and specialized weather and ice forecasts to various users.

The centers communications network includes two teletype receivers, a facsimile and an auto-answer telecopier system. The communications also includes a dedicated National Weather Service marine circuit that links six National Weather Service offices and the Toronto Weather Office in Canada, with the Center in Cleveland and five commercial marine radio stations.

Much of the data used by the center for operational purposes as well as most of the data used for scientific and engineering research studies are archived at the Lake Survey Center. This data includes ice thickness measurements dating from the early 60's as well as water temperature data, visual ice reports, photomosaics and ice charts. Much of this data is disseminated on an individual request basis.

DATA USES

The uses of ice information are many and varied, ranging from the prediction of freeze-up in selected harbors to the determination of vessel routes through the ice. Ice extent and coverage information is currently serving as an input into heat budget and evaporation studies on Lakes Erie and Ontario. The data are also being used to study erosion rates and processes on the lakes and connecting channels as well as serving as baseline data to evaluate the environmental impact of ice breaking and commercial vessel navigation on the lakes and connecting channels.

Forecast techniques are being developed and tested to forecast freeze-up, break-up, ice growth rates and maximum ice coverage. Ice information currently being collected as well as that from the historical data base serves as a primary input. In addition the associated water temperature data will serve to monitor the heating and cooling cycles of the lakes and provide an input into other Great Lakes research studies.

On the operational side the information is used by the National Weather Service in the preparation of daily ice messages and 30-day outlook charts showing the expected ice cover on the lakes. The Coast Guard uses the data for daily ice summaries and for the scheduling of ice breaking operations. The ice information aids commercial navigation interests in scheduling their spring outfittings as well as winter layups and in planning extended season vessel transits through critical areas of the system are likely to have adverse environmental effects such as ice james, flooding and damage to shore structures.

ICE INFORMATION NEEDS

Because of the heavy cloud cover over the lakes during the winter season an active rather than passive sensing system is needed to provide continuous synoptic ice information. Ideally the system should be capable of sensing both the areal extent and thickness of the ice cover. A limitation of the current passive sensing by satellite is that it is incapable of "seeing through" the heavy cloud cover which extends over the lakes. A possible system might consist of an active sensor, such as SLAR, with high resolution, such as ERTS, and have frequent passes over the lakes. At the present time the thermal sensors on the NOAA-2 will provide valuable information for monitoring the cooling cycle of the lakes. This, along with the DAPP system will be used this winter to monitor the ice cover whenever possible. The ERTS satellite although limited by its low return interval does provide for accurate assessment of various ice characteristics.

A good starting point in the application of remote sensing to the lakes has been NASA's successful SLAR program on Lake Erie Most of the lakes are of the size that a single pass down the lake axis yields a complete synoptic view of the ice cover, regardless of weather conditions. The extension of the program to the other Great Lakes is being investigated for inclusion in the Winter Navigation program for the coming winter. If successful, it could supplant most of the visual aerial coverage of the lakes.

A second application of remote sensing now undergoing development is an accurate sensor to measure ice thickness. Currently NASA is working on a pulse radar system for this purpose. A radar pulse is partially reflected from the ice surface and from the ice-water interface. The difference in the reflections indicates the ice thickness. The sensor would be used to measure ice thicknesses throughout the lakes but with particular emphasis on vessel routes, connecting channels and harbors. It should be adaptable for measuring both normal ice thicknesses and ice ridges and dams up to about 25 feet thick.

An all weather system for determining ice movement would also be a valuable tool of ice information. This could be either shore based or use a satellite as a platform and should be capable of monitoring ice movements in all types of weather conditions.

With the improvement and development of remote sensing systems capable of sensing the extent and thickness of the Great Lakes ice cover and its movements will come a better understanding of the physical processes involved in forming the ice cover and their impact upon the lake processes. This, along with providing a more efficient real time system will lead to a more effective use of the water resources of the Great Lakes.

REFERENCES

Assel, R.A., 1972. Great Lakes Ice Cover, Winter 1970-71. Detroit, Mich: Lake Survey Center, NOAA.

Lake Survey Center, and Artic Institute of North America. 1971 Annoted Bibliography on Freshwater Ice. Detroit, Mich: Lake Survey Center, NOAA. 425p.

Rondy, Donald R.,1968. Great Lakes Ice Atlas. Detroit, Mich: Lake Survey Center, NOAA.

MODERN DEMANDS ON THE CANADIAN ICE ADVISORY SERVICE

W.E. Markham
Department of The Environment, Canada

ABSTRACT

Canada has developed a comprehensive sea ice reconnaissance and forecasting program over the past fifteen years which extends from the Great Lakes and Grand Banks in winter to the fringes of the Arctic Ocean in summer. Initial visual mapping and message relay from short range aircraft has given way almost exclusively to teams of observers aided by an array of remote sensors passing data directly to ice-breakers and from airport to forecast office using facsimile equipment.

At the same time, ice forecasting requirements have become more and more involved, their period has lengthened and seasonal outlooks are now required in most areas.

This evolution has also extended to the data store for northern and offshore petroleum developments, planning and design of larger ice-breakers and strengthened cargo vessels are continually raising new questions concerning the nature of the ice pack, of its ridges, the size, numbers and paths of icebergs that tax the capability of an all-purpose survey system.

Considerable improvements in observing capability and in data handling are required and steps are being taken to explore and exploit these as national economics permit.

INTRODUCTION

Canada has conducted an ice reconnaissance survey on an annual basis since 1940, but until 1956 it was a limited operation concerned only with the opening of marine navigation through Gulf of St. Lawrence to Quebec and Montreal or through Hudson Strait and Hudson Bay to Churchill. These initial surveys were conducted by Marine Services Ministry of Transport and concentrated primarily on reporting existing conditions to ships in the area.

In the mid-fifties responsibility for ice reconnaissance and forecasting was transferred to the Meteorological Branch and support was soon provided for Arctic resupply missions as well as the first winter traffic into the Gulf of St. Lawrence. Gradually the service has expanded in area, in times of operation and in the complexity of the ice data obtained and distributed.

When ice forecasting began in 1958, the raw data was obtained by observers in DC-3 aircraft which were on short term charter and thus had no sophisticated aids for navigation or data relay. Information was passed by teletype message from the aircraft base to Ice Central and forecasts were relayed to ships through marine radio stations and a facsimile broadcast of charts from Halifax or from certain bases in the Arctic during the summer.

Progress since that time has been significant in some areas and rather minor in others so that the present situation is rather imbalanced in many ways.

DATA COLLECTION

From the short-term charter of two engined aircraft, the ice service progressed gradually to larger aircraft and longer contracts. We have recently completed a 6 year charter of 2 DC-4's and are now in the second year of a long contract for 2 Lockheed Electras. For ice reconnaissance these aircraft cruise at about 200 kts, have a 12 hour range and carry both inertial and Omega navigational systems. From the original observers position in the co-pilot's seat (except in take-off and landing) we have progressed to a special observing canopy on the top of the aircraft just behind the pilots with additional blisters on the sides that provide a straight down view of the ice.

As sensors, the progression has gone from pure "eyeball" through various radar installations to a high resolution ground mapping set. Lately, a laser profilometer and an infra-red line scanner have been added along with tri-metrigon aerial cameras, but full exploitation of this array is just being developed.

We have, thus, a long range reconnaissance aircraft that can obtain ice data which is positionally accurate anywhere in our offshore waters, providing the weather is reasonable. This point is one on which pressure for improvement is now developing, as I will discuss more fully later.

Even though they were not designed for ice surveillance, the TIROS and ESSA satellites have been a valuable source of ice data ever since the orbital inclination was increased to 50 degrees. Recognizing that satellites have their limitations, mainly because of cloud and darkness, it is clearly apparent that the VHRR imagery of NOAA-2 and the multi-spectral capabilities of ERTS-A have a distinct place in the synoptic scale surveillance of ice in North American waters. As the resolution has improved and the scale has increased, the operational use of satellite imagery appears to be a distinct possibility in the near future in certain circumstances. At present, though, the climatological aspect is strongest with some indirect operational benefit in directing the aerial reconnaissance.

DATA RELAY

The original teletype messages describing an ice map were subject to error in preparation and in replotting in Ice Central. A number of years ago, a facsimile relay over long distance telephone lines to Ice Central was tested and this procedure spread later from the Eastern Seaboard area to include Hudson Bay and the Arctic by 1971. The most recent addition in this regard has been telecopier relay via satellite to and from the Arctic during the past summer.

Prompt relay of data from the observers to Ice Central and a scheduled broadcast of ice data and forecasts is one of the essentials of a thorough national service, but since the aim of the programme is to assist shipping, it is pertinent to have a capability of relaying data directly from the reconnaissance aircraft to ships in the area. With this in mind, a facsimile transmitter was installed in the DC-4's and regular broadcast of data was soon standardized. This is of considerable assistance to the ships in the reconnaissance area on the day of a flight, but it does not provide the broad area picture that is available from Ice Central. Ships also need updated information when the aircraft is not in the area.

From the time the programme began, a facsimile broadcast from Halifax has been conducted twice daily for up to 40 minutes at a time. Coverage on the Eastern Seaboard is quite adequate, but reception in the Arctic is only marginal. For the first few years of the programme, broadcasts over low powered transmitters in the Arctic were conducted, but the coverage provided by them did not have adequate range. The next step was to provide a broadcast from a more powerful station in Edmonton, and for 2 or 3 years, a staff of forecasters worked in that office during the summer. More recently, the work has been done in Ice Central and charts are passed by land line to Edmonton for broadcast. The total time available for ice data is almost 2 hours per day on this station, but again, reception in the Arctic is subject to interference. This is also one of the areas where pressure for improvement in the service is growing. It is, in fact, the weakest link in the whole system.

FUTURE DEVELOPMENTS

The limitation imposed by weather conditions often arises at very critical times for winds are strongest and ice is most mobile during the passage of migratory storms. The fact that reconnaissance is impossible at this time is a serious limitation on the Eastern Seaboard during the winter. In northern waters, although storms are less frequent, the occurrence of fog is more common and reconnaissance is often limited for that reason. In the far north, there is also the limitation imposed by light level in October and November, particularly, and in addition reconnaissance data concerning the age of snow-covered ice is sometimes inaccurate. After considerable testing, the Atmospheric Environment Service is budgeting for Sideways Looking Airborne Radar (SLAR) in 1975-76. This will give us capability of consistent ice reconnaissance up to 30 miles from the aircraft track regardless of weather, snow cover, and lighting conditions. It will, however, raise considerable problems with regard to data handling.

Our present system uses a chart scale of 1:2 million, although air to ship relay is electronically increased to 1:1 million during transmission. Once SLAR is available, the observers in the aircraft will have to have a capability of real time interpretation; the data must be passed to the ships in the area and it is likely that the scale used will vary according to the ice conditions. It is quite conceivable that in severe ice conditions the ships will progress along their intended track by dodging certain floes and navigation will be by ice conditions rather than along a predetermined track. This facet of a SLAR oriented programme does not appear to present any major problems, although transmission time from the aircraft will likely increase.

Once the aircraft reaches its base, the interpreted SLAR imagery will have to be transmitted to Ottawa, but the present channels are likely to be adequate. Telecopier relay is marginal in this case, because of its size limitation, but the larger facsimile equipment can likely be provided at most bases.

Data relay from Ottawa to the ships presents a major problem for extra time is simply not available on the high powered broadcasts that are now in use.If the scale of the charts being relayed is changed only from 1:2 million to 1:1 million a 4 time expansion of broadcast time is required.

In difficult ice situations, a scale of 1:500,000 may be necessary and this will require 8 times as long for relay of the same area. Teams are at work on this problem, but no solution has yet been provided. When the reliability of receipt of the data in the Arctic is also to be improved it becomes a real problem in technology.

One other concept is also worth considering in discussing a SLAR oriented programme. If a ship is proceeding through severe ice conditions and if there are particularly large floes of old ice to be avoided, relay of a detailed chart from the aircraft to the ship will be particularly valuable. However, the aircraft cannot be in all areas every day and later vessels will have considerable trouble in identifying these large old floes in proceeding down channel, for their later motion will depend on both wind and current. Tidal effects are completely ignored in present ice forecasts and advisories, but these effects too will have to be considered in cases such as this. It is apparent that some sort of markings device, perhaps a transponder, would be very useful if one could be dropped by the aircraft to identify a particular floe. This would assist the ships in avoiding it as they approach and will also provide additional information about ice motion during repeated reconnaissance flights.

This area has not been explored fully as yet, but a transponder capable of withstanding an airdrop is available and further effort in this area will likely be useful.

Other remote sensors may also be developed which will provide useful information in the future. A prototype system for airborne measurement of the thickness of fresh-water ice is already available I understand, and considerable effort is being directed towards development of one for salt-water ice. Without wanting to be a "Doubting Thomas", I suspect that such a system is several years down the road for when one considers the actual variations in thickness of ice over even a 100 meter square and propose to measure this from one to two thousand feet at about 200 kts, the task is obviously a gigantic one. Probably a helicopter-borne system will be the first stage of the development and even that will not be available next year.

Passive and active radiometers are also being developed and tested, but as in the case of the thickness sensors, time is needed to identify their capabilities and limitations before employing them in an operational manner.

On the observational side then, there is considerable pressure for all-weather reconnaissance and increasing signs that surveillance of Arctic ice conditions will soon be required during the winter night. The major problems at present are in acquiring an adequate number of dollars for the required equipment and the identification of a reliable relay method for passing data from Ottawa to ships at sea. This, too, relates in part to expense.

On the forecasting side there has not been any major changes in either procedure or content. Ice forecasting is limited in two ways - knowledge of water motions (both tidal or residual currents) is skimpy at best, and the effect of one floe on another - the so called internal stress is not well known. Besides these two items, the concept of a detailed forecast of ice conditions in an area such as the Gulf of St. Lawrence over even a 24 hour period would require so much information that the length of the bulletin would make it nearly useless.

As a result, our forecasts concentrate on a forecast of the wind induced ice drift and an indication of where pressure is expected to develop. They are, besides, limited to the shipping track in actual use. The ship captain should be able to interpret this information on the basis of his own position and intended motion, but although icebreakers are capable in this regard, many commercial vessels have limited knowledge of ice and its behaviour.

The output from Ice Central is largely concerned with tactical forecasts covering up to the third day from time of issue. Longer range forecasts based on the National Weather Service 30 Day Outlook are also provided for distribution by mail.

Seasonal Outlooks of conditions on the Eastern Seaboard and in Arctic Waters are issued as appropriate. The one for the Gulf of St. Lawrence is based on an oceanographic survey and considerations of future heat losses during the freeze-up period. It appears that there is a persistence of weather patterns in this area during the winter which permits extension of the estimate through the entire growth cycle and this type of forecasting has been moderately successful. In the Arctic areas, the Outlook is closely related to ice conditions just prior to freeze-up combined with information on wind drift and formation of polynyas in the spring. Its accuracy is clearly lower than that of the Outlook for the Eastern Seaboard.

Offshore developments such as drilling rigs on the Grand Banks and in Hudson Bay will present a need for spot forecasting rather than route forecasting and considerable effort is being devoted to development of this capability. Of particular concern is the coastal sector of the Beaufort Sea where active drilling is expected by 1975. For areas like Hudson Bay, there is a clear need for a 14 day forecast for this is the lead time required for entry of the drilling rig and for its escape during the freeze-up period. At present there is no known meteorological basis for such forecasts, but a close examination of past ice records may make a computer-based approach regarding rates of disintegration a useful guide.

In southern waters there is an increasing requirement for an ability to forecast freeze-up in the St. Lawrence Seaway. The St. Lawrence Study Office has made some progress in this regard using present and past records of water temperature, but the range required for a useful forecast - 60 days - is a real challenge and one which will require a statistical type of approach. For forecasting, then, there is a need for detailed spot forecasting in several areas and for an increasing capability in the long range forecasting field

PROCESSED ICE DATA

Over the past 4 or 5 years, there has been a marked increase in queries related to coastal and offshore marine ventures. New harbour installations, offshore drilling, improved ship design, and feasibility of longer operating seasons in the north are some typical examples. In this regard, our data bank is being pressed to the utmost, but some useful information can now be extracted. One of the programmes developed in Ice Central soon after it was established was to prepare a weekly map of ice conditions in as broad an area and in as great a detail as possible. As a result, we now have a series of these charts going back 10-12 years in all the major areas. This period is not particularly long, of course, but study and interpretation of the charts in the light of the meteorological regime makes it possible to extend the effective coverage to 30 years or more in the Arctic and 70-80 years on the Eastern Seaboard.

This is one of the major advantages of using meteorologists for ice studies.

Although northern ice data is limited to the break-up period, a digitizing programme to place much of this data on punch cards has already been started and preparation of detailed ice atlases will soon be possible. Major progress is being made on the southern areas from the Gulf of St. Lawrence to Hudson Bay and these will be the first to appear. The card design can be extended into Arctic Waters, but before that programme is commenced, a review of latest advances in this field should probably be undertaken. In view of the knowledge that has been acquired in the past 15 years concerning ice conditions in North American waters, all existing atlases are hopelessly outdated. Advances in ship design and powering methods have accented this deficiency. In summary, then, the data banks which have been developed are being computerized, but the questions raised regarding ice occurrence, thickness, floe size, details of ridge structure and other questions are pushing our capabilities to the limit, and sometimes beyond it.

In all the foregoing discussion, I have concentrated mainly on the occurrence and movement of sea ice, but at the same time, glacial ice is a matter of real concern. Traditionally, the International Ice Patrol is the authority for iceberg data on the Eastern Seaboard, but, in fact, their terms of reference are clearly related to the Grand Banks only. Other data concerning berg numbers, drift rates, size, etc., have been gathered by the Patrol, but these are an irregular bonus rather than their true mission. A great many questions are now being asked regarding drift of bergs, the depth and frequency of scour on the sea bottom in Labrador, Davis Strait and Baffin Bay, and about the occurrence of ice islands and their fragments in the Arctic Archipelago. Information in this regard is almost non existent, and in many cases special surveys using sonar equipment are necessary on the part of the surveying engineers. Pipeline construction is concerned with these answers and will also be affected by the ice regime in the coastal zones where the lines go below the sea. Specific surveys in this regard are under way, but there is no data bank to assist in such studies.

SUMMARY

At the present moment, the Canadian ice program is approaching a major challenge because of the increased demands placed upon it. Conversion to an all-weather system will increase its operational cost by about one-third, but there is a major technological problem to be solved in the relay channel to ships at sea.

The data bank available is being "massaged" to provide guidance to marine developers and a series of ice atlases will be produced. Many questions still remain to be answered, but new sensors and special ground based studies have been initiated to seek the answers needed.

SUMMARY OF DISCUSSION
(Papers 3.1 to 3.7)

Concern was expressed that present methods of dealing with the spatial distribution of snow and ice are based on empirical methods rather than physical models. It appears that this is due to the lack of manpower and funding limitations. In other situations, it may be due to the inaccessibility of particular regions, or to the dangers involved in obtaining data in certain areas (e.g.; crevasse areas of glaciers).

It was agreed that more efficient methods of collecting data are necessary, and that one technique that can be used to good advantage is stratified sampling such as that used by Stepphun and Dyck. It was pointed out that this method may alleviate the problem of getting to regions of difficult accessibility year after year. Measurements in such areas may only have to be carried out over a few years in order to actually get representative values for that particular type of area. Some experienced surveyors are using a form of intuitive stratification based on their past knowledge of which sample areas give the best results.

A discussion of precipitation gauge versus ground survey water equivalents quickly developed into a question of "What happens to the alpine snowpack during the winter?" Several participants mentioned instances of loss from the alpine pack under non-melt conditions. This was also pointed out for instances of snowfall in the Antarctic. There was some disagreement as to whether or not the snow sublimated, and if so, whether sufficient energy was available, or the snow was redistributed. The consensus was that under blowing conditions the airborne snow particles could undergo rapid sublimation.

SUMMARY OF DISCUSSION
(Papers 3.8 to 3.12)

Papers 3.8 to 3.10

There was a general consensus that avalanche research has not received adequate attention in the U.S., particularly in the area of avalanche forecasting techniques. For example, the release mechanism involved in avalanche initiation is not fully understood, and adequate data exists for only a few locations in the U.S., so knowledge of the condition of a snowpack prior to, during, and after an avalanche is very poor. Compared to other natural hazards, such as floods or tornadoes, avalanches cause relatively less damage to life and property in the U.S., and therefore avalanche research is generally not given a great amount of attention or funding. However, many of the participants felt that if the field of avalanche research is to be allowed to make necessary, and overdue, advances, funding for studies of the basic physical and mechanical properties of snow must be given a high priority.

Avalanche forecasting also suffers from a lack of accurate mountain weather information. If this were available, current forecast schemes could provide a public warning of the chance of avalanches in a given region.

Instrumentation, such as the profiling snow gauge and the 12-foot snow pillow, was mentioned. It was pointed out that these have been tested and are in use where applicable. The problem still remains of developing instrumentation to be placed in or near potential avalanche starting zones to measure the physical and mechanical properties within these zones.

Papers 3.11 and 3.12

In an exchange of comments concerning active and passive sensing systems, it was suggested that a passive system could be used effectively to observe open water, but that radar shows ice better. Passive imagery from the ERTS satellite, however, has been disappointing, for in studies in the Great Lakes Region it seemed that whenever the satellite passed overhead (every 18 days) there was almost always some cloud cover. An application of passive microwave imagery to detect ice cover was suggested, and a slide was shown of an image obtained from an altitude of 3 km.

It was noted that the NOAA-2 satellite provides good data on cloud-free days. Quinn (paper 3.11) responded that he needed both more frequent data, and to be able to depend on getting data through any weather.

A consensus developed that various kinds of remote sensing data needed to be pulled together and provided as a package to operational users.

Some of the difficulties in interpreting side-looking airborne radar (SLAR) imagery were discussed. Three points were brought out: One was that the fidelity of the imagery could not be counted on to be consistently good (or consistently bad). Another was that interpretation of the imagery depended on the ability of the interpreter. The last was that the quality of the imagery depended a lot on the skill of the SLAR operator during acquisition of data.

Another discussion involved the problems of measuring ice thickness in salt water. Doubt was expressed that a practical device could be developed that would be able to collect useful data from an airdraft flying at 200 knots. However, such a device might be successfully mounted on a helicopter whose mission was to seek out the best path for its parent icebreaker. One participant noted that a prototype of such a device has been developed. It is mounted on a helicopter and is flown at a constant altitude over the ice, using a computer aboard the helicopter to process immediately the impulses received.

Session IV

Radar Techniques

Chairman:

W. I. Linlor, Ames Research Center, NASA

General Reporter:

H. L. Ferguson, Atmospheric Environment Service,
Environment Canada

USE OF RADAR TECHNIQUES FOR SEA ICE MAPPING

S. K. Parashar, R. K. Moore, A. W. Biggs
The University of Kansas Remote Sensing Laboratory

ABSTRACT

Radar scattering data from sea ice at 400 MHz and 13.3 GHz were analyzed for an April flight near Point Barrow, Alaska. Seven categories of sea ice were identified on aerial photographs. The radar backscatter cross-section (σ^o) is compared for different categories of sea ice at various incident angles. Multi-year ice (ice more than 180 cm thick) gives the strongest return at 400 MHz. Water can be differentiated at both the frequencies. Some ambiguity exists between radar returns from the thinnest ice layer (under 18 cm thick) and a moderately thick layer (90 to 180 cm thick) at 13.3 GHz. Although 400 MHz is not as satisfactory for ice identification as 13.3 GHz, combining a 400 MHz and 13.3 GHz system eliminates the ambiguity regarding the very thin ice. These results may be important in design of ice-mapping imaging radars.

INTRODUCTION

A new interest has been developed in the Arctic Ocean with the hope that it can be utilized more for navigational purposes. Better ice navigation systems and methods for determining surface features and thickness of sea ice are needed in order to facilitate routing of ships. This requires mapping large areas of sea ice. The all-weather, day-night operational capability of radar systems is particularly useful in the regions such as the Arctic where light and weather conditions are uncertain most of the time.

The ability of radar systems to discriminate sea ice types has been demonstrated to some extent (Anderson, 1966; Rouse, 1969; Johnson and Farmer, 1971). There is a need for a design of an optimum system which can specifically be used as a sensor for discriminating ice types. Before optimum parameters for such a system can be specified, the effect of frequency, polarization, angle, and resolution in discriminating sea ice types has to be understood. This information can be achieved in part from the radar scatterometer data. Radar scatterometers permit more quantitative and detailed observation of radar scattering behavior than radar imagers (Moore, 1966).

It should be emphasized that, at spacecraft altitudes, a scatterometer will have a poor resolution, whereas a radar imager can be made to operate with fine resolution from both spacecraft and aircraft altitudes. Hence, one of the major purposes of conducting scatterometer measurements over sea ice is to determine parameters required for radar imagers.

RADAR SCATTEROMETER

The radar scatterometers used here are instruments designed to measure the radar scattering coefficient σ^o (radar cross-section normalized to the illuminated area) as a function of the illuminated angle θ (angle measured from the vertical). The 13.3 GHz scatterometer transmits a vertically-polarized CW (Continuous Wave signal) and collects return from an illuminated region which corresponds to $\pm 60^o$ along the flight line (along-track) and 3° wide (across-track).

The physical dimensions of the antenna beam for the 400 MHz scatterometer are 7.5° in the across-track and $\pm 60^{\circ}$ in the along-track direction. Reflectivity measurements are made for transmitted signals having horizontal and vertical polarizations by the 400 MHz scatterometer. In addition, cross-polarized reflectivity measurements for each transmitted polarization are also made.

Because of the aircraft motion in the along-track plane, the radar returns at the illuminated incidence angles are coded by Doppler shift frequencies displaced from the radar center frequency. The return is processed through Doppler filters to obtain the scattering coefficient σ^{o} at each of several discrete angles within the beam ($\pm 60^{\circ}$). Since σ^{o} is measured at different times for different incidence angles, the time must be "compressed" to yield a σ^{o} vs θ plot for a particular patch on the surface. In addition to the above, σ^{o} vs time plots (time-history plots) can also be constructed for each forward and aft angle. The results presented here correspond to forward-beam measurements only.

SCATTEROMETER AND GROUND-TRUTH MEASUREMENTS

The results presented here were obtained from the analysis of 13.3 GHz and 400 MHz scatterometer data for NASA Earth Resources Aircraft Program Mission 126. This mission was conducted in April 1970, off the coast of Pt. Barrow, Alaska. The analyzed data were obtained from the experiment designated Phase I, Part C. To minimize the effects of aircraft drift angles, parallel lines were established for the scatterometer runs with an upwind and downwind orientation. The results given here correspond only to Site 93, Flight 3; scatterometer lines 20, 3, 4, 5, 6 and 7 (NASA designation). The test site was located approximately 40 kilometers northwest of Pt. Barrow and was selected on the recommendations of a Naval Research Lab ice observer (R. Ketchum).

The purpose of distinguishing different types and thicknesses of sea ice from radar backscatter can be achieved if a category of sea ice and its thickness can be associated with radar backscatter. The category of sea ice and its thickness as viewed by the scatterometer at a particular instant in time can be obtained from the "ground truth" information, but unfortunately no "ground truth" was collected for this mission. Hence, this information could only be obtained from interpreting the aerial photographs. The following seven categories of sea ice were identified and classified on the aerial photographs by visual interpretation on the basis of thickness, age, tone, and texture (Anderson, 1970):

Category 1: Open Water

It gives a very black tone on the photograph.

Category 2: New Ice (0 to 5 cm thick)

It gives a black tone on the photograph. It is very thin, new ice. These open-water or thin-ice areas will normally occupy parts of larger thicker ice, or will occur as cracks and leads between very thick ice.

Category 3: Thin Young Ice (5 to 18 cm thick)

The upper limit of 18 cm was chosen in this category because this is generally the point when sea ice can begin to support a dry snow cover. It can preserve rafting. It is dark grey in tone, but where rafting takes place it appears as light grey because of air trapped in the overlapping layers.

Category 4: Thick Young Ice (18 to 30 cm thick)

It can be identified by the presence of finger rafting patterns. For the ice above 30 cm thick the pattern begins to become more sinuous than angular.

Category 5: Thin First-Year Ice (30 to 90 cm thick)

One indication of this type of ice is that ice floes begin to become rounded and their interior areas are smooth.

Category 6: Thick First-Year Ice (90 to 180 cm thick)

It has a light tone and appears severely marked by pressure ridges. It surrounds well-rounded multi-year ice floes.

Category 7: Multi-Year Ice (180 to 360+ cm thick)

It is probably easiest to identify on the photographs. It gives a bright tone and floes are round and have a smooth surface. Since this type of ice has gone through at least a summer melt, it can also be characterized by the pressence of small, irregular dark tones within the white background of a floe; these dark tones indicate the smooth ice of the frozen puddles.

The photo mosaic was then aligned with the time-history plot for each scatterometer line. The boundaries, corresponding to boundaries on the photo mosaic, were established on the time-history plots. The scatterometer data were then separated for each category.

For both 13.3 GHz and 400 MHz scatterometers the dimension of the resolution cell in the along-track direction varies as a function of incident angle. For an average aircraft ground speed of 90 m/sec at an altitude of 900 m, the along-track resolution for the 13.3 GHz scatterometer varies from 30 m at 5° to 125 m at 60°. For the same aircraft ground speed and altitude the along-track resolution for the 400 MHz scatterometer varies from 30 m at nadir to 250 m at 60°. Since there is one point on the time-history plot of σ° (radar scattering cross-section) at every 1.3 and 1.1 seconds for 13.3 GHz and 400 MHz scatterometer respectively, there is a point corresponding to about every 90 m on the ground. Because of this there is a certain amount of overlap between consecutive resolution cells on the ground. This overlap is the maximum at 60° from the vertical and decreases as the angle decreases.

To ensure that scatterometer data corresponded to homogeneous patches of each category of sea ice the points at the boundaries had to be deleted when taking averages. Because of overlap between consecutive resolution cells, at least three points had to be deleted at the boundaries. To make results more meaningful and accurate, two points were deleted from the beginning and two points from the end of each category. This means that four points were deleted at the boundaries which corresponded to about 360 m on the ground. This ensured us that return from only one category was being averaged, so the averaged data only corresponded to patches of sea ice which were at least 360 m along the flight line. Even after deleting four points at the boundaries enough data points were left to carry out a meaningful statistical analysis.

ANALYSIS OF 13.3 GHz SCATTEROMETER DATA

First all data were averaged for each scatterometer line, forward angle, and sea ice category. There are twelve fore angles: 5, 10, 15, 20, 25, 30, 35, 40, 45, 50, 55 and 60°. Since the value of σ° (radar scattering cross-section) was given in dB on the time history plot, each value of σ^o was converted to absolute value before the average of all the points was taken. The resultant averaged value was then converted back into dB. Averaged σ^o vs θ curves were produced for each category of sea ice and scatterometer line. This was done to see the consistency of scatterometer results from one line to another. Each scatterometer line did not have every type of sea ice. The averaged value of radar return for each category was relatively consistent from line to line (Parashar, 1973).

The average of all these lines was taken for each category at each forward angle and results are shown in Figure 1. It was considered appropriate to average the data from all the scatterometer lines because the maximum difference between two scatterometer lines was 5 km. The lines were normally 30 km long and the area thus covered by the scatterometer was approximately 30 km by 30 km. The physical properties and other parameters of each category of sea ice should thus be reasonably constant.

It can be seen from Figure 1 that Category 1 (open water) gives strongest return at near-vertical angles. This is expected because water is comparatively smooth and has a high dielectric constant. Apart from vertical, Category 7 (multi-year ice) gives more return than water and all other types of sea ice for all angles. Category 6 (first-year thick ice, 90 to 180 cm thick) gives more return than all other types of ice except Category 7. Sea ice less than 18 cm thick (Categories 2 and 3) gives more return than thick young ice and thin first-year ice (Categories 4 and 5).

The reason for separating the sea ice into different categories according to thickness was that the backscatter could later be associated with the ice thickness. On this basis σ^o vs thickness curves for various fore angles were produced and are shown in Figure 2. The major changes in the shape of σ^o vs thickness curve occur at the angles 5, 20, 25, 30 and 50° from the vertical. The really thin ice gives more return than moderately thicker types of ice. As can be seen from this curve, the shape of σ^o vs thickness curve is very much dependent on the angle. There is a change of 8 dB in the value of σ^o when ice thickness changes from 30 to 270 cm.

ANALYSIS OF 400 MHz SCATTEROMETER DATA

The analysis of 400 MHz scatterometer data was carried out in a way similar to the 13.3 GHz data analysis. The data for 400 MHz were available for all the four polarizations (VV*, HH, VH and HV), whereas data for 13.3 GHz were available for only one polarization (VV). Thus the effect of polarization in discriminating sea ice could be studied from the 400 MHz data analysis.

The data for each category, angle and polarization were averaged for each scatterometer line. There are twelve fore angles: 0, 5, 10, 15, 20, 25, 30, 35, 40, 45, 50 and 60°. The resultant σ^o vs θ curves give indication of the consistent operation of scatterometer. The data looks to be consistent from one line to another as far as the slope of the σ^o vs θ is concerned. The value of σ^o varies from one line to another but the relative variation in σ^o

---

*Vertical transmit-Vertical receive

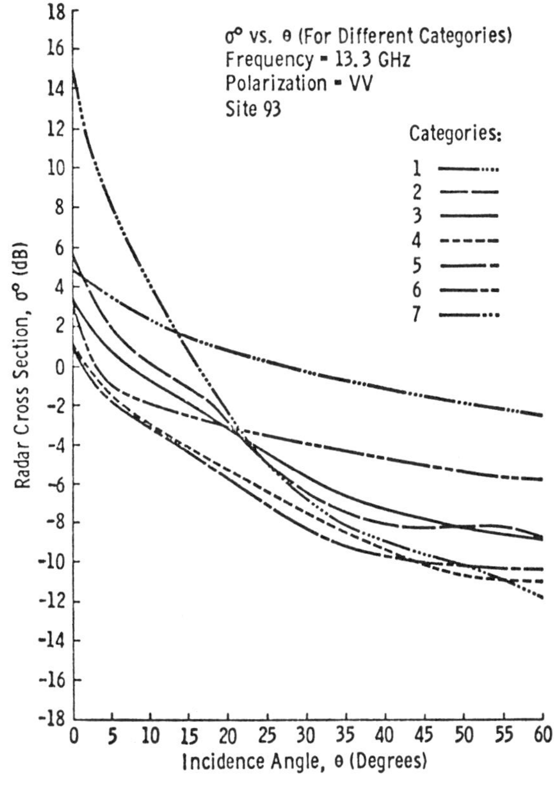

Figure 1.

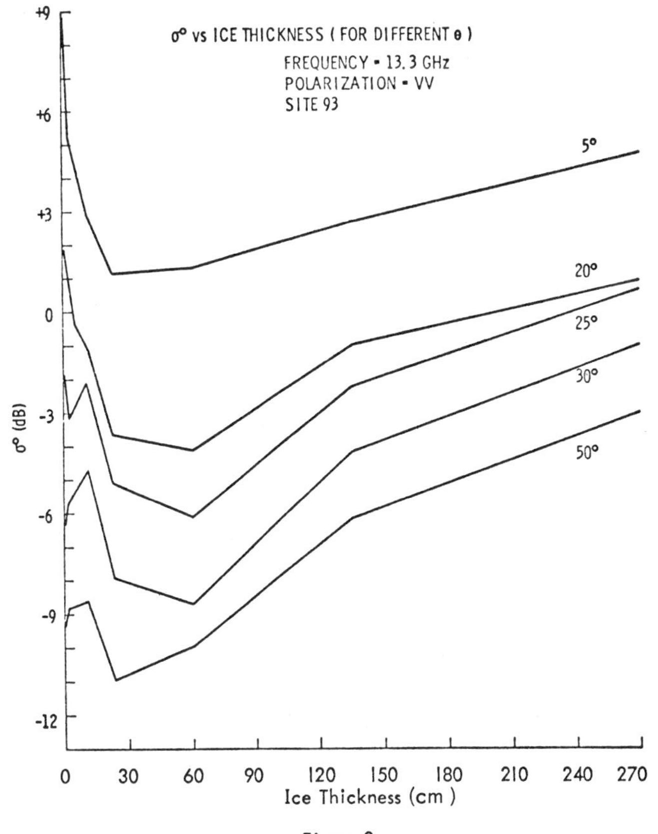

Figure 2.

vs θ for sea ice categories within each scatterometer line is reasonably consistent from one line to another line. After this, data from all the scatterometer lines were combined to give mean σ^o vs θ curves for each polarization and sea ice category. The resultant mean σ^o vs θ curves for each category and three polarizations are shown in Figures 3, 5 and 7.

It is interesting to note that Category 1 (open water) gives strongest return at all angles for VV polarization. Categories 5 and 6 (ice 30 to 180 cm thick) give more return than Category 7 (ice 180 to 270 cm thick) but less than Category 1. Categories 2 and 3 (ice less than 180 cm thick) give minimum return at all angles.

For HH polarization, Category 1 (open water) gives strongest return for all angles less than 20°. Categories 5 and 6 (ice 30 to 180 cm thick) give strongest return for angles greater than 20°. Even for this polarization Categories 2 and 3 (ice less than 18 cm thick) give minimum return at all angles.

Category 7 (multi-year ice) gives more return than Categories 2 and 3 but less than Categories 1, 5 and 6. For cross-polarization (VH), Category 1 (open water) still gives strongest return for almost all the angles. Categories 5 and 6 give strongest return for some of the angles. Categories 2 and 3 give minimum return in this case also. The σ^o vs θ curves for each category are very similar for the two cross-polarizations (VH and HV).

The wind speed at which these measurements were made was 25 kts. The σ^o vs θ curve for water was compared with a σ^o vs θ curve from NRL North Atlantic measurements at wind speeds of 25 kts. It was seen that the value of σ^o obtained from NRL data was the same as the value obtained from these data. This indicates that the results obtained by this analysis are consistent with previous observations (Parashar, 1973).

As in the 13.3 GHz analysis, σ^o vs ice thickness curves were produced for each polarization at different angles, as shown in Figures 4, 6 and 8. Sea ice less than 18 cm thick is seen to give minimum return and water maximum return. Thick first-year ice gives more return than multi-year ice. The difference in σ^o between return from Category 3 and Category 5 is about 6 dB for the like polarization VV at 55°. The difference in σ^o between the two categories is about 10 dB at the same angle for cross-polarization VH. This shows that there is more spread in σ^o between thin and thick first-year ice at cross-polarization than like-polarization.

The data were normalized to the value of σ^o at 0°. The effect of normalization is that the spread of the σ^o vs θ lines for different categories increases. Water and thick first-year ice still give maximum return. For HH polarization water could not be differentiated for unnormalized data, but water can easily be differentiated on the normalized curves.

DISCUSSION OF RESULTS AND CONCLUSIONS

As can be seen from some of the results given above, the scatterometer has potential as a sensor in discriminating sea ice. The results given here will help in specifying optimum parameters for an imaging radar and in better understanding the nature of radar backscatter from sea ice. Water can be differentiated from both normalized and unnormalized data for both 13.3 GHz and 400 MHz. There is a reversal in backscatter from thin ice at the two frequencies.

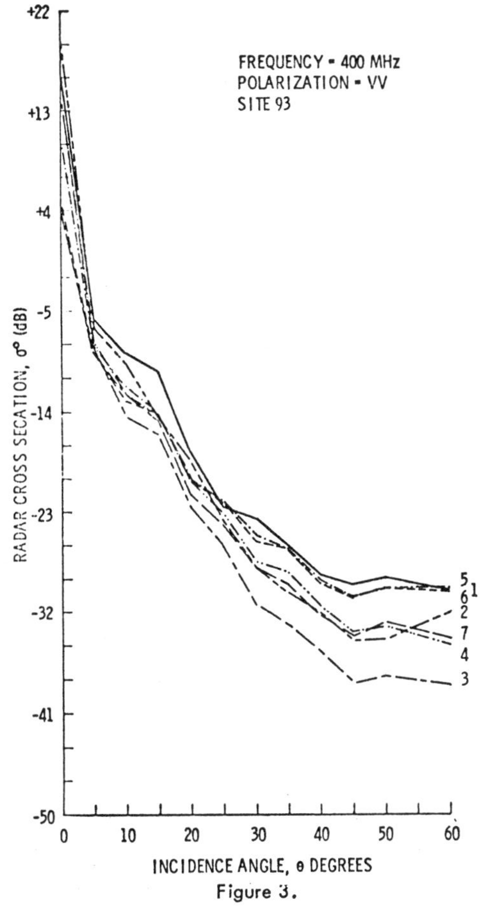

Figure 3.

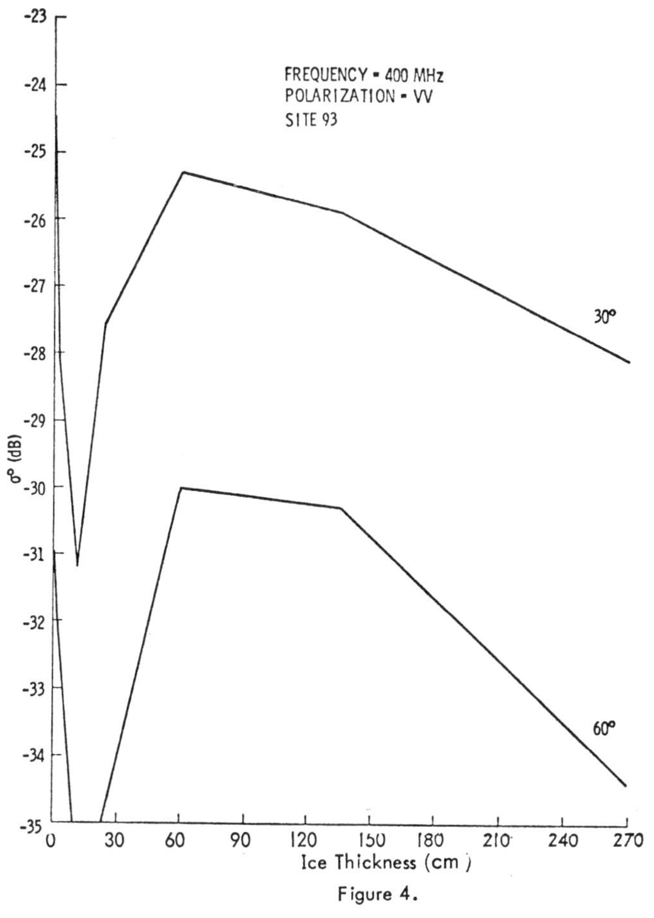

Figure 4.

4.2

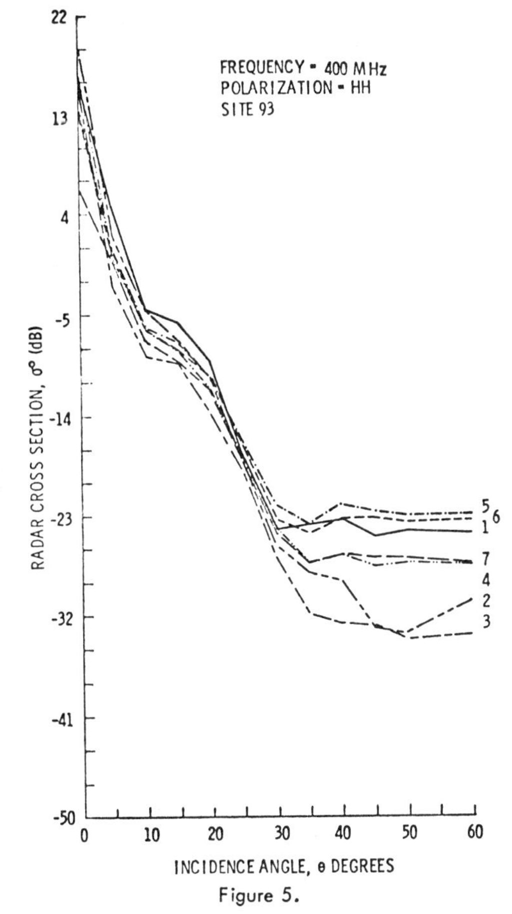

Figure 5.

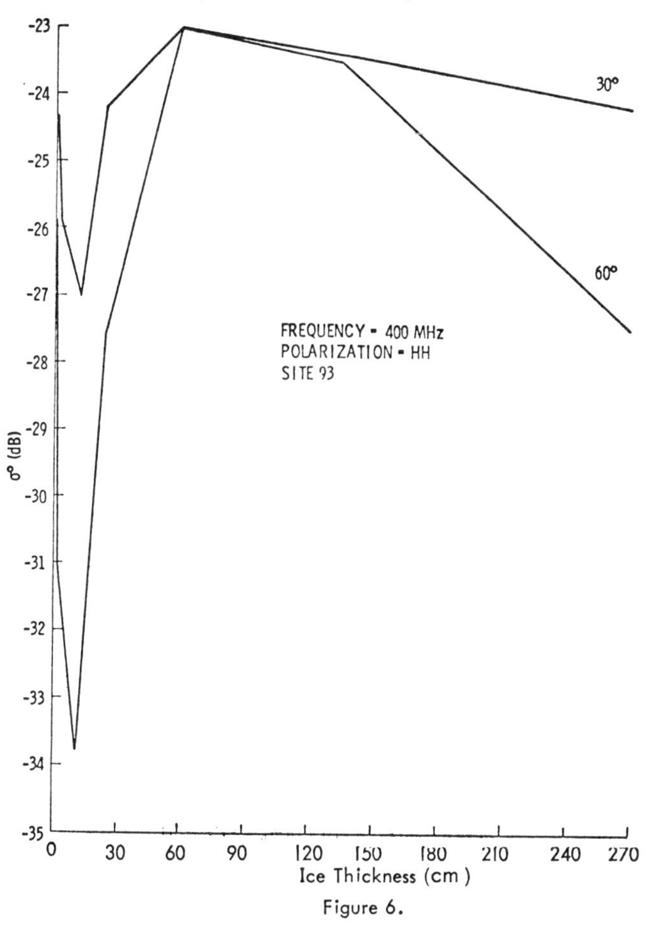

Figure 6.

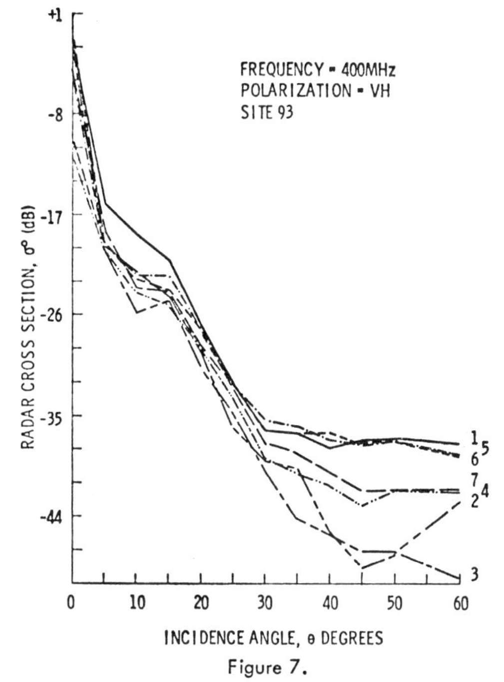

Figure 7.

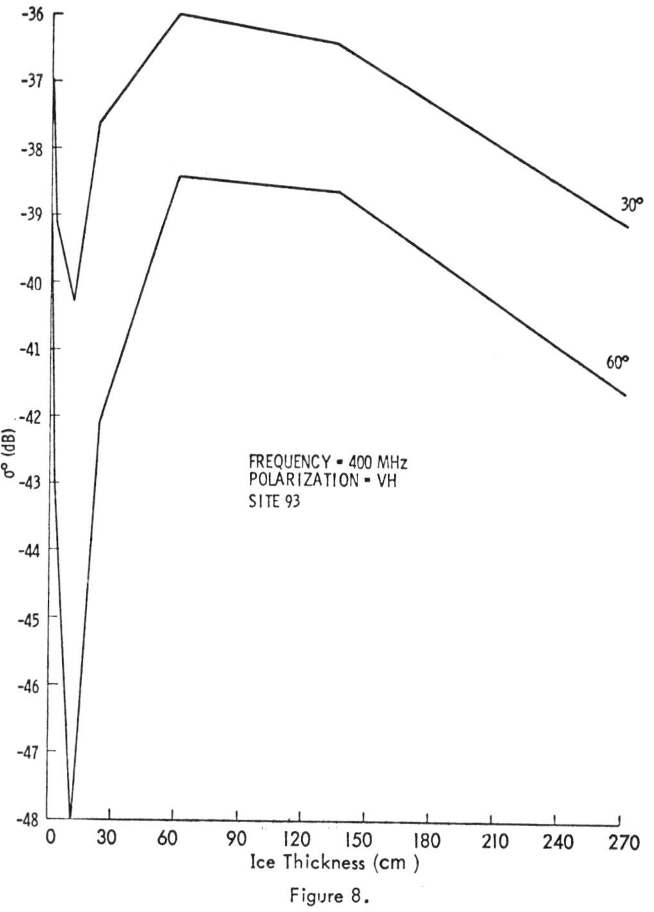

Figure 8.

At 13.3 GHz thin ice (Categories 2 and 3) gives return comparable with moderate ice (Category 5), so it is hard to differentiate between the two. The thinner ice gives minimum return at 400 MHz, so by combining the data from two frequencies, thin ice can be discriminated easily. Multi-year ice gives maximum return at 13.3 GHz at near-grazing angles. First-year thick ice (Categories 5 and 6) gives maximum return (less than water) at approximately all angles at 400 MHz, so both multi-year ice and thick first-year ice can be discriminated.

It is evident from above that more research is needed before final conclusions can be drawn. There is a need to collect more data, especially the cross-polarized data at Ku-band. Only then could it be said categorically which frequency and polarization is better for sea ice discrimination. It is conceivable that a system operating at two frequencies will be better in discimінating sea ice thickness.

The fact that all these data were obtained under spring conditions at Pt. Barrow must be remembered. Whether the results shown here would apply when the surface was colder must await further experiment.

ACKNOWLEDGEMENTS

The authors are grateful to Mr. Vern Anderson of the Photographic Interpretation Corporation, Hanover, N.H., (formerly with Cold Regions Research and Engineering Laboratories [CRREL], Hanover, N.H.) for his expert guidance and advice in identification and categorization of sea ice types on the aerial photographs. The authors would also like to acknowledge support from the University of Kansas Computation Center, NASA Contract NAS9-10261, and U.S. Navy Contract N60921-70-C-0221.

REFERENCES

Anderson, V. H., 1966. High altitude, side looking radar images of sea ice in the Arctic. Proc. Fourth Sym. on Remote Sensing of Environment, University of Michigan, Ann Arbor. 845-857.

Anderson, V. H., 1970. Sea ice pressure ridge study: an air-photo analysis. Photogrammetria. Amsterdam: Elsevier Pub. Co.

Johnson, J. D., and L. D. Farmer, 1971. Use of side-looking airborne radar for sea ice identification. J. Geophy. Res. 76:2138-2155.

Moore, R. K., 1966. Radar scatterometry—an active remote sensing tool. University of Kansas Center for Research, Inc., CRES Technical Report 61-11.

Parashar, S. K., 1973. Investigation of radar discrimination of sea ice. University of Kansas Center for Research, Inc., CRES Technical Report 185-13.

Rouse, J. W., Jr., 1969. Arctic ice type identification by radar. Proc. IEEE. 57:605-614.

MEASUREMENT OF SNOWFALL BY RADAR

James W. Wilson
The Center for the Environment and Man, Inc.

ABSTRACT

An S-band and C-band radar were used to estimate the water equivalent of snow above approximately 75 gages near Lake Ontario for the winter of 1972—73. Thirteen of the gages were specially sited to provide the maximum amount of natural protection from the adverse effects of wind on the gage catch. Results of radar to gage comparisons indicate that radar can be used to measure snowfall with approximately the same accuracy as for rainfall provided an empirically derived range correction is applied to the radar estimates. The maximum radar range of useful snowfall estimates is less than for rainfall because, for snow, precipitation rates are lower and precipitation growth occurs at lower altitudes.

The accuracy of the radar estimates varies with storm type. The highest accuracies were associated with large scale storms which are characterized by relatively high altitude snowfall growth; the lowest accuracies were associated with storms where orographic lifting of the air and lake-generated showers predominated. There was evidence to suggest that the radar reflectivity was affected by the type of snow crystal.

The root mean square error in the radar estimates can be reduced up to 50% through the use of a reference gage to adjust the radar snowfall estimates.

Based on the above results, recommendations are given concerning the siting, characteristics and operation of weather radar for snowfall measurement.

INTRODUCTION

With the advent of mini-computers and reliable equipment to automatically digitize the radar reflectivity from precipitation, it is now possible to utilize weather radar for the operational measurement of precipitation. In fact, the National Weather Service is presently conducting field experiments at several radar sites, utilizing automated procedures to collect and process radar data for the purpose of rainfall measurement. While there have been numerous studies directed at the use of weather radar to measure rainfall, only a few limited studies have been made of the measurement of snowfall by radar.

Precipitation measurements by radar are based on a relationship between the radar reflectivity factor Z_e (measured by radar) and the precipitation rate R given by the equation $Z = aR^b$, where a and b are coefficients dependent on the size distribution of the raindrops or snowflakes. A complicating unknown factor with snowflakes is that the shape and structure of the flakes may affect the reflection.

Kodaira and Inabe (1955) concluded, from measurements of backscattered power from falling snow, that large variations in the Z-R relationship occur, these variations being associated with changes in snow crystal type and size distribution. Ohtake and Hemni (1970) have reported

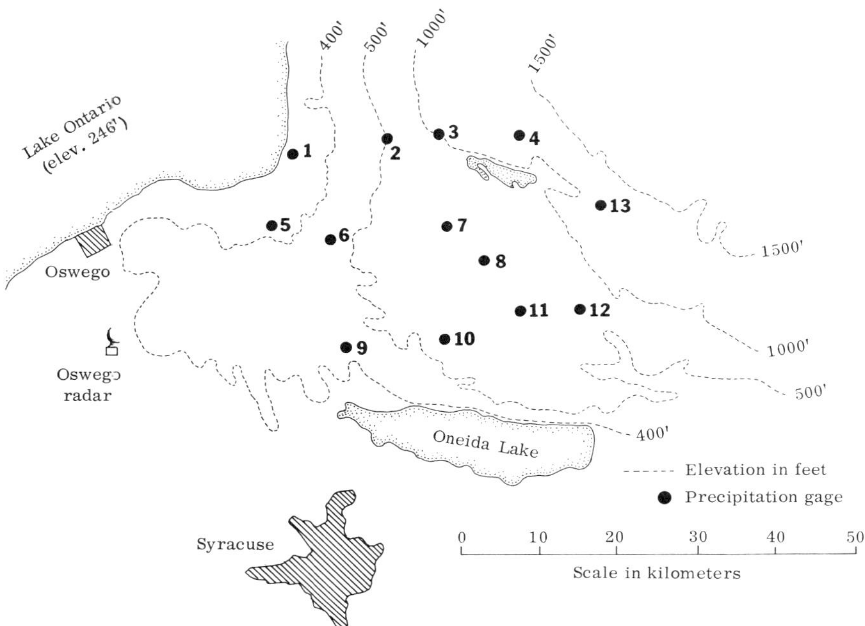

Figure 1 Location of Oswego radar and precipitation gages.

large variations in Z-R relationships for six types of crystals. However, Sekhon and Srivastava (1970) have pointed out inconsistencies in the derivation of Z-R relationships made by other investigators. After reanalysis of available snow particle data, they report the relationship $Z = 1780R^{2.21}$ fits the data quite well. These data are from numerous geographical regions with varying snow crystal types. From actual measurements of snowfall by radar for one snow storm by Carlson and Marshall (1972) and for 67 hours from nine storms by Jatila (1973), it would appear that the accuracy of measurements in snowfall are similar to those found in rain storms by Huff (1966) and Wilson (1970). Jatila (1973) showed that the radar measurements during snow were improved if a reference gage is used to calibrate the radar. Wilson (1970) demonstrated the success of a similar technique for improving rainfall measurements by radar.

The purpose of this paper is to present comparisons of radar and gage measurements of snowfall for the vicinity of Lake Ontario in New York for the winter of 1973—73. These data were acquired in conjunction with the International Field Year for the Great Lakes (IFYGL) in which an intensive program was made to measure the precipitation falling into the Lake Ontario watershed for a one-year period. In addition, recommendations are presented concerning radar characteristics, siting considerations and collection procedures suitable for snowfall measurement.

DATA

The primary data for this study was obtained from a C-band radar (Vitro MR 782) located 15 km south of Oswego, New York. The radar had a beam width of 1.7 deg and a transmitted peak power of 250 kw. The radar was equipped with an automatic digitizer and tape recorder built and installed by the National Severe Storms Laboratory. The signal processing and recording system has been described by Sirmans, Watts and Horwedel (1970). A radar scan was recorded on magnetic tape every 10 minutes for each mile in range from 8 to 120 mi (15—222 km) and for each 2 deg of azimuth. The data are quantized into 16 levels (3db intervals). Accumulated precipitation was obtained for each 2 deg by 1 mi increment by summing the products of the radar derived precipitation rate and the time interval between observations (10 min). The radar estimates during snow were based on the relationship $Z = 1780R^{2.21}$ reported by Sekhon and Srivastava (1970). Rainfall measurements were based on the relationship $Z = 200R^{1.6}$. When comparing radar amounts with gage amounts, the radar estimate is for an area with a radius of 2 mi centered on the gage. No allowance was made for drift of snowflakes by the wind.

Surface snowfall measurements were obtained from 13 Universal weighing-recording precipitation gages. The gages were located between 26 and 65 km to the E and NE of the Oswego radar (see Figure 1). Each gage site was carefully chosen to provide the maximum natural protection possible from the wind. In general, the gages were located in small clearings in coniferous forests. The surface snowfall measurement network is described in detail by Peck, et al. (1973). It is concluded by Peck that the catch by these gages is close to the "true" snowfall, certainly closer than that measured by most climate gages. In addition, there were approximately 50 high school and junior high school students living within the gage network who measured snow depths and water equivalents. They also obtained some limited information on the type and size of the snowflakes.

Comparison of radar and gage measurements were made for the period 26 November 1972 to 19 March 1973. During this period the special 13 gage network measured between 15 and 23 inches (38—58 cm) of water. Approximately one half fell as snow—a relatively snowless winter for the area.

In addition to the Oswego radar, data was collected by the National Weather Service WSR-57 radar at Buffalo, New York, which is located 185 km to the west of the Oswego radar. The Buffalo radar is an S-band radar with a beam-width of 2.0 deg and a peak transmitted power of 410 kw. Similar to the Oswego radar data, the data from the Buffalo radar was automatically digitized and archived on magnetic tape. However, there was not a surface network to provide reliable snowfall measurements near the Buffalo radar. It was then necessary to make comparisons with existing National Weather Service climatological gages. Because appropriate precautions are generally not taken in locating these gages in sites well protected from the wind, their catch during snowfall is frequently low. Therefore, only a few general results are presented for the Buffalo data.

RADAR-GAGE COMPARISONS (SEASON)

Comparison of the gage and radar estimates of the melted snow-water equivalents shows the radar estimates are consistently lower than the gage measurements and the underestimate increases with increasing distance from the radar. The results of these comparisons are shown in Figure 2 for the Oswego radar where the ratio between the gage measure-

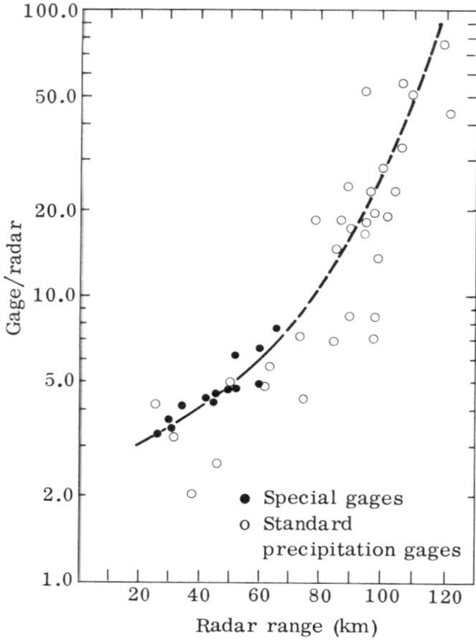

Figure 2 Comparison of G/R ratios of snowfall measurement as a function of radar range.

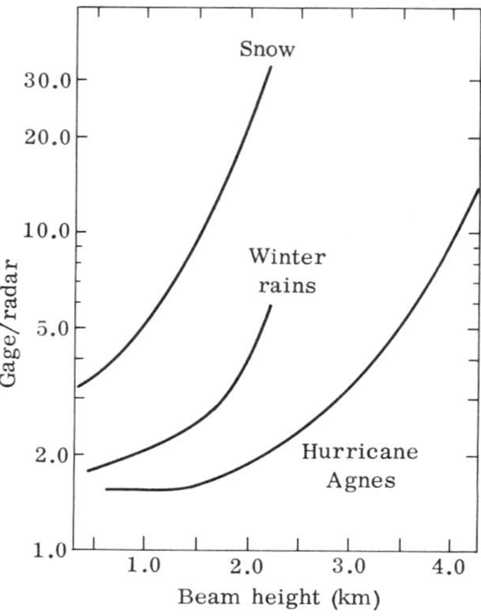

Figure 3 Comparison of G/R ratios as a function of radar beam height above the ground for snow storms, winter rain storms and Hurricane Agnes as measured by the Oswego radar. The snow and winter rain curves are averaged for all storms during the 1972–1973 winter.

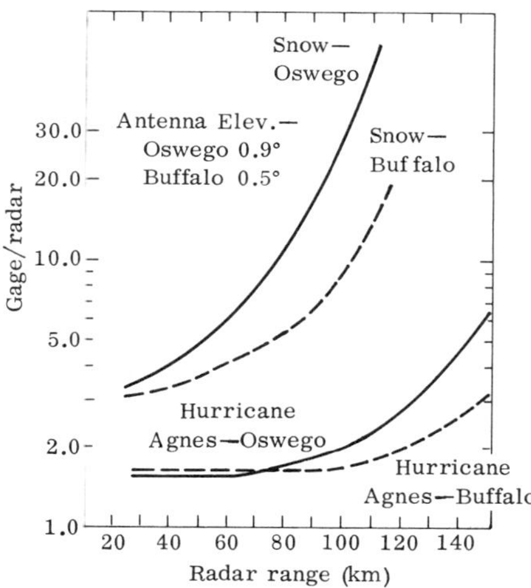

Figure 4 Comparison of G/R ratios as a function of range for Buffalo and Oswego radars for snowstorms and Hurricane Agnes.

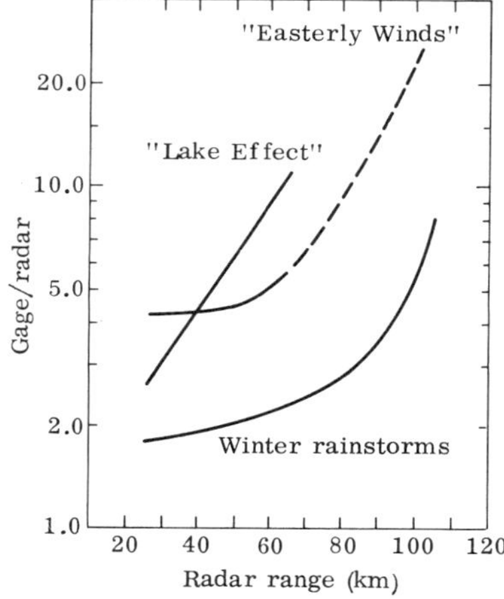

Figure 5 Comparison of G/R ratios for "Lake Effect" snowstorms, "Easterly Wind" snowstorms and winter rainstorms.

ment (G) and radar precipitation estimate (R) over the gage are shown as a function of radar range. The G/R ratios are based on the total snow measured between 30 November 1972 and 19 March 1973. In addition to the 13 special gages, comparisons are also shown for existing climatological stations. Because of the low reliability of snowfall measurements from the climate gages, the data for these gages are used only as an indication of the nature of the curve beyond 60 km. Note the larger scatter of the G/R ratios about the curve for the climate gages.

Earlier results of radar estimates of rainfall from Oklahoma thunderstorms (Wilson, 1970) also showed the estimates to be consistently less than the observed rainfall. For the Oklahoma thunderstorms, the G/R ratio averaged 1.7 to a radar range of approximately 100 km. It is not certain why the radar are consistently low and particularly why the underestimate is much larger for snow than rain. However, until the reason for the low estimates by radar is appropriately accounted for the radar estimates can be corrected using the empirically derived adjustment factor.

The increase of G/R with range is due to the increasing height and width of the radar beam with increasing range. The magnitude of the effect then depends on the vertical profile of the precipitation. When cloud tops are low and there is considerable low-level precipitation growth, G/R ratios will increase rapidly with range; this is frequently the situation during snowstorms. To illustrate this phenomena, G/R as a function of radar beam height for snowstorms, winter rainstorms and a large summer rainstorm (Hurricane Agnes) are shown in Figure 3. All these data were obtained from the Oswego radar during the IFYGL. Snowstorms characteristically have low cloud tops and considerable low-level precipitation growth where summer rainstorms have high cloud tops and relatively little growth at low levels, with winter rainstorms generally falling in between. To more closely sample the precipitation that reaches the ground, it is then desirable to keep the radar beam width narrow and sample as close to the ground as possible. Practical problems, such as obstructions, make it necessary to frequently elevate the radar beam when collecting data. The Oswego data was collected with an antenna elevation angle of 0.9 deg. Thus, if the data had been collected at a lower angle, the G/R ratio would have increased at a slower rate with increasing range. This can be illustrated by comparing the G/R ratios from the Buffalo and Oswego radars, since the Buffalo data were collected with an antenna elevation of 0.5 deg and Oswego at 0.9 deg. This comparison is given in Figure 4 for the average of all snowstorms and Hurricane Agnes. If the ordinate in Figure 4 were beam height instead of radar range, the Oswego and Buffalo curves would nearly coincide.

COMPARISONS BETWEEN STORM TYPES

It was expected and found that between snowstorms there would be considerable variability in the G/R ratio as a function of range, the reason being that the "lake-effect" snowstorms that frequent the region are known to have low cloud tops, generally below 3 km (Jiusto and Holroyd, 1970) with considerable low-level precipitation growth. Snowstorms associated with synoptic scale lows and fronts generally have higher cloud tops. Storms were classified into three groups: "lake-effect," "easterly wind" and "westerly wind." "Lake-effect" storms occur during Arctic air outbreaks when the cold air flows over the relatively warm lake waters. "Easterly wind" storms were associated with synoptic scale low pressure storms. Both "lake-effect" and "westerly wind" storms generally have winds from the west or southwest, the primary difference in their classification being the vertical temperature lapse-rate in the lowest 1.5 km. To be classified as a "lake-effect" storm, the difference between the lake temperature and 850 mb temperature had to exceed the adiabatic lapse-rate. With winds from the southwest, there is orographic lifting of the air over the network. It was quite apparent that this lifting was responsible for increased snowfall at gages 3, 4 and 13. Four storms were not classified because of their variable nature. Of the 27 snow periods, 12 were classified as "lake-effect," six as "easterly wind," and five as "westerly wind" storms.

For "lake-effect" storms the G/R ratio increased very rapidly with range. To illustrate this difference, G/R ratios are compared in Figure 5 for "lake-effect" and "easterly wind" snowstorms. The G/R ratios for the "westerly wind" storms approximated the curve for the "lake-effect" storms. Essentially all the points in Figure 1 above the curve are associated with high elevation stations that receive above average snowfall. Much of the additional snowfall for these stations occurs during "lake-effect" and "westerly wind" storms. Since these type storms have steeper G/R-range curves, it is apparent why these stations are above the curve in Figure 1.

The curve for the "easterly wind" storms (Figure 5) is level within 50 km of the radar, thus indicating there is little additional snowfall generation below about 2 km. The curve for the "lake-effect" storms has a constant slope indicating that, even for the closest gages, there is precipitation growth within and below the radar beam. Thus, for all gages, it would be expected that for a given snowfall rate measured at the ground the radar would measure less snow in a "lake-effect" storm than an "easterly wind" storm. However, Figure 5 shows that this is not the case for radar ranges less than 40 km. This indicates that the reflection from snowflakes in the "lake-effect" storms is greater than from those in the "easterly wind" storms. Ohtake and Henmi (1970) have reported from snowflake size distribution data and empirical fall speed equations that snowflakes consisting of spatial dendrite crystals should give higher reflectivities than other crystal forms and that plate and column crystals should give the lowest reflectivities. Jiusto and Holroyd (1970), from study of "lake-effect" storms, suggest that dendrite aggregates predominate in these storms and plate and columns are rare.

While there was only a limited number of snow crystal observations from the students living within the gage network, their data indicate that spatial dendrites were frequently observed during the "lake-effect" storms. There is also indirect evidence from the student observations that spatial dendrites were common during "lake-effect" storms. Jiusto and Kaplan (1972) have suggested that the predominant snow crystal type within a storm can be inferred from the snow depth to water equivalent ratio. Their data show that spatial dendrites have the largest snow depth to water ratio.

Observations of snow depth to water ratios by the student observers show the "lake-effect" storms had the highest ratios averaging between 15 and 27 per storm with an overall average of 21. The students made observations only during January, February and March. Therefore, they observed only one "easterly wind" storm, the average ratio for the storm being 7.5. The snow depth to water ratios for the "westerly wind" storms averaged between 9 and 13 with an overall average of 10. Thus, there is strong evidence that the predominant crystal type differs between "easterly wind" and "lake-effect" storms and that spatial dendrite crystals predominate in "lake-effect" storms and that these crystals have a higher reflectivity than others.

Z-R RELATIONSHIP

Scatter diagrams of hourly water equivalents measured by gage and radar were made to determine if the assumed value of 2.21 for the exponent b in the Z-R relationship was realistic. Because of the naturally occurring limited range of snowfall rates, primarily between 0.2 and 4 mm/hr, combined with the scatter in the data points, a change in b of two or three tenths has little affect on the radar estimates. Thus, it can only be said there is no detectable difference in the value of b between storms and that a value of b between approximately 1.9 and 2.4 would have little affect on the accuracy of the radar estimates.

STORM-TO-STORM COMPARISONS

As found above, there are large differences within and between snowstorms in the G/R ratio as a function of range. These differences were partially dependent on storm type and indicate the need to use different relationships for converting radar reflectivity to precipitation for different type storms. In addition to differences between storm types, there are variations in the G/R ratio within storm types. The extent of this variability is shown in Table 1 for three gages. The G/R ratio is given for each of 27 snow periods for gage 5, the closest gage to the radar (26 km), gage 8 located 48 km from the radar, and gage 13, the farthest gage from the radar (64 km). The radar estimates in Table 1 were based on the relationship $Z = 1780R^{2.21}$ and were then range-adjusted for each storm type using the appropriate G/R-range curve given in Figure 5. The notation R_s is used for these estimates to distinguish between those that are labled R and based solely on $Z = 1780R^{2.21}$. The "lake-effect" curve was used for the "westerly winds" and the average snow season curve given in Figure 2 was used for the four unclassified storms. Examination of the G/R_s ratios in Table 1 indicates there is variability within storm types, the "easterly wind" storms have the least variability and the "lake-effect" storms the greatest.

From Table 1 some general statistics on the accuracy of the radar estimates is possible. For gage 5, 89% of the radar estimates are within a factor of 2.0 of the gage measurements, compared to 84% for gage 8 and 83% for gage 13. For a factor of 1.5 the figures are 63%, 60% and 44%, respectively, for gages 5, 8 and 13. Similar statistics are obtained when just the average range-G/R curve of Figure 2 is used to adjust the radar estimates. It should be noted that the figures for gage 5 are higher than those obtained from Oklahoma thunderstorms and those for 13 are about equivalent for data from Oklahoma thunderstorms (Wilson 1970). There were 19 rain periods during the 1972—73 winter over the snow network. Examination of the variability of the storm G/R_S ratio shows that the accuracy of radar estimates between winter rain and snow are quite similar.

Table 1: Comparison of gage (G) and radar (R_S) measurements of water equivalents for gages 5, 8 and 13, located, respectively, 26, 48 and 64 km from the radar.

Storm Type	Date	GAGE 5		GAGE 8		GAGE 13	
		Gage Amt (mm)	G/R_S	Gage Amt (mm)	G/R_S	Gage Amt (mm)	G/R_S
E	26 Nov	17.5	0.81	—	—	—	—
E	30 Nov-1 Dec	13.0	0.65	15.7	0.63	—	—
E	4- 5 Dec	8.2	1.16	8.9	1.15	9.2	1.02
E	8 Dec	5.8	0.85	6.3	0.82	6.9	0.76
E	15-16 Dec	17.3	1.17	23.4	1.23	22.4	0.91
E	14-15 Feb	14.5	1.09	19.3	1.36	20.0	1.30
L	28-29 Nov	4.7	0.63	—	—	—	—
L	1- 5 Dec	5.8	1.18	9.1	0.74	—	—
L	16-17 Dec	3.3	0.78	14.0	1.84	7.9	2.19
L	1- 2 Jan	5.1	1.27	8.1	1.07	6.9	1.04
L	5- 6 Jan	1.0	0.47	4.1	0.48	4.3	0.66
L	9-10 Jan	14.2	1.17	11.9	1.12	12.2	1.06
L	11 Jan	3.9	1.74	2.3	0.86	1.0	0.62
L	12 Jan	1.7	0.65	1.0	0.64	0.8	0.58
L	30-31 Jan	3.3	0.84	6.4	0.69	5.9	0.52
L	10 Feb	3.9	1.34	1.0	0.53	1.3	1.92
L	15-16 Feb	4.1	0.30	3.8	21.50	3.8	*
L	18 Feb	4.6	1.32	1.0	2.86	0.0	*
W	14-15 Jan	3.9	1.58	3.3	0.72	5.9	0.99
W	24 Jan	1.5	0.75	5.6	0.65	6.6	0.77
W	3- 4 Feb	4.8	1.09	9.9	0.81	7.4	1.87
W	8- 9 Feb	12.5	1.78	8.6	1.25	9.2	1.74
W	18-19 Mar	19.0	1.23	17.8	0.73	16.2	1.70
U	27-28 Dec	14.2	1.42	17.8	1.21	16.3	0.88
U	29 Jan	4.8	0.49	5.3	0.52	7.6	0.53
U	20-25 Feb	1.3	0.33	2.3	0.71	2.5	5.33
U	26-27 Feb	2.5	0.85	1.5	0.43	2.0	0.47

E - "Easterly Wind" W - "Westerly Wind" — Not operational
L - "Lake Effect" U - Unclassified * Radar measured 'none'

REFERENCE GAGE FOR RADAR ADJUSTMENTS

Comparison in Table 1 of the G/R_S ratios between gages for the same storm indicates a positive correlation, particularly for the "easterly wind" storms. This suggests that the use of a reference gage to adjust the radar estimates would improve the radar estimates at other gages. This was tested by using gage 7, the most centrally located gage, as the reference gage. The procedure was to obtain the G/R_s ratio for gage 7 for each storm and then use this ratio to adjust the R_s estimates at the other 12 gages. The resulting estimates, denoted as R_{sg} to indicate both gage and storm type adjustment, were then compared to the R_s estimates. Comparisons were based on the root-mean-square (rms) error of the logarithmic difference between the gage catch and radar estimate. Taking all storms together, the rms error of the R_{sg} estimates in comparison to the R_s estimates varied from being reduced 48% at gage 8, the closest gage to the reference gage, to being increased by 40% at gage 4. The R_{sg} estimates were improved at all gages except the most northern gages (gages 1, 2, 3 and 4). However, the usefulness of the calibration gage varies geographically with storm type. For "easterly wind" storms there is general improvement everywhere; the average reduction in the rms error is 40%. For "westerly wind" storms, the reduction in the rms error averages 25% over the western three-fourths of the network. For the eastern quarter of the network (gages 4, 13, 11 and 12) the R_{sg} estimates were slightly worse than the R_s estimates. For "lake-effect" storms, the rms error was reduced an average 38% over the southeast half of the network (gage 8, 9, 10, 11, 12 and 13); for the northwestern half of the network the rms error increased an average 27%.

The cause for the variability in the usefulness of the reference gage between storm types is not yet understood. However, it is likely linked to spatial variations in crystal type and degree of low-level snowflake growth. These two factors are likely to be affected by orographic lifting and low level moisture, both of which are highly variable in space during "lake-effect" and "westerly wind" storms.

CONCLUSIONS AND RECOMMENDATIONS

Radar can be used to estimate water equivalents of snowfall with approximately the same accuracy as for rainfall. However, the maximum radar range of useful snowfall estimates is more limited than that for rainfall. The maximum useful range for snowfall is between 80 and 100 km providing the radar beam width is not over 2.0 deg and the antenna elevation angle is 0.5 deg or less. The accuracy of the radar estimates varies with storm type, the highest accuracy occurring with large scale storms where the primary snowfall production is above 2 km and the lowest accuracy occurring with "lake-effect" storms that are characterized by small horizontal and vertical extent. It is important to use an empirical range correction for the radar estimates of snowfall.

The root-mean-square error in the radar estimates can be reduced up to 50% through the use of a reference gage to adjust the radar snowfall estimates. The reduction depends on distance from the reference gage, geographical location and storm type. It is felt that variations in the accuracy of the radar estimates is caused by variations in snow crystal type and degree of low-level snowfall growth. These factors are likely affected in an unknown manner by orographic lifting and moisture injection from the lake.

The characteristics, operation and siting of the radar for snowfall measurement are more important than for rainfall measurement because, for snow, the cloud tops are lower, precipitation rates are lower and precipitation growth generally extends to lower altitudes. To obtain good quantitative measurements during snowfall, the radar beam width should be less than 2.0 deg. To help achieve a narrow beam width at relatively low cost the use of C-band radar is appropriate and even X-band, if rainfall measurements are not desired. The radar beam should be kept as low to the ground as possible; thus the antenna should not be elevated above 0.5 deg. This means the radar site must be carefully chosen so that the antenna is sufficiently above the ground to minimize screening of the radar beam by objects close to the horizon while still being located close enough to the surface to minimize ground-return echoes.

The inverse range-square normalization of the radar returned power that is usually accomplished electronically within most weather radars should not be used during snowfall. The use of such range normalization circuits increases the minimum detectable precipitation rate at closer ranges and thus will result in the non-detection of light snowfall. Range normalization can be easily accomplished by the computer program when deriving the precipitation totals.

ACKNOWLEDGMENTS

Many people were responsible for collecting the high quality data that made this study possible. In particular the author would like to acknowledge with gratitude: Professor Robert Sykes and Norman Knox of the State University College at Oswego and the Oswego radar staff for the collection of the Oswego radar data and Oswego snow network data; Dr. Eugene Peck and Dr. Lee Larson and Mr. Donald Quick of National Weather Service for selecting the sites for the gages; Mr. George Bosworth of the New York State University at Albany for his assistance in training the students in snowflake observations; Messrs. Paul Cardinali, Kenneth May, Walter Kling, James Munger and William Besaw and their respective students who made the snow observations within the Oswego network; Mr. Kenneth Wilk and Mr. Walter Watts of the National Severe Storms Laboratory, and Mr. Arthur Hansen of NOAA, for the installation of the Oswego radar and radar processor; and Mr. Stephen Vigeant and Mr. Dean Gulezian for assistance in the data analysis. Special thanks are extended to Dr. Edwin Kessler, Director of the National Severe Storms Laboratory and Dr. Eugene Peck of the National Weather Service Office of Hydrology for their special efforts to arrange funding and to acquire the gages for this project.

The work was supported by the NOAA-IFYGL Project Office under Contract 2-35353. The computer computations were performed on the IBM 360/65 at the University of Connecticut Computer Center, which is partially supported through the National Science Foundation Grant GJ-9.

REFERENCES

Carlson, P.E., and J.S. Marshall, 1972: Measurements of snowfall by radar. J. Appl. Meteor., 11, 494—500.

Huff, F.A., 1966: The adjustment of radar estimates of storm mean rainfall with rainguage data. Reprints of papers, Twelfth Radar Meteor. Conf., Norman, Okla., Amer. Meteor. Soc., 198—203.

Jatila, Erkki, 1973: Experimental study on the measurement of snowfall by radar. Geophysica, 12:2 (Helsinki).

Jiusto, J.E., and E.W. Holroyd, III, 1970: Great Lakes snowstorms—Part 1. Tech. Rpt. Atmospheric Sciences Research Center, State University of New York at Albany, 142 pp.

Jiusto, J.E., and M.L. Kaplan, 1972: Snowfall from "lake-effect" storms. Mo. Wea. Rev., 100, 62—66.

Kodaira, N. and M. Inaba, 1955: Measurements of snowfall intensity by radar. Papers, Meteor. Geophys. (Japan), 6, 126—129.

Ohtake, T., and T. Henmi, 1970: Radar reflectivity of aggregated snowflakes. Reprints of Papers, Fourteenth Radar Meteor. Conf., Tucson, Ariz., Amer. Meteor. Soc., 209—210.

Peck, E.L., L.W. Larson, and J.W. Wilson, 1973: Lake Ontario snowfall observational network for calibrating radar measurements. Interdisciplinary Symposium on Advanced Concepts and Techniques in the Study of Snow and Ice Resources, Monterey, Calif.

Sekhon, R.S., and R.C. Srivastava, 1970: Snow size spectra and radar reflectivity. J. Atmos. Sci., 27, 229—307.

Sirmans, P., W.L. Watts, and J.H. Horwedel, 1970: Weather radar, signal processing and recording at the NSSL. IEEE Trans. Geosci. Electron., GE-8, No. 2, 88—94.

Wilson, J.W., 1970: Integration of radar and rain gage data for improved rainfall measurement. J. Appl. Meteor., 9, 489—497.

APPLICATION OF SLAR FOR MONITORING GREAT LAKES TOTAL ICE COVER

R. J. Jirberg
R. J. Schertler
R. T. Gedney
H. Mark
Lewis Research Center, National Aeronautics and Space Administration, Cleveland, Ohio 44135

ABSTRACT

A series of X-band SLAR images is presented showing the development and disintegration of the entire ice cover on Lake Erie during the winter of 1972-1973. Simultaneous ground truth observations and ERTS-1 photography establish accurate correlations of radar responses with ice conditions. The all-weather, broad areal mapping capability of SLAR is seen to be the means for obtaining the repeated coverage needed for winter navigation on the Great Lakes.

INTRODUCTION

The success of current efforts to demonstrate the feasibility of extending the Great Lakes shipping season into the winter months hinges on a number of factors, not the least of which includes improvements in ice information techniques. To facilitate navigation by optimizing shipping routes through ice-bound waters, it is imperative that quick, accurate, and comprehensive appraisals of the ice cover be made available to shippers on a timely basis. In view of the tremendous areal extent of the ice cover and the need for repeated, in some cases daily, reconnaissance, it is clear that automated surveillance is the only viable approach. Among the various imaging devices under consideration for such an automated system, side-looking airborne radar (SLAR) offers two important advantages. One is the ability of the microwave radiation to penetrate all but the most severe weather, day or night. The other is its ability to map much broader swaths from aircraft altitudes than is possible with optical systems.

There have been a few investigations concerning the use of SLAR for ice monitoring beginning early in the 1960's mostly dealing with Arctic ice. From the onset it was apparent that the interpretation of the radar image in terms of the ice types and thickness would be the critical difficulty. Image analysis has been hampered both by a general lack of "ground truth" and by the absence of an accurate analytical model for the scattering of microwave radiation by the ice. Thus a continuing effort involving coordinated ground truth operations and SLAR overflights was launched to firmly establish the connection between the radar image and ice types. Through the cooperation of helicopter teams attached to the 9th District U.S. Coast Guard, it was possible in the first year of this effort to make some of the needed on-the-ice observations and thickness measurements in Lake Erie and in the Whitefish Bay area of Lake Superior. In the course of the study it became obvious that some of the previously held interpretations of radar imagery as to

4.4

ice type and thickness made without benefit of ground truth were inaccurate.

The SLAR used in this study was the Motorola AN/APS-94C system developed for the U.S. Army and flown aboard a Grumman Mohawk OV-1B. It operates in the X-band at 9.2 GHz on a real aperture principle with horizontal transmit and received polarization. Two previous SLAR studies of fresh-water ice have been reported, one by the U.S. Coast Guard (1972) using the Ku-band AN/DPD-2 system, and the other by Larroew et al. (1971) of Michigan's Willow Run Labs (now ERIM) using an X-band synthetic aperture system. The unique feature of the AN/APS-94C system among these and other present day SLAR systems is its extremely broad range of coverage--a capability that makes it feasible to repeatedly map areas as extensive as the Great Lakes.

LAKE ERIE TIME-LAPSE IMAGES

The broad mapping capability of the SLAR is illustrated in the series of images (Figures 2, 3, and 4) showing the entire ice cover on Lake Erie for the period from February 22 through March 5, 1973. This series of SLAR flights was preceded by a rare cloud-free look at the area by the ERTS-1 satellite as shown in Figure 1. Although four days had elapsed between the SLAR flight on the 22nd and the ERTS-1 pass, the correlation of features is striking. The influence of the winds is clearly evident. Note for example that northerly winds prior to the 18th had pushed the ice pack against the southern shore, carrying a relic pattern of the northern shoreline to near mid lake. Between the 18th and the 22nd, southwesterly winds then had carried the ice against the bottleneck at Long Point and into the eastern basin. In the process an extensive network of pressure ridges developed which are seen on the radar image as the dominant lacy pattern.

The flight of the 23rd was terminated midway down the lake, but is important in showing the development of large leads in the western basin due to the continued westerly winds. It also shows the fracturing of the new ice pinned against Pelee Point. A shift of the winds to the northeast during the 25th to the 27th is seen to have shoved the entire ice pack westward, shearing it off at Long Point and at Pelee Point. On the 28th, southerly winds developed shoving the ice pack northward, producing numerous leads as shown in the March 1 image. Also about this time, temperatures rose above freezing. Consequently by the 5th the disintegration of the ice cover had become well advanced.

To the image interpreter these repetitive views can reveal in a way not possible to aerial observers or from aerial photographs the intricate dynamics of the ice. The history is needed to infer thickness information because ground truth observations verified that tonal relations in the image do not directly convey thickness information as sometimes implied in the visible. This can be illustrated by comparing the differences in response to the visible and microwave radiation for the various ice types and thicknesses shown in the pair of images in Figure 5. The various regions of smooth unbroken ice, though varying in thickness from 31 to 41 cm., produce similar dark radar imagery, but in the visible are distinctly different. The radar image, however, is more clearly able to distinguish the rough pancake ice north of Pelee Island which had become layered to a depth of 115 cm. Other bright return areas in

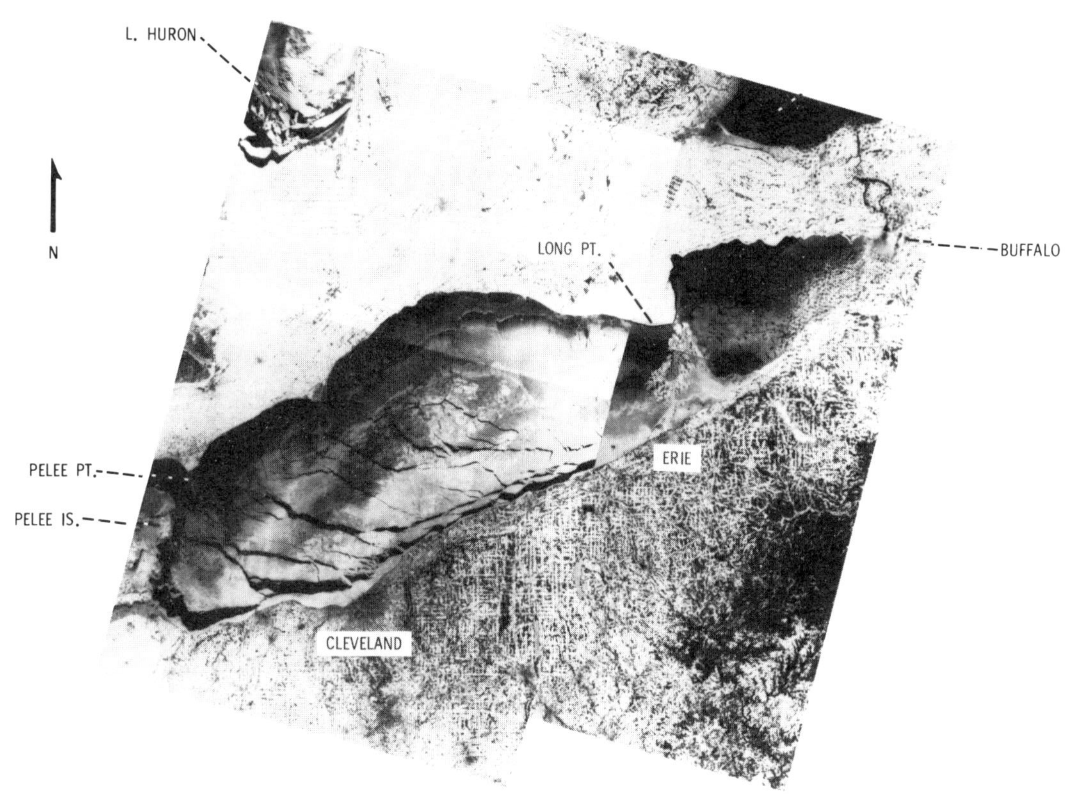

Figure 1. Photomosaic of ERTS-1 satellite imagery from band 4 showing Lake Erie ice cover on Feb. 17 (eastern basin) and Feb. 18, 1973 (central basin). Note the ice movement that occurred in one day along the southern shore.

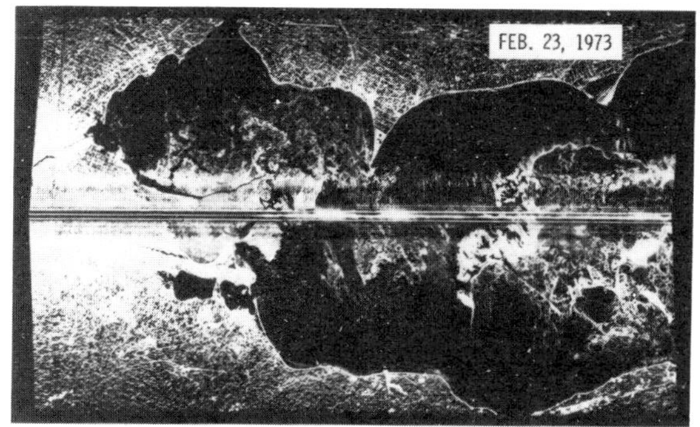

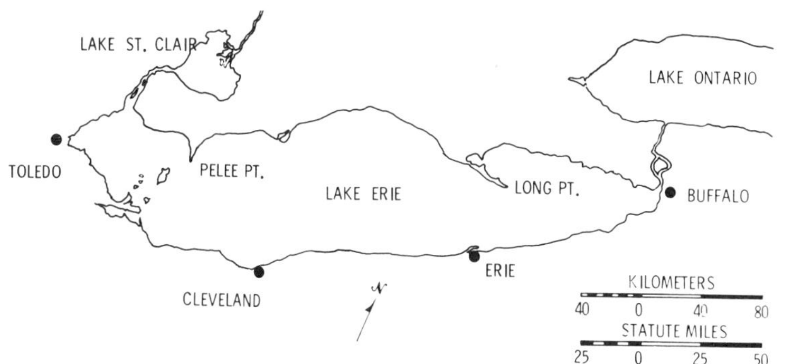

Figure 2. Radar images of Lake Erie ice cover on February 22 and 23, 1973. Time of flight over the lake was approximately 70 minutes. The center line of the image is the blind ground track of the aircraft.

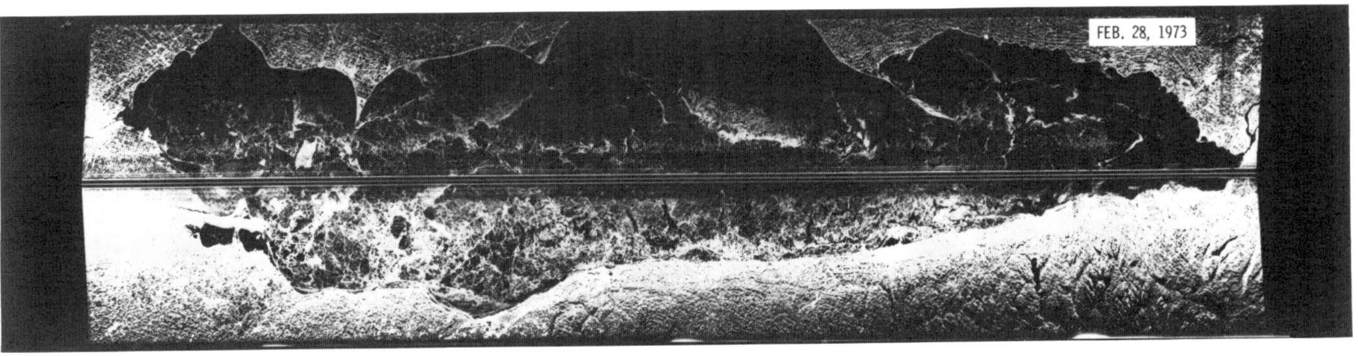

Figure 3. Radar images of Lake Erie ice cover on February 26 and 28, 1973. The compression of the image on the 28th resulted from an uncertainty in ground speed. The distorted eastern end of the lake was due to a slight deviation in course.

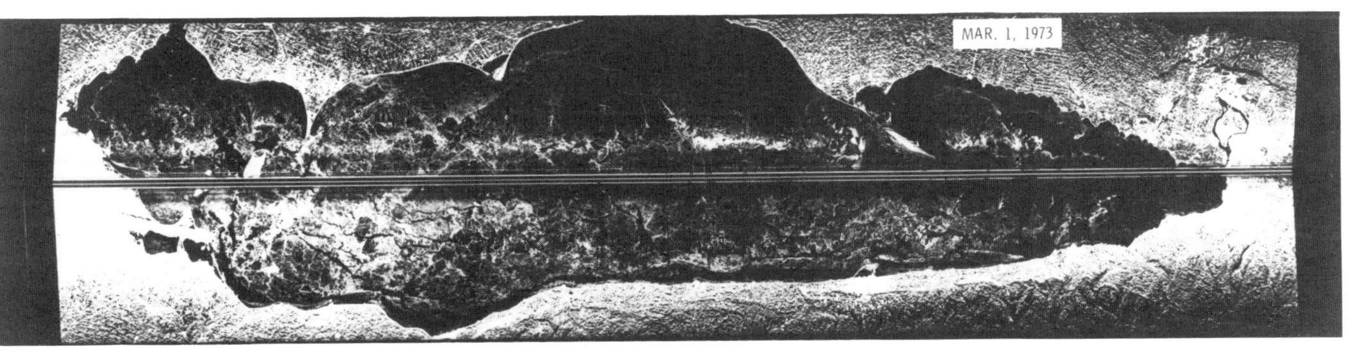

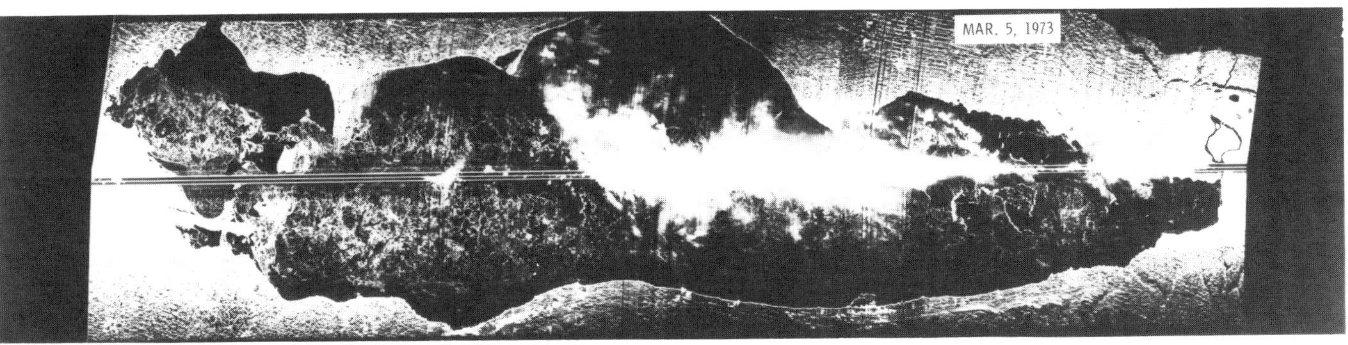

Figure 4. Radar images of Lake Erie ice cover on March 1 and 5, 1973. Rain and hail - filled clouds on the 5th which partially mask the ground were the only weather limitations encountered. Buffeting of the aircraft by strong gusty winds on the 5th produced some smearing of the image.

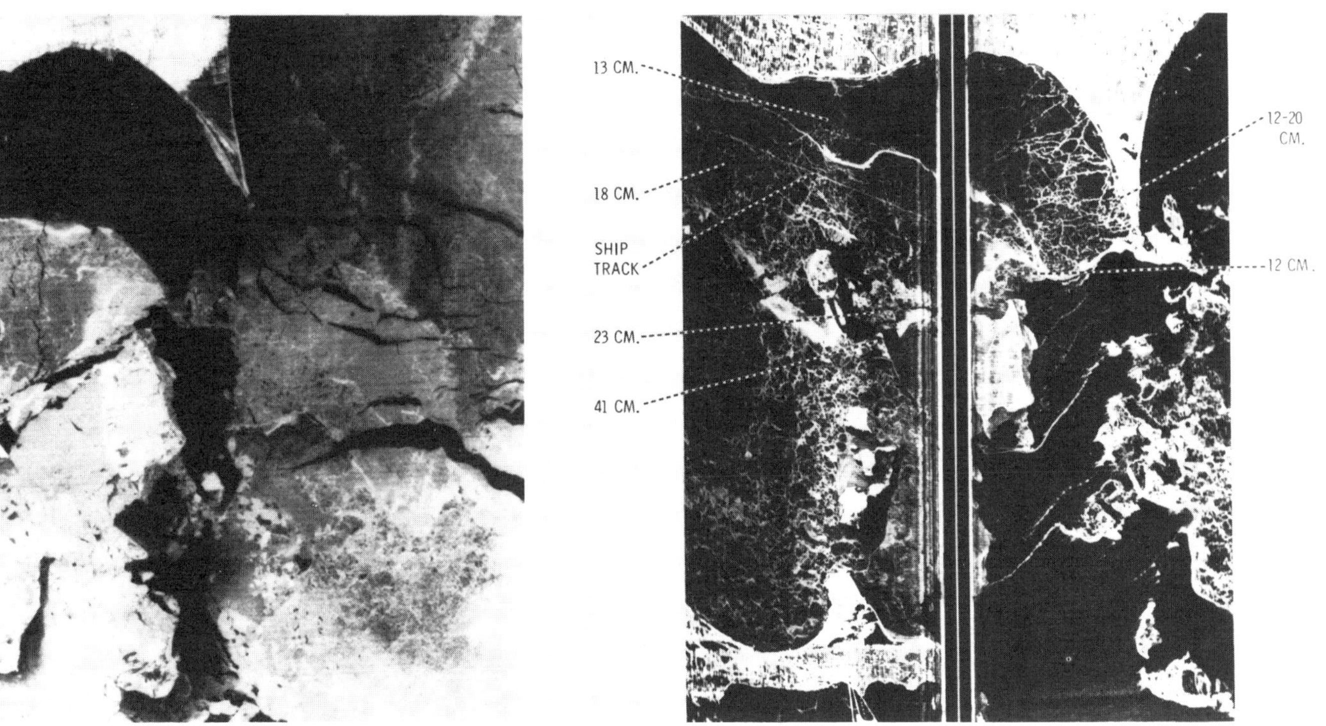

Figure 5. Western Lake Erie as seen from ERTS-1 in band 5 on Feb. 18, 1973 (left) and by SLAR (right) one day later. Pelee Island is prominent in the center with Pelee Point jutting down from the top.

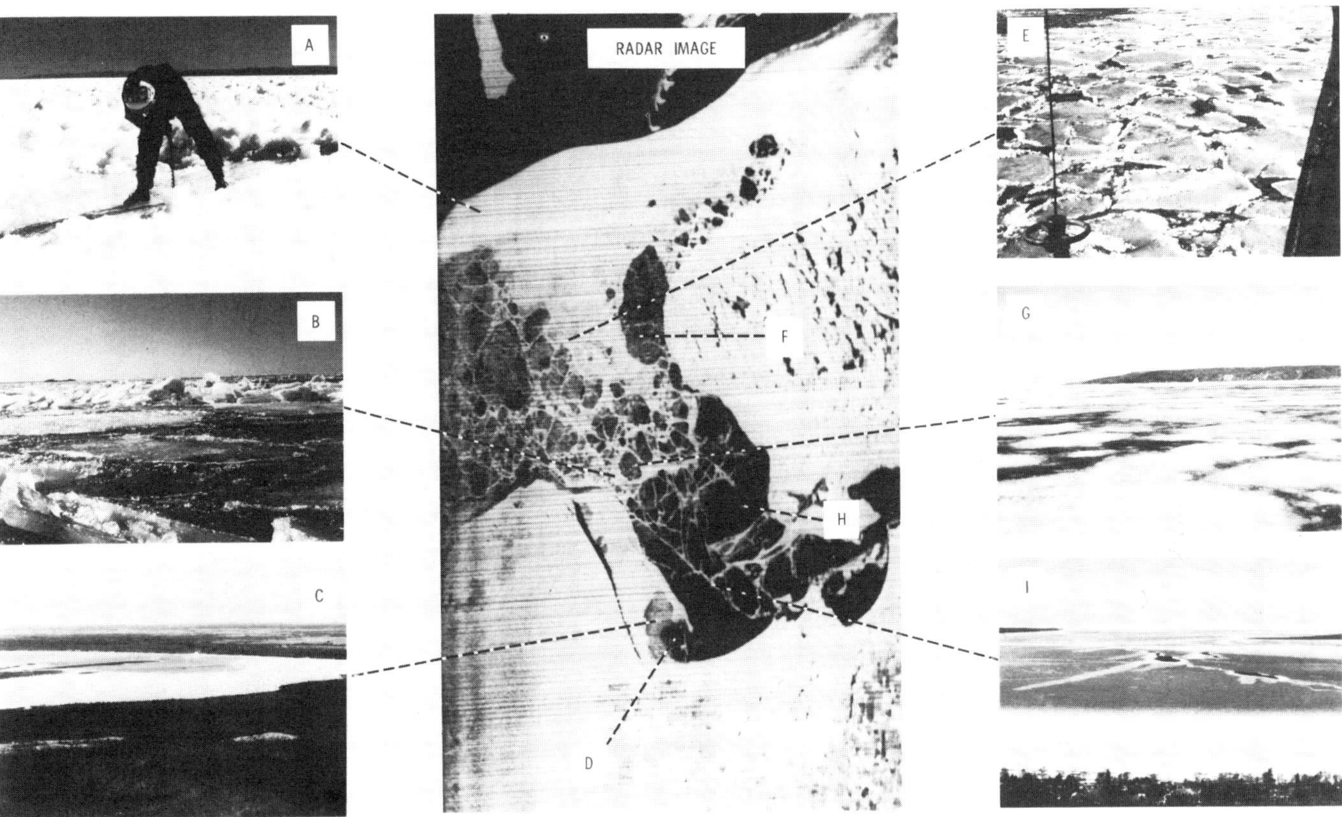

Figure 6. Whitefish Bay (L. Superior) ice cover on March 20, 1973 as mapped by SLAR and photographed at various ground truth test points.

the SLAR image such as that off the tip of Pelee Point originate from regions of consolidated brash which, though deceptively smooth on the surface, had a stepped bottom profile which served as an effective radar reflector. This ice off Pelee Point was 15 cm. thick brash refrozen in a 10 cm. thick matrix.

WHITEFISH BAY GROUND TRUTH

Further indications of the nature of the microwave scattering by the ice were obtained in Whitefish Bay. In Figure 6, the medium-gray area at point C was found to be fast ice, 41 cm. thick, with a 1 or 2 cm. surface ripple. To the eye this area was milky white. The difference in returns between this ice and the adjacent ice at D is believed to be associated with the inclusions present in the fast ice that gave it its milky appearance rather than with the surface ripple. Unfortunately no detailed examination of this milky ice was made, but one possibility is that tiny voids (bubbles) are responsible for its appearance. The ice at D was similarly rippled, but being newer (10 cm. thick), was dark blue in color and unbroken except for fine hairlike cracks. Since this newer ice contained no appreciable scattering sites, it appears no different from the large open water area at H.

The long linear feature at I was a major pressure ridge. Although not an especially striking feature among the surrounding surface clutter when viewed at higher altitudes, its rafted edges make it a good radar target. The thickness of the surrounding ice sheet varied between 35 and 55 cm.

The high return area in and around point B was an area that had undergone considerable crushing, forming a ridged ice zone. The large cellular areas such as F and G were brash ice 44 to 50 cm. thick smoothly refrozen in a 28 cm. thick darker ice matrix. Here again, these cells were difficult to locate among the surface detail from altitude, but because of the microwave scattering off the stepped bottom profile, they clearly stand out in the radar image.

The area that includes point E was found to be composed of rafted angular plates of 32 cm. thick white ice, each on the order of 1 to 5 meters across, refrozen in 24 cm. thick dark ice. Ice build up on the edges of the plates gave the area a high degree of vertical roughness. The roughest ice was found at the edge of the ice pack at point A. Here, due to the constant lapping of the waves and the wind-driven spray, the surface had accumulated a globular foam-like coating of ice 40 to 60 cm. thick. Beneath was a layered accumulation of brash ice extending to depths in excess of 1 meter.

CONCLUDING REMARKS

Except for the case of fast ice (shore ice), it has become apparent that the radar "sees" a world of edges rather than one of bulk forms as we are used to in the visible. In the radar image, ice types such as brash, pancake, and related forms are often the brightest simply due to the large vertical cross section presented by their edges. Strong returns are also produced by ice types with a surface roughness of 10 cm. or more such as occur when edges are built up by wave and spray action, or as occur when ice debris is accumulated along pressure ridges. Similarly, ice types with a comparably rough bottom profile such as consolidated brash produce strong returns due to penetration of the ice by the microwave radiation and its subsequent backscatter off ice-water corner interfaces. Only in the case of fast ice have we seen any indication of volume scattering such as gives us our visual impressions.

Further work is warranted to determine the microwave scattering cross sections for the various ice types and thus allow a better fix on the dynamic range of signals bearing useful information. With the sensitivity and video gain characteristics as designed into the AN/APS-94C system, it was not possible to discriminate unfractured ice from open water directly simply because smooth clear ice lacks sufficient defects to backscatter enough radiation to be detected. Also since clear ice is so transparent at X-band, the dominant sources of backscatter are the edges and ice-water corners rather than volume scatterers. This together with the fact that the density of scatterers can vary widely from one ice type to another, and may not always be correlated to ice thickness, hinders the discrimination of varied ice thicknesses. It remains to be determined to what extent a radar system calibrated in terms of the ratio of backscattered to incident radiation can resolve differences in returns from the various ice types and thicknesses. An additional source for thickness information may be in the degree of depolarization of the radiation with thickness of the ice as noted by Bryan and Larson (1973). Other clues as to ice thickness may be found in the way in which the ice yields to wind stress and in the way ice types develop with time.

Further effort is also warranted to correlate the impeding forces and other difficulties actually encountered by vessels transiting the ice to the information available in the radar image. It is significant that the rougher ice and especially brash ice which is among the most difficult to transit is also that which gives the strongest returns in the radar image. Feedback from shippers in this regard will be valuable.

REFERENCES

Bryan, M. L., and R. W. Larson, 1973. Application of dielectric constant measurements to radar imagery interpretation, Presented at the Second Annual Remote Sensing of Earth Resources Conference, University of Tennessee Space Institute, March 26-28, 1973.

Larrowe, B. T. et al., 1971. Fine-resolution radar investigation of great lakes ice cover. Final Report, Willow Run Laboratories, University of Michigan, January 1971.

U.S. Coast Guard, 1972. Analysis of SLAR imagery of Arctic and lake ice. Final Report, DOT-CG-14486-A, April 1972. Interpretation of winter ice conditions from SLAR imagery. Final Report, DOT-CG-14486-1A, February 1972.

LAKE ONTARIO SNOWFALL OBSERVATIONAL NETWORK FOR CALIBRATING RADAR MEASUREMENTS

Eugene L. Peck, National Weather Service
Lee W. Larson, National Weather Service
James W. Wilson, Center for the Environment & Man, Inc.

ABSTRACT

A select network of weighing-recording precipitation gages was installed to provide improved measurements for evaluating the use of radar for measuring snowfall. The locations of the gages were chosen to provide a uniform distribution for one sector of the radar coverage. Each gage site was selected to provide the maximum amount of natural protection to minimize the adverse effects that wind movement has on gage measurement of snowfall. The importance of having reliable snowfall measurements for evaluating other measurement techniques is demonstrated by comparison of point and areal values among those from the special network, the radar estimates and regular climatic stations in the same area.

INTRODUCTION

Nearly all research, management, or operational activities in the field of snow and ice hydrology require for some reason or other an estimate of solid precipitation values. Unfortunately, it is generally agreed that most point estimates of solid precipitation based on precipitation gage readings are deficient due mainly to wind and its associated adverse effect on gage catch. The worth of areal precipitation estimates based on these point measurements is questionable. A possible technique for obtaining better areal estimates of solid precipitation is through the use of radar. The techniques and procedures which must be developed in order to utilize radar to measure snowfall have only recently begun to be investigated.

The National Weather Service (NWS), in cooperation with the International Field Year for the Great Lakes, has recently been involved with a program to obtain meteorological data by radar. A C-Band 5.3 cm MR-782 Weather Service Radar was installed south of Oswego, New York to make precipitation observations, especially solid precipitation observations in the high snow area just east of Lake Ontario. Storm precipitation values were determined from the radar observations and the appropriate Z-R relationships.

For comparison with the precipitation values determined from the radar equations, it was decided to install a special snow measuring network of 13 Universal weighing-recording precipitation gages. Each gage site in this network was carefully chosen to provide the maximum natural protection possible. The ultimate goal of the network was to provide the best possible input to the radar study.

The network was utilized to investigate the importance of gage location and the improvement in precipitation measurements which may be made by proper site selection and gage location. Point and areal comparisons were made between the special network, the radar estimates, and other stations in the area.

The University of New York at Oswego maintained and operated the network and radar. In addition, the University recruited special student volunteer observers who made snowfall observations during selected periods of time for comparison with the special network.

NETWORK DESCRIPTION

The snowfall observational network is located north of Syracuse, New York and east of Lake Ontario and covers an area of roughly 350 miles^2 (906 km^2) (Figure 1). The elevation generally increases from west to east and south to north across the network and varies from about 250 feet MSL (82 m) to 1300 feet MSL (427 m). The area is a combination of farmland and gently rolling woodlands.

An attempt was made to place all of the 13 network gages in what would be considered "well-protected" locations. In general, this would mean that each gage was sheltered in all directions by coniferous forest subtending angles of 30 - 45° from the gage orifice with the forest of sufficient depth to minimize eddy effects (Peck, 1972). However, since the gages were placed in natural openings, the exposure did vary somewhat from site to site.

OROGRAPHIC INFLUENCES

The seasonal precipitation during the period of observation ranged from near 14 inches (35.5 cm) over the western portion of the study area to near 22 inches (54.2 cm) on the east (Table 1). Most of the variation can be explained by elevation (Figure 2). An analysis of the individual departures from the elevation-precipitation line of Figure 2 indicated that there is a major influence due to the location of Lake Ontario. The maximum lake effect occurs slightly below the level of maximum elevation, but the lake and elevation effects are interdependent. This accounts to some extent for the high correlation (r = .93) for the best fit line in Figure 2. This dual effect may also be seen in the seasonal isohyet analysis shown in Figure 1.

TABLE 1
OSWEGO SNOW NETWORK
PRECIPITATION TOTALS
2 DECEMBER 1972 - 26 MARCH 1973

Gage Number	1	2	3	4	5	6	7	8	9
SUM (CM)	35.53	39.01	52.25	54.20	35.86	36.78	46.76	45.44	39.62
MEAN (CM)*	.37	.41	.55	.57	.38	.39	.49	.48	.42
STD. DEV.	.51	.54	.67	.68	.52	.54	.66	.64	.57

*N = 95

TABLE 1
(continued)

Gage Number	10	11	12	13	Camden	Mallory	Bennetts Bridge	Selkirk Shores
SUM (CM)	42.93	45.24	39.07	50.52	39.88	35.84	40.05	30.00**
MEAN (CM)	.45	.48	.41	.53	.42	.38	.42	.67
STD. DEV.	.59	.64	.54	.72	.67	.55	.58	.61

**N = 45 (Selkirk Shores only, total from 1 Dec.)

GAGE COMPARISONS

Of primary interest were comparisons between gages already located in the network area, mostly NWS climate gages, and the special network gages. The assumption was that most gage locations in the past were not primarily chosen to provide good gage exposure. Instead, the gages were generally placed at locations where people were willing to be volunteer observors. It was felt that gage sites which were specifically chosen for good exposure would probably provide a better estimate of point and areal precipitation for the radar study than would pre-existing gages from the same general area.

Table 2 tends to support the assumption that good exposure is essential for reliable snowfall measurements. Network gage #9 compared to the climate gage at Mallory for snow and mixed storms shows a monthly deficit for Mallory of -5 to -28% and a final four-month deficit of -14%. This is in spite of the fact that the observer at Mallory is extremely capable and conscientious. An inquiry as to this particular observer's observational techniques revealed that he goes to great lengths to provide the best solid precipitation data he possibly can. His system utilizes a standard eight-inch gage, two snowboards plus snowstakes. After each snowfall, a comparison is made between the gage reading and the snowboard readings with the most representative value being logged. The gage itself has little natural protection, one snowboard has natural protection from the east while the other snowboard has natural protection from the west. This particular example shows quite clearly that the most reliable observer will be deficient in his measurements of solid precipitation if the initial site selection was poor.

The comparison of catch at Bennetts Bridge with the average of gages #3, #4, and #7 showed a monthly deficit of -2 to -36% and a four-month deficit of -26%. The comparison of total catch at Selkirk Shores (a Lake Survey gage) with special network gage #1 showed a monthly deficit of -7 to -74% with a final deficit of -23%.

The climate gage at Camden when compared to network gage #12 showed no significant difference. This can perhaps be explained by several factors. First, the climate station at Camden is apparently fairly well protected being located in a barnyard with buildings on the north, south, and west. Second, network gage #12 site is the least protected location in the special network. Initial comments on this site prior to gathering any data was that protection to the east and south was marginal.

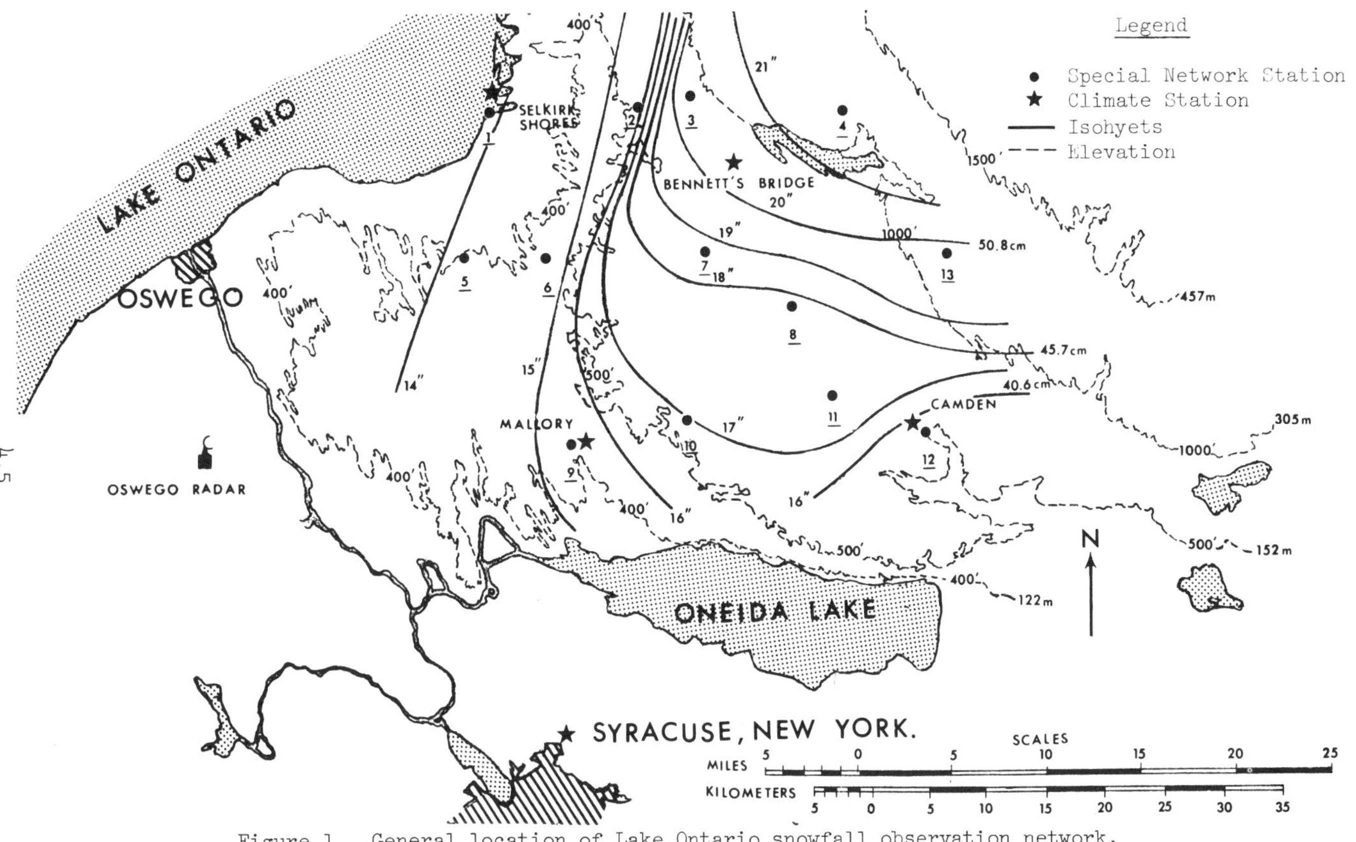

Figure 1. General location of Lake Ontario snowfall observation network.

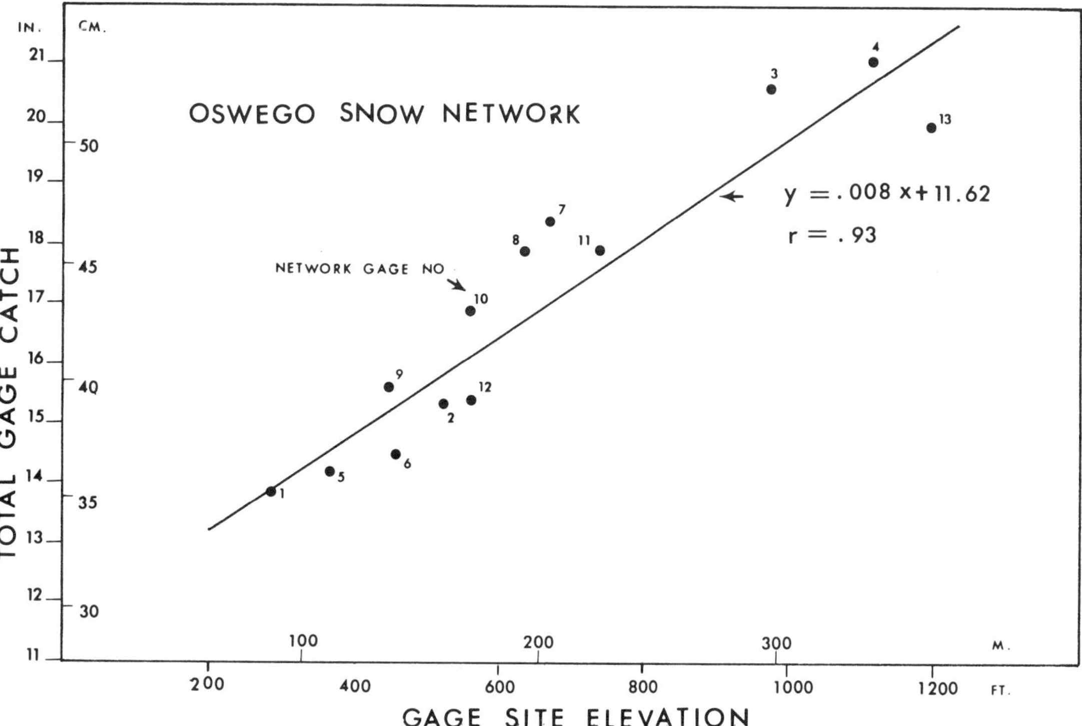

Figure 2. Gage catch versus gage site elevation.

TABLE 2
COMPARATIVE PRECIPITATION GAGE CATCHES
SNOW AND MIXED PRECIPITATION ONLY

Mallory Versus #9

Month	Nov.		Dec.		Jan.		Feb.		Mar.		Total	
Gage	9	M	9	M	9	M	9	M	9	M	9	M
Catch*	.81	.64	10.92	9.40	5.21	4.37	5.87	5.49	2.11	2.01	24.92	21.89
Diff.		-.17		-1.52		-.84		-.38		-.10		-3.03
% Diff.		-28		-16		-19		-7		-5		-14

Bennetts Bridge Versus Mean of (#3, #4, #7)

Month	Nov.		Dec.		Jan.		Feb.		Mar.		Total	
Gage	3,4,7	BB	3,4,7	BB	3,4,7	BB	3,4,7	BB	3,4,7	BB	3,4,7	BB
Catch*	4.34	3.28	16.48	12.12	7.98	6.63	8.38	6.68	4.70	4.62	41.88	33.32
Diff.		-1.06		-4.36		-1.35		-1.70		-.08		-8.56
% Diff.		-33		-36		-20		-25		-2		-26

Camden Versus #12

Month	Nov.		Dec.		Jan.		Feb.		Mar.		Total	
Gage	12	C	12	C	12	C	12	C	12	C	12	C
Catch*	2.67	2.69	11.76	12.17	5.11	4.93	6.63	7.04	2.18	2.29	28.35	29.11
Diff.		.02		.41		-.18		.41		.11		.76
% Diff.		1		3		-4		6		4		3

Selkirk Shores Versus #1

Month	Nov.		Dec.		Jan.		Feb.		Mar.		Total	
Gage			1	SS	1	SS	1	SS	1	SS	1	SS
Catch*			12.24	11.44	8.86	5.09	7.75	6.61	8.13	6.86	36.98	30.00
Diff.				-.80		-3.77		-1.14		-1.27		-6.98
% Diff.				-7		-74		-17		-18		-23

* (cm)

AREAL ESTIMATES OF PRECIPITATION

Areal estimates of storm and total precipitation values for the network were made using the Thiessen method. Comparisons were made between precipitation estimates based on the entire 13-gage special network, the four-gage climate network (i.e. Mallory, Camden, Bennetts Bridge, and Selkirk Shores), and the four-gage special network composed only of gages close to climate gages (i.e. #1, #9, #12, and average of #3, #4, #7). The total period (four months) areal precipitation values, based on the 13-gage special network was 17.5% greater than the result based on the climate gages. (If the analysis is based only on the four-gage special network, the result is that the special network produced an estimate of areal precipitation 16.2% greater than that based on the climate network.)

An estimate of sampling errors within the special network was made utilizing the assumption that the areal estimate based on all 13 network gages was correct. By reducing the number of gages in the special network and comparing the areal estimates with those from the 13-gage network for monthly data, the following results were obtained. For a network of eight to 10 gages and monthly precipitation of four to six inches, an error of 2% in an areal estimate of precipitation was obtained. The estimated sampling error for a five to seven gage network increased to 3% while the sampling error for a one to four gage network was about 8%. The relatively small increase in sampling error as the number of gages in the network was reduced can perhaps be explained by the fact that care was taken to insure that the remaining gages in each case were uniformly distributed throughout the network area and that each gage site in the network was carefully chosen to provide a uniform type of exposure from site to site.

EXPOSURE

Wind is the major cause of error in precipitation gage measurements. This error increases with wind speed and is larger for solid precipitation than for liquid precipitation (Kurtyka, 1953). A generally accepted theory is that, in addition to site turbulence, much of the total measurement error is a result of turbulence and increased wind speed in the vicinity of the gage orifice resulting from the obstacle of the gage itself to the windstream (Robinson and Rodda, 1969).

An analysis of gage catch deficiencies versus wind speed was performed for the climate sites. The gage catch deficiency was found by comparing the climate gage catch to the corresponding network gage catch. An example of the results from Bennetts Bridge are that at 10 mph (16.1 km/hr) the ratio of climate to network gage is .80 and at 20 mph (32.2 km/hr) the ratio is .65. This would indicate that wind has a greater adverse effect on the climate gage than on the network gage and therefore the climate gage exposure is probably inferior to that of the network gage.

It has been shown that an Alter shield will increase gage catch in windy situations (Warnick, 1956). Within reasonable limits, up to 25 mph (40.2 km/hr) (Weiss, 1961), the windier the gage site, the greater will be the difference in catch between a shielded and an unshielded gage. Conversely, if a gage site is well enough protected so as to eliminate or minimize horizontal wind speeds, then the Alter shield will make little difference in gage catch. In the 13-gage special network, five uniformly distributed gages (#1, #4, #7, #9, and #12) were installed with Alter shields. It is interesting to note that the average total catch for these five gages is 17.0 inches (43.1 cm) while the average total catch for the other eight unshielded gages is 17.1 inches (43.5 cm). This would indicate that within the network, gage sites are at least well enough protected so as to eliminate the beneficial effect of the Alter shield.

An excellent example of gage catch difference due to exposure is Selkirk Shores and gage #1. The Selkirk Shores gage is located very near the waters edge of Lake Ontario. Gage #1 is located nearby, but in a forest clearing. From December through March, gage #1 caught nearly 23% more precipitation due to better gage exposure.

The performance of each gage in the network as a function of wind direction was investigated. Unfortunately, sufficient snow data was not obtained to make definite conclusions as to the relationship between individual gage sites and wind direction. It did seem, however, that the sites which were initially evaluated as being weak in exposure in a given direction generally had lower relative catches from that direction.

ADDITIONAL OBSERVATIONS

Some 55 student observers were recruited throughout the network to provide additional input data to this study. One of their functions was to provide solid precipitation measurements through the use of snowboards. During the months of January and February (some 14 individual observation periods), the snowboard results were compared to the 13-gage network average and were found to be about 17% higher. The precipitation during these 14 observation periods represented about 70% of the total precipitation which fell during these two months.

Although snowboards are assumed to represent only new and actual snowfall, it should be recognized that the exposure of the snowboard site is critical. In some locations, blowing snow may add or detract from the snowfall measurement.

RADAR MEASUREMENTS

A thorough analysis of the radar results from this study are available in a companion paper by Wilson (Wilson, 1973). Some interesting results applicable to this paper are presented in Figure 3. In this figure, the ratios of total snowfall by gage catch and radar estimate are plotted against distance from the radar.

The lower line represents the best fit curve for the climate gages in the area while the upper line is the best fit curve for the 13-gage special network. It may be seen that the data for the special network results in a curve with a higher correlation coefficient (.90 to .64) and also with a smaller standard error (.10 to .28). The better fit of the data from the special network supports the contention that well-protected sites are essential for proper evaluation of radar measurements of snowfall.

CONCLUSIONS

The importance of gage exposure in the measurement of precipitation, expecially solid precipitation, cannot be overemphasized. In this particular study, properly exposed gages from the special snowfall measuring network averaged 16% greater catch than existing climate gages. Except for unusual situations, it is generally assumed that the larger the catch, the closer it represents the amount of precipitation which actually fell at the site (Brown and Peck, 1962). Wilson has shown that precipitation measurements from a properly protected gage provide a better index to areal precipitation than if the gage is poorly exposed (Wilson, 1954). It seems obvious therefore that both point and areal estimates of precipitation obtained from the special network in this study are more reliable and closer to "true" precipitation than similar data obtained from normal climate stations in the same area. It also seems obvious that the use of precipitation gages to calibrate or evaluate the effectiveness of radar in measuring precipitation (especially solid precipitation) must first be preceded by a careful analysis and evaluation of the gage exposure of existing gages and/or a careful site selection process for new gages.

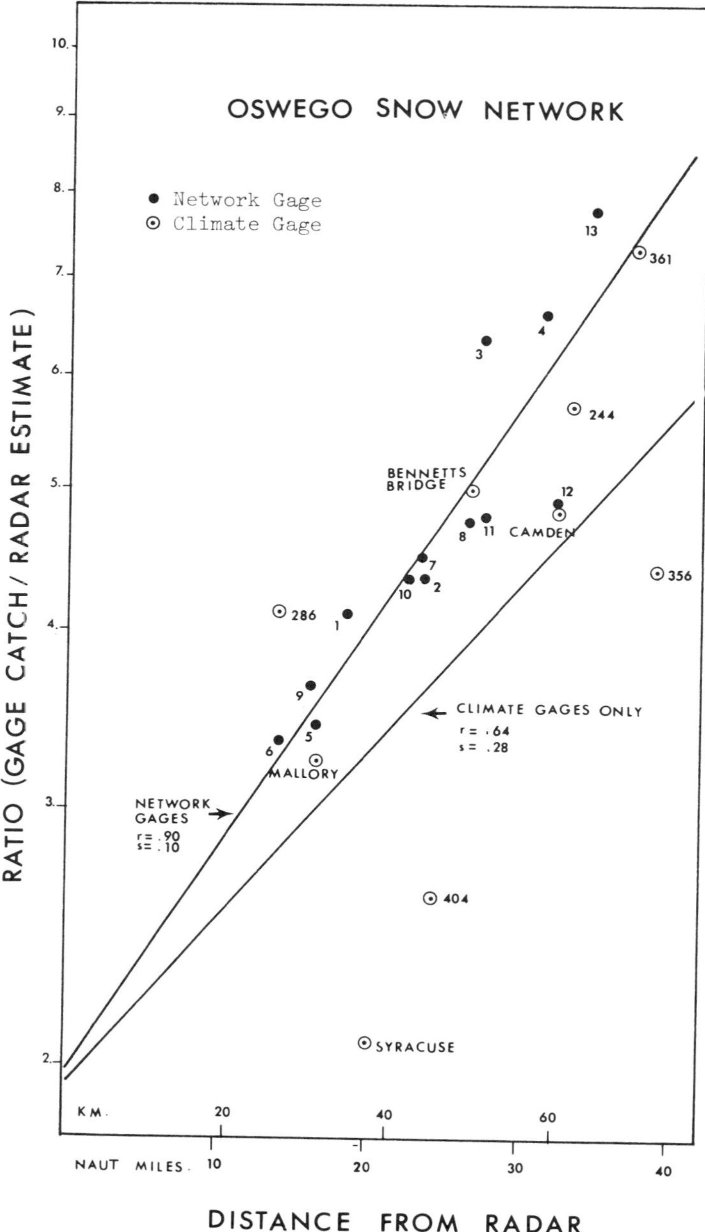

Figure 3. Gage catch/radar estimate versus distance from radar.

REFERENCES

Brown, M. J., and E. L. Peck, 1962. Reliability of Precipitation Measurements as Related to Exposure. Journal of Applied Meteorology. Vol. I, No. 2.

Kurtyka, J. C., 1953. Report of Investigation Study No. 20. State Water Survey Division. Urbana, Illinois.

Peck, E. L., 1972. Snow Measurement Predicament. Water Resources Research. Vol. 8, No. 1.

Robinson, A. C., and J. C. Rodda, 1969. Rain, Wind and the Aerodynamic Characteristics of Rain-Gauges. Meteorological Magazine. London. 98.

Warnick, C. C., 1956. Influence of Wind on Precipitation Measurements at High Altitude. Univ. of Idaho. Eng. Exp. Sta. Bulletin No. 10.

Weiss, L. L., 1961. Relative Catches of Snow in Shielded and Unshielded Gages at Different Wind Speeds. Monthly Weather Review. Pages 397-400.

Wilson, J. W., 1973. Measurement of Snowfall by Radar. Interdisciplinary Symposium on Advanced Concepts and Techniques in the Study of Snow and Ice Resources. Monterey, Calif.

Wilson, W. T., 1954. An Analysis of Winter Precipitation Observations in the Cooperative Snow Investigations. Monthly Weather Review. Pages 183-195.

AIRBORNE PROFILING OF ICE THICKNESS USING A SHORT PULSE RADAR

R.S. Vickers
Dept. of Electrical Engineering
Colorado State University

J.E. Heighway
R.T. Gedney
NASA/Lewis Research Center

ABSTRACT

The acquisition and interpretation of ice thickness data from a mobile platform has for some time been a goal of the remote sensing community. Such data, once obtainable, is of value in monitoring the changes in ice thickness over large areas, and in mapping the potential hazards to traffic in shipping lanes.

This paper describes measurements made from a helicopter-borne ice thickness profiler of ice in Lake Superior, Lake St. Clair and the St. Clair river as part of NASA's program to develop an ice information system.

The profiler described is a high resolution, non-imaging, short pulse radar, operating at a carrier frequency of 2.7 GHz. The system can resolve reflective surfaces separated by as little as 10 cm. and permits measurement of the distance between resolvable surfaces with an accuracy of about 1 cm. Data samples are given for measurements both in a static (helicopter hovering), and a traverse mode. Ground truth measurements taken by an ice auger team traveling with the helicopter are compared with the remotely sensed data and the accuracy of the profiler is discussed based on these measurements.

INTRODUCTION

Techniques for the remote measurement of ice thickness have been studied for many years. In some cases, the thickness is inferred from the surface temperature as in the case of thermal infrared scanning, or from the microwave brightness temperature (Meier and Edgerton, 1971). A more direct method, that of profiling with an active radar system was used in 1966 by the U.S. Army (Rinker, 1966) to map the thickness of the Greenland Ice Cap. This technique proved successful in terms of penetration through a great thickness of ice, but the resolution was not suitable for other applications involving relatively thin ice due to the small (14 MHz) bandwidth employed.

In this paper, a radar system is described which has been used to profile lake ice down to a thickness of 10 cm. Data from the system both in a static and mobile mode are given. The platform used was a Sikorski H-52 helicopter furnished through the cooperation of the U.S. Coast Guard stations at Selfridge Field and Traverse City.

4.6

MICROWAVE PROPAGATION THROUGH ICE

A summary of the microwave properties of ice and snow is given by Holter and Lowe (1966) based on data published by Von Hippel (1954) and Cummins (1952). At 3 GHz the value given for the loss tangent of ice is 0.9×10^{-3}, corresponding to an attenuation of 0.44 db/meter for ice with a dielectric constant of 3.2. This figure assumes no losses from scattering processes.

The losses indicate that a plane parallel layer of ice could easily give rise to multiple returns when illuminated by a radar pulse, due to energy being multiply reflected within the layer with minimal absorption. This phenomenon was noted in some of the data obtained over smooth ice. One advantage of the low absorption is the attendant high penetration capability of the radar. Using a dielectric approximation for the calculation of the reflection coefficient, the values to be expected at 3 GHz are as follows:

Air/Ice interface	28%
Air/Snow interface	10%
Snow/Ice interface	19%
Ice/Water interface	67%

The presence of surface roughness would of course modify these figures, as would also the presence of surface films of water in the liquid phase. The calculations do indicate however that in an ideal case of horizontally stratified snow/ice/water, a substantial reflection would result from each boundary. Previous work (Vickers and Rose, 1972) on layered snow and ice supports this contention.

SYSTEM DESIGN

A diagram of the system is given in Figure 1. The generator is a solid state source capable of tuning from 2.5 to 2.9 GHz. The frequency was adjusted to 2.7 GHz to
minimize leakage from the mixer. The C.W. signal from the oscillator was then switched with a double balanced mixer driven by a 1 ns pulse to give a short burst of 2.7 GHz carrier. This signal was amplified to a peak power of 1W with a Travelling Wave Tube amplifier (TWT) and then transmitted through a double ridged horn antenna. The TWT also had an amplitude modulation capability, thereby allowing range dependant gain to be employed if necessary.

The receiver chain consisted of two solid state amplifiers with a total gain of 70 db feeding an envelope detector. The output from this detector was then displayed on a sampling oscilloscope and recorded on either polaroid film (static mode) or on continuous strip film (traverse mode).

Options available to the operator included triggering the display on the first return, which removed helicopter altitude variations from the data, and an intensity modulation display mode.

The system was designed to operate at altitudes around 40m above the ice, although the gain was sufficient for greater altitudes if necessary. The detected return pulse, shown in Figure 2, had a half width of 1.3 ns. A photograph of the completed system appears in Figure 3.

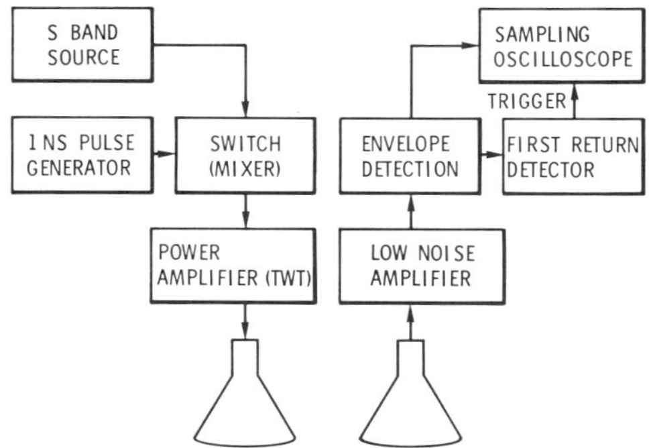

Figure 1. - Block diagram of ice profiler.

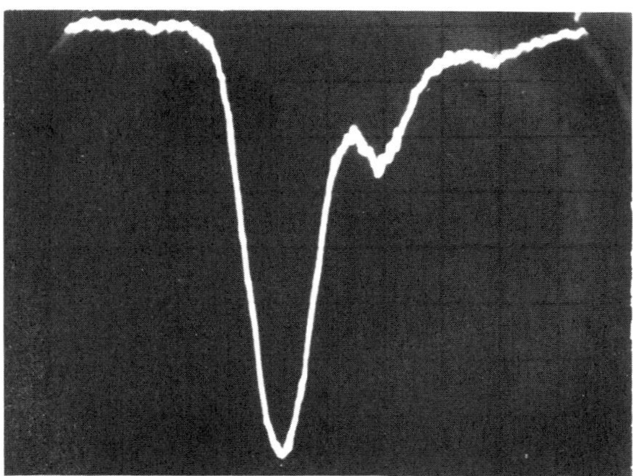

Figure 2. - Waveform (1 ns per division) as passed through complete system, with receiving horn facing the transmitting horn.

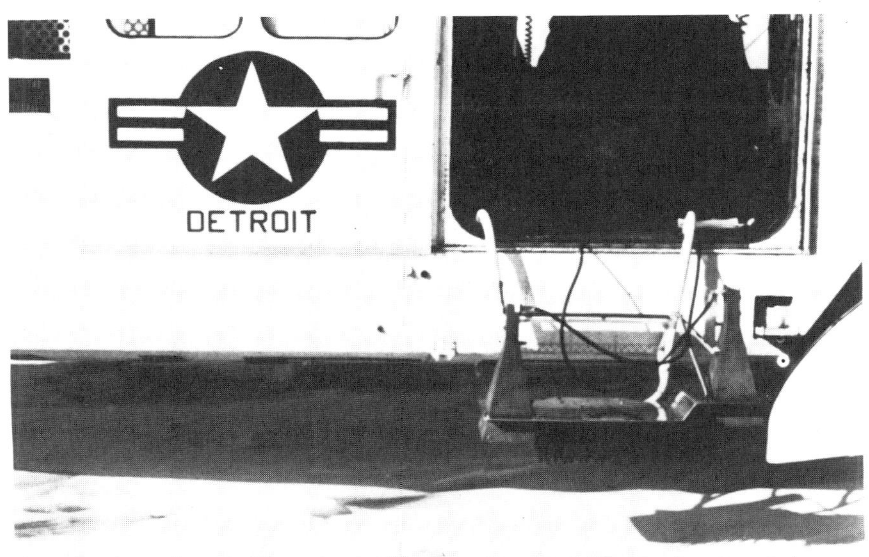

Figure 3. - Radar profiler installed in Sikorsky H-52 helicopter.

4.6

A protype of the equipment, with some minor differences, had already operated for one field season in the measurement of snow depth in the Colorado Rockies (Vickers and Rose, 1972). The present system was therefore given brief preliminary testing by using it to view a vertical ice wall, of adjustable thickness built from commercial ice blocks. This experiment was used to determine the minimum measurable thickness of ice, under relatively controlled conditions. This minimum was less than 10 cm. Also evaluated were the effects of having the first surface (air/ice interface) covered with water to simulate melting ice conditions. The thin film of water retained on the vertical face of the ice wall had no appreciable effect. An effect was observed during these tests which later gave some experimental difficulty, namely the beamwidth of the antennas (45°) was large enough to allow returns from a number of paths to interfere with that from the line-of-sight path from the target. This interference effect between the returns caused significant amplitude modulation of the total signal as the range was varied, and made data interpretation difficult. The same effect was later noted in the helicopter installation where the return coming from the ice interfered with a reflection from the vehicle landing gear. It was eventually removed by lowering the antennas. An example of the test data from a 19 cm. thick wall of ice is given in Figure 4.

EXPERIMENTAL PROCEDURE AND RESULTS

After a number of trial flights over Lake St. Clair, the system was taken over Whitefish Bay in Lake Superior for demonstration. The procedure employed was to fly the system to a suitably thick region of ice and land the helicopter. An ice auger team then took a sounding of the ice thickness, and the radar system was subsequently flown over the area at an altitude of 20 to 30 m. A typical return from these tests is shown in Figure 5 where the two pulses correspond to the upper and lower surfaces of ice 30 cm. in thickness. The sites were photographed from the ground and from the helicopter for later correlation with the data. The radar data itself was recorded on polaroid film. In addition some traverse data was recorded on strip film.

An example of data taken in the traverse mode is given in Figure 6, where each excursion of the trace to the left indicates a target. An alternative and more powerful presentation in terms of visual impact is to detect the minima in each trace, and show these bright points in an intensity modulated display as shown in Figure 7. Visual separation of the ice stratigraphy from random returns due to roughness is then more easily achieved.

It was found that the ice thickness could be accurately measured in the case of a single unbroken layer thicker than 10 cm. and smooth on both its upper and lower surfaces. The accuracy of measurement was about 2 cm. for such sheets of ice. Quite often, however, one encounters ice that exhibits roughness on one or both of its surfaces, or possesses a layered structure. Surface roughness can be caused by 1) the freezing of slush formed from the melting of snow, or 2) the freezing of spray and droplets generated in the leads between storm-tossed floes, or 3)the upthrusting of edges in pressure ridges.

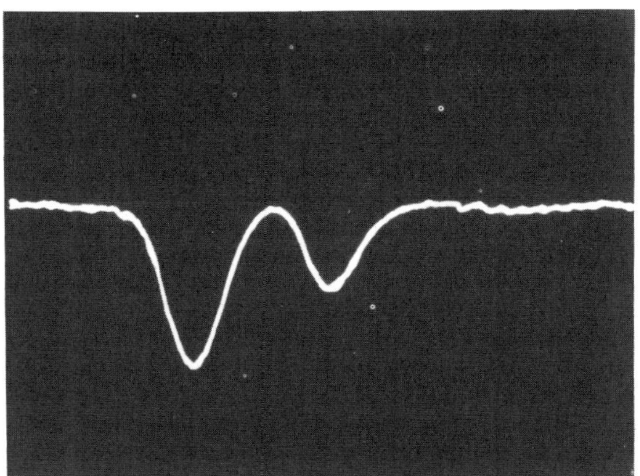

Figure 4. - Radar return from 7 1/2'' of block ice (1 ns per division).

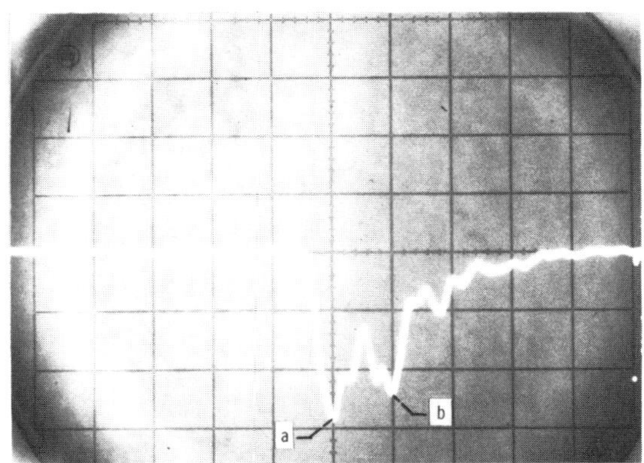

Figure 5. - Radar return from old ice cemented by newer thin ice (5 ns per division). a: Return from air-ice interface; b: return from ice-water interface.

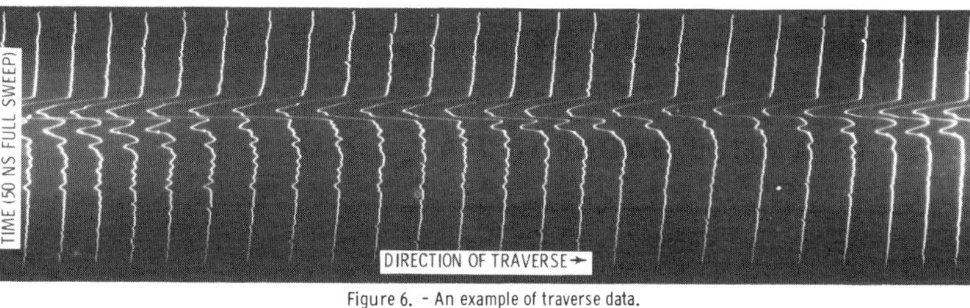

Figure 6. - An example of traverse data.

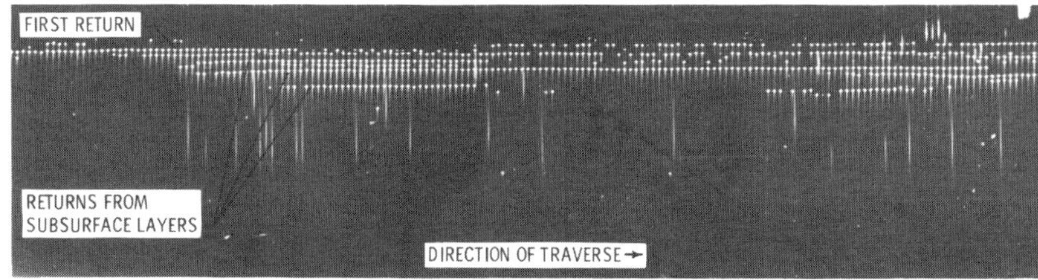

Figure 7. - An example of intensity modulated traverse data showing stratification.

Lower surface roughness is observed in pressure ridges, of course. Besides this, one frequently encounters a "stepped" lower surface in re-frozen brash in which relatively thick floes are held together by a matrix of thinner ice.

Upper surface roughness causes difficulty because signals reflected from bumps located off the axis can interfere with the signal of primary interest, namely, that reflected from the ice-water interface. An example of such a case is given in figure 8.

The "stepped" lower surface encountered in refrozen brash seems to produce a complex return as off-axis signals are efficiently returned from surfaces forming corner reflectors, as it were. These reflections make interpretation of the return difficult. If the most straightforward interpretation is used, (identifying the largest peak after the first peak as that reflected from the ice-water interface) significant errors are made.

The problem of interference by off-axis reflections is judged to be the most serious impediment to the development of an operational radar ice thickness system.

Because of the large attenuation one would expect that if ice sheets are separated by more than a few millimeters of liquid water, lower sheets would be undetectable. No data bearing upon this issue was taken.

In the case of smooth sheets of ice, it was found that multiple reflections were often observable. While these multiple reflections caused no difficulty in inferring ice thickness for ideal ice sheets, they obviously increase the difficulty of interpreting the returns from more complex ice structures. This problem is considered to be minor as compared with that of off-axis reflections.

Figure 9 presents data from 14 sites comprising a variety of ice types. The R.M.S. error of these measurements is about 15%. Previous measurements on ice and snow show that with perfectly smooth surfaces an accuracy of better than 10% is easily achievable (Vickers and Rose, 1972).

CONCLUDING REMARKS

In summary, it has been demonstrated that a short pulse radar system can make accurate measurement of the thickness of simple, smooth ice sheets, and at least give an indication of the structure of rough-surfaced or broken ice. Furthermore, the data indicate that a system capable of making thickness measurements of nearly all types of ice is feasible, provided that interference from off-axis reflections can be overcome. The most natural attack on this problem is to reduce the beamwidth of the system. For the 1973-74 season, an array of four receiving antennae will be used, reducing the beamwidth from 45 degrees to somewhat less than 20 degrees. In addition, the use of a higher carrier frequency (up to 5 GHz) will be studied to determine whether the higher attenuation will reduce the amplitude of multiple reflections.

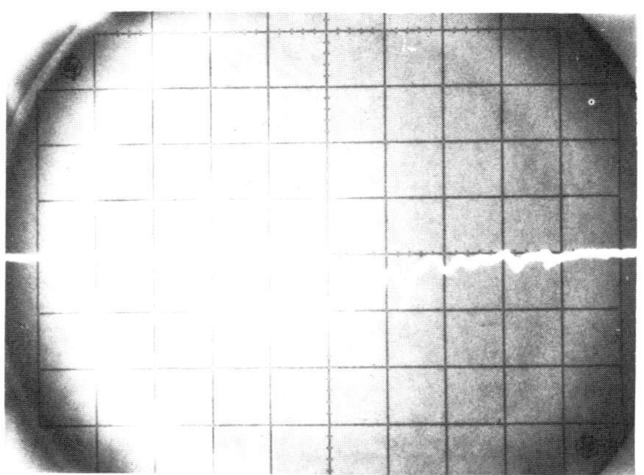

Figure 8. - Radar return from 1 meter ice with 30 cm of debris, 5 ns per division.

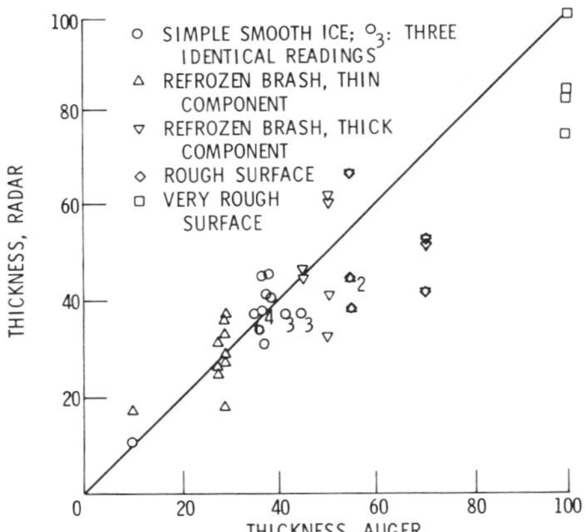

Figure 9. - Comparison of radar derived thickness (cm) with ice auger readings.

4.6

ACKNOWLEDGEMENT

The authors wish to acknowledge the generous cooperation of the United States Coast Guard Stations at Selfridge Field and at Traverse City.

REFERENCES

Cummins, W.A., 1952. The Dielectric Properties of Snow and Ice at 3.2 cms. J. Appl. Physics, vol. 23, No. 7 p. 768.

Holter, M.R. and Lowe, D.S. ed, 1966. Peaceful Uses of Earth Observation Spacecraft. vol. III, Report No. 7219-1-F (III). Willow Run Labs., University of Michigan.

Meier, M. F. and Edgerton, A.T., 1971. Microwave Emission from Snow, a Progress Report. Proc. 7th Int. Symp. on Remote Sensing of Environment, University of Michigan. pp. 1155-1163.

Rinker, J.N., 1966. Radio Ice Sounding Techniques, Proc. 4th Int. Symp. on Remote Sensing of Environment. University of Michigan.

Vickers, R.S. and Rose, G.G., 1972. High Resolution Measurements of Snowpack Stratigraphy using a Short Pulse Radar. Proc. 8th Int. Symp. on Remote Sensing of Environment. vol. 1, University of Michigan.

Von Hippel, A.R., 1954. Dielectrics and Waves, John Wiley & Sons, Inc.

CONTINUOUS SEA AND FRESH WATER ICE THICKNESS PROFILING USING AN IMPULSE RADAR SYSTEM

K. J. Campbell
A. S. Orange
Geophysical Survey Systems, Inc.

ABSTRACT

An impulse radar system that provides a continuous profile of sea and fresh water ice thickness was put into operational use in early 1973. This system can be considered the electromagnetic equivalent of single-trace acoustic profiling systems used in marine sub-bottom profiling. For the radar system, an electromagnetic pulse is generated on the ice surface and the reflections from the surface and the ice/water interface are displayed on a continuous strip-chart recorder. Travel times of the reflected pulses can be converted directly to ice thickness. System calibration, accuracy and repeatability have been investigated under varying conditions of first-year and multi-year sea ice and fresh water ice. Structural features within the ice including both "wet" and "dry" cracks as little as 4 to 5mm wide, can be detected with the system. Operational surveys have been performed for oil companies and geophysical contractors at several locations in the Canadian Arctic to insure the safety of on-the-ice operations and to contribute to the more economical utilization of personnel and equipment. It is likely that the system also can be applied to thickness measurement of glacial ice and to crevasse detection.

INTRODUCTION

For several years Geophysical Survey Systems, Inc. has been using an impulse radar system as a shallow subsurface exploration tool for engineering applications. Recently this system has been applied to the measurement of ice thickness both on sea and fresh water ice. In the course of a feasibility study performed in December 1972, the basic operating parameters and limitations of the tool when operated on ice were determined. Following the feasibility study, operational surveys were performed totalling approximately eleven crew-months and covering in excess of 1500 km of survey route at several locations in the Canadian Arctic.

The technique is known as Electromagnetic Subsurface Profiling (ESP) and can be considered the electromagnetic equivalent of the single-trace acoustic profiling methods used for marine sub-bottom profiling. In practice, ice-thickness profiling is done by towing a sled-mounted antenna behind a tracked vehicle containing the impulse system (Figure 1). Real-time profile data is displayed graphically on a strip-chart recorder. The data may also be recorded on magnetic tape for later processing and playback.

Figure 1. Sled-mounted impulse radar antenna, towed behind all-terrain vehicle.

PRINCIPLES OF ELECTROMAGNETIC SUBSURFACE PROFILING

A. Introduction. The physical principle upon which the method is based is the reflection that occurs when an electromagnetic wave is incident upon an interface where a change in electrical properties occurs (Kraichman, 1970). In the case under discussion, if an electromagnetic wave is incident upon an ice-water interface, part of the energy will be reflected from the interface. In the impulse radar system an extremely short electromagnetic pulse is generated at the surface and radiated into the ice. The "echo" or reflection from the ice/water interface is received and the travel time from the surface to the interface and back measured. Once the velocity of propagation within the ice is known, either from a knowledge of the electrical properties or direct mechanical calibration by measurement of ice thickness, then the travel time can be converted directly to ice thickness.

B. System Description. The block diagram of the impulse radar system is shown in Figure 2. The power supply feeds a regulated D.C. voltage to the transmitter, which, when triggered, generates a pulse of the order of 2 nanoseconds long. It is important to note that this is a voltage impulse, not pulsed cw (continuous wave or carrier) as in the conventional navigation radar. The pulse is shaped to quasi-gaussian form and then radiated via the antenna into the ice. Frequency content of the signal is broadband, with a bandwidth of the order of 300 Mhz. The pulse repetition frequency of the transmitted impulse is 20 Khz. The reflected signals are received on the same antenna used for transmitting and are fed into the receiver portion of the system. The receiver design is based on sampling techniques that involve conversion of the input waveform, a transient that occurs in nanoseconds of time, to audio frequencies.

This allows data processing and recording utilizing relatively simple techniques. The sampling frequency is a variable but is of the order of 16 hz. That is, a complete "scan" of the input transient requires 62.5 milliseconds in which time 1250 successive transmitted pulses have been sampled to yield the composite scan. It should be noted that at the normal profile speed (4-5 km/hr) the antenna has covered a distance of the order of 25 cm in this time. Once the audio signal has been generated, it is filtered and recorded.

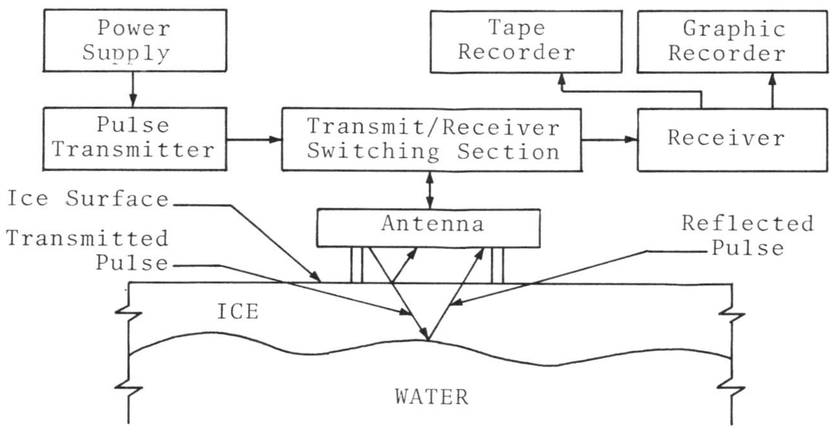

Figure 2. Block diagram of impulse radar system.

C. Data Presentation. The ice thickness profile data is displayed on a graphic recorder identical to those used in marine bottom and sub-bottom profiling. The input waveform to the recorder, sketched in Figure 3A, consists of the transmit pulse, the reflection from the surface of the ice and the reflection from the ice/water interface. Note that while the transmitted pulse input to the antenna during the transmit cycle is a short, monopolarity pulse, the filtering effect of the finite-bandwidth antenna and the subsurface media itself results in the reflected signals appearing as damped sinusoidal in shape.

The recorder is an intensity-modulated device, with the sweep of the stylus across the chart paper snychronized with the pulse transmitter trigger. Signal amplitudes above a pre-set threshold level are printed as black while weaker signals remain white. The paper is moved under the stylus as the stylus is swept, resulting in the profile or "section" as shown in Figure 3B. The dark bands occur at signal peaks (both positive and negative peaks are recorded), the narrow white lines being the "zero crossings" between peaks. The chart paper is calibrated in terms of nanoseconds of time, and thus can be also calibrated in terms of ice thickness once the travel time/thickness relationship is known. However, the profile displays only total ice thickness. It does not necessarily display the true relationship between ice surface topography and ice bottom topography since time zero remains constant regardless of surface topography.

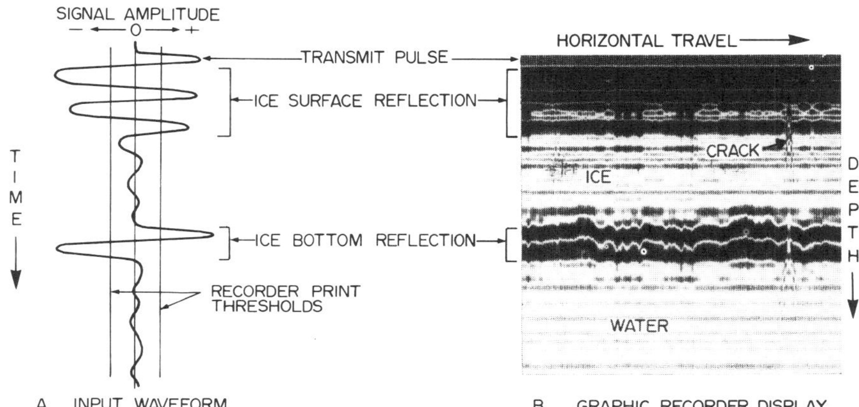

Figure 3. Explanation of ice thickness profile data.

D. Calibration and Interpretation. While an estimate of the electrical properties of ice (dielectric constant, conductivity) can be used to approximate the pulse velocity and thus ice thickness, an accurate interpretation requires more precise calibration. This calibration is accomplished by augering one or more holes in the ice and mechanically measuring ice thickness where profile data is also recorded. The calibration results in a scale or conversion factor by which ice thickness is read directly on the recorder chart as the ice is profiled. Since the electrical properties of sea ice are a complex function of chemical, physical, and temporal relationships (Hoekstra and Cappilino, 1971), calibration is necessary each time an operation is begun in a "new" area. In addition, the calibration is checked at least once each day and more frequently in areas of complex ice conditions. The electrical properties of fresh water ice generally do not vary significantly and thus calibration is not required as often.

EXPERIMENTAL STUDIES

The initial experimental work was carried out near the southern tip of Amund Ringnes Island in the Canadian Arctic during December 1972. Areas of both first-year sea ice about one meter thick and multi-year sea ice up to 3 meters thick were available for study. The results presented are based largely on these experiments and to a lesser extent on experimental work done under widely varying ice conditions until June 1973.

A. Position of electrical interface between ice and sea water. It was evident from the start of the experimental program that when the system was operated on sea ice a strong reflection was received. While it was felt that this reflection was related to the ice/water interface, it was not known at what point the reflection was actually taking place. A simple experiment was designed to provide this information. A small hole was augered through smooth first-year sea ice 95 cm thick, and a length of 3.8 cm diameter steel pipe inserted under the ice. The pipe was held tightly against the ice bottom with a rope. The location was scanned by the profiling system immediately after pipe installation and then periodically afterwards during the course of a continuing field program. As the ice grew downward with the pipe remaining at a constant depth, the location of the "pipe signal" relative to the "bottom signal" was determined.

No evidence of the pipe was seen immediately after installation. This was not surprising since the bottom few centimeters of sea ice (referred to as the "skeleton layer" by Assur, 1958) are characterized by numerous voids open to the sea water below. Thus, when first installed, there was in effect a layer of sea water between the pipe and the "electrical" ice bottom. The pipe signal was first detected on the eighth day following installation. At that time the ice had grown 11 cm to a thickness of 106 cm.

The results of the above experiment indicate that the interface as observed by the impulse radar system is of the order of 8 cm above the mechanical ice/water boundary. In actuality, this electrical interface is probably not a distinct plane but a diffuse zone extending over a vertical distance of several centimeters. The ability to interpret the profile data in terms of ice thickness to an accuracy of centimeters over distances of kilometers is an indication that the position of the electrical interface did not vary greatly at the time of the study.

B. Anisotropic nature of sea ice (ice polarization). It was discovered early in the experimental program that the amplitude of the reflected signal from the sea ice/water interface commonly varied markedly (from a clearly identifiable signal to no apparent signal at all) with antenna azimuthal orientation on first-year sea ice. On more than one occasion it was observed that the orientation of maximum reflected signal from undeformed first-year sea ice remained nearly constant for distances of several kilometers. However, in areas of first-year ice characterized by rotated plates refrozen along the boundaries, azimuthal orientation of maximum signal often varied from plate to plate. Although the effect is not always observed, and is usually less pronounced than on first-year ice, it has also been noted on multi-year sea ice. It is of interest to note that fresh water ice does not exhibit this effect. This was demonstrated during a lengthy survey on the Mackenzie River north of Inuvik,

N.W.T., early in 1973.

The variation in signal strength, or apparent anisotropy or polarization of the ice as the linearly-polarized antenna was rotated, was unexpected based on published reports concerning ice structure (Pounder, 1965). This phenomenon often made effective profiling difficult when using the original antenna. The problem was solved with a new antenna design that allowed the simultaneous scanning of the orthogonal polarizations.

In an attempt to understand this polarization phenomenon, an oriented ice core was taken from a recently refrozen lead covered with 25 cm of ice. A detailed examination was made of the microstructure, especially of platelet orientation at the core bottom. A definite intercrystal preferred orientation of the platelets was observed. It was found that the maximum bottom reflection occurred when the antenna radiating elements were oriented perpendicular to the trend of platelets, and the minimum where the antenna radiating elements were parallel to the platelet trend.

If this apparent relationship between aligned platelets and the polarized antenna (over relatively large distances) is real, it is suggested that platelet alignment results from currents beneath the ice during crystal growth or from systematic mechanical flexure of the ice. Platelet alignment may also have important ramifications with respect to ice strength (Peyton, 1963). Clearly a complete understanding of electrical anisotropy of sea ice must await further experimental work.

C. Dielectric constant of sea ice. The velocity of propagation of an electromagnetic wave in a medium other than free space is a function of the dielectric constant and electrical conductivity of that medium (Kraichman, 1970). For the electrical properties of sea ice and the frequency range of interest of the impulse radar system, it has been assumed for the sake of simplicity that the dominant factor determining the velocity is the dielectric constant. Thus a knowledge of the velocity, obtained from pulse travel time and measured ice thickness, leads to a computation of dielectric constant.

The technique used was to simply measure the round trip travel time for the pulse at points of known ice thickness (from augered holes) and compute an apparent relative dielectric constant, ε_r. The dielectric constant thus obtained is a <u>bulk</u> value, or an average value which the ice would have if of uniform dielectric constant throughout its total "electrical" thickness (i.e., the ice thicknesses used for the computations were equal to the distance from the ice surface to the electrical interface detected by the system. This interface was about 8 cm above the physical ice/water interface.) It is well known (Hoekstra and Capillino, 1971) that the actual electrical properties of sea ice vary with depth into the ice, but it is the average or bulk dielectric constant that describes the observed data most simply and is the value that can be computed from the recorded information and thus used to compare ice from varicus locations.

The bulk dielectric constant was computed for several locations on both first-and multi-year ice and the results plotted on Figure 4 as a function of "electrical" ice thickness. The data is further broken down into the interval during which each measurement was made. System timing inaccuracies, especially the definition of exact t_o (or ice surface) could result in a consistent bias of these data such that ε_r could be high by as much as 20%. However, relative accuracy of the data is estimated on the basis of repeatability to be of the order of 95% or higher.

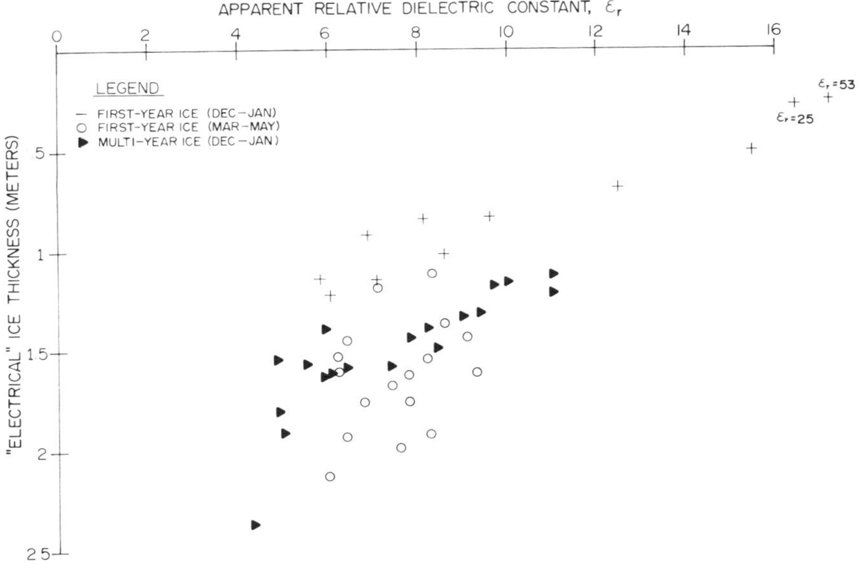

Figure 4. Relationship between apparent relative dielectric constant and "electrical" ice thickness.

Figure 4 shows that ε_r is a weak function of ice thickness. The relationship appears to be somewhat stronger for the multi-year ice than the first-year ice until it is noted that no measurements of ε_r were made on multi-year ice between March and May. In general, it would be expected that ε_r would decrease as ice thickness (and time since initial freeze-up increased) and the brine slowly drained. At a test site in first-year ice it was shown that ε_r of the first meter of ice decreased from 6.9 to 5.2 during a month's time in December/January. However, it is presumed that this trend would be reversed as the over-all ice temperature rose due to increasing ambient air temperature. This may explain the two fairly distinct populations the first-year ice can be separated into based on the time of measurement. Measurements of

ε_r between March and May do not show a significant decrease in ε_r as compared with those made during December and January although average ice thickness had increased considerably. The amount of scatter in these values, aside from any experimental error, may be related to differential snow cover and differences in ambient air temperature over several hundred miles of latitude. The apparent relationship between ε_r for measurements made on first-year and on multi-year ice during December and January was not expected. It is not clear why ε_r as measured for much of the multi-year ice is greater than for the first-year ice although thermal relationships may be in part responsible.

It is emphasized that the data shown in Figure 4 was taken in several different areas, on sea ice both with and without snow cover, and on ice presumably having distinctly different temperature profiles (the data was collected between December and May). As indicated earlier, the electrical properties (including ε_r) of sea ice are a complex function of chemical, physical and temporal relationships. Thus the full significance of Figure 4 cannot be determined at this time. It is planned that several selected sites on both first-year and multi-year ice will be instrumented to provide detailed temperature and salinity profiles as a function of time. This data will be used as control for the apparent relative dielectric constant measurements to be made periodically at these sites with the impulse radar system.

D. Cracks. In addition to ice thickness, profile data commonly displays a characteristic pattern that has been correlated with cracks in the ice (see Figure 3B). "Wet" and "dry" cracks as little as 4-5 mm wide have been detected in all types of ice both with and without concealing snow. In some cases the cracks intersect the ice surface while in others they appear to be restricted to the bottom half of the ice. At this time it is not possible to differentiate between "wet" and "dry" cracks using the profile data.

E. The effect of snow cover. Profile data has been obtained on both plowed and unplowed ice in order to determine the effect of snow on system capabilities. To summarize the results, the presence of snow did not inhibit the reception of the ice bottom reflection unless the snow was greater than about 1.5 meters thick. Snow may, however, introduce an error in the interpretation of ice thickness. This factor must be taken into account in the course of a survey, usually by assuming an average snow depth based on spot measurements (snow depth in the open is rarely in excess of 30 cm or so) and applying a correction factor uniformly. Since the velocity of pulse propagation in the snow is faster than in the ice, the error introduced in thickness determination by assuming no snow cover is generally of the order of 20 cm for each 30 cm of snow cover.

F. Fresh water ice. Operation on fresh water ice posed fewer problems because of its comparative electrical simplicity. The distinct electrical interface between ice and water, corresponding to the physical bottom of the ice, consistently reflected strong signals. No anisotropic effect was observed and less frequent calibration was required than on sea ice.

It was also possible to profile river and lake bottom topography through the ice to water depths of 7 meters. Profiling sea bottom topography is not possible because of the high conductivity of sea water.

OPERATIONAL EXPERIENCE

In the course of roughly eleven crew-months of survey work last season, ice thickness data was obtained throughout the Canadian Arctic Islands and the Mackenzie River Delta areas. Under virtually all conditions of fresh water ice and first-year and multi-year sea ice a clearly recognizable ice/water interface was observed. Figure 5 is a representative section of data obtained in the course of an operational survey on sea ice. Accuracy of ice thickness determination using the calibration procedures previously described was of the order of 5-10% for fresh water ice and undeformed first-year ice, and 15-25% for complex, rafted, or multi-year ice. The data were repeatable to within the timing and recording accuracy of the system, of the order of 1%.

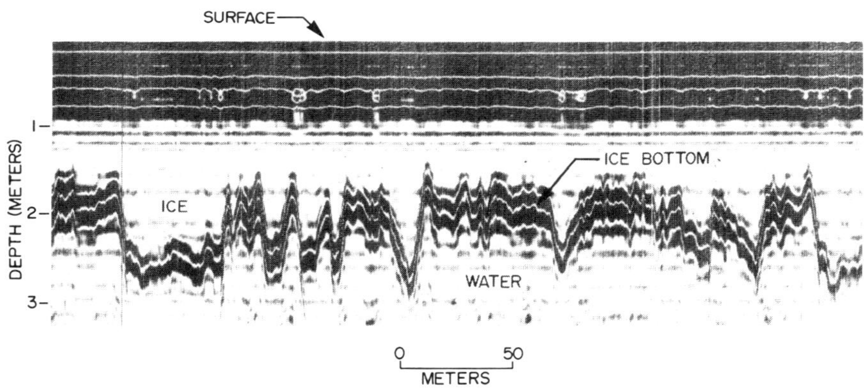

Figure 5. Ice Thickness Profile of Multi-Year Sea Ice.

Operational problems concerning the anisotropic nature of the sea ice bottom reflection were encountered early in the work, as discussed earlier, and were solved by equipment redesign. Later in the season a more serious problem was recognized. In the vicinity of some pressure ridges on sea ice, sea water from attendant cracks intruded between ice and snow cover, resulting in a slush layer impenetrable to the signal. When this occurred it was necessary to determine ice thickness by mechanical measurement. The problem became more acute as the season progressed and the ambient temperature rose above freezing. Weak or discontinuous bottom signals were also received from highly deformed or extensively rafted ice and from areas where a steeply dipping ice/water interface was common. However, the total mileage involved due to all of these problems was a small fraction of the total surveyed.

The typical ice thickness survey was performed using a lightweight, all-terrain vehicle towing the antenna. Survey speed was of the order of 4-5 km/hr and was a function of surface roughness and the amount of detail required by the client. A normal day's total survey would encompass 10-30 km of route, the distance depending on weather, visibility, and surface roughness. The full range of high-arctic weather conditions was encountered. Several surveys were performed in mid-winter total darkness.

Profiling was done to insure the safety of on-the-ice operations ranging from laying out ice routes for rig moves to preceeding seismic trains traveling across the ice to determining ice thickness for a through-the ice drilling operation. An extensive ice thickness reconnaissance survey was undertaken as one phase of a study to determine feasibility of inter-island pipeline laying. In a somewhat different but related application, river bottom topography was profiled through the ice to measure water depth for locating suitable barge landing sites.

Hazards to on-the-ice operations encountered in the course of the work varied from ten-centimeter-wide cracks to refrozen leads a hundred meters wide and kilometers long. In some cases strong inter-island water currents resulted in anomalously thin ice, usually with no surface indication of the thin zone. When hazards were encountered the procedures varied from removing the snow cover (to promote rapid ice growth) to attempting to locate alternate routes.

POTENTIAL APPLICATIONS

In addition to use for ice thickness determination, other potential applications for the ESP technique are glacier profiling and crevasse detection. Although depth of signal penetration in glacial ice is unknown at this time, it is expected that the system will be particularly useful for detailed profiling of small valley glaciers. Crevasse detection is especially promising because of the distinct electrical boundary presented by the ice/air or snow/air interface.

ACKNOWLEDGEMENTS

The authors would like to acknowledge the Sun Oil Company who, acting as operator for a group consisting of Sun, Gulf Oil, and Global Marine Canada, supported the initial experimental work. Our success is due in large part to their support and active assistance in the field.

REFERENCES

Assur, A., 1958. Composition of sea ice and its tensile strength. Arctic Sea Ice, National Academy of Sciences, National Research Council Publication 598, P. 106-138.

Hoekstra, P. and Capillino, P., 1971. Dielectric Properties of sea and sodium chloride ice at VHF and microwave frequencies. J. Geoph. Res. 76, 20:4922-4931.

Kraichman, M.B. 1970. Handbook of electromagnetic propagation in conducting media. Document NAVMAT P-2302. U. S. Government Printing Office.

Peyton, H.R., 1963. Some mechanical properties of sea ice. Ice and Snow, Cambridge, Mass.: M.I.T. Press 107-113.

Pounder, E.R., 1965. The Physics of ice. New York: Pergaman Press, 151p.

SUMMARY OF DISCUSSION
(Papers 4.1 to 4.7)

Some concern with the catch of special network gauges used in the studies by Wilson (paper 4.3), and by Peck, et al (paper 4.5), was expressed. It was noted that the special gauge network caught 17% less snow than the amounts measured on the ground with snowboards. ("Lake effect" storms contribute significantly to total snowfall in the Oswego, New York, area. In response to a question, Wilson stated that winds accompanying these storms are frequently in the range of 15-30 knots.) Smith observed that he had found significant undercatches by snow gauges even in sheltered areas. Storr mentioned a paper by Weiss* in which he suggests that the forested clearings used for the special gauges (subtending angles of 30° to 45° from gauge to treetops) were too large.

The problem of discriminating between open water and clear ice on side-looking airborne radar (SLAR) images was discussed (see discussion of papers 3.11 and 3.12). Jirberg noted that certain clues were sometimes available in the surrounding imagery or in historical data that could assist in the analysis. Also, he said, SLAR is best employed in a system of multi-spectral techniques, a method that helps in overcoming problems of single-channel ambiguity. For example, simultaneous thermal imagery, as described in paper 5.8, and short pulse radar, as described in paper 4.6, would greatly assist in the interpretation of SLAR imagery.

The bulk of the discussion period was devoted to paper 4.7, with R. Morey representing the authors. The question was raised as to what interface the pulse radar actually observes. Wilheit pointed out that theory indicates that the penetration of the radar pulse into the sea ice would be limited to one quarter the effective wave length, and suggested that the signal return identified as the bottom of the ice layer might be a reflection from the gradient in salinity just below the ice surface. Morey replied that the metal marker test described in the paper disproved this. Assur noted the importance of the dielectric constant in this study. Linlor, elaborating on this point, observed that the dielectric constant is dependent on both frequency and salinity.

Weeks noted that there would be a systematic bias in the data in proceeding from a melt pond to a hummock because of significant changes in the vertical gradients of salinity below these features. However, a simple relationship exists between these variations and ice thickness, so it should be possible to devise a correction procedure to improve the thickness estimates. Weeks also noted that the findings on polarization were extremely interesting, but that further studies might be necessary because the crystal sizes could be large in relation to the diameter of the present 3-inch diameter core sampling.

Assur stated that the return signal would be affected by ambient temperatures and brine pockets in the ice as well as ice thickness. He noted that the technique appears to have promise but, to make it acceptable, good synoptic salinity and temperature information would have to be collected and the effects of these on the thickness estimates well documented. Since the paper suggests shortcomings in the present theory of radar pulse penetration into sea ice, the results of further studies on the system will be followed with much interest.

---

*Weiss, L. L., 1963. Securing more nearly true precipitation measurements. Proc. ASCE, Hydraulics Div. 89(Hy2), pp 11-18.

Session V

Remote Sensing Techniques

Chairman:

R. J. P. Lyon, Department of Applied Earth Sciences,
Stanford University

General Reporters:

P. S. Eagleson, Department of Civil Engineering, M.I.T.

E. P. McClain, National Environmental Satellite Service,
NOAA

THE ROLE OF SATELLITES IN SNOW AND ICE MEASUREMENTS

Donald R. Wiesnet
NOAA, National Environmental Satellite Service

ABSTRACT

Earth-orbiting polar satellites are desirable platforms for remote sensing of snow and ice. Geostationary satellites at very high altitude (35,000 km) are also desirable platforms for many remote sensors, particularly for communications relay, flood warning systems, and telemetry of unattended instrumentation in remote, inaccessible places such as the Arctic, Antarctic, or mountain tops. Optimum use of satellite platforms is achieved only after careful consideration of the temporal, spatial and spectral requirements of the environmental mission. The National Environmental Satellite Service will maintain both types of environmental satellites as part of its mission. The NOAA-2 satellite currently in use provides twice-a-day earth coverage in a near-polar orbit at an altitude of 1500 km. The NOAA-2 Very High Resolution Radiometer (VHRR) is a new, significantly improved dual channel scanner (visible, .6-.7μm; infrared 10.5-12.5μm) that has been used experimentally for detailed snow mapping. Basin mapping of percent of snow cover is believed to be accurate to 5% in either mountainous terrain or flatland basins greater than 5,000 km^2 when the visible channel is used on cloudfree days. Cloud contamination and thick coniferous forests are problems that can and do affect such measurements.

Although snow thickness and water equivalent cannot as yet be determined from satellite data, under certain special conditions visible band reflectance seems to be related to snow depth. Research directed toward determining snow cover in regions of heavy forest canopy, such as the Adirondacks would be highly beneficial to all investigators.

Satellite detection of melting snow and melting lake ice using visible and near-IR has been achieved using ERTS-1 data as well as Nimbus-3 data. Multispectral techniques have permitted identification of melting snow, but the relation of spectral reflectance of snow to type, age, water equivalent and metamorphosis of snow is poorly known and rarely studied.

NASA's Nimbus-5 satellite launched in December 1972, carries an Electrically Scanning Microwave Radiometer (ESMR) capable of measuring the earth's brightness temperature--hence its snow cover--through almost any cloud cover. Ground resolution of the instrument ranges from 25 x 25 km to 45 x 160 km, thereby limiting the usefulness of this instrument--at least for snow mapping--to large-area studies.

Many properties of snow cannot be measured today by satellites: snow thickness, water equivalent and snow density are foremost among them. On the other hand, determining the percent of snow cover in basins is easily possible today if the need is demonstrated. The percent of the snowpack melting at the surface cannot be accomplished with present NOAA sensors, but may be possible in a future generation of sensors.

INTRODUCTION

Today, a wide variety of environmental satellites equipped with various sensors are transmitting data on snow and ice to Earth at unbelievable rates. These data streams are, for all practical and operational purposes, virtually ignored by the hydrologic community. Our capability for collecting data on snow and ice far exceeds our ability to analyze the data. Why is this so? What can be done to alleviate the situation? What is already scheduled for future generations of satellites and sensors? Is fieldwork obsolete? What kind of satellite snow and ice data are available to the private sector of research and to university researchers? These and other questions should be asked of NASA and NOAA/NESS, the agencies responsible for developing and maintaining experimental and operational environmental satellites, respectively. The EROS staff of the Department of Interior is also importantly and properly concerned in view of the spectacularly successful ERTS satellite and its hydrologic applications.

ORBITAL CONSIDERATIONS

The NOAA-2 satellite represents the third generation of environmental satellites in the National Operational Environmental Satellite System. The orbit is near polar and sun synchronous so that the satellite always crosses the equator at the same local solar time, in this case 0930 and 2130. This orbit is a typical ITOS orbit, providing a twice-daily IR image and one-time visible band image of an area. Orbital characteristics of NOAA-2 are given in Table 1.

This type of orbital coverage permits 12-hour change detection for dynamic snow and ice events. Cloudiness commonly reduces these observations, but in most of the USA it is possible to secure at least one cloudless view per week. The primary sensor for hydrologic use aboard NOAA-2 is the Very High Resolution Radiometer (VHRR), a new, significantly improved dual channel scanner (visible, .6-.7μm; infrared, 10.5-12.5μm).

Table 1.--Nominal orbit parameters[1/]

Parameters	Values
Altitude (above Earth surface).....	790 n.mi. (1464 km)
Apogee and Perigee.................	790 ±25 n.mi. (1464 ±46km)
Inclination........................	101.7°
Nodal Period.......................	115.14 min
Spacecraft Sun Angle...............	Varies with month of launch. Should be kept between 30° and 60° during the active lifetime of the spacecraft.
Precession of Nodes................	0.9857° per day (for complete sun synchronism)
Equator crossing time.............. (southbound)	0851 local solar time

1/ From Schwalb (1972)

Orbital characteristics of ERTS-1 are shown in Table 2. This unique satellite has a very different orbit from NOAA-2; it provides an 18-day revisit cycle. Uncooperative cloud cover may, at times, result in 36- or 54-day cycles of actual observation of the Earth's surface. For snow and ice studies in temperate zones, a data gap of 36 days is intolerable, but in Arctic regions it might be tolerated. The ERTS orbit, however, does not include areas north of 81°N or south of 81°S. Winter sun angles in polar areas further restrict the areas and periods that can be imaged.

Table 2.--ERTS-1 orbit parameters [1/]

Parameter	Actual Orbit
Semi-major axis....................	7285.82 km
Inclination........................	99.114 deg
Period.............................	103.267 min
Eccentricity.......................	.0006
Time at descending node........... (equatorial crossing)	9:42 a.m.
Coverage cycle duration...........	18 days
Distance between adjacent......... ground tracks (at equator)	159.38 km at equator

1/ From General Electric (1972).

Geostationary orbits---NASA's Advanced Technology Satellite (ATS) has demonstrated the value of a 35,000 km geosynchronous orbit, in which the satellite appears to hover "motionless" over a point on the earth. The advantages of this type of orbit are:

1. The viewing station is constant.
2. Almost 1/6 of the Earth may be observed almost synoptically.
3. Observations may be more frequent, e.g., every 15 or 20 minutes.
4. Pointable "telescopic" observations of those areas where high resolution is required for detailed observations can be made.
5. Time-lapse imagery of storms, floods, snow cover etc. can be prepared to study the genesis and dynamic aspects of important hydrologic events.
6. The satellite can collect and relay data in real time from instruments located at remote inaccessible sites, upon command, 24 hours a day. Furthermore, these data readouts can be programmed to coincide with scheduled detailed imagery if desired.
7. Processed data products can be retransmitted from central processing and analysis centers via satellite to local forecast/warning centers in near real time.

SNOW MAPPING

Unquestionably, snow-extent mapping of basins as small as 5,000 km^2 is feasible as an operational technique using NOAA-2 VHRR visible band imagery (Wiesnet and McGinnis, in press). To be of greatest value to the hydrologic community, snow-extent mapping must be done on some type of river basin map at a suitable scale. As NOAA-2's

VHRR transmits an unrectified image, it must be rectified prior to mapping. A Bausch and Lomb Zoom Transfer Scope (ZTS) is an acceptable optical method of rectification, and the SRI console appears to be acceptable electronic/optical method (Evans and Serebrey, 1973). Sophisticated computer programs for automated production of snow-extent basin maps are possible and desirable, but none is known to be in use at this time.

The 1000-meter resolution or "spot size" of the VHRR system, tends to eliminate the problem of distinguishing the "true" snowline from intermittent patches of snow interspersed with bare ground (Figures 1 and 2). The VHRR tends also to integrate the snow areas near the line of continuous snow cover and to eliminate the small snow patches which lie outside of the main body of continuous snow cover. The shadow effect, in which a low sun angle can cause shadows that decrease the reflectivity of the snow in a steep-sided valley and the vastly better resolution of ERTS are probably the reasons for VHRR snow extent measurements being consistently lower than ERTS-1 measurements (Figure 3). This effect will be more severe at higher latitudes. A third problem, and probably the most vexing, is the difficulty of detecting snow in heavily wooded areas, particularly in the coniferous forests.

The ERTS-1 satellite provides superb, sharp (100 m ground resolution) multispectral images for snow extent mapping (Figure 4). With these ERTS-1 images of unsurpassed cartographic accuracy one can quickly and easily map snow extent in basins as small as 272 km^2 (Meier, 1973). Basins larger than 10,000 nmi^2 (34,000 km^2) would exceed the size of one ERTS frame and present some mapping problems. For example, on a cloudless day over a basin the first ERTS overpass would secure part of the desired imagery. The following day might be overcast resulting in no data on snow.

Other factors limiting the usefulness of ERTS imagery for snow studies are: 1) the 18-day revisit time; 2) the long wait (30-60 days) from collection of data to receipt by the investigator. A quick-look capability, such as developed by the Canadians at their ERTS data collection site, would eliminate this second problem.

NASA's Nimbus-5 satellite launched in December 1972, carries an Electrically Scanning Microwave Radiometer (ESMR) that is capable of measuring the Earth's brightness temperature--hence its snow cover--through almost any cloud cover. Ground resolution of the instrument ranges from 25 x 25 km to 45 x 160 km, thereby limiting the usefulness of this instrument--at least for snow mapping--to large-area studies.

In summary, plans for operational snow-extent mapping by satellite for certain selected river basins are being implemented for the 1973-74 snow season at NOAA/NESS. The ZTS and the improved resolution of the NOAA-2 VHRR makes this possible. ERTS-1 has been and is extremely useful as a calibration tool. It has by far the best resolution of any environmental satellite. Snow extent mapping by satellite is today a viable technique of monitoring snow packs. As this fact is increasingly appreciated by hydrologists, we may expect to see a great deal of activity in developing improved and automated techniques.

RECENT SNOW AND ICE SATELLITE OBSERVATIONS

Barnes and Bowley (1968) demonstrated that snow cover can be

Figure 1. Enlarged VHRR-VIS image of the central Sierra Nevada 1713 GMT, orbit 1728 (recorded data). Figure 2 was prepared from this distorted image. Scale is about 1:4,000,000.

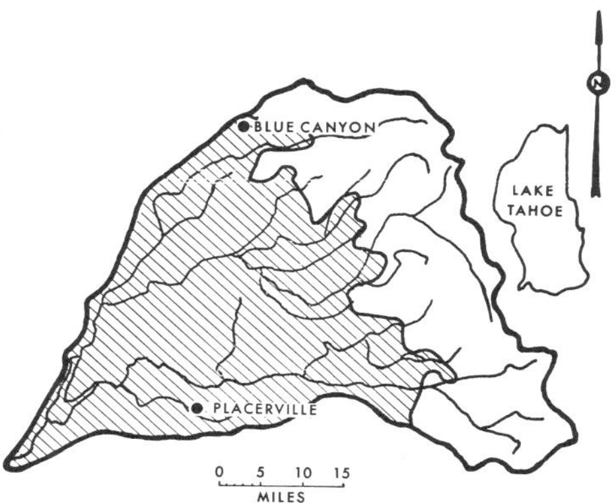

Figure 2. Snow-extent map for the American River basin, Calif., prepared from NOAA-2 VHRR imagery shown in figure 1. Patterned area is snowfree.

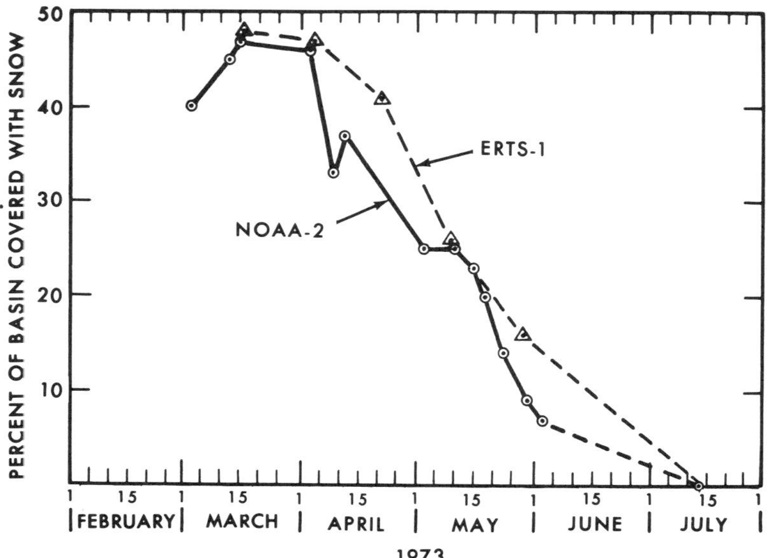

Figure 3. Comparison of snow-extent mapping of the American River basin, California, by NOAA-2 and ERTS-1 satellites.

reliably identified from (ESSA) satellite images with snow depths of 2-3 cm or more, commonly appearing as continuous snow cover. Snow lines were estimated to be accurately located to within about 20 km in nonmountainous terrain. Basins as small as 1400 km^2 could be mapped.

Clouds have been a constant problem, as their reflectance commonly matches that of snow; but of course cloud patterns are not stable from image to image. McClain and Baker (1967) used a computer program to produce 5-day CMB (Composite Minimum Brightness) charts, which in effect "removed the clouds" by computer selection of minimum brightness.

McClain (1973) in a comprehensive summary of satellite snow survey methods points out the chief problem is very often the location of the entire area to be mapped (the basin) rather than the extent of the snowline.

The new NOAA-2 VHRR images have vastly improved our capability for mapping snow. This new operational environmental satellite launched Oct. 15, 1972, has a 1 km resolution, i.e., 4 times better than the previous series of satellites in the visible band and 10 times better in the IR. By use of the Zoom Transfer Scope and the NOAA-2 imagery (Figure 2), snow cover was mapped within several river basins (Wiesnet and McGinnis, 1973) as a pilot study of snow-extent mapping in the American River basin of California in cooperation with the National Weather Service/Office of Hydrology.

The plan was to prepare a map using the ZTS as soon as possible

Figure 4. Mosaic of ERTS-1 images of Lake Erie showing ice coverage on Feb. 18-19, 1973. Air temperatures are below freezing everywhere. Ice cracks have occurred during past 24 hours.

after the imagery was received, then measure areal extent of the snow; and lastly, using the NWS telecommunication network, send the percent of basin covered by snow to the cognizant River Forecast Center (RFC). In this way, the RFC would get the data usually about 30 hours after the satellite passed over the basin. Our goal was to furnish data no more than 36 hours after satellite transit of the basin. Fourteen VHRR snow cover maps (see Figure 2) and snow-extent measurements of the American River basin were made in the spring of 1973 (Figure 3).

About two-man hours were required to map the American River basin (5,601 km^2) snow extent. The same amount of time was required to map the Red River of the North, which is a 104, 120 km^2 basin. In mapping the Red River of the North basin, the coniferous forest canopy problem affected the snow-extent maps, although the problem is more acute in the Adirondack Mountains, which lie partly in the Lake Ontario basin and in Canada. By carefully mapping the forested areas, which lay chiefly at the eastern edge of the basin, it was possible to extrapolate the snow extent into the forest. Conversely, determining whether the snow persists under the canopy subsequent to melting in the open fields is conjectural. Research directed toward devising methods of determining snow extent in regions of heavy forest canopy would be welcomed by all investigators.

COST/ COMPARISON ESTIMATE OF SATELLITE SNOW MAPPING

At the request of the Director of the National Environmental

Service, a cost/comparative figure was calculated for satellite (NOAA-2) measurement of Sierra Nevada snow cover vice conventional aircraft measurements. Assuming 1) 20 basins were of interest and 2) a simple altimeter survey by a light plane, at least 40 hours would be required at a total cost of at least $20,000. Using satellite data the entire Sierra block could be mapped in two man days for a direct cost of about $100. This is a comparative/cost ratio of 200:1.

ICE STUDIES OF THE GREAT LAKES

The ambitious International Field Year on the Great Lakes (IFYGL) has caused many hydrologists to realize that satellite data is perhaps the only way to gather synoptic environmental data over a body of water as large as Lake Ontario. The VHRR thermal scanner aboard NOAA-2 could provide a thermal map of the lake every 12 hours. Cloud cover reduces the amount of useful data, but once-a-week cloud-free views are quite common. Upwelling water is easily detectable; snow extent can be mapped; and ice can be observed and measured. By using VHRR digital data and known water surface temperatures at a few locations, thermal maps can be prepared to assist in evaporation calculations over the lake surface. Ice extent likewise may be measured to determine the lake's contribution to snow storm effects on the lee side. Ice conditions are also of great economic importance to Great Lakes shipping interests.

The mosaic of ERTS-1 images in Figure 4 shows the ice on Lake Erie near Cleveland on February 18, 1973; and near Buffalo on February 17, 1973, in MSS band 7. (.8-1.1μm). Air temperatures are below freezing but a moderate (10 knot) wind has begun to break up the ice cover by moving it to the east.

MELTING SNOW AND ICE

Satellite detection of melting snow and melting lake ice using visible and near IR (.8-1.1μm) data jointly has been achieved using ERTS-1 data as well as Nimbus-3 data (Strong et al., 1970; Wiesnet, 1972). Although these multispectral techniques have permitted identification of melting snow, the relation of spectral reflectance of snow to type, age and water equivalent, is poorly known in general and rarely studied. Recent (not yet published) work by H.W. O'Brien and Richard Munis performed at CRRL (under contract to NOAA/NESS) will present several spectral reflectance measurements taken of dry, melting, and refrozen snow. The need for basic studies of snow and ice reflectance and emission in various bands of the electromagnetic spectrum cannot be overstressed. Mistakes in interpretation and analysis of snow scenes cannot be avoided unless we know the emission and reflection characteristics of snow and ice under varying conditions.

There is considerable interest in developing the capability of measuring the percent of the snowpack that is melting at the surface. Whether this technique can be made operational and whether a sensor should be developed specifically for this application remain speculative questions.

SNOW THICKNESS

Brightness or reflectivity of snow has been related to snow thickness (Barnes and Bowley, 1968; McGinnis and Wiesnet, 1973) but this technique has limitations. Vegetation canopy, variable photo processing, urban effects of buildings, age and condition of snow, refreezing, etc. are factors that influence reflectivity determinations and render most correlations speculative at best. Ground- or aircraft-based research on the "behavior" of snow reflectivity as snow accumulates, ages, and melts is too sparse and should be encouraged.

WATER EQUIVALENT OF SNOW

A very dynamic parameter of ripening snowpacks is the water equivalent of the snow. It is also one of the most meaningful measurements that the operational hydrologist can have. Despite some encouraging progress in passive microwave experiments (Meier and Edgerton, 1973), no satellite technique for determining water equivalent is yet available. Peck, et al. (1971), however, using low flying aircraft (90 meters) have made areal surveys using sensitive scintillometers to measure water equivalent values. Bissell and Peck (1973) described a ground-station instrument that could measure water equivalents and telemeter or relay that measurement via a geostationary satellite (GOES) upon command. Ten such instruments in the Sierra Nevada would go a long way toward ameliorating the lack of data from remote sites.

CONCLUDING REMARKS

Earth-orbiting polar satellites are desirable platforms for remote sensing of snow and ice, not only as relay stations for telemetering commands and data, but also for data collection on spectral reflectivity, as well as infrared and microwave emissions. Hydrologists are, in general, unfamiliar with satellite data, and are waiting to be convinced of the utility and reliability of satellite data. Yet, satellites have a demonstrated capability for mapping snow-extent in small ($<1,000$ km^2) basins and mesoscale ($<10,000$ km^2) basins. During the melt season, weekly snow-extent measurements by VHRR are feasible in many areas. ERTS data have been (Meier, 1973), and VHRR data ultimately will be related to meltwater runoff curves. This relationship will benefit all water users as well as flood forecasters. The ERTS and NOAA-2 satellites truly present a golden opportunity to the hydrologic community to plan and carry out an improved program of satellite snow and ice observations. The capability for snow mapping exists; only the commitment is lacking.

ACKNOWLEDGEMENTS

I would like to acknowledge the work of D.G. Forsyth, who prepared the original snow maps of the American River basin, D.F. McGinnis, who is actively assessing the reflectivity snow-thickness question, Mrs. Olivia Smith for typescript preparation, and E.P. McClain for his continuing interest and professional support of our satellite hydrology research program. The work on ERTS-1 data was funded by NASA under Contract S-70246 AG.

REFERENCES

Barnes, J.C. and C.J. Bowley, 1968. Snow cover distribution as mapped from satellite photography. Water Resources Research, 4:257-272.

Bissell, V.C., and E.L. Peck, 1973. Monitoring snow water equivalent by using natural soil radioactivity. Water Resources Research, 9:885-890.

Evans, W.E., and S.M. Serebreny, 1973. Analysis of ERTS imagery using special viewing/measuring equipment. NASA SP-327, 1:1211-1218.

General Electric Corp., 1972. Data users handbook. NASA Goddard Space Flight Center Document No. 71SD4249.

McClain, E.P., 1973. Snow survey from earth satellites. World Meteor. Organiz. No. 353, Geneva, Switzerland.

McClain, E.P., and D.R. Baker, 1967. Experimental large-scale snow and ice mapping with composite minimum brightness charts, ESSA Tech. Memo. NESTCM 12. 19 p.

McGinnis, D.F., and D.R. Wiesnet, 1973. Snowline mapping using NOAA-2's very high resolution radiometer [Abs] Trans. Amer. Geophys. Un., 5:272.

Meier, M.F., 1973. Evaluation of ERTS imagery for mapping and detection of changes of snowcover on land and on glaciers, NASA SP-327, p. 863-875.

Meier, M.F. and A.T. Edgerton, 1973. Microwave emission from snow--a progress report, Proc. 7th Internat. Sympos. on Remote Sensing of Environment, May 1971, Ann Arbor, p. 1155-1163.

Peck, E.L., V.C. Bissell, E.B. Jones, and D.L. Burge, 1971. Evaluation of snow water equivalent by airborne measurement of passive terrestrial gamma radiation. Water Resources Research, 7:1151-1159.

Schwalb, A., 1972. Modified version of the improved TIROS Operational satellite (ITOS D-G). NOAA Tech. Memo. NESS 35, 48 p.

Strong, A.E., E.P. McClain, and D.F. McGinnis, 1971. Detection of thawing snow and ice packs through combined use of visible and near-infrared measurements from earth satellites. Monthly Weather Rev. 99:828-830.

Wiesnet, D.R., 1973. Detection of snow conditions in mountainous terrain, Proc. Earth Resources Technol. Satellite Symposium, Sept. 29, 1972, Goddard Space Flight Center, X-650-73-10.

Wiesnet, D.R., and D.F. McGinnis (in press). Hydrologic Applications of the NOAA-2 very high resolution radiometer, Proc. Amer. Water Resources Assoc. Symposium on Remote Sensing and Water Management, June (1973), Burlington, Ontario, Canada.

SOME NEW SATELLITE MEASUREMENTS AND THEIR APPLICATION TO SEA ICE ANALYSIS IN THE ARCTIC AND ANTARCTIC

E. Paul McClain
NOAA, National Environmental Satellite Service

ABSTRACT

The NOAA-2 operational environmental satellite carries a new Very High Resolution Radiometer (VHRR) capable of one kilometer ground resolution in the visible and thermal infrared portions of the spectrum. This resolution is significantly higher than that previously available from NOAA operational satellites or from NASA Nimbus satellites, and the areal coverage and frequency of coverage is much greater than that afforded by NASA's Earth Resources Technolcgy Satellite (ERTS-1).

The VHRR data have greatly improved the ease and reliability of satellite image interpretation for the detection and monitoring of ice pack features and condition, even during the months of polar night darkness. Operational applications of VHRR images are already being made, and research exploring the extraction of quantitative information from these data is underway also.

Visual and infrared observations of sea ice from satellites are often severely limited by cloud cover. The microwave imager on the Nimbus 5 satellite does not suffer this constraint.

INTRODUCTION

The ice cover in the waters of the Arctic and Antarctic, which is variable in extent both seasonally and on shorter and longer time scales, has been receiving increasing attention in recent years because of its importance, among other things, on weather and climate--even global climate. Untersteiner and Maykut (1969) have stated that the surface albedo and its variations, which are almost totally a function of the presence or absence of ice and the type or condition of the ice, are probably the most important regional factors affecting the heat and mass budgets of the Arctic Basin in summer. Similar statements have been made about the Southern Ocean by Fletcher (1969) and others. Ice is the major factor in polar regions that controls the exchange of heat and momentum between ocean and atmosphere in winter. The heat flux into the atmosphere from open water in areas such as leads and polynya is much greater in magnitude than it is from the surrounding ice.

In order to monitor the seasonal patterns of surface heat exchange over the polar oceans, one should systematically observe the location and extent of pack ice, its concentration, and the extent and position of open leads and polynya. Such surveillance could also serve as an aid to navigation in polar waters (Vasil'ev, 1968). Although special ice reconnaissance aircraft are being used by several nations for this purpose, such ice missions cover relatively limited areas on an infrequent basis, and they are quite expensive. The vast areas involved, their general remoteness, and the need for repetitive coverage, makes comprehensive or frequent monitoring of sea ice impractical or impracticable except by Earth-surveillance satellites.

EARTH SATELLITE DATA APPLICABLE TO ICE ANALYSIS

Pictures from the early TIROS weather satellites were immediately recognized for their value in ice analysis (Wark and Popham, 1962). Improved operational versions of the TIROS research satellites, the ESSA series, and the Nimbus research satellites, generated better and more complete data for ice analysis. Daily global coverage was provided by the Advanced Vidicon Camera System (AVCS) on ESSA-1, 3 ...9, whereas local coverage was provided by direct readout to simple ground stations by the APT (Automatic Picture Transmission) system on ESSA-2, 4, ...8. AVCS and APT systems were carried by the early Nimbus satellites also, which were equipped with High Resolution Infrared Radiometers (HRIR) as well.

Next came the Improved TIROS Operational Satellites (ITOS). ITOS combined the AVCS and APT functions in a single spacecraft, and it also carried the first dual-channel (visible and thermal infrared) Scanning Radiometer (SR) for global and local direct readout. ITOS-1 was followed by NOAA-1, identically equipped.

On later Nimbus satellites the AVCS and APT were replaced by an IDCS (Image Dissector Camera System). A daytime channel sensitive in the reflective infrared band was added to the HRIR on Nimbus 3, and the HRIR was replaced by the THIR (Temperature-Humidity Infrared Radiometer) on Nimbus 4 and 5.

Nimbus 5 is equipped with an SCMR having three detectors in the following wavelength bands: 0.8-1.1, 8.3-9.2, and 10.5-11.3 micrometers (NASA, 1972). Ground resolution at nadir was designed to be 0.8 km. A limited number of high-quality images were obtained from this sensor before it became inoperative a few weeks after launch. The ESMR on Nimbus 5 operates at a frequency of 19.35 GHz (1.55 cm) with a nominal ground resolution of 30 km. The special usefulness of microwave sensors will be discussed briefly in a later section.

The launch of the first Earth Resources Technology Satellite (ERTS-1) took place in July 1972. The approximately 100-meter ground resolution of the Return Beam Vidicon (RBV) and Multispectral Scanner (MSS) systems on ERTS-1 is superior to anything heretofor available from satellites for ice analysis (General Electric, 1972). Moreover, two bands in the visible (0.5-0.6 and 0.6-0.7μm) and two in the reflective infrared (0.7-0.8 and 0.8-1.1μm) are provided.

All of the satellites above are in near-polar, sun-synchronous orbits at altitudes from 925 to 1500 km. A more complete summary of Earth satellites and satellite sensor systems suitable for sea ice studies is given in a recent article by McClain (1973c).

THE NOAA-2 SATELLITE, WITH EMPHASIS ON THE VHRR

NOAA-2 is equipped with a number of environmental sensors, most of which are primarily for meteorological usage, but two of these--the SR and the VHRR--are also useful for oceanographic and hydrologic purposes, including sea ice monitoring and studies (Schwalb, 1972). The Scanning Radiometer (SR) is a two-channel device designed both for cloud mapping twice daily on a global scale and local direct readout by simple ground stations. The visible band detector of the

SR is sensitive to reflective solar radiation in the 0.5-0.7μm wavelengths, and it has a nominal ground resolution of 3.5 km at nadir. The thermal infrared detector is filtered to 10.5-12.5μm and thus measures thermal radiation emitted by clouds and Earth's surface. This band is in the "cleanest" portion of the atmosphere's 8-13μm "window", which minimizes atmospheric attenuation of the signal. Water vapor corrections in the relatively cold and dry polar atmosphere are small, generally being less than 2-3°C.

The Very High Resolution Radiometer (VHRR) is designed primarily for direct readout use, but the required ground station is considerably more costly than that needed for local readout of SR data. NOAA presently is able to read out VHRR data at its two command and data acquisition stations, one at Wallops Island, Virginia and the other near Fairbanks, Alaska. A third NOAA VHRR data acquisition station will go into operation early in 1974 near San Francisco. The area that can be covered in the direct readout mode is rather large, viz. a strip of about 2200 km wide and more than 5000 km long for the satellite pass most nearly overhead. The VHRR also has a stored-data mode, but its capacity is very limited, viz. eight minutes (about 8%) per orbit. This yields an image about 2200 km wide and 2200 km long.

Fairly complete VHRR coverage of the North American side of the Arctic from Greenland and Labrador westward to beyond the Chukchi and Bering Seas was obtained by direct readout during this past year.

Special stored VHRR data coverage was obtained of the Ross Sea area of Antarctica from November 1972 through January 1973 to assist U.S. Navy ice forecasters in their support of resupply missions. Similar coverage of the Palmer Peninsula and Weddell Sea area is being provided during recent months. Special coverage of several other areas was arranged for during shorter periods of the spring and summer: North Water area, Foxe Basin, East Greenland Current, Iceland, and northern Scandanavia. Detailed experimental ice analyses based on VHRR imagery were produced by a NOAA/NESS ice analyst and transmitted to the Alaska area for guidance to the National Weather Service there.

TECHNIQUES OF SATELLITE DATA APPLICATION FOR SEA ICE ANALYSIS

Employment of satellite vidicon photographs for detection and mapping of sea ice in the Arctic or Antarctic has emphasized photo-interpretation techniques to delineate the ice boundaries and features. One of the problems most encountered is the discrimination between ice, with or without a snow cover, and clouds, for all often have reflectances in the same general range. Taking advantage of the relatively conservative behavior of the ice fields compared with the often fast changing cloud masses, careful study of the area for several successive days generally enables differentiation of ice-covered areas from cloudy ones. Also helpful is that ice often has characteristically different patterns, shapes, or texture from clouds. Furthermore, clouds can cross coastlines, cast shadows, or are partially transparent. The better the spatial resolution of the sensor, the more effectively are these factors employed in the interpretation process. Spatial resolution also affects the ability to deduce pack ice type and concentration (Potocsky, 1970; Swithinbank, 1970).

Cloud filtering and suppression from visual satellite images can also be achieved by computer manipulation of the satellite brightness values or reflectances to generate what are termed Composite Minimum Brightness, CMB, charts (McClain and Baker, 1969). McClain (1973a) developed a procedure to calibrate the CMB values externally and found that characteristic brightness values correspond to ice pack concentration and snow cover or puddling conditions. Wendler (1973) used this procedure to study ice movements in a small area of the Arctic Ocean off the shore of northern Alaska and to derive monthly mean albedo maps.

Thermal infrared imagery and measurements from the HRIR and THIR on Nimbus satellites and the SR and VHRR on the NOAA satellites are seeing increased operational and research application for sea ice surveys, e.g. the studies by Barnes et al. (1972). These investigations have shown that the 8 km resolution IR imagery provides a means of mapping gross ice boundaries during periods of polar darkness, although it is generally inferior to the 3.5 km resolution visual imagery that can be used during periods of sufficient solar illumination. During those times when visible and thermal IR can be used jointly, preliminary results indicate that additional ice information, such as ice concentration and possibly thickness, can be inferred. An analysis of a small sample of daytime and nighttime IR data gave evidence that diurnal variations are up to three times greater over pack concentrations of 7/10 or more than those over 6/10 and less.

Thermal contrasts between ice and water are not as large as reflectance contrasts, therefore IR imagery generally requires some special treatment, i.e., enhancement, in order to make optimum use of it in ice analysis. When a thermal infrared image is displayed for general meteorological purposes, it appears rather "flat" because the 16 or so distinguishable gray tones have been spread over the entire meteorological temperature range of over 100°C. Figure 1 is a VHRR IR image of the North Water area in northern Baffin Bay that has been enhanced to make it more useful in ice studies. After digitization of the original analog data tape, the data were redisplayed such that the available gray tones were spread only over the temperature range of primary interest, which was less than 30°C.

High latitudes in winter are often characterized by strong low-level temperature inversions, and this can easily result in the tops of low cloud layers being warmer than the underlying pack ice. Therefore, the ice analyst should be aware that clouds can appear either brighter (colder) or darker (warmer) than ice cover in the thermal infrared imagery as it is conventionally displayed.

The about 5 km ground resolution (at nadir) of the visible range vidicons on the TIROS, ESSA, ITOS-1, NOAA-1, and early Nimbus satellites, and in particular the 8 km resolution of the thermal infrared channel of the HRIR, THIR and SR, have tended to limit the application of these satellite data to fairly gross delineation of main pack ice boundaries and detection of large polynya. The newly-available VHRR visible and thermal infrared data, whose one kilometer ground resolution yields a fourfold improvement over most previous visible observations and a tenfold improvement over previous IR measurements, represents a substantial gain in ice mapping capability from space. Cloud discrimination techniques are applied more effectively with the higher resolution images. Furthermore, details of floe, lead, and fracture patterns stand out clearly in the new data, the improvement in ice information content being especially marked in the infrared imagery. DeRycke (1973) of NOAA/NESS used VHRR imagery of the Ross Sea area of Antarctica to study ice motions

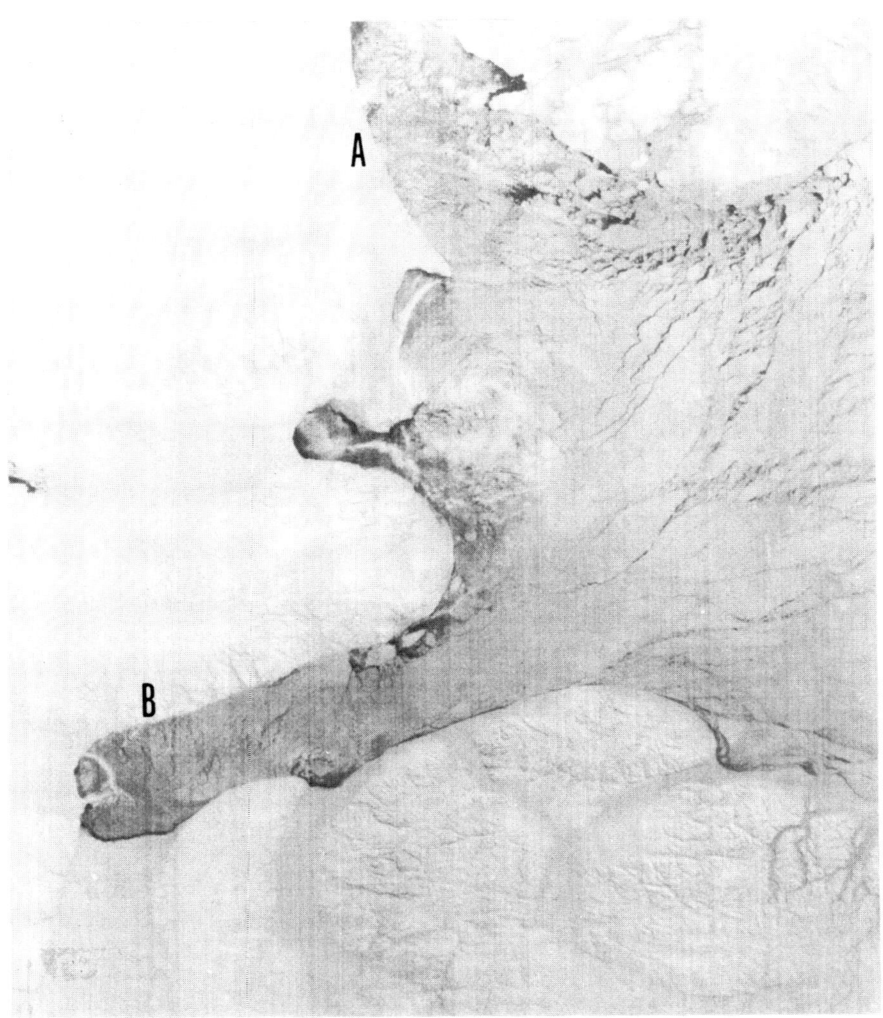

Figure 1. Enhanced VHRR-IR image, 22 April 1973, in northern Baffin Bay, including North Water (A) and Lancaster Sound (B). Darker tones are warmer radiating surfaces.

(Figure 2). He found it relatively easy to track individual floes and tabular icebergs for periods of up to a week, and even the actual calving of large icebergs was observable. Rather highly variable short-term ice movements were found, ranging from nearly zero to almost 30 km per day. The tabular icebergs, as might be expected from their much greater mass, exhibited a much more sluggish response to the wind.

A similar study of ice motions in the Labrador coast area is underway in NOAA/NESS. Preliminary results indicate much more

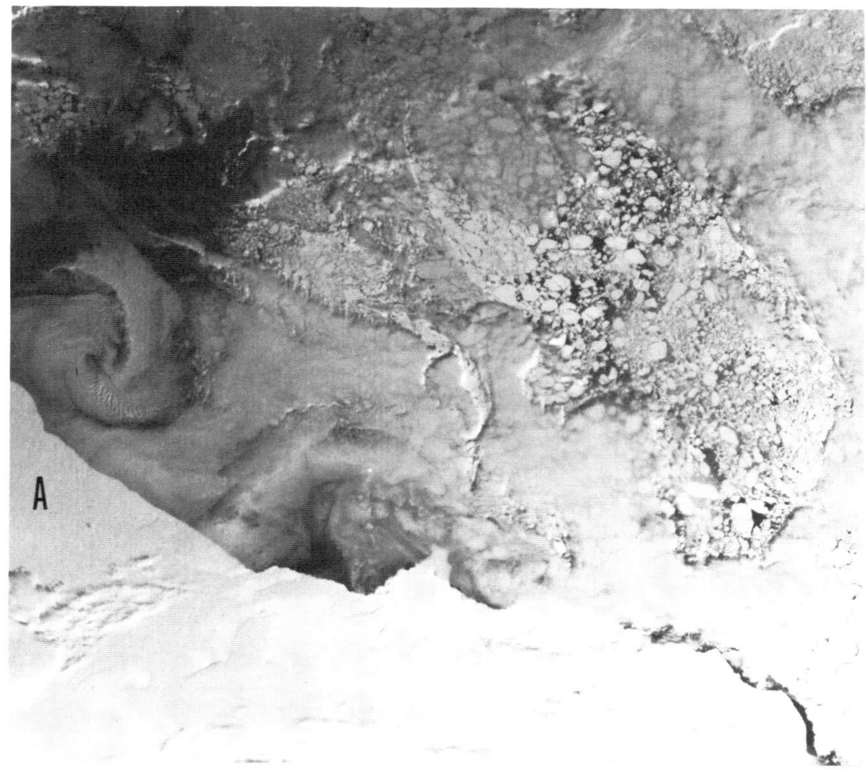

Figure 2. VHRR-VIS image, 12 December 1972, in Ross Sea area of Antarctica. Visible in this scene are a part of the Ross Ice Shelf (A), pack ice, open water, and clouds.

systematic floe movements than in the Ross Sea, generally toward the south and southeast with the prevailing flow of the Labrador Current. Even in this region, however, short-term motions were found to be rather variable, ranging from a few kilometers per day to as much as 45, and with directions occasionally even perpendicular to the general current flow. Figure 3 is a VHRR view of a part of the Labrador ice stream. Not only is considerable ice concentration and structural detail available in such an image, but systematic gradations in the reflectance of the ice can provide information on such processes or conditions as snow cover or puddling, and the change in compactness or thickness resulting from ice motion toward or away from the coast. In the latter case, several stages of refreezing of near inshore waters is evident in Figure 4.

Figure 4 shows VHRR views of the northeastern Greenland area five weeks apart during the early summer ice melt period. The retreat of the seasonal snow line in the coastal region and the decay of the ice pack offshore clearly demonstrate the value of

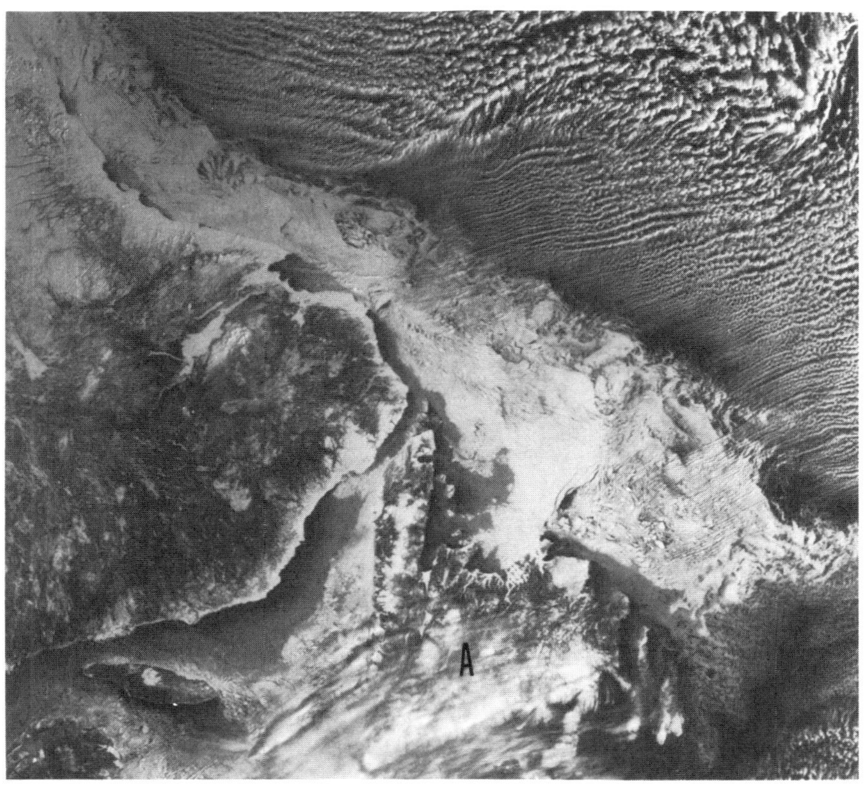

Figure 3. VHRR-VIS image, 3 March 1973, along lower Labrador coast and upper Gulf of St. Lawrence, including Newfoundland (A). Note offshore movement of main pack under influence of northwesterly winds.

higher resolution satellite observations for sea ice surveys and research.

Research using digitized VHRR data has been initiated recently within NOAA/NESS to further development of ice enhancement methods and other investigations, but digitization of VHRR data is limited to small areas. Rectification and mapping, which are requirements for production of CMB charts from visible range data or some type of similar composite chart from IR data, are precluded for the near future.

The 100-meter resolution of the experimental ERTS data permits the most detailed and precise interpretation and mapping of ice features ever possible from an Earth satellite, but the coverage provided by the 185 km wide swath and the 18-day return cycle precludes most operational use, especially when cloudiness and data processing delays are taken into account. Overlapping ERTS coverage at high latitudes sometimes permits views of restricted areas for several successive days, and it has been possible at times to track specific ice features over short periods to gain information on wind, current,

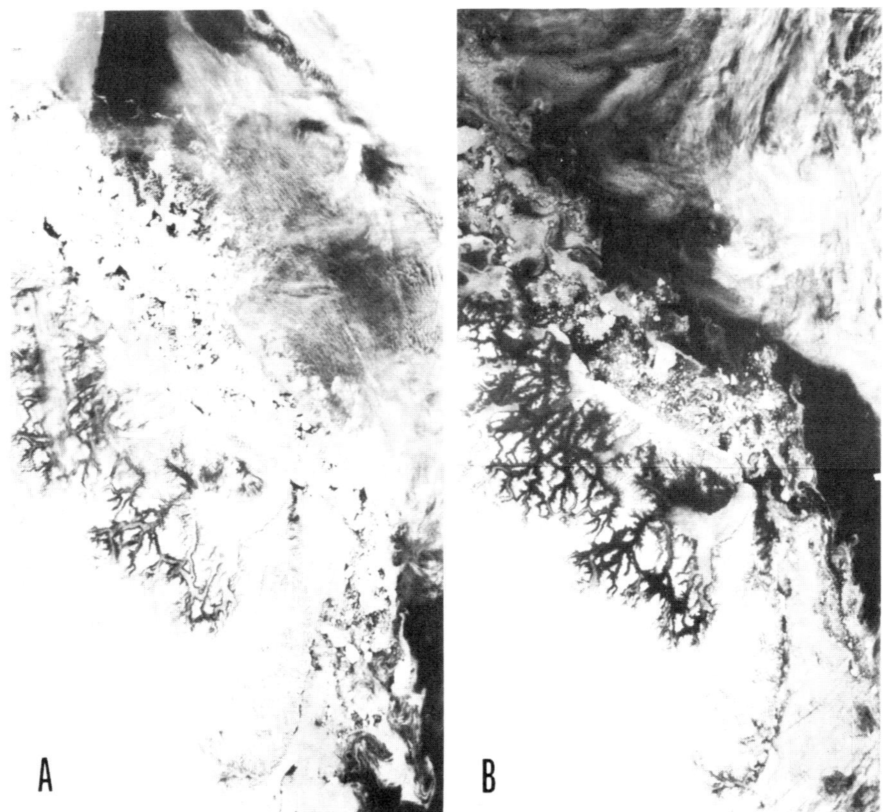

Figure 4. VHRR-VIS images of northeastern Greenland coastal area: (A) 11 June 1973; (B) 17 July 1973. Note snow melt at lower elevations, drop in reflectance of fast ice, and decay of the pack ice.

and ice interactions (Campbell, 1973). ERTS data can be used to detect and delineate areas of pack ice where melting has occurred and thus the extent of puddling or melt water pools (McClain, 1973a). This capability, which was first demonstrated using the visible and near-infrared imagery from Nimbus 3 (Strong et al., 1971), derives from the circumstance that the reflectivity of ice (or snow) diminishes sharply in the wavelengths from 0.7-1.3μm but much less at visible wavelengths when its surface becomes wet.

The success of the Electrically Scanning Microwave Radiometer (ESMR) on Nimbus 5 represents a significant advance in our capability to monitor sea ice throughout the year and in virtually all weather (Campbell et al., 1973). The greatest single limitation on satellite sensing of sea ice at visible and infrared wavelengths is cloud cover, but natural emission from Earth's surface at microwave wavelengths is attenuated only slightly by clouds typical of polar regions. Large brightness temperature contrasts, which are due to emissivity rather than physical temperature differences, are found at ice/water boundaries and these are readily observed through clouds. The ice surface itself has been observed to have emissivity variations, first-year ice being somewhat more emissive (higher brightness temperature) than multi-year ice (Gloersen et al., 1973). The only drawback to the use of ESMR imagery for some types of ice studies is the relatively coarse spatial resolution of the data, viz. 30 km. Better resolution than this will require the orbiting of substantially larger antennas.

CONCLUDING REMARKS

The sensor complement of the NOAA series of polar-orbiting operational satellites will remain unchanged through about 1977, except that the visible channel will be opened up to 0.5-1.0μm on some of the Scanning Radiometers. The application of SR and VHRR measurements, especially the latter, to sea ice problems is likely to increase greatly as the number of ground stations equipped to receive these data continues to grow and as better equipment and techniques for processing, displaying, and applying these observations are developed. The next in the Nimbus series, sometime in mid-1974, will also carry sensor systems of interest to ice researchers: a THIR and an improved ESMR (0.8 cm wavelength). ERTS-2, presently scheduled for 1976, will have the same RBV and MSS sensor systems as ERTS-1 except that a fifth channel will be added to the MSS. This extra detector will provide measurements in the thermal infrared with a nominal resolution at nadir of 250 meters.

Satellite data are used most effectively in sea ice investigations when they can be calibrated and controlled with aircraft remote sensing and surface-based observational programs, such as in AIDJEX (Arctic Ice Dynamics Joint Experiment). Such an approach would be valuable, if not essential, to the upcoming POLEX (Polar Experiment) program. The potential results of such coordinated investigations will have application not only to the solution of sea ice problems, but to short-term weather and long-term climatological forecasting over areas distant from the Arctic and Antarctic.

REFERENCES

Aber, R.G. and E. Vowinckel, 1972: Evaluation of North Water spring ice cover from satellite photographs. Arctic 25:263-271.

Barnes, J.C., D.T. Chang, and J.H. Willand, 1972. Application of ITOS and Nimbus infrared measurements to mapping sea ice. Final Report on NOAA/NESS Contr. No. 1-36025 by Allied Research Assoc. 86 p.

Campbell, W.J., P. Gloersen, W. Nordberg, and T.T. Wilheit, 1973. Dynamics and morphology of Beaufort Sea ice determined from satellites, aircraft, and drifting stations. Goddard Space Flight Center, Greenbelt, Md. Preprint X-650-73-194. 27 p.

DeRycke, R.J., 1973. Sea ice motions off Antarctica in the vicinity of the Eastern Ross Sea as observed by satellite. J. Geophy. Res. 78. (in press)

Fletcher, J.O., 1969. Ice extent on the Southern Ocean and its relation to world climate. Memorandum RM-5793-NSF, The Rand Corporation. 107 p.

General Electric Company, 1972. ERTS Reference Manual. General Electric Co., Space Division, Philadelphia, Pa. 184 p.

Gloersen, P., W. Nordberg, T.J. Schmugge, and T.T. Wilheit, 1973. Microwave signatures of first-year and multi-year ice.

J. Geophys. Res. 78:3564-3572.
McClain, E.P., 1973a. Quantitative use of satellite vidicon data for delimiting sea ice conditions. Arctic 26:44-57.
McClain, E.P., 1973b. Detection of ice conditions in the Queen Elizabeth Islands. Proceed. of Earth Resources Technology Satellite-1 Symposium, NASA Goddard Space Flight Center, Greenbelt, Md. p. 127-128.
McClain, E.P., 1973c. Earth satellite measurements as applied to sea ice measurements. Proceed. of Symposium on Approaches to Earth Sciences Through the Use of Space Technology, COSPAR Working Group 6, Konstanz, Germany, May 1973 (in press).
NASA, 1972. Nimbus 5 User's Guide. Goddard Space Flight Center, Greenbelt, Maryland. 162 p.
Potocsky, G.J., 1968. Use of satellite photographs to supplement aerial ice information. Informal Report No. 68-72, U.S. Naval Oceanographic Office. 13 p.
Schwalb, A., 1972. Modified version of the Improved TIROS Operational Satellite (ITOS D-G). NOAA Technical Memorandum NESS-35. Washington, D.C. 48 p.
Strong, A.E., E.P. McClain, and D.F. McGinnis, 1971. Detection of thawing snow and ice packs through combined use of visible and near-infrared measurements from Earth satellites. Mon. Wea. Rev. 99:828-830.
Swithinbank, C., 1970. Satellite photographs of the Antarctic peninsula area. Polar Record 15:19-24.
Untersteiner, N. and G.A. Maykut, 1969. Numerical prediction of the thermodynamic response to Arctic sea ice to environmental changes. Memorandum RM-6093-PR, The Rand Corporation. 173 p.
Vasil'ev, K.P., 1968. Use of meteorological satellite data as a navigation aid. Trudy No. 36, Problems of Satellite Meteorology, Hydrometeorological Research Center of the USSR (Translation TT-70-50001 for ESSA and NSF, 1970, p. 46-57).
Wark, D.Q. and R.W. Popham, 1962. The development of satellite ice surveillance techniques. Proceed. of First International Symposium on Rocket and Satellite Meteorology, p. 415-418.
Wendler, G., 1973. Sea ice observation by means of satellite. J. Geophy. Res. 78:1427-1448.

APPLICATION OF SATELLITE VISIBLE & INFRARED DATA TO MAPPING SEA ICE

James C. Barnes
Clinton J. Bowley
David T. Chang
James H. Willand
Environmental Research & Technology, Inc.
Lexington, Massachusetts 02173

ABSTRACT

This paper describes the use of the high-resolution, multispectral data from ERTS-1 (Earth Resources Technology Satellite) for mapping arctic sea ice. Ice features mapped in the eastern Beaufort Sea and the Bering Sea during the 1973 spring season are discussed. The results demonstrate that ERTS imagery has a high potential for monitoring ice conditions during the time of maximum ice extent and the beginning of the ice break-up season. Examples of the application of satellite thermal IR data for detecting ice are also discussed. Evidence is found that the combined use of IR and visible data may provide more information on ice conditions than can be deduced from either type of data alone, and that during periods of alternating daylight and darkness a relationship exists between the diurnal variation in IR temperature and ice thickness and/or amount.

INTRODUCTION

Nearly a decade ago, relatively low resolution video imagery from earth satellites was already being used to acquire useful information on arctic sea ice distributions. In fact, it was at the time that Fletcher (1966) stated that the "observational barrier" in the Arctic was crumbling primarily under the impact of satellite observation systems. Now, ERTS-1 is providing multispectral data at a resolution that has not previously been available. Techniques being developed using ERTS data will lead eventually to the use of operational satellite systems to collect ice data on a far more complete and economical basis than has heretofore been possible.

ERTS-1 was placed into a near-polar, sun-synchronous orbit at a height of 900 km on 23 July 1972. Except for a brief period in early August of 1972, the MSS (Multi-Spectral Scanner) sensor system has operated continuously, returning approximately 188 data scenes each day, 44 of which are over the United States. The MSS views a swath 180 km wide, with coverage of the exact same area occurring every 18 days. Because of orbital overlay at the higher latitudes, coverage of arctic areas can occur as often as three or four consecutive days during each 18-day cycle. Features greater than about 70 m in width can be identified in ERTS imagery and can be mapped accurately to a scale of at least 1:250,000. The MSS senses in four spectral bands: MSS-4 (0.5 to 0.6 μm), MSS-5 (0.6 to 0.7 μm), MSS-6 (0.7 to 0.8 μm), and MSS-7 (0.8 to 1.1 μm).

Despite the tremendous potential of ERTS, the MSS spectral bands provide useful information only during periods of adequate solar illumination. Monitoring of sea ice, however, is also required during the winter dark period; for example, knowledge of the wintertime distribution of open water is needed for heat balance studies. A means for observing ice during the winter dark period is through use of satellite thermal IR sensing.

The use of satellite data, including IR, for ice survey has been summarized by McClain and Baliles (1971). Specific techniques to use thermal IR data for mapping sea ice distributions have been developed in studies by Barnes, Chang, and Willand (1970, 1972a, 1972b). The results of these studies have shown that principal sea ice boundaries can be reliably mapped from nightime IR imagery. Image enhancement experiments have also shown that the information content of the IR images can be increased by selecting a discrete number of gray levels. The gray levels must be readily distinguishable, however, since the purpose of the display is to reduce ambiguity by matching the information content of the interpretation. In a series of subjective tests, a five gray-scale display was found to be optimum in that more accurate interpretations of the temperatures could be obtained and the depiction of significant features enhanced.

USE OF ERTS IMAGERY TO MAP SEA ICE

Analysis of ERTS-1 imagery indicates that sea ice can be identified in all of the MSS spectral bands because of its high reflectance. It has been found, however, that the photographic processing of prints from the original 70 mm negatives can be important for ice detection, since exposures selected to retain detail in bright, snow-covered land areas may result in the loss of significant ice features. Although ice and clouds may have similar reflectances, ice can be distinguished through various interpretive keys (Barnes and Bowley, 1973).

Ice features have been mapped for several arctic areas including the Bering Sea, the eastern Beaufort Sea, parts of the Canadian Archipelago, and the Greenland Sea. Tentative identification of ice types has been made, based on variations in reflectance and the shapes and sizes of the patterns. Differences in reflectance between the shorter wavelength bands (MSS-4 and MSS-5) and the near-IR band (MSS-7) have been found very useful for differentiating ice types and for determining characteristics of the ice surface. For example, in some of the earliest ERTS images from late July 1972, ice in Crozier Channel and Kellett Strait along the coasts of Prince Patrick and Melville Islands appears very dark in the MSS-7 band. Because of the differences in reflectance in the MSS-4 and MSS-7 images, these areas are believed to be areas where the ice surface is covered by melt water. As pointed out by McClain (1972), even a thin film of water on the ice can cause a sharp drop in reflectance in the near-IR portion of the spectrum, whereas in the visible the reflectance does not change significantly. Daytime temperatures during this period at Mould Bay, located on Prince Patrick Island, are reported to be well above freezing. Several smaller, very dark spots can also be detected in the ice in M'Clure Strait; these spots are dark in the MSS-4 imagery as well, so are deduced to be areas that contain open water rather than melt water on the ice.

Ice distribution mapped from the ERTS data during the late summer and early fall of 1972 are discussed in detail in the paper by Barnes and Bowley (1973). Some ice features mapped from data collected during this past spring season are discussed in the following sections.

DETECTION OF LEADS IN THE BEAUFORT SEA

ERTS imagery from several passes during the period from the first of April through mid-June 1973 covering the general area of the 72-80°N and 126-150°W has been examined. This area contains compact (10/10) and very close (9/10 to less than 10/10) pack ice, but even in the early April data extensive leads can be seen, some being as much as 180 km or more in extent. In these early spring images, complex patterns are evident, with newer leads often intersecting older refrozen leads. As the spring season progresses, the pack ice undergoes deformation from large ice fields into smaller ice floes.

In the imagery of 1 April, a large lead that is about 50 km wide at the widest point exists near 74°N, 140°W. Distinct tonal variations are evident within the lead. The differences in tone can be distinguished in both the visible and near-IR spectral bands, and are presumably representative of the gradual refreezing of the lead through formation of grey and grey-white ice. Twenty days later, on 21 April, the same lead can be seen, although it has narrowed considerably to a maximum width of only 20 km. On this date, the entire lead appears to contain grey-white ice. Within the grey-white ice, small shearing leads are visible; the dark tone of these features indicates either open water or nilas ice. Finally, in an image of 7 May, the lead can barely be distinguished as only a very small difference in reflectance is detectable between the refrozen lead and surrounding ice.

Sequential Observation of the Eastern Beaufort Sea. Ice movement and deformation in the eastern Beaufort Sea near Prince Patrick Island and the mouth of M'Clure Strait can be mapped throughout the spring season through use of ERTS imagery. Early in the spring, in imagery on 8, 10, 11, and 28 April, a quasi-permanent flaw lead can be observed just west of Prince Patrick Island. The lead separates fast ice along the coast and in M'Clure Strait from pack ice to the west. In the April imagery, all land and ice surfaces are snow covered and, therefore, appear smooth.

Significant changes in the configuration of the lead can be mapped from the imagery. On the 8th, the maximum width of the lead is 4 km; two days later, it has widened to a width of 10 km near Prince Patrick Island and almost 30 km at M'Clure Strait; one day later, on the 11th, the maximum width is as much as 35 km. Later in the month the widening continues, so that on 28 April the lead is as much as 50 km wide. As the lead opens, new ice formation is also evident. On the 10th and 11th, the low reflectance indicates the formation of grey ice, whereas on the 28th the higher reflectance in the western portion of the lead indicates that much of the ice has become grey-white. On each date the area of new ice formation is along the immediate eastern edge of the lead. Ice floes, judged to be either first-year or multi-year ice because of their high reflectance, are also embedded in the newer ice within the lead.

The same area can be observed again in imagery on 2 and 3 May. During the short interval since the previous observation (28 April), ice conditions have changed dramatically. The lead has now closed in again and is full of ice floes, with very little grey or grey-white ice evident. Changes continue to take place, and by the next ERTS cycle (17 May) the lead has once again reopened in the area south of

Prince Patrick Island. In these three observations, the formation of a giant floe (more than 150 km across) can also be traced. The floe develops from fractures that can be detected on the 2nd and 3rd of May, and can be identified in later data on 20 May and 6 June. By the last date, the floe has become surrounded by a broken ice field containing much smaller floes, and surface features can now be seen on the floe.

Monitoring Ice Deterioration in Amundsen Gulf and M'Clure Strait. Through repetitive ERTS coverage during mid-June, ice deterioration can be monitored in Amundsen Gulf and M'Clure Strait. On 10, 12, and 13 June the eastern part of Amundsen Gulf is observed, and on 16, 18, and 20 June, M'Clure Strait is observed. In both areas the ice appears to be in a state of rapid deterioration. The ice surface is covered with various patterns indicative of puddling and thawing, and ice breakup is occurring in Amundsen Gulf and at the western end of M'Clure Strait. Shore polynyas exist along Banks Island, particularly at the mouths of the rivers. In Amundsen Gulf, the movement of ice floes over the 4-day period can be observed. Surface weather charts indicate that temperatures throughout the Canadian Northwest rose well above freezing during the month of June.

Farther north, in the area that includes Crozier Channel and the Kellett and Fitzwilliam Straits, the ice appears similar to that observed the previous summer. In both years, much of the ice has a very low reflectance in the near-IR, indicative of surface melt water. The differences in the reflectances in the visible and near-IR images can be seen in Figure 1. In these areas of low reflectance, the ice is intersected by bright linear patterns, many of which can also be identified in the imagery from the previous summer; the interpretation of these patterns is that leads or cracks, too small to be resolved by the satellite, have permitted the water on the sea surface to drain away.

Correlation Between ERTS Imagery and Observations Made on the Bering Sea Expedition (BESEX). The Bering Sea Expedition (BESEX), a joint effort between the USA and USSR, was conducted during the period from mid-February through early March 1973. The primary area of interest extended southward from St. Lawrence Island to beyond the edge of the pack ice. During the experiment some 14 flights were made by the NASA CV-990 aircraft, operating out of Elmendorf Air Force Base in Anchorage; on all flights numerous on-board experiments were conducted including nearly continuous vertical viewing aerial photography.

During the BESEX period ERTS data were collected on a total of four passes over the Bering Sea. On two dates, 27 and 28 February, the ERTS passes crossed the easternmost part of the Bering Sea in the Nunivak Island area; on 6 and 7 March, the passes crossed St. Lawrence Island and the prime BESEX area of interest. In the ERTS imagery of 6 March, shown in Figure 2, an area of low reflectance, apparently grey ice, extends southward from St. Lawrence Island. Along the immediate south coast of the island, a band of open water exists. Over the open water small streaks of stratus cloud can be seen; the ice to the south, however, appears to be cloud-free. To the north and to the southeast of St. Lawrence brighter ice floes exist, surrounded by the grey ice. Numerous fractures and leads cut through the bright (first-year and multi-year) ice and the surrounding grey ice. Differences between the apparent grey ice and the older ice are especially noticeable in the MSS-7 imagery.

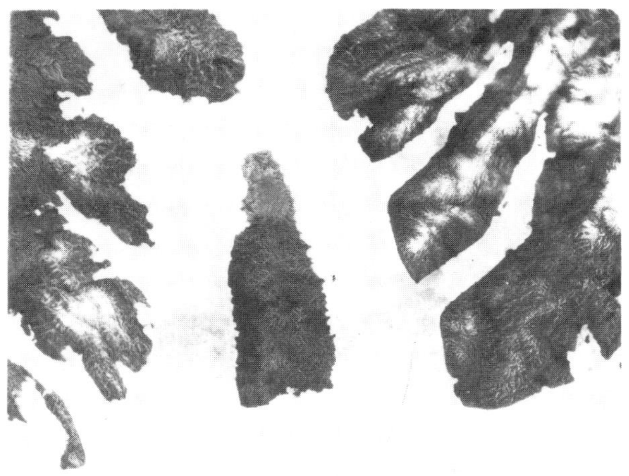

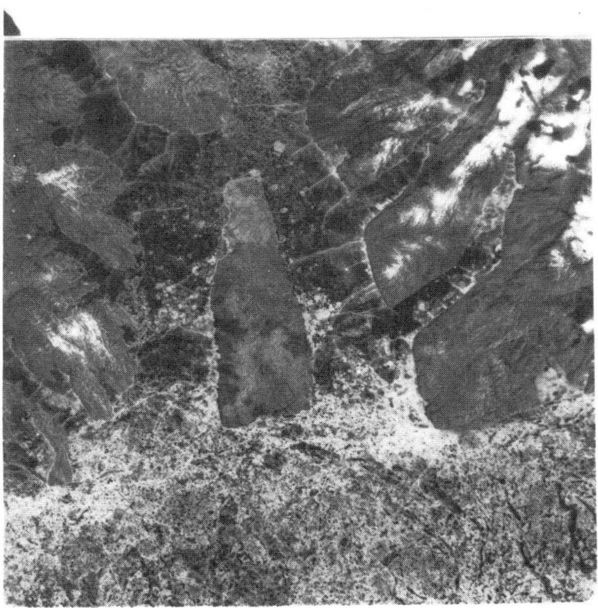

Figure 1. ERTS-1 imagery showing ice in Crozier Channel and Kellett Strait just north of M'Clure Strait, 18 June 1973. Prince Patrick Island is at the left, Melville Island at the right. Top: MSS-4 visible band; Bottom: MSS-7 near-IR band. Dark patterns in the near-IR image are related to surface melt water (ERTS ID 1330-20064).

Figure 2. ERTS-1 imagery showing sea ice in the vicinity of St. Lawrence Island in the Bering Sea, 6 March 1973. The scenes shown are the MSS-7 near-IR band (ERTS ID 1226-22171 and 22174).

The commentary by the ice observer onboard the CV-990, given in the Navigational Flight Data report, is in remarkably good agreement with the ice conditions deduced from the ERTS data, even though the flight is a day earlier. To the north of St. Lawrence the observer reports generally 80% first-year and multi-year ice and 20% grey ice. Where the observer reports simple shearing leads, leads are detectable in the imagery. Also, the north coast of the island is reported as having compacted ice; in the imagery, the ice in that area does appear solid, quite different from the ice south of the island.

As the flight approaches the western end of St. Lawrence Island the commentary indicates that the amount of first-year and multi-year ice is decreasing and that the ice ahead is mostly grey ice; the location of the ice boundary separating these different ice types in the ERTS imagery corresponds almost exactly with the location where the above comment was made. After crossing the island, the observer reports all grey ice a few days old, some of which has undergone deformation in the form of stretching to the southwest. This observation confirms the ice type deduced from the ERTS data; furthermore, leads indicative of stretching of the ice can be seen in the imagery.

The ice conditions displayed in the ERTS imagery on 7 March are very similar to those of the previous day. On the 7th, however, the stratus streaks have increased, obscuring some of the grey ice south of St. Lawrence Island. Farther south some ice features can be detected, but much cloudiness exists. One segment of the CV-990 flight on this day follows the southern boundary of the grey ice, an area that is cloud-free. The observer reports a vast expanse of grey ice, and then reports running into a stratus deck; the comment is made that the "stratus streaks are due to open water along the south shore of St. Lawrence." Thus, this commentary verifies exactly the ice and cloud conditions apparent in the ERTS imagery.

DETECTING SEA ICE IN THERMAL IR DATA

Techniques to detect and map sea ice in thermal infrared data are described in detail in reports such as Barnes, Chang and Willand (1970, 1972a, 1972b). Two examples illustrating application of thermal IR data are briefly discussed in the following paragraphs.

Comparison Between Visible and Thermal IR Imagery. An ITOS-1 DRSR (Direct Readout Scanning Radiometer) image received at the Fairbanks station on 11 March, 1971 was compared with a corresponding AVCS (Advanced Vidion Camera System) photograph taken 12 hours earlier on the same date. The ground resolution of the Scanning Radiometer, which operates in the 10.5-12.5 μm spectral band, is 7.4 km; the AVCS resolution is about 4 km. In both the IR and visible images the Alaskan Coast and the Aleutian Islands can be identified, indicating that the area is essentially cloud-free except for low stratus over the ocean. Just south of the Aleutians, "wakes" in the stratus exist as a result of the wind flow over the individual islands.

Despite the fact that the DRSR image contains considerable noise, several ice features are detectable. The ice edge mapped in the Bering Sea is in close agreement with the ice edge mapped from the photograph, and many smaller scale features within the ice pack in the photograph can also be detected. In the eastern part of the

Bering Sea, a well-defined thermal gradient extends from St. Lawrence to Nunivak Island and then westward toward St. Matthew Island. Because this gradient is detectable in DRSR imagery on the following two days as well, it is almost certainly associated with a real ice feature. In the corresponding visible data, however, no significant change in brightness is evident, as is the case in the area along the southern side of St. Lawrence Island. Thus, the thermal gradient is deduced to be associated with the edge of thicker ice or the edge of ice with a greater amount of snow cover, but not with a boundary separating different ice concentrations.

Measurements of Diurnal Variations in Ice Surface Temperature. Thermal IR data from the Nimbus Satellite have been used to examine the magnitude of the diurnal difference in surface temperatures for various ice features. The purpose of this investigation was to determine the feasibility of using these differences as an additional key for the objective identification and mapping of ice features. In one case, digitized data obtained over the Barents and Kara Seas area of the USSR during both the daytime and nighttime passes on 18 April, 1970 were analyzed. For these dates the region under investigation did experience alternating periods of sunshine and darkness; at 65-70°N, approximately 15 hours of daylight exists during mid-April and the maximum solar elevation is about 30°. On 18 April, the area over Novaya Zemlya and the adjacent seas of the USSR had essentially clear sky conditions. These conditions are depicted in the IDCS (Image Dissector Camera System) photograph taken on the daylight pass over this region.

Full resolution digitized THIR (Temperature-Humidity Infrared Radiometer) data were used in the analysis. The data are in the 10.5-12.5 μm spectral band, and the ground resolution of the radiometer is about 7 km. Mean values of the effective IR temperatures for the daytime and nighttime passes were computed for segments of the individual scan lines crossing geophysical features identified in the corresponding photograph. From these mean values, representative temperature values were computed.

The results of the analysis show a number of interesting features. The first is that the effective temperatures are fairly uniform, both during the day and at night, over the uniform sea ice targets. More importantly, the data show an unambiguous difference in the values computed for each of the seas. Based on interpretation of the ice coverage using the visible data, it may be tentatively concluded that the difference in the diurnal effective temperature is due to the differences in ice type. Over the compact pack ice of the Kara Sea, the surface temperatures of the ice respond strongly to the alternating cycle of daytime solar heating and nighttime radiative losses under clear sky conditions resulting in a diurnal range of 7 to 10°K. Over the thinner ice, or what is more probably open pack ice, of the Barents Sea, the large heat capacity of the exposed water surface prohibits large diurnal variations in surface temperature.

CONCLUSIONS

The results of the analysis of data collected during the spring of this year demonstrate that ERTS imagery has a high potential for monitoring arctic sea ice conditions during the time of maximum ice extent and the beginning of the ice-breakup season. In the eastern

Beaufort Sea area, the combination of ERTS orbital overlap and a high incidence of cloud-free conditions during the spring assures a high frequency of repetitive satellite coverage. With this repetitive coverage, the deformation and movement of ice features can be mapped throughout the early April to late June period. Ice features that can be identified in ERTS imagery include the following: the development of fractures leading to the formation of distinct ice floes; the growth and deterioration of leads; evidence of shearing movements of ice masses; the formation of new grey ice within leads; the distinction between grey, grey-white, and older forms of ice; and the deterioration of the ice surface evidenced by the formation of puddles, thaw holes, and drainage patterns. Features as small as 80-100 m wide can be detected, and ice types can be identified most reliably through analysis of both the visible and near-IR spectral bands.

Data from ERTS passes crossing the Bering Sea in early March have been correlated with ice observations collected in the Bering Sea Expedition (BESEX). On two flights of the NASA CV-990 aircraft, the ice conditions in the vicinity of St. Lawrence Island reported by the onboard observer are in remarkably close agreement with the ice conditions mapped from the corresponding ERTS imagery. The ice identified in ERTS imagery and substantiated by the aerial observer include the locations of boundaries between areas consisting of mostly grey ice and of mostly first and multi-year ice, the existence of shearing leads, and the occurence of open water with the associated development of stratus cloud streaks.

Thermal IR measurements in the 10.5-12.5 μm spectral interval provide a means for mapping gross ice boundaries during periods of polar darkness when observations in the visible portion of the spectrum are restricted. Moreover, evidence was found indicating that the combined use of thermal IR and visible data can provide more information on ice conditions than can be deduced from either type of data alone. In observations of the Bering Sea, some areas were observed to have higher IR temperatures combined with lower visible reflectances; these areas are deduced to consist of lesser ice concentrations. On the other hand, a higher IR temperature was measured in another area where no significant decrease in reflectance was observed in the visible data; because of the high reflectance, this area is deduced to consist of either thinner ice or ice with less snow cover, but not a lesser ice concentration.

In analyses performed from a rather limited data sample, the diurnal difference in effective blackbody temperature appears to be correlated with the thickness of the ice and/or the percentage of ice cover, or more likely a combination of both of these important aspects of the ice field. Specifically, it was found that for close pack ice the diurnal range of temperature was of the order of 9° to 10°K, a value characteristically different from the value for thin and broken pack ice, which was of the order of 3°K. Although these results are obviously geographically and seasonally dependent, the observed temperature differences may have use as signatures to aid in mapping ice features.

REFERENCES

Barnes, J.C., D.T. Chang, and J.H. Willand, 1970: Satellite Infrared Observation of Arctic Sea Ice, AIAA Paper No. 70-301, AIAA Earth Resources Observations and Information Systems Meeting, Annapolis, Maryland, March 1970.

Barnes, J.C., D.T. Chang, and J.H. Willand, 1972a: Image Enhancement Techniques for Improving Sea Ice Depiction in Satellite Infrared Data, Journal of Geophysical Research; Oceans and Atmospheres Edition, 77(3), pp. 453-462.

Barnes, J.C., D.T. Chang, and J.H. Willand, 1972b: Application of ITOS and Nimbus Infrared Measurements to Mapping Sea Ice, Final Report under subcontract No. 19333 to ARA, Inc. (NOAA/NESS Contract No. 1-36025), ERT, Inc.

Barnes, J.C. and C.J. Bowley, 1973: Mapping Sea Ice from the Earth Resources Technology Satellite, Arctic Bulletin, 1 (1), pp 6-13.

Fletcher, J.O., 1966: Forward, Proc. of Symposium on the Arctic Heat Budget and Atmospheric Circulation, Memorandum RM-5233-NSF, The RAND Corporation.

McClain, E.P., 1972: Detection of Ice Conditions in the Queen Elizabeth Islands, Proc. of Symposium on Earth Resources Technology Satellite-1, September 29, 1972, National Aeronautics and Space Administration.

McClain, E.P., and M.D. Baliles, 1971: Sea Ice Surveillance from Earth Satellites, Mariners Weather Log, 15 (1), pp 1-4.

SNOW STUDIES USING VISIBLE & INFRARED MEASUREMENTS FROM EARTH SATELLITES

James C. Barnes
Clinton J. Bowley
David A. Simmes
Environmental Research and Technology, Inc., Lexington, Mass. 02173

ABSTRACT

This paper describes recent studies of the applications of ITOS and Nimbus thermal IR measurements and of ERTS-1 multispectral imagery for detecting and mapping snow extent. The results of these investigations demonstrate that thermal IR measurements can be a useful tool for mapping gross snow-cover patterns, at least in relatively flat terrain; in IR imagery, snow-covered areas are usually depicted by significantly brighter tones (lower temperatures) than the surrounding snow-free terrain. The analysis of ERTS imagery for two test sites, the Salt-Verde Watershed in central Arizona and the southern Sierra Nevada in California, indicate that snow extent can be mapped from ERTS in more detail than is depicted on aerial survey snow charts. Moreover, it appears that although small details in the snowline can be mapped better from higher-resolution aircraft photographs, boundaries of the areas of significant snow cover can be mapped as accurately from the ERTS imagery as from the aircraft photography.

INTRODUCTION

Snow cover was perhaps the first of the earth's water resources to be observed from space. In some of the initial pictures taken by TIROS-1 in April 1960, snow could be identified in northeastern Canada. Since then, techniques to map snow have been developed using later TIROS, ESSA, ITOS, and Nimbus data (Barnes and Bowley, 1968, 1970; McClain, 1970). These studies have shown that valuable information on snow extent can be derived from spacecraft observations.

Now, ERTS-1 is providing multispectral data at a resolution not previously available. ERTS-1 was placed into a near-polar, sun-synchronous orbit at a height of 900 km on 23 July 1972. Except for a brief period in early August of 1972, the MSS (Multi-Spectral Scanner) sensor system has operated continuously, returning approximately 188 data scenes each day, 44 of which are over the United States. The MSS views a swath 180 km wide, with coverage of the exact same area occurring every 18 days. Features greater than about 70m in width can be detected in the ERTS imagery; this compares with a resolution of about 4 km for the earlier meteorological satellite photography. The MSS senses in four spectral bands: MSS-4 (0.5 to 0.6 μm), MSS-5 (0.6 to 0.7 μm), MSS-6 (0.7 to 0.8 μm), and MSS-7 (0.8 to 1.1 μm).

Although satellite thermal IR radiometers have not provided data with the spatial resolution of sensors such as the ERTS multispectral scanner, the IR data have been shown to have various geophysical applications, including the measurement of sea surface temperature and the mapping of sea ice. With regard to snow, IR sensors have the potential for measuring the surface temperature of the snow pack, an important parameter for predicting snowmelt. Moreover, the requirement for adequate solar illumination restricts the use of video or visible-spectrum radiometer systems to daytime operation, seriously limiting data acquisition at high latitudes. With the continued economic exploitation of the Arctic, as evidenced by projects such as the proposed trans-Alaska oil pipeline, the application of remote sensing to monitor snow cover and other geophysical phenomena during daytime and nighttime will become increasingly important.

ANALYSIS OF THERMAL IR MEASUREMENTS

Physical Considerations. The detection of a geophysical feature in thermal infrared measurements depends on the existence of a temperature contrast between the "target" and its surroundings, assuming that both surfaces have emissivities near unity. In laboratory experiments using an emissivity box, Shafer and Super (1971) found the emissivity of snow in the 8 to 14 μm interval to range from 0.966 to 0.989, with the average value for all types of snow surfaces examined to be 0.978. From aircraft measurements, Griggs (1968) found a mean emissivity of 0.99 for a melting snow surface. Thus, in the thermal IR spectral interval snow can be considered to have essentially blackbody characteristics. The emissivities of various ground and vegetation surfaces that might exist adjacent to snow cover have been reported by Griggs (1968) and others to have values generally of 0.90 or higher. Whereas these surfaces may not approximate a blackbody as closely as a snow surface, the blackbody assumption is still valid within the accuracies of the satellite measurements used in this study.

Assuming that both the snow surface and surrounding terrain can be considered blackbodies, then the snow surface - to be detectable in IR measurements - must have a temperature different from the surrounding snow-free terrain. Analyses by Estoque and Yee (1963) of the surface temperature as a function of albedo, have shown that a surface with a high albedo absorbs a smaller amount of solar radiation than is absorbed by surfaces with lower albedoes; therefore, a black surface can be expected to have a higher temperature and greater diurnal temperature range than a white surface. Snow cover, of course, does have a high albedo. In measurements from aircraft, Griggs and Marggraf (1967) found an overall mean value of 0.74 for snow surfaces that were from one to three weeks old. Newly fallen snow would likely have an even higher albedo, perhaps approaching 0.90 in some instances. Hence, snow-covered areas would be expected to have lower temperatures than the surrounding snow-free areas.

When melting, snow also has an upper temperature limit of 273°K (0°C). On a warm day, this limiting temperature may be many degrees lower than that of nearby snow-free surfaces. On the other hand, because of a high content of melt water, a melting snow pack could remain warmer at night than the bare ground, which undergoes more rapid radiational cooling. In any event, under most conditions during winter or spring, a snow surface would theoretically have a temperature significantly different from that of surrounding snow-free surfaces.

Data Sample. Infrared measurements from the Nimbus-4 THIR (Temperature Humidity Infrared Radiometer) and the ITOS-1 SR (Scanning Radiometer) during 1970 and 1971 were used in this study (Barnes and Bowley, 1972). The thermal IR channels of the THIR and SR sensors operated in the same spectral interval (10.5 to 12.5 μm).

Since these were the first high-resolution radiometers to operate in the 10.5 to 12.5 μm channel, Nimbus-4 and ITOS-1 were the first satellites to provide both nighttime and daytime thermal IR data. At orbital altitudes, the ground resolution at nadir of the THIR is about 6.7 km and of the SR about 7.4 km. The output of the THIR and SR is specified in terms of an effective blackbody temperature (T_{bb}). For the 10.5 to 12.5 μm channel, the T_{bb} measured over a cloud-free region provides a first estimate of the actual surface temperature. For a surface with an emissivity of unity in this spectral region and in the absence of any atmospheric attenuation, T_{bb} is, in fact, the surface temperature. However, since the emissivity for all surfaces is less than or equal to one and the atmosphere above the surface attenuates the emitted radiance even in the window channel, the measured effective temperature, T_{bb}, will, in general, be slightly lower than the actual surface temperature.

From a purely theoretical point of view, the THIR and SR are identical. However, the interpretation of the measurements from these two radiometers over the same target areas may be quantitatively different resulting from differences in the sensor characteristics and calibration procedures. In fact, a consistent difference in the temperatures measured by the two satellites was noted. An external calibration of the Nimbus radiometer using ART (Aircraft Radiation Temperature) surveys and ship reports showed that during February 1971, the Nimbus IR temperatures are 4 to 5°K lower than the corresponding sea surface temperature. Similar calibration of the ITOS data shows the measured IR values to be consistently 2 to 4°K higher than corresponding sea surface temperatures during the same month. A direct comparison of the measurements from the two satellites indicated the ITOS IR temperatures to be 6 to 8°K higher than the Nimbus IR temperatures.

Summary of Results of Thermal IR Data Analysis. Analyses were performed using both daytime and nighttime Nimbus, THIR, and ITOS SR imagery and digitized data. The data sample comprised passes from the winter and spring seasons over both mountainous and relatively flat terrain. The ITOS imagery consisted almost entirely of direct readout (DRSR) at the Wallops station. The principal results are summarized below:

- In Nimbus daytime IR imagery during the spring of 1970, the snow cover retreat at higher elevations in New England and New York from mid-April to early May can be monitored. In these film strips, the snow-covered higher terrain appears distinctly colder (brighter tone) than the non-snow covered areas. In several ITOS DRSR images across the Northeast during February 1971, considerable small-scale detail in the temperature patterns measured over snow cover is evident; however, the detailed patterns are not consistent from image to image. Of interest is that a complete reversal in the thermal patterns is observed over the Northeast between the daytime Nimbus imagery during spring and the nighttime ITOS imagery during winter. In the former, water areas are colder than the land, and the higher terrain areas are colder than the river valleys. In the latter, the lakes and coastal waters are warmer than the land, and the higher, thickly forested Adirondacks area is not as cold as the surrounding lower terrain.
- In an ITOS DRSR image showing the south-central part of the country in early February 1971, areas of distinctly lower temperature (brighter tone) are in very close agreement with areas of

recent snowfall. The digitized data for this case show that the snow covered areas have temperatures about 10°K lower than non-snow covered areas; the mean IR temperature for the snow covered areas is 260°K, with the lowest values being 252-256°K, whereas for the non-snow covered areas the mean temperature is 270°K, with the highest values being 277-280°K.

- Although snow-covered areas appear colder in daytime Nimbus film strips over the upper Mississippi-Missouri Basins region, the edge of the snow is poorly defined. In data from five passes during December, however, an area in the eastern Dakotas in which an inch or less of snow is reported appears noticeably warmer (2-5°K) than the surrounding region in which greater snow amounts exist.
- In Nimbus daytime film strips for the western United States, the Sierra Nevada and other mountain ranges appear distinctly colder than the terrain at lower elevations. However, the extent of the coldest (brightest) areas seems to be associated as much with the higher terrain as with the snow extent. In the Sierra Nevada area, agreement between the thermal patterns and the snow extent seems better in the late spring than in the early spring. The agreement is also better in the southern Sierras where elevation differences are more distinct than in the north.
- In the sample of Nimbus daytime passes covering the northeastern part of the country during April 1970, IR temperatures measured over snow covered areas are as high as 290°K. The data analyzed were on days during which air temperatures in the same areas were well above freezing (273°K). Nevertheless, the measured IR values are as much as 17°K higher than would be expected even over a melting snow surface, the temperature of which would remain near 273°K (see, for example, Geiger, 1965). Similarly, Nimbus IR temperatures measured over the Sierra Nevada during April and May 1970 and ITOS IR temperatures measured over the same area during April 1971 are as high as 288°K and 282°K, respectively. Since snow survey data indicate considerable snow cover still remaining in the high Sierras on the dates of the satellite passes analyzed, these measured values are also considerably higher than would be expected.

The apparent discrepancies of as much as 15°K cannot be accounted for solely by calibration problems or atmospheric attenuation effects. However, measurements by Edgerton *et al.* (1968) show that on a clear day, the surface temperature of bare soil increases rapidly and may exceed the surface air temperature by many degrees. In experiments using a high resolution aircraft radiometer, Shafer and Super (1971) measured temperatures across a northwest slope to be as much as 10°K higher than those over the adjacent valley floor (approximately 268°K versus 258°K); they conclude that the higher temperatures are due to exposed south-facing rock ledges heated by solar radiation, and point out that the high temperatures of the rock ledges represents a significant error source if the type of surface viewed is not considered. One may conclude from these measurements that the surface of bare soil or rocks within a snow field can exceed 280°K and may even reach 300°K, while the temperature of the snow remains at about 273°K. On a warm day, trees can also be expected to have temperatures higher than the snow, with some of the heating being due to the absorption of solar radiation being reflected off the surrounding snow surface.

Computations demonstrate that if the field-of-view of the radiometer is 50 percent snow and 50 percent a snow-free surface with a temperature of 300°K, then the effective temperature measured by the radiometer exceeds 287°K. At the resolution of the Nimbus and ITOS radiometer (about 7 km), the amount of vegetation or bare rock could approach half of the field of view and on a warm day the temperature of these snow-free surfaces could approach 300°K. Thus, it is concluded that the relatively high IR temperatures mapped over snow during the spring are due primarily to the presence of bare surfaces in the melting snow field.

MAPPING SNOW EXTENT USING ERTS IMAGERY

An investigation is currently in progress to evaluate the application of ERTS data for mapping snow cover in the mountainous areas of the western United States. In this study, which has been concentrated in the southern Sierra Nevada in California and the Salt-Verde Watershed in central Arizona, the snow extent mapped from the ERTS imagery has been correlated with standard snow measurements, aerial survey snow charts, and aerial photography (Barnes and Bowley, 1973).

Examination of the ERTS data has shown that contrast between snow covered and snow free terrain is greatest in the MSS-4 (0.5 to 0.6 μm) and MSS-5 (0.6 to 0.7 μm) spectral bands. The MSS-5 data appear to be the more useful of the two bands for snow mapping, because in some of the MSS-4 images snow covered areas are near saturation, causing a loss of detail in the snow pattern. In the longer wavelengths, especially the MSS-7 near-IR band (0.8 to 1.1 μm), snow cover is more difficult to detect. However, the near-IR band, which has been found extremely useful for studies of glaciers and sea ice in the Arctic, may provide information for certain purposes such as detecting melting conditions. A limited sample of color composite data have also been examined. Although a thorough evaluation of the advantages of the color product must await additional data, the initial examination indicates that color may have some advantages for detecting and mapping snow.

Snow cover can be identified in the MSS-5 data because of its greater reflectance than the surrounding snow free terrain. Although snow and clouds have similar reflectances, mountain snow cover can be differentiated from cloud primarily because the configuration of the snow fields is very different from cloud patterns and can be instantly recognized. The snow boundaries are also sharper than typical cloud edges, and snow fields usually appear with a more uniform reflectance than do clouds, which have considerable variation in texture. Furthermore, cloud shadows are usually visible, especially with cumuliform clouds, and various terrestrial features can be recognized in cloud-free areas. Because of the high-resolution of the ERTS data, numerous terrestrial features that are not visible in lower-resolution meteorological satellite photographs can be recognized. In addition to natural features, such man-made features as roads, power line swaths, and cultivated fields are detectable. In the heavily forested areas of the Cascades, timber cuts are clearly visible.

ANALYSIS OF DATA FOR SALT-VERDE WATERSHED

The 1972-73 winter season produced a record snowpack accumulation in the Salt-Verde Watershed in central Arizona. Precipitation for the October through April period was much above normal, and the snowpack at its maximum in early April was estimated to be as much as 500 percent of normal in the Verde River Watershed and 300 percent in the Salt. In April the runoff forecast for the Verde was nearly 400 percent of the 1953-1967 average and for the Salt more than 350 percent. Obviously, throughout this past winter-spring season, snow hydrology was a vital concern in Arizona water management programs.

Snow extent for at least a portion of the Salt-Verde Watershed could be mapped from imagery for seven of the ten ERTS cycles between mid-November 1972 and early May 1973. Maps comparing the snow extent mapped from ERTS and as depicted on aerial survey charts indicate that more detail in the snow line can apparently be mapped from the ERTS data than can be mapped by the aerial observer. In nearly all areas in which a discrepancy occurs, the aerial survey chart depicts a greater snow extent than is mapped from the ERTS imagery. In some cases this difference can be explained by melting that occurred during the interval between the observations. The snow extent in mid-February mapped from ERTS and from aerial survey is shown in Figure 1.

To obtain a quantitative evaluation of the differences between the ERTS data and aerial survey charts, the total Salt-Verde Watershed was divided into three areas, and the percentage of snow cover computed using a planimeter. The mean overall difference in the percentage of the Watershed snowcovered is 8 percent. In the two cases with the greatest difference (18 percent), weather charts indicate that a significant change in snow cover probably occurred between the time of the ERTS observation and the aerial survey. In the comparison between ERTS imagery and aerial photography, detailed snow features that cannot be detected in the ERTS imagery can be seen in the aircraft photography. However, it appears that all significant snow cover (i.e., substantial snow cover, not small amounts such as might be found along small topographic features) can be detected in the ERTS imagery.

ANALYSIS OF DATA FOR SOUTHERN SIERRA NEVADA

Similar analyses were conducted for four river basins in the southern Sierra Nevada: the Kings, Kaweah, Tule and Kern Basins. The snowpack in these basins was also above normal this past season. As was true for the Arizona test site, a major part of the southern Sierras area was sufficiently cloud-free to be mapped on seven of the ten ERTS cycles between early December and late May.

For each river basin the snow line elevation was determined directly by comparing the snow map derived from the ERTS image with superimposed elevation contours. In addition, the snow line elevation was determined by measuring the percentage of the basin snow-covered and referring to the area - altitude curve for the particular basin. In an earlier study using meteorological satellite photography, the snow line elevation for the Kings Basin was determined in this way (Barnes and Bowley, 1970). Recently, the snow line determined from areal snow extent, or the equivalent snow line altitude (ESA), has been discussed further with regard to ERTS data (Meier, 1973).

The snow line elevation determined from a direct comparison with a contour chart varies considerably within each river basin. The Kings River Basin in the southern Sierras, for example, was divided into three sections, and the mean elevation for each section was determined from a large number of data points. For three cases during the spring season, the mean difference between the section with the highest snow line elevation and that with the lowest is of

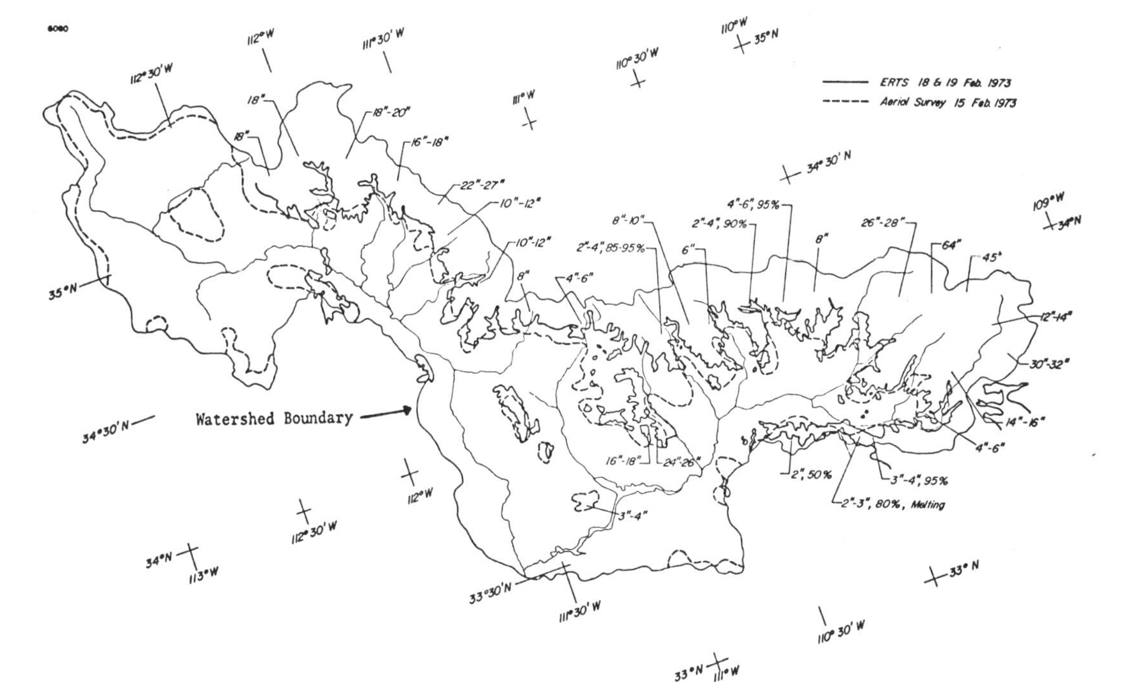

Figure 1. Comparison between snow line mapped from ERTS MSS-5 images (ID 1210-17273 and 1211-17332) and that depicted on aerial survey snow chart, Salt-Verde Watershed, Arizona, mid-February. The eastern part of the area (east of about 110°30'W) was covered on 18 February; the extreme western part was not covered in the ERTS imagery.

the order of 400 m. Because of the observed variation in the snow line elevation measured directly, which can be influenced significantly by small mapping errors, it is believed that the equivalent snow line altitude (ESA) is a more meaningful measurement with regard to the application of satellite data to snow mapping. As was pointed out in the discussion of the Arizona data, it appears that considerably more detail in the snow line can be mapped from ERT imagery than is depicted on aerial survey snow charts.

The snow extent mapped from ERTS data was compared with the snow extent depicted on aerial survey charts for three spring cases. Because of cloud cover, the Kern Basin was not mapped on two of these dates. The results show that in every case except one, the difference in percentage snow cover is less than 10 percent. The mean difference for the Kings Basin for the three cases is 5 percent, for the Kaweah 12.5 percent, and for the Tule, 5.5 percent. The mean differences in the equivalent snow line altitude are 160, 360 and 210m respectively. For each case analyzed, the percentage of snow cover determined from ERTS data is greater than that of the aerial survey chart; thus, the ESA determined from ERTS is lower than the ESA shown in the aerial survey chart.

Similar aircraft data as were collected over the central Arizona mountains were collected by the NASA/ARC Earth Research Aircraft Project (ERAP) over the southern Sierra Nevada on 20 February. Three segments of the flight cross areas covered in the ERTS imagery of 25 February; parts of two of the segments are within the four-basin area, whereas the third is just northeast of that area. The segment northeast of the Kings River Basin crosses Mono Lake and the Owens River in the vicinity of Bishop. In both the aircraft and ERTS data, the snow line can be identified in the area north of Bishop indicated on the topographic chart to be an area of volcanic tableland essentially unvegetated. The snow line appears to be at about the 1500m level, with little change having occurred during the five-day interval between 20 and 25 February. More detail in the snow line and same patchy snow south of the edge of the solid snow cover can be mapped from the aircraft data. However, the edge of the area of significant snow cover can be mapped as precisely from ERTS as from the aircraft photography.

CONCLUSIONS

Based on the results of the completed analysis of ERTS-1 data, it is concluded that the amount of information in ERTS imagery with practical application to snow mapping is substantial. Moreover, for two mountain areas in which snow hydrology is a major concern, the Salt-Verde Watershed in Arizona and the southern Sierra Nevada in California, useful snow cover information could be derived from ERTS data on 70 to 80 percent of the cycles during the past winter and spring seasons. Thus, in these two areas, cloud obscuration does not appear to be a serious deterrent to the use of satellite data for snow survey.

The results of the analysis of ERTS imagery for the Arizona and California test sites indicate that the extent of the mountain snowpacks can be mapped from ERTS data in more detail than is depicted in aerial survey snow charts. In nearly all cases in which significant discrepancies occur, the differences can be explained by changes in snow cover during the interval between the ERTS observation and the aerial survey. In addition to comparative analysis with aerial snow charts, the ERTS data have also been compared with high-altitude aircraft photography. The results of the comparative analysis indicate that although small details in the snow line that cannot be detected in the ERTS data can be mapped from the higher resolution aircraft data, the boundaries of the areas of significant snow cover can be mapped as accurately from the ERTS imagery as from the aircraft photography.

The costs involved in deriving snow extent maps from ERTS imagery appear to be very reasonable in comparison with current data collection methods. For example, the flight time to survey the Salt-Verde Watershed is approximately five hours, with another hour or so needed to compile the snow chart. On the other hand, the snow extent can be mapped from an ERTS image covering nearly the entire Watershed area by an experienced analyst in about two hours. Eventual machine processing can be expected to reduce this time considerably.

The results of these studies demonstrate that spacecraft acquired thermal IR measurements (10.5 to 12.5 μm spectral interval) can also be a useful tool for mapping gross snow cover patterns, at least in relatively flat terrain regions. In IR imagery, snow-covered areas are usually depicted by significantly brighter tones (lower temperatures) than the surrounding snow-free terrain. Analysis of corresponding digitized data shows that areas with snow cover have IR temperatures, on the average, 5 to 10°K lower than areas with no snow.

The results indicate, however, that certain difficulties are inherent in the interpretation of thermal IR data. The spatial resolution of the Nimbus and ITOS-1 radiometers is marginal for mapping snow with the accuracy desired for many hydrologic applications, and in mountainous areas, IR temperature measurements are influenced by terrain elevation regardless of the snow distribution. Also, IR temperatures measured over snow during the spring months can be considerably higher than 273°K (0°C), the value that would be expected for a melting snow surface. Since an external correction of the satellite data indicates that calibration problems alone cannot account for the apparent discrepancy, the relatively high IR temperatures are believed to be primarily the result of the radiometer viewing warmer surfaces, such as trees or bare rock, in addition to the melting snow field. Sensors such as the current NOAA-2 VHRR (Very High Resolution Radiometer) have greatly improved spatial resolutions, and, thus, can be expected to provide a significantly greater amount of operationally useful information.

REFERENCES

Barnes, J.C., and C.J. Bowley, 1968. Snow cover distribution as mapped from satellite photography. Water Resources Research, 4(2). pp. 257-272.

Barnes, J.C., and C.J. Bowley, 1970. Measurement of mountain snow extent using satellite data. Paper presented at National Fall Meeting of AGU, San Francisco, California, Dec. 1970.

Barnes, J.C., and C.J. Bowley, 1972. Snow studies using thermal infrared measurements from earth satellites. Final Report under subcontract No. 19334 to ARA, Inc. (NOAA/NESS Contract No. 1-35350), ERT, Inc.

Barnes, J.C., and C.J. Bowley, 1973. Use of ERTS data for mapping snow cover in the western United States. Proceedings of symposium on significant results obtained from ERTS-1, NASA-Goddard Space Flight Center.

Edgerton, A.T., et al., 1968. Passive microwave measurements of snow, soils and snow-ice-water systems. Technical report no. 4 (SDG 8-29-6), contract NOnr 4767 (oo) NR 387-033 to Office of Naval Research, Space Divison, Aerojet-General Corporation.

Estoque, M.A. and W.S. Yee, 1963. Air-earth interface characteristics at the atmospheric boundary layer. Report No. 36, Contract No. AF 19 (604) - 7484 to Air Force Cambridge Research Laboratories, Hawaii Institute of Geophysics.

Geiger, R., 1965. The climate near the ground. Harvard University Press, Cambridge, Massachusetts. 611 p.

Griggs, M., 1968. Emissivities of natural surfaces in the 8- to 14-micron spectral region. JGR, 73 (24). pp. 7545 to 7551.

Griggs, M., and W.A. Marggraf, 1967. Measurements of cloud reflectance properties and the atmospheric attenuation of solar and infrared energy. Final Report, Contract No. AF 19 (628)-5517, to Air Force Cambridge Research Laboratories, Convair Division of General Dynamics.

McClain, E.P., 1970. Applications of environmental satellite data to oceanography and hydrology. ESSA Technical Memorandum NESC TM-19, Environmental Science Services Administration.

Meier, M.F., 1973. Evaluation of ERTS imagery for mapping snowcover. Proceedings of symposium on significant results obtained from ERTS-1, NASA-Goddard Space Flight Center.

Shafer, B.A., and A.B. Super, 1971. Infrared temperature sensing of snow-covered terrain. Final Report under Contract No. DAHC04-67-C-0058 to Dept. of Army (Durham), Montana State University.

DETECTION OF BURIED GLACIAL AND GROUND ICE WITH THERMAL INFRARED REMOTE SENSING

Ray Lougeay
State University of New York at Geneseo

ABSTRACT

Thermal remote sensing techniques are demonstrated in the detection and mapping of moraine-covered glacial ice and shallowly buried ground ice. It is often difficult to determine the total extent of an active glacier, or rock glacier, when much of the ablation zone and surrounding area are covered by bare detrital material. This is especially true in aerial surveys, when photo interpretation is strongly dependent upon varying surface albedo values (albedo determining the amount of solar energy which is reflected back to the camera).

Ground level observations have shown that surface temperature, and thus the emitted terrestrial radiation, is closely correlated to the thickness of detrital mantle which covers an ice core. Various types of ice-cored features are discussed which show strong thermal contrast, in the thermal infrared regions of the spectrum, but similar surface configuration and albedo values.

Thermal remote sensing of buried ground ice is dependent upon the magnitude of thermal contrast between various surfaces. Field observations of thermal contrast over various ice-cored surfaces are discussed. The temporal and spatial patterns of thermal contrast indicate optimum conditions for detection of buried ground ice, using thermal remote sensing techniques.

The field methods discussed in this paper are useful for preliminary reconnaissance to locate large masses of ice which would be hidden from the air photo interpreter. Also, these field methods have proved essential in determining optimum conditions for thermal remote sensing surveys related to an inventory of buried ice.

The importance of ice-cored features in the alpine/periglacial landscape is poorly understood. Ablation moraines on active glaciers, ice-cored moraines (lateral, terminal and recessional), and rock glaciers can all have similar morphology and exhibit surfaces of bare detritus. These ice-cored landforms are often found adjacent to other geomorphic features which do not contain ice within them, such as talus slopes, alluvial fans, and areas of glacial outwash. The usefulness of aerial photography is limited in the detection of buried glacial ice. This is due to the fact that the loose rock and rubble surfaces of all these features have similar albedo values and are portrayed on the aerial photograph in similar tones of gray (or brown, in the case of color photography). Photographic film is sensitive to solar energy which is reflected from the surface as a function of surface albedo. The photograph is, thus, a spatial display of varying surface albedo values, portraying reflected sunlight in the visible (400 - 700 nm) and near infrared (700 - 1,000 nm) portions of the spectrum. An infrared thermal remote sensing device responds to varying amounts of electromagnetic energy in the thermal infrared portion of the spectrum (3,000 - 20,000 nm) incident upon the detector. Slight changes in surface temperature will cause different amounts of radiation to be emitted from the earth's surface. Thus, an infrared detector responds to varying surface temperatures. By understanding the controls upon the varying surface temperatures within an area one can understand the thermal patterns obtained by a thermal scanning system. In this study both surface contact thermometers and infrared radiometers were used to monitor changing thermal patterns. In this way, the potential usefulness of thermal infrared remote sensing for detection of buried glacial ice was tested.

It was hypothesized that thermal infrared remote sensing would be able to detect buried glacial ice and to distinguish between various features containing buried glacial ice. Surface temperature contrasts might exist between these ice-cored surfaces as a function of the individual energy budget of each environmental surface and the thickness of detrital mantle over an ice core, producing different soil heat flux rates. Assuming their emissivities to be approximately equal, these surfaces would then emit infrared radiation at differing rates. A thermal remote sensing system could detect the thermal contrasts between these surfaces by monitoring their respective rates of infrared radiation (Lougeay, 1972).

Detection of an object or feature on remote sensing imagery requires an image density contrast between the object and its background. In the case of photographic imagery (i.e., aerial photography) the change in image density across the boundary at the edge of the imaged subject is a product of differing amounts of solar energy reflected from the subject and its background. Image contrasts on thermal infrared imagery are recorded only when there is a difference in the amounts of infrared radiation emitted from the subject and its background. This thermal contrast may result from a difference in surface temperatures or differing emissivities of target surfaces. One purpose of ground truth is to monitor these radiant emission patterns to predict the occurrence of radiant thermal contrasts so that there is some assurance that the subject will be detected on airborne imagery.

Most environmental surfaces observed in this study exhibited an emissivity value of approximately 1.0 in the thermal infrared spectral band. However, glacier ice was found to have an emissivity of 0.985. This produced an apparent radiometric temperature of -1.0°C for melting glacial ice known to have a surface temperature of 0.0°C (Figure 1). Varying emissivity values must be accounted for when comparing radiometric and contact surface temperature measurements. Actual contrast measurements, as detected by thermal remote sensing, must include emissivity ("e") in the relationship $R = e \sigma T^4$; where R is the radiant energy emitted from the surface in ly min^{-1}, σ is a constant = 8.132×10^{-11}, and T is the absolute surface temperature. (One ly min^{-1} is equivalent to 0.698 kJ m^{-2} s^{-1}.)

Observations of both actual and radiant surface temperatures and related environmental parameters were recorded at selected study sites. A transect was established across various environmental surfaces at the terminal margins of the Donjek Glacier in the St. Elias Mountains, Yukon Territory, Canada (61° 09' North,

Blackbody infrared radiance of a surface at $0^{\circ}C$ (emissivity = 1.0)

$$\sigma\, 273^{\circ} K^4 = .452 \text{ ly min}^{-1}$$
$$(.3154 \text{ kJ m}^{-2} \text{ s}^{-1})$$

Blackbody infrared radiance of a surface at $-1^{\circ}C$ (emissivity = 1.0)

$$\sigma\, 272^{\circ} K^4 = .445 \text{ ly min}^{-1}$$
$$(.3106 \text{ kJ m}^{-2} \text{ s}^{-1})$$

Infrared radiance from surface at $-1^{\circ}C$ is 98.5% of that at $0^{\circ}C$

$$.445 \div .452 \times 100 = 98.5\%$$

Infrared radiance of a surface at $0^{\circ}C$ with an emissivity value of 0.985

$$0.985 \times \sigma\, 273^{\circ} K^4 = .445 \text{ ly min}^{-1}$$
$$(.3106 \text{ kJ m}^{-2} \text{ s}^{-1})$$

Figure 1. Effect of emissivity upon radiant temperature.

139° 21' West). These surfaces included bare glacial ice, thin ablation moraine (5 - 15 cm thick on the active glacier), ice-cored moraine (1 - 3 m detrital mantle, covering a core of clean ice over 30 m thick), and morainic outwash containing no apparent buried ice. These sample areas were representative of much of the alpine/periglacial region in the foothill region bordering the high peaks and massive St. Elias ice field. This region of rugged terrain, totaling over 10,000 square kilometers, is almost totally characterized by surfaces of bare detritus with only small patches of alpine tundra and some spruce forest at lower elevations. An analysis of aerial photographs revealed that perhaps 20 percent of this region may be underlain by buried glacial ice, with another 20 percent indistinguishable on aerial photographs.

To determine the potentials of thermal remote sensing in this region, surface temperatures of the sample areas were measured radiometrically under varying weather conditions and at all hours of the day. This provided knowledge of both temporal and spatial patterns of thermal contrast. Strong thermal contrast between various moraine types indicated that thermal remote sensing could easily determine the extent of active glacier covered by ablation moraine. Also, stable ice-cored moraines could be distinguished from adjacent non ice-cored terrain (Figure 2 and Figure 3).

Thermal contrasts between environmental surfaces may vary with the time of day and changing meteorologic conditions because various ground surface materials heat and cool at different rates. At the Donjek Glacier study area the diurnal trend of infrared emission held relatively constant throughout the day (Figure 2 and Figure 3). At midday, the ice-cored moraine surface remained a few degrees cooler than the non ice-cored morainic outwash material. Ablation moraine remained considerably cooler than both outwash and ice-cored moraine, but was always warmer than the glacial ice. These significant thermal contrasts indicate that thermal remote sensing would be quite useful in distinguishing

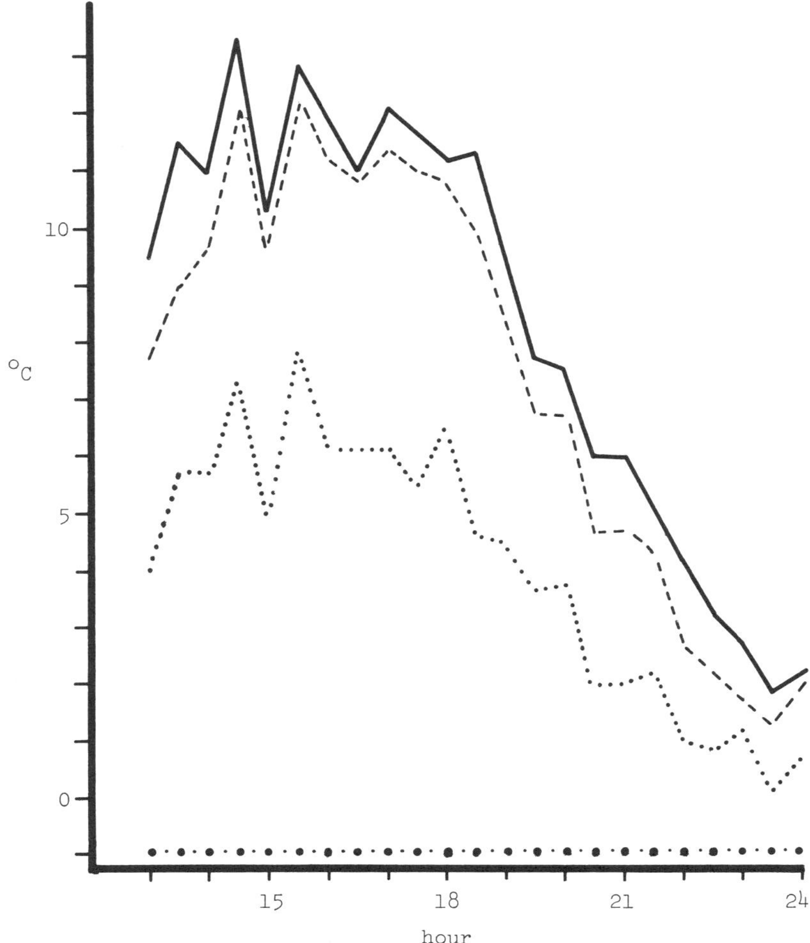

Figure 2. Thermal contrast patterns for one afternoon (June 15) with cloud cover of 3/10 cirrus and 2/10 alto cumulus. Surfaces are: glacier ice (alternating large and small dots), thin ablation mcraine (dotted line), ice-cored terminal moraine (dashed line), and morainic outwash (solid line).

different types of moraine and glacier ice surfaces. A major controlling factor of surface temperature, and the amount of solar energy incident on a given site, is slope and exposure relative to the angle of direct beam incoming solar radiation. The Donjek Glacier moraine study sites were chosen so that variations in slope and exposure characteristics would be sampled. In this case there was still a significant degree of thermal contrast between environmental surface types (e.g., ice-cored moraine and veneer ablation moraine), so that variations within one environmental surface type did not mask the thermal contrast between different surface types. In general, strongest thermal contrast between all surfaces can be expected at mid afternoon when differential heating due to localized slope exposure is minimized.

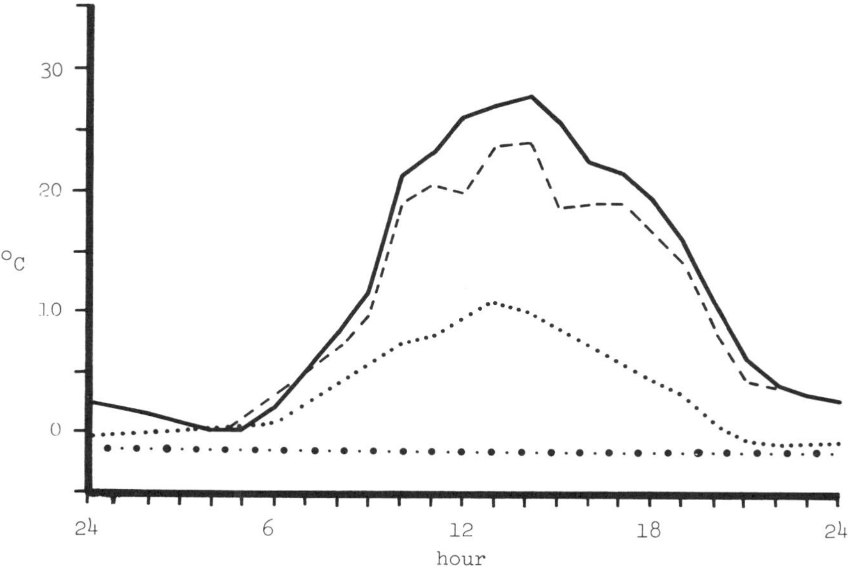

Figure 3. Mean diurnal pattern of radiant temperature under clear skies. Surfaces are: glacier ice (alternating large and small dots), thin ablation moraine (dotted line), ice-cored terminal moraine (dashed line), and morainic outwash (solid line).

The relative pattern of thermal contrast held constant under all meteorologic conditions; however, with increased overcast and decreased atmospheric transmissivity, the magnitude of thermal contrast between surfaces was markedly decreased. Correlation of the infrared radiant temperatures with incoming solar energy indicated a rather strong dependency of surface radiant temperature upon solar insolation. Incoming short wave radiation is used here as a measure of atmospheric transmissivity (i.e., cloud cover and density). There is a direct causal relationship between emitted infrared radiation (i.e., the radiant temperature) and surface temperature ($R = e \sigma T^4$). The statistical analysis of these data, for the observed surfaces, revealed high correlation of surface temperature with infrared radiant temperature (Figure 4). Thus, it is possible to predict the radiant temperature from known surface temperature. A regression analysis, comparing infrared radiant temperature to surface temperature, for each environmental surface studied, indicated a very high relationship.

Errors in monitoring surface temperature with contact thermometers, such as the problem of sampling vegetation and soil simultaneously, have resulted in a rather large standard error of estimate. This problem is alleviated, however, when using an infrared thermometer (i.e., radiometer). The methods used here seem adequate since this study was primarily concerned with the thermal contrast between environmental surfaces and not the absolute surface temperature.

Surface type	Multiple correlation coefficient	Regression equation (y=mx+c)	Standard Error of estimate °C
Morainic outwash	0.86	y=1.38x+(-5.65)	5.03
Ice-cored moraine	0.92	y=1.04x+0.27	3.37
Veneer ablation moraine	0.91	y-1.27x+0.35	2.75
Glacial ice	-	-	-

Figure 4. Regression analysis of infrared radiant temperature and surface temperature.

The nature of the ground truth needed for a given study is dependent upon the objectives of the study and the surfaces to be imaged. In general, the lesser the thermal contrast between target surfaces the greater the need for more extensive ground truth. When thermal contrasts are small there may be few periods of the day when one can safely predict that airborne thermal remote sensing imagery will be able to show a contrast between these surfaces. With similar detrital materials mantling a core of glacial ice, and the mantle thickness approaching or exceeding the soil heat flux damping depth (Outcalt, 1971, 1972), thermal contrasts between ice-cored and non ice-cored surfaces will be slight. However, with thinner mantles of detrital overburden, buried glacial ice can easily be detected. These thermal contrast patterns can be readily obtained through ground truth observations.

Since σ is a constant in the equation $R = e \sigma T^4$, only differences in surface temperature or emissivity can produce changes in surface radiant emission power and apparent thermal contrasts. A relatively small emissivity difference can produce thermal contrasts on the order of a degree or two between various surfaces, even where surface temperatures may be equal. This is shown to be the case with glacier ice (Figure 1). However, if one has a knowledge of the emissivities of surface materials he is working with, or the thermal contrasts are sufficient to mask the effects of emissivity differences, he can then rely on ground truth measurements of surface temperature to indicate thermal contrast which would be depicted on infrared imagery. In this case surface temperature probes, a contact pyrometer, or simple thermometers could yield valuable ground truth data at a minimum of financial expense. The surface temperature data plotted on a diurnal curve could provide information on the time of day when thermal contrasts would be such that airborne imagery could provide a broad regional survey of ice-cored features and buried ice.

For finer accuracy in ground truth observations, however, it is necessary to deal with the question of emissivity. The emissivity of surface materials can be obtained by radiometric and contact measurement of surface temperatures. Emissivity can be expressed as follows: $e = R_{surf} / (\sigma T^4)$; where e is emissivity, R_{surf} is the radiant emission of the surface, and T is the absolute surface temperature.

Absolute emissivity values, as measured in this process, however, need not concern the researcher who is interested only in thermal contrasts for the flight planning or airborne infrared imagery. The question of surface emissivity and the problems of accurately measuring surface temperature can all be eliminated by the use of an infrared radiometer. The radiometer responds to infrared radiation emitted from the ground surface in the same way as the airborne infrared detector. The radiometer integrates all surface temperatures within its field of view. This is important when one deals with varied surfaces such as vegetation over exposed soil, or coarse gravel where there are both sunlit and shady patches. On surfaces such as these, temperature will vary markedly and it is difficult to integrate contact temperature measurement. The radiometer, however, does this automatically.

In summary, these ground level observations have shown that the thermal remote sensing system is capable of detecting buried glacial ice and distinguishing between various ice-cored geomorphic features. Such broad-scale mapping of buried glacial ice is not possible with conventional aerial photography. The use of thermal remote sensing may in the future provide knowledge of ice resources previously unknown and not inventoried. It should be noted that thermal contrasts are most pronounced with relatively thin detrital overburden. When the thickness of the detrital mantle exceeds three to four meters, thermal contrasts between environmental surfaces will be quite small, requiring very sophisticated remote sensing systems and analysis for accurate interpretation.

REFERENCES

Lougeay, R., 1972. Infrared radiometric temperatures in the alpine/periglacial environment as related to thermal remote sensing. Icefield Ranges Research Project Scientific Results, III. American Geographical Society and Arctic Institute of North America, New York.

Outcalt, S., 1971. A numerical surface climate simulator. Geographical Analysis 3:379-392.

Outcalt, S., 1972. The simulation of subsurface effects on the diurnal surface thermal regime in cold regions. Arctic 25:305-307.

HYDROLOGIC CHARACTERISTICS OF SNOW-COVERED TERRAIN FROM THERMAL INFRARED IMAGERY

Ambrose O. Poulin
U. S. Army Engineer Topographic Laboratories

ABSTRACT

Snow cover, in addition to its own contribution to the hydrological state of the terrain, often conceals other hydrologic features or changes in their conditions. The differing subsurface thermal regimes of features such as streams, frozen lakes, and areas of deeper snow often produce surface temperature differences that are sufficient for the production of thermal images of those features. Measurements at a frozen arctic shoreline indicate that the part of these temperature differences resulting from systematic variations of the energy balance components is sufficiently large that it can be distinguished from the part due to random variations. This suggests that there is a degree of predictability to the information available in thermal infrared imagery of snow-covered terrain. However, it is often necessary to consider topographic relationships and short term environmental phenomena in order to understand some of the transient, but useful, image features. A phenomenon, which was given the name "cold fringe effect," was discovered that should allow the identification of areas in which the ice formed on bodies of water is frozen to the bottom. A qualitative model was developed for the diurnal variation of a shoreline temperature differential throughout the winter season, and studies are in progress to evaluate it using measurements from both aircraft and spacecraft.

INTRODUCTION

Hydrologic or hydrologically-related features are one of the major classes of terrain elements affected by the masking influence of snow. Depressions get filled in, and the steep slopes of river banks, shorelines, and terrace faces become ramp-like. Where the terrain is treeless and has little relief even when it is snow-free, such as in some arctic coastal plain areas, snow cover can make the land areas and water bodies visually indistinguishable, thereby resulting in possible unanticipated problems in travel on or across sea or lake ice. The lack of recognizable terrain features hampers one's ability to correlate points of interest on a map with their position on the ground. Another important cold weather problem is the location of water supplies. Although a snow cover helps to protect some water sources from total freeze-up, it also makes them difficult to find.

These observations lead to the conclusion that conventional (summer) topographic maps are often inadequate for use under snow-covered conditions. The U. S. Army Topographic Command (now Defense Mapping Agency Topographic Center) has produced a few prototype maps from winter aerial photography of areas in Alaska, and these appear to be an improvement. However, research in thermal infrared sensing of snow-covered terrain performed over the past 15 years has shown that this method holds promise for significant enhancement of winter

5.6

maps of arctic areas. This work was done by the Photographic Interpretation Division of the Army Engineer Topographic Laboratories (formerly of the Cold Regions Research and Engineering Laboratory) in cooperation with the Canadian Defence Research Board and other Canadian agencies. Figures 1 and 2 demonstrate some of these capabilities of thermal infrared imagery in the spectral range of 8-14 micrometers.

Figure 1 shows concurrent infrared imagery and aerial photographs of a proglacial pond and its outlet channels near Cape Discovery, Ellesmere Island. Although a hint of the outlet channels is apparent in the photographs, the pond may be identified only in the thermal infrared imagery. The pond's location was confirmed on summer aerial photographs.

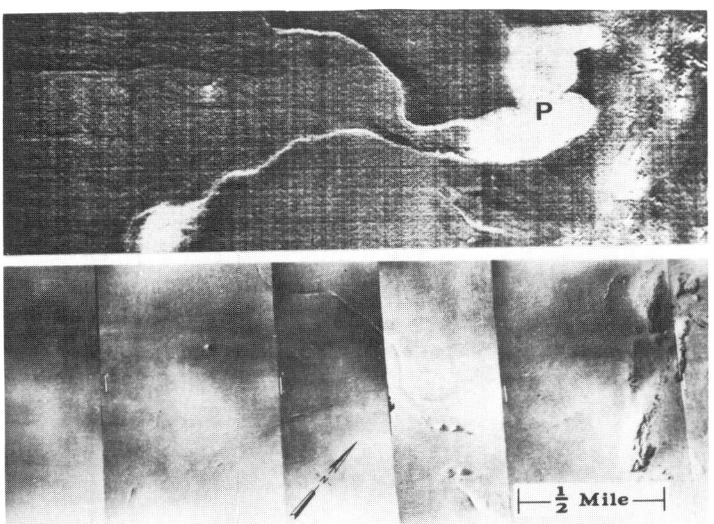

Figure 1. Thermal infrared image, 8-14 micrometer, (top) and concurrent aerial panchromatic photographs of a proglacial pond, P, near Cape Discovery, Ellesmere Island. April 19, 1963.

Figure 2 is composed of four segments from a strip of thermal imagery of the Ruggles River, Ellesmere Island obtained January 26, 1964. Segment 1 depicts the river as a warm signal related to the relatively large flow above the stream bed as it leaves Lake Hazen (the water surface no doubt being at least partially covered by ice and snow). In Segment 2, it appears that the flow had gone underground in the gravels of the stream bed at least enough so that any remaining above-surface flow was largely masked by snow and ice. At the upper part of Segment 3 the water had apparently surfaced again, probably as a result of the reduced gradient of the river valley as it approached base level, and had subsequently spread out, either to be frozen or again to filter into the river bed before it could reach the head of the fiord, at the middle of Segment 4. The white arrows

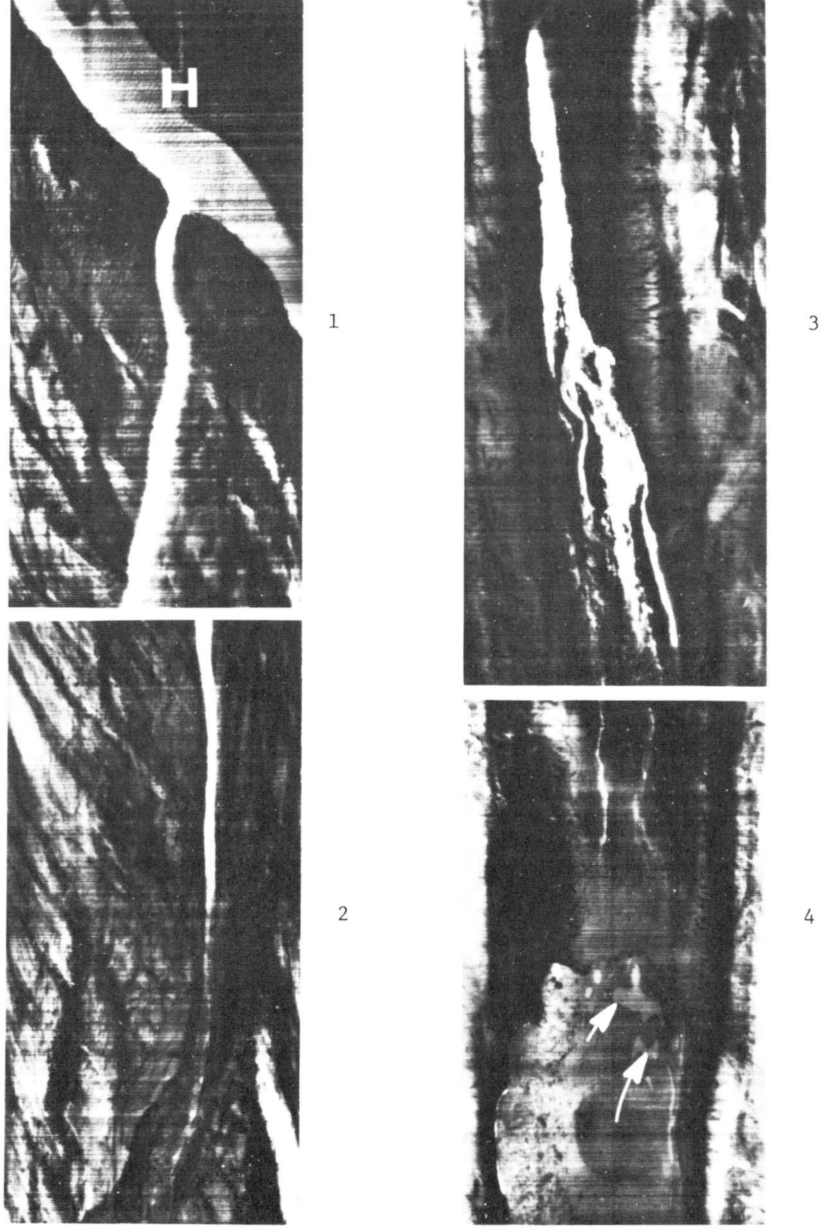

Figure 2. Four segments from a strip of thermal infrared imagery of the Ruggles River, Ellesmere Island. January 26, 1964.

in Segment 4 point to warm areas on the ice cover of the fiord. These warm areas probably resulted either from flow onto the ice earlier in the winter, from upwelling of fresh water from the river with the consequent thinning of the ice, or from upwelling through cracks in the snow-covered ice as a result of either storm or tide activity.

It is interesting to note that in a comparison of the imagery with the best available map (Lady Franklin Bay, 1:250,000), the lower 6 to 8 kilometers of the river valley were found to have been mapped as part of Chandler Fiord. This error was possibly due to the inability to delineate the shoreline in the mapping photographs, either because it was frozen and snow-covered, or because the lower part of the valley was flooded by a high runoff volume.

These examples are impressive, but they raise questions concerning the origin and variability of surface thermal signals and of their detectability. Since 8-14 micrometer emissions are a surface or near surface phenomenon, are they really a dependable source of information or is their variability due to changing surface temperatures resulting from transient weather conditions so great as to preclude any degree of predictability on the basis of such as seasonal radiation balance changes, freezing index or other relatively long-term phenomena? What magnitude of temperature differences can be expected from a surface covered by such a good insulator as snow? To what extent does the atmosphere, including cloud and haze, actually interfere with the detection of infrared radiant energy? And, with the advent of such instruments as NOAA's "very high resolution radiometer," to what extent will satellite imagery be useful for terrain studies? Although only partial, qualitative answers to these questions have thus far been attained, some of then have been encouraging.

THERMAL NATURE OF SNOW-COVERED TERRAIN

Fundamental considerations that help in understanding the thermal nature of snow-covered terrain involve the relative thermal and radiation properties of snow, ice and soil. Although this subject deserves considerable discussion, only a few general observations will be made here.

Figure 3 shows, on a logarithmic scale, the ranges of thermal conductivities for snow, sea ice, fresh ice, frozen fine-grained soils, and frozen coarse-grained soils. It shows that there is an overall range of two orders of magnitude in the thermal conductivity of these materials and that, while values for soil and ice could be about the same, there is very little chance that either will have the same conductivity as snow. Such an occurrence would be limited to very dense snow and a light, dry soil. Limiting consideration to more common snow densities, on the order of 0.4 gm/cm^3 or less, largely precludes an overlap between snow and other natural materials under common conditions. The significance of this observation is that a snow cover, even a very thin one, strongly influences the conducted heat flux and surface temperature. Moreover, this influence is a function of both depth and density, thereby complicating its evaluation in thermal imagery.

Other factors that have a strong, and often dominant, influence on the differential thermal responses of snow-covered terrain are the relative radiation properties of snow, ice, and soil. Specific properties to be considered are spectral reflectance, emissivity, and transmissivity, for both shortwave and longwave radiant energy. But, due to the transient nature of incoming radiant energy and the complexity of the interaction of radiant energy with snow, the relative influences of these phenomena on thermal imagery are often hard to assess. Nonetheless, there are times when even an ordinarily low order phenomenon, such as the differential surface heating of snow by longwave (atmospheric) radiant energy, can be identified in thermal imagery.

The fact that snow and ice transmits shortwave radiant energy, while soils do not, has an important differential effect between the snow covers on ice and land, particularly in the spring of the year under conditions of incipient melting. The energy that penetrates the snow over land is either absorbed or reflected at the snow/soil interface, while at a snow/ice interface a significant portion of the energy continues to be transmitted into and through the ice. Thus, the snow cover over land is warmed faster than that on an ice-covered lake, as evidenced by the earlier disappearance of snow over land. In an earlier stage, prior to the onset of large scale melting of the snow, there is evidence that the normal condition of warmer surface temperatures on frozen water bodies is reversed for some period during sunny days. Weller and Schwerdtfeger (1967) indicate that a significant shortwave radiant energy flux occurs to depths exceeding 4

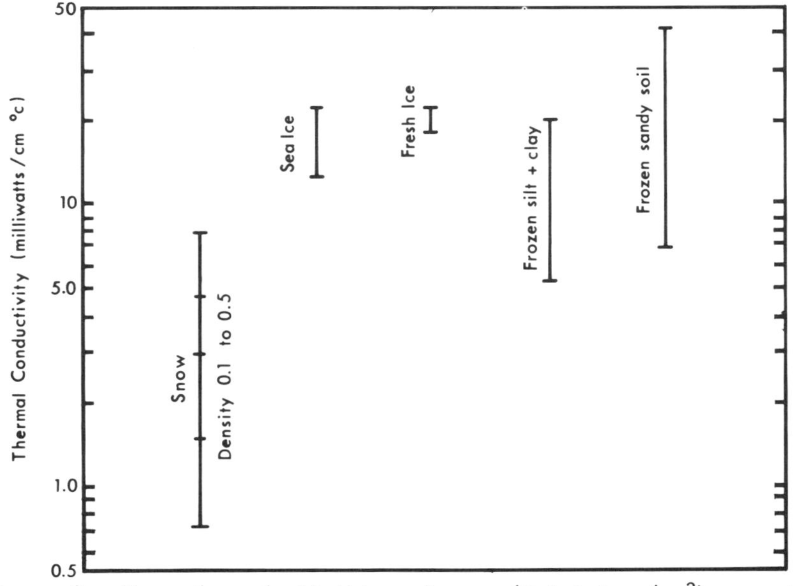

Figure 3. Thermal conductivities of snow (0.1-0.5 gm/cm^3), sea ice, fresh ice, frozen fine-grained soil (1-2 gm/cm^3, 10-40% moisture), and frozen coarse-grained soil (1.2-2 gm/cm^3, 5-25% moisture). Soil moisture is percent of dry weight. Extreme values for soil are not included. Sources for these values are from the listed references.

meters in ice. Their value for the extinction coefficient ($0.043 cm^{-1}$) for windblown snow of density 0.42 gm/cm^3 indicates that approximately one percent of the energy entering the snow would penetrate to a depth of one meter. Thus, for ice of one to two meters thick covered with 10 to 20 cm of snow (a common condition) a significant portion can be expected to penetrate into and through the ice and be unavailable for heating the snow cover.

In short, the point to be stressed here is that the thermal nature of snow-covered terrain and the different surface temperature responses to the various components of the surface energy balance equation are very complicated phenomena. Therefore, it is necessary to develop an understanding of these phenomena in order to optimize use of the imagery.

OBSERVATIONS OF SHORELINE TEMPERATURE DIFFERENCES

In an attempt to obtain some quantitative information regarding the predictability and magnitudes of thermal signals from snow-covered arctic terrain, an experiment was conducted at Tuktoyaktuk, on the Mackenzie Delta. The main effort in the experiment was to make aerial measurements of the temperature differences at shorelines in that area and to correlate these measurements with subsurface temperatures at one of these locations over an extended period of time (March to May) during the cold period and to the start of melting of the snow cover.

Subsurface temperature profiles were obtained from thermocouple strings placed to depths, measured from the bottom of the snow cover, of 130 centimeters in the soil just above the beach, 110 centimeters in the zone where the ice was frozen to the bottom, and 167 centimeters in the floating fast ice. Surface temperature (horizontal) profiles across shorelines were obtained with a Barnes PRT-5 radiometer from an aircraft.

Figure 4 shows the results obtained from approximately 160 aerial radiometric measurements of shoreline temperature differences made during the period from March 20 through May 2. Each point is an average value for from 5 to 36 measurements. On the basis of time and solar altitude, the points divide into two groups: (1) those for solar altitudes less than about 25 degrees in the morning and less than about 5 degrees in the afternoon and (2) those for solar altitudes greater than 25 degrees in the morning and greater than 5 degrees in the afternoon. This asymmetry about midday is the result of the time lag between maximum irradiation by solar radiant energy and the attainment of maximum surface and subsurface temperatures. The figures of 5 and 25 degrees are nominal values determined from visual inspection of the data, and the solid straight lines are least-square fits for the two groups of points intended to show the trends. This systematic separation of the trend lines and their convergence to zero at about the same date (which date is in agreement with the measured subsurface temperatures) indicates that a significant systematic component exists for surface temperature differences at snow-covered shorelines, thus encouraging the idea that they have a degree of predictability.

Figure 5 shows a sequence of three sets of temperature profiles representative of the subsurface temperature trend observed at Tuktoyaktuk during the warming period of early spring. Note: (1) the

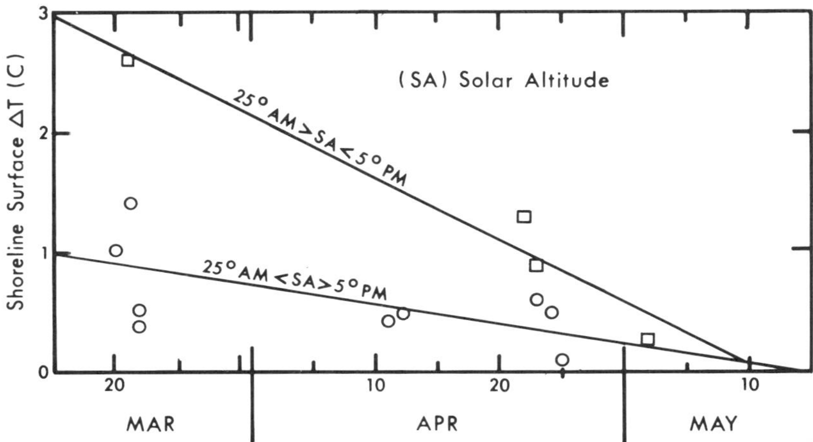

Figure 4. Average values and linear least square curves for 160 aerial measurements of shoreline temperature differences in the Mackenzie Delta coastal area, 1969.

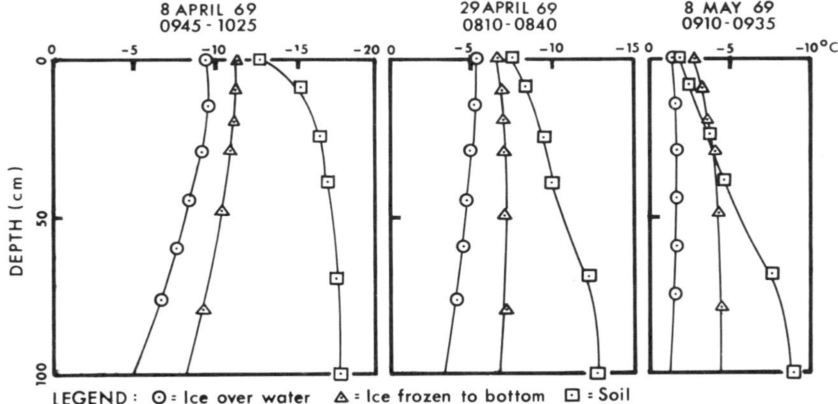

Figure 5. Progression of temperature profiles in the upper meter of ice and soil at a shore during early spring. Tuktoyaktuk, Northwest Territories, 1969.

rapid time rate of soil warming with respect to both of the ice temperatures, and (2) the continued lag of the temperatures of the ice frozen to the bottom with respect to those of the floating ice. Later in the day, the soil temperatures near the surface also became warmer than the ice over water. These subsurface temperature relationships suggest the possibility of two basic deviations from the normal surface temperature difference of the ice being warmer than the land. The first possibility is a simple reversal of surface temperature (ice colder than land). The second possibility is that of both the floating ice and the land appearing warmer than the zone where the ice is frozen to the bottom. Aerial observations showed that these conditions actually do develop.

A simple temperature reversal was observed on only one occasion, early in the afternoon of April 24, when four measurements showed land areas ranging from 0.2 to 0.5 degrees centigrade warmer than the ice. Cold zones at shorelines were observed a total of 34 times on three difference occasions--the mornings of April 25 and May 2 and the evening of May 2. These 34 surface temperature difference observations had a minimum value of 0.1 and maximum value of 1.5 degrees with an average of 0.5 degrees centigrade. The cold area was in the transition zone between the land and floating ice. In a thermal image such a surface temperature variation would cause frozen water bodies to be outlined by a cold zone. Therefore, this phenomena was given the name "cold fringe effect."

Four reasons seem to explain the persistence of this cold zone: (1) the insulating effect of a deeper snow cover which commonly exists along shorelines as a result of the geometric discontinuity where the upper ice surface contacts the beach, (2) the colder ice temperatures here with respect to ice over water, (3) the greater thermal inertia (product of the thermal conductivity and the volumetric specific heat) of the underlying, frozen, saturated soils with respect to both ice and the unsaturated upper soils on the beach, and (4) the heat input to the beach soils by radiant energy, as described above in the discussion of the differential shortwave transmission phenomenon.

A MODEL FOR THE SEASONAL VARIATION OF SHORELINE TEMPERATURE DIFFERENCES

Factors that influence the shoreline signal may be classed as systematic or random. The systematic factors include soil temperature, seasonal air temperature trends, ice thickness, aging of the snow cover, and the daily trend of shortwave irradiation. Random factors include short term fluctuations of air temperature, cloud cover, haze, precipitation, snow depth, and wind. The most favorable conditions for maximum signals would be clear, calm, cold nights during a period of sharply falling temperatures when there is a minimal cover of relatively old snow on thin ice. The opposite, of course, would be true for minimum signals. Since the basic signal is warm over frozen water bodies and cold over land, the signal differential would tend to increase with decreasing soil temperatures, which reach a minimum in late winter, but this parameter is in conflict with the effect of ice thickness, which tends to reduce the differential and which is also maximum in late winter. Moreover, the coldest and clearest periods are usually also those during which ice and snow cover are the thickest. Consequently, one might reasonably expect at least two major maxima in the seasonal trend of shoreline signal magnitudes, such as shown in Figure 6. It is estimated that the greatest maximum would occur prior to or early in the dark season for the following reasons: (1) ice thickness and snow cover are of first order importance and are minimum early in the winter; (2) very cold and rapidly dropping temperatures do occur early in the winter; (3) clear, calm conditions might occur at anytime even though they are most common in late winter and early spring; (4) the soil in the active layer of permafrost freezes rapidly; and (5) the time required for the temperature response of thin ice with minimal snow would be relatively short, precluding the necessity of a long duration for optimum environmental conditions. A second maximum might occur during the coldest and clearest conditions, probably late winter or very early spring.

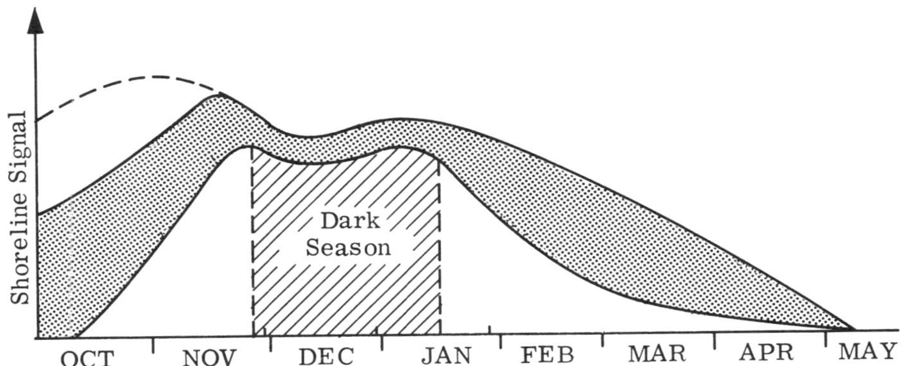

Figure 6. A qualitative model for shoreline thermal signals under nominal weather conditions, i.e., exclusive of storms and heavy haze or fog. Ordinate difference of the envelope represents systematic diurnal variations. Dark season is based on 70° N. latitude.

Other factors dictating the shapes and ordinate differences in Figure 6 are: (1) before the dark season the signal should have a diurnal range which diminishes from a maximum at the start of freezing to a minimum at the start of the dark season, with possible reversals (land warmer than water area) early in the period resulting from an intermittent or melting snow cover; (2) during the dark season the daily air temperature range is diminished and there is no surface heating by solar irradiation. Thus surface temperature variations and the diurnal range are minimized; (3) after the dark season the daily variation increases and subsequently decreases to zero when the snow cover reaches the melting point. Reversals, as was shown earlier, occur late in the season. Figure 6 does not include consideration of the cold fringe effect.

The significance of a model such as this lies in its potential application as an indicator of how other primary thermal signals may vary and, consequently, as a guide to flight planning for the acquisition of thermal imagery. It would also assist in the development of criteria for seasonal maps--maps, for instance, that depict the location and condition of hydrologically significant terrain features. However, a considerable amount of research will be needed for the realization of such benefits on a dependable basis.

USE OF SATELLITES

Considering the extent and remoteness of many snow-covered areas, satellites are obviously a preferred vehicle for obtaining thermal imagery. However, it is doubtful that the present NOAA satellites, with their noise equivalent temperature differentials (NEΔT) of approximately 2°C, will be useful for this purpose. Transmission coefficients of typically hazy arctic atmospheres can be expected to be relatively low, probably less than 0.5, and surface temperature differences commonly are less than 2 to 4°C. An atmospheric transmission of 40 percent and a surface temperature difference of 1°C would require an NEΔT better than 0.4°C; on many occasions, appreciably lower values would be required. Design criteria for such a satellite system should probably consider an NEΔT on the order of 0.1 or better.

CURRENT STUDIES

Efforts are currently in progress to evaluate the model depicted in Figure 6 through a program of aerial measurements of shoreline temperature differences at many different locations in the arctic at various times during the winter. This is being done in cooperation with the Canadian Defence Research Board and the Atmospheric Environment Service. Also, attempts will be made to correlate aerial measurements with comparable data obtained from satellites in cooperation with the Environmental Sciences Group, National Environmental Satellite Service, in order to determine atmospheric transmission values.

REFERENCES

Dorsey, N. E., 1940. Properties of ordinary water substance. Reinhold Publishing Co., N. Y.

Kersten, M. S., 1949. Laboratory research for the determination of the thermal properties of soil. Engineering Experiment Station, University of Minnesota, Minneapolis, Minnesota.

Mantis, H. T., 1951. Review of the properties of snow and ice. SIPRE Report 4, Institute of Technology, University of Minnesota, Minneapolis, Minnesota.

Mellor, M., 1964. Properties of snow. Part III, Section A, Cold Regions Science and Engineering, U. S. Army Cold Regions Research and Engineering Laboratory, Hanover, New Hampshire.

Pounder, E. R., 1965. Physics of ice. Pergamon Press, London.

Schwerdtfeger, P., 1963a. The thermal properties of sea ice. Journal of Glaciology, vol. 4, no. 36, 789-807.

Schwerdtfeger, P., 1963b. Theoretical derivation of the thermal conductivity and diffusivity of snow. International Association Sci. Hydrology, General Assembly, Berkeley, Publication No. 61.

Weller, G. and P. Schwerdtfeger, 1967. Radiation penetration in antarctic plateau and sea ice. Polar Meteorology, World Meteorological Organization, Technical Note No. 87, Geneva, Switzerland.

REMOTE SENSING APPLICATIONS IN CANADIAN ICE RECONNAISSANCE

Henry G. Hengeveld
Atmospheric Environment Service,
Department of the Environment
Canada

ABSTRACT

Remote sensing techniques are being investigated and integrated into the Canadian airborne ice reconnaissance program in order to improve the program to an all weather, day and night capability. Two Electra aircraft being used for a routine visual and radar ice observing program have been equipped with an infrared line scan system, a laser profilometer system and a tri metrigon Vinten camera system. Applications and interpretation problems of Side Looking Airborne Radar imagery over ice of all ages and types are also being investigated in preparation for future installations of SLAR systems. Real time data availability for immediate inflight interpretation and analysis is being emphasized. The formats of the real time as well as permanent data records, the interpretation methods, and the analysis of the available information are discussed. Proposed methods of data reduction and coordination with visual and radar data for use in direct marine tactical support are presented.

INTRODUCTION

All of Canada's navigable waters, with the exception of the Pacific Coast and a small part of the Atlantic Coast, are congested with floating ice for at least part of the year. Much of the coastal areas are ice encumbered for more than nine months each year, and in the high Arctic, extensive areas are ice covered year round. Ice, therefore, plays a very important role in the economy, the environment, and to a certain extent the social cycle of Canada as a nation and a people. With this role comes a vital need for accurate information on the whereabouts, conditions and movements of ice.

Canada has been active in operational ice reconnaissance and forecasting since the mid 1950's. Working from the south and east in winter to the north in summer, highly skilled observers ply their trade in routine reporting and charting of ice parameters throughout her navigable waters, providing the data to marine vessels for tactical support, to forecasters for preparation of summaries and advisories, and to the climatologist for preparation of a statistical data bank for consulting and climatological applications. This reconnaissance program, managed by the Atmospheric Environment Service, Department of the Environment, has been, and still is, primarily one of visual observations at low altitudes (300 to 1500 metres

ASL) by specially trained personnel. Supplementary information on ice edges and large leads through cloud or beyond range of visibility is provided by airborne search radar. The observing platforms, currently two Lockheed 188C Electra aircraft, are equipped with up to date inertial and omega navigation systems to ensure accurate location of ice features at all times. A program depending predominantly on visual observations, however, is also subject to the limitations of the human eye -- limitations due to range, cloud, darkness, lack of resolution of detail of ice surfaces, etc. Such a program lacks the capability of acquiring information through cloud or during darkness conditions or to provide detailed information of the ice surface characteristics, such as quantitative roughess data, ice pressure information, or ice thickness.

Although the above limitations have in the past not been critical, the increasing need for a more detailed, all weather, and day and night ice reconnaissance capability in support of marine interests in Canada has been apparent for many years. The rapid escalation of the economic development of the north during recent years and the resultant increase in activity as well as duration of the Arctic shipping season is making this requirement of increasing importance. Similar requirements are also apparent, to the south during the winter ice season, where increasing marine activities in the Gulf of St. Lawrence create a pronounced impact on the country's economy. With this in mind, the Atmospheric Environment Service has since 1969 been involved in a program for the development of an airborne remote sensor package capable of updating the ice reconnaissance program to an operation independent of weather or darkness conditions over the reconnaissance regions.

REMOTE SENSING APPLICATIONS

The development of a remote sensing package for ice reconnaissance has required extensive testing and evaluation of various sensors as to their operational characteristics and the applicability of their data output in a reconnaissance operation. A limited sensor package, including a laser profilometer system, an infrared line scan system and a trimetrigon type 70 mm camera system, has already resulted. Operational plans call for inclusion of Side Looking Airborne Radar (SLAR) by 1975. Real time, hard copy sensor data display to allow for immediate analysis for tactical support applications is being stressed. A data acquisition system for storage of data in a coordinated and computer compatible form is also proposed.

A. Laser Profilometer System

The laser profilometer system consists of a Spectra Physics Geodolite 3A Laser Profiler, an Accudata 174 preamplifier, and a Honeywell 2206 Visicorder. The 25 milliwatt laser beam, generated by a helium-neon laser within the Geodolite, is amplitude-modulated at selected precision frequencies and transmitted to the ice surface. The phase delay of the reflected signal, received from the surface by an eight inch receiving telescope, is precisely measured with respect to the modulation signal on the transmitted light. It is then converted to an analog voltage output which increases from 0 to 10 volts as the phase delay changes from 0 to 360°. At the highest selectable modulation frequency of 49mHz, each complete phase change during the time of light beam transit represents a measuring distance of 10 feet (3.05 metres). Hence for a total range of 10N + x feet, the output signal will be x volts. Lower modulation frequencies, stepped down by factors of ten, can be selected to scales of 100, 1000, 10,000 and 100,000 feet (30.5, 305, 3050, and 30,500 metres).

The 0 - 10V analog output signal from the Geodolite is fed through the Accudata 174 preamplifier to the Honeywell recorder. The Visicorder is a direct recording light beam oscillograph. The trace produced on the light sensitive chart paper in the recorder is visible within seconds after passing under a small latensifier, and can be analyzed almost immediately.

In an ice reconnaissance application, the Geodolite is normally used in a 10 foot full scale mode. The signal recorded indicates the change in air to ground distance and thus consists of two components--aircraft altitude perturbations and the ice surface profile. The two can be distinguished from eachother easily by subjective analysis since altitude perturbations have a much lower frequency pattern than ice surface profiles. Trials to test the application of a pressure port calibrator to remove the altitude component gave negative results. Frequency filters show possibilities for this application, but more promising are the results of tests by Defence Research Establishment Ottawa (DREO) in double integration of an accelerometer signal to obtain an altitude trace. Should these tests prove successful, this technique for removal of the altitude parameter will likely be applied.

Presently the laser data is analyzed in flight for two predominant surface features - ridge heights and frequency of ridging greater than 2' (0.61 metres) in height per mile of flight. In the past, ice charts indicated ice roughness characteristics in terms of the number of tenths of the ice surface covered by ridging, hummocking and rafting. This has now been revised to include the above data. Additionally, frequency characteristics of each particular ice type are readily identifiable in the laser trace. For example, new ice will show a relatively smooth trace, as shown in Figure 1, while second year ice is characterized by large, weathered ridges and a generally rough topography. Grey white ice appears relatively smooth, but shows an occasional spike representing a small ridge. Particularly under low ambient light conditions, these characteristic frequencies are very useful in identification of ice types present on the sea surface.

B. Infrared Line Scan System

According to Planck's radiation laws, intensity of radiation emitted from the earth's surface is a function of surface emissivity and the fourth power of the surface temperature. Imagery of the earth's surface recorded through the 8-14u infrared atmospheric window, in which region the earth's radiation spectrum also happens to peak, is therefore a presentation of the distribution of the apparent surface temperatures. Such a thermal display has an important practical application, particularly under darkness conditions, in a reconnaissance operation over ice covered waters.

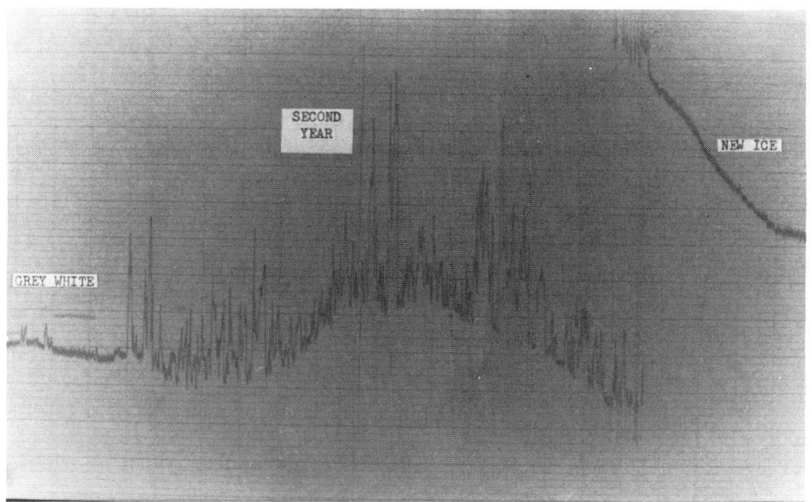

Figure 1. A sample laser trace of sea ice in Lancaster Sound, October 1972, showing characteristic traces of three different ice types. Dark horizontal and vertical lines indicate one foot and one second intervals respectively. Note 6 foot (1.8 metres) ridge in centre of trace.

During the period of sea ice formation and growth, which in the Arctic extends from late summer to early spring, temperatures of the surface air layers are normally well below those of the sea surface. The resultant upward heat flux through any ice formation present causes a variation in ice surface temperatures directly related to the thickness of the ice. Thermal imagery will show the warmest surfaces, in this case open water, as "hot" spots, with ice surfaces of increasing thickness appearing increasingly "cold". Thus relative ice thicknesses up to 3 feet (0.9 metres) or more can be identified through variations in the emitted radiation intensity. Additional characteristic features such as rafting and ridging, open water areas, and presence of frozen puddles on ice surfaces can be associated with image density to identify ice types being observed. Recent ridging, implying presence of ice pressure, can also be distinguished from older fractures because initially the broken ice is moist and "warm".

In the summer melt season, surface conditions change considerably. Ice surfaces begin to melt at about $0^{\circ}C$, usually warmer than the surrounding sea water. Characteristic puddling and ridging (zones of no puddling) patterns are now used to identify ice types. Sea surface thermal patterns, due to presence of warm melt water from floes as well as cold upwelling from beneath moving floes, can offer further information on ice movement directions.

The instrumentation for recording the thermal imagery consists of a Bendix LN-2 thermal mapper, a Honeywell 1856 fibre optics real time chart recorder, and a signal monitoring oscilloscope. The mapper, equipped with a detector sensitive to the 8 - 14u infrared region, scans the ice surface beneath the aircraft through a 120° sweep and records the signal, by means of a glow modulator, onto a 70mm film. Real time application, however, is essential. The signal, therefore, is also fed into the fibre optics recorder to be recorded onto a light sensitive chart. The image which appears within seconds is somewhat limited in contrast (only 4 to 5 grey levels) but, as is apparent in the sample images in Figures 2 and 3, is sufficient for identification of predominant ice types and features.

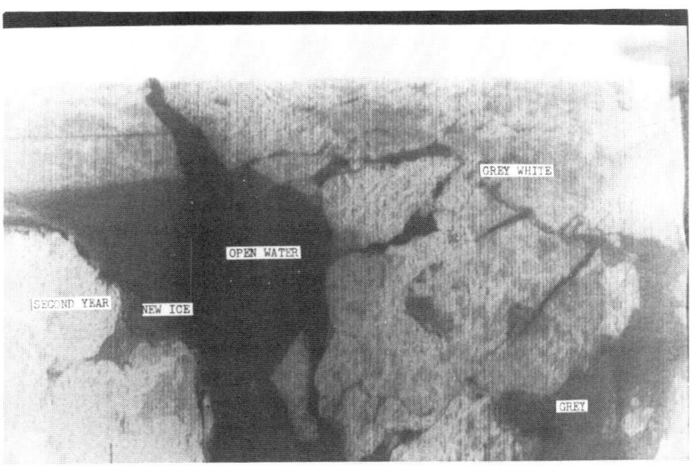

Figure 2. Copy of a real time infrared image recorded in Lancaster Sound, October 1972. Four ice types can be identified.

Continued exposure of the paper to strong light will, however, degrade the image. This type of record is, therefore, impractical for permanent archives. Other real time displays of the screen display type have also been evaluated, but duration of observation time was generally found to be insufficient for accurate and detailed analysis.

Although the infrared data is as yet not being incorporated into ice charts on a routine basis, they are presently the prime source of detailed ice information under darkness conditions available on the reconnaissance platform. With increasing requirements for day and night reconnaissance, the role of this information will become an essential part of the data acquisition program.

C. Side Looking Airborne Radar

Although the previous two sensors discussed play an important role in a day and night reconnaissance operation, neither has a cloud penetration capability. Yet an all weather collection capability is of significant strategic, as well as economic, importance to shipping and would greatly reduce the amount of unproductive reconnaissance, estimated as high as 30%, currently

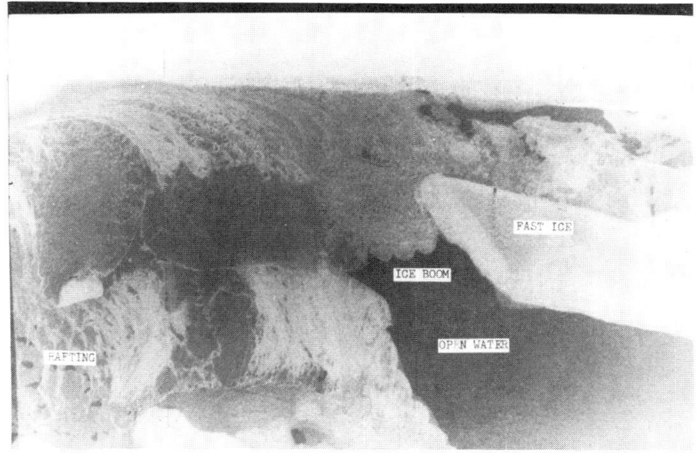

Figure 3. Real time infrared image recorded over mostly grey white ice on the St. Lawrence River in the vicinity of Prescott, Ontario, January, 1973.

flown over cloud covered ice. Considerable effort in investigating the sensor market for a solution to this problem has, therefore, been undertaken. Side Looking Airborne Radar (SLAR) equipped with a real time capability appears to provide the only realistic answer. These systems, having wide area coverage, high resolution and unique imaging characteristics, are effective under almost all weather and visibility conditions. Action for procurement of a SLAR system for each of the platforms is already underway, with current operational plans providing for acquisition and installation into the Electras by 1975. That is still more than one year away. Yet we must remain aware that this will not immediately solve all our data acquisition problems. Full utilization of SLAR imagery in a real time mode will require extensive periods of training, experimentation and development of interpretation techniques. It is, therefore, fortuitous that the Canadian Armed Forces, who have already equipped one of their military aircraft with a SLAR system, have agreed to cooperate in the advance collection of SLAR over ice covered waters to aid in preliminary development of applications and analysis procedures prior to actual SLAR implementation into the ice program. Several projects for SLAR data collection of ice in areas such as the St. Lawrence River (1969), the Gulf of St. Lawrence (1972) and the Canadian Arctic (1973), have already been undertaken.

The types of information required by a ship in ice congested waters include mesoscale and synoptic scale ice patterns and ice movements for selection of optimum route direction, larger scale information to identify features such as areas of minimum resistance versus areas of solid ice within an ice pack, and finally close in detail of ice in the immediate vicinity of the vessel for navigation around solid floes. SLAR imagery shows very useful applications in all of these areas.

Small scale, long range SLAR images (eg. 1:2,000,000) show excellent examples of zones of convergence or divergence of ice, shear lines, and general changes in characteristics from, for example, large, smooth floes to generally heavy broken ice. Identification and location of interconnecting areas of looser ice conditions or open water is possible over large areas. Ice edges can be located well beyond visual range to distances of 50 miles (80 km).

On the intermediate scale (eg. 1:500,000), various characteristics of ice within a particular ice field can be identified. Areas of fast ice can be recognized, although care must be taken not to confuse with open water. Total concentration of ice within the area can be estimated. Location of large and old floes are relatively easy, and lighter ice conditions among old floes can be outlined. Icebergs show prominent returns, with spot size varying with physical size of the bergs. At low altitudes, shadowing of bergs is apparent.

In the large scale (1:250,000) imagery, details of floes, such as ridging, relative roughness and characteristic tone, allow identification of predominant ice types. Detail of berg population is also enhanced.

A number of the above features are illustrated in the image shown in Figure 4. This image was recorded over Arctic

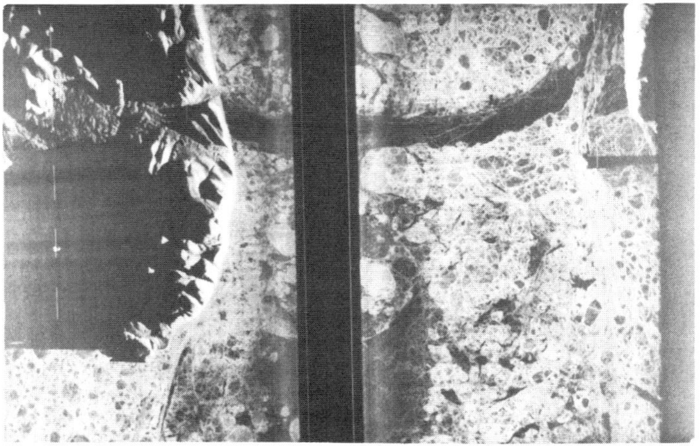

Figure 4. SLAR image recorded in 1:250,000 scale over ice in Nares Strait, January, 1973. Greenland lies on the right side of the image. Flight altitude was 2,000 feet (610 metres) above sea level.

ice in the winter of 1973. Note the darker appearance of the smoother first year ice patches in comparison with the much rougher multi-year ice. Ridging is prominently displayed both on the first year and multi-year ice. Shear lines are apparent on each side. A long shadow appears behind the island on the Greenland coast.

Identification of features on SLAR imagery is, however, frequently subjective and requires considerable experience to obtain acceptable credibility. Even then, certain ambiguities remain. For example, smooth fast ice may show the same low returns as open water. Possible clues such as ragged boundaries or fragments of ice often associated with open water must be used. On the other hand, heavily broken ice gives a strong return regardless of age. It is in these and similar cases that other sources of data -- visual, infrared, or laser -- become extremely useful as a key for interpretation of SLAR. Through cloud, the observer must rely entirely on his skill.

The information collected by SLAR must be made available to the ship as soon as possible. Real time displays (wet processing films) are already available and operational with some SLAR units. Dry processing displays are currently under development. Because of the relatively small scale of the imagery, and thus slow change in the area that can be observed for analysis at any one instance in flight, the observer has sufficient time to carefully interpret the features in the image. Pertinent information, such as open water areas, regions of low ice concentrations or young ice, outline and position of large old floes, location of icebergs, etc., can be traced onto a blank chart, the scale and two latitude/longitude positions can be indicated on the chart, and it can then be transmitted directly to any ship in the region by airborne facsimile to be used immediately for tactical navigation. This information can also be included on the regular scale ice charts for transmission to the central forecasting office at the end of the flight. Alternatively, communications facilities are available for transmission of the radar signal directly to a receiver/recorder on board marine vessels by data-link. This approach is more costly and would require receivers and interpreters on board each marine vessel requiring information.

D. Other Sensors

Each of the reconnaissance aircraft is also equipped with a tri-metrigon type camera array. The array consists of three model 1815 Vinten cameras, each equipped with automatic iris control. Although the cameras cannot provide real time information, they provide a source of hard copy "ground truth" data useful in remote sensing trials and for ice climatological applications.

There remains one sea ice parameter of which little has as yet been said -- quantitative sea ice thickness. Current remote sensing technology has, as yet, not come up with a practical solution to this data collection problem. Research in techniques using passive microwave systems is, however, showing considerable promise. The development of a fully operational sensor capable of providing reliable thickness measurements over ice of varying salinity and temperatures may be closer to realization than we think. Such a development would unquestionably be very valuable to our reconnaissance operation.

SUMMARY

In conclusion, the Canadian ice reconnaissance program has already developed a limited capability of night time data acquisition. By 1975, a complete sensor package capable of providing detailed ice information over ice congested waters regardless of weather or light conditions will be available. With the development of skills in interpretation and application of this data and the implementations of improved communication facilities for transmission of the data to the users, a vastly improved level of support to marine navigation within Canada's ice congested waters will become available.

APPLICATION OF THERMAL IMAGERY TO THE DEVELOPMENT OF A GREAT LAKES ICE INFORMATION SYSTEM

R.J. Schertler
C.A. Raquet
R.A. Svehla
LEWIS RESEARCH CENTER, NATIONAL AERONAUTICS AND SPACE ADMINISTRATION, CLEVELAND, OHIO 44135

ABSTRACT

Recent measurements and analysis have shown that thermal infrared imagery (wavelength, 8-14μm) can be employed to delineate the relative thicknesses of various regions of freshwater ice, as well as, differentiate new ice from both open water areas and thicker (young) ice. Thermal imagery was observed to be generally superior to visual (0.4 - 0.7μm) and our SLAR (3.3 cm) imagery for estimating relative ice thicknesses and delineating open water from new ice growth. In a real-time Great Lakes Ice Information System, thermal imagery can not only provide supplementary imagery but also aid in developing interpretative methods for all-weather SLAR imagery, as well as, establishing the areal extent of spot thickness measurements.

INTRODUCTION

Previous studies in the arctic by Ketchum, 1966 and Polin, 1966 have demonstrated the potential of such thermal imagery for determining ice distributions and relative ice thicknesses in adjacent areas. However, the geographic location and expanse of the Great Lakes Basin is such that storms, winds, currents, upwellings, and other hydrometeorological parameters impart important changes in the ice cover in a matter of one day. The Great Lakes ice situation is much different from the Arctic and therefore warrents new considerations in the application of thermal imagery systems for ice information.

Interpretative keys involving the size, shape, tone, and texture associated with the various ice types and ice surface topographic features are generally used in classifying ice types. These various ice types, in turn, are arranged into a hierarchy corresponding to ice age and development, which categories are then associated with ranges of ice thickness. This classification system, developed primarily for Arctic type ice identification, has many shortcomings when applied to the classification of the freshwater ice encountered in the Great Lakes. Notably, visual clues are confusing and sometimes totally lacking which would allow new ice, in an early stage of development, to be delineated from open water areas as well as from young ice which ranges between 10 and 30 cm thick. Slush ice layers which form especially as a result of the freezing of water saturated snow help to obliterate many of the visual clues associated with new and young ice formations. Frequently, surface topographic features and patterns associated with finger rafting and pressure ridging are obscured by a clutter of relict slush patterns, snow filled cracks, and wind streaks.

5.8

This paper will review some of the preliminary findings associated with the applicability of thermal imagery in providing ice information and discuss its overall role in a Great Lakes Ice Information System.

THERMAL SCANNER

A single channel Bendix airborne line scanner, mounted in the NASA Lewis C-47 aircraft, was used in these investigations. Radiation from the ground was scanned by a single surfaced, flat mirror (IFOV of 2.5 milliradians). The energy within a 120 degree FOV, centered beneath the aircraft, was collected and focused by reflective optics through an optical filter (8-14μm transmission) onto a photovoltaic detector cooled with liquid nitrogen.

An internal blackbody reference source allowed the imagery to be calibrated with respect to absolute temperatures. This system had a capability of detecting scene temperature differences as small as 0.1°C.

VISUAL - THERMAL IMAGERY COMPARISON

Some of these visual ice interpretative problems that were mentioned previously will now be illustrated using visual and thermal imagery. Figure 1 allows the information available from photographic imagery in the visible region of the spectrum to be contrasted with that obtained from the scanner in the thermal infrared region. These images encompass an area of approximately 7 square kilometers in Lake St. Clair. In the visual image, the various photographic tones are determined by variations in reflectivity, while the tones in the thermal image result from variations primarily in surface temperature. The radiation imaged by the thermal scanner is associated with the surface "skin" temperatures of the upper fraction of a millimeter of the ice and water surface due to the large absorption coefficients for both water and ice which range from 500 cm^{-1} at 8μm to over 1000 cm^{-1} at 14μm. The emissivity of water and ice is generally very high (>0.97) in the 8 - 14μm region and for angles of incidence less than approximately 50 degrees.

In the positive thermal image the lighter toned areas correspond to warmer temperatures while the darker toned areas indicate colder temperatures. Very light toned, open water areas are readily delineated from the darker tones of the newly frozen ice in the thermal image while such distinctions are very difficult, if not impossible, to discern in the visual image. Close examination of the visual image will reveal hairline type, white streaks in some of the seemingly open water areas which are characteristic of water and snow filled cracks in new ice formations. Generally such hairline crack features are difficult to detect because their very narrow width falls below the detectable resolution element in such wide angle, high altitude imagery. The relative temperature variations associated with the various tonal contrasts within the thermal image are indicated on the figure. The very dark toned, ribbon like regions in the thermal image, a characteristic of areas of colder, thicker ice, correlate with the light toned

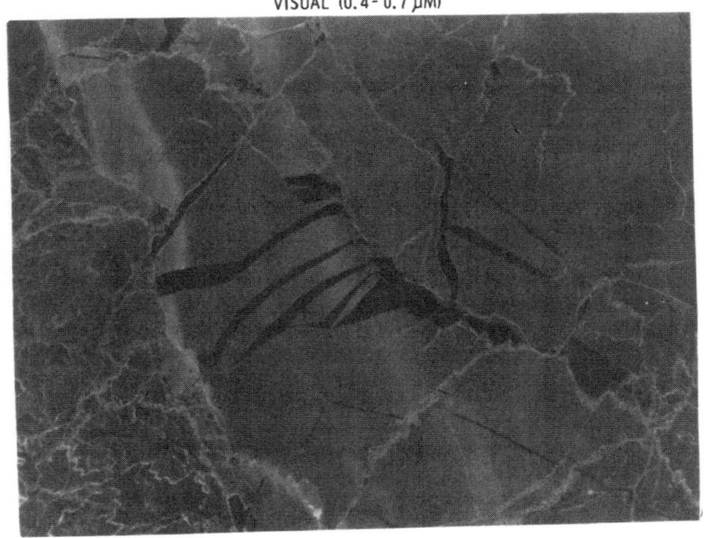

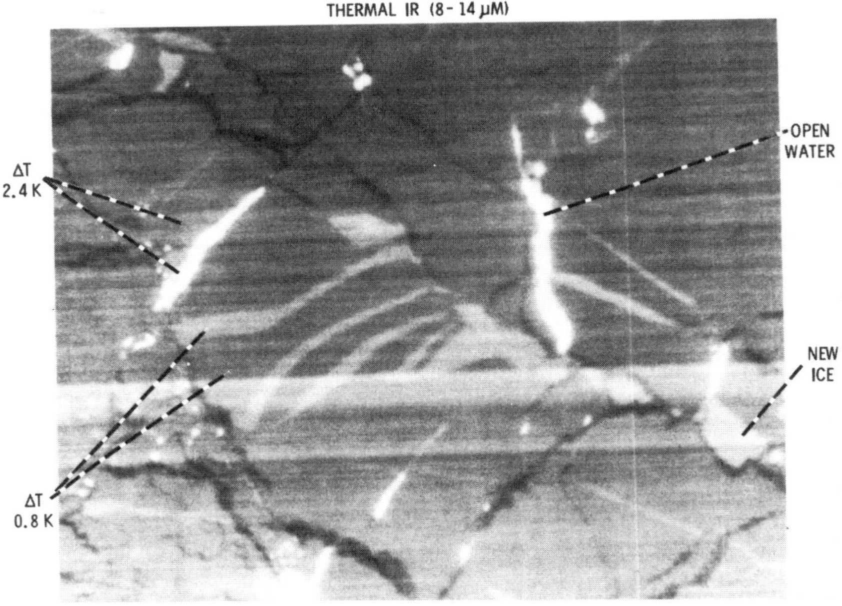

Figure 1. An example of ice conditions in Lake St. Clair as seen in the visible (upper) and thermal IR (lower).

regions in the visual image. These mound-like features often result when cracks and leads in the ice become saturated with snow and refreeze forming a thick interface between adjacent pieces of ice. The overall, uniform tonal shading of the thermal image indicates that the ice cover is at the same surface temperature and therefore is of generally the same relative thickness.

It is evident from these above comparisons that thermal imagery can provide a much stronger capability in delineating various regions of ice, especially those associated with new ice development, than visual imagery.

ICE THICKNESS

The surface temperature of freshwater ice, bounded by relatively warm and constant temperature water beneath and colder surface air temperatures above, reflects an equilibrium between the heat conducted from the water through the ice to the surface, the heat absorbed by the ice from the incident solar radiation, the heat radiated from the surface, and the heat convected from the surface by the surrounding air. Since ice is a fairly good insulator, the thicker the ice the less observable is the warming influence of the underlying water and the closer the upper surface temperature of the ice will approach the outside air temperature. To adequately model the thermal environment of the ice, parameters such as thermal conductivities, heat transfer coefficients, water and air temperature, local wind velocity, average surface, roughness, type of ice, amount of snow cover, incoming solar radiation, percentage of cloud cover, and other hydrometeorological factors must be considered. These complexities make it difficult to obtain thickness measurements on an absolute bases; however, relative temperature measurements and their corresponding correlation to ice thicknesses can be provided.

Using tonal contrasts associated with open water areas, finger rafting patterns, and shore ice formations in thermal imagery, a range of actual ice thicknesses, correlated with relative surface temperatures, may be established. Open water areas could establish the upper temperature, zero thickness, limit; while shore ice formations could provide a low temperature, thick ice, lower limit.

Thrust structure patterns, frequently observed in ice sheets on the Great Lakes and characteristic of an intermediate range of ice thicknesses, could become the key interpretative clues in relating surface temperature to ice thickness using thermal imagery.

Figure 2 is used to illustrate this technique. An example of the fairly well-defined pattern associated with a finger rafting type thrust structure in young ice is shown in this figure.

Finger rafting consists of an alternating series of parallel rectilinear overthrusts and similar shaped underthrusts, producing a square-wave type pattern. This pattern easily identified in the visual image is seen as a ribbon like dark toned, colder temperature feature in the thermal image. An enlargement of a thermal image of this same pattern, taken two days earlier and as a lower altitude is also shown in this figure. Newly formed thin ice will produce a jagged pattern with irregular and poorly defined edges due to the inability of the ice to transmit the

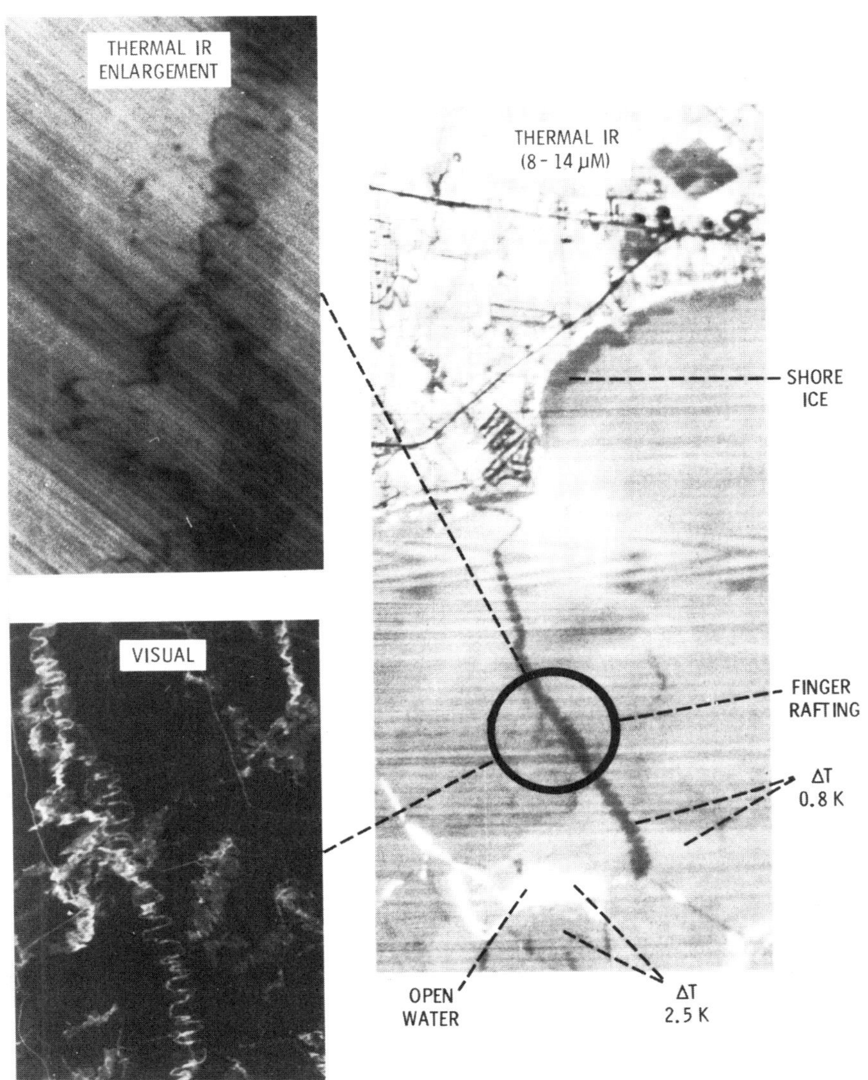

Figure 2. An example of a finger rafted thrust structure in the ice cover on Lake St. Clair.

structural forces. As the ice becomes thicker, such patterns are characterized by more clearly defined rectilinear edges. Well defined square wave patterns are associated with young ice approximately 15 - 18 cm thick. Ice thicker than approximately 20 cm will tend to form pressure ridges. Finger rafting patterns are also frequently detectable even under a snow cover because of the tendancy of water within the snowpack to soak into and delineate the edge patterns.

The rounded edges, shown in these images, are characteristic of young ice approximately 10- 12cm thick. The darker tones, colder temperatures, are associated with the over- and under-thrusted ice sheet which is twice as thick as the adjacent ice sheet. By measuring the relative surface temperatures of the shore ice rafted regions, and open water area and obtaining information regarding the ice thickness along the shore line it is possible to establish a relationship between relative surface temperature and ice thickness for this ice region and its adjacent area.

VISUAL - RADAR - THERMAL IMAGERY COMPARED

The distinctively different responses of ice in widely separated regions of the spectrum and the importance of thermal imagery for accurate interpretation are illustrated in Figure 3. Note that only in the thermal IR image is the open water area near the tip of Pelee Point clearly discriminated. The basic difference that sets the thermal IR apart from both the visual and the SLAR is that it responds to emitted rather than scattered radiation by the ice. In the case of the SLAR microwave radiation it is primarily the backscatter off the edges of the ice at ice-water corner interfaces which produces the image as described by Jirberg, et. al. (1973). Since the new unfractured ice along the west shore of Pelee Point does not contain sufficient scatters, it is not differentiated from the adjacent open water area. Similarly in the visible, the lack of defects in the new ice sheet that scatter the solar illumination precludes it being separated from the dark open water area.

Further comparison of thermal IR to visual, in this case ERTS-1 satellite imagery, is made in Figure 4.

At the time of the satellite pass, the area was under a high (above 6 Km) cirrus cloud layer. In an enlarged portion the ERTS imagery near the southern shore of Lake Erie off Lorain, Ohio, a rather large distinctive light tone feature is observed with a dark toned background. The thermal image, as well as, simutaneous acquired oblique angle visual imagery, reveals that this entire area is covered with a layer of skim ice which has been broken up by wind and wave action. The distinctive pattern highlighted in every channel of the visual and near infrared imagery, is not discernable in the thermal imagery, apparently possessing no distinctive thermal contrast with the adjacent areas. The mottled light gray pattern of the thermal imagery evidences the honeycombed pattern of the wind broken, skim ice. Fast ice development along the shore can also be distinguished in both ERTS and thermal imagery. In another area of the ERTS image, near

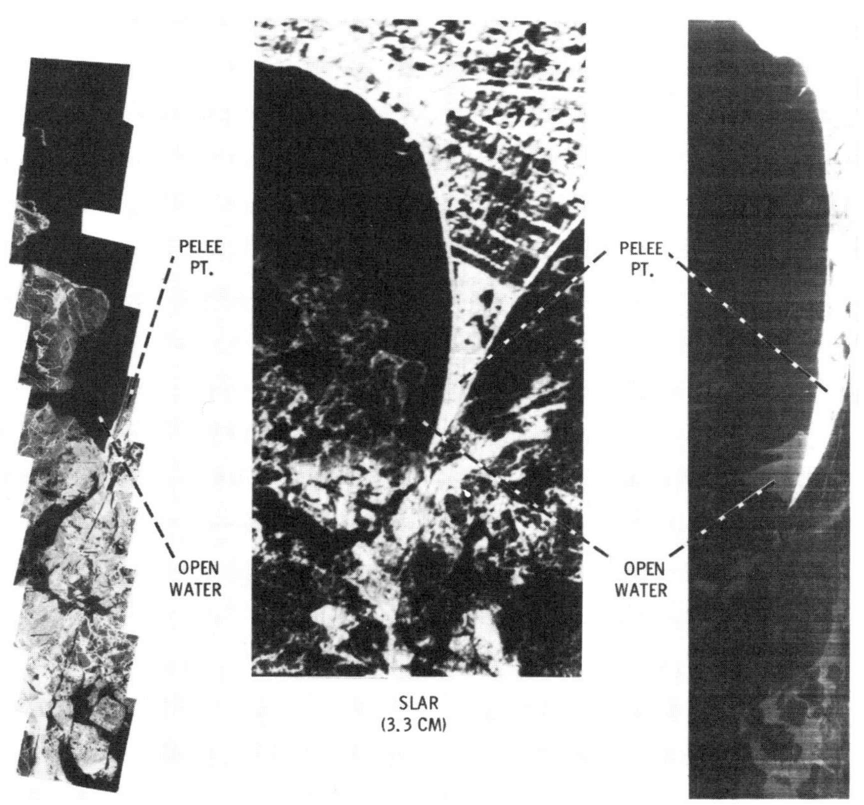

Figure 3. Thermal IR, radar, and visual imagery of ice conditions in Lake Erie near Pelee Point.

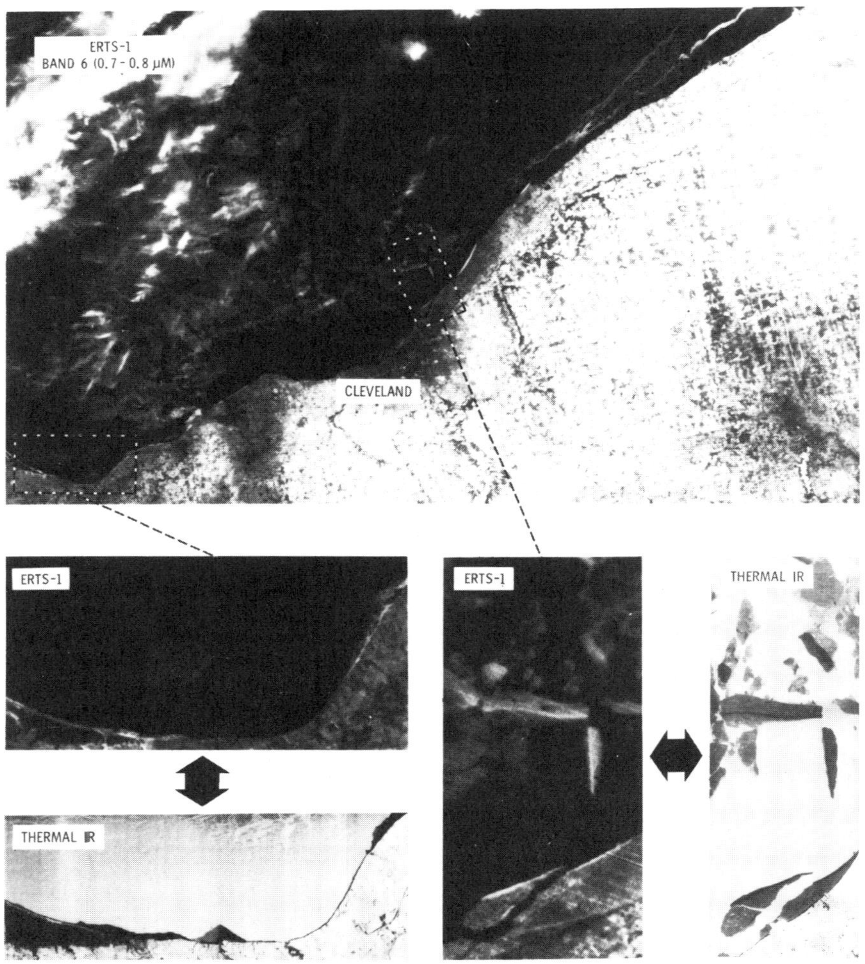

Figure 4. Thermal IR and ERTS-1 Satellite Imagery of Ice conditions in Lake Erie on January 13, 1973.

Cleveland, tonal features in the broken ice floes correlate in both the visual and thermal spectral regions. The light tones of the satellite imagery corresponds to the dark tones, colder temperature of the thermal imagery indicating relatively thicker ice.

GROUND-TRUTH DATE SUPPORT

Thermal imagery obtained in conjunction with either ground truth thickness measurements or with data from a non-imaging, pulsed radar ice thickness system being developed at Lewis (Vickers, et.al.,1973) can aid in determining the areal extent of these spot thickness measurements to the surrounding ice sheet by delineating regions of relative uniform thickness adjacent to such spot measurement locations. An analyses of thermal imagery in advance of planning ground truth measurements will aid in determining locations where ground truth will be most valuable in supplying a comprehensive sampling of the present stage of ice development.

CONCLUDING REMARKS

Preliminary analysis has shown that tone contrasts in thermal infrared imagery (8-14μm) allows various regions of similar ice thickness to be established thereby permitting new ice to be readily delineated from open water areas and thicker young ice found within the Great Lakes. The complexities involved in establishing a realistic thermal radiation model for the ice surface may preclude the use of thermal imagery for establishing absolute ice thicknesses. However, by making use of the tonal characteristics of open water areas, shore ice formations, and surface topographic features such as finger rafting, a range of ice thicknesses can be established for adjacent areas from thermal imagery. Future experience in interpreting such imagery should allow even more definitive characteristic to be established for various ice features and types which in turn may provide additional clues for determining ice thicknesses. Thermal imagery was generally superior to both visual and SLAR imagery for estimating relative ice thicknesses and differentiating regions of new ice in open water areas. Thermal imagery with its rather limited areal coverage and dependence on cloud-free conditions will play only a secondary role in a total Ice Information System for the Great Lakes. By providing interpretative keys for SLAR imagery evaluation and delineating the areal extent of pulsed microwave spot thickness measurements, thermal imagery will occupy a vital supplementary role in ice reconnaissance.

REFERENCES

JIRBERG, R.J. et al., 1973. Application of SLAR For Monitoring/ Great Lakes Ice Conditions, paper 4.4, Interdisciplinary Symposium on Advanced concepts and techniques in the study of snow and ice resources, Monterey, California, Dec. 1973.

REFERENCES (continued)

KETCHUM, R. D., and WITTMAN, W. I., 1966. Infrared Scanning Of Arctic Park Ice Proc. Fourth Symposium Remote Sensing of the Environment, Ann Arbor, Michigan, pg. 635-656.

POLIN, A.O., 1966. Infrared Imagery In Arctic Under Daylight Conditions. Proc. Fourth Symposium Remote Sensing of the Environment, Ann Arbor, Michigan, pg. 231-241.

VICKER, R. S., et al., 1973. Airborne Profiling of Ice Thickness Using A Short Pulse Radar, paper 4.7, Interdisciplinary Symposium on Advanced Concepts and Techniques in the Study of Snow and Ice Resources, Monterey, California, Dec. 1973.

BREAK-UP CHARACTERISTICS OF THE CHENA RIVER WATERSHED, CENTRAL ALASKA

Gerd Wendler
Robert Carlson
Doug Kane
University of Alaska

ABSTRACT

The snow melt for a small watershed (5130 km^2) in Central Alaska was successfully monitored with ERTS imagery. Aerial photography was used as supporting data for periods without satellite coverage. Comparison both with actual measurements and with a computer model showed good agreement.

INTRODUCTION

Of the many water resource problems faced by the State of Alaska, floods are the most widespread; the damage is large in small population centers, and statewide the floods cause annually a great amount of damage. The art of the prediction of floods, either on a day-to-day operational basis or as a part of a long term study, has advanced to a rather high state of knowledge. There are available a wide number of computer oriented, graphical, analytical, and empirical methods which are used successfully with experience.

However, regardless of the degree of knowledge, the engineer or forecaster is limited by the completeness of the measured input. In the case of rainfall, the provision of adequate input data, although not always easy, is quite straightforward. Snowmelt, on the other hand, provides an input, to the runoff system which is, in itself, the output of a rather complex process as the snowmelt depends on a variety of factors; some of a permanent physiographic nature and others of a more transient environmental nature, (Wendler, 1967). Because of the complexity of the process, snowmelt measurement on a pointwise basis and extrapolation of point measurements to even a relatively small basin become very difficult. This is especially true of the Chena River in particular and Alaska in general, where precipitation gauge densities are extremely small (Searby, 1968).

Because of Alaska's rather severe cold climate the surface hydrology in most areas of the state is only active during four to six months. As a result, the spring snow and ice break-up dominate the cycle and the resulting snowmelt runoff contributes most of the stream flow and causes many of the floods.

Major floods have occurred in the Fairbanks area in 1930, 1937, 1948, 1960, and 1967.

To overcome the shortcomings of conventional point measurements (Snow Surveys, 1973), satellite imagery has been applied (Barnes and Bowley, 1968a,b, 1969; McClain and Baker, 1969). However, up till now the resolution of the satellite imagery was insufficient for small watersheds. The launch of ERTS 1, having a resolution of 100m together with supporting aerial photography has made it possible to monitor the snow melt for much smaller basins such as that of the Chena River (5130 km^2).

DESCRIPTION OF THE CHENA RIVER BASIN

The Chena River Basin lies in the interior of Alaska, about 100 miles (160 km) south of the Arctic Circle. The climate is continental sub-polar, with cold winters and relatively warm summers; the precipitation is relatively low with a long-term annual mean of 11.3 inches (287 mm) (Searby, 1968). The elevation of the basin ranges from very small areas in excess of 5000 ft (1525 m) to 440 ft (134 m), at the confluence of the Chena with the Tanana River (Figure 1). The area is hilly, mostly lying below 2500 ft (760 m) and less

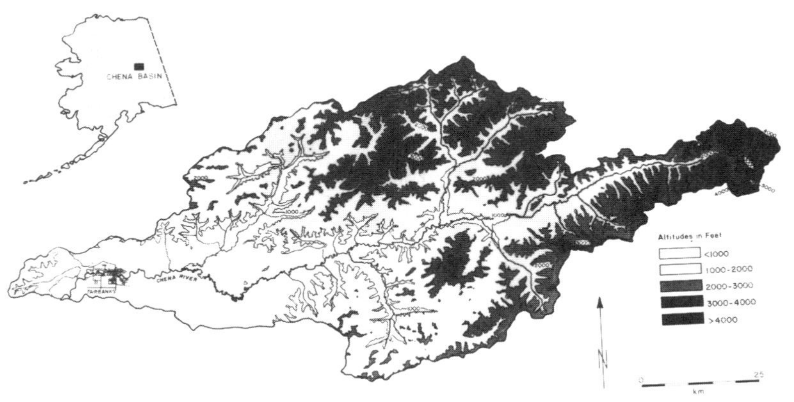

Figure 1. Topographic map with Chena River Basin Boundary outlined.

than 20% has an elevation above this level (Figure 2). The basin has a size of 1980 sq. miles (5130 km^2). Major tributaries of the Chena are the Little Chena River, West, East, and South Forks and Munson Creek. The precipitation increases with altitude, as can be seen from the snow course data (Snow Survey, 1970).

METEOROLOGICAL CONDITIONS DURING BREAK-UP

The climatological data for Fairbanks, the lowest point in the basin, are shown in Figure 3 (NOAA, 1973). On 28 March 1973 the average daily temperature rose above the freezing point for the first time in the season. However, positive temperatures of any longer duration were observed only after 6 April 1973. During 16-27 April a cold spell occurred, with temperatures around the freezing point. After this period the temperature remained above freezing, and most of the melting occurred. The dew point remained below the saturation point for water until 7 May 1973, thus indicating that evap-

CHENA RIVER DRAINAGE BASIN

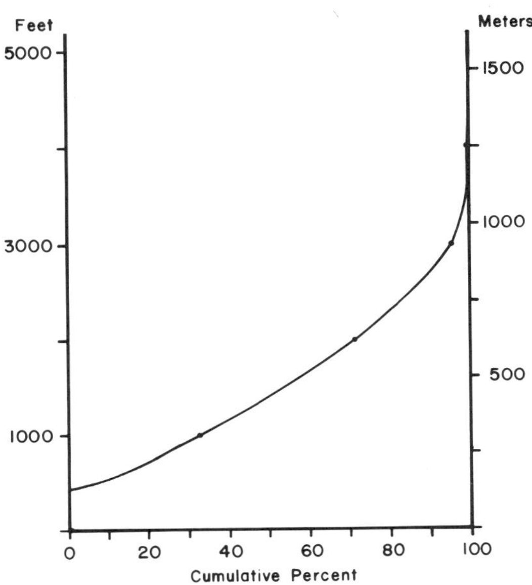

Figure 2. Elevation distribution for the Chena River Basin.

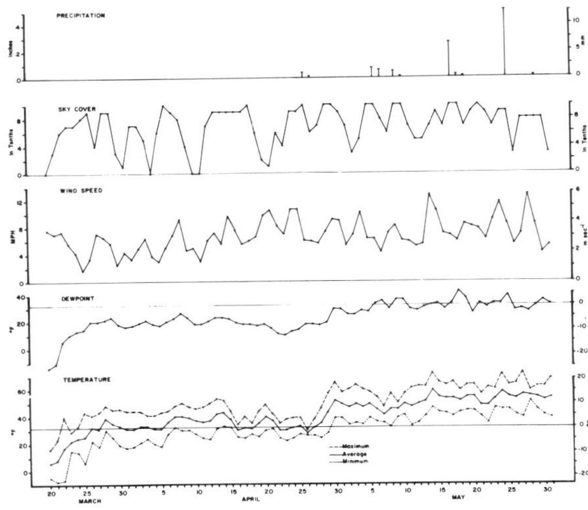

Figure 3. Climatological conditions at Fairbanks (temperature, dew point, wind speed, cloud cover and precipitation) during break-up (20 March to 31 May 1973) on the Chena River.

oration occurred, which is typical for the early part of the break-up (Wendler, 1967). Thereafter, values around the saturation value for melting snow were observed, showing that the condensation balanced the evaporation. The wind speeds were relatively low (around 3 m sec^{-1}), which is typical for the interior of Alaska (Searby, 1968). The cloudiness with a mean of 6/10 was average for the spring period. No appreciable precipitation was observed during the break-up until the latter part of May 1973.

SNOW MELT MODEL

Because of the difficulties of comparing satellite imagery with spot measurements of the snow cover, a snow melt model which had been developed by Carlson and Norton (1973) was used. This model is based on calculation of energy fluxes at the surface and within the snowpack. Cloudiness is used to estimate the radiative fluxes, temperature and wind speed to estimate the sensible heat flux, and dew point and wind speed to calculate the latent heat flux. Also the process of melting at the surface and refreezing at deeper layers in the snow pack is considered. This model is tested against actual measurements for the spring of 1973 (Figure 4). It can be seen that

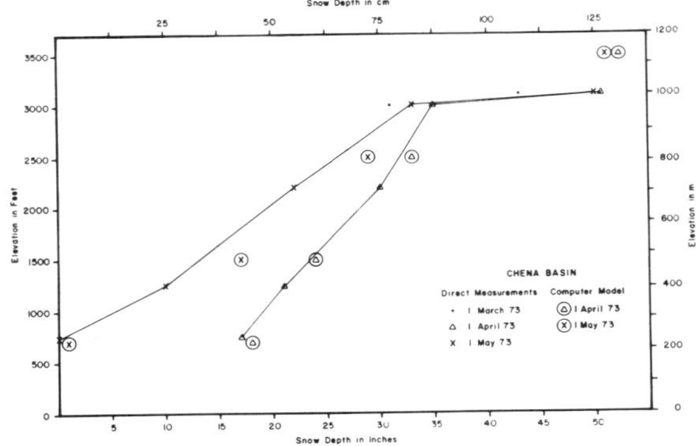

Figure 4. Snow depth - actual measurements and model calculations in relation to altitude for particular times during break-up on the Chena Basin, 1973.

the agreement is very good, especially if one considers the variability of the snow pack for point measurements. Hence, data obtained from this model might be compared directly with the snow cover extent of the basin.

ERTS IMAGERY AND AERIAL PHOTOGRAPHY

Cloudfree ERTS imagery during break-up was obtained on 12 and 30 April and 2 May 1973. Later in May, it was cloudy at the time the satellite passed. Therefore, two aerial flights (height about 12000 ft (3660 m), stereo photography) were carried out on 11 May and 20 May over part of the basin. Altitude lines were transferred from topographic maps onto the photographs with a zoom transfer scope (Bausch and Lomb), and areas were obtained with a planimeter. Patches of snow were found down to the lowest lying areas on north slopes on 11 May, while no snow was found on a slope with southerly exposure below 1250 ft (380 m) (Figure 5). The percentage of snow-covered area

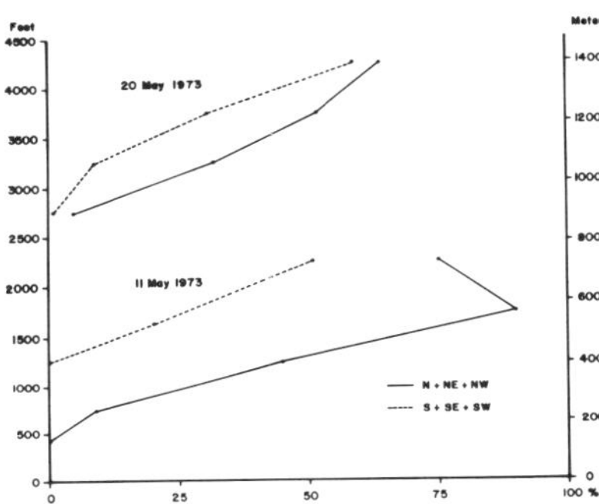

Figure 5. The amount of snow cover in percent as function of altitude for northerly and southerly exposure on two occasions during break-up.

increased with height. A similar percentage cover was attained at heights some 750 ft (230 m) higher on the south slopes than on the north slopes on 11 May and 400 ft higher (120 m) on 20 May. The decrease in snow cover for the greatest altitude range on 11 May 1973 is at first sight astonishing. However, when the photographs are analysed, it can be seen, that it is real. The area photographed did not exceed 2500 ft (780 m), and hence this area represented the top of the hills. As no forest is present on these hill tops and they are thus more freely exposed to the wind, some of the areas were snow free even before the melting started. An example is given in Figure 6. Further, it can be seen from Figure 5, that the snow line ascended about 2000 ft (610 m) in the 9 days from 11 to 20 May 1973. This rapid ascent of the snow line was caused by the high temperature (see Figure 3) during this period.

The Chena Basin was outlined on the ERTS imagery for the three dates, at which pictures were obtained. It was found, that channel 5 (0.5 - 0.6 μm) was the best suited to distinguish the snow cover from snow free areas. An example is given in Figure 7, showing the Chena Basin at two times during break-up. It can be seen, that at the later data, most of the snow in the lower lying areas (compare with Figure 1) has melted. A VP-8 analyser was used to obtain quantitative results. This instrument displays a grey scale in color, which makes it easier for the eye to distinguish different levels. Also, an automatic built in planimeter makes it possible to rapidly measure areas with different snow coverage. Such a task using a hand planimeter is a very time consuming task. Using this instrument the

Figure 6. Aerial photography of an area north of Fairbanks. (Note that the southerly and westerly slopes of Haystack Mountain (2525 ft or 770 m) are snow free, but at altitudes below 2000 ft (610 m) snow is observed.

following values were obtained:

Percent of Snow Cover Determined from ERTS Imagery
Chena River Basin
(1980 sq. mi.)

	<50% snowfree or snow patches	>50% broken or continuous snow	
12 April 1973	18	82	%
30 April 1973	28	72	%
2 May 1973	46	54	%

COMPARISON OF THE STAGE OF BREAK-UP

The areal percentage of the basin, which was snow free, could be obtained from both the ERTS imagery and aerial photography. These values are compared with data from the snow melt model, from which the melt could be calculated for various altitude intervals, and by knowing the altitude distribution (Figure 2), the percentage of the area, which was snow free, could be calculated. One can see (Figure 8), that the agreement is very good. To appreciate such an agreement, one should notice, that a snow melt model never reflects nature in every detail, and further, that measurements of the snow cover by satellite or aircraft are hindered by vegetation and the shadowing effects of mountains.

In addition, the amount of discharge, measured in Fairbanks, is shown in Figure 8. This curve is not totally comparable with the other ones, as there is a lag between the time of melting, and the time at which this water arrives in Fairbanks. A lag time of about a week was found, which is probably somewhat longer than might be expected.

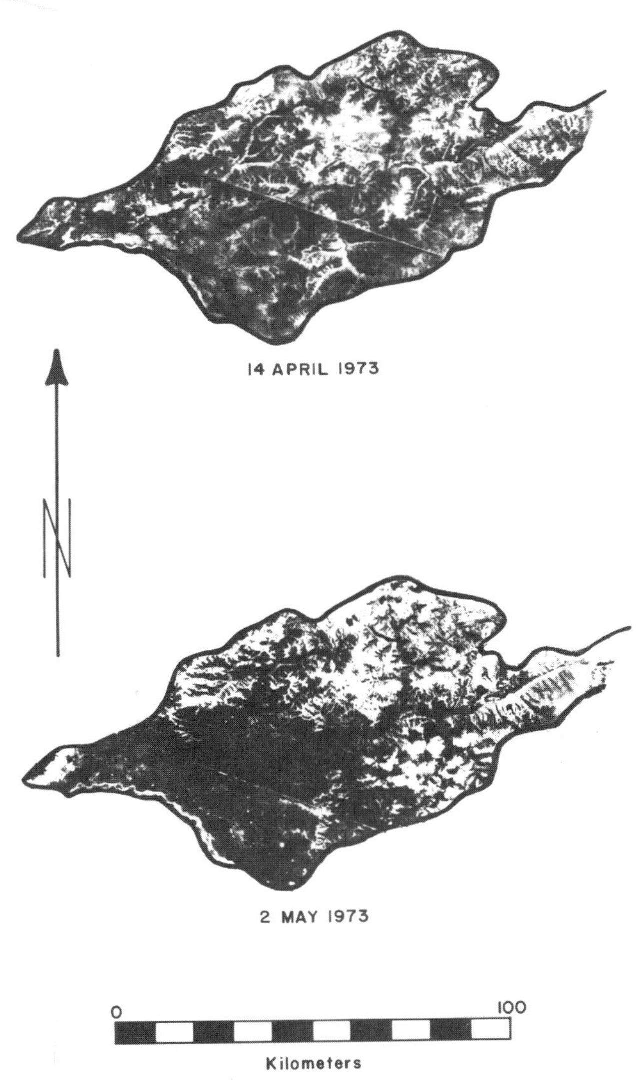

Figure 7. Imagery of the Chena River Basin as seen from ERTS.
7a. 14 April 1973 (Note that most of the area is snow covered.)
7b. 2 May 1973 (Note that the lower lying areas are snow free.)

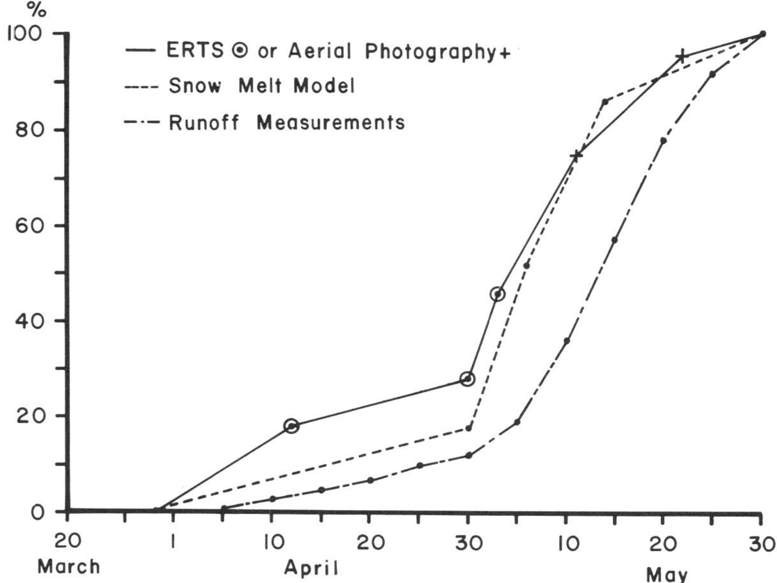

Figure 8. Variation in snow free area with time for the Chena Basin during the break-up of 1973. Values obtained from ERTS and aerial photography, the snow melt model, and direct run off measurements.

CONCLUSION

It has been shown that ERTS imagery can successfully be used to monitor the snow melt in a relatively small watershed. Comparison with actual measurements as well as with a computer model showed good agreement. These same techniques can therefore be used to study the snow melt behavior in watersheds where little or no hydrological or climatological data are presently being collected. Such information will make it possible to draw preliminary conclusions about the most dynamic portion of the hydrologic cycle in the Arctic, namely the spring break-up.

ACKNOWLEDGEMENT

This work was supported by NASA, Grant NAS5-21833. We are thankful to J. McDougall who directed the computer model and to K. Ahlnas, J. Leap and T. McClung, who helped in reducing the data.

REFERENCES

Barnes, J.C., and C.J. Bowley, 1968a. Snow Cover Distribution as Mapped from Satellite Photography. Water Resources Research, 4:2, p. 257.

Barnes, J.C., and C.J. Bowley, 1968b. Operational Guide for Mapping Snow Cover from Satellite Photography. ESSA Contract No. E-162-67(N), Allied Research Associates, Concord Mass., May, 116pp.

Barnes, J.C., and C.J. Bowley, 1969. Satellite Surveillance of Mountain Snow in the Western United States. Final Report, ESSA Contract No. E-196-68, Allied Research Associates, Concord Mass., June, 78pp.

Carlson, R., and Norton, W., 1973. Modeling Snowmelt Runoff in an Arctic Coastal Plain. in press.

McClain, E.P., and D.R. Baker, 1969. Experimental Large-Scale Snow and Ice Mapping with Satellite Composite Minimum Brightness Charts. ESSA Technical Memorandum NESCTM-12, Dept. of Commerce, Washington D.C., 16pp.

NOAA, 1973. Fairbanks Climatological Data. U.S. Department of Commerce, March - May.

Searby, H.W., 1968. Climates of the States, Alaska. ESSA, Climatology of the United States, No. 60-49, U.S. Weather Bureau, 23pp.

Snow Survey, 1970. Summary of Snow Survey Measurements for Alaska. 1951-1970, Federal-State-Private, Cooperative Snow Surveys, U.S. Department of Agriculture, Soil Conservation Service, and Alaska Soil Conservation District, p. 1-101.

Snow Survey, 1973. Cooperative Snow Surveys for Alaska. Federal-State-Private, U.S. Department of Agriculture, Soil Conservation Service, and Alaska Soil Conservation District, Feb. - May.

Wendler, G., 1967. The Heat Balance at the Snow Surface during the Melting Period (March-April 1966) near Fairbanks, Alaska. Gerlands Beiträge zur Geophysik, 76, Heft. 6, p. 453-460.

TECHNIQUES FOR DETERMINING AREAL EXTENT OF SNOW IN THE SIERRA NEVADA MOUNTAINS USING HIGH ALTITUDE AIRCRAFT AND SPACECRAFT IMAGERY

Donald T. Lauer
William C. Draeger
Center for Remote Sensing Research
University of California, Berkeley, California

ABSTRACT

Sequential aerial photography, Earth Resources Technology Satellite (ERTS-1) imagery and ground data were used to: (1) develop an image interpretation key, to be used as an aid in interpreting areal extent of snow in forested areas, (2) develop an efficient manual photo interpretation technique for estimating acreages of snow cover, and (3) test the level of accuracy of the interpretation technique as applied to a 1.5 million acre (600,000 ha) area within the Feather River watershed in northern California. The results showed that the interpretation key and the interpretation technique, when applied in a test situation, resulted in an estimate of snow cover with an accuracy of 96.3 percent.

INTRODUCTION

It has been recognized by many investigators that areal extent of snow could be a valuable parameter for improving the effectiveness of existing stream flow forecasting models (Baker, 1971; Barnes, 1969; Barnes and Bowley, 1969; Leaf and Haeffner, 1971; Tarble and Burnash, 1971; U.S. Army Corps of Engineers, 1956). However, this parameter has rarely been used because it has been nearly impossible to economically acquire such data for large areas at the accuracy required. However, it appears that Earth Resources Technology Satellite (ERTS-1) imagery (which is obtained over any particular location every 18 days, weather permitting), used in conjunction with the appropriate supporting aerial photography can provide a means for obtaining accurate and efficient estimates of areal extent of snow cover during critical periods of the melt season.

Consequently, a study was carried out which resulted in the following: (1) sequential ERTS-1 imagery, U-2 high flight photography and ground data were used to develop a suitable reference document, in the form of an image interpretation key, which could effectively be used as training material for the determination of areal extent of snow in forested areas; (2) an efficient manual analysis technique was developed for estimating acreages of snow. This technique capitalizes on the human's ability to amalgamate information on the appearance of snow as seen on satellite photos and the type, density and distribution of the vegetation/terrain within which the snow occurs and which greatly influences the appearance of snow; and (3) within the Feather River watershed in northern California an estimate of areal extent of snow was made over a 1.5 million acre (600,000 ha) area using ERTS-1 imagery. The level of accuracy of the estimate was verified through the use of sample vertical aerial photos, taken from a light aircraft, showing in detail the actual snow cover conditions.

PROCEDURES AND RESULTS

During the 1970-71 melt season, small scale (1:100,000) 70 mm black-and-white photographs of a 170,000 acre (68,000 ha) test area within the Feather River watershed were taken on five different occasions. During the 1971-72 season, the NASA U-2 aircraft stationed at the Ames Research Center, Mountain View, California flew four missions, each of which covered the entire Feather River watershed. In addition, during the 1972-1973 melt season, ERTS-1 imagery was acquired over the Feather River watershed every 18 days with clear weather prevailing on April 4, April 22, May 10 and May 28. Consequently, for three successive melt seasons, a sufficient amount of imagery was available for study.

To correlate interpretations made of the appearance of snow as seen on the small scale aerial or space imagery with ground conditions, it was necessary to collect data on true snow cover conditions. During the 1971-72 season, on three occasions coincident with U-2 overflights, a field crew visited a series of plots established within the Spanish Creek watershed. The field crews traveled by snowmobile and on snowshoes to each plot location. Data on extent and condition of snow, including depth, were collected and recorded, consistent with California Cooperative Snow Survey procedures.

Data collection on the ground, however, was very tedious and time-consuming. Furthermore, only a few samples could be collected before so much time had elapsed that further sampling became meaningless with respect to a specific high flight mission. Consequently, the primary method employed for estimating true snow cover conditions during the 1973 season was through the use of relatively large scale vertical color aerial photographs (see Figure 1). Coincident with each ERTS-1 overpass, as many as many as 150 35 mm photos were taken from a single-engine aircraft along preselected flight lines which transected the entire watershed. These photos proved to be of excellent quality, and the presence or absence of snow on the ground was easily interpreted on them, even in areas where a dense forest canopy obscured much of the snow. Supporting aircraft flights were flown in 1972 on January 10, January 31, March 28, and May 2 and in 1973 on April 6, May 11, May 29, and June 4.

Development of an Image Interpretation Key

A comprehensive image interpretation training and reference key was prepared using 1971-72 melt season U-2 photography and ground data. The key is for evaluating snow pack conditions as seen on small scale synoptic view imagery. The primary value of this key is that it documents, in the form of word descriptions and photo illustrations, the appearance of snow when influenced by a variety of vegetation/terrain conditions. A selective type of key was prepared based on eight vegetation/terrain categories -- dense conifer forest, sparse conifer forest, dry site hardwood forest, brushland, meadow or rangeland, urban land, water or ice, and rock or bare ground. Within each category, examples of different conditions of elevation, steepness of slope and direction of slope (aspect) were chosen. For example, within dense conifer forests, snowpack conditions are described for (1) high elevation and steep north slopes,

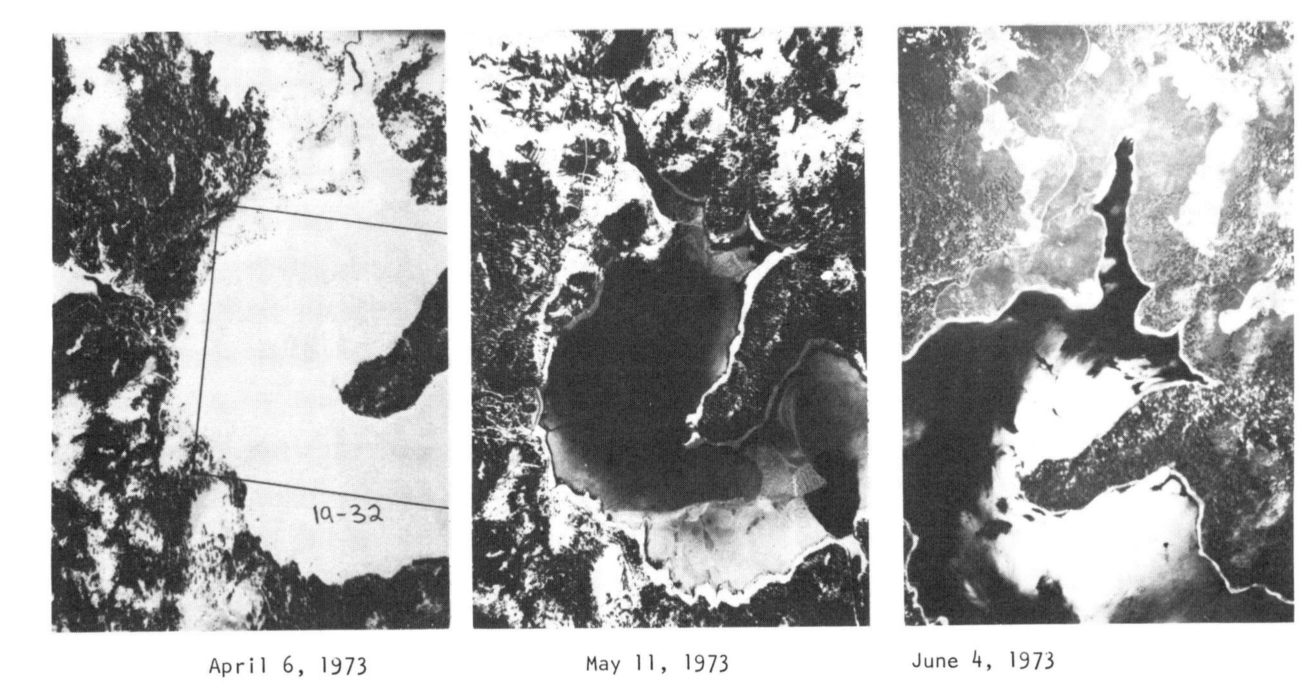

Figure 1. These photos were taken over Bucks Lake, California from a single engine aircraft flying at 17,000 feet (5,180 m) above sea level, using a 35 mm camera with a 24 mm lens. The original photos (in color) were used to make accurate estimates of actual snow conditions.

(2) high elevation and moderate north slopes, (3) high elevation and gentle south slopes, (4) medium elevation and flat areas, and (5) low elevation and steep north slopes. Each one of these specific examples represents an entire page in the key and consists of five illustrations -- one before the snow season, two at the height of snow accumulation and two during the depletion period. In the completed version of the key the eight vegetation/terrain type categories are illustrated with a total of twenty-two specific examples.

Possibly the most important value of the image interpretation key is that it allows the interpreter to become cognizant of the fact that for any given area, snow may be present on the ground but may not be visible on U-2 or satellite imagery. This situation often occurs within a dense coniferous forest in which a deep snowpack can be completely obscured by the crown canopy. It has been emphasized in the key, however, that the presence of snow in a dense forest usually can be deduced by examining the appearance of adjacent vegetation/terrain types. For example, if a dense stand of timber is surrounded by meadows or brushlands, and snow can be detected within these adjacent types, the interpreter can safely predict that the dense, heavily shaded timbered area also contains snow.

Development of an Image Interpretation Technique

It was found in developing the interpretation key that four environmental factors greatly influence the appearance of snow as seen on aerial photographs, viz., elevation, slope, aspect and vegetation/terrain type. For example, in certain areas a snow boundary may appear to follow a line of equal elevation but will drop down in elevation considerably on north facing slopes. In addition, it was found that the presence or absence of snow was easily detectable in meadows and bare areas, sometimes, but not always, detectable in sparse coniferous forest, and nearly impossible to detect in dense coniferous forest. However, once an interpreter is properly trained to recognize various combinations of environmental conditions, and is aware of the relationships among these various conditions and the appearance of snow associated with them, it was hypothesized that he could effectively detect the presence or absence of snow at any location, even on very small scale imagery such as that obtained by ERTS-1.

Consequently, an interpretation technique was developed which allows a trained interpreter to accurately estimate areal extent of snow, is suitable for the interpretation of very large, complex forested regions, and is fast and inexpensive to implement. This technique capitalizes on the ability of the interpreter to amalgamate several kinds of information and quickly arrive at a decision. By examining two images simultaneously, one taken during a snow free period and the other during the melt season, an interpreter can concentrate on both the appearance of snow, and on the vegetation/terrain condition. Thus reliable estimates of areal extent of snow can be derived.

The procedure is simple and consists of the following steps:

1. Using a standard mirror stereoscope, the interpreter observes two images simultaneously -- one taken during the summer season showing vegetation density and terrain conditions, and one taken during the snow melt season showing snow conditions.

2. Through minor adjustments to the stereoscope and image viewing height, the two images are brought into common register, i.e., the images as seen by the interpreter appear to be superimposed on one another.

3. By winking the right or left eye, the interpreter can quickly examine the same area both with and without snow cover.

4. A grid of known dimensions is placed over the winter image and the total percent area within each grid cell that is covered by snow is estimated.

5. Each cell is coded as follows:

Code	Snow Cover Class
1	No snow present
2	0-20 percent of ground covered by snow
3	20-50 percent of ground covered by snow
4	50-98 percent of ground covered by snow
5	98-100 percent of ground covered by snow

6. Areal extent of snow is calculated for any given watershed by calculating the total acreage in each cover class (number of cells in each class x average number of acres per cell), and then multiplying this gross acreage value in each class by the percentage midpoint of the respective class.

Testing the Interpretation Technique

The interpretation technique described above and the image interpretation key were used in an interpretation test applied to the west half of the Feather River watershed. Two color composite ERTS-1 images, one taken on August 31, 1972 and the other on April 4, 1973 (See Figure 2), were used, with a grid size of approximately 980 acres (392 ha) per cell. In addition, vertical 35 mm color aerial photos taken on April 6, 1973 were used to help train the interpreter (using 24 cells) and to evaluate the accuracy of the interpretation results (using 52 cells).

The interpreter spent 3 hours training himself to interpret the ERTS-1 data and in 6 hours classified the 1,592 cells comprising the test area. By applying the appropriate weighting factors (i.e., using midpoints of the snow cover classes), the areal extent of snow for the 1,560,160 acre (624,064 ha) area was estimated to be 899,670 acres (359,868 ha). This estimate could have been easily broken down by any sub-units of the large test area that were desired.

To evaluate the accuracy associated with this estimate, the interpretation results for 52 of the cells were compared with estimates made on relatively large scale (1:45,000) color aerial photos. These large scale estimates were felt to be very accurate, and truly indicative of the actual ground condition. Figure 3 shows the results of this comparison. Note that only 5 cells out of 52 were incorrectly classified, resulting in an overall "correctness" of 90.4 percent, based on number of cells. Furthermore, 4 of the 5 cells were misclassified by only one class. In addition, when the areal extent of snow was calculated on an acreage basis, the image interpretation estimate of snow cover made using ERTS-1 imagery was

August 31, 1973

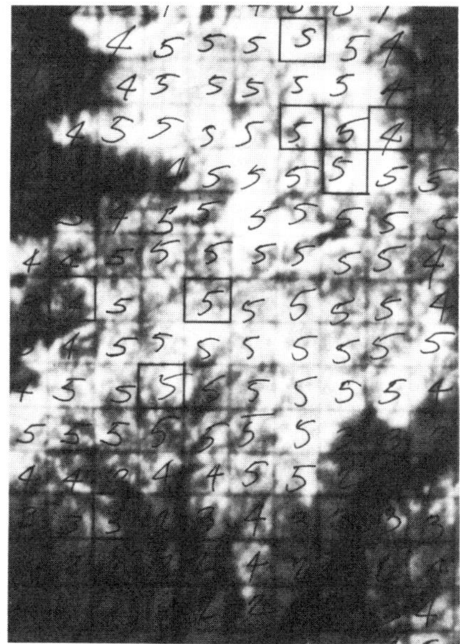

April 4, 1973

Figure 2. The photos above illustrate the two types of ERTS-1 imagery used to estimate areal extent of snow. The interpreter views both the summer and winter image simultaneously and then makes an estimate of percent snow cover in each cell on the grid. The interpreter is able to amalgamate information seen on the summer image relating to vegetation density and terrain type with information seen on the winter imge pertaining to snow cover. Bucks Lake, California appears near the center of each of these two images, and also appears in each of the photos in Figure 1.

		LARGE SCALE PHOTO DATA							
		Snow Cover Classes					Total # Cells	Total Acreage	Total Acreage of Snow
		1	2	3	4	5			
ERTS IMAGE INTERPRETATION — Snow Cover Classes	1	6	1				7	6,860	0
	2		9		1		10	9,800	980
	3			4	1		5	4,900	1,715
	4				10	1	11	10,780	7,977
	5				1	18	19	18,620	18,434
	Total # Cells	6	10	4	13	19	52	50,960	29,106
	Total Acreage	5,880	9,800	3,920	12,740	18,620	50,960		
	Total Acreage of Snow	0	980	1,372	9,428	18,434	30,214		

Overall percent "correct" (based on cell count) = 90.4%
Overall percent "correct" (based on acreage count) = 96.3%

Figure 3. Results of the comparison between estimates of snow cover made on ERTS-1 imagery and that made on 1:45,000 vertical color aerial photos. Note that "percent correct" was greater than 90% on both a cell and an acreage basis.

29,106 acres (11,642 ha), and the "correct" snow cover estimate made using the large scale 35 mm photos was 30,214 acres (12,086 ha), resulting in an overall "correctness" based on acreage of 96.3 percent.

SUMMARY AND CONCLUSIONS

Sequential aerial photography, Earth Resource Technology Satellite (ERTS-1) imagery and ground data were used to develop an image interpretation key to be used as an aid in training photo interpreters to estimate the areal extent of snow in forested areas, and as inputs to rapid and relatively simple manual analysis technique for estimating acreages and extent of snow cover on satellite imagery. This technique is based on the interpreter's ability to amalgamate information on the appearance of snow as seen on aerial or ERTS-1 imagery with the type, density and distribution of the vegetation and characteristics of the terrain within which the snow occurs and which greatly influence the appearance of a snowpack as seen from above. By examining two images simultaneously, one taken during a snow free period and the other during the melt season, an interpreter can concentrate on both the appearance of snow, and on the vegetation/terrain condition. Consequently, reliable estimates of areal extent of snow in large complex forested areas can be derived.

A test of the interpretation technique was conducted over a 1.5 million acre (600,000 ha) area in the Feather River watershed in northern California. Using percent snow cover estimates made on relatively large scale color aerial photographs of sample plots within the test area and taken coincident with the ERTS overpass as a basis for comparison, it was found that the interpretation of ERTS-1 imagery yielded estimates of snow-covered acreage with a 96.3 percent accuracy.

ACKNOWLEDGEMENTS

The research reported here was conducted under the sponsorship and financial assistance of the National Aeronautics and Space Administration, Earth Resources Survey Program, NASA Grant NGL 05-003-404. In addition, two members of the Center for Remote Sensing Research staff, Edwin F. Katibah and Steven J. Daus contributed a great deal of time and effort to the work reported herein.

REFERENCES

Anderson, Henry W., 1956. Forest covers effects on snowpack accumulation and melt. _In_ Transactions of the American Geophysical Union. Central Sierra Snow Lab. June.

Anderson, Henry W., 1958. Snow in forest openings and forest stands. _In_ Proceedings of the Society of American Foresters.

Anderson, Henry W., 1967. Snow accumulations as related to meteorological, topographic, and forest variables in central Sierra Nevada, California. International Association of Scientific Hydrology. Publication No. 78.

Baker, D., 1971. Remote sensing of snow fields from earth

satellites. In International Workshop on Earth Resources Survey Systems. National Aeronautics and Space Administration. NASA SP-283. May 3-14.

Barnes, James C., 1969. Operational snow mapping from satellite photography. In Proceeding 26th of the Eastern Snow Conference.

Barnes, James C. and C. J. Bowley, 1969. Satellite photography for snow surveillance in western mountains. In Proceedings of the 37th Annual Meeting of the Western Snow Conference. Salt Lake City.

Finnegan, W. J., 1962. Snow surveying with aerial photographs. Photogrammetric Engineering. 28(5). November. pp. 782-790.

Hannaford, Jack, R. H. Bush and R. E. Barsch, 1970. The development of a hydrologic model as an operational tool. In Proceedings of the 38th Western Snow Conference.

Leaf, Charles F., 1969. Aerial photographs for operational stream flow forecasting in the Colorado Rockies. In Proceedings of 37th Annual Western Snow Conference. Salt Lake City.

Leaf, Charles and J. Washichek, 1969. Report on distribution and measurement of snow cover. In Proceedings of the Workshop on Snow-Ice Hydrology.

Leaf, Charles F. and Arden D. Haeffner, 1971. A model for updating streamflow forecasts based on areal snow cover and a precipitation index. In Proceedings of the 39th Annual Meeting of the Western Snow Conference. Billings, Montana. April 20-22.

Popham, R. W., 1963. Progress report on snow and ice observations from Tiros satellite. In Proceedings of the 20th Eastern Snow Conference.

Popham, R. W., A. Flanders and H. Neiss, 1966. Second progress report on satellite applications to snow hydrology. In Proceedings of the 23rd Eastern Snow Conference.

Smith, James L., 1966. Progress report. California Coop Snow Management Research. USFS Pacific Southwest Forest and Range Experiment Station. Berkeley, California.

Tarble, R. D., 1963. Areal distribution of snow as determined from satellite photos. International Association of Scientific Hydrology. Publication 65.

Tarble, R. D. and R. J. C. Burnash, 1971. Directions in water supply forecasting. In Proceedings of the 39th Annual Meeting of the Western Snow Conference. Billings, Montana. April 20-22

U.S. Army Corps of Engineers, 1956. Snow hydrology.

POLAR SEA ICE OBSERVATIONS BY MEANS OF MICROWAVE RADIOMETRY

P. Gloersen
T. C. Chang
T. T. Wilheit
NASA Goddard Space Flight Center

W. J. Campbell
U.S. Geological Survey, Tacoma

ABSTRACT

Principles pertinent to the utilization of 1.55 cm wavelength radiation emanating from the surface of the Earth for studying the changing characteristics of polar sea ice are briefly reviewed. Recent data obtained at that wavelength with an imaging radiometer on-board the Nimbus 5 satellite are used to illustrate how the seasonal changes in extent of sea ice in both polar regions may be monitored free of atmospheric interference. Within a season, changes in the compactness of the sea ice are also observed from the satellite. Some substantial areas of the Arctic sea ice canopy identified as first-year ice in the past winter were observed not to melt this summer, a graphic illustration of the eventual formation of multiyear ice in the Arctic. Finally, the microwave emissivity of some of the multiyear ice areas near the North Pole was found to increase significantly in the summer, probably due to liquid water content in the firn layer.

BACKGROUND

The imaging microwave radiometer utilized in these studies has been described elsewhere (Wilheit 1972, Gloersen et al. 1973). Briefly, The Electronically-Scanned Microwave Radiometer (ESMR) operates at a wavelength of 1.55 cm. In this wavelength region, the Rayleigh-Jeans approximation applies to all targets of interest, and the microwave power received from a radiating target depends linearly on the brightness temperature of that target. This simplifies calibrating the output of the radiometer directly in terms of the brightness temperature of the target. In general, the observed brightness temperature depends on the physical temperature and emissivity of the viewed surface, and both direct and reflected components from the intervening atmosphere. The atmospheric components at a wavelength of 1.55 cm result principally from the liquid water and water vapor present. Since these quantities are small in the Arctic and Antarctic atmospheres, they have been neglected in determinations of surface brightness temperatures reported here. The stratus clouds usually obscuring the polar regions consist mostly of ice crystals, which are transparent to 1.55 cm radiation.

The observed variation of the brightness temperatures result from the combination of seasonal, diurnal, and climatic changes in the physical surface temperatures amounting to differences of about 40° K, and from significant variations in the emissivity of the surface, as illustrated in Figure 1, which translate into brightness temperature contrasts of 20 K and 120 K for multiyear sea ice and open water vs. first-year sea ice, respectively. The emissivity differences shown for first-year sea ice, multiyear sea ice, and open water result in microwave image patterns that are readily discernable in spite of variations in the physical temperature of the surface, since the latter vary sufficiently slowly in space and time compared to the emissivity differences. It should be noted that the emissivity of the central Arctic ice canopy, where most of the multiyear ice is located, has been observed to increase markedly over the values shown in Figure 1 in the summer months, probably due to a liquid water content in the firn layer. (Edgerton et al. 1971, Meier 1972) When the instantaneous field-of-view of the ESMR contains a mixture of the three polar sea components, intermediate brightness temperatures are observed. This may be used to infer compactness of sea ice, as illustrated in Figure 2.

The curve in Figure 2 is simply a linear interpolation between the cases of completely open water ($T_B = T_W = 131°K$) and completely consolidated first-year sea ice with a surface temperature, T_s, of 260°K, typical of areas near the edge of the ice pack, especially in Spring and Fall. The shading indicates the climatic variation of T_s expected, in addition to the uncertainty of the brightness temperature measurement. This leads to an average uncertainty in the determination of the fraction of open water of about 4% from the observed brightness temperature. It has been observed that measurable open water usually occurs only in areas where little or no multiyear ice is present. Therefore, the curve fits most cases of interest; some downward adjustments to T_s would be required for sea ice well-removed from its outer boundary in Antarctica during Austral Winter.

SEA ICE COMPACTNESS

The curve of Figure 2 has been used to analyze the sea ice cover around Greenland and Antarctica, as shown in Figures 3 and 4, respectively. The distribution of compactness shown on these particular days has been observed to change significantly from one polar projection to another over time intervals as short as 10 days, the minimum time interval utilized so far in the data processing. It is evident that such determinations, combined with records of wind stress, would provide useful data for the study of ice dynamics. Such studies are currently in progress. In addition to the areas discussed so far here, other areas of considerable ice dynamic activity, as evidenced by the data on hand, are the Hudson Bay and the Bering, Chuckchi, East Siberian, Laptev, and Beaufort Seas.

SEA ICE BOUNDARIES

Nimbus 5 ESMR data are now available for determining the near minimum and maximum sea ice boundaries for both polar regions in 1973. Such data are illustrated in Figures 5 and 6, illustrating quantitatively what is already

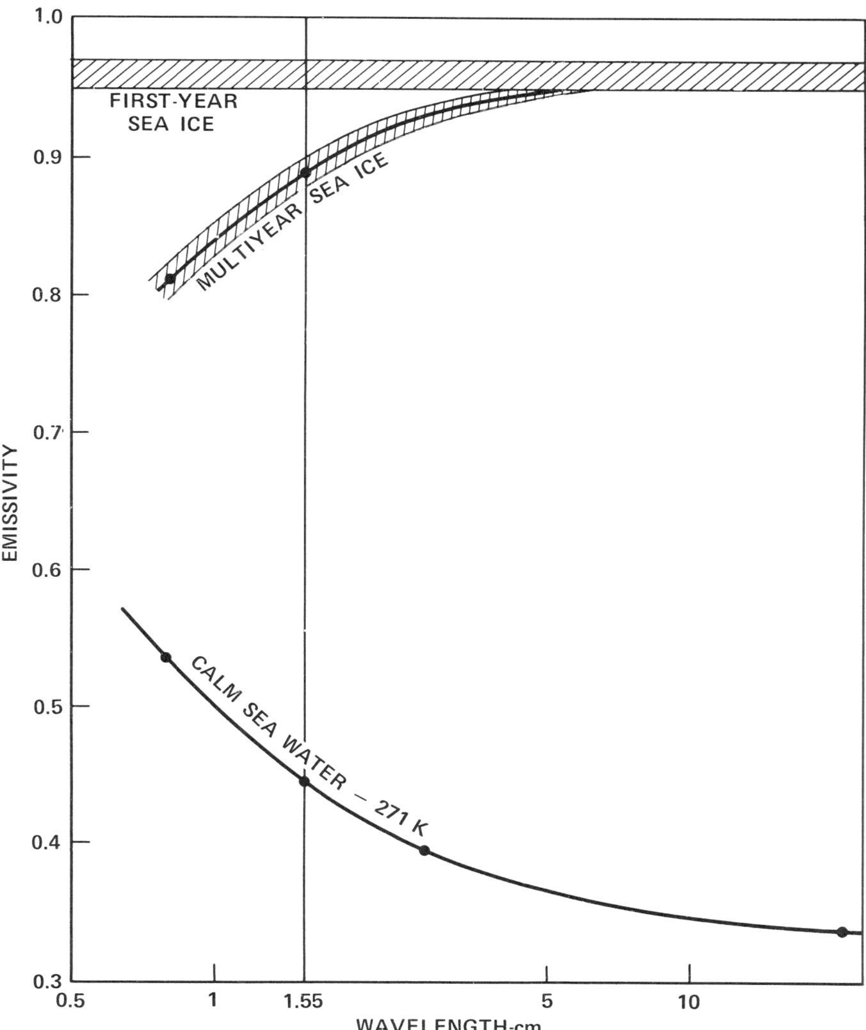

Figure 1. Microwave Emissivities of the Polar Seas

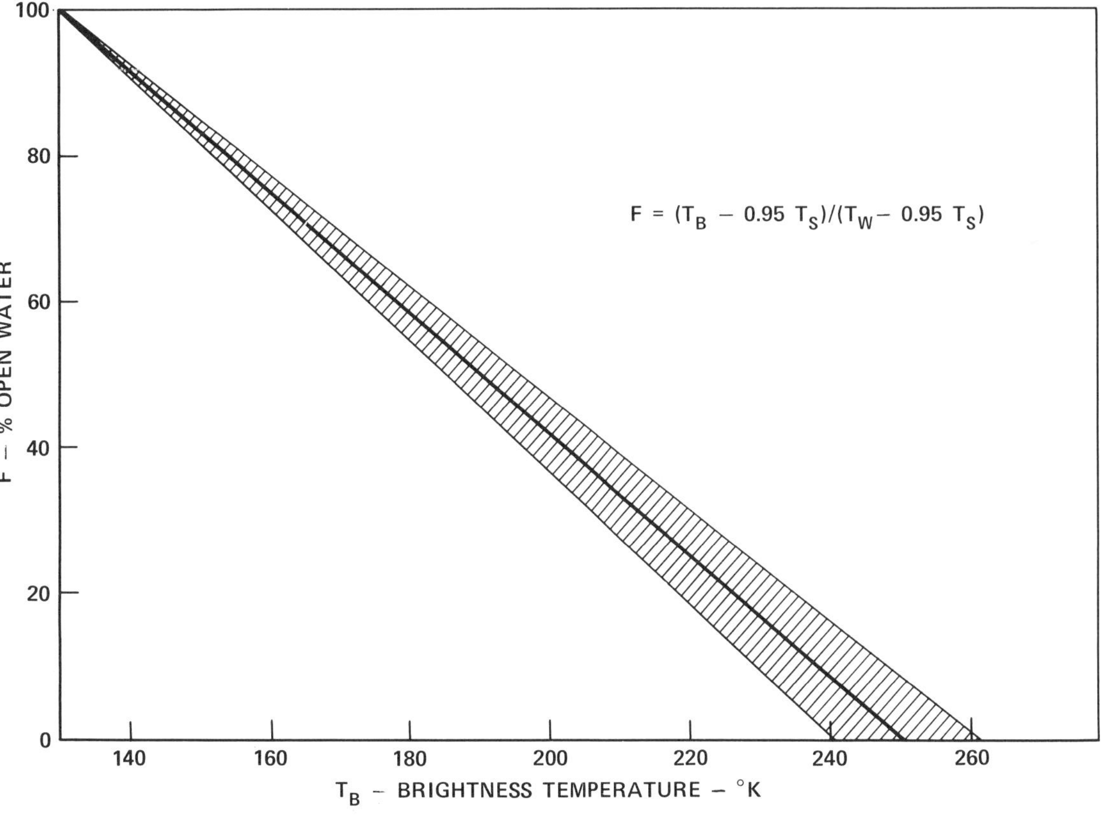

Figure 2. Determination of Unresolved Open Water in Areas Containing Sea Ice from the Microwave Brightness Temperature at $\lambda = 1.55$ cm

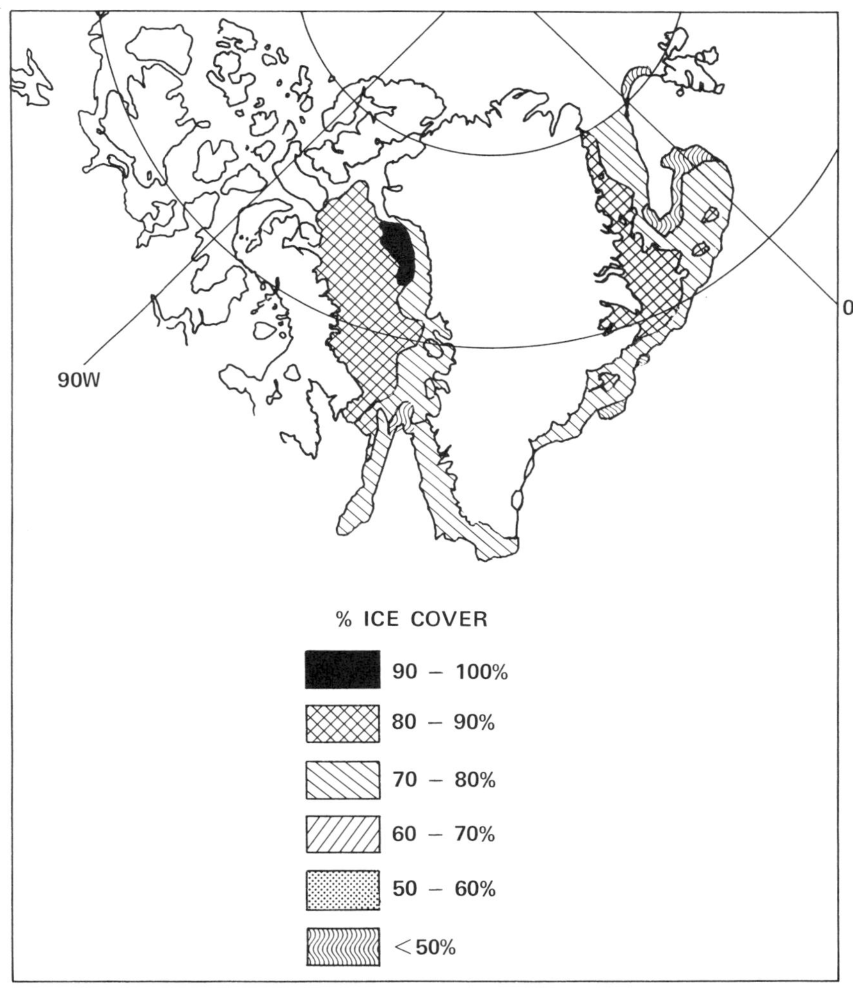

Figure 3. Compactness of Sea Ice Around Greenland on February 26, 1973.

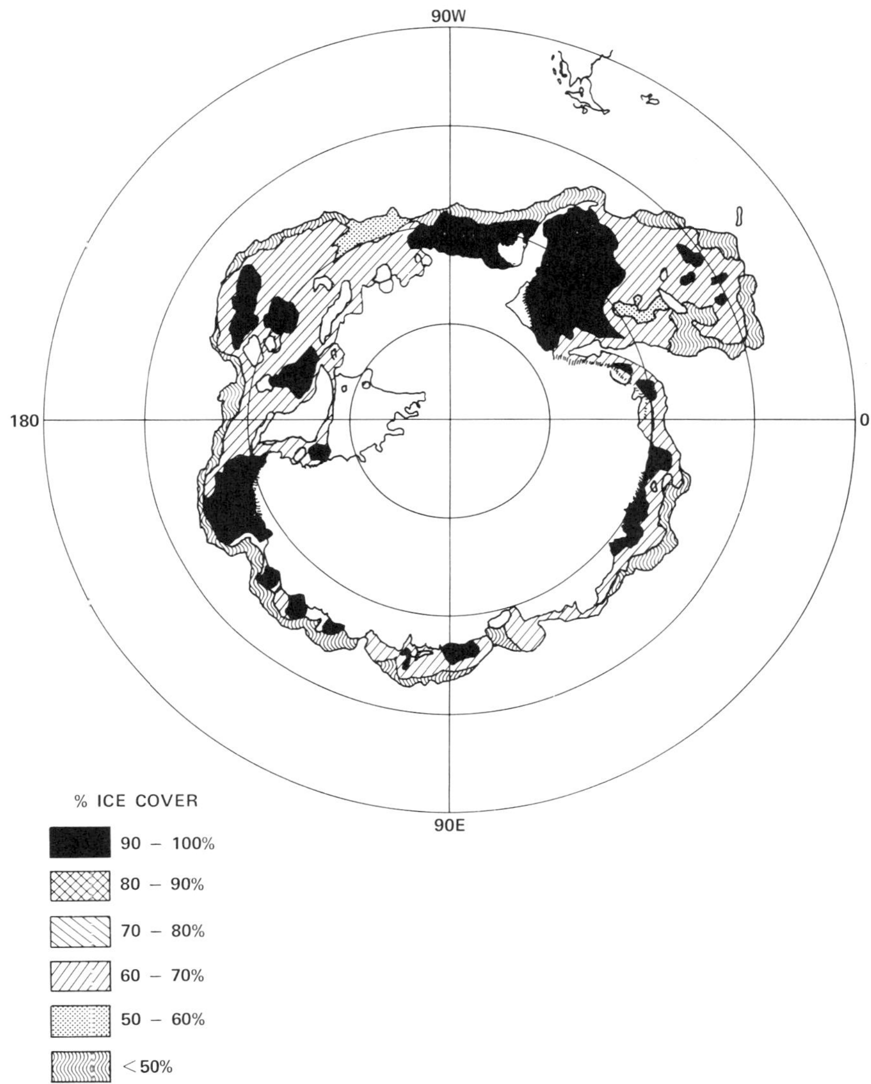

Figure 4. Compactness of Sea Ice Around Antarctica on December 26, 1972.

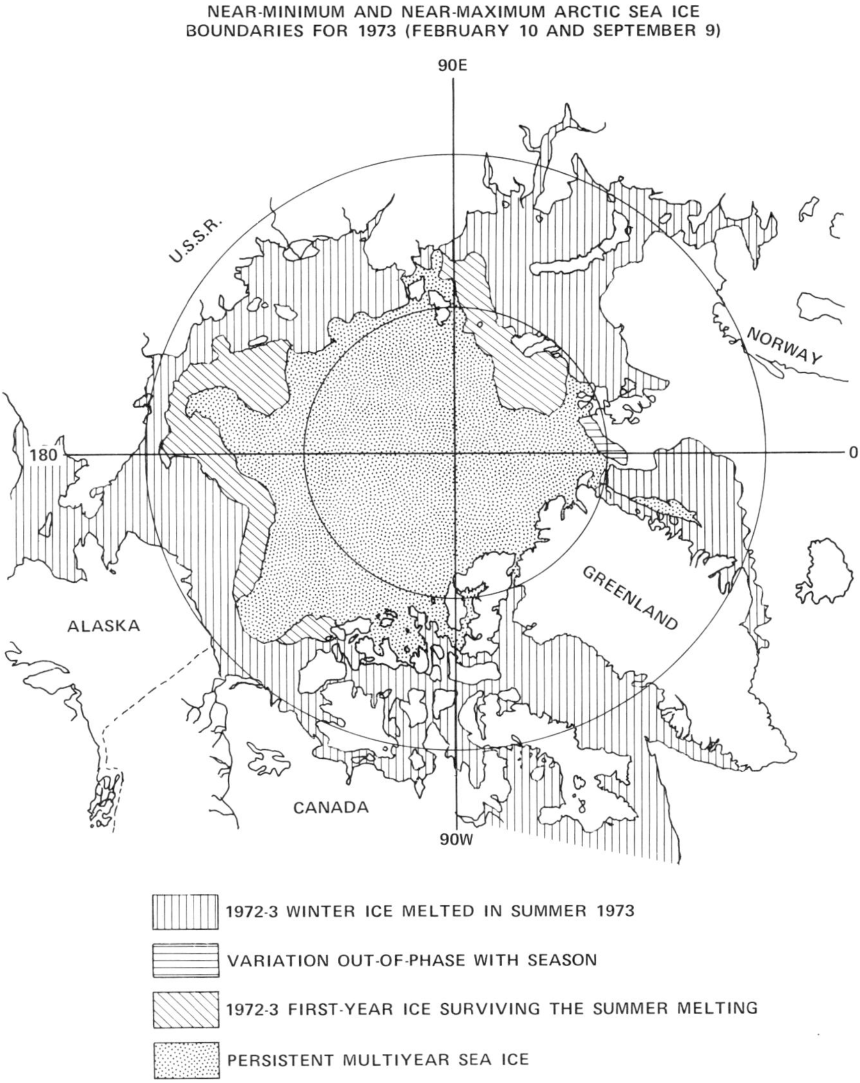

NEAR-MINIMUM AND NEAR-MAXIMUM ARCTIC SEA ICE
BOUNDARIES FOR 1973 (FEBRUARY 10 AND SEPTEMBER 9)

90E

U.S.S.R.

NORWAY

180

0

ALASKA

GREENLAND

CANADA

90W

1972-3 WINTER ICE MELTED IN SUMMER 1973

VARIATION OUT-OF-PHASE WITH SEASON

1972-3 FIRST-YEAR ICE SURVIVING THE SUMMER MELTING

PERSISTENT MULTIYEAR SEA ICE

Figure 5. Seasonal Variation of Sea Ice in the Northern Hemisphere.

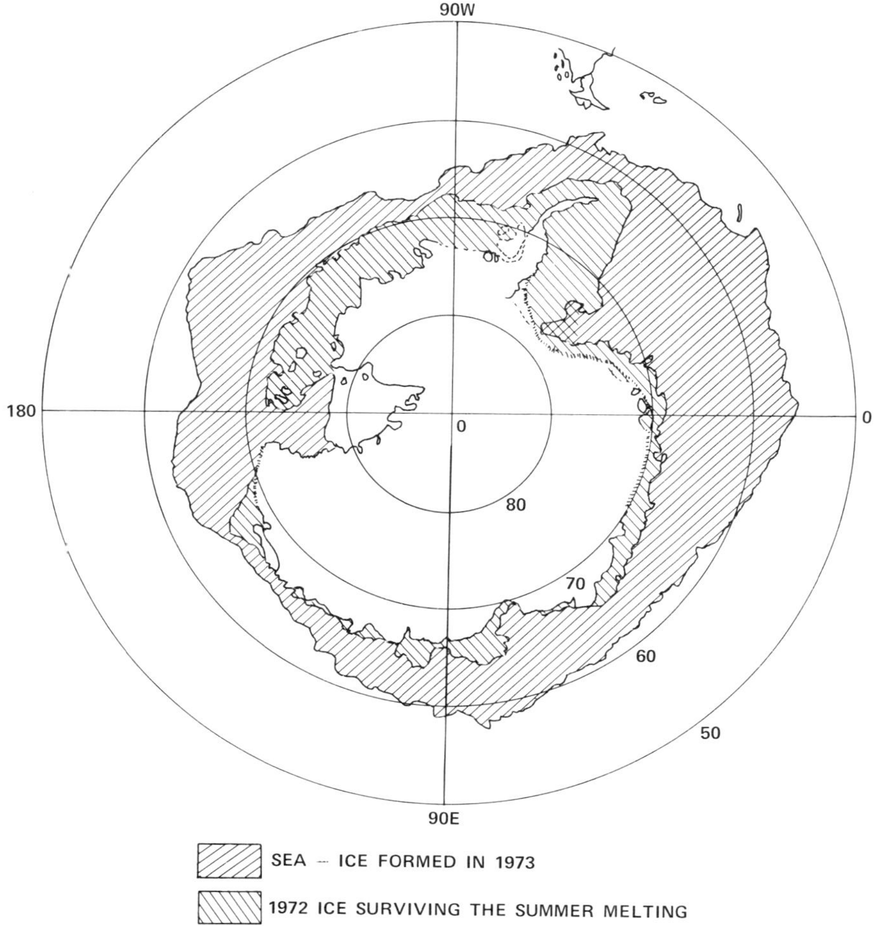

Figure 6. Seasonal Variation of Sea Ice in the Southern Hemisphere.

known in a qualitative way, i.e. that the amount of ice cover changes by about a factor two in both polar regions during a given year.

In Antarctica (Figure 6), the observations might seem to indicate the formation of multiyear ice in some regions, since not all of the ice formed in 1972 melted during Austral Summer. On the other hand, these same areas had a distinct first-year signature in Austral Spring 1972 and Austral Winter 1973 indicating, at least, that it was not ice formed in previous years. This, coupled with the knowledge obtained by observers in these areas indicating high ice dynamic activity and the absence of multiyear sea ice leads to the conclusion that the sequential images are representative of a steady-state phenomenon, i.e. the areas of persistent sea ice cover are accumulation regions in a system of steady depletion and replenishment.

In the Arctic region (Figure 5), there are a number of noteworthy points to be made. First, the near-minimum boundary of the sea ice illustrated will be the basis for following the changing distribution of multiyear ice in the winter of 1973-4, giving valuable insight into the macroscale ice dynamics in the north polar region. It was also observed that not all of the sea ice identified as first-year in Winter 1972-3, particularly in the Kara, northern Chukchi, and Beaufort Seas, melted in Summer 1973. Thus, observing the changes in brightness temperature of these areas during Winter 1973-4 may result in new knowledge of the emissivity of sea ice vs. age at some intermediate points. In addition, such observations will give insight into the sources of multiyear ice in the Arctic. Finally, there is clear evidence that the sea ice northwest of Svalbard is strongly affected by ocean currents, since the boundaries are observed to change well out-of-phase with expectations based on seasonal variations alone.

APPLICATIONS

Since the amount of open water in the polar regions has a profound impact on global circulation patterns of the atmosphere (Fletcher 1972), the detailed information of extent and location of open water in the polar regions is a valuable input for computer models of the atmosphere and climate.

In terms of shipping, the Fleet Weather Facility of the U.S. Navy at Suitland, Maryland is utilizing Nimbus 5 ESMR data obtained in near real-time to produce weekly ice limit reports for military, commercial, and international use. Plans for substituting an automated system for the present hand-analysis from quick-look ESMR records are under consideration.

The nature of the persistent radiometric patterns that have been observed in the continental ice sheets on Greenland and Antarctica have been described elsewhere (Gloersen et al. 1973). These patterns may be related to snow accumulation rates on these continents, but the correlation in Greenland is not high, based on the available contours drawn through relatively few measured points on the basis of a climatic model (Mock 1967). If subsequent surface measurements give more substantial support to a correlation with accumulation rates, it may be possible to monitor long-term changes in such rates with this instrument.

Although no such occurrances have been observed to date, it may be possible to monitor the calving of shelf ice of sufficient size from sources such as the Ross Ice Shelf. Similar studies of such enormous glaciers as the Lambert Glacier in Antarctica may also be possible, but the spatial resolution capability of the instrument would be stressed.

REFERENCES

Edgerton, A. T., A. Stogryn, and G. Poe, 1971. Microwave radiometric investigations of snowpack. Final Report No. 1285 R-4 for U.S.G.S. contract No. 14-08-001-11828, Aerojet-General Corp., Microwave Division, El Monte, California.

Gloersen, P., T. T. Wilheit, T. C. Chang, W. Nordberg, and W. J. Campbell, 1973. Microwave maps of the polar ice of the earth. NASA/GSFC Report No. X-652-73-269.

Meier, M. F., 1972. Measurement of snow cover using passive microwave radiation. Proc. Intl. Symposia on the Role of Snow and Ice in Hydrology, Banff, Canada.

Mock, S. J., 1967. Accumulation patterns on the Greenland ice sheet. USA/CRREL Report No. 233. Also, 1973. Priv. Comm.

Wilheit, T. T., 1972. The electronically scanning microwave radiometer (ESMR). The Nimbus 5 User's Guide, pp. 59-104, U.S. Government Printing Office 1972-735-963/259.

MICROWAVE SIGNATURES OF SNOW AND FRESH WATER ICE

T. Schmugge, T. T. Wilheit, and P. Gloersen
NASA/Goddard Space Flight Center
Greenbelt, Maryland 20771

M. F. Meier and D. Frank
Water Resources Division
U.S. Geological Survey
Tacoma, Washington

I. Dirmhirn
Utah State University
Logan, Utah

ABSTRACT

During March of 1971, the NASA Convair 990 Airborne Observatory carrying microwave radiometers in the wavelength range 0.8 to 21 cm was flown over dry snow with different substrata: Lake ice at Bear Lake in Utah; wet soil in the Yampa River Valley near Steamboat Springs, Colorado; and glacier ice, firn and wet snow on the South Cascade Glacier in Washington. The data presented indicate that the transparency of the snow cover is a function of wavelength. False-color images of microwave brightness temperatures obtained from a scanning radiometer operating at a wavelength of 1.55 cm demonstrate the capability of scanning radiometers for mapping snowfields.

INTRODUCTION

In order to obtain data on the microwave signature of snow, flights were made over 3 snow covered sites by the NASA CV-990 Airborne Observatory in March 1971. The sites were Bear Lake on the Utah-Idaho border, the Yampa River Valley south of Steamboat Springs, Colorado, and the South Cascade Glacier and Lake in Washington. Ground truth measurements of snow depth and density were obtained at each of the three sites.

Ground based measurements (Edgerton et al. 1971, Meier and Edgerton, 1971) have indicated that for dry snow over frozen soil the effective microwave emissivity decreases as the snow pack increases. For example, at a wavelength of 8 mm and at a look angle of 45° the emissivity decreases from near unity for frozen soil to about 0.78 as a snow pack of density 500 kg/m^3 is built up to a thickness of 80 cm. At longer wavelengths, the decrease was less for the same snow pack. The emissivity calculated for a very thick, uniform snow pack from the Fresnel equations (Jackson, 1962) is 0.97, so the bulk dielectric properties of snow cannot explain the low emissivity that was observed. The effect of liquid water in the snow on microwave emission from snow is striking. The emissivity at a wavelength of 8 mm increases from 0.78 to near unity (Edgerton et al., 1971) as the liquid water content of the surface layer increases from 0 to 1%. These effects are similar for both the vertical and horizontal polarizations.

The objective of these flights was to determine over what wavelength range the emissivity of dry snow is abnormally low and at what wavelengths the emission from a snow pack can be described by the bulk dielectric properties of snow. We also wish to determine the wavelengths for which the emissivity increases when liquid water becomes present in the snow.

The radiometers on-board the aircraft are listed in Table 1. The 1.55 cm radiometer scans ±50° in a direction perpendicular to the aircraft's flight path. Thus, the radiation from a swath whose width is approximately twice the altitude of the aircraft above the ground is mapped. The remaining radiometers were non-scanning and, with the exception of 0.8 cm radiometer were nadir viewing. The 0.8 cm radiometer looked to the rear of the aircraft at an angle of 45° and received both horizontally and vertically polarized radiation. Surface temperatures were obtained with an infrared radiometer operating in the 10-12 micrometer band. In addition, color photography was obtained over each target and was used to determine the flight path of the aircraft.

SUMMARY OF RESULTS

The brightness temperatures for all the radiometers from each target area are presented in Table 2. At Bear Lake and Steamboat Springs the values are averages for the total time over each target. The values from the Cascade Glacier are the averages for the periods (2-3 seconds) that the aircraft was over each specific area.

The results from the 0.8 cm and 1.55 cm radiometers exhibit the least amount of variation among the 3 target areas. This may result from the fact that the surface snow at each site had similar characteristics. The variation increases with wavelength indicating that these radiometers are responding to the variation in the substrata. The details of these variations will be discussed for each site individually in the following paragraphs.

STEAMBOAT SPRINGS

Snow depth and density measurements were obtained along a 4 mile N-S track in the Yampa River Valley south of Steamboat Springs Colorado. The snow depths ranged from 75 to 90 cm and the snow density ranged from

Table 1

Radiometer Characteristics

Freq. GHz	Wavelength cm	Pointing Relative to Nadir	3 db Beam Width	RMS Temp. Sens.	Reference
1.42	21	0°	15°	5 K	Edgerton et al. (1971)
2.69	11	0°	27°	0.5 K	Gray et al. (1971)
4.99	6.0	0°	5°	15 K	Edgerton et al. (1971)
10.69	2.8	0°	7°	1.5 K	Wilheit et al. (1972)
19.35 H	1.55	SCANNER	2.8°	1.5 K	Oister and Falco (1967)
37 V	0.81	45°	5°	3.5 K	Wilheit et al. (1972)
37 H	0.81	45°	5°	3.5 K	Wilheit et al. (1972)
INFRARED	10 microns	14°	<1°	<1 K	Kuhn et al. (1971)

Table 2

Observed Brightness Temperatures, in Kelvins

	Snow Thickness (m)	21 cm	11 cm	6 cm	2.8 cm	1.55 cm	0.8 cm Horiz.	0.8 cm Vert.	IR 10-12μ m
Bear Lake Alt. 1805 m	.15	123	156	163	193	206	198	231	268
Steamboat Springs Alt. 2070 m	.8	212	235	246	243	215	211	235	266
South Cascade Glacier									
South Cascade Lake[1] Alt. 1610 m	.5	222	252[2]	242	250	232	232	255	271
P-0 - below firn line Alt. 1770 m	4.9	247	258	244	241	205	208	235	271
P-1 - above firn line Alt. 1890 m	6.8	242	256	238	231	202	214	238	268
P-3 Alt. 2040 m	8.4	239	252	231	224	194	206	235	268
Chickamin Glacier Alt. 2310 m		234	246	231	235	210	218	238	263

1) Values at 21 and 11 cm are for the lowest altitude pass only.
2) Field of view includes areas of dry snow on slopes bordering lake.

260-330 kg/m^3. The ground beneath the snow was not frozen and the soil moisture was generally greater than 35%, by weight.

The results presented in Table 2 are from a flight at an altitude of 4000 m above the ground on March 3, 1971. The ground was partially obscured during the overflight because of cloudy conditions.

The most interesting result is the low brightness temperature of 212 K at a wavelength of 21 cm. This temperature is about that seen for very moist agricultural fields (Schmugge et al., 1973) and thus it appears that the snow is transparent at this wavelength. The brightness temperatures at 0.8 cm (horizontal polarization) and 1.55 are as low as those observed in the ground measurement of Edgerton et al. (1971). The results at the intermediate wavelengths (11, 6, and 2.8 cm) are approximately the results one expects from the bulk dielectric properties of the snow.

BEAR LAKE

The surface of Bear Lake is at an altitude of 1800 m. The ground truth site was selected to be representative of the whole lake. It was located 700 m from the Limnological Laboratory of Utah State University. The depth of snow was measured over a 12,000 m^2 (3 acre) site and ranged from 13 to 17 cm, having an average of 15 cm. The average density of the snow was 200 kg/m^3. This snow fell about one week before the flight. During the intervening period the snow temperature stayed below freezing and there was little or no wind which would have packed the snow. The albedo of the snow was high (0.86) indicative of new snow. The ice thickness at this location was 25 cm.

Figure 1 is the 1.55 cm microwave image of the pass over Bear Lake at an altitude of 3400 m above the terrain along with a map of the area on which the flight line is plotted. The pass was from north to south at a speed of 185 m/sec. The eastern edge of the lake has a steep slope with areas of exposed rock having a higher brightness temperature, 245 K, than the snow covered lake. In general, the brightness temperature of the lake was 194-210 K except for a region with a value of about 190 K between 18:51:30 and 18:52:00 which corresponded to an area with a large number of cracks in the ice.

The stripchart results from all the radiometers for this pass over Bear Lake are shown in Figure 2. The plot for the 1.55 cm radiometer is the average of the center 5 beam positions which simulates a radiometer beam width of 10°. The much lower brightness temperatures observed by the longer wavelength radiometers over the lake can be understood quantitatively using a straightforward layered model.

The reflectivity of a layered dielectric may be calculated as a boundary value problem according to the principles found in electromagnetic theory texts such as Jackson (1962). Assuming a uniform thermodynamic temperature this result can be expressed as a brightness temperature. In this case, we treat four layers: air, snow, ice and water. The dielectric constants for snow and ice are given by Evans (1965) and for water by Lane and Saxton (1952). Using these data and assuming a density of 200 kg/m^3 for the snow and a temperature of 273 K for the water, we get the dielectric constants given in Table 3.

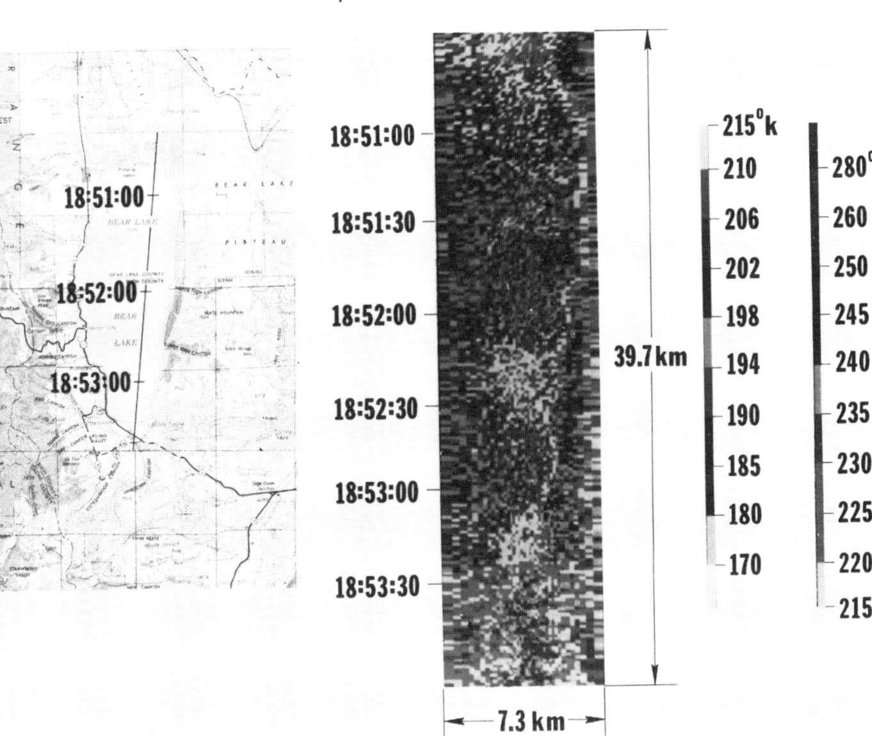

Figure 1. Map and false color 1.55 cm microwave image of Bear Lake.

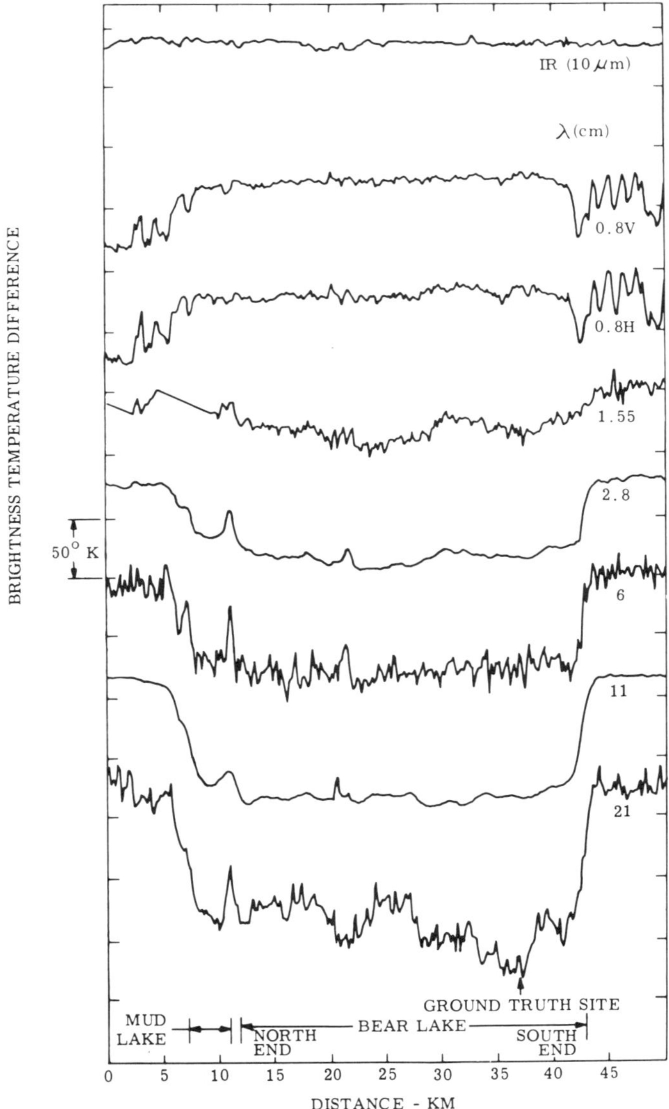

Figure 2. Multispectral data obtained over Bear Lake. The H and V refer to the horizontal and vertical channels of the 0.8 cm radiometer which viewed the surface at an angle of 45°. The remaining radiometers were nadir viewing. Average values for the brightness temperatures over the lake at each frequency are given in Table 2.

Table 3

Dielectric Constants For Bear Lake Calculation

Layer	Dielectric Constant at 21 cm	Dielectric Constant at 11 cm
AIR	1.00 + 0 i	1.000 + 0 i
SNOW	1.32 + .0002 i	1.32 + .0002 i
ICE	3.2 + .004 i	3.2 + .004 i
WATER	85.4 + 13.3 i	80.3 + 23.6 i

Since the imaginary components for the ice and snow layers are small, the calculation is not sensitive to the exact value nor to the temperature of the layer; we have accordingly assumed a temperature of 273 K for all layers. The results of such a calculation are shown in Figure 3 as contours of constant brightness temperature for the 21 cm channel as a function of ice thickness and of snow thickness. The contours are 120 K, 160 K and 200 K. The observed brightness temperature at 21 cm for the lake, 123 K, is in good agreement with the calculated value from the measured snow and ice thickness as indicated by the black square in the figure.

As indicated in the figure, the calculated brightness temperature is a strong oscillatory function of the ice thickness with a weaker snow thickness dependence modulating the amplitude of the oscillations, as is to be expected on the basis of interference of the reflected components from the various surfaces. The separation between adjacent brightness temperature maxima is 6 cm on the ice thickness scale. Thus there may be brightness temperature gradients of about 20-30 K per cm of ice thickness. The observed variation of the 21 cm brightness temperature is about 40-50 K which implies ice thickness variations of between 1 and 2 cm. These are consistent with thickness variations that can be expected for a lake of this size.

A similar calculation was performed for the 11 cm channel. Again there was good agreement between the calculated and the observed values of brightness temperature for the lake. As expected for the shorter wavelength, the separation between adjacent brightness temperature maxima is reduced. This would imply that there should have been larger variations of the observed 11 cm brightness temperature than those observed at 21 cm. However, this was not the case; the observed variation at 11 cm was 15 K or about one third the variation observed at 21 cm. We believe that this is due to the averaging effects of the larger beamwidth of the 11 cm radiometer (see Table 1) and its shorter wavelength.

SOUTH CASCADE GLACIER

At South Cascade Glacier, the snow surface was very smooth with only small snow dunes. Snow at the surface was soft and light with a density of $250 kg/m^3$, increasing with depth. Ground truth measurements had been made one week before the overflight and air temperature, precipitation and runoff were recorded through the time of overflight. Three relatively level points on the glacier and on South Cascade Lake north of the glacier were selected for

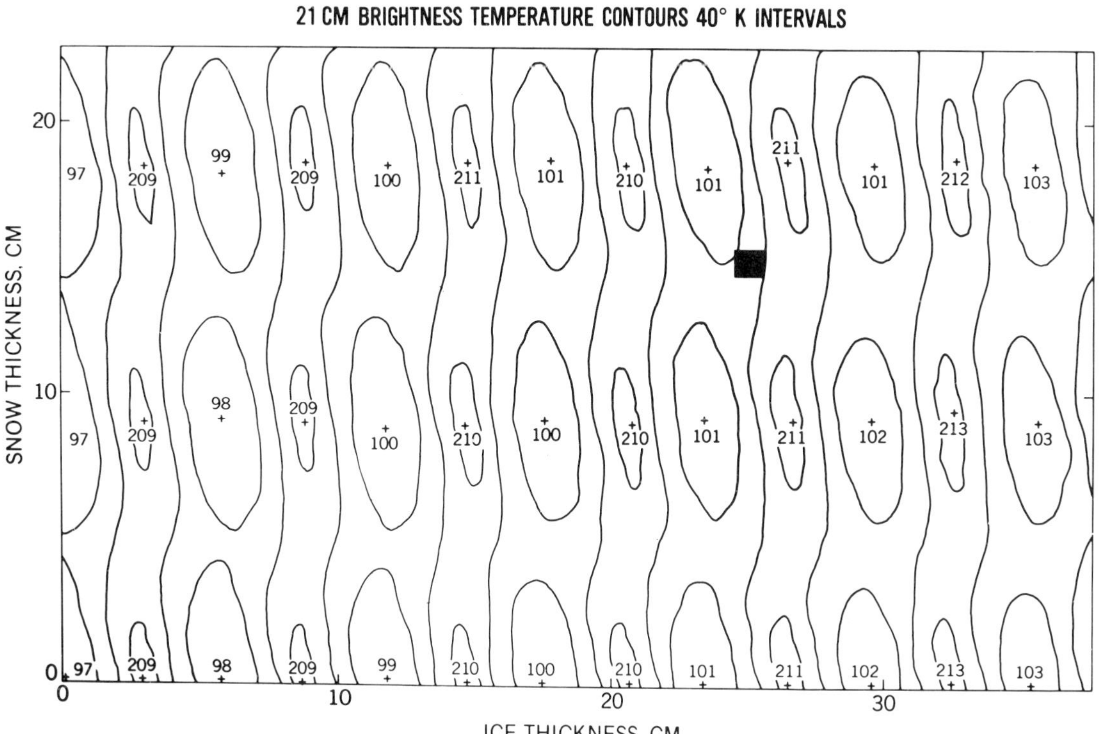

Figure 3. Calculated brightness temperature contours for the 21 cm radiometer as a function of snow and ice thicknesses. The 120 K, 160 K, and 200 K contours are drawn. The black square is the range of observed snow and ice thicknesses at Bear Lake.

analysis. The total snow depth at P-0, altitude 1770 m, a point below the firn line near the terminus of the glacier, was 4.9 m over the ice. At P-1, altitude 1890 m, a point slightly above the firn line near the center of the glacier, the snow depth was 6.8 m over firn. The total snow, firn and ice depth was over 200 meters at this point. A third location, P-3, altitude 2040 m, had a snow depth of 8.4 m over the firn layer. The temperature records indicate that there was no liquid water in the top meter of snow at any point on the glacier.

At the fourth ground truth site, South Cascade Lake, there was a transition layer, approximately 0.5 m thick, of light snow containing some liquid water. This layer was below about 0.2 m of light dry snow and above 3 m of wet snow, slush and water. This was the only site with any wet snow present. The Chickamin Glacier 3 km southeast of the South Cascade Glacier was also over-flown. While no ground truth was obtained at this glacier it did provide another high altitude (2300 m) snow target with steeper slopes, thicker and colder snow and more surface roughness than the South Cascade Glacier.

The aircraft made three passes over the glacier at altitudes of 3330 m, 3420 m, and 4830 m above sea level. The results from the three passes varied by no more than the instrument noise. The values presented in Table 2 are the averages for these passes.

In Figure 4, the 1.55 cm microwave image of the South Cascade Glacier area from the 4830 m pass, the blue and dark green areas are those with the lowest brightness temperatures and these areas approximately correspond to the glaciers. Some of the glaciers which can be located on the microwave image are the South Cascade, Chickamin, LeConte, Dome, Dana and Spire. Generally the brightness temperatures of the glaciers are below 205 K. The snow-covered mountain slopes, e.g., the area between the Cascade and Chickamin glaciers, have brightness temperatures in the range 210-225 K, indicated by yellow and light green in the microwave image. The high brightness temperature areas, violet and red (240-255 K), are forested areas where there is a foliage canopy above wet snow.

South Cascade Lake shows a somewhat higher brightness temperature, about 230 K, than that of the glacier. A 15 to 20 K rise was also observed in the horizontal and vertical channels of the 0.8 cm wavelength radiometer. These increases in the brightness temperature of the snow are possible indications of the presence of liquid water in the snow, in agreement with the surface observation. The 21 cm radiometer has a lower brightness temperature for the lake which may be indicative of its ability to penetrate a sufficiently thin wet snow layer.

On the glacier itself the shorter wavelength radiometers (1.55 cm and 0.8 cm) had the lowest brightness temperatures while the 21 and 11 cm wavelength radiometers had the highest brightness temperatures. The results at the longer wavelengths can be understood in terms of the same layered dielectric model used with the Bear Lake results. In this case, the observed snow density was used to calculate its dielectric constant using the Weiner mixing formula (Evans, 1965). At P-0 the snow density increased linearly from 250 kg/m^3 at the surface to 500 kg/m^3 at a depth of 4.9 meters at which point the ice begins. For a surface temperature of 270 K the calculated brightness temperature is 265 K which is in good agreement with the 258 K value observed by the 11 cm radiometer which was the most accurate of the

1.55 CM MICROWAVE IMAGE OF
SOUTH CASCADE GLACIER AREA
MARCH 8, 1971

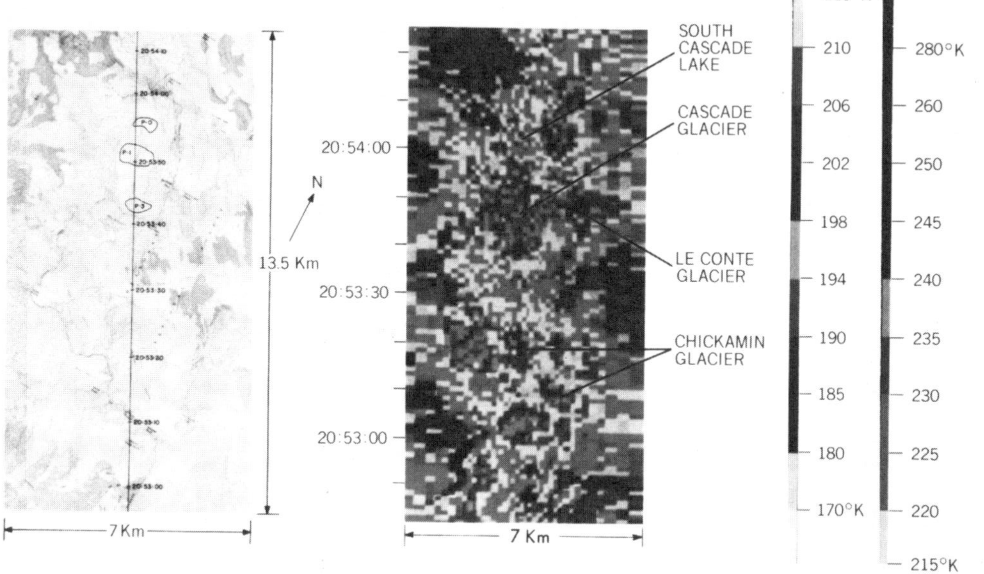

Figure 4. Map and false color 1.55 cm microwave image of South Cascade Glacier.

radiometers. Instrument uncertainties preclude determining how well the 21 and 6 cm results agree with the calculation. At the shorter wavelengths the brightness temperatures are substantially below the calculated results. At 2.8 cm, the brightness temperatures range from 241 K to 224 K as the altitude increases on the glacier. The brightness temperatures at wavelengths of 1.55 cm and 0.8 cm (horizontal polarization) are below 210 K yielding emissivities of 0.8 or less. This is essentially the same spectral response as that observed for the emissivity difference between first year and multi-year sea ice (Gloersen et al., 1973, Wilheit et al., 1972), i.e., there was no difference observed in the 6 cm, 11 cm and 21 cm results, while the difference increased as the wavelength decreased from 2.8 cm to 0.8 cm. As was the case with the multi-year ice, we expect that the low emissivities result from volume scattering occurring in the snow and firn layers on the glacier. The scattering centers in this case are suspected to be the ice grains. The strength of the scattering depends on the wavelength in ice, increasing as this wavelength becomes comparable to or less than, the size of the scattering centers. At wavelengths much larger than these centers e.g., 11 and 21 cm, volumn scattering is unimportant and the medium can be described by its bulk dielectric properties.

Since we did not observe any difference in brightness temperature when crossing the firn line on the glacier, we conclude that either the snow cover is sufficiently thick so that we would not expect to see such a transition or that the emission from the firn is not significantly different from the emission from new snow.

Finally, it can be seen from the 0.8 cm data in Table 2 that the scattering cross-section appears to be polarization-dependent. However, we are not certain about the significance of this result due to uncertainties about the relative calibration of the two channels.

CONCLUSIONS

These results indicate that the effects of volume scattering in dry snow and firn become noticeable for free space wavelengths shorter than about 3 cm as had been observed earlier for multi-year sea ice (Wilheit et al., 1972 and Gloersen et al., 1973). When liquid water is present, the effective loss tangent for the snow increases to wash out the effects of volume scattering. This is the same mechanism that produces the higher brightness temperature at these wavelengths in first year sea ice when the scattering centers, brine cells, are filled with liquid.

At the longer wavelengths, 11 and 21 cm, the results at Bear Lake and the dry snow areas of Cascade Glacier are in good agreement with those calculated using the bulk dielectric properties of ice and snow, indicating that scattering may not be an important mechanism at these wavelengths.

These results are in reasonable agreement with the earlier ground-based measurements of microwave properties of dry and wet snow (Edgerton et al., 1971) and airborne observations at 1.55 cm (Meier, 1972). The rise in brightness temperature for wet snow indicates that it may be possible to detect the onset of snow melting by looking at brightness temperature differences between day and night passes over snow fields.

ACKNOWLEDGEMENTS

The authors wish to acknowledge the assistance of A. Edgerton of Aerojet-General Corporation, and J. C. Blinn III of the Jet Propulsion Laboratory, in calibrating, installing, and operating various instruments aboard the NASA CV-990 (see Table 1). The authors wish to thank G. Hidy of North American Rockwell Corporation, and P. Kuhn of the NOAA Environmental Research Laboratories for the use of the data obtained with their instruments, Dr. Eugene Peck of the NOAA Hydrologic R&D Laboratory for providing the ground truth data at Steamboat Springs, and Earl Petersen and the CV-990 crew of the Ames Research Center for their support.

REFERENCES

Edgerton, A. T., A. Stogryn and G. Poe, 1971. Microwave radiometric investigations of snowpack. Final report no. 1285 R-4 for U. S. G. S. contract no. 14-08-001-11828, Aerojet-General Corp., Microwave Division, El Monte, California.

Evans, S., 1965. Dielectric properties of ice and snow - a review. Journal of Glaciology 5:773.

Gloersen, P., W. Nordberg, T. J. Schmugge, T. T. Wilheit and W. J. Campbell, 1973. Microwave signatures of first year and multi-year sea ice. J. Geophys. Res. 78:3564.

Gray, K. W., W. F. Hall, W. N. Hardy, G. M. Hidy, W. W. Ho, A. W. Love, Mr. J. Van Melle, and H. Wang, 1971. Microwave measurement of thermal emission from the sea. Proc. Seventh Intl. Symp. on Remote Sensing of the Environment 3:1827.

Jackson, J. D., 1962. Classical electrodynamics. New York: John Wiley & Sons, Inc.

Kuhn, P. M., M. S. Lojko, and E. V. Petersen, 1971. Water vapor: stratospheric injection by thunderstorms, Science 174:1319.

Lane, J. A., and J. A. Saxton, 1952. Dielectric dispersion in pure polar liquid at very high radio-frequencies. Proc. Roy. Soc. A214:400.

Meier, M. F., 1972. Measurement of snow cover using passive microwave radiation. Proc. Intl. Symposia on the Role of Snow and Ice in Hydrology, Banff, Canada.

Meier, M. F., and A. T. Edgerton, 1971. Microwave emission from snow—a progress report. Proc. Seventh Intl. Symp. on Remote Sensing of Environment, pp. 1155-1163.

Oister, G., and C. V. Falco, 1967. Microwave radiometer design and development. Final report contract no. NAS5-9680.

Schmugge, T., P. Gloersen, T. Wilheit and F. Geiger, 1973. Remote sensing of soil moisture with microwave radiometers. J. Geophys. Res. (to be published).

Wilheit, T., J. Blinn, W. Campbell, A. Edgerton, and W. Nordberg, 1972. Aircraft measurements of microwave emission from Arctic sea ice. Remote Sensing of Environment 2:129.

MESOSCALE DEFORMATION OF SEA ICE FROM SATELLITE IMAGERY

William K. Crowder
Harlan L. McKim
Stephen F. Ackley
William D. Hibler III
Duwayne M. Anderson
U.S. Army Cold Regions Research and Engineering Laboratory
Hanover, N.H.

ABSTRACT

Sequential mesoscale movement and deformation in the pack ice approximately 90 km northeast of Point Barrow, Alaska, have been observed in the ERTS-1 multispectral imagery of 19 to 22 March 1973. At this latitude, sidelap of adjacent ground tracks of daily overpasses is about 75%. This sidelap, together with the coincidence of five cloud-free days and a major westward movement of the pack in the Beaufort Sea Gyre, permitted observation of drift and deformation in an area of about 1.4×10^4 km^2.

Strain calculations using several 10-point arrays yielded shear and divergence rates as large as 1.3×10^{-6} sec^{-1} (0.5% per hour). Continuous deformation measurements through the fast ice/pack ice boundary indicated a sharp change in the sign of the vorticity as the shear zone was crossed. Measured drift velocities varied from 0.24 m/sec to 0.4 m/sec (0.9 to 1.4 km/hr).

These results indicate that detailed deformation and movement data can be obtained from sequential ERTS-1 images. Such data are useful for determining scaling effects in the ice velocity field and for testing existing mathematical models of the response of sea ice to meteorological and hydrodynamic stresses.

INTRODUCTION

There has been considerable interest recently in quantitatively measuring sea ice drift and deformation. Such measurements (Hibler *et al.*, 1973a, b; Thorndike, 1973) are critical in the formulation of realistic models for the drift of the arctic ice pack and in the development of an accurate internal sea ice constitutive law applicable to pack ice on a geophysical scale of tens of kilometers. Two particular types of data have been especially in demand: 1) data on ice deformation across the shear zone region where the moving pack grinds against the stationary shore-fast ice, and 2) sequential deformation measurements on a variety of spatial scales to determine the spatial nature of the ice velocity field.

Although it has been considered feasible that satellite imagery might be utilized for quantitative deformation measurements, such studies have not been performed to date. The principal reason for this is that prior to ERTS-1, satellite sensors have not had sufficiently high resolution; thus, investigations utilizing meteorological satellite data have been primarily qualitative. Even with this limitation, however, it has been possible to detect gross ice boundaries, calculate albedo maps on synoptic scale, and improve our knowledge of the heat balance of the Arctic Ocean (Wendler, 1973).

ERTS-1 provides, for the first time, sufficiently high resolution multispectral imagery to perform detailed ice studies. Preliminary analysis using these data has shown that ice types such as fast ice, pack ice of various concentrations, brash ice, rotten ice, leads, fractures, puddled areas, and flooded ice can be identified (Barnes *et al.*, 1973) and first year floes may be differentiated from multiyear floes (Campbell *et al.*, 1973). Despite such progress, it has not been demonstrated to date that satellite imagery, or other remote sensing imagery, can be used to quantitatively measure the velocity of grid points on the ice pack.

In this paper such a quantitative study has been carried out. By using high resolution multispectral ERTS-1 imagery of sea ice, it has been possible to quantitatively determine sequential positions of grid points on the ice pack. The data analyzed represent a four-day sequence of deformation across the shear zone northeast of Point Barrow, Alaska. The resulting deformation calculations have relevance to the boundary conditions used for drift calculations.

METHODOLOGY

A location map showing the coverage of the satellite imagery used in this investigation is shown in Figure 1. The available imagery consisted of four sequential ERTS-1 multispectral images taken during the period 19 to 22 March 1973. Due to favorable orbital parameters, the imagery obtained during daily coverage swaths at this latitude has sidelap of about 75%.

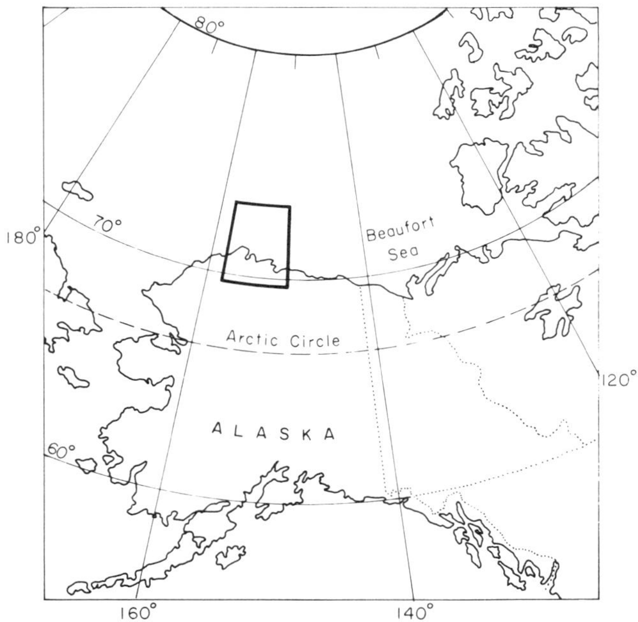

Figure 1. Study area.

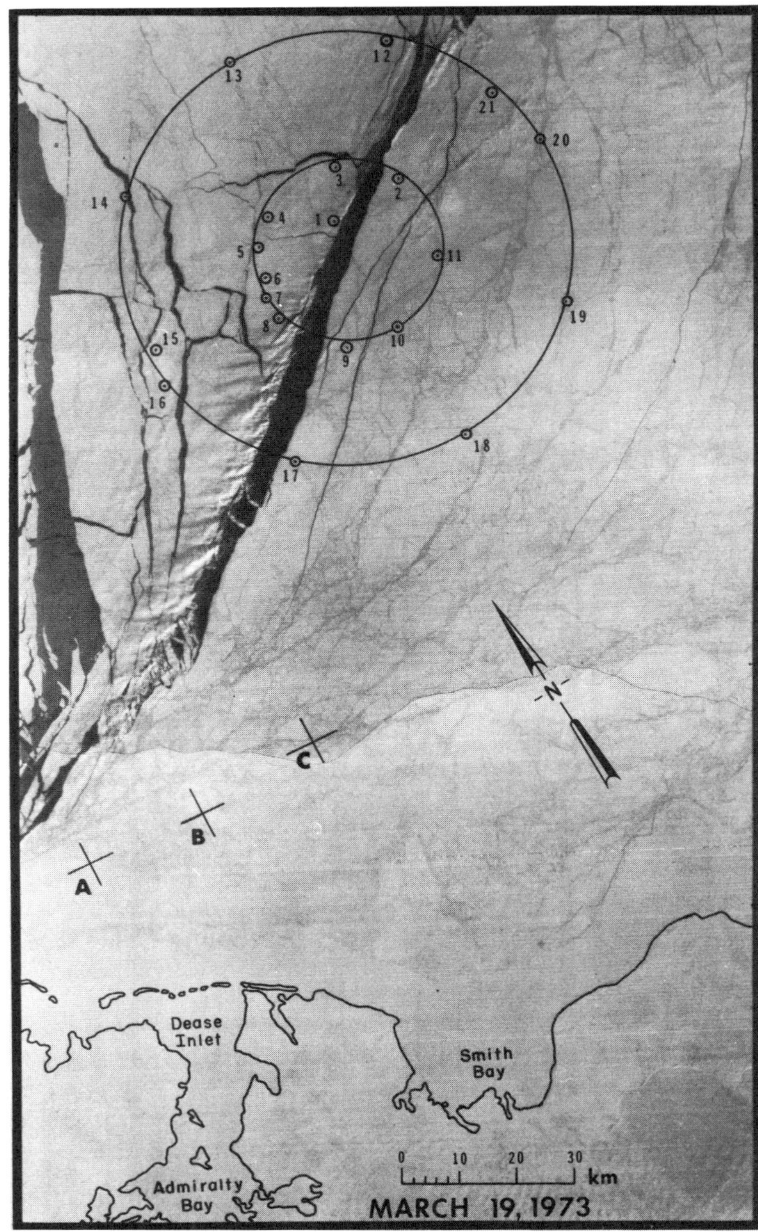

Figure 2. ERTS-1 multispectral images 1239-21461-7 and 1239-21463-7 (band 7, near-infrared) taken 19 March 1973, showing circular strain arrays on the arctic ice pack northeast of Point Barrow, Alaska.

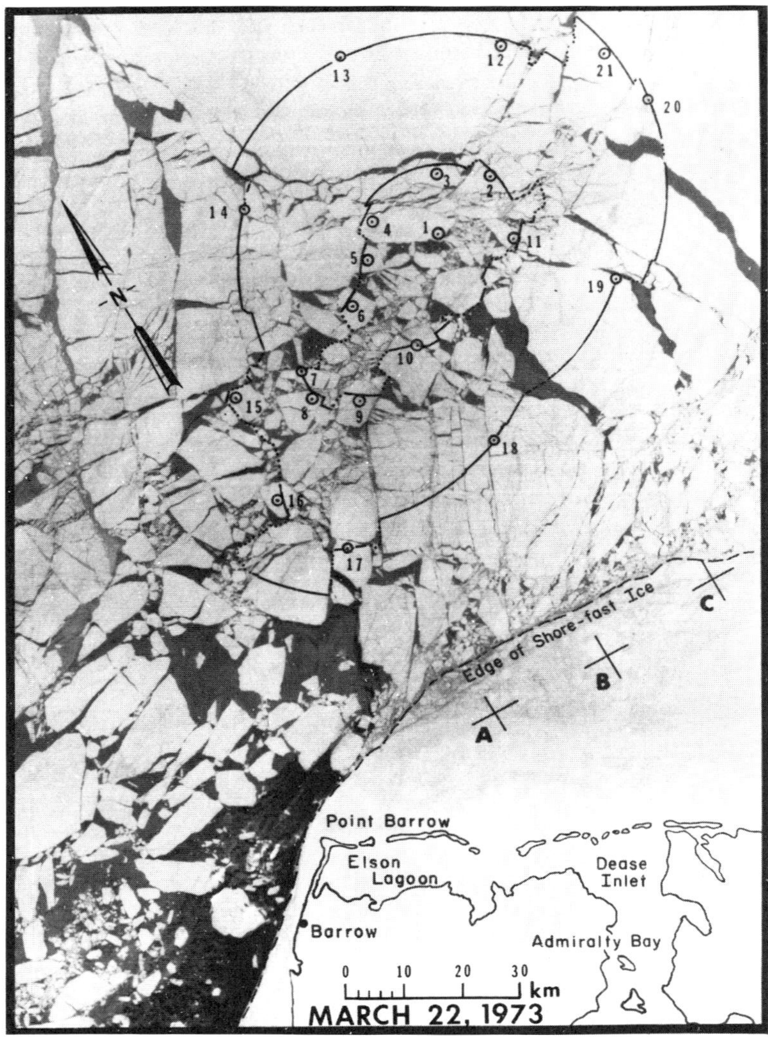

Figure 3. ERTS-1 multispectral image 1242-22032-7 (band 7, near-infrared) taken 22 March 1973, showing the deformation of the arctic pack ice northeast of Point Barrow, Alaska.

On 19 March 1973, deformation along the shear zone was commencing as is shown in Figure 2. Also shown in Figure 2 are two nested strain arrays (starting as circles on the first day) consisting of approximately 20 points and 3 registration points in the apparent shore-fast ice labeled A, B and C. Figure 3 shows the deformation of the nested strain arrays and the drift of the pack ice in relation to the north coast of Alaska as of 22 March 1973. The position of each point on the imagery was followed (with an accuracy of about ± 100 m) throughout the four-day period of deformation relative to a constant fiducial position.

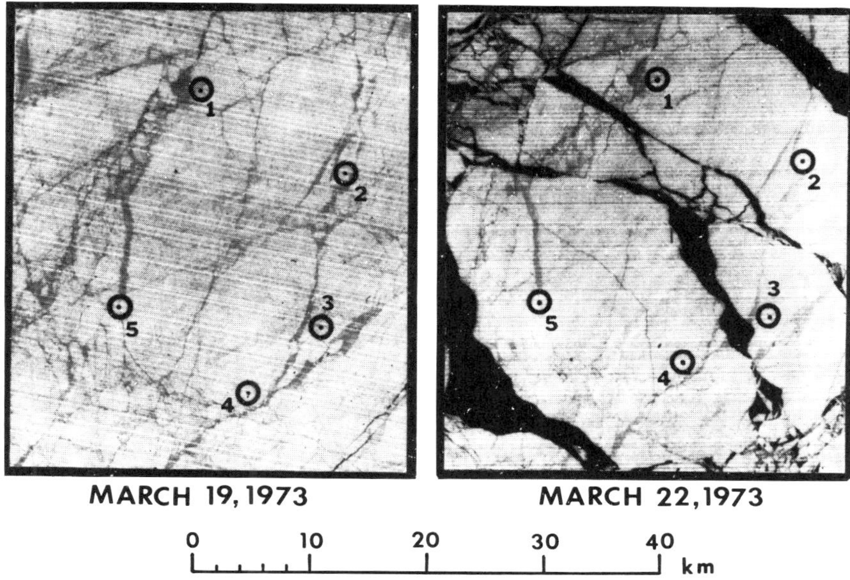

Figure 4. Enlarged portions of ERTS-1 images 1239-21461-7 and 1242-22032-7, showing fracture patterns utilized in transferring strain array points.

The following technique was employed to accomplish the transference of point positions. First a point was established on a given floe or within the apparent shore-fast ice on the initial positive transparency. Then the positive transparency for the next day was placed over it on a light table and positioned so that the fracture pattern of the floe came into registration; the point was then transferred. Fracture patterns are refrozen leads recognized by tone gradations in the imagery (Figure 4). This procedure was used to establish all points on the strain arrays for each successive day. It was done both for drifting floes and for the fixed shore ice so that the relative position of all the selected points could be determined. The width of a typical point was about 0.1 mm, which is equivalent to 100 m at the working scale of 1:1 million.

Using grids of successive position points, the first derivatives of the velocity field (i.e. the strain rate and vorticity) were calculated using a least squares procedure. The grid positions were first converted to a series of strain lines over which linear strain and rotation were measured. These measurements of linear strain and rotation were then used according to the equations discussed by Hibler *et al.* (1973b) which yielded least squares strain rates and vorticity as well as inhomogeneity errors due to the fact that the ice velocity field is not homogeneous. To understand the meaning of the strain field, it is useful to visualize a contour plot of the x (or y) velocity component of the ice. The computer program used to characterize strain in essence fits a planar surface through the contour plot with the slope of the plane being the strain rate. Since the actual velocity components will deviate, possibly in a random distribution, from a perfect plane, there will be some uncertainty in the slope angles of the plane. This uncertainty is represented by the inhomogeneity error.

DEFORMATION RESULTS

A number of deformation calculations were performed using the above techniques. The calculations discussed in this paper were designed with two primary goals in mind: 1) to compare the deformation rate magnitudes and spatial scaling behavior with results obtained from measurements in the interior of the pack (Hibler *et al.*, 1973b), and 2) to examine the variation and the time dependence of the deformation across the shear zone. The nature of this shear zone deformation is important for determining boundary conditions for drift calculations.

Time dependence of deformation and scaling effects

To examine the time dependence of the deformation we used two 10-point arrays consisting of the two circles shown in Figures 2 and 3 and a center point. Ten strain lines were constructed for each array. In Figure 5 we show the four first derivatives of the ice velocity field, calculated from both the inner and outer arrays, together with inhomogeneity errors. The drift rate of the central point is also shown. Since the vorticity Ω, defined by $\Omega = 1/2[(\partial u_y/\partial x) - (\partial u_x/\partial y)]$, **is a measure of the rotation of the ice, we have represented** the off diagonal velocity components by Ω and the shear term $\partial u_x/\partial y$. The x axis is west to east. Note that the shear zone is essentially east-west and hence a positive value of $\partial u_x/\partial y$ indicates a shearing motion with the ice nearer the shore moving *faster* than the ice further north. The strain rates in Figure 5 are three to five times larger than strain rates typically measured in the interior of the pack (Hibler *et al.*, 1973b). The maximum shear and divergence rates are of the order of 0.5% per hour.

From Figure 5 we can make the following observations. The deformation initially begins with a compression of ice in the y direction and an extension in the x direction as the ice moves away from the shore. The vorticity and shear become larger with time (both were small initially) indicating that the pack is generally rotating clockwise with the ice nearer the shore moving more rapidly than the ice further away. These results suggest that the pack is attempting to rotate essentially as a whole, but cannot do so until the bond to the shore is broken. Note also that the deformation rates for the inner array (~ 15-km radius) and the outer array (~ 40-km radius) are quite similar, with the inhomogeneity error generally larger for the inner array, as expected (see Hibler *et al.*, 1973b).

Dependence of deformation on distance from shore

To examine the dependence of the deformation on the distance from the unmoving shore ice, a series of position measurements were made across the shear zone. A schematic array of the position grid is shown in Figure 6. A least squares strain analysis was performed for each four-point box. Of particular interest is the nature of the shear, $\partial u_x/\partial y$, as one proceeds across the shear zone (Rothrock, 1973). Consequently, in Figure 7 we plot the shear rate $\partial u_x/\partial y$ versus distance

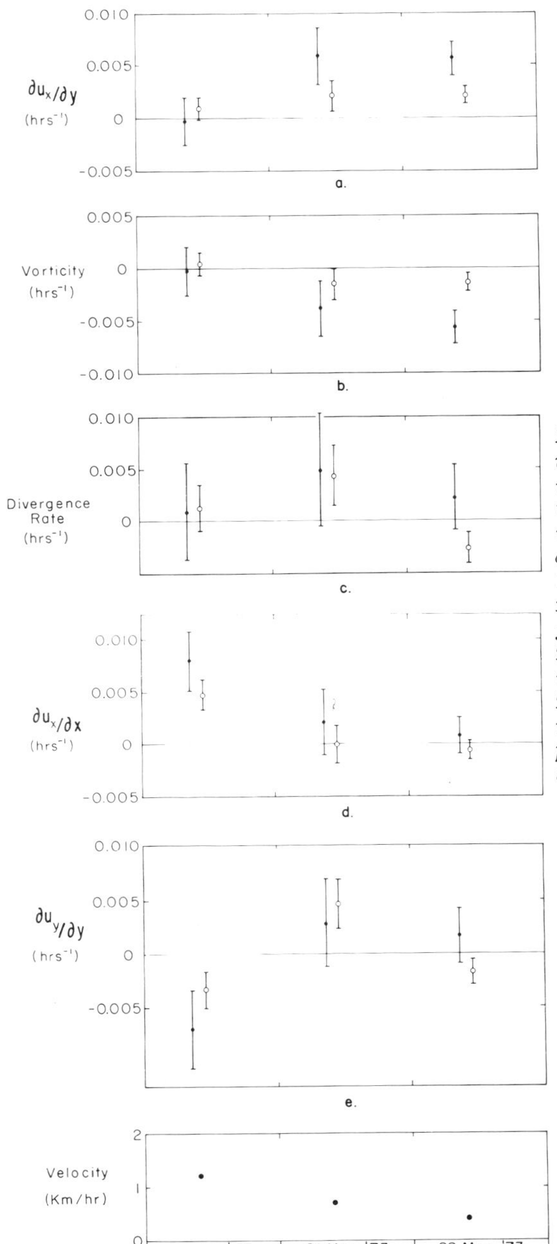

Figure 5. Strain rates, vorticity and ice velocity calculated using the two 10-point arrays shown in Figure 2. The solid data points were calculated from the inner array and the open circles from the outer arrays. The error bars represent inhomogeneity errors. The ice velocity is the velocity of the center point of the two arrays shown in Figure 2.

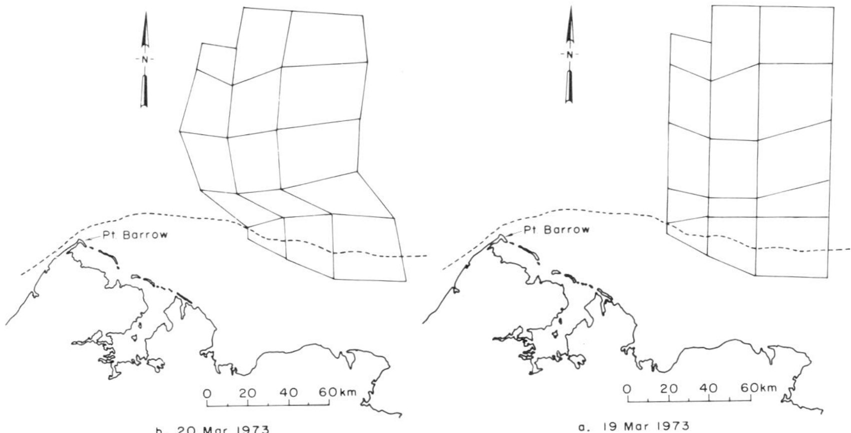

Figure 6. Schematic array of grid points used to study the spatial variation of the strain field. The dashed line marks the boundary between the stationary shore ice and the moving pack.

from the stationary shore ice. In this figure, array 1 refers to deformation measurements obtained using the two right-most lines of the array, array 2 the two inner lines, and array 3 the two left-most lines.

In all cases the shear dips to a large negative value before becoming positive about 60 km away from the shear zone. Also the magnitude of the shear rates is about ten times larger at the shear zone boundary than in the pack. These results generally support the view that the pack as a whole is rotating clockwise and slipping within a narrow region (~ 50 km) at the boundary. This view is in opposition to a model where the pack deforms continually with no slip at the boundary.

Relevance to drift calculations

The deformation results in Figures 5-7 indicate that if a viscous model for ice drift is used, one must either allow slip at the boundary or allow the viscosity to become much smaller in the active shear zone area. This basically follows from the fact that for a viscous model with no slip at the boundary, the slope of the shear rate versus distance curve (Figure 7) would be approximately constant or at least slowly varying near the boundary. This is an important point, because Rothrock (1973) has argued, on the basis of a viscosity model with no slip at the boundary, that the viscosity must be less than ~ 10^{11} g/sec. This value is relatively small compared to viscosities (~ 10^{15} g/sec) required to obtain the best agreement with mesoscale strain measurements (Hibler, 1973c) and to obtain the most desirable average arctic circulation patterns using a steady state viscous model (Campbell, 1965). Our results indicate that the assumption of no slip at the boundary coupled with a viscous model is probably not tenable.

The slope of the shear rate curve away from the shear zone may be used to estimate a shear viscosity for sea ice (Figure 7). Assuming

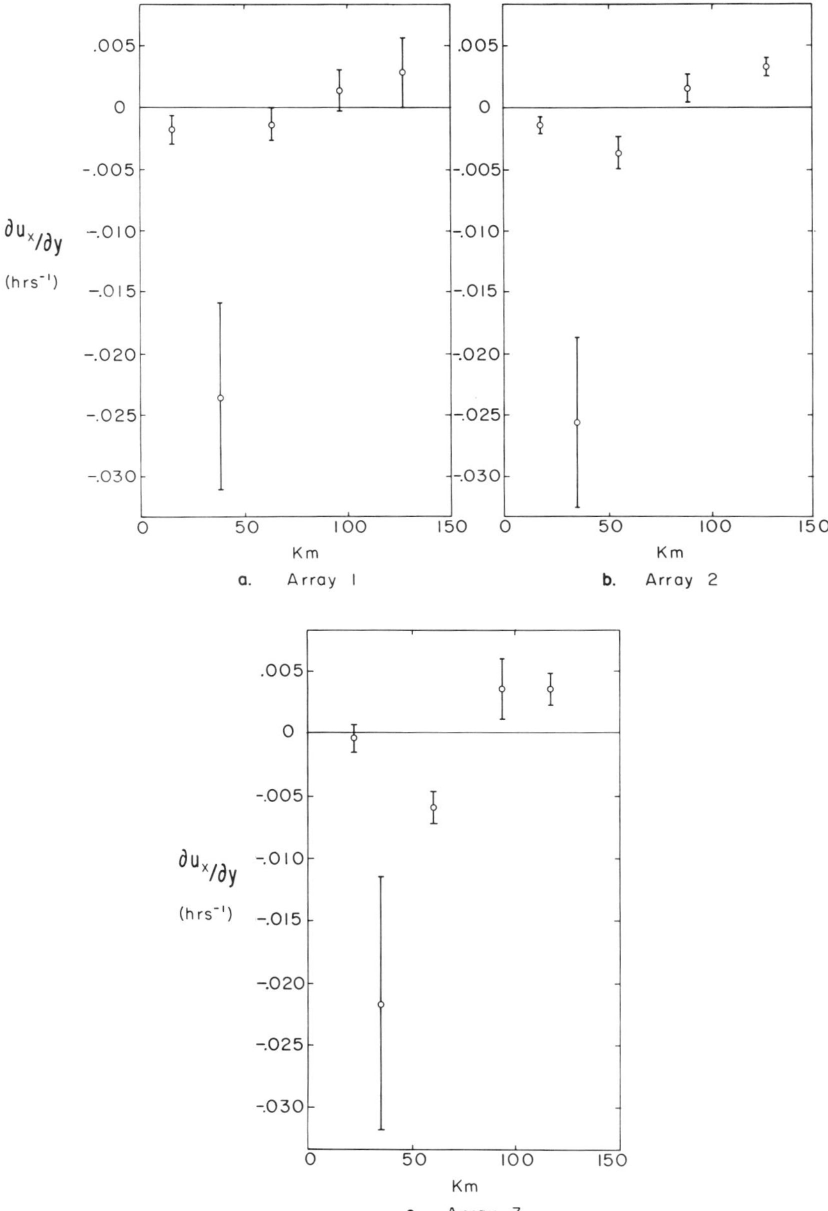

Figure 7. Shear rate as a function of distance from the stationary shore ice. Array 1 refers to the two right-most lines in Figure 6, array 2 the middle two lines, and array 3 the two left-most lines.

a flow law of the form $F_x \sim \eta\partial^2 u_x/\partial y^2$ (Glen, 1970) where η is a shear viscosity, and assuming a force on the ice of ~ 10 dynes/cm^2 due to external forces (wind, water), we have $\eta \sim 5 \times 10^{13}$ g/sec, where we have estimated $\partial^2 u_x/\partial y^2$ to be ~ 2×10^{-13} (cm sec)$^{-1}$ from Figure 7. Although this is somewhat smaller than 10^{15} g/sec, it represents a relatively large viscosity as compared to 10^{11} g/sec.

CONCLUSION

This study indicates that with certain limitations satellite imagery may be used to quantitatively measure deformation and drift of arctic pack ice. One such limitation is the difficulty of obtaining sequential cloud-free imagery coincident with the occurrence of significant deformation.

An important advantage in utilizing satellite images is that the synoptic view permits detailed spatial studies, such as that illustrated here, to be performed. In particular, this investigation of the shear zone suggests that the pack is behaving as a relatively cohesive mass (i.e. highly viscous) with slippage over a narrow region (~ 50 km) at the boundaries. This indicates that the assumption of no slip at the boundary coupled with a single viscosity model is probably not tenable for the arctic ice pack.

REFERENCES

Barnes, J.C. and C.J. Bowley, 1973. Mapping sea ice from the Earth Resources Technology Satellite. Arctic Bulletin 1(1):6-13.

Campbell, W.J., 1965. The wind-driven circulation of ice and water in a polar ocean. J. Geophys. Res., 70(14):3279-3301.

Campbell, W.J., P. Gloersen, W. Nordberg and T.T. Wilheit, 1973. Dynamics and morphology of Beaufort Sea ice determined from satellites, aircraft, and drifting stations. NASA Rpt. No. X-650-73-194.

Glen, J.W., 1970. Thoughts on a viscous model for sea ice. AIDJEX Bulletin No. 2, October 1970, p. 18-27.

Hibler, W.D. III, W.F. Weeks, S.F. Ackley, A. Kovacs and W.J. Campbell, 1973a. Mesoscale strain measurements on the Beaufort Sea pack ice (AIDJEX 1971). J. Glaciol., 12(65):187-206.

Hibler, W.D. III, W.F. Weeks, A. Kovacs and S.F. Ackley, 1973b. Spatial and temporal variations in mesoscale strain in sea ice. AIDJEX Bulletin No. 21, p. 79-114. (Also submitted to J. Glaciol.)

Hibler, W.D. III, 1973c. Comparison of mesoscale strain measurements with linear drift theory predictions. AIDJEX Bulletin No. 21, p. 115-138. (Also submitted to J. Glaciol.)

Rothrock, D.A., 1973. The steady drift of an incompressible arctic ice cover, AIDJEX Bulletin No. 21, p. 49-78.

Thorndike, A.S., 1973. Analysis of position measurements, AIDJEX, 1972. EOS, 53(11).

Wendler, G., 1973. Sea ice observation by means of satellite. J. Geophys. Res., 78(9).

SUMMARY OF DISCUSSION
(Papers 5.1 to 5.6)

Three items dominated the discussion of this session.

1) A discussion of the role of cost/benefit analyses in research and development was an outgrowth of the General Reporter's criticism of one author's use of the term "cost/benefit analysis" when speaking of a cost comparison between satellite and aircraft sensing of snow extent. On the one hand, there was a feeling that some of the proponents of satellite-based remote sensing were promoting a technique which either: (a) had not yet reached the state of development required by the recognized uses for the information; or (b) could provide good measurement of parameters for which there was as yet no recognized need. On the other hand, it was felt that research should proceed wherever it leads us, unencumbered by measures of effectiveness, and that the need will develop once the capability has been demonstrated.

The general feeling coming out of the discussion seemed to be that we may no longer be able to afford to pursue all research leads without regard to cost. We should, however, be careful not to make cost-based decisions governing the direction of research too early in the research and development process. Later on, cost/benefit analyses should be used as a guide when branch points in the process are reached.

As an example of the value of remote sensing, a speaker noted that the State of California could gain about $40 million/year from a good water information system.

2) It was pointed out that many investigators are collecting data such as ice thickness on an "ad hoc" basis, but the data are not being made available to the general research community. A plea was made for a special purpose data center to receive this information, catalog it, and make it available to users. In reply, a speaker suggested that both the U.S. Army Cold Regions Research and Engineering Laboratory (CRREL) and the Naval Hydrographic Office would be happy to receive ice thickness data from any investigator.

3) A speaker pointed out that the mode of scientific investigation seems to have changed with the advent of satellite-based remote sensing. Formerly a natural anomaly would be noted, and this would spawn an observational program designed to develop an understanding of what had happened. Now, we are getting images from space, studying them, noting anomalies and then going out in the field to see what happened. It was also pointed out that analytical model-builders and remote-sensing specialists need to coordinate their work more closely. Each of these fields should advance in parallel with the other in an interlocking and mutually-reinforcing fashion. Advances in model-building should suggest which variables need to be measured, and the resolution required; advances in sensing technology should be planned to reveal the details of phenomena in a way that allows improvement of the models.

SUMMARY OF DISCUSSION
(Papers 5.7 to 5.13)

There was considerable interest in Gloersen's false-color composite (shown at the Symposium, but not contained in paper 5.11), generated from the Nimbus 5 microwave imager, of the Arctic and Antarctic near the time of maximum sea ice extent. In particular, there was discussion of the fact that multi-year ice in the Weddell Sea exhibited brightness temperatures associated with first-year ice. It was suggested that perhaps the so-called "bottom ice", apparently unique to this area, has different microwave emission characteristics than the conventional multi-year ice. Another anomaly, as yet unexplained, is the trend toward higher brightness temperatures with increasing elevation on the Greenland ice cap, in contrast to the decrease with height found (as expected) over interior Antarctica.

Wendler (paper 5.9) was asked about his use of a false-color density slicer with an electronic planimeter in his snow extent study. He explained the advantages of its use for rapid and repeatable measurements of the snow-covered or snow-free areas of a basin from ERTS images.

Crowder (paper 5.13) was complimented on his use of ERTS images for strain calculations. It illustrated how quantitative data can be obtained from satellite images and then related to models.

Lauer's paper (5.10) came in for considerable discussion also. Part of this dealt with the validity of the test for accuracy of the method for determining snow extent from ERTS images. A speaker commented that errors of 10% or more can stem from the great irregularity of the snow boundary due to bare spots, snow patches, etc. In response to the question of whether further refinement or automation of the technique was justified by the need for such data in current hydrologic models, the consensus of those present was definitely in the affirmative.

Schertler (paper 5.8) was asked to elaborate on his findings with regard to the use of infrared imagery (aircraft) for estimating ice thickness on the Great Lakes. He showed a graph (not in his paper) illustrating the relationship between surface temperatures and ice thickness based on a simple model (no snow on ice). It showed little dependence of the temperature on the thickness for thicknesses greater than about 20 cm. A participant remarked that the relationship in question is greatly complicated by snow conditions and by whether the ice is forming or decaying.

In general discussion, not specifically related to any particular paper, it was observed that a ratio of the MSS-5 to MSS-7 channels on the ERTS satellite is a useful quantity in snow analysis, particularly when the detection of wet snow is involved.

A speaker asked if anyone is investigating the use of spectral reflectance analysis for removing clouds from snow images. Purdue University's Laboratory for Applied Remote Sensing (LARS) has experimented with a scheme using ERTS channels, but with no real success. A similar attempt using Skylab data apparently has had some success, but the snow was probably wet, which causes a drop in the reflectance of the snow but not of the clouds when near-infrared bands are compared with visible bands.

Finally, it was pointed out that hydrologists and other potential users of space data should let NASA know what their requirements are that are not being met by present systems so that the requirements can be taken into account in the design of future satellite systems.

Session VI

Nuclear Techniques

Chairman:

E. J. Langham, INRS/EAU, Université du Quebec

General Reporters:

R. E. Isaacson, Advanced Process Development, Atlantic
Richfield Hanford Co.

Zolin Burson, E G & G, Inc.

AN EXPERIMENTAL GAMMA-RAY SPECTROMETER SNOW SURVEY OVER SOUTHERN ONTARIO

R.L. Grasty
Geological Survey of Canada
Department of Energy, Mines and Resources
Ottawa, Ontario

H.S. Loijens
Environmental Management Service
Department of the Environment
Ottawa, Ontario

H.L. Ferguson
Atmospheric Environment Service
Department of the Environment
Toronto, Ontario

ABSTRACT

In the winter of 1972-1973, four gamma-ray spectrometer surveys, each 1850 km long, were flown over Southern Ontario at 150 m, using a 50,000 cm^3 sodium iodide detector system. Total radioactivity and potassium information was used to calculate a snow-water equivalent for 16 km sections along each flight line. The airborne results are compared with ground data from 10 snow courses established along the flight lines.

A root mean square deviation of 1.2 cm water equivalent was found between the ground and potassium-airborne results, whereas a deviation of 1.7 cm water equivalent was calculated between data from the ground and those from the total radioactivity information.

Soil moisture corrections from measurements at selected sites were found to decrease the calculated snow-water equivalent an average of 1.7 cm. Errors in the soil moisture measurements and background variations encountered along the flight lines were found to be more important than statistical errors.

INTRODUCTION

Since 1967 the Geological Survey of Canada has been mapping surface concentrations of potassium, uranium, and thorium using a high sensitivity airborne gamma-ray spectrometer. In the winters of 1970-1971 and 1971-1972, encouraging results were obtained over a uniformly radioactive test strip near Ottawa, which was established to measure the water equivalent of a snowpack by monitoring the absorption of natural gamma rays emitted by potassium-40, thallium-208, and bismuth-214 (Grasty, 1973). The measurement technique used was initially developed in the U.S.S.R. (Kogan *et al.*, 1965), where it is now in general use (Dmitriev *et al.*, 1971). Similar research has been carried out in Norway (Dahl and Odegaard, 1970) and in the U.S.A. (Peck *et al.*, 1971).

In 1972, the Department of the Environment, with the Geological Survey of Canada initiated a cooperative project to determine the feasibility of measuring the spatial variation of snow-water equivalent over a large area. Southern Ontario was selected because of the immediate need of the Water Planning and Operations Branch for more information in forecasting the inflow from snowmelt into the Great Lakes, and because the large number of snow courses already being surveyed would make possible a comparision of results.

DATA COLLECTION

During the fall and winter of 1972-1973, four airborne surveys were carried out, each 1850 km long and divided into 18 distinct flight lines for ease of navigation. Flight lines varying in length from 10 km to 190 km, were established in cooperation with the agencies operating the existing snow courses. The flight lines were laid out approximately perpendicular to the average snow-water equivalent isolines, which are strongly influenced by the Great Lakes. In addition, eight parallel flight lines, approximately 20 km long and one kilometer apart, were established over the Wilmot Creek Research Basin (265 km^2). The locations of the flight lines and snow courses are shown in Figure 1. In late October and early November 1972, two surveys were flown under very moist soil conditions. These presnow surveys were expected to determine the reproducibility of the radiation measurements and to give some indication of the accuracy of the airborne technique. Lack of ground information for the November presnow survey, however, and the presence of newly fallen snow over part of the area, made a comparison difficult. Airborne snow surveys were flown in January and February 1973. All surveys were flown with a nominal terrain clearance of 150 m at approximately 190 km/hr. An inventory of all snow courses intercepted by the

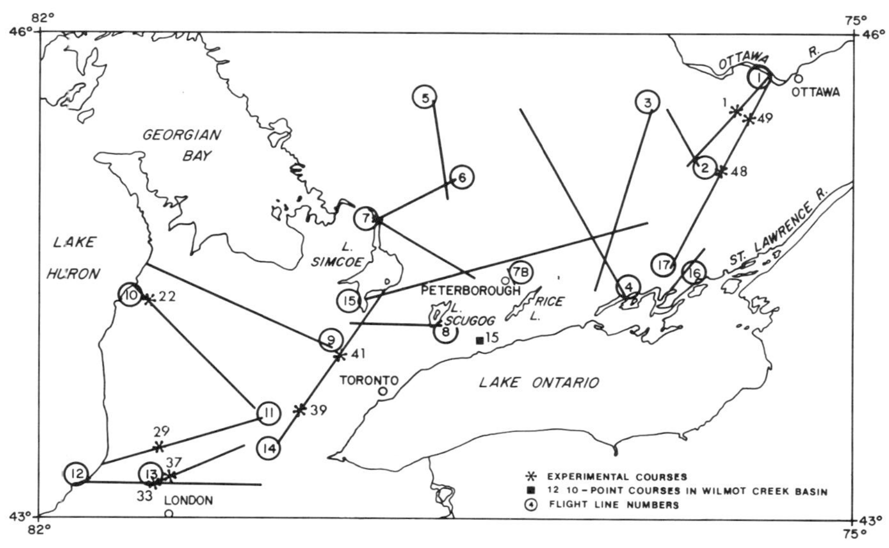

Figure 1. Flight line and snow course locations

6.1

airborne survey and of those in the immediate area are presented in Loijens and Grasty (1973).

After the flight plan was firmly established, 10 experimental courses were set up under the flight track. These courses were generally 2-7 km long, compared to a maximum of 900 m for the existing courses (Figure 1). Snow-depth soundings and density measurments were generally made at equidistant points along the experimental courses, with 10 times as many depth measurements as density measurements. The Meteorological Service of Canada (MSC) snow sampler was used to measure snow density. Details of the ground survey for these experimental courses are presented in Table 1.

TABLE 1 INFORMATION ON EXPERIMENTAL SNOW COURSES

COURSE NUMBER	NAME OF COURSE	LENGTH (M)	NO. OF SNOW SAMPLE POINTS	NO. OF SOIL SAMPLES
1	Almonte	2500	42	9
15	Wilmot Creek	3100	120	12
22	Paisley Traverse	4800	260	5
29	Farquhar	6500	65	7
33	Granton	7050	47	8
37	Willburn	6400	16	8
39	Blue Springs Creek	5300	12	4
41	Cold Creek	3000	172	11
48	Perth	1400	28	5
49	Carleton Place	2900	45	6

Because almost all existing courses are surveyed on the first and the fifteenth of each month, the airborne surveys were planned to coincide with these dates. Weather conditions delayed several flights, however, and ground and airborne surveys were carried out on the same day on few of these existing courses. Corrections of the ground-survey data to that of the airborne survey were difficult, because of the mild and erratic winter weather. In most cases the experimental courses were surveyed on the day on which the gamma-spectrometer survey was flown, because these ground surveys were operated on a stand-by basis. In this paper, we shall discuss only results for the experimental courses and the Wilmot Creek Basin, which had detailed ground coverage. The results for all existing courses are contained in a report by Loijens and Grasty (1973).

Because radiation emitted from the ground depends on soil moisture content, soil samples were taken on the majority of the snow courses. For a homogeneous soil, 72 percent of the gamma radiation at the surface is emitted from the top 10 cm and 17 percent from the next 10-20 cm (Zotimov, 1968). Soil samples were taken to a depth of 15 cm, using an Oakfield soil sampler kit which contains a tube for unfrozen soil and an auger for frozen and heavy soils. Each sample was sealed in a can and shipped to Ottawa for analysis; the sampling scheme was designed to obtain a representative value of soil moisture along each course. Three samples per vegetation class per course were considered to be the minimum, but the actual number varied from four to twelve (Table 1). Soil samples were collected for each survey except the second presnow cover flight in November (Table 2).

TABLE 2 MEAN SOIL MOISTURE (percent of dry weight) AND STANDARD ERROR

COURSE NUMBER	OCTOBER	JANUARY	FEBRUARY
1	30 ± 2	48 ± 7	64 ± 8
15	31 ± 5	45 ± 4	73 ± 10
22	33E	42E	47 ± 3
29	50E	64 ± 13	76 ± 11
33	42E	54 ± 6	60E
37	57E	73 ± 21	56 ± 5
39*	15 ± 2	17 ± 1	18 ± 2
41	30 ± 3	44 ± 3	57 ± 10
48	26 ± 4	44 ± 7	39 ± 6
49	35 ± 2	60 ± 9	67E
AVERAGE	28 ± 3	57 ± 7	

E estimated

* samples analysed by the University of Guelph

THEORY

The relation between the detector count rate N and flying height H of the aircraft for an infinite homogeneous half-space emitting mono-energetic gamma radiation is given by Godby et al. (1952).

$$N = N_o \int_1^{\infty} \frac{e^{-\mu_a Hx} dx}{x^2} = N_o E_2 (\mu_a H) \qquad (1)$$

E_2 is known as the exponential integral of the second kind; μ_a is the linear absorption coefficient of air corresponding to that particular energy, and N_o is the count rate at the surface of the ground, which depends on the radioactive concentration and absorption coefficient of the ground and various physical parameters of the detector system.

The count rate N_o at the surface is dependent on soil moisture content and is given by Kogan et al. (1969).

$$N_o = \frac{100 \ N_D}{100 + k\omega} \qquad (2)$$

where ω is the soil moisture content expressed in percentage of dry weight, and N_D is the count rate at the surface for dry rock. k is the ratio of the number of electrons in 1 g of water to the number of electrons in 1 g of rock. Because most soils and rock material are made up predominantly of oxygen, silicon, and the lighter elements, k is approximately 1.11.

For a presnow flight, equation 1 can be written as

$$N_{ps} = \left[\frac{100 \ N_D}{100 + k\omega_{ps}}\right] E_2 \ (\mu_a H) \qquad (3)$$

where the subscript ps is used to designate presnow conditions and all other terms are as previously defined.

For ground covered by a snow layer with water equivalent D, the detector count rate N_s is given by a similar expression

$$N_s = \left[\frac{100\ N_D}{100 + k\omega_s}\right] E_2\ (\mu_a H + \mu_w D) \qquad (4)$$

where μ_w is the linear absorption coefficient of water and ω_s is the soil moisture content for the snow flight. If a course is flown both with and without snow cover, the count rate N_{ps} before snow, and N_s with snow, corrected to the same temperature, pressure and height, will then be related by

$$\frac{N_{ps}}{N_s} = \left[\frac{100 + k\omega_s}{100 + k\omega_{ps}}\right] \left[\frac{E_2(\mu_a H)}{E_2(\mu_a H + \mu_w D}\right] \qquad (5)$$

The linear absorption coefficient of air for the integral, potassium, uranium, and thorium channels were experimentally determined by flying over a uniform test strip near Ottawa at different altitudes (Grasty, 1973). The linear absorption coefficient of water was calculated from these experimental values for air, using the relationship that water is 1.11 times as effective as the same mass of air in absorbing gamma rays of the energies concerned (Davisson, 1965). This factor 1.11 is the ratio of the number of electrons in 1 g of water to the number of electrons in 1 g of air. These experimental values for air and water are presented in Table 3.

TABLE 3 EXPERIMENTAL LINEAR ABSORPTION COEFFICIENTS

	Air (0°C, 76 cm Hg) m^{-1}	Water m^{-1}
Potassium	0.005352	4.60
Uranium	0.002183	1.87
Thorium	0.004296	3.68
Integral	0.003733	3.21

TABLE 4 SPECTRAL WINDOW WIDTHS

Element	Isotope	Photon energy	Window width
Potassium	K-40	1.46 MeV	1.37 - 1.57 MeV
Uranium	Bi-214	1.76 MeV	1.66 - 1.86 MeV
Thorium	Tl-208	2.62 MeV	2.41 - 2.81 MeV
Integral			0.41 - 2.81 MeV

THE AIRBORNE SYSTEM

The detector system consists of twelve 22.86 x 10.16 cm sodium iodide (thallium activated) crystals, maintained at 38°C in thermally insulated boxes to minimize spectral drift. Pulses from these detectors are fed into a 128-channel analyzer, from which four energy windows can be preselected by a plug-in program card and the pulses accumulated in four scalers. The energy windows representing integral count, potassium, uranium, and thorium are shown in Table 4, with the radioelement and photon energy. The accumulated counts from the scalers are recorded digitally on magnetic tape with details of height, position, and manually inserted operational information. Integral count information is recorded every 0.5 seconds and the three elements every 2.5 seconds.

The spectrometer is flown in a Short Skyvan aircraft, operated by the Geological Survey of Canada. The aircraft is fitted with a Doppler navigation system, for monitoring along and cross track position and a radar altimeter. The output from the radar altimeter, recorded on magnetic tape, is the average terrain clearance over each 2.5 second counting period. The image from a TV camera, fitted with a wide-angle lens, and mounted in the nose of the aircraft, is relayed onto a 12.7 cm monitor screen in front of the equipment operator and is used for confirmation of track. The along-track Doppler information is also displayed on a corner of the TV screen as a navigational aid.

The spectrum is calibrated on the ground by a program card, switched into the system manually, with three windows bracketing the potassium-40 peak. This prominent potassium-40 peak is present in any gamma-radiation spectrum from the ground. By adjustment of the high-voltage supply to the photo multiplier tubes, the spectrum can be shifted so that the window centered on the potassium peak has maximum count rate. A storage scope display of the spectrum and window positions can also be used for in-flight monitoring of the calibration. Normally no adjustment of the spectrum is necessary during the course of a flight.

DATA-PROCESSING PROCEDURE

Before the count rates for the presnow and snow-cover flights can be used to calculate snow-water equivalent, the data must be corrected for background radiation, variations in flying height, temperature, and Compton scattering in the crystals. Coordinate information for each data point must also be supplied. The computer techniques for the automatic data processing of the magnetic tapes to produce final corrected count rates and coordinate information has been documented by Grasty (1972).

The count rates in each window must first be corrected for background radiation originating from cosmic radiation, contributions from the aircraft and its equipment, and variable amounts of bismuth-214, a decay product of radon present in the air. One background value was used for each line from measurements taken over a large lake, normally flown before the start of each flight, and from other lakes found along the line.

Because of the finite size of the detectors, a certain fraction of high-energy thallium-208 photons that would be detected in the thorium window are incompletely absorbed and appear as counts in the lower-energy uranium and potassium windows. Similarly, incompletely absorbed bismuth-214 photons can appear as counts in the lower-energy potassium window. The fraction of counts in a high-energy window that appear in a lower energy window because of incomplete absorption is known as a stripping ratio or Compton scattering coefficient. The Compton coefficients α (uranium counts/thorium count = 0.35), β (potassium counts/thorium count = 0.33) and γ (potassium counts/uranium count = 0.56) have been experimentally determined using uniformly radioactive slabs of known composition, constructed at Uplands Airport, Ottawa, Ontario (Grasty and Darnley, 1971). These calibration slabs can be considered to be effectively infinite in size for a detector located at or very close to the surface.

If K_B, U_B, Th_B are the potassium, uranium and thorium background count rates, and K, U and Th are the total counts detected, then the corrected count rates K_C, U_C, and Th_C are related by the following equations:

$$Th_C = Th - Th_B \qquad (6)$$

$$U_C = U - U_B - \alpha\, Th_C \qquad (7)$$

$$K_C = K - K_B - \beta\, Th_C - \gamma\, U_C \qquad (8)$$

The detector count rates are also dependent on the mass of air between the aircraft and the ground, and consequently on the altitude of the aircraft and the air density. The effective flying height H at 0^oC and 76 cm mercury was calculated from the general relation

$$H = H_T \frac{273}{T} \frac{P}{76} \qquad (9)$$

in which H_T is the flying height, T the absolute air temperature (^oK) and P atmospheric pressure (cm Hg). No correction was made for atmospheric pressure changes: equation 9 consequently reduces to

$$H = H_T \frac{273}{T} \qquad (10)$$

The air temperature used was read by the pilot on a thermometer mounted outside the aircraft. All count rates were corrected to a height of 150 m at 0^oC, using data presented in Table 3. Each data point was given a coordinate position on a 1:250,000 map, based on distances indicated by the Doppler navigation system. Profiles of the corrected potassium, thorium, and integral count were plotted together with aircraft altitude for visual checking of the data.

The ratio of the count rates before snow and with snow cover (equation 5), can now be used to calculate the water equivalent of the snow pack. However, because of statistical fluctuations in the count rate associated with the natural gamma radiation, the ratio cannot be used reliably to calculate the water equivalent for each data point. By accumulating counts over 16 km sections, the conflicting requirements of resolution, necessary for comparing the ground and airborne results, and good counting statistics were best satisfied. The average count rates for potassium, thorium, and the integral over the same 16 km section were determined for both the snow and presnow flights. These average count rates were used to calculate the average water equivalent for that section. For the Wilmot Creek Research Basin, an average snow-water equivalent was determined for all eight flight lines.

Background radiation measurements over large lakes were normally taken at the beginning of each flight and as frequently as possible during the survey. Previous work indicated that variations in the integral and potassium channels were directly related to variations in the bismuth-214 content of the air as measured by the uranium channel. The high-energy thorium count rate has remained constant over the previous four summers' surveys, suggesting that the cosmic background shows little variation with time. This situation has also been observed by other authors (Zotimov, 1965; Dahl and Odegaard, 1970). The thorium values obtained during this winter's survey were appreciably higher and exhibited a much greater variation than those found previously. These variations were sufficiently large that the measurements of snow-water equivalent, by monitoring the absorption of the thorium channel, were considered unreliable and they are therefore not presented. The high and variable thorium background has not yet been explained, although there always remains the possibility of equipment malfunction.

GROUND AND AIRBORNE RESULTS

For experimental courses 29, 33, 37, and 39 (Table 1), a water equivalent measurement was taken at each sample point. For these courses the arithmetic mean snow-water equivalent and associated standard error were calculated. For the remainder of the courses, a different number of density and depth measurements were taken. An average density was determined for each vegetation class and thereafter, the water equivalent calculated for each point. The mean snow-water equivalent and standard error were then calculated for the entire line (Table 5).

TABLE 5 GROUND AND AIRBORNE MEASUREMENTS OF SNOW-WATER EQUIVALENT (CM)

		GROUND RESULTS	AIRBORNE RESULTS				
	COURSE NUMBER		NO SOIL-MOISTURE CORRECTION INTEGRAL	NO SOIL-MOISTURE CORRECTION POTASSIUM	THEORETICAL SOIL-MOISTURE CORRECTION INTEGRAL	THEORETICAL SOIL-MOISTURE CORRECTION POTASSIUM	EXPERIMENTAL SOIL-MOISTURE CORRECTION INTEGRAL
JANUARY	1	14.1 ± 0.7	19.6	17.0	16.7	14.9	17.4
	15	4.1 ± 0.2	4.5	5.4	2.4	3.8	2.9
	29	5.6 ± 0.4	7.1	7.3	5.3	5.9	5.7
	33	3.9 ± 0.4	5.4	5.9	3.7	4.6	4.1
	37	4.0 ± 0.5	8.1	4.8	6.0	3.4	6.4
	39	6.1 ± 0.4	8.3	7.3	7.9	7.1	8.1
	41	6.3 ± 0.3	5.9	6.2	3.7	4.6	4.2
	48	10.7 ± 0.4	11.5	12.6	8.7	10.5	9.3
	49	12.0 ± 0.3	11.8	13.0	8.1	10.3	8.9
FEBRUARY	1	13.1 ± 0.6	16.0	15.7	10.8	11.8	12.0
	15	1.4 ± 0.2	2.9	3.7	0.0	0.0	0.0
	22	1.5 ± 0.1	2.7	3.3	0.6	1.8	1.1
	37	1.8 ± 0.3	3.2	1.7	3.3	1.8	3.3
	39	1.7 ± 0.3	2.8	4.1	2.3	3.8	2.5
	41	1.9 ± 0.1	3.1	4.8	0.0	1.9	0.2
	48	10.8 ± 0.7	13.4	14.8	11.8	13.2	11.8
	AVERAGE	6.2	7.9	8.0	5.7	6.2	6.2

The mean soil moisture content expressed as percent of dry weight and the standard error for each course in the October presnow flight and the two snow surveys are presented in Table 2. A large variation in the soil moisture content was found along most courses, but for some courses, soil moisture data were not available. At four sites soil samples were not collected during the October flight. In these circumstances, a soil moisture value was estimated from the average proportional soil moisture increase of 28 percent, found between the October and January measurements of samples obtained at all existing and experimental courses. An average proportional increase of 11 percent in soil moisture was found between the January and February flights. Values estimated from these increases are indicated in Table 2.

The variation of count rate with soil moisture (equation 2) assumes that the ground is homogeneously radioactive. If the radioactivity is concentrated toward the surface, it is evident that soil moisture will have less effect on the count rate. Equation 2 also assumes that any gamma-ray photon that loses energy through a Compton collision caused by increasing soil moisture will be lost to the detector. This assumption is not valid because the scattered photon, in many circumstances, still has sufficient energy to be detected in the integral window. A small fraction of photons in the gamma-ray spectrum

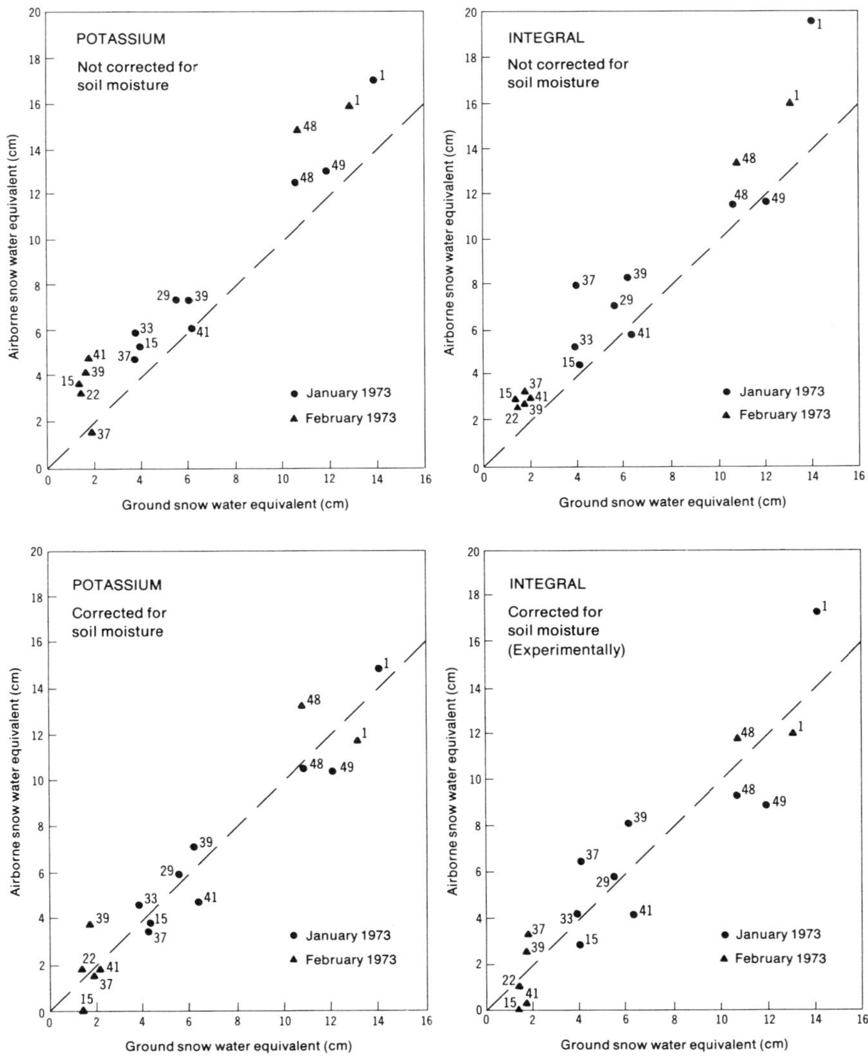

Figure 2. Ground and Airborne Results

6.1

have an energy exceeding that of the potassium window. Only a small percentage of these high-energy photons will be scattered into the potassium window with the majority appearing in the integral window. It is therefore to be expected that the theoretical equation 2 would hold closely for potassium but may not hold for the integral count rate.

Table 5 lists the snow-water equivalent obtained from ground and airborne measurements. The airborne measurements, using both potassium and integral counts, are given without correction for soil moisture and with soil moisture correction based on the theoretical equation 5. The uncorrected integral and potassium data, together with the corrected potassium data are also presented in Figure 2. The results in Table 5 show that the theoretical soil moisture correction, when applied to the airborne potassium measurements, shows an average decrease of 1.8 cm in the indicated snow-water equivalent. The average of the corrected potassium values is the same as that of the ground values, and indicates that the theoretical equation 2 for soil moisture effect is valid for potassium. However, the average of the airborne integral values without soil moisture correction is 1.7 cm higher than the ground value. With the theoretical correction, the airborne integral average is 0.5 cm lower than the ground value, suggesting that the theoretical correction is too large in the integral case.

Previous workers have made various estimates of the effect of soil moisture on integral count levels. Present results suggest a reduction of the soil moisture correction for integral count measurements. Equation 2 was modified to include a reduction factor f and is expressed by the relation

$$N_o = \frac{100\ N_D}{100 + k\omega f} \qquad (11)$$

where N_o, N_D, k and ω are the same parameters shown in equation 2. The factor f, equal to 0.69, was calculated so that the average ground and airborne measurements would have the same value (Table 5).

From the above relationship, a soil with a 30 percent soil moisture content that is increased by 10 percent to a 33 percent soil moisture content will show a 2.0 percent decrease in the integral count rate. From data presented by Zotimov (1965), a 2.7 percent decrease in the integral count rate can be calculated for this 10 percent increase in soil moisture, whereas Peck *et al*. (1971) assumed a reduction of 3 percent. The theoretical equation 2 gives a 2.5 percent decrease in the potassium count rate for this same increase in soil moisture. Based on an increasing radioactivity towards the surface, Peck *et al*. calculated a 5 percent decrease in the potassium count rate.

From the experimental soil moisture corrections of equation 11, new airborne snow-water equivalents, determined from the integral data, are presented in Table 5 and Figure 2. The root mean square deviation between the ground results and these new airborne values is 1.7 cm compared to 2.2 cm for the uncorrected values. The potassium airborne results indicate a root mean square deviation of 1.2 cm compared to 2.1 cm for the uncorrected values.

DISCUSSION OF ERRORS

The airborne results using the potassium window have proved superior to those obtained from the integral information. If only statistical errors caused by the random nature of the radioactive process were to be considered, the integral results would be expected to be more reliable because of the greater number of photons detected. An important consideration, however, is the variability of background along each flight line especially because the automatic processing procedure allowed only one background for each line and because of the considerable length of some lines. The average background radiation and its variation along all survey flight lines is shown in Table 6 with the average ground radiation.

In addition to errors caused by counting statistics and background variations along a flight line, errors are also associated with the different experimental conditions for the two flights necessary for this survey. These operational errors are related to equipment stability and duplication of the flight track. Duplication of the flight track is particularly important in areas of variable ground radioactivity, such as the northern part of the survey area where there are many lakes and swamps, and varying amounts of outcrop and overburden typical of the Canadian Shield. Drift of the spectrum and changes in equipment sensitivity and resolution will also result in operational errors. The magnitude of these errors is difficult to evaluate directly, but the error can be minimized by adequately stabilizing the equipment and choosing easily navigable flight lines that pass over areas of uniform ground radioactivity.

Because the count rates have to be corrected for soil moisture changes between the two flights, it is important to obtain reliable soil moisture measurements representative of the area viewed by the detectors. From Table 2 it is apparent that large variations are associated with these ground measurements, particularly those carried out in January and February.

It is useful to evaluate the relative importance of the various errors on the calculation of snow-water equivalent. The calculation depends on the ratio of the count rates obtained for the presnow and snow survey, and errors associated with both flights have to be considered. Errors in the average count rate for a 16 km section of line flown at 190 km/hr were calculated using average count rates for the entire survey and these are presented in Table 6. For the integral count technique, the statistical error is equal to the square root of the total number of photons detected in the 5-minutes flying time for

TABLE 6 AVERAGE GROUND AND BACKGROUND RADIATION

SURVEY	AVERAGE GROUND RADIATION (COUNTS/MIN)		AVERAGE BACKGROUND AND VARIATION (COUNTS/MIN)	
	INTEGRAL	POTASSIUM	INTEGRAL	POTASSIUM
PRESNOW	47400	4250		
			35760 ± 1260	2832 ± 94
SNOW	35900	2750		

the 16 km section. This error is expressed as a percentage of the average ground radiation in Table 7. In this calculation, no errors were assumed in the background measurements. The statistical errors for the potassium window are somewhat complicated by the Compton scattering corrections from the higher-energy uranium and thorium windows carried out according to equations 6, 7, and 8. Background errors from data presented in Table 6, are also expressed as a percentage of the ground radiation. The effect on the corrected count rates of errors in the measurement of soil moisture can be computed using equations 2 and 11, by increasing the average soil moisture value by an amount equal to the average error presented in Table 2; a percentage error in the ground radiation can then be found. Average errors in the count rates for the

TABLE 7 ERRORS IN AIRBORNE TECHNIQUE

	INTEGRAL			POTASSIUM		
	STATISTICAL	BACKGROUND	SOIL-MOISTURE	STATISTICAL	BACKGROUND	SOIL-MOISTURE
PRESNOW COUNT RATE (%)	0.3	2.7	1.9	0.9	2.2	2.5
SNOW SURVEY COUNT RATE (%)	0.3	3.5	3.8	1.3	3.4	4.9
WATER-EQUIVALENT ERROR (cm)	0.1	0.8	0.8	0.2	0.6	0.8

presnow survey and the two snow surveys can then be used to determine an average error in the snow-water equivalent. It is interesting to note that because of the higher potassium absorption coefficient, the same percentage error in the integral and potassium count rates gives a smaller snow-water equivalent error for the potassium window.

A significant fraction of the errors in the airborne measurements are believed to be caused by navigational problems in duplicating the flight track. By flying over a test strip 37 times in the Mont Laurier area of Quebec, the operational errors associated with the integral and potassium measurements were estimated to be 3.5 and 5.1 percent, respectively (Grasty, 1973). The airborne results, however, were not corrected for soil moisture or temperature, and these values could very well be reduced. In this type of survey, it is difficult to lay out a flight track with distinguishable landmarks at sufficiently frequent intervals. If a much smaller area were covered, roads, hydro lines, and so forth could be used as landmarks. From Figure 1, we see that navigation is complicated because all the lines must be flown with different magnetic headings. More than 90 1:50,000 map sheets, some of them produced from aerial photographs taken in the 1920's were used for the survey. An attempt was made, using the gamma-spectrometry data, to establish when the pilot was off track. This was done by dividing each 16 km section into 3.2 km sections and by calculating a water equivalent for these shorter sections. When a large variation in the water equivalent was found for the short sections, the operator monitoring the flight path from the TV camera in the nose of the aircraft noted that the plane was deviating from its intended flight track (Loijens and Grasty, 1973). By flying parallel lines with the same magnetic headings and using reliable aerial photographs or detailed topographic maps, these operational errors could undoubtedly be reduced.

No atmospheric pressure corrections were applied to the data, and consequently any pressure variation between the presnow and snow survey will result in an apparent increase or decrease in the calculated water equivalent. From data obtained at the Ottawa weather office, the maximum errors incurred are not expected to exceed 0.3 cm water equivalent. To minimize all possible sources of error, it is recommended that the atmospheric pressure be monitored aboard the aircraft and the airborne data corrected according to equation 9.

Background variations along lines introduced considerable sources of errors, particularly for the snow-survey flight when the count rates are reduced (Table 7). Backgrounds should be updated frequently from values obtained over lakes on or close to the flight line, otherwise, similar errors would occur. In most of the area surveyed, lakes are fairly abundant; if this type of survey were carried out in other areas, obtaining accurate backgrounds could well be a serious problem.

The difficulty in obtaining reliable soil moisture data has also been a source of error in the airborne technique. If accurate measurements of the water equivalent of the snow pack are required, considerable effort should be made to obtain representative soil moisture samples over the survey area. In operational hydrology, a distinction between water storage above the soil surface and in the top-soil may not be required.

Most of the errors in the technique arise from soil moisture correction procedures, background, and operational errors; there is consequently little advantage in flying with a large volume detector for this type of survey. Reducing the detector volume by a factor of 10 will increase the statistical errors by the square root of 10, with little effect on the integral count method and with statistical errors in the potassium technique comparable to errors caused by background and soil moisture. A much smaller and cheaper detector system, flown in a single-engined aircraft, could thus be used to considerably reduce the costs per line kilometer except perhaps where deeper snow packs are encountered; statistical errors could then become important.

When the magnitude of the errors associated with the airborne technique is taken into consideration, the excellent agreement between the ground and potassium airborne results suggests that the potassium experimental linear absorption coefficient applies over all the experimental courses. The integral results, expected to be almost as good as the potassium results, suggest that this may not be the case for the integral absorption coefficient. Experiments over areas with abnormal relative concentrations of uranium and thorium have shown that the absorption coefficients for all four windows can vary appreciably (Grasty, private communication). In addition, the radioactivity of the overburden in the Canadian Shield area is known to be generally lower than the underlying bedrock when the bedrock is highly radioactive, and higher when the underlying bedrock is low in activity (Charbonneau <u>et al</u>. 1973). In these situations the experimental linear absorption coefficients would vary, because they are dependent on the variation of radioactivity with depth. The absorption coefficients may vary over any survey area, and calibration flights at different altitudes are therefore recommended.

SUMMARY

The results of the gamma-ray spectrometer survey over Southern Ontario indicate that with the equipment and procedures described, the average water equivalent of the snowpack over 16 km sections can be measured to an accuracy of 1.2 cm using the potassium count information, and to 1.7 cm using total radioactivity.

The largest errors arose from background variation along the flight lines, and from lack of accurate soil moisture data. Because only one background was used for each flight line, background errors could well be reduced by updating the background whenever the flight lines pass over large bodies of water. Care should be taken when laying out the flight track to ensure that adequate background information will be available. Accurate soil moisture measurement is difficult to obtain over a large area, and present experience suggests that a few selected sites supplying reliable data would be the best alternative.

For the range of snow-water equivalent encountered (up to 14 cm), and by averaging the airborne data over 16 km sections, statistical errors were small. For shallow snowpacks, operational costs could be reduced by flying a smaller volume detector system in a single-engined aircraft. Calibration flights at different altitudes to determine the absorption coefficients would be necessary for any new detector system and if any new area is surveyed.

ACKNOWLEDGEMENTS

The assistance of individuals from the following agencies and universities is gratefully acknowledged: The Ontario Ministry of the environment, the Ontario Ministry of Natural Resources, Ontario Hydro, Trent University, McMaster University, University of Guelph, Canada Department of the Environment and Canada Department of Indian Affairs and Northern Development.

This experimental survey was carried out using the Geological Survey of Canada spectrometry system with financial support from the Department of the Environment and the Ontario Ministry of the Environment.

We should also like to thank Peter Holman, the equipment operator; Mrs. Barbara Elliott, Dave Underwood, and Terry Beck for assistance in data reduction, and we are grateful to D.F. Witherspoon, St. Lawrence-Great Lakes Study Office, Cornwall, for his interest in and support of the project, especially during the planning phase.

REFERENCES

Charbonneau, B.W., K.A. Richardson, and R.L. Grasty, 1973. Airborne gamma-ray spectrometry as an aid to geological mapping, Township 155, Elliot Lake Area, Ontario. Geol. Surv. Can., aper 73-1, Pt. B.

Dahl, J.B., and H. Odegaard, 1970. Aerial measurements of water equivalent of snow deposits by means of natural radioactivity in the ground. Isotope Hydrology 1970, 191-210, IAEA, Vienna, Austria.

Davisson, C.M., 1965. Gamma-ray Attenuation Coefficients. α, β, γ -RAY Spectroscopy. Vol. 1, pp. 827-843, ed. Siegbahn, K., North Holland, Amsterdam.

Dmitriev, A.V., R.M. Kogan, M.B. Nikiforov and Sh.D. Fridman, 1971. Aircraft gamma-ray survey of snow cover. Nordic Hydrology 2, 47-56.

Godby, E.A., S.H.G. Connock, J.F. Steljes, G. Cowper, and H. Carmichael, 1952. Aerial prospecting for radioactive minerals. Atom. Energy Rept. 13.

Grasty, R.L., and A.G. Darnley, 1971. The calibration of gamma-ray spectrometers for ground and airborne use. Geol. Surv. Paper Canada 71-17.

Grasty, R.L., 1972. Airborne gamma-ray spectrometry data-processing manual. Open File 109, Geol. Surv. of Canada, Ottawa.

Grasty, R.L., 1973. Snow-water equivalent measurements using natural gamma emission. Nordic Hydrology 4, pp. 1-16.

Kogan, R.M., I.M. Nazarov, and Sh.D. Fridman, 1965. Determination of water equivalent of snow cover by method of aerial gamma-survey. Soviet Hydrology: Selected Papers, No. 2, AGU, pp. 183-187.

Kogan, R.M., I.M. Nazarov, and Sh.D. Fridman, 1969. Gamma Spectrometry of Natural Environments and Formations. Israel Program for Scientific Translations, No. 5778, 337 pp., Jerusalem, 1971.

Loijens, H.S., and R.L. Grasty, 1973. Airborne measurements of snow-water equivalent using natural gamma radiation over Southern Ontario. 1972-73. Scientific Series No. 34, Inland Waters Directorate, Dept. of Environment, Ottawa.

Peck, E.L., V.C. Bissell, E.B. Jones, and D.L. Gurge, 1971. Evaluation of snow water equivalent by airborne measurement of passive terrestrial gamma radiation, Wat. Res. Research, Vol. 7, No. 5, 1151-1159.

Zotimov, N.V., 1965. A surface method of measuring the water equivalent of snow by means of soil radioactivity. Soviet Hydrology: Selected Papers No. 6, AGU, pp. 537-547.

Zotimov, N.V., 1968. Investigation of a method of measuring snow storage by using the gamma radiation of the earth. Soviet Hydrology: Selected Papers No. 3, AGU, pp. 254-266.

AREAL SNOWPACK WATER-EQUIVALENT DETERMINATIONS USING AIRBORNE MEASUREMENTS OF PASSIVE TERRESTRIAL GAMMA RADIATION

E. Bruce Jones, PE*
Allen E. Fritzsche
Zolin G. Burson
Dale L. Burge
EG&G, Inc., Las Vegas, Nevada

ABSTRACT

Airborne measurement of the areal average of the water equivalent of snow by the terrestrial gamma attenuation method is accurate to 1.2 cm or 10% whichever is greater. Airborne survey data from study areas in the midwest show that airborne radon daughters, soil moisture and gamma counting statistics are the predominant sources of error. The method has good potential for large, rapid areal measurements of water equivalent over flat or rolling terrain.

INTRODUCTION

The measurement of depth and water equivalent of snow have generally been on a point-by-point basis, the results of which were assumed to be representative of fairly large areas. Errors have been found to exist in this assumption, primarily because of the non-uniform distribution of the snow cover and because of the difficulty in accounting for all the variations in snowpack that can occur on various portions of a given watershed (Peck, 1972; Richards, 1972; Kagan, 1972; and Bray, 1973). Measurement of the water equivalent of a snowpack on an areal basis is required for a more accurate runoff prediction.

During the mid-1960s, interest was expressed in measuring the water equivalent of a snowpack from an aircraft using the principle that the mass of the snowpack (water, ice and water vapor) will cause an attenuation of the natural gamma radiation emitted from the surface of the earth. An operational approach to this technique was reported by Kogan et al., (1965). The initial results were encouraging; but, more important, it began to show a way to measure snow-water equivalent on other than a point-by-point basis.

---

*Hydrologic consultant to EG&G, M. W. Bittinger and Associates, Inc., Fort Collins, Col.

6.2

General descriptions of the concepts as well as equipment and results have been given by Dahl (1970); Peck *et al.* (1971); Dmitrieyev *et al.* (1970); Kogan *et al.* (1971); Vershinina and Dimaksyon, *eds*, (1971); and Grasty (1972).

EXPERIMENTAL STUDIES

The current research effort in the USA was initiated in 1969.+ Data acquisition was made by the ARMS* aircraft which utilizes a package of fourteen 10x10 cm NaI(Tl) detectors. The gamma ray data were recorded through multi and single channel analyzers onto punched paper tape, along with radar altitude, aircraft position and meteorological parameters.

Two primary snow courses were initially chosen for the study. The Luverne, Minnesota, snow course is located between the Iowa-Minnesota state line and Luverne, Minnesota, along US Highway 75, at an elevation of approximately 1,450 ft (442 m) above sea level. It is 8-1/2 miles (13.7 km) long. The course is fairly flat with little or no trees or rock outcropping nearby. It was chosen because: (1) of these topographic conditions, (2) fairly uniform deposits of snow were expected, (3) historic hydrologic data are available and (4) convenience of obtaining ground truth data (water equivalent of snow, percent soil moisture and ground cover).

The Steamboat Springs survey site is located in a north-south direction extending from 4 to 8 miles (6.4 to 12.8 km) south of Steamboat Springs, Colorado. The center line of this area lies about 7,000 ft (2148 m) above sea level. This valley site has many of the same advantages as the Luverne site, but receives more snow.

Both summer and winter surveys were conducted at several altitudes. Ground based measurements of the snow water equivalent and/or soil moisture were obtained along the flight line on each survey. Gross gamma counts above 50 keV were recorded each second while a complete gamma energy spectrum was recorded for each pass over a course. The gamma data were computer processed by summing gross count over 1 mile (1.6 km) intervals, as well as the whole flight line, and by computing spectral photopeak areas. These were plotted against the equivalent air mass separating detector and ground. The plots were then evaluated for curve shape and standard error.

+Sponsored by the Office of Hydrology, National Weather Service, National Oceanographic and Atmospheric Administration (NOAA) in cooperation with the U S Atomic Energy Commission (USAEC).

*ARMS refers to the Aerial Radiological Measuring System owned by the USAEC and operated by EG&G (ARMS, 1973).

Net count data (gross counts less background) taken over a 3-year period are shown in Figure 1 for the Luverne snow course. Even though the attenuation curves are not precisely exponential, the experimental data show that the use of a simple negative exponential to represent the total gamma flux transmission is both convenient and accurate. This agrees with the Russian work (Kogan et al., (1971).

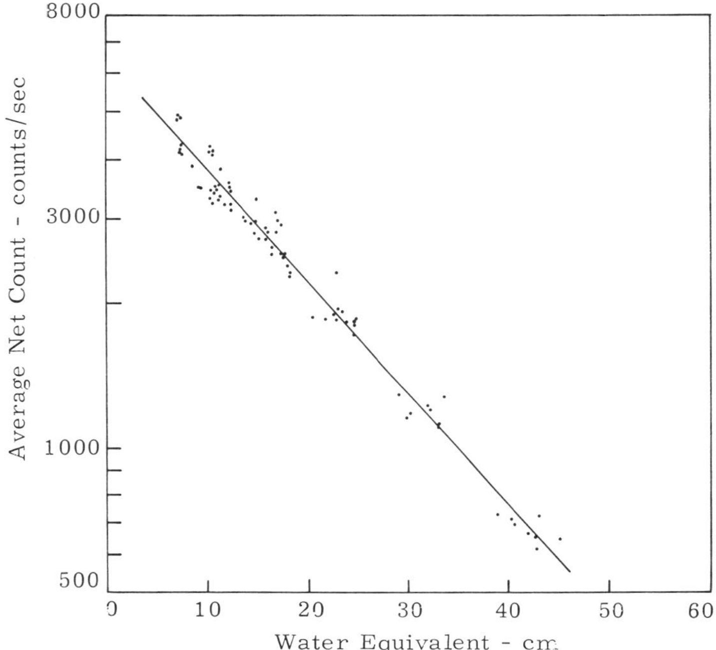

Figure 1. Net count survey line averages versus water equivalent at Luverne, Minnesota.

Twenty-eight survey lines in the Ontario basin of New York State were flown in summer and winter as part of the International Field Year for the Great Lakes (IFYGL) program. Three of these lines were chosen as calibration courses for comparison with ground-based snow tube and soil moisture measurements. The resulting water equivalent estimates for 1 March 1973 for both the gamma method of airborne surveys and that obtained from snow sampling tubes are given in Table 1. About 1/3 of the uncertainty in the gamma data is due to soil moisture correction and the other 2/3 is due to background correction and counting statistics. Uncertainties in the snow tube data as applied to line averages are based on standard deviations and number of sample points.

Table 1. Water Equivalent (cm) of Snow Cover on IFYGL Calibration Lines

	Gamma Method+				Snow Tube+
Line	Net Count	^{40}K	^{208}Tl	Weighted Average	
R 180	2.9 ± 1.3	3.4 ± 1.0	3.6 ± 1.4	3.3 ± 1.0	3.3 ± 0.3
S 030	1.6 ± 1.3	1.6 ± 1.4	1.3 ± 2.0	1.5 ± 1.3	2.0 ± 0.5*
S 050	9.8 ± 2.6	9.1 ± 1.0	9.3 ± 1.9	9.4 ± 1.0	9.3 ± 1.1

+Errors shown are one standard deviation.
*Estimated. Data were taken adjacent to the flight line.

SYSTEM PARAMETERS AND ERRORS

The gamma signal strength or flux is dependent on the natural isotope concentrations in the top few centimeters of the soil. The concentration of ^{40}K, ^{238}U, ^{232}Th are generally a few parts per million. The ^{40}K specific signal strength (gammas per ppm) is the largest and ^{232}Th the smallest.

The sum of the gamma fluxes due to all three isotopes is generally called the gross flux. This flux consists of both discrete and continuous energy components from less than 50 keV to 2.62 MeV. The discrete components are due to source gammas that have not interacted with the air or ground. The continuous energy components are due to the inelastic collisions of discrete energy gammas with atomic electrons in the ground, air and aircraft.

The NaI(Tl) crystal efficiency ranges from 20% to nearly 100% as gamma photon energy decreases from 2.62 MeV to 50 keV. The energy distribution or spectrum of the gamma flux is perturbed by the NaI(Tl) crystal in the interaction process. Figure 2 shows a typical gamma energy spectrum obtained from the ARMS system. The area under a peak in the spectrum is proportional to the discrete gamma flux incident on the crystal. Likewise, the total area under the spectrum curve is proportional to the gross gamma flux.

Theoretically, the discrete energy flux which is proportional to the spectrum peak area from a terrestrial source is given as:

$$\varphi = (1/(2\mu_g))(E_2(\mu_a h + \mu_w WE)) \quad (1)$$

where:

E_2 = exponential integral of the second kind
μ_g = mass absorption coefficient of soil (cm^2/g)
μ_a = mass absorption coefficient of air, (cm^2/g)
h = air mass, (g/cm^2)
μ_w = mass absorption coefficient of water, (cm^2/g)
WE = water equivalent, (g/cm^2)

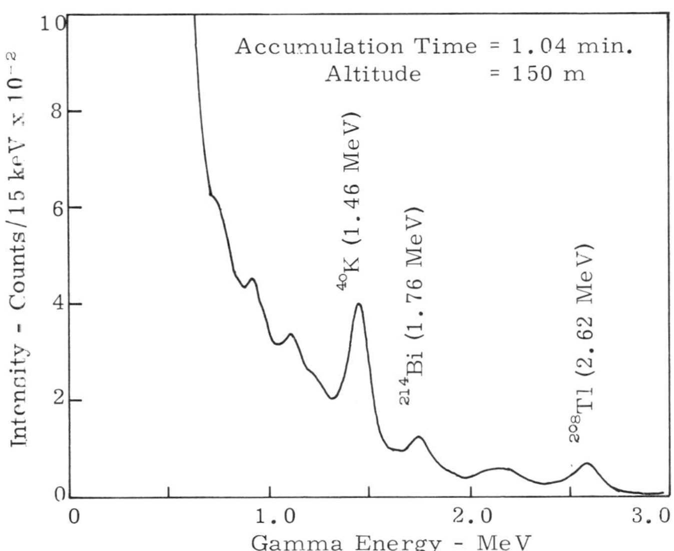

Figure 2. NaI (Tl) energy response to natural sources in the soil.

Empirically, the discrete energy flux has been found to be:

$$\varphi = (K/(2\mu g))\exp(-\alpha WE) \quad (2)$$

where:

K = terrestrial source strength
α = attenuation coefficient (cm^2/g)

The gross flux can be written in the same form. The attenuation coefficients (α) are 0.076 for ^{40}K, 0.064 for ^{208}Tl, and 0.059 for gross flux.

Also, it is possible to a good approximation to write

$$\mu_g = \mu_{go}(1 + M) \quad (3)$$

where M is the commonly measured soil moisture i.e., soil moisture weight divided by soil dry weight, and μ_{go} is the dry soil coefficient.

If aerial surveys are made on different dates of the same area, the ratio of the count rates, whether peak area or gross count, is:

$$N_1/N_2 = ((1 + M_2)/(1 + M_1))\exp(-\alpha(WE_1 - WE_2)) \quad (4)$$

Equation (4) is used to obtain a change in water equivalent between two surveys of the same site. For the case when $M_1 = M_2$ (no soil moisture change):

$$WE_1 - WE_2 = (1/\alpha)\ln(N_2/N_1) \quad (5)$$

when the moisture changes, however:

$$WE_1 - WE_2 = (1/\alpha)(\ln(N_2/N_1) + \ln((1 + M_2)/(1 + M_1))) \quad (6)$$

Thus, change in water equivalent cannot be determined by the airborne gamma technique alone without knowing the change in soil moisture.

The soil moisture effect has proven to be one of the larger uncertainties in an airborne water equivalent determination. Soil moisture changes from 30% to 50% (by weight) from summer surveys to winter surveys have been noted. This contributes about 2 cm to the airborne measurement of water equivalent if unaccounted for.

The theoretical accuracy in water equivalent due to counting statistics of gross count and peak area can be obtained from the empirical flux transport equations. Typical computed counting errors are listed in Table 2.

Table 2. Typical Counting Errors in Water Equivalent Units for 5 cm and 20 cm Total Water Equivalents

Data Type	Accumulation Time	Spatial Resolution	5 cm Counting Error	20 cm Counting Error
^{40}K	3 min	15 km	.4 cm	.6 cm
^{208}Tl	3 min	15 km	.9 cm	1.2 cm
Gross Flux	20 sec	1.5 km	.2 cm	.2 cm

The radon perturbation is the predominant source of uncertainty in the gross count measurement technique. The ^{208}Tl peak area is not affected, while the ^{40}K peak area is slightly affected. The radon gas, ^{222}Rn, in the ^{238}U radioactive chain has a half life of 3.8 days which allows it time to diffuse out of the ground into the atmosphere. Its short lived daughters, ^{214}Bi and ^{214}Pb, appear the same to a gamma detector as these same daughters in the ground. Since these daughters contribute a significant fraction of the total gamma flux, they perturb airborne detection of terrain radiation and introduce errors in water equivalent estimates. Their contributions range from nearly zero to over 50% of the total gamma flux. The immediate air pressure history and current atmospheric mixing are the principle determinants of ^{222}Rn concentration. Errors in water equivalent in the gross count technique in the IFYGL surveys ranged from .5 cm (R 180 and S 030) to 2.5 cm (S 050).

Several methods are used to account for the radon perturbation in the gross flux. These are: (1) flying a survey line twice at 2 altitudes; (2) examining spectrum shape; (3) correlating radon daughter activity in air flow filters with radon induced gross count; (4) using dual detector systems where one detector is shielded; and

(5) flying over nearby lakes. The dual altitude method has a radon measurement uncertainty of about 25%. The filter technique has an uncertainty of 25% or less. The dual detector technique demands system weight be 3 times a single detector system if acceptable counting statistics are to be obtained. Time varying radon distributions in the atmosphere further diminish the practicality of this method. The spectrum shape method is under study, and relies on the difference in shape of ground source spectra as opposed to spectra from airborne emitters. Flying over a nearby lake eliminates terrestrial radiation and is satisfactory when good atmospheric mixing has occurred.

Spatial changes in water equivalent and soil activity cause small errors in the estimated water equivalent (Fritzsche and Burson, 1971) because the detection system performs a nonlinear, weighted average. This error has been less than 5% for the surveys to date.

The cosmic ray contribution and aircraft background in both the gross flux and peak area data are significant. However, multiple flights over lakes on low radon days have provided an accurate profile of this contribution versus pressure, and it contributes negligible error to the water equivalent determination if accounted for.

The total error in the airborne water equivalent determination can be stated in terms of the standard error of data about the exponential fits. These are: 1.2 cm water equivalent for the ^{40}K peak area technique, 1.5 cm for the gross flux, and about 2 cm for the ^{208}Tl peak area for water equivalents from 1 to 10 cm. For water equivalents from 10 to 25 cm, the error is a minimum of 10%. No data have been obtained at greater depths.

The effective area sampled on the ground by a detector in an aircraft is large. For example, at an altitude of 150 meters, the area is 500 meters in diameter which is 2 billion cm^2 as compared to a few cm^2 sampled by a conventional snow tube. Direct comparison of results of aerial surveys to results of ground surveys, then, is not trivial. One must guard against the tendency to obtain a few ground-level measurements, average them, and apply this average as the "true" average for the entire area of investigation. For example, for line S 050 in Table 1, (IFYGL Flight Program) even though 183 individual snow tube measurements were taken, the resulting average still contained a 12% uncertainty of the true average because of large deviations among the individual data points -- again a common dilemma of the snow hydrologist. To minimize errors in air-to-ground comparisons, as many ground measurements as operationally practical are obtained for each aerial survey over calibration sites.

OPERATIONAL PARAMETERS

An airborne gamma radiation measuring system includes the aircraft, navigation instruments, radiation instruments, and data reduction hardware. The size and weights of all these components are inter-related. The ARMS system parameters are listed in Table 3.

Table 3. ARMS System Parameters

Aircraft type	Twin Beech Bonanza
Aircraft speed and range	150 mph, 600 mi (241 kmph, 966 km)
Survey altitude	500 ft (152 m)
Crew	Pilot, navigator, instrument operator
Detector size and weight	700 in^3, 160 lb (11480 cm^3, 73 kg)
Electronics weight	565 lb (256 kg)
Nominal gross flux rate	3500/sec
Nominal ^{40}K peak area	2500/min
Nominal ^{208}Tl peak area	500/min

More sensitivity can be obtained by adding more detectors. Eliminating the instrument operator can be achieved by automating the electronics system. Punched paper tape serves as the medium between the data collection system and computer-aided data reduction. Magnetic tape may be substituted with no weight penalty for higher density data output. Experience at IFYGL surveys has shown that flight data can be reduced to average water equivalents in a few hours using a time sharing computer system.

Operation logs of surveys show that the flight factor (fraction of each 8-hour day that an aircraft can be airborne) averages about 0.4. Thus, the expected flight time over snow fields is 3.2 hours per day. The difference (4.8 hours) is required for maintenance and waiting for clear weather. The expected watershed coverage then is about 500 linear miles (805 km) per day.

CONCLUSIONS AND FUTURE APPLICATIONS

The current airborne survey technique, when employed over flat or rolling terrain, can determine the water equivalent of snow cover to within ± 1.2 cm for a spatial resolution of 3 miles (4.8 km) using the photopeak area technique. The gross flux technique (± 1.5 cm) is recommended for smaller spatial resolution because of its larger signal.

The chief errors are soil moisture effects in both the peak area and gross flux techniques, counting statistics in the peak area, and airborne radon perturbations in the gross flux. These are currently under study and results should improve the accuracy for water equivalent determinations, although errors are not likely to be less than 0.5 cm (Kogan et al., 1971).

The areal coverage of the airborne gamma system is a major advantage over ground-based point measurements. An entire watershed of hundreds of square kilometers can be effectively surveyed by an airborne system in a single day.

The present data from snow tube and other point methods for predicting water equivalent can be greatly extended by the use of airborne gamma radiation surveys. The existing ground-based reporting network employed by the National Weather Service can be used to corroborate the airborne system measurements. Total snow-water inventory in large areas can then be obtained from the airborne data. Timely water equivalent results are possible immediately from an airborne minicomputer or in a few hours from an earthbound computer.

The airborne gamma system, as described in this paper when used over relatively flat or rolling terrain, appears to be a significant part of the answer to the hydrologist's dilemma of quickly and efficiently estimating the water equivalent of a snowpack on an areal basis. Conceptually, the hydrologist may wish to use the system on an operational basis for various purposes: (1) to assist in the delineation of areas of abnormal water content, particularly in those cases when major flooding is a concern; (2) to follow the buildup of water equivalent throughout the snow accumulation season in certain preselected or indexed watersheds; (3) as a tool to assist the research hydrologist in the study of snowpack behavior. The system appears to be well suited to these tasks within the operational limits previously set forth.

One could speculate on the future use of this system, but there is one application that is immediately apparent: that is, when a flood potential exists in the snowpack in the north-central United States, large-scale areal water equivalent information could be obtained on a daily or weekly basis.

As the airborne gamma survey system becomes operational, consideration could be given to other measurements. As an example, photographic instrumentation could be added to aid in interpreting water equivalent data from partial snow cover. Assuming the aircraft has the capability, various types of microwave systems could also be considered.

The existing airborne gamma system has shown evidence of being a valuable tool in the field of snow hydrology, and when combined with other tools, the combination will become even more valuable.

6.2

REFERENCES

Bray, Dale I., 1973. A Report on the Variability of Snow Water Equivalent Measurements, Proceedings of the 1973 Eastern Snow Conference, pp 20-31.

Dahl, J. B., 1970. Aerial Measurements of Water Equivalent of Snow Deposits by Means of Natural Radioactivity in the Ground. IAEA/SM-129/13: Symposium on Isotopes in Hydrology, Vienna.

Dmitriev, A. V., R. M. Kogan, M. V. Nikiforov, and Sh.D. Fridman, 1971. Principles of Aircraft Gamma-Ray Survey of Snow Cover. XV General Assembly of the IUGG, Moscow.

Fritzsche, A. E. and Z. G. Burson, 1971. Water Equivalent of Snow Measurements Using Natural Terrain Radiation. EG&G Report No. 1183-1533, EG&G, Inc., Las Vegas, Nevada.

Grasty, R. L., and P. B. Holman, 1972. The Measurement of Snow Water Equivalent Using Natural Gamma Radiation. First Canadian Remote Sensing Symposium, Ottawa, Canada.

Kagan, R. L., 1972. Snow Cover - Statistical Principles, IN Casebook on Hydrological Network Design Practices, WMO - No. 324, World Meteorological Organization, Geneva. Switzerland, pp. 3.2-1 through 3.2-9.

Kogan, R. M., I. M. Nazarov, and Sh.D. Fridman, 1971. Gamma Spectrometry of Natural Environments and Formations. Translated from Russian, Israel Program for Scientific Translations, TT70-50092, NTIS, Springfield, Va., 337 p.

Peck, E. L., V. C. Bissell, E. B. Jones, and D. L. Burge, 1971. Evaluation of Snow Water Equivalent of Airborne Measurement of Passive Terrestrial Gamma Radiation. Water Resources Research, 7, 5:1151-1159.

Peck, E. L., 1972. Snow Measurement Predicament, Water Resources Research, 8, 1:244-248.

Richards, T. L., 1972. Snow Cover - North Americal Experience, IN Casebook on Hydrological Network Design Practices, WMO - No. 324, World Meteorological Organization, Geneva, Switzerland, pp. 3.1-1 through 3.1-4.

Vershinina, L. K. and A. M. Dimaksyan, eds., 1969. Determination of the Water Equivalent of Snow Cover. Translated from the Russian Israel Program for Scientific Translations, TT70-50093, NTIS, Springfield, Va., 142 p.

MEASUREMENT OF SNOW AT A REMOTE SITE: NATURAL RADIOACTIVITY TECHNIQUE

V. C. Bissell
E. L. Peck
NOAA, National Weather Service
Silver Spring, Maryland 20910 U.S.A.

ABSTRACT

The use of natural gamma radiation from the soil as a basis for snow water equivalent measurements at remote sites has been under investigation by the National Weather Service since 1970. Results to date indicate that measurements with about five percent error in the five to forty centimeter water equivalent range can be obtained in periods uncomplicated by precipitation or considerable change in soil moisture. Periods of active melt can be subject to serious errors.

A new natural radioactivity method is also proposed. The use of highly penetrating cosmic radiation appears to have excellent potential for point snow water equivalent measurement in extremely deep snow.

INTRODUCTION

The accurate measurement of snow water equivalent at remote locations is in many parts of the country the most critical element in the formulation of water supply forecasts and continuous river stage forecasts during the spring melt season. Within the last two decades a great deal of work has been done developing automated snow measurement systems to supplement manned snow surveys. A fairly recent nuclear counting method involves the use of naturally occurring radioactive isotopes in the soil as the sole radiation source (Zotimov, 1968; Kogan, et al., 1971; Peck, et al., 1971). The feasibility of the technique for automated measurement at fixed remote locations has been tested since 1970 by the National Weather Service at the NOAA-ARS cooperative snow study site near Danville, Vermont. Test results during the 1970-71 snow accumulation period have been given by Bissell and Peck (1973). This paper provides an update on publication of test results as well as suggesting yet another nuclear counting method which appears to have excellent potential for measuring the water equivalent of deep snow.

EXPERIMENT AND RESULTS

In the autumn of 1970 a gamma radiation detector with a small (2.54 cm diameter by 3.81 cm length) NaI scintillation crystal was suspended about two meters above the ground and count rates during the snow season were monitored for several hours on the days the station was attended. Data were not obtained during the 1971-72 snow season until just prior to spring runoff. At that time an hourly recorder was hooked up to the detector. For the 1972-73 snow season a second (identical) detector was placed just above the first with a 2.5 cm thick lead slab between the two. This dual detector system was placed with the hope of differentiating between atmospheric- or cosmic-derived radiation and radiation coming from the ground. Separate hourly counting and recording systems (different from that used in spring 1972) monitored the count rates from the two detectors. An initial analysis of the dual detector system showed that its discriminating ability in improving water equivalent measurements is secondary to other error sources encountered. No further detailed analysis has yet been done. The water equivalent of the snow cover at the site during all three seasons was measured by snow tube on days the station was attended (most weekdays) at three snow courses, each within twenty meters of the detector. At least two cores were taken in each course area.

Time variation of the count rate within a given day is clearly shown by the hourly data obtained. The hour-by-hour count traces for three different three-day periods in 1973 are given in Figure 1. A marked diurnal effect is strongly evident in both the no-snow period (August 14-16) and during the period with greatest snow cover (February 24-26). This effect is attributed to the dynamic nature of radon, a radioactive noble gas which originates in the soil and whose daughter products ^{214}Pb and ^{214}Bi are major sources of the gamma radiation seen by the detector. During daytime hours the atmospheric mixing at the surface of the earth allows a great deal of the radon diffusing from the soil and through the snow to be transported beyond the range of the detector. When evening comes, cooling at the earth's surface weakens the transport mechanism and the radon concentration in and near the surface of the earth (snow or soil, as the case may be) increases. The dynamic nature of radon has long been known and is further discussed in another paper in this symposium [Bissell, paper 6.4]. Points to be made here are: (1) during daylight hours count rate fluctuations due to radon effects are reduced, (2) surface buildup of radon during the night is highly sensitive to slight meteorological effects and hence may disallow nighttime use of natural gamma radiation for snow measurement, and (3) the snow cover under study inhibited the diffusion of radon from the soil into the atmosphere, but not completely. The count rate trace for the March 14-16 period in Figure 1 shows the diurnal effect to a lesser extent, but contains two other points of interest as well. The first is the deposition of radioactive aerosols on the snow surface by precipitation, noted previously by Bissell and Peck (1973). The anomalously high count rates produced during precipitation require about three or four hours after the end of precipitation to decay back to their nominal values, as seen in the two events of March 15. A second point of interest is the considerable change in daytime count rate between March 14 and March 15 during a period of constant snow cover. More will be said of this event subsequently.

The hourly count rates obtained through the 1970-71 and 1972-73 snow seasons are shown plotted against water equivalent in Figure 2 and Figure 3, respectively. The earlier report by Bissell and Peck (1973) was based on points in Figure 2 which are indicated as observations during the snow accumulation period. To eliminate precipitation-induced error, only those observations which had no precipitation during or immediately prior to the counting period are shown in Figure 2 and Figure 3. The 1970-71 counts were generally obtained during the afternoon, while the 1972-73 plot includes average counts from 0900 to 1200 and from 1200 to 1600. Other hours from the 1972-73 data were not included because of the diurnal effect

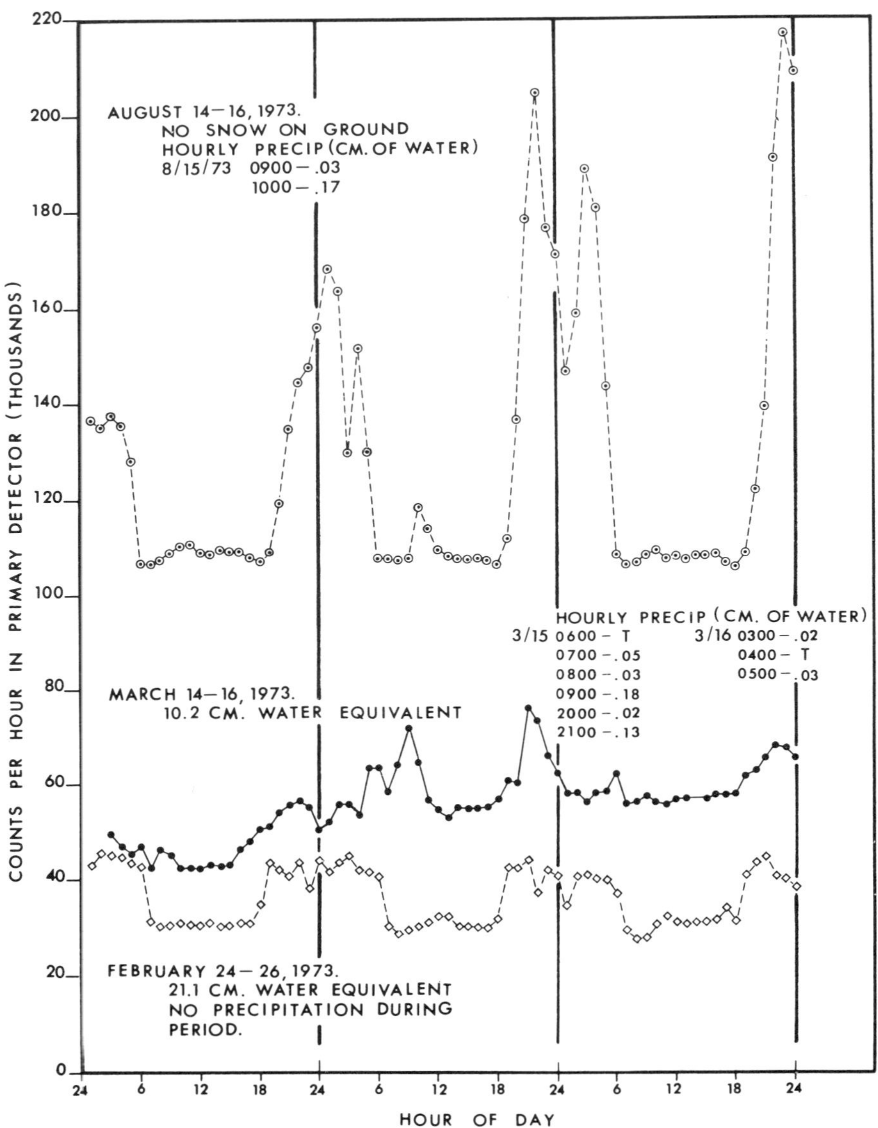

Figure 1. Hourly count rate traces in primary detector during February 24-26, March 14-16, and August 14-16, 1973.

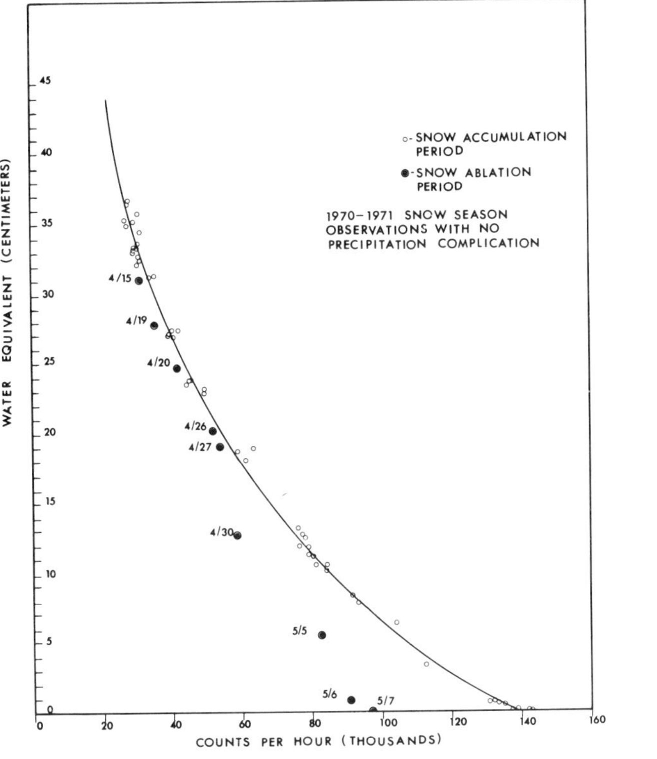

Figure 2. Count rate versus water equivalent relationship during 1970-1971 snow season at townline station.

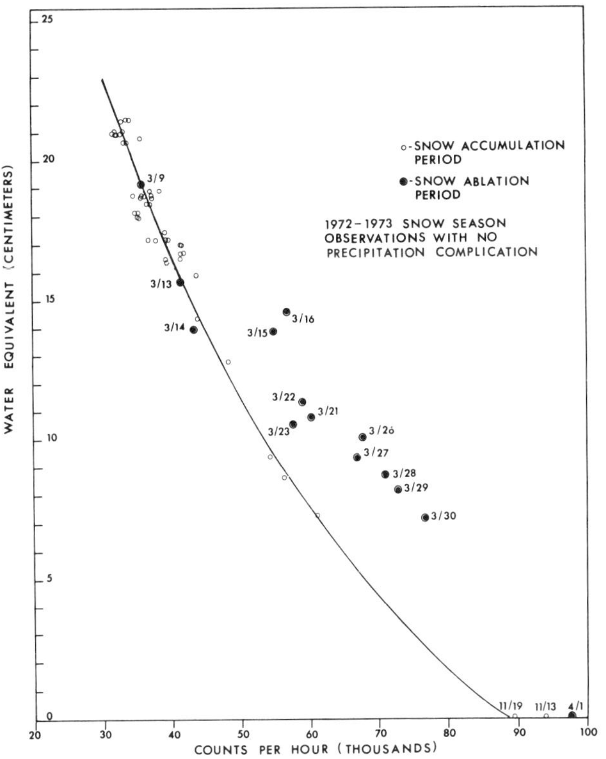

Figure 3. Count rate versus water equivalent relationship in primary detector during 1972-1973 snow season.

mentioned earlier. During periods of snow accumulation, the consistency of the counts-water equivalent relationship is evident. The root-mean-square deviations from best-fit lines through these points are given in Table 1.

Table 1. Root-mean-square deviations (cm) from best-fit lines during snow accumulation.

Range (cm)	1970-71 (precipitation events excluded)	1972-73 (precipitation events excluded)	1972-73 (coef. of var. less than 0.04)
5-13	0.58	0.25	0.91
13-25	1.14	0.89	1.03
25-40	1.27	no observations	no observations

In an operational framework separation of precipitation-complicated events could be difficult since it would require telemetering of site hourly precipitation data. Since the count rate is generally highly variable during occurrence of significant precipitation events, the coefficient of variation of the count rate is suggested as a screening criterion. The coefficient of variation of hourly counts was computed for the periods 0900 to 1200 and 1200 to 1600 during the 1972-73 snow accumulation period. The efficacy of two levels of application of this criterion in reducing error is given in Table 2. The effect of the coefficient of variation screening at high C_v levels is to eliminate only the most anomalous events. More restrictive values of the coefficient of variation would not appear to give additional benefit.

Table 2. Effect of coefficient of variation as screening a criterion, 1972-73 snow accumulation period

	Root-Mean-Square (cm)	Percent of observations
All points	1.27	100
C_v less than 0.04	1.02	80
C_v less than 0.015	1.02	50

Much of the above discussion about accuracy during snow accumulation must be tempered with one very critical consideration - the effect of soil moisture. An increase in soil moisture will produce a decrease in count rate which is difficult to separate from snow change effects. For a specific site and detector there will be an entire family of count rate versus water equivalent relationships, each characterized by a parameter relating to soil moisture. The apparently consistent relationships during snow accumulation in Figure 2 and Figure 3 could well range through a wide range of soil moisture values rather than representing a single curve in the family of relationships. If this is the case then the historical relationships would have diminished value in providing a basis for measurement in subsequent years unless proper allowance is made for soil moisture during both the calibration and operational periods. Unfortunately, use of different detector configurations and counters during each of the three years of this study disallows direct comparison of count rates between any two years.

Turning to the periods of snow ablation, inspection of Figure 2 and Figure 3 reveals that the observations during melt depart significantly from the accumulation period relationships between count rate and water equivalent. Further, the 1971 and 1973 departures are in opposite directions. The depressed count rate in the 1971 melt period is ascribed to some combination of the following two reasons: (1) Snow accumulation began in the fall of 1970 over a rather dry soil which received very little additional moisture from rain or melt events during the remainder of the fall and early winter. An estimated 0.025 cm per day melt due to ground heat at the site [Anderson, 1973] would have added roughly four centimeters of water to the soil by the time twenty centimeters water equivalent had accumulated in mid-February. It is important to note that much of this ground melt water may have been held in the upper soil layers by the temperature gradient in the soil [Peck, 1973]. Water percolating into the soil after active melt began in mid-April would then give only a small additional contribution to the top five or ten centimeters of soil from which the great bulk of terrestrial gamma radiation comes. (2) Water percolating through a rapidly melting snowpack and into the soil will carry with it some radon gas. This "washing out" of radon from the snow and upper soil levels would reduce the radiation source strength and hence further depress the count rate. This phenomenon is further discussed elsewhere in this symposium [Bissell, paper 6.4].

The 1973 melt period count trace in Figure 3 is somewhat of a mystery. A sharp departure from the accumulation relation occurred between March 14, 1973 and March 15, 1973. After this time count rates continued considerably higher during ablation than during snow accumulation days with similar water equivalents. The hourly count rate trace during March 14 and March 15 is shown in Figure 1 as mentioned earlier, and is well corroborated by the independent hourly trace (not shown) of the secondary detector. Due to a very wet fall it would be expected that the ablation observations would be reasonably aligned with the accumulation period observations. The two following considerations are put forward as possible reasons for the differences: (1) A common influence may have caused similar changes in both the primary detector and the secondary detector. The most suspect in this category would be the possibility of an earlier heater failure in the box housing both detectors. No indication of heater repair, however, was found in the station log. Further, previous calibration tests in a cold room with the primary detector yielded only a two percent reduction in count rate over the temperature range a heater failure would have produced. (2) The freezing rain which occurred the morning of March 15 may have produced a thin ice seal on the snow surface, producing a buildup of radon concentration in the snowpack and thus increasing the radiation source strength. A heavy rain (3.20 cm) March 17 followed by snow showers and five consecutive days during which air temperature never got above freezing produced an ice layer which was covered by about six cm of new snow. This ice layer was noted in the station log on March 22 and March 23, and could have provided a stronger and more enduring seal than the small precipitation event of the 15th. There are inconsistencies with the radon "capping" possibility however. First warm air temperatures on March 16 would have quickly removed a light crust which may have been formed on the 15th. This would have allowed the radon buildup in the snowpack to deplete and count rates during the afternoon of the 16th should have been back down to previous levels rather than remaining high as they did. Second, with the onset of heavy melt about March 26 it is questionable that the sealing would have been sustained through March 30 as Figure 3 would indicate.

A final observation regarding Figure 3 is that the post-melt count rate over bare ground on April 1 would be more in line with the accumulation relation. The lower no-snow count rates on November 19 and November 25 could be due to soil moisture. Although the soil was quite wet both in the fall and immediately following melt, the soil freezing in the fall could account for additional moisture in the top layers.

It is hoped that continuation of the study may provide some additional insights into the March 15-March 16, 1973 event since such a sustained anomaly would be especially difficult to detect in a remote measurement setup.

Now let's look at the 1972 spring melt. Six-hourly water equivalent values inferred from the radiation count rate are shown in Figure 4. These are "hindsight" measurement values since snow tube water equivalent measurements taken during the period were used to generate a count vs. water relationship which in turn was used to go back and produce the continuous water equivalent trace. Precipitation-complicated measurements are distinguished in the figure by asterisks. The point to be made here though is found in Figure 5, which provides better detail of the April 28-May 1, 1972 heavy melt period. Under ideal conditions of heavy melt and low radon interference, the gamma method appears to have good potential for measuring hourly decreases in snowpack water equivalent, especially as noted on April 29 and May 1, 1972. Snowpack runoff appears to have begun around noon both days. The heavy melt during the afternoon of both days appears to end in the evening around 1900 hours. Anomalies in the water equivalent trace during the late morning and early afternoon of April 30 magnify the obvious fact that hourly count traces would require careful screening and review before application to compute hourly runoff.

USE OF COSMIC RADIATION TO MEASURE DEEP SNOW

At this point we would like to depart somewhat from the main thrust of this paper to document a proposal for a natural radiation method of point snow measurement which may find application in extremely deep snow. In the preceding discussion, cosmic radiation was regarded as a noise component in the count rate. A slight realignment of viewpoint, however, allows cosmic radiation to be viewed as signal rather than noise. Cosmic radiation is extremely penetrating as demonstrated by the fact that ionization intensity from cosmic radiation decreases by only a factor of ten between the surface of a water body and at 50 meters depth (Wilson, 1952). Cosmic radiation, however, has several components and the actual attenuation of detector response by water depends on the detector's efficiency of response to the various components. To get some idea of what a NaI scintillation crystal response might be to cosmic radiation under various depths of water we have considered the count rates between three and six MeV at various altitudes over Lake Meade reported by Burson and Fritzsche (1972). Essentially all counts above 3 MeV are due to cosmic

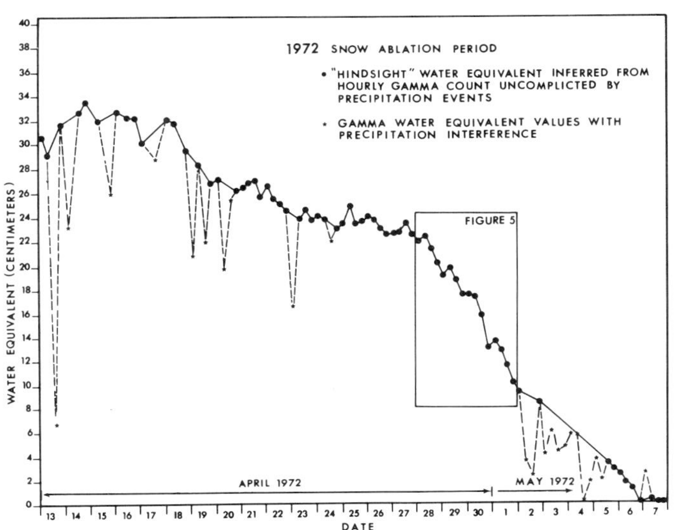

Figure 4. Six-hourly water equivalent trace inferred from gamma counts during 1972 snow ablation period.

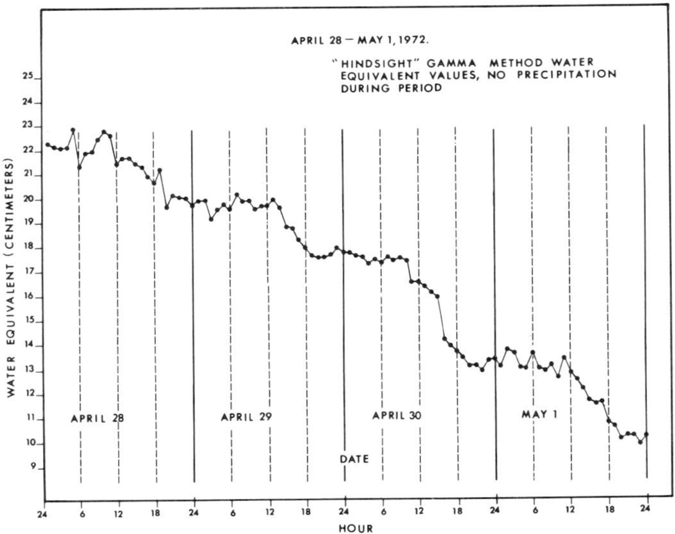

Figure 5. Hourly water equivalent trace inferred from gamma counts April 28-May 1, 1972.

radiation and hence would be free of terrestrial radiation interference. The data show that a single 10 cm diameter by 10 cm length NaI crystal at 1800 meters above mean sea level would produce roughly 6500 counts per hour between three and six MeV energy. A single discriminator counting all detector pulses above three MeV would produce about twice this number (Burson, 1973, personal communication), or 13000 counts per hour. A cursory analysis of the data shows that if the attenuating effect of a unit mass thickness of water can be roughly equated to the effect of a unit mass thickness of air (similar atomic weights make this reasonable) a 100 cm thick layer of water overlying the detector would reduce the 13000 counts per hour to around 6500 counts per hour. If two detectors were placed one at ground level and one above the snow surface, daily measurement of a 100 cm water equivalent by the ratio of counts in the two detectors would have a standard deviation of roughly 0.5 centimeter due to random statistical fluctuations of count rate alone. The feasibility of such a method would depend on cost, availability of required technology (e.g., automatic gain stabilization on detectors), and most of all on what the attenuation of cosmic count rate in a NaI scintillation detector due to snow water equivalent really looks like.

SUMMARY AND CONCLUSIONS

The natural terrestrial radiation method has both advantages and disadvantages in comparison with other unmanned snow measurement methods. Among the advantages are: (1) Since the radiation source is spread out in the soil over the surface of the earth, the measurement is representative of a larger area than other methods with remote capability; (2) Since the output of a nuclear counter is digital in form, the analog-to-digital conversion and attendant calibration required for weighing devices is bypassed; (3) In contrast to the artificial radiation method, there is no hazard from radiation source transport and placement, nor is continuous exercise of safety precautions necessary; (4) The source strength may be increased by mixing into the top soil mantle some material with a higher radioisotope concentration.

Among the method's disadvantages are: (1) Measurement accuracy decreases as water equivalent increases since the signal-to-noise ratio is diminished. Forty centimeters is probably a reasonable upper limit of measurable water equivalent; (2) Measurement accuracy is strongly affected by soil moisture changes. Some estimation of soil moisture in an operational framework will be required; (3) Due to the transport of radon gas, the radiation source is a dynamic one subject to both meteorological and hydrologic processes; (4) The strength of the natural background radiation and the way it is attenuated by snow differ from site to site. Supplemental control measurements taken at periods of different snow cover would be required initially to formulate a relationship between count rate and water equivalent at each site; (5) Precipitation events sharply perturb gamma radiation levels near the ground. The frequent precipitation in most areas where remote snow measurement is needed presents a problem because there could be long periods during which no measurements free of this effect could be obtained.

Overall, the use of terrestrial gamma radiation for low and mid-range snow water equivalent measurements appears quite promising, especially during periods of snow accumulation. For maximum accuracy counting periods should be limited to daylight hours to allow stabilization of radon effects, and simulated soil moisture values should be used to account for the attenuating effect of soil water. Due to the magnitude of other effects, initial analysis of the dual detector for atmospheric radiation delineation was inconclusive.

REFERENCES

Anderson, E. A., 1973. National Weather Service River Forecast System - Snow Accumulation and Ablation Model. NOAA Tech. Memo. NWS HYDRO-17. U.S. Department of Commerce, NOAA, National Weather Service, Washington, D.C. 86 pp.

Bissell, V. C., and E. L. Peck, 1973. Monitoring Snow Water Equivalent by Using Natural Soil Radioactivity. Water Resources Research, 9 (4). pp. 885-890.

Burson, Z. G., and A. E. Fritzsche, 1972. Snow Gaging by Airborne Radiological Surveys - Status through September 1972. EGG-1183-1565, Tech Rep. L-1078. E G & G, Inc., Las Vegas, Nevada. 48 pp.

Kogan, R. M., I. M. Nazarov, and S. D. Fridman, 1971. Gamma Spectrometry of Natural Environments and Formations. Translated from Russian, TT70-50092, Israel Program for Sci. Transl., Jerusalem. 337 pp.

Peck, E. L., V. C. Bissell, E. B. Jones, and D. L. Burge, 1971. Evaluation of Snow Water Equivalent by Airborne Measurement of Passive Terrestrial Gamma Radiation. Water Resources Research, 7 (5). pp. 1151-1159.

Peck, E. L., 1973. Soil Moisture - A Critical Factor in Snowmelt Runoff?. Paper presented at the Fifty-Fourth Annual Meeting of the American Geophysical Union, April 16-20. Washington, D. C. 17 pp.

Wilson, J. G. ed., 1952. Progress in Cosmic Ray Physics. North-Holland Publishing Company, Amsterdam. 510 pp.

Zotimov, N. V., 1968. Investigation of a Method of Measuring Snow Storage by Using the Gamma Radiation of the Earth, Soviet Hydrology Selected Papers 3:254-266.

NATURAL GAMMA SPECTRAL PEAK METHOD FOR SNOW MEASUREMENT FROM AIRCRAFT

Vernon C. Bissell
NOAA, National Weather Service
Silver Spring, Maryland 20910 U.S.A.

ABSTRACT

The accuracy of airborne measurements of snow water equivalent over large areas using the ^{40}K gamma radiation spectral peak is limited both by the accuracy of soil moisture estimates and by variation in general radon distributions between missions. The accuracy of the ^{208}Tl spectral peak method is constrained only by soil moisture estimate accuracy and is therefore preferred over large areas. The use of an optimal linear combination of the ^{40}K and ^{208}Tl method water equivalent measurements significantly improves accuracy over small basins where only one or two flights are made.

To complement current literature on gamma snow surveys, dynamic aspects of radon in both soil and atmosphere are reviewed.

INTRODUCTION

Recent studies have shown that airborne surveys of terrestrial gamma radiation can provide measurement of snow cover water equivalent with operational accuracy (Peck et al., 1971; Grasty, 1973; Burson and Fritzsche, 1972; Vershinina and Dimaksyan, 1969). Due to the difficulty of obtaining ground-based point measurements in windblown drifted snow fields, this method may be extremely valuable in obtaining water equivalent measurements more representative of areal conditions. The United States National Weather Service, in cooperation with EG&G, Inc., and the U.S. Atomic Energy Commission, has been conducting research in the method since 1969. Research missions have been conducted primarily at three lines. The first two are located just south of Luverne, Minnesota, while the third study line is in a level mountain valley near Steamboat Springs, Colorado. Another paper in this symposium (Jones et al., paper 6.2) reports many of the program results to date. This paper complements those results by discussing the composition of spectral peak errors, as well as reviewing some of the dynamic aspects of the natural radiation environment.

THE DYNAMIC RADIATION ENVIRONMENT OF SNOW SURVEYS

The principal sources of count rates obtained with the ARMS detector system at 500 feet (152 m) altitude and their relative magnitudes are: (1.) radioactive decay of ^{40}K in the ground and of ^{238}U and ^{232}Th and their decay products in ground and air (about 75%); (2.) cosmic radiation (about 15%); and (3.) the background of the aircraft and detector system itself (about 10%). The cosmic radiation flux is nearly constant in time at a given latitude, most variations being due to atmospheric pressure changes and time within the 11-year solar cycle. Neither the cosmic flux (slight variability) nor the aircraft and detector background (nearly constant) present major problems in the radiation method of snow surveys.

6.4

A major difficulty arises, however, from the variability of ^{214}Pb and ^{214}Bi, major gamma emitters in the ^{238}U chain. These radioisotopes are decay products of ^{222}Rn, a noble gas with a sufficiently long half-life to allow its transport from soil into the atmosphere in considerable quantities. While the name "radon" is generally reserved for the isotope ^{222}Rn, there are two other isotopes of radon as well. The isotope ^{219}Rn ("actinon") is relatively insignificant, whereas the isotope ^{220}Rn ("thoron") is a parent atom of ^{208}Tl, which produces the important gamma spectral peak at 2.62 MeV. The distance a radon atom may travel is limited by its life time. Radon gas (^{222}Rn) has a half-life of about 3.8 days. Its range in the earth environment therefore is much greater than that of thoron, the half-life of which is only 52 seconds. The decay products of radon and thoron are heavy metals and are therefore immobile unless produced by decay in the atmosphere. Over 90% of the natural radioactivity in the atmosphere results from radon and its short-lived daughters (Shafrir et al., 1967).

The dynamics of radon distribution in the natural radiation environment involve three basic processes: (1.) transport of radon from soil particles into the soil water or soil air; (2.) transport of radon by and through the soil water or soil air; and (3.) the transport of radon and its subsequent decay products in the atmosphere. To illustrate the importance of these processes, Faul (1954) points out that the majority of gamma activity from uranium minerals comes from trapped radon and its decay products. The variable radon content in the upper soil layer then profoundly affects the gamma radiation flux over porous rocks and soil.

One of the most important steps of radon transport within a soil grain occurs at the instant of its formation (Tanner, 1964). All isotopes of radon are formed by the alpha decay of a radium isotope. When this occurs, the recoil atom's kinetic energy is dissipated along its path until it comes to rest, after which it either decays within the soil grain or diffuses through the grain into the soil interstices. Comparison of recoil path lengths in air, soil, and water points up two important phenomena: (1.) only a small fraction of the radon atoms will terminate their recoil paths in soil air, since the distance traveled per kinetic energy expended is much larger in air than in soil; (2.) moisture in the soil interstices will increase the fraction of recoil atoms escaping the soil particles, since the recoil range in water is much less than in air. Even if a decay atom comes to rest in a water-filled pore, its chances of escape into the atmosphere are greater than if it had stopped within a soil grain crystal because of the difference of diffusion constants in these materials (Tanner, 1964).

The total fraction of radon atoms which escape the mineral grains by either recoil or diffusion is known as the "emanating power" of the rock or soil. Emanating powers range from nearly nonexistent to 100% depending on the material with most measurements from 10 to 30 percent (Faul, 1954; Tanner, 1964).

Migration of radon isotopes through soil interstices can occur either by diffusion or convection. Diffusion is dominant in unsaturated undisturbed soil or unfractured rock, while in fractured rock or disturbed soil convection by soil air will dominate (Tanner, 1964). Also, radon can be convected to deeper soil layers by infiltrating water. Since the equilibrium ratio of radon concentration in water at 0°C to its concentration in air is roughly twice that at 25°C (Faul, 1954), this effect would be more important during the percolation of snowmelt water than during summer rain. This phenomenon, combined with the attenuating effect of additional soil water, may cause spurious results from gamma snow measurements during periods of active melt.

The convection of radon by soil air can be produced by several phenomena: (1.) at the outset of a rain or snowmelt event the wetting surface of the percolating water moves downward, displacing radon-rich air upward as it goes (Tanner, 1964); (2.) a drop in barometric pressure results in expansion of radon-rich soil air, providing an upward displacement (Kraner et al., 1964); (3.) gusting winds provide a "pumping" action which draws radon-rich air out of the soil by a pressure gradient. As the gust dies down, diluted air replaces the soil air drawn out (Kraner et al., 1964).

Soil water and ice can significantly inhibit diffusion and convection of radon from the soil. A "capping" effect can occur in which precipitation (or snowmelt), freezing, or snow cover tend to seal the ground, thereby causing a buildup of radon in the uppermost soil layer. Perhaps the most revealing study of hydrologic and meteorological effects on radon in soil gas was conducted by Kovach (1945), who found the following: (1.) frozen ground produced the highest radon concentration. In fact, the depletion profile is practically eliminated during long periods of ice and snow coverage, causing the radon content at all depths to approach the same value; (2.) snow on the ground enhanced the content of radon in soil gas, but was not a greater influence than frozen ground; (3.) prolonged dry spells produced a very consistent profile of radon concentration from 25 to 150 cm depth; (4.) ground temperature changes apparently had no great effect on radon concentration in soil gas.

Capping by freezing depends not only on the fact that the ground is frozen, but also on the structure of the ice formed. Kraner et al. (1964) found that ground frozen to six inches (15.24 cm) depth or more still exhaled radon at 60% of the rate observed during summer. This fact might have important future application to river forecasters, to whom the permeability of frozen ground to rain or snowmelt water is extremely critical. Atmospheric radon levels may be a potential indicator of soil ice struature.

Radon exhalation rates encountered over agricultural areas will probably be larger than those in areas of undisturbed soils. Styra et al. (1970) found that thoron (^{220}Rn) exhalation from a ground surface plowed to a twenty-centimeter depth is two or three times greater than that from smooth, bare, undisturbed ground.

Once radon escapes the soil surface, major influences producing its distribution in the atmosphere are the rate of atmospheric diffusion, transport by turbulent mixing (convection), transport by wind (advection), and "washout" by precipitation. Since radon comes from the ground, its concentration will be maximum at ground level. Transport to higher levels takes place mainly through eddy diffusion, which varies widely according to wind variation and atmospheric stability (Suschny, 1968). Under the normal decrease of temperature with altitude, transport of radon to higher levels is achieved quite easily. The presence of an inversion, however, blocks this transport and causes a buildup of radon below the inversion. This effect is magnified by the presence of snow cover. Peck and Bissell (1973) have pointed out this phenomenon as a major source of error in airborne gamma snow surveys.

The most marked characteristic of atmospheric radon is its diurnal fluctuation. The formation of low-level inversions during nocturnal cooling of the earth's surface causes a buildup of radon which is depleted the following morning by increased convection and eddy diffusion as the inversion "burns out." An example of this effect is found in another paper in this symposium (Bissell and Peck, paper 6.3). The diurnal phenomenon would indicate that midday and afternoon aerial gamma snow surveys are preferred to early morning flights.

Work by Kirichenko (1970) perhaps best illustrates the variability of atmospheric radon. A ten-day sequence of vertical concentration profiles of radioactive aerosols showed sharp variations, most resulting from inversions at various altitudes. The natural radioactivity around noon at 200 m (roughly the altitude at which United States snow reconnaissance missions are expected) varied by a factor of ten for the ten days. The highest activity was observed under a strong inversion at 250 meters, while another extreme followed the passage of a warm front.

Another notable characteristic of atmospheric radon is its seasonal variation. Moses et al. (1963) found typically high values in summer and fall three feet (91 cm) above the ground at Argonne, Illinois. The highest monthly average for early morning concentration appeared in the late fall, probably due to the dry ground and reduced convection. Reduced exhalation from the soil due to occasional frozen ground and snow cover and from gradually increasing soil moisture caused atmospheric radon concentrations to decrease as fall progressed into winter. The minimum concentrations occurred in spring due to the combination of wet ground and strong gusty winds which rapidly mixed radon to higher levels. These results indicate that bare ground calibration flights for total-count snow reconnaissance should not be conducted during summer or dry fall weather.

The advection of radon-rich or radon-poor air into an area is a process which can produce rapid changes in local radon concentration (Kirichenko, 1970). Aircraft flights at 100 meters passing through cold fronts demonstrated that markedly high radioactivity is usually observed just ahead of a cold front. After the cold front passes, the radioactivity may be only about half that observed ahead of the front. This phenomenon was considered due to advection of radon-rich air away from the area, while rising pressure following the passage of the front inhibited exhalation of additional radon from the soil. Kirichenko also noted a second frontal passage effect in increased radioactivity following the passage of a warm front. Kirichenko suggested this phenomenon was due to radon "pumped" from the soil by the fluctuating unstable warm air mass behind the front.

All of the foregoing demonstrate that a radon-complicated gamma radiation spectrum is complicated indeed. Several methods for removal of airborne radon contributions to total count in snow surveys have been tried with varied degrees of success (Burson and Fritzsche, 1972), and research is continuing.

DETERMINATION OF SPECTRAL PEAK METHOD PARAMETERS

The ^{40}K spectral peak is only moderately affected by radon daughter activity, while the ^{208}Tl peak is unaffected. To use these peaks for water equivalent measurements then requires removal of cosmic and aircraft background from both peaks and removal of (radon daughter) ^{214}Bi contributions from the ^{40}K peak. On the basis of theoretical considerations, and with a correction for soil moisture attenuation of gamma radiation (Fritzsche and Burson, 1970), the spectral peak count rate as a function of water equivalent is taken to be

$$N(W) = \frac{N_o}{1+KS} E_2(\alpha W) \quad (1)$$

where W = effective water equivalent, including air blanket (g/cm^2);
α = mass attenuation coefficient in water (cm^2/g);
S = soil moisture (fraction of dry weight);
K = ratio of gamma attenuation cross section in water to cross section in air, taken to be unity;
N_o = no-snow count rate with zero soil moisture;
E_2 = second order exponential integral function.

The actual peak count rates due to soil isotopes are computed from spectral window counts as

$$N_K = C_K - B_K + \beta_1(C_B - B_B) + \beta_2 C_C$$

and

$$N_{Tl} = C_{Tl} - B_{Tl} + \gamma C_C$$

where C_K = count rate in ^{40}K spectral peak window;
B_K = aircraft background in ^{40}K window;
C_B = count rate in 1.76 MeV ^{214}Bi peak window;
B_B = aircraft background in 1.76 MeV peak window;
C_{Tl} = count rate in ^{208}Tl spectral peak window;
B_{Tl} = aircraft background in ^{208}Tl peak window;
C_C = count rate in high energy cosmic window;
β_1, β_2, γ = "stripping coefficients."

The stripping coefficients allow removal of contributions to spectral peaks from radiation flux of higher energy, and are characteristic of the detector system. The mass attenuation coefficients are theoretically constants determined by the energy of the spectral peak, but in snow survey practice can vary somewhat between sites due to different radiation source geometry. Values of the mass attenuation coefficients α and no-snow count rates N_o at each of the three survey lines were determined by a combination direct-search and regression which optimized the least-squares total of gamma-measured versus ground-measured water. The values of the detector stripping coefficients were determined simultaneously in the same optimization, and were in reasonable agreement with physical considerations. Removal of the ^{208}Tl contribution from the ^{40}K peak was not attempted since warping of the ^{40}K calibration curve to include this effect seemed preferable to the loss of statistical resolution which would result from such an attempt. For comparison, an exponential curve was used in a similar optimization with only slightly poorer results. The critical consideration in selection of a curve type, however, is which will perform best when water equivalent values are encountered outside the range of those used in the calibration period. For this reason, the E_2 curve was chosen for further discussion on the basis of its theoretical justification. Table 1. summarizes the ^{40}K method results using an E_2 curve.

Table 1. Optimization of results of E_2 calibration curve values for ^{40}K peak.

Location	N_o (cts/min)	α (cm^2/g)	R.M.S. error (cm of water)
Steamboat Springs	14658	.0390	1.24
Luverne A	14256	.0428	0.56
Luverne W	15339	.0457	0.67

The estimated standard deviation in the data set, allowing eight degrees of freedom for parameter determinations, was 0.95 cm using the ^{40}K peak and 1.03 cm using the ^{208}Tl peak.

Table 2. Spectral stripping coefficients obtained by optimization of water measurements.

Curve type	β_1	β_2	γ
E_2	-1.16	-1.45	-1.75
Exponential	-1.14	-3.16	-2.55

The stripping coefficients determined are given in Table 2. The consistency of the ^{214}Bi correction to the ^{40}K peak between the two curve types is excellent. The poorer consistency of the cosmic stripping coefficients probably reflects both the use of the two different curves and the poor statistical quality of the cosmic window data used to derive the coefficients.

IDENTIFICATION OF RESIDUAL ERRORS

The determination of N_o and α values ("calibration") of operational flight lines could take several years. Archiving water equivalent values with over-snow gamma counts would allow the "best" calibration curve parameters to be approached as time progresses. In the following error analysis curve-fitting errors are assumed negligible (this is reasonable since a large number of missions were used in parameter determination) and the method accuracy is taken to represent the long-term performance achievable. This approach complements the short-term (one calibration flight) error analysis given by Jones et al. (paper 6.2, this symposium).

A model II analysis of variance was used to separate the "within groups" errors from "between groups" errors. "Within groups" errors (those producing differences in measurement results between flights on the same day) were attributed to statistical count fluctuations in the spectral windows, rapid fluctuations in radon distributions, instrument calibration drift, flight navigation errors, and air mass calculation errors. The "between groups" errors (those producing a bias in all measurements on a given day, but a different and statistically independent bias on a different mission date) were attributed to soil moisture measurement errors, "ground truth" water equivalent measurement errors, changes in general radon distribution, and (since cosmic count rates were averaged over a mission date) the error in determining the cosmic radiation contribution. The resultant "within day" and "between day" errors are summarized in Table 3.

Table 3. Errors within days and between days from analysis of variance.

Measurement peak	"Within" error (cm)	"Between" error (cm)
^{40}K	.64	.67
^{208}Tl	.86	.53

A first order approximation using Equations 1.-3. of the water measurement variance due only to statistical count fluctuation showed this to be the dominant source of "within day" error. The relatively small remaining "within day" error (0.25 cm) was attributed, for convenience, to air blanket measurement errors. This assumption does not affect the error analysis in a substantive way. The breakdown of operational "within day" errors given in Table 4. assumes a detector volume identical to the ARMS detector, but assumes a cosmic spectral window wide enough that cosmic averaging over a day would not be necessary.

Table 4. Within-day errors ($^{40}K/^{208}Tl$) with ARMS detector volume, wide cosmic window, and 2.5 minute counting time.

Effective water equivalent (snow plus air) in cm	Counting statistics (except cosmic) cm	Cosmic statistics cm	Air mass & others cm	Total cm
5-15	.20/.35	.01/.08	.25/.25	.32/.44
15-25	.37/.54	.03/.15	.25/.25	.45/.60
25-35	.63/.80	.04/.24	.25/.25	.68/.87
35-45	.98/1.17	.07/.39	.25/.25	1.01/1.26

The "between-day" errors encountered using the operational detector system assumed above would be only due to soil moisture estimate errors and general changes in the radiation source, and are given in Table 5.

Table 5. Between-day errors ($^{40}K/^{208}Tl$)

Std. dev. of soil moist. estimate	Soil moisture estimate (cm)	Radon and thoron variations (cm)	Total error (cm)
.02	.25/.23	.37/0	.45/.23
.05	.62/.58	.37/0	.72/.58
.07	.87/.81	.37/0	.95/.81
.10	1.17/1.12	.37/0	1.23/1.12

The effect of soil moisture error on water equivalent measurements was determined from Equation 1. The radon and thoron variation errors were determined as residuals after evaluating the errors common to both methods by correlating the ^{40}K and ^{208}Tl "between day" errors. The information in Tables 4. and 5. is combined in Figure 1. for the ^{40}K peak method. The relative standard error (coefficient of variation C_V) is shown in Figure 1. as well.

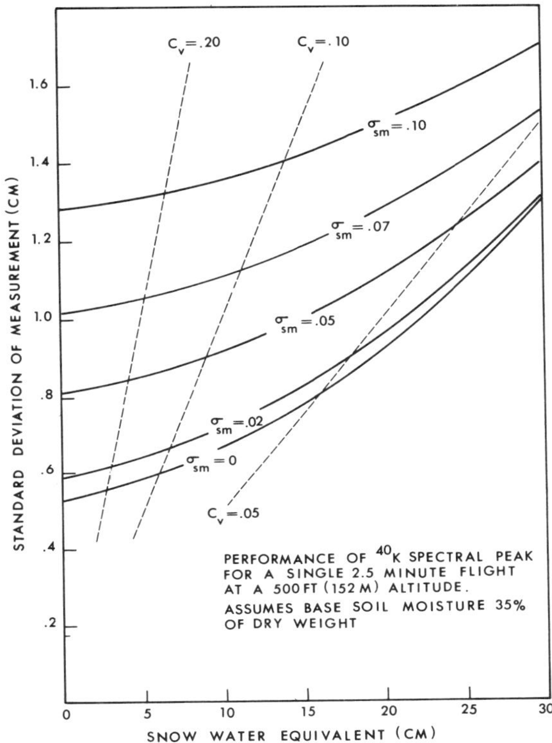

Figure 1. Single-flight accuracy of ^{40}K peak. Assumes ARMS detector volume, wide cosmic window, and previous long-term calibration.

It is important in application to make best use of the water equivalent measurements in the two peaks. A weighted water equivalent estimate can be written as

$$W_F = F\, W_{^{40}K} + (1-F) W_{^{208}Tl} \tag{4}$$

An expression for the variance of W_F in terms of F, the errors determined previously and the number of flights over the line or basin on a given day was differentiated with respect to F and set to zero to determine the optimal linear combination. The optimal F values and resulting combination accuracy are shown as a function of the number of flights in Figure 2.

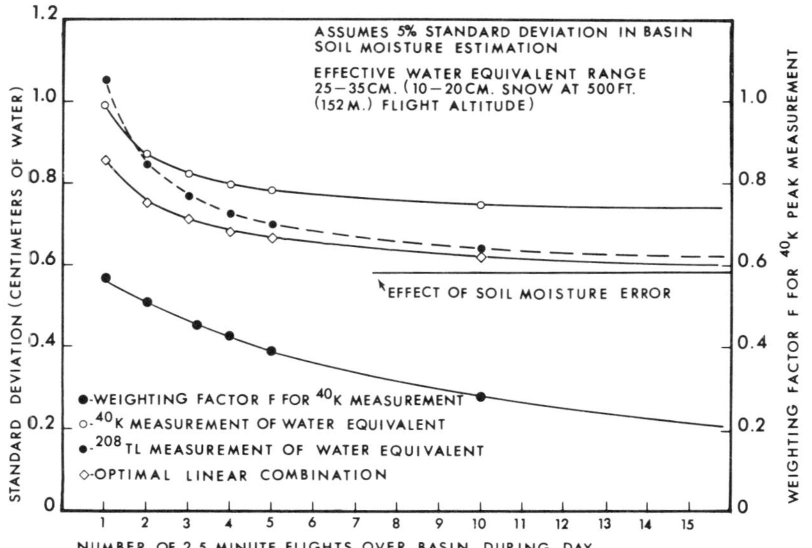

Figure 2. Performance of ^{40}K and ^{208}Tl spectral peaks.

The statistical quality of the ^{40}K peak measurement dominates when only a few flights over a line or basin are made, but as the number of flights increases, the statistical error disappears. When many flights are used the ^{208}Tl peak performance is superior to that of the ^{40}K peak because it is subject only to basin soil moisture estimate error. The ^{40}K peak is subject to general radon level bias as well. Thus, the optimal linear combination accuracy tends to the ^{208}Tl peak accuracy which is ultimately constrained by surficial soil moisture estimate accuracy.

Finally, it should be noted that the above analysis only treats errors in snow measurement over selected lines. Basin water equivalent estimates are also subject to errors due to nonrepresentativeness of selected lines. Treatment of these network errors was not within the scope of this report.

REFERENCES

Burson, Z. G., and A. E. Fritzsche, 1972. Snow Gaging by Airborne Radiological Surveys - Status Through September 1972. EGG-1183-1565, Tech. Rep. L-1078. EG&G, Inc., Las Vegas, Nevada. 48 pp.

Faul, H., ed., 1954. Nuclear Geology, John Wiley and Sons, Inc., New York. pp. 138-143.

Fritzsche, A. E., and Z. G. Burson, 1970. Water Equivalent of Snow Surveys. EGG 1183-1495. EG&G, Inc., Las Vegas, Nevada. 112 pp.

Grasty, R. L., 1973. Snow-water Equivalent Measurement Using Natural Gamma Emission. Nordic Hydrology, 4. pp. 1-16.

Kovach, E. M., 1945. Meteorological Influences Upon the Radon-content of Soil-gas. Trans. Am. Geophys. U., Vol. 26, No. II. pp. 241-248.

Kraner, H. W., G. L. Schroeder, and R. D. Evans, 1964. Measurements of the Effects of Atmospheric Variables on Radon-222 Flux and Soil-gas Concentrations. The Natural Radiation Environment. Adams, J.A.S., and W. M. Lowder, eds. Publ. for William Marsh Rice Univ. by U. of Chicago Press. pp. 191-215.

Moses, H., H. F. Lucas, Jr., and G. A. Zerbe, 1963. The Effect of Meteorological Variables upon Radon Concentrations Three Feet Above the Ground. Air Pollution Control Assoc. J., 13:12-19.

Peck, E. L., V. C. Bissell, E. B. Jones, and D. L. Burge, 1971. Evaluation of Snow Water Equivalent by Airborne Measurement of Passive Terrestrial Gamma Radiation. Water Resources Research, 7(5). pp. 1151-1159.

Peck, E. L. and V. C. Bissell, 1973. Aerial Measurement of Snow Water Equivalent by Terrestrial Gamma Radiation Survey. Bull. of Int. Assoc. of Hydrol. Sci., XVIII, 1. pp. 47-62.

Shafrir, U., G. Assaf, and M. Rindsberger, 1967. Possible Uses of Radioactive Isotopes for Meteorological Instrumentation. Scient. Rep. No. 1, ESSA Agreement No. E-94-67(n), U.S. Dept. of Commerce. 32 pp.

Styra, B. I., T. N. Nedveckaite and E. E. Senko, 1970. New Methods of Measuring Thoron (Radon 220) Exhalation. J. Geophys. Res.; 75(18). pp. 3635-3638.

Suschny, O., 1968. The Measurement of Atmospheric Radioactivity. Tech. Note No. 94, World Meteorological Org., Geneva. 109 pp.

Tanner, A. B., 1964. Radon Migration in the Ground: A Review. The Natural Radiation Environment. Adams, J. A. S., and W. M. Lowder, eds. Publ. for William Marsh Rice Univ. by U. of Chicago Press. pp. 161-190.

Vershinina, L. K., and A. M. Dimaksyan, eds., 1969. Determination of the Water Equivalent of Snow Cover. Translated from Russian, Israel Program for Sci. Transl., TT70-50093, NTIS, Springfield, Va. 142 pp.

UTILITY OF ISOTOPE PROFILING SNOW GAGE FOR WATER MANAGEMENT

Fred A. Limpert
Bonneville Power Administration
U.S.D.I.

James L. Smith
Pacific Southwest Forest and Range Experiment Station
Forest Service, U.S.D.A.

ABSTRACT

The electric power needs of the Pacific Northwest at the present time are satisfied primarily by hydropower projects. Three basic ingredients are embodied in a determination of the hydroelectric power potential of a river basin: streamflow and associated head, storage, and power plant installation.

The broad storage component is examined in some detail. Storage is generally viewed as water in a reservoir behind a dam; however, the groundwater pool and the snowpack are also a part of the storage component.

Reservoirs can be controlled, and to a much lesser degree so can the groundwater component; but except for some very specialized situations, there is virtually no control over the snow component of storage. The snow component amounts to nearly 3/4 of the average runoff of the Columbia River.

Experience to date with the isotope profiling snow gage indicates this instrument will be a valuable adjunct to the hydrometeorological network. The profiler system, a method for translating field measurement data to management use, and existing and proposed utilization of the profiler data for management purposes, are described.

The waters of the Columbia River Basin are used and reused by man for power, industry, navigation, municipal water supplies, irrigation, recreation, fish and wildlife, and many other purposes. The use of these waters for power production and the role of the isotope profiling snow gage for management of these waters is our subject.

Three basic ingredients are embodied in the determination of the hydroelectric power potential of the Columbia River Basin: first, streamflow and associated head; second, storage; and third, the hydroelectric power plant installation. An examination of the basin as a whole provides a better understanding of the basic ingredients.

The Columbia is the fourth largest river in North America on the basis of average runoff. The Mississippi River is the largest, the St. Lawrence River next, and the Mackenzie River third. As a producer of energy, the Columbia River is second to none. The Columbia River Basin contains more than 259,000 square miles and includes most of Washington, Oregon, and Idaho; that part of Montana west of the Rocky Mountains; small areas of Wyoming, Utah, and Nevada; and the southeastern corner of British Columbia.

Runoff in the Columbia River Basin results from precipitation in the form of rain or snow. The average annual precipitation over the basin is about 28 inches. Of this amount, 12 inches is returned to the atmosphere by evapotranspiration, and 1 inch is consumed, leaving 15 inches for runoff. This amount is equivalent to 185 million acre-feet of water annually, or a flow of 255,000 cubic feet per second.

Two large variations in runoff patterns in the basin have a major effect on hydropower production, seasonal variation and long-term annual variation. The large seasonal variation is caused by differences in precipitation patterns, snowmelt, temperature, geology, and topography. Coastal streams have their high runoff during the winter months when precipitation in the form of rain generally occurs. Conversely, the main stem of the Columbia and its inland tributaries generally experience high runoff in the summer months.

About one-half of the runoff from the interior basin streams occurs during the three summer months of May, June, and July, primarily from snow melt. The remaining half of the runoff takes place during the nine months from August through April. In general, precipitation in the form of snow occurs during the months of November through March. A large portion of this water in the form of snow remains stored in the snow pack until temperatures are high enough to melt the snow and release the water in the summer. A portion of the melting snow as well as a part of the rainfall is added to the groundwater pool. Flow from the groundwater pool occurs throughout the year as base flow. The base flow sustains the flow of the stream when there is no precipitation contributing to the runoff. The remainder of the precipitation in the form of rain provides for the balance of the direct runoff.

Basin runoff records also show significant annual long-term variations. The Geological Survey's report to BPA shows three major long-term trends that were remarkably consistent throughout the region. An above-average runoff trend occurred for a 12-year period ending about 1904 or 1905. Below-average runoff began about 1922 and ended about 1942 to 1945. Above-average runoff began about 1945 and persisted through 1959 to 1960.

The records at The Dalles, Oregon, cover a period of over 93 years. During this time the flows have ranged from a low of 35,000 cubic feet per second on January 12, 1937, to a high of 1,240,000 cubic feet per second on June 6, 1894. The high flow is about six times the long-term average of 194,000 cfs at The Dalles, and the low flow is about one-sixth the average.

We have noted that, in general, the high flows in the basin occur in the summer months. The area power loads on the other hand, show the opposite pattern. Peak power requirements come in the winter months during the low runoff. It would seem then, from a hydropower standpoint, that area runoff would be in excess of power needs in the summer and deficit in the winter. This imbalance has been offset a substantial amount by large storage projects in the basin.

Existing power storage in the basin amounts to slightly more than 40 million acre-feet. There is a total of about 50 million acre-feet that is economically feasible to develop in the basin. This is about one-third the mean annual streamflow at the mouth of the Columbia River. Installation of more hydropower units at existing projects will further offset the amount of runoff not being used for generation. Further adjustment to the imbalance is achieved by the large interties between the Pacific Northwest and Pacific Southwest.

Associated with runoff is the available head for developing hydropower. The main stem of the Columbia River in the United States has a gross head of 1290 feet starting at Grand Coulee Dam. All of this head has been or is under development with the exception of the 42 feet in the vicinity of Pasco, Washington.

On the tributary streams in the United States, some undeveloped head under active consideration remains near the mouth of the Flathead River, on the Wenatchee River, and on the Snake River. A small amount of head remains for consideration in the Canadian portion of the basin.

The final item is the hydroelectric power plant installations. The current installed capacity is about 20,000 megawatts. The additional installation of units would increase the total to 30,000 megawatts.

Our ability to forecast the seasonal volume runoff and day to day runoff, particularly during the summer melt season, is of vital concern in the multiple purpose management of the hydro system of the Pacific Northwest. As previously stated, about one-half of the runoff comes from snow melt in the interior basins. General forecasting of streamflow is done by the Columbia River Forecast Service, sponsored by the National Weather Service, U.S. Corps of Engineers, and Bonneville Power Administration. One of the tools used by this group is the Streamflow Synthesis Reservoir Regulation model (SSARR), which is being continually developed and improved for operational river forecasting and river management activities (U.S. Army Corps of Engineers, North Pacific Division, 1972).

Snow melt in the SSARR model can be calculated by the temperature index method or by the use of a generalized snow melt equation which provides an energy budget approach to compute snow melt for a basin. The temperature index method is generally used for daily forecasting because of the lack of energy budget data on a real-time basis.

The generalized snowmelt equation is subject to several problems affecting accuracy of the forecasts. The relation between snowmelt and the climate is dependent upon snow albedo, snow temperature, cloud cover, wind movement, etc., and is affected by overhead cover, rainfall, etc. Most models use some form of snowmelt-temperature relationship (e.g. degree-day) because of the relative abundance of temperature data. However, temperatures are not normally available from snow fields and must be adjusted from valley locations.

While the temperature index method or the energy budget approach can be used successfully to predict melt under most conditions, it is difficult to determine when this melt water will leave the pack (Halverson 1971). When surface melt occurs, it is absorbed within the pack until the free water deficit is satisfied. After this, it begins to leave the pack. The amount of water the pack can hold varies, dependent upon pack morphology, prior melt and water absorption and other factors (Smith and Halverson 1969); (U.S. Army C.E., NPD, 195

The snowpacks contributing to flow of the Columbia River can be considered as uncontrolled headwater reservoirs. There are two problems inherent in their operation in relation to more efficient operation of the downstream reservoirs which they feed. These are: (1) determination of amount of water in storage, and (2) time of release of water from these uncontrolled reservoirs.

Both amount of water in a snowpack and its interior density profile can be determined with the profiling isotope snow gage, developed by Smith et al. (1965,1972). The gage operates on the principle of gamma transmission. It consists of a separated radioactive source and a scintillating photo multiplier detector which are concurrently profiled through two tubes set 26" (66 cm.) apart and extending from above the snow-air interface to some distance within the soil. Associated electronics may be located at or away from the measuring site.

Data from the gage provides: (1) total snow depth; (2) snow density at 0.5-inch vertical intervals and average pack density; (3) total water content; (4) changes in water content in the pack and where they occur; (5) amount of new snow since the last measurement; (6) amount and intensity of rainfall on the snow until the pack begins to discharge water; (7) melt rate between measurements when melt is occurring; (8) and changes in soil moisture content.

From time sequence profiles, Smith and Halverson (1969) were able to predict the ripeness of snowpacks and predict the amount of water that snows of various density and morphology could absorb before release of water from the pack.

Water routing through the pack could be monitored. Thus, the effect upon water movement of ice lenses, different snow density layers, etc., upon time of water yield could be determined.

The Smith (1972) gage was modified to provide remote telemetered operation, and five of these gages were installed throughout the West (Randolph et al. 1972). Two of the gages were placed in the Columbia basin for evaluation by user groups in forecasting and water management. These gages, located on Mt. Baldy and on Mt. Hood, are operated via telephone lines from the U.S. Atomic Energy Commission's National Reactor Testing Station at Idaho Falls, Idaho. Teletype printouts are then mailed at about weekly intervals to the various users. There is a one to two days lag from the time the data is collected at the remote gage until its receipt by the user, and another two days delay in key punching and plotting the snow depth, density, and cumulative water content. Plots are printed on transparent film, making possible the overlay of a number of films to observe the progressive changes with time that take place in the snowpack, Figure 1.

Location of these two test gages and length of time between data taking and availability of the data to the users restricts the utility of the data.

Data from the snow profilers are used subjectively at the present time. We have found these data helpful in assessing the "ripeness" of the snowpack for determining start of runoff and for selecting proper melt rate and for more accurate determination of daily snowmelt for use in the SSARR model.

A better understanding of the snowmelt, rainfall, runoff relationship and the movement of rainfall through the snowpack, has been possible from study of these profiles. We feel there is a greater potential to be realized from use of the profiling snow gage systems. However, to realize the fullest potential will require a more intensive network of strategically placed snow gages yielding real-time radio-relayed data to be used in daily operation of downstream water reservoirs.

The first use of the snow gage should be to increase the accuracy of the estimate of the amount of water in storage in the snowpacks at any point in time.

In snow surveying, we have historically collected snow water content at monthly intervals beginning about January 1 of each year. From a few sample points in each subbasin, a prediction is made of the streamflow to be delivered in the spring and summer. The method relies on correlation analysis. The historical record of snow water content on a particular date and the streamflow in subsequent months

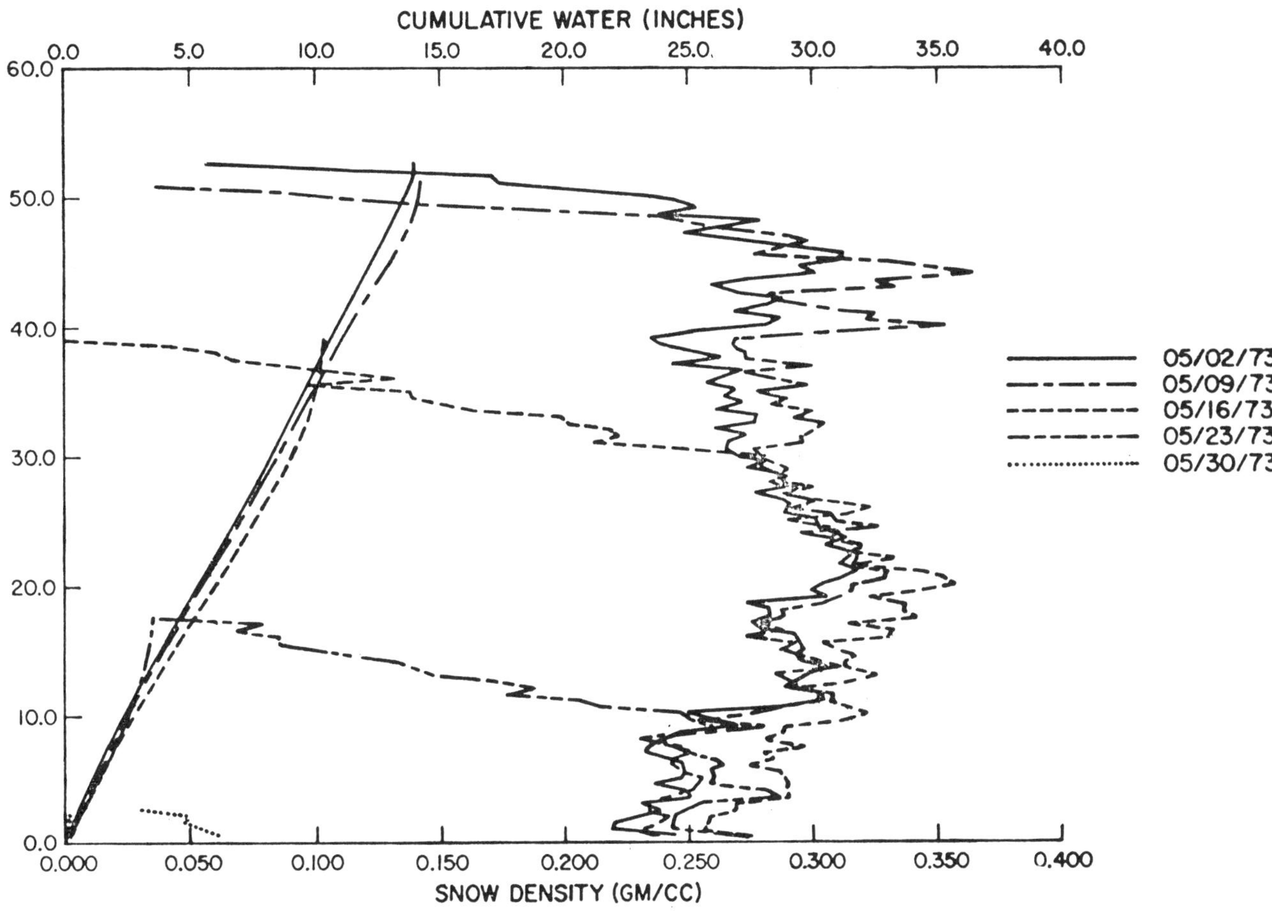

Figure 1. Snow profile - Mt. Baldy, Idaho

is correlated for all the years of record. From this, one can then estimate future streamflow.

This system depends upon a long record for best accuracy. If massive forest fires or timber harvest changes the melt relationship of the snow, and snowmelt timing changes, the whole relationship changes. Accuracy is least in deficient snow years and in massive snow years. These are times when the greatest accuracy is needed.

If one desires to use the current method for weekly or daily values of streamflow, the cost for data collection is exorbitant and accuracy is poor! As an example, in the Columbia River Basin, the total runoff in this area is prepared on the basis of 150-200 forecasts for the entire drainage network. The average forecast error for the streamflow period April 1 - September 30 is 8 percent to 14 percent. Occasionally, forecasts miss by 40 percent - 50 percent. The deviation of the streams amounts to 12 percent - 20 percent. Thus, by our forecasts, the error has been reduced by only half. Yet, plans for managing the Columbia require accuracy of forecast down to just a very few percent and on a daily basis. It is doubtful the current method of snow survey will meet the need.

The profiling snow gage offers an opportunity for development of a new concept in determining the amount of water in a snowpack and in determining the rapidity with which snow melts under the influence of warm rain or atmospheric-solar heat. It should offer a more accurate method for predicting streamflow levels from snow surveys.

The suggested new method for snow surveys would be based upon estimation of the amount of water in the snowpack and the rapidity of melt of snow from the various facets of a watershed and its movement through the snow to the streams.

It is visioned that one or more telemetered profiling snow gages, a precipitation gage, temperature sensors, and other equipment would be placed in a central location within each subbasin. Auxiliary and less expensive equipment for measuring snow water amount would be placed in other well-selected sites covering the major altitudinal, aspect, and cover conditions within each subbasin. A study would be made over a period of time, correlating the amount of snow and the melt found at any time at the central location to the amount found at other locations within the subbasin. All combinations of aspect, slope, and cover would be sampled. The melt rate for each combination of cover slope and aspect and altitude would be related to that from the central location. A computer program would be written, in which the acreage, location, aspect, and cover of each area would be listed along with its mathematical relation to the central location. When the cover was changed on any part of the subbasin, the mathematical effect of this change would be fed to the computer. Daily melt, measured at the snow gage site, would be adjusted in the computer to yield a basin-wide figure for use in the SSARR model. The melt rate-temperature regime for the SSARR model would thus encompass all elements of the watershed in relation to the percent of the area they cover. With such a program, one would be able to determine the actual amount of water on the watershed.

From study of the profiles taken with the centrally located profiling snow gage, one could determine how much water from rain or melt could be held in the pack prior to initiation of runoff at that point. By correlations developed between the snow reaction at the central location, and that for the various satellite locations, the melt relation for the basin should be determinable.

From a long-range weather forecast, the future heat input to the pack could be estimated, and, hopefully, more accurate predictions developed for addition of this data to the current SSARR model. The prediction "lead time," before snowpack outflow reached downstream reservoirs, and the accuracy of prediction of total outflow amounts, should be increased as a result of use of such a system. This should result in more efficient reservoir operation and increased water storage. In turn, this should increase power generation.

Addition of aerial snow coverage obtainable from satellite photography should further improve the accuracy by providing real-time data on snow line location from day to day, as a further estimate of melt rate and total aerial contribution to melt.

With the combination of satellite photography and the proposed profiling snow gage system, it may be possible to improve the accuracy of streamflow forecasts from snowmelt from statutory wilderness areas. If a few strategically located snow gages can be placed at sites near wilderness areas at representative elevational, aspect, and cover conditions, it appears probable that correlations between the measured sites and the adjacent wilderness sites can be developed. From these, predictions of melt rate and water flow from the wilderness areas may be developed. Aerial snow coverage would be obtained from satellite photographs.

Our subjective use of the data from the Mt. Baldy and Mt. Hood gages and study of research data obtained from snowpacks at Central Sierra Snow Laboratory in California utilizing the snow gages, suggest the profiling snow gage can furnish an important new dimension to forecasting streamflow eminating from snowmelt. This, in turn, can lead to more efficient operation of the multi-purpose reservoirs in the Columbia River system.

REFERENCES

Halverson, Howard G., 1971. Seasonal Snow Surface Energy Balance at the Central Sierra Snow Laboratory. Dissertation University of Arizona, Tucson, Arizona. 107 p.

Randolph, P. D., R. A. Coates, E. W. Killian, L. O. Johnson, R. L. Heath, 1972. A Network of Telemetered Profiling Isotope Snow Gages. Paper presented at International Symposium on the Role of Snow and Ice in Hydrology, Banff, Canada.

Smith, James L. and Howard G. Halverson, 1969. Hydrology of Snow Profiles Obtained with the Profiling Snow Gage. Proceedings Western Snow Conference. April 1969. pp 41-48.

Smith, James L., Howard G. Halverson, and Ronald A. Jones, 1972. Central Sierra Profiling Snow Gage: A Guide to Fabrication and Operation. TID-25986. U.S. Atomic Energy Commission. National Technical Information Service, U.S. Department of Commerce.

Smith, James L., Donald W. Willen, and Michael S. Owens, 1965. Measurement of Snowpack Profiles with Radioactive Isotopes. Weatherwise 18: 246-251.

U.S. Army Corps of Engineers, North Pacific Division, 1972. Program Description and User Manual for SSARR Model.

U.S. Army Corps of Engineers, North Pacific Division, 1956. Summary Report of the Snow Investigations, Snow Hydrology.

NUCLEAR TECHNIQUES FOR SNOW AND ICE STUDIES IN CANADIAN SUBPOLAR REGIONS (DEVON ISLAND)

F. A. Prantl
Canada Centre for Inland Waters
Burlington, Ontario

R. M. Koerner
Polar Continental Shelf Project
Department of Energy, Mines and Resources
Ottawa, Ontario

E. Robertson
Atomic Energy of Canada Limited
Chalk River Nuclear Laboratories
Chalk River, Ontario

ABSTRACT

In a cooperative multidisciplinary programme, snow and ice samples have been collected on a traverse from the south-east to the north-west across Devon Island, North-West Territories, and analyzed for stable and radioactive isotopes: deuterium, oxygen-18, tritium, and total β-activities originating from fission product debris of atmospheric nuclear weapons testing. The snow samples consisted generally of one annual layer. The ice under the annual snow cover was sampled, and, additionally samples were taken from pits at various locations, comprising up to three annual layers of net accumulation. The purpose of the study is to demonstrate to what extent it is possible by use of nuclear methods in Canadian subpolar regions under melting and drifting conditions to identify origin and seasonal patterns of snow deposits, date the layers, and, determine the net accumulation rates for mass balance studies.

INTRODUCTION

Over the past fifteen years extensive programmes have been conducted to investigate the natural resources of the Canadian Arctic, its climate and the history of snow accumulation. In recent times nuclear techniques have become powerful tools for snow and ice studies in polar and Alpine regions. This investigation was directed at examining their usefulness and limitations in the Canadian Arctic.

Two isotope programmes were started in 1972 on the Devon Island ice cap. The purpose of the first was to determine the characteristics of variations of stable and radioactive isotopes within the snow deposits of the past two decades, and to find out if melting and drifting altered them significantly. This information was required also for the second programme, intended to measure and interprete the isotopic changes of glacier ice over thousands of years from 3 deep cores drilled down to bedrock in the summers of 1971 to 1973, and several large volume samples of old ice.

The paper gives results of the first isotope programme, illustrates the potentialities and limitations of isotope techniques in identifying the origin and seasonal patterns of recent snow deposits, dating the annual layers and determining net accumulation rates.

DESCRIPTION OF ISOTOPE TECHNIQUES

Two different techniques have been examined in this study. One is based on the analysis of snow for the heavy stable isotopes of the water molecules, the other on activity measurements of tritium and of fission products attached to particles in the snow. The basic principles of the two techniques are reviewed briefly.

Water molecules are combinations of the hydrogen isotopes ^{1}H, ^{2}H, ^{3}H and oxygen isotopes of which ^{16}O and ^{18}O are the most abundant in nature. All are stable and of natural origin except ^{3}H which decays with a half-life of 12.4 years emitting weak β-radiation. The maximum β-energy is 18 keV. Since about 1954 the tritium concentration in precipitation reflects the artificial production during nuclear weapons tests and by other man-made sources rather than the seasonal variations of the natural stable isotope ratios that will be described in the following. Therefore, tritium will be discussed under the second isotope technique.

The isotopic species of the water molecule differ in their diffusivities and vapour pressures. Therefore, each change of phase during the hydrologic cycle is accompanied by isotope fractionation which changes the relative isotopic abundances. The magnitude of this change depends on the temperature at which the process has occurred. This means the atmospheric temperature at which precipitation was formed can be inferred from studies of the isotopic ratios of D/H or $^{18}O/^{16}O$ (Merlivat et al., 1967). The seasonal variations of atmospheric temperatures are reflected by higher D/H and $^{18}O/^{16}O$ ratios during summer than during winter and in this way seasonal accumulations can be distinguished, accumulation rates determined, and the annual layers can be dated by counting the number of seasonal isotope variations from the surface downwards (Epstein et al., 1963, 1965; Johnson et al., 1972).

The second isotope technique consists in analyzing the snow for tritium or for fission products attached to particles in the snow and can be applied only to glaciological mass balance studies over the past two decades (Crozaz et al., 1966; Ambach and Dansgaard, 1970; Prantl et al., 1972). During this period isotopic composition and radioactivity levels of the snow reflect sizes, frequencies and types of nuclear weapons tested in the atmosphere so that characteristic features of the radioactivity profiles of the snow can be ascribed to the time scale of the bomb test history. In this way, annual accumulations can be dated absolutely without referring to the surface as must be done by glaciological and stable isotope techniques.

The usefulness of both isotope techniques is limited by the degree and rate of modification of the isotopic stratification after the deposition of the snow. Isotopic homogenization occurs by melting, refreezing, vapour transport, drifting and radioactive decay. All of these processes, except of course, radioactive decay, are related to the specific conditions of the locality under study.

SITE DESCRIPTION AND SAMPLING

Devon Island is one of the group of the Queen Elizabeth Islands of the North-West Territories separated from Greenland by Baffin Bay. Its ice cap of 15,568 km^2 (figure 1) covers about 1/3 of the island's area and is among the largest in Canada. Glaciological studies since 1961 have shown that for at least the past 12 years the ice cap has generally had a negative annual balance, i.e. ablation has exceeded accumulation (Koerner, 1970). Accumulation rates on the south-east side (45 g cm^{-2} y^{-1}) have been higher than on the north-west side of the ice cap (12 g cm^{-2} y^{-1}). Over the same 12 year period the elevation of the equilibrium line, where accumulation equals ablation, has varied between 600 and 1500 m above mean sea level on the north-west side of the ice cap. In some years the ice cap encompasses all the

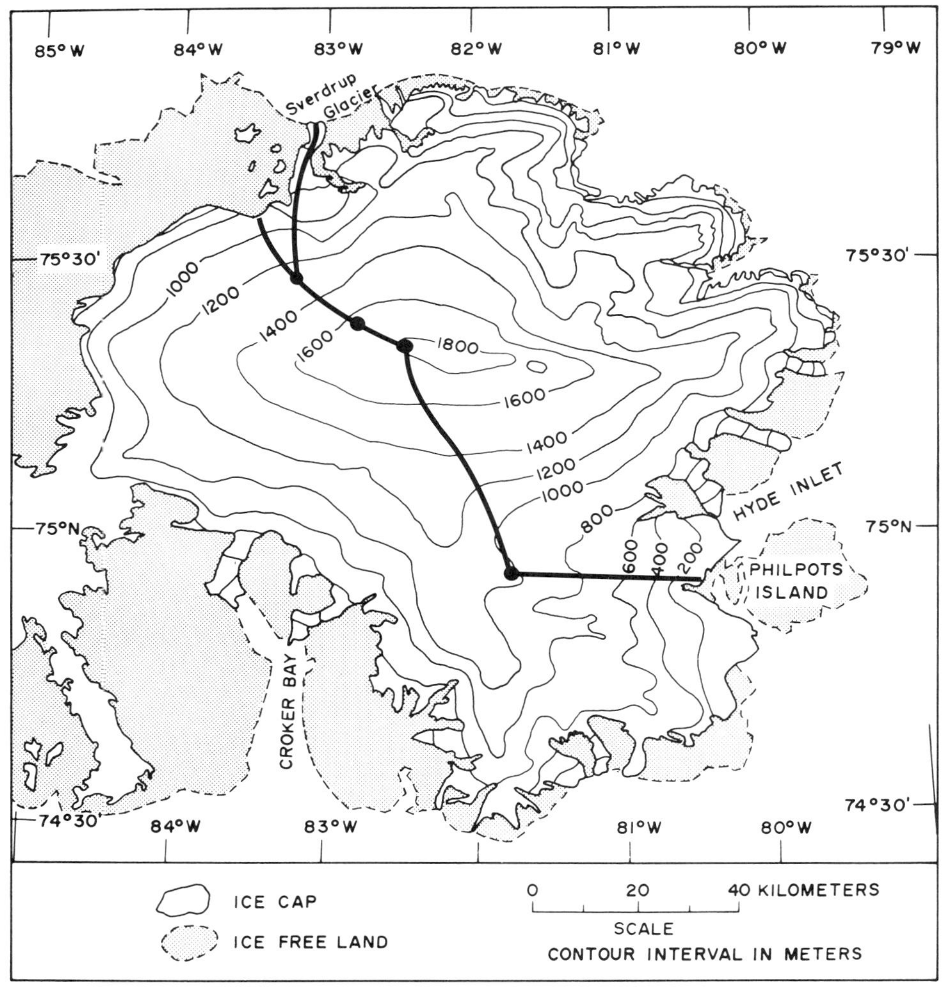

Figure 1. Devon Island ice cap showing the traverse route and the sites of the detailed samples from pits and cores (heavy dots).

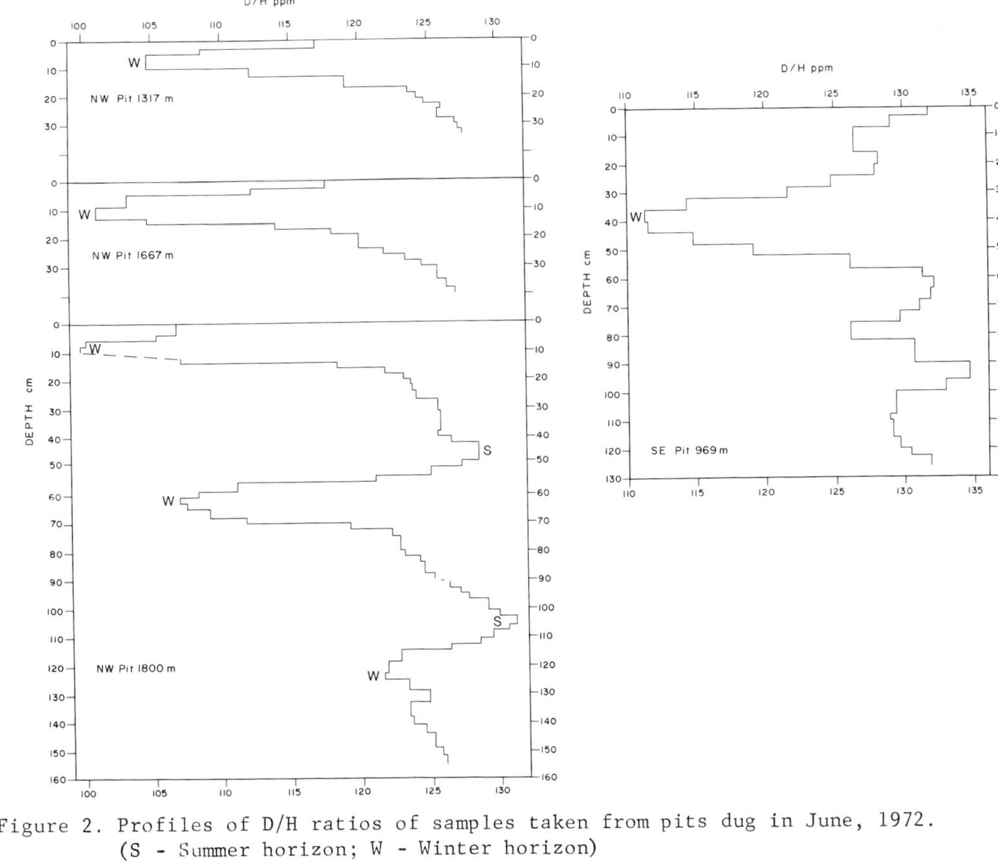

Figure 2. Profiles of D/H ratios of samples taken from pits dug in June, 1972.
(S - Summer horizon; W - Winter horizon)

6.6

genetic snow zones from dry to soaked deposits (Benson, 1961). Generally the top of the ice cap is in the percolation zone, i.e. the meltwater refreezes within the current balance year layer.

The isotope programme was designed to encompass various ablation conditions. Therefore, two types of samples were collected at the locations shown in figure 1:

Firstly, detailed samples from pits and cores were taken between surface and 10 m depth at 4 different altitudes on the north-west and south-east side of the ice cap. Stratigraphy and density were recorded. The temperature at 10 m depth was -23°C. Four sets of samples have been analyzed for stable isotopes and cover from one to three consecutive annual layers. The first, most recent of 1972/71 was unaffected by melting, the second of 1971/70 and the third of 1970/69 had been affected by melting. Two more sets of samples cover in total two decades of annual net accumulations and are being analyzed for tritium and fission products attached to particles in the snow.

Secondly, current annual snow cover samples were taken at about 60 locations at different altitudes on a four day south-east to north-west traverse across the ice cap in June, 1972. There was no precipitation during this time and the snow cover had not been affected by melting. Each sample comprises the accumulation from August, 1971 to June, 1972 and was collected in a stainless steel tube pushed vertically into the snow to the hard firn or superimposed ice forming the boundary between the current and the previous balance year deposit. Additionally, several ice samples were taken mainly in the ablation zone and in most cases represent snow deposited originally at higher elevations in the accumulation zone. Water equivalents and stable isotope ratios of the samples were determined.

All samples were stored in sealed polyethylene bags or screw-cap bottles, shipped to the laboratory and kept frozen until ready for analysis. D/H was determined by mass-spectrometry at the Chalk River Nuclear Laboratories of Atomic Energy of Canada Limited, and $^{18}O/^{16}O$ at the University of British Columbia. In this paper we refer to D/H ratios only, not to $^{18}O/^{16}O$ ratios. The two are related linearly and, therefore, for the purpose of this study provide similar information. Tritium concentrations were determined by liquid scintillation counting, total β-activities by low-level β counting at the Radiation Protection Bureau of the Department of National Health and Welfare.

RESULTS

Isotope profiles

The purpose of the detailed samples from pits and cores analyzed for stable isotopes and fission products was to determine the characteristics of isotope variations within the annual deposits and to find out if they could be used to determine the accumulation rates under the varying melting conditions encountered on the ice cap.

The results of the four sets of samples analyzed for stable isotopes are illustrated by the four pit profiles in figure 2. The typical seasonal variation characteristic of stable isotopes can be recognized in all deposits. However, in the 1800 m profile this variation diminishes in the lower layers. Here, in particular the shift of the low winter values towards higher values should be noted as it must be related at least in part to the scale of melting. Summer melting causes isotopically heavy melt water to percolate through the snow pack where it has refrozen at depth. In this way the winter layer has become isotopically heavier whereas the mass of the summer layer has been reduced but isotopically it was not altered significantly. The mean shift of the three consecutive winter values was about 12 ppm y^{-1}, the largest difference of D/H values in this profile about 35 ppm. Assuming similar conditions for the previous balance years it would mean that the characteristics of the isotope variation could disappear in about three years at this location. However, no general predictions can be made because conditions can vary from year to year. At the same location analysis of the detailed samples between 3-5 m for tritium which as a water isotope undergoes similar processes of isotopic homogenization as stable isotopes, have shown a peak in a strata dated by snow stratigraphy and isotope techniques as the deposit of 1963/62 when the tritium levels from nuclear weapons testing were at their highest. This peak was preserved largely because of a lack of melting in the ensuing summers of 1963 to 1965 at this altitude. There is a high probability that the tritium peak can be found at similar elevations on other ice caps of the Canadian Arctic and can be used as a glaciological dating mark and reference horizon for accumulation studies.

The major conclusion to be drawn so far from the measurements of total β-activities of fission products attached to particles in the snow was that the profiles have not been affected significantly by isotopic homogenization processes and can be used to determine the accumulation rates over the past two decades.

Accumulation rates

The accumulation rates are computed by using the isotope profiles and associated density measurements. From previous glaciological studies already mentioned here, we knew that the annual accumulation was about four times greater in the south-east than in the north-west of the ice cap. The D/H profiles in figure 2 emphasize the difference between the locations of the south-east and the north-west. Furthermore, they allow the determination of seasonal accumulation rates indicating that in spring and fall accumulation was approximately 3 times higher at the south-east location, but in winter only twice as high as at the north-west locations. The accumulation rates were 18.1 g cm^{-2} for 1972/71 and 23.8 g cm^{-2} for 1971/70 at 1800 m.

The accumulation rate computed by using the tritium profile was 22.0 g cm^{-2} y^{-1} for the period from 1963 to 1972. By using the total β-activity profile an accumulation rate of 21.7 g cm^{-2} y^{-1} was obtained as a preliminary result for the period from 1953 to 1972.

Results of the traverse study

The purpose of analyzing the traverse samples from the current annual snow cover for stable isotopes was to locate the source of moisture and to examine the role of scouring and drifting. The results are plotted in figure 3 and figure 4. The general trends of decreasing D/H ratios with decreasing amount of precipitation and with increasing distance from the south-east coast should be noted as it

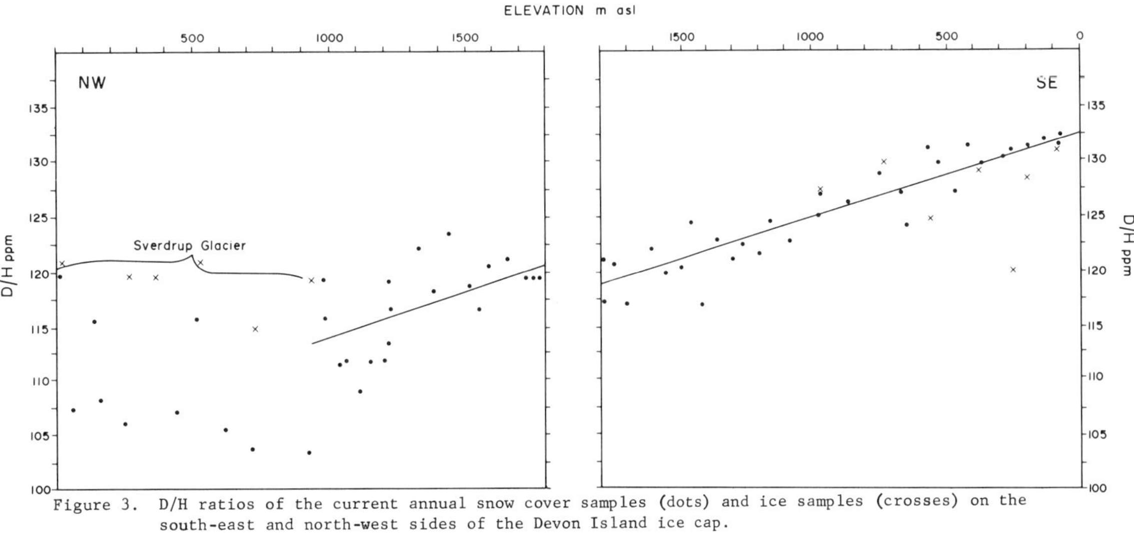

Figure 3. D/H ratios of the current annual snow cover samples (dots) and ice samples (crosses) on the south-east and north-west sides of the Devon Island ice cap.

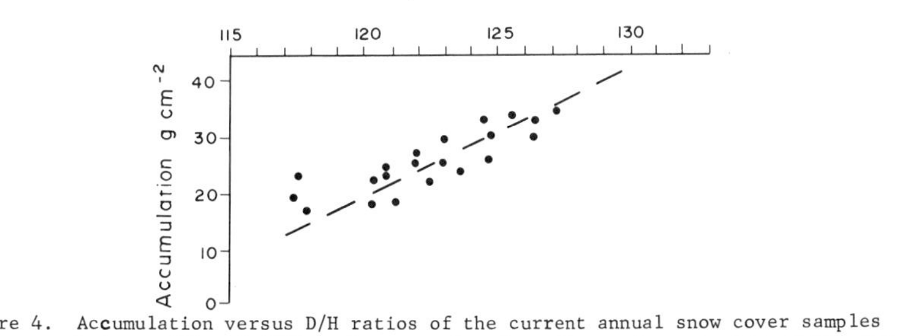

Figure 4. Accumulation versus D/H ratios of the current annual snow cover samples between 900 and 1800 m above sea level on the south-east side of the ice cap.

indicates that the prevailing source of moisture was located in Baffin Bay. The air masses were depleted gradually of their moisture by precipitation as they crossed the island. Along this passage each subsequent condensation was accompanied by isotope fractionation. As the air mass rose to higher altitudes of generally lower temperatures the preferential loss of deuterium from it was enhanced during condensation. The resulting D/H effect calculated by a least square fit of the data in figure 3 was 0.8 ppm/100 m change of altitude. This preferential loss of deuterium from the air mass with increasing altitude of the condensation level persisted after the air mass had crossed the top of the ice cap but there is a much greater scatter of the points on the north-west side. This result agrees with that of similar studies in the Sierra Nevada area where the greater scatter of the D/H values on the generally lee-ward side of the mountain crest was attributed to additional turbulence as the air mass moved across the top ridge (Friedman, 1970). On the Devon Island ice cap katabatic winds cause scouring and downslope drifting of the snow (Holmgren, 1971). There is a zone of scouring between about 1200 and 1600 m above sea level in the north-west where the isotopically light winter snow has been removed and deposited downslope in the zone of drifting between about 1000 and 1200 m above sea level. As a result the isotope values in the zone of scouring were too high and those in the zone of drifting were too low. The existence of a systematic relationship between D/H values and accumulations in figure 4 suggests that there was very little modification of the current annual snow cover by downslope drifting in the south-east. No systematic trend was found on the north-west side.

The mean annual values of D/H calculated for the 1972/71 deposit from the 4 profiles in figure 2 fit into the general trend of the traverse samples in figure 3. However, the lowest values of this deposit, the winter values, seem to suggest that the winter layers could originate from moisture carried by air masses from the north-west as well as from Baffin Bay.

The results of the ice sample analysis for deuterium were added to figure 3. The values seem to be distributed at random, and cannot be explained from our present knowledge of the conditions involved. Studies are continuing on this subject.

CONCLUSIONS

Two isotope techniques were examined. The first is based on analysis of the snow for stable natural isotopes, the second on radioactivity measurements of fission products from nuclear bomb tests. Deuterium studies were useful to identify Baffin Bay as the major source of precipitation on the ice cap. The use of deuterium profiles for determining the accumulation rates was limited by the varying degree of melting of the deposits. This limitation applies in general also to the use of tritium profiles. However, at an elevation of 1800 m a characteristic tritium peak in the 1963/62 deposit was preserved because of lack of melting at this altitude during the summers of 1963 to 1965. Reference to this horizon allows the determination of the accumulation rate over the last decade. Total β-activities of fission products attached to particles in the snow were not so affected by melting. Therefore, this is a promising method for determining the accumulation rates over the past two decades.

ACKNOWLEDGEMENT

We gratefully acknowledge cooperation of Mr. W. M. Thurston, Atomic Energy of Canada Limited, Chalk River Nuclear Laboratories, who provided the deuterium analyses, and of Dr. H. Taniguchi, Radiation Protection Bureau, Department of Health and Welfare, who supervised the tritium and β-activity measurements. Field work and part of the sample preparation was supported by the Polar Continental Shelf Project, Department of Energy, Mines and Resources. We thank all who rendered technical assistance. Our appreciation is due to the Department of Environment for allowing Dr. F. A. Prantl to continue this work at the Canada Centre for Inland Waters.

REFERENCES

Ambach, W., and W. Dansgaard, 1970. Fallout and Climate Studies on Firn Cores from Carrefour, Greenland. Earth Plan. Sci. Letters 8: 311-316.

Benson, C. S., 1961. Strategraphic Studies in the Snow and Firn of the Greenland Ice Sheet. Fol. Geogr. Dan. Tom. 9:13-37.

Crozaz, G., C. C. Langway, E. Picciotto, 1966. Artificial Radioactivity Reference Horizons in Greenland Firn. Earth Plan. Sci. Letters 1:42-48.

Epstein, S., R. P. Sharp, I. Goddard, 1963. Oxygen Isotope Ratios in Antarctic Snow, Firn and Ice. Journ. Geology 71, 6:698-720.

Epstein, S., R. P. Sharp and A. J. Gow, 1965. Six Year Record of Oxygen and Hydrogen Isotope Variation in South Pole Firn. Journ. Geophys. Res. 70, 8:1809-1814.

Friedman, I. and G. I. Smith, 1970. Deuterium Content of Snow from Sierra Nevada Area. Science 169:467-470.

Holmgren, B., 1971. Climate and Energy Exchange on a Sub-Polar Ice Cap in Summer, Arctic Inst. North Am. Devon Island Expedition 1961-1963, Part A. Physical Climatology Met. Inst. Uppsala Univ., Meddelande 107:83.

Johnson, S. J., W. Dansgaard, H. B. Clausen, C. C. Langway, Jr., 1972. Oxygen Isotope Profiles through the Antarctic and Greenland Ice Sheets. Nature 235:429-434.

Koerner, R. M., 1970. The Mass Balance of the Devon Island Ice Cap, Northwest Territories, Canada, 1961-66. Journ. Glaciol. 9, 57: 325-336.

Merlivat, L., C. Lorius, M. Majzoub, G. Nief and E. Roth, 1967. Etudes isotopiques en profondeur d'un glacier en Antarctique. Proc. Symp. Isotopes in Hydrology. Vienna, 1966:671-679.

Prantl, F. A., W. Ambach and H. Eisner, 1973. Alpine Glacier Studies with Nuclear Methods. Proc. Intern. Symp. Role of Snow and Ice in Hydrology, Banff, 1972 (in print).

USE OF THERMOLUMINESCENT DOSIMETERS FOR STUDIES OF THE SNOW COVER

W.E. Grummitt and F.A. Prantl*
Biology and Health Physics Division
Atomic Energy of Canada Limited
Chalk River, Ontario, Canada

ABSTRACT

A preliminary study has shown the feasibility of using thermoluminescent dosimeters (TLD's) to determine the 'water equivalent' of the snow cover by gamma ray attenuation measurements. Small lithium fluoride detectors were placed at a convenient height of 1-2 metres above ground at the beginning of the observation period, and collected afterwards to be read in the laboratory. Attenuation studies of gamma-radiation from point sources and natural background were carried out. The latter require TLD exposure times of several weeks. Point sources of 10 to 500 uCi located at ground level allow much shorter exposure times. In general the source strength is chosen according to height of snow cover and length of observation period. Advantages of this system are: cheapness, flexibility, easy monitoring, simple field installation. Therefore it can serve a large number of observation stations.

Several methods have been proposed for the measurement of snow depth by the change in intensity of transmitted gamma radiation. Both artificial (Ferronskiy 1968) and natural (Dahl 1970) sources have been used and measurements have been made in the field at ground level and from aircraft (Grasty 1972).

The limitations of the various schemes are now well understood. Methods using natural background suffer from errors due to partially covered outcrops of rock and the variable background from Radon and its daughter products. Counting equipment is usually complex, expensive and difficult to maintain under field conditions. Some of these problems can be eliminated through use of artificial sources and passive detectors.

---

*Present address, Physical Sciences Labs., Canada Centre for Inland Waters, Burlington, Ontario.

6.7

The proposed method employs thermoluminescent dosimeters (TLD's) for detection of the radiation. These have the advantages of low initial cost, ease of handling in the field, insensitivity to low temperatures and lack of maintenance. When used with an artificial source there are almost no site or snow depth restrictions, though there is some inconvenience associated with the handling of exposed sources. Also, in most countries, there are regulations governing the use of radioactivity which normally apply to quantities larger than 10 μ Ci.

METHOD

Dosimeters

The principal difference between a TLD and other radiation detectors is that the radiation effect is stored in the TLD crystal and is subsequently released and measured in the laboratory. Dosimeter exposure times with natural background are relatively long (days), but many TLD's can be placed in the field at one time because the initial cost is small.

Lithium fluoride (TLD-100*) is frequently chosen for the measurement of environmental gamma radiation. It has the advantage of long term stability (~ 5% loss per year) and energy-independent response. Its sensitivity is moderate; an exposure of 10 mR can be determined with a standard deviation of $\pm$ 1.0 mR.

The thermoluminescent properties of most crystals can be altered by the incorporation of foreign elements into the lattice. Calcium fluoride:manganese has 3 to 10 times greater sensitivity than TLD-100 so correspondingly smaller exposure times are required. This is a very useful dosimeter for snow measurements even though it does not have energy-independent response or long term stability. Lithium fluoride was used in the present investigation because it was the only dosimeter available at CRNL at the time. Calcium fluoride or sulfate would have been a better choice from the standpoint of sensitivity to radiation.

Before use all TLD's were annealed by heating to 400°C for one hour and 100°C for two hours. They were

*Harshaw Chemical Co.

then individually packaged in paper and sealed in plastic to exclude moisture and light. Finally they were stored in a shielded lead castle. Dosimeters could be removed for exposure in a shaded location at any time provided they were returned to the shield following exposure. Controls (usually eight TLD's) remained shielded until the end of the experiment at which time half were exposed to one Roentgen. This provided the calibration and permitted direct comparison of gamma intensity with and without snow cover. Details of the procedure used for reading the TLD's has been given elsewhere (Jones 1973).

Radiation Sources

The absorption of gamma rays from a point source of monoenergetic radiation is given by:

$$I = I_o e^{-\mu x}$$

where I = gamma intensity with absorber
I_o = original gamma intensity
x = thickness of absorber
μ = linear absorption coefficient
ρ = density.

For water μ/ρ, the mass absorption coefficient is equal to μ the linear absorption coefficient.

Terrestrial gamma radiation can be used with TLD's at medium snow depths to provide data for calculation of water equivalent thickness. It is convenient to consider this source as an infinite slab of material. Carmichael et al. (1951) developed an expression for this geometry where:

$$I = I_o E_2^{(\mu x)}$$

Values of the function $E_2(x)$ are tabulated by Placzek (1953). The expression is valid only for monoenergetic radiation which has a characteristic value of μ. To circumvent the problem arising from the complex nature of the gamma rays from natural background, some investigators have used a semi-empirical formula (Dahl 1970) (Zotimov 1968). However, the above expression will fit the absorption of terrestrial gamma radiation in air down to values of I/I_o = 0.15 (Grasty 1973), so it should be equally valid for snow.

When an artificial source is used for snow depth measurements it is normally placed at ground level with the detector vertically above at some fixed distance (normally one to two metres). This arrangement, i.e. point source geometry, offers several advantages: 1) the terrain can be uneven with outcrops of rock, 2) an independent calibration can be made using water as the absorbing medium, 3) greater snow depths can be measured by using larger sources, 4) pre-season measurements are not required, and 5) background corrections are small so there is greater inherent accuracy.

Cobalt-60 has a convenient energy for this type of measurement. The half-value layer, $0.693/\mu$, is about 20 cm water equivalent so snow depths even five times that thickness can be measured (3% transmission). A TLD-100 placed one metre above a 100 μCi source at ground level requires exposure times of about 5 days for an accurate measurement. Greater source-detector distances require a more sensitive TLD or correspondingly larger sources.

The method was tested by making absorption measurements in a shallow body of water with the source resting on the bottom and the TLD's one metre above the source position. A rectangular aluminum (Dexion) frame, one metre cube, with a thin aluminum floor provided the support with minimum backscattering. The frame could be moved into water of various depths. A 10 mCi cobalt-60 source was used with an exposure time of one to two days. This is a larger source than normal but it provided accurate data for calibration. (Figure 1.)

RESULTS

The results for water are given in the Figure. Each point is the mean of four TLD measurements at a particular depth. Because of the 'build up' of scattered radiation, the linear portion of the curve intersects $I/I_0 = 1$ at a depth of 2 cm water instead of at zero. A positive correction of 2 cm should therefore be applied when calculating snow depths from the data in Table 1.

Mean values of μ/ρ the mass absorption coefficient are given in Table 1 for water, polystyrene and styrofoam*. The measurements were made with TLD-100 using a point source of cobalt-60. Polystyrene is somewhat less absorbing

*Expanded polystyrene foam plastic, $\rho = 0.0332$.

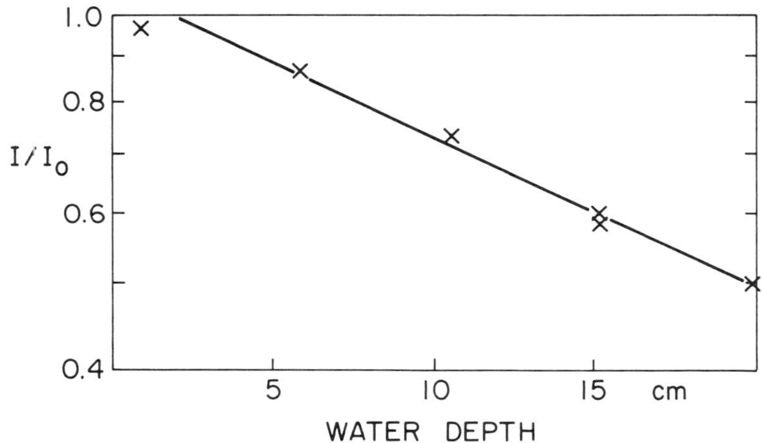

Figure 1. The absorption of cobalt-60 gamma rays by water.

than water (3.3 %) due to its slightly lower electron density. This effect can be seen in the results for μ/ρ in Table 1.

Table 1. Mean Values of the Mass Attenuation Coefficient

Absorbent	Source	Collimation	μ/ρ
Water	^{60}Co	No	0.0385 cm^2g^{-1}
Water	^{60}Co	Yes	0.037
Polystyrene	^{60}Co	Yes	0.0358
Styrofoam	^{60}Co	Yes	0.0363
Snow	Natural γ	-	0.032

Styrofoam was selected to test the method experimentally because of its similarity to snow. The thickness of a pile of this material was determined using μ/ρ measured for polystyrene sheet (ρ = 1.04). The results in Table 2 indicate that the thickness of a low density material such as uncompacted snow can be satisfactorily determined from a mass absorption coefficient obtained with water placed close to the source.

Table 2. Comparison of Calculated and Measured Thicknesses of Styrofoam

% Transmission	Mass Thickness	Calculated Thickness	Measured Thickness
95.0	1.43	43.0 cm	42.7 cm
90.0	2.94	88.6	87.9
86.5	4.05	122	120
82.8	5.27	159	155

A mean absorption coefficient was also measured for natural background and used to estimate snow depths in the field. The area chosen was located on the Laurentian Shield near CRNL. Because natural gamma radiation is a distributed source, direct calibration with water is not feasible, but both air and snow have been used previously (Grasty 1973) (Dahl 1970). In the present case μ / ρ was measured for snow using an energy-compensated Geiger counter (Jones 1971). This instrument and TLD-100 have almost energy-independent response and give very similar absorption curves with cobalt-60 and water. The error in using this sort of calibration for estimating snow depths is small relative to other factors. These are discussed in a later section.

Early last spring four TLD-100 dosimeters were exposed to natural background radiation during a period of relatively stable snow conditions. In April and May when the snow had melted, a second series gave data from which I/I_o could be calculated and water equivalent snow depths could be estimated. The results are given in Table 3 together with the amount measured gravimetrically. The mean depth of 19.4 cm obtained by gamma ray absorption agrees well with the gravimetric result.

Table 3. Net Accumulation of Snow

I/I_o	Nuclear Method 'Water Equivalent'	Gravimetric 'Water Equivalent'
0.30	17.1 cm	19.2 cm
0.24	21.4	19.2
0.25	20.6	19.2
0.28	18.4	19.2

DISCUSSION

When using natural background radiation to measure snow depth, several factors must be carefully considered in order to obtain reliable results. This applies to all types of detectors including TLD's. The site must be large, relatively smooth and free from areas which are thinly snow covered. Accurate measurements require a separate determination of I_o for each site surveyed. Soil moisture may be important because the measured intensity increases by 3% for each 10% decrease in soil moisture (Peck 1971).

Radon is known to affect the total count when measurements are made from aircraft. Extrapolating to ground level, it would appear that at some sites the effect can be significant at equivalent snow depths of 20 cm or more. A correction can be applied by exposing the controls at the centre of a large, frozen lake rather than in a lead castle. Additional information is required to assess the magnitude of this effect and to relate it to the uranium content of the overburden or to other sources of Radon.

There are fewer errors associated with artificial sources of radiation and much less care is involved in the choice of sites. Snow measurements can be made in forested areas and on relatively steep slopes. If a calibration curve is available over the whole range of snow depths, a graphical approach will reduce errors to the minimum. Natural background can be corrected for, if required, by exposing controls at the same time and location at a suitable distance from the source. With these precautions and with TLD's which can measure 1 ± 0.1 mR it should be possible to use a 10 μCi cobalt 60 source to estimate snow depths to 20 cm 'water equivalent'. Such a source might conveniently be placed at one end of a probe with a TLD holder at the other end so that the source could readily be inserted in the snow or moved to a new site. Care would have to be used in design to avoid shielding between source and TLD. Ten μCi is the maximum strength source which can be used in Canada without a permit from the Atomic Energy Control Board.

With the foregoing precautions one could expect the results to show a standard deviation of ± 10% for natural background and ± 5% for the point source exposures. The standard error will of course be lower if four or five dosimeters are exposed at each site and the results averaged.

A combination of methods is advisable when measurements are required at a variety of sites or over a range of snow depths. Such a compromise will reduce the effort to a minimum and produce results of optimum accuracy.

SUMMARY

A preliminary investigation into the use of TLD's for estimation of snow depth has indicated the feasibility of this type of measurement. It is suggested that the major part of the snow survey be done with natural background radiation and a sensitive TLD. Artificial sources are more inconvenient to use and should be restricted to difficult terrain or very large snow depths. They may also be used as a check on the natural background method. At the present time it requires a minimum of 10 days for a determination but the development of more sensitive TLD's will allow shorter exposure times to be used.

REFERENCES

Carmichael, H., G. Cowper, E.A. Godby, G.S. Levy and J.F. Steljes, 1951. Aerial prospecting for radioactive deposits in snowfalls. National Research Council (Canada) report CRR-477. Also available as Atomic Energy of Canada Limited report AECL-64, Chalk River, Ontario.

Dahl, J.B. and H. Ødegaard, 1970. Areal measurements of water equivalent of snow deposits by means of natural radioactivity in the ground. Symposium on Isotope Hydrology, International Atomic Energy Agency, Vienna.

Ferronskiy, V.I., A.I. Danilin, V.T. Dubinchuk, V.S. Goncharov, V.A. Polyakov, Yu. B. Seletskiy and N. Ya. Flekser, 1968. Radioisotope investigative methods in engineering geology and hydrogeology. Atomizdat, Moscow. Translated by addis Translations International, P.O. Box 4097, Woodside, California 94062.

Grasty, R.L. and P.B. Holman, 1972. The measurement of snow water equivalent using natural gamma radiation. First Canadian Symposium on Remote Sensing. (In press).

Grasty, R.L., 1973. Snow water equivalent measurement using natural gamma emission. Nordic Hydrology. (In press).

Jones, A.R., 1971. A gamma monitor for measuring environmental gamma doses and dose rates. Atomic Energy of Canada Limited report AECL-3989, Chalk River, Ontario.

Jones, A.R., 1973. Measurement of low level environmental gamma dose with TLDs and Geiger counters. IEEE Nuclear Science Symposium, San Francisco.

Peck, E.L. and V.C. Bissell, 1971. Aerial measurement of snow water equivalent by terrestrial gamma radiation survey. XV General Assembly of the IUGG, Moscow.

Placzek, G., 1946. The functions $E_n(x)$. National Research Council (Canada) report 1547. Also available as Atomic Energy of Canada Limited report MT-1, Chalk River, Ontario.

Zotimov, N.V., 1968. Investigations of a method of measuring snow storage by using the gamma radiation of the earth. State Hydrol. Inst. (Trudy GGI) 152: 114.

STABLE ISOTOPES IN THE STUDY OF SNOW AND ICE RESOURCES

H. Roy Krouse
Department of Physics, The University of Calgary
Calgary, Alberta, Canada T2N 1N4

ABSTRACT

Stable isotopic techniques have elucidated many problems in the study of snow and ice resources. Emphasis has been placed on the measurements of the most abundant isotopic species of water, $H_2{}^{16}O$, $HD^{16}O$, and $H_2{}^{18}O$. Their abundances are altered markedly in the hydrologic cycle because of their different vapor pressures. Potential uses and accomplishments of this technique include: differentiation of freshwater and sea ice; relating snow accumulations to meteorology, particularly in remote areas; elucidating water movement in snowpacks; evaluating snow and glacier runoff (stratified lakes); elucidating the history and age of ice caps. Field studies, mainly from the Arctic, will be used to illustrate these concepts. The information obtainable will be related to the instrumental requirements, cost of analyses, and time factors involved.

INTRODUCTION

Many researchers have used abundances of stable isotopic species of water as an effective tool in studying snow and ice. Regrettably, references to many excellent studies are omitted in this brief presentation.

The fundamental concept underlying stable isotope research is that isotopes of an element differ in their masses. Many natural processes are mass dependent and hence the relative abundances of isotopes are altered. In the case of water, the major factor which changes the relative abundances of the isotopic species is their differences of vapor pressures. Transport in the vapor, liquid, and solid phases further alters the isotopic distribution.

Isotopic compositions are usually measured with a mass spectrometer. The most important species of water, $H_2{}^{16}O$, $HD^{16}O$, and $H_2{}^{18}O$ occur with average abundances in the ratios 10^4:3:20. Absolute abundances can be rarely measured to an accuracy of better than 0.1%. However, the relative isotopic abundances of two samples can be routinely compared with precisions of better than 0.01% in a given laboratory. Therefore, isotopic abundances are expressed on a 'del' scale. For oxygen isotopes, this scale is defined as

$$\delta^{18}O_{sample} \text{ in ‰} = \left[\frac{[{}^{18}O/{}^{16}O]_{sample}}{[{}^{18}O/{}^{16}O]_{SMOW}} - 1 \right] \times 1000.$$

The scale expresses the deviation in parts per thousand (‰) of the relative numbers of ^{18}O's to ^{16}O's in a sample to a standard. The standard SMOW (Standard Mean Ocean Water) does not exist physically but represents an average value for ocean waters (Craig, 1961). A positive δO^{18} value means the sample is enriched in O^{18} as compared to SMOW, whereas negative δO^{18} values express O^{18} depletions relative to SMOW. A similar δD scale is defined for H/D abundances. A given sample may be assigned a range of δ values among laboratories, dependent upon preparation techniques, instrumental biases, and the extent to which necessary corrections are recognized. Consistency among laboratories of the world has been improved by provision of standard samples by the International Atomic Energy Agency in Vienna.

GLOBAL DISTRIBUTION OF ISOTOPIC SPECIES OF WATER

One cannot interpret isotopic data from snow and ice without an understanding of isotopic fractionation in meteorology. The world acts as a complex 'still' of isotopic species of water with net evaporation at the equator and net precipitation at the poles. For evaporation, the process may occur so slowly that near isotopic equilibration conditions are attained between the liquid and vapor phases. In this situation, if the liquid phase had isotopic composition SMOW, at 20° C $\delta^{18}O$ vapor $\approx$ -9‰ and δD vapor $\approx$ -79‰ (Merlivat et al, 1963; Reisenfeld and Chang, 1936). Under conditions of rapid evaporation, kinetic isotope effects may predominate whereby the isotopic rates of evaporation $R_{H_2{}^{16}O} > R_{HD^{16}O} > R_{H_2{}^{18}O}$. $R_{H_2{}^{16}O}/R_{H_2{}^{18}O}$ values as high as 1.025 have been reported (Dansgaard, 1961; Bakr, 1971).

In very early studies, the oceans were found to be fairly uniform isotopically except near freshwater tributaries (Epstein and Mayeda, 1953). This implies that the oceans of the world are well mixed, a fact verified by sulfur isotope abundance studies (Thode et al, 1961). With increasing latitude, continental precipitation is generally depleted in ^{18}O and D with $\delta^{18}O$ values in the range -50‰ found at the South Pole (Epstein and Sharp, 1959). $\delta^{18}O$ and δD values become generally more negative with increasing altitude (Sharp et al, 1960; Friedman and Smith, 1970). In the case of Mt Logan (6100 m elevation), $\delta^{18}O$ values as low as -45‰ were found (West and Krouse, 1972).

In 1961 the International Atomic Energy Agency and the World Meteorological Organization set up a worldwide survey of $\delta^{18}O$ and δD variations. Dansgaard (1964) has summarized and interpreted many of these data. Dansgaard and Tauber (1969) have mapped iso-$\delta^{18}O$ lines for the northern hemisphere based on annual mean δ values. For a number of stations in the northern hemisphere, δD and $\delta^{18}O$ were related by $\delta D = (8.1 \pm 0.1)\delta^{18}O + (11 \pm 1)$‰. Further, for many stations the δ values had temperature dependence given by (t in °C): $\delta^{18}O = 0.69_5 t - 13.6$‰; $\delta D = 5.6t - 100$‰. These relationships are consistent with evaporation from SMOW under conditions slightly removed from isotopic equilibrium and condensation isotopically equilibrated with the vapor phase (Dansgaard, 1964). A similar relationship between $\delta^{18}O$ and δD was found for data from Mt Logan and vicinity in the St Elias Mountain Ranges (West and Krouse, 1972). Although temperature is the most important parameter in determining the isotopic compositions, attempts have been made to correlate additional variables. In studies of the Andes, Moser et al (1973) related the deuterium content of precipitation to the average temperature and the distance of sampling points from the coast. Many locations in the world, particularly at low latitudes, do not exhibit such simple relationships between isotopic composition and other parameters. In the Sierra Nevadas it is not unusual to have rain with greater $\delta^{18}O$ depletions than snow which has been deposited at lower temperatures (Krouse and Smith, 1973).

SNOWPACKS AND THEIR DYNAMICS

As the temperature gradient between the source and precipitation area varies with season, the δ-value of precipitation is expected to reflect seasonal trends, the more negative δ-values corresponding to the colder events. As discussed above, this temperature dependence holds for a number of stations at higher latitudes. Thus, in remote areas an isotopic examination of the snow accumulation in early spring provides a record of the winter's

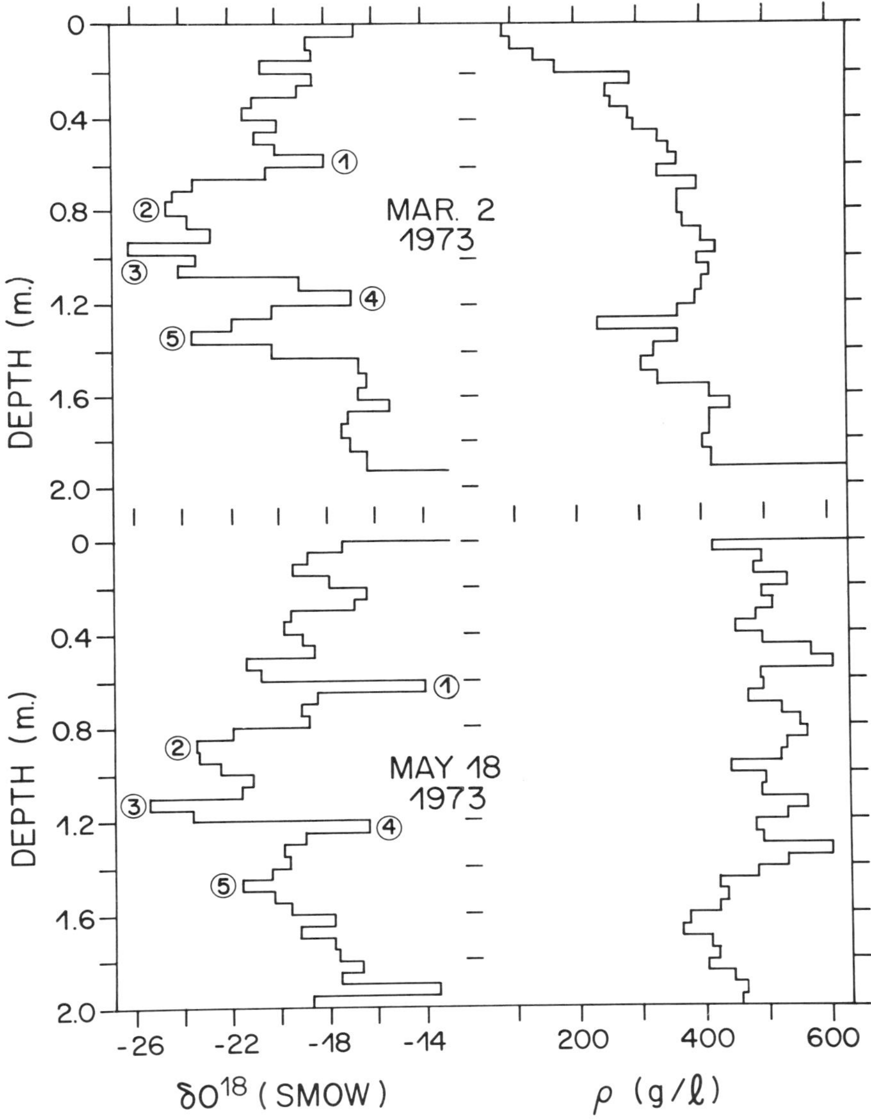

Figure 1. $\delta^{18}O$ and density profiles from two pits in the vicinity of Peyto Glacier. Hislop and Krouse (1973)

meteorology. Mixing by high winds may obscure the seasonal record. This was believed to be the case in a snow pit sampled on Mt Logan at elevation 5250 m during the summer of 1968 (West and Krouse, 1972). Even this situation provides information about the previous winter conditions.

As we progress from spring to summer, melting alters the isotopic composition of a snowpack. The extent of alteration depends very much on location. In some situations the snow entirely disappears, while in other cases it represents accumulation for a glacier or ice cap. Melting processes generally lead to a smoothing of the isotopic profile and a trend towards enrichment in ^{18}O and D (Sharp et al, 1960; Deutsch et al, 1966; Macpherson and Krouse, 1967; Arnason et al, 1973).

Recently, Hislop and Krouse (1973) isotopically monitored snow accumulation in the vicinity of Peyto Glacier (51° 40'N, 116° 33'W). Figure 1 shows data from two pits approximately 18 km apart and at a time interval of 2.5 months. Features of the winter's meteorology can be identified at both locations as denoted by the circled numbers. The Figure also shows that although processes such as compaction and water movement have increased the mean density and reduced its variation, the isotopic record still retains easily recognizable winter events.

Krouse and Smith (1973) encountered a more dynamic situation in their studies of snowpacks at 1900 m elevation in the central Sierra Nevadas of California. In this area the storm systems are complex. Precipitation does not arrive from the southwest as a solid front but rather as cells of rain and/or snow of varying intensity. Consequently, no isotopic minima could be identified with the coldest winter events as in drier locations. The isotopic composition of marked strata in the pack changed markedly with water movement. The $\delta^{18}O$ change could be positive or negative depending on the isotopic composition of the water involved. It was further necessary to evaluate the water retention abilities of layers in the pack as well as maintaining a good record of the water budget in order to interpret the isotopic changes. In particular, frequent in situ profiles with a nearby radioactive isotope profiling snow gauge (Smith et al, 1965; 1969; 1972) contributed immensely to the elucidation of the stable isotope behavior.

RUNOFF FROM SNOWPACKS AND GLACIERS (STRATIFIED LAKES)

In considering water resources and flood control, it is of interest to evaluate how much water infiltrates the soil as compared to surface runoff.

In a mountain watershed in northern Czechoslovakia, Dincer et al (1970) effectively combined ^{18}O data with tritium concentrations to conclude that about two-thirds of the meltwater infiltrated the soil to fill a subsurface storage estimated to be $2.6 \times 10^6 m^3$. The mean residence time of the subsurface water was approximately 2.5 years.

In a study of the Colorado Front Range, Meiman et al (1973) showed that the deuterium content of snow was sufficiently different from base streamflow and rain to permit an evaluation of the snow input to the hydrologic system. Snowmelt also underwent considerable mixing with subsurface water before appearing as runoff.

Recently, Holecek and Krouse (1973) examined an interesting situation in the Spring Creek Basin (54° 53'20"N, 117° 51'40"W). A number of holes drilled to various depths (down to 250 feet) were logged and these revealed the presence of two underground water systems. One was shallow and localized above a relatively impervious layer, while the deeper storage embraced a much larger area. Isotopic analyses of waters in several standpipes of different depths in the fall of 1972 provided two clusters of ^{18}O data, one at -15‰ and one at -19‰ corresponding to the deep and shallow systems respectively. Three rivers in the area gave $\delta^{18}O$ values -14.7, -15.3 and -18.5‰, the former two being related to the deep system and the latter to the shallow system. Snow accumulations were monitored in the area from January to April 1973. The mean value decreased from -25‰ in January to -23‰ in April. This resulted from warmer snows being progressively added to the pack and homogenization processes. During April a shallow well continued to display a $\delta^{18}O$ value of -18.5‰, as found the previous fall. This is explained by the frozen ground not permitting runoff to penetrate to the shallow system. The isotope studies verify that the shallow system is small and easily depleted. Its contents at any time represent precipitation of a few months previously. The deeper, larger system has an isotopic composition more consistent with the mean for annual precipitation.

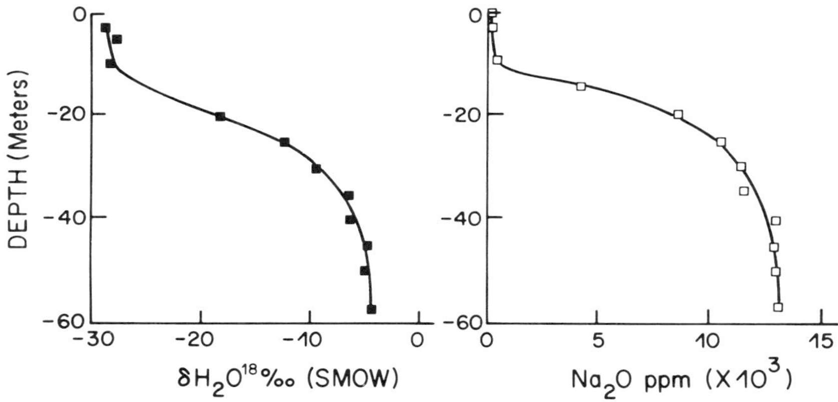

Figure 2. Stratified lake, northern Ellesmere Island

On Ellesmere Island (83° N, 75° W), Hattersley-Smith et al (1970) found lakes that formed as a result of seawater trapped either through glacier advance or postglacial rise of land. Stratification of a number of parameters arose as the surface received freshwater contributions from glacier runoff. Figure 2 shows $\delta^{18}O$ data which we obtained on Lake 'A' of their report. These clearly show that the trapped seawater has not effectively mixed with fresh surface waters over a time span of the order of 10^3 years.

ICE

Ice is found in different forms on the earth's surface. In order to evaluate this resource one should know the origin (fresh or seawater) and the age. Isotopic fractionation by freezing is relatively small. Ice contains about 2% more deuterium than the associated water. Thus, ice formed from seawater has δ-values slightly positive to SMOW, as demonstrated with data from Woods Hole, Massachusetts and Hopedale Bay, Labrador (Friedman et al, 1964). In the spring of 1973 we found $\delta^{18}O$ values for large sea ice chunks which had drifted on the shore of Hudson Bay near Ft Churchill to be -0.4‰, while a lake in the same area and the water supply system from Ft

Churchill had values of -10.8 and -13.2‰ respectively. In contrast, we found $\delta^{18}O$ values for icebergs sampled during August 1971 by the CCGS Louis S. St Laurent (76° 49'N, 73° 00'W) in the range -25 to -28‰, which is characteristic of precipitation at higher latitudes. These data support a concept which was recognized many years ago: namely, that sea ice and ice of meteorological origin could be readily distinguished at higher latitudes on the basis of ^{18}O or D content.

Permafrost is abundant in Canada's Arctic, but its existence beneath the sea floor has been firmly established rather recently. For example, icy sediments in boreholes beneath the southern Beaufort Sea were first recovered in the spring of 1970 (Mackay, 1972). The presence of permafrost on land or offshore has importance in regard to future activities in the Arctic. If the ice content is high and irregular in its distribution, thermal disturbances are undesirable. Interpretations in geophysical exploration are also difficult. Stable isotopes can potentially elucidate the origin and dynamics of the permafrost-ground ice system. Mackay (1972) presents a few data which suggest that in two boreholes beneath the Beaufort Sea, ice samples originated from rain and snow, and not the sea.

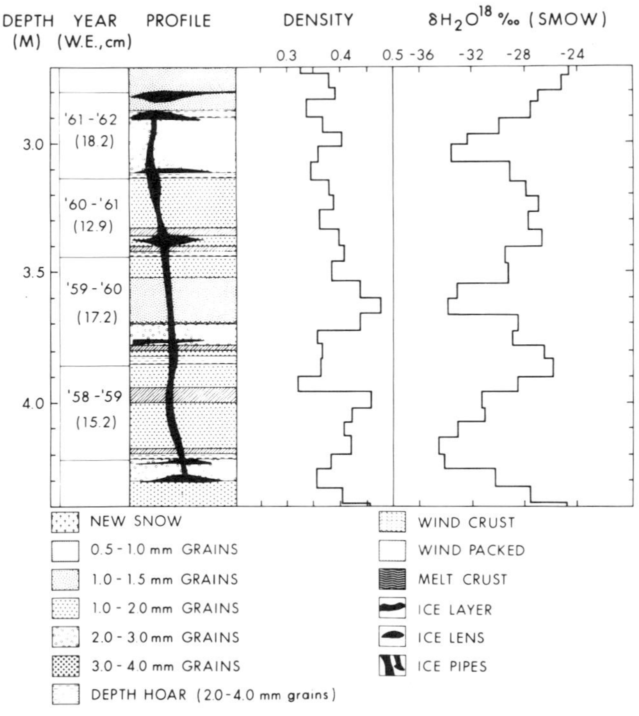

Figure 3. Snow-pit, northern Ellesmere Island. West (1972)

The isotopic record in glaciers and large ice caps of the world is particularly informative. In relatively dry conditions, seasonal isotopic trends may be preserved for many years. In more temperate glaciers or wet conditions, seasonal variations may be erased. An intermediate situation is presented in Figure 3. These data are from an ice cap northwest of Tanquary Fiord, northern Ellesmere Island (West, 1972). It is interesting to note that lens and pipe formations are localized phenomena which do not mask seasonal trends in the snow. Seasonal trends were reasonably well preserved in this ice cap, although the $\delta^{18}O$ values fluctuations decreased with depth. Figure 3 also shows an inconsistency between seasons determined by stable isotopes and by stratigraphy.

Discrepancies among glaciological and isotopic methods of determining the rate of snow accumulation were also reported by Picciotto _et al_ (1968) at the USSR station situated at the Pole of Relative Inaccessibility (eastern Antarctic Plateau). Whereas the 1955 radioactive fallout horizon and the ^{210}Pb method agreed with stake measurements, oxygen isotope variations gave accumulation values high by a factor of two, while stratigraphy was difficult to interpret.

Stable isotope variations over time spans of centuries have identified major ice-age events in Greenland (Dansgaard _et al_, 1969), Antarctica (Epstein _et al_, 1970), and more recently the Devon Ice Cap (Paterson, 1973).

In contrast, Koerner _et al_ (1973) examined a complete ice core from surface to bedrock from the Meighen Ice Cap (80° N, 100° W). This was a stagnant ice cap whose accumulation was almost entirely superimposed ice with occasional years of firn accumulation. In such a situation it is next to impossible to decipher the complex relationship between the δ-value of the superimposed ice layer and the original snow. Further, much of the record is lost when a number of annual accumulations undergo ablation. Despite the difficulties in isotopic interpretation, some interesting correlations were found and it could be established that the ice cap was considerably less than 10,000 years old.

BENEFITS VERSUS COSTS

Scientific information which relates to basic research cannot be assigned an absolute value. The cost of obtaining information might be compared for a number of techniques. Information which relates to a product which in turn has a price, may be more readily evaluated. However, when one is considering snow and ice resources and their many uses, the problem of pricing the products is very complex. How much is a glass of water, or the recreation provided by lakes or ski slopes, worth?

Information might be classified into the urgent or dynamic and the more casual type. The former category includes items like the dynamics of water in a snowpack and flood forecasting, while the latter embraces such questions as the age of a major ice sheet. The degree of urgency is sometimes taken as a measure of the applied aspect of the research.

It is clear that complex natural situations require frequent measurements of many parameters. For example, the monitoring of meteorology and processes in the snowpacks of the central Sierra Nevadas requires many more data than those required to describe drier accumulations at higher latitudes. Thus, location may dictate to a large extent the costs and feasibility of various techniques.

One must ask if other techniques will adequately provide the necessary information. The stratified lake of Figure 2 can be described by the depth dependence of its salinity, density, or a number of chemical parameters (such as $[Na^+]$ plotted in Figure 2). Icebergs can be monitored by photographic means from source to final dissipation, so one questions the value of isotopic measurements. Sea ice can be differentiated from freshwater ice on the basis of salinity (sea ice may have typically 10% salinity as compared to 33% for ocean water).

It is important to realize that stable isotopes provide an internal label which tells the history of water itself. Techniques such as chemical analysis or dirt layer assessment in ice really describe the history of materials co-existing with water. Therefore, certain types of information are best obtained with stable isotopes. Although a random piece of freshwater ice might be so identified by other means, isotopic measurements might reveal its origin. The decline of the Wisconsin ice-age may be readily detected in an isotopic profile of a deep ice core. The presence of the two groundwater reservoirs in the Spring Creek Basin study was readily detected with stable isotopes.

Success in unravelling a complex system depends on utilizing many techniques and correlating many parameters. Stable isotope data are most meaningful if related techniques have been carried out well. From the basic research viewpoint, it may be that one cannot afford to neglect stable isotope measurements in a given situation. It is also scientifically inefficient to omit any reasonably priced technique if one is undertaking a major costly exercise such as coring through an ice cap.

Stable isotopes can in principle be used as applied rather than natural tracers for smaller systems. For example, the fate of a spray of distilled seawater (isotopic composition near SMOW) on a snowpack at high latitude could be readily traced. This technique is devoid of criticisms which can be directed at radioactive tracer techniques.

The costs of stable isotope analyses can be estimated, assuming that the mass spectrometer is the measuring instrument. In recent years, small permanent magnet analyzers (AEI MS20; Atlas Mat GD150; Micromass) have become available in the $20,000 range. Additional monies are required for preparation lines. In terms of productivity, up to 40 sample preparations and $\delta^{18}O$ analyses have been obtained in the author's laboratory with two technicians working a 10-hour day. There is no doubt that automation can provide more analyses with less labor, as witnessed by H/D techniques at the Chalk River Laboratories (Thurston, 1970). Commercial quotes for analyses depend to a large extent on the number of samples involved. If a figure of the order of $25 per sample is considered, and a researcher requires only a few hundred analyses per year, investment in a laboratory is impractical unless the instrumentation will be used for other purposes. In principle, other techniques can be utilized for stable isotope determinations. If the same effort is expended on these as has been the case with mass spectrometry, the cost factor may decrease considerably.

ACKNOWLEDGMENTS

Financial support for the author's research over the years has been provided by the National Research Council of Canada, Defence Research Board of Canada, Atmospheric Environment Services of Canada, and the Boreal Institute of the University of Alberta. Field assistance has been provided by the Arctic Institute of North America, Defence Research Board and the Department of Energy, Mines and Resources of Canada.

REFERENCES

Arnason, B., Th. Buason, J. Martinec, and P. Theodorsson, 1973. Movement of water through snowpack traced by deuterium and tritium. Internat. Symp. on Role of Ice and Snow in Hydrology, Banff, Canada, 1972.

Bakr, A.A., 1971. Isotopic studies of retardation of evaporation of water by cetyl alcohol films. MSc Thesis, University of Alberta.

Craig, H., 1961. Standard for reporting concentrations of deuterium and oxygen-18 in natural waters. Science, 133:1833-34.

Dansgaard, W., 1964. Stable isotopes in precipitation. Tellus, 16:436-468.

Dansgaard, W., and H. Tauber, 1969. Glacier oxygen-18 content and Pleistocene ocean temperatures. Science, 166:499-502.

Dansgaard, W., S.J. Johnsen, J. Møller, and J.R.C.C. Langway, 1969. One thousand centuries of climatic record from Camp Century on the Greenland Ice Sheet. Science, 166:377-378.

Deutsch, S., W. Ambach, and H. Eisner, 1966. Oxygen isotope study of snow and firn on an alpine glacier. Earth and Planet. Sci. Lett., 1:197-201.

Dincer, T., B.R. Payne, T. Florkowski, J. Martinec, and E. Tongiorgi, 1970. Snowmelt runoff from measurements of tritium and oxygen-18. Water Resources Res. 5:110-124.

Epstein, S., and T. Mayeda, 1953. Variations of the O^{18} content of waters from natural sources. Geochim. et Cosmochim. Acta, 4:213-224.

Epstein, S. and R.P. Sharp, 1959. Oxygen-isotope variations in the Malaspina and Saskatchewan Glaciers. J. Geol., 67:88-102.

Epstein, S., R.P. Sharp, and A.J. Gow, 1970. Antarctic ice sheet: Stable isotope analysis of Byrd Station cores and interhemispheric climatic implications. Science, 168:1570-1572.

Friedman, I., A.C. Redfield, B. Schoen, and J. Harris, 1964. The variation of the deuterium content of natural waters in the hydrologic cycle. Rev. Geophys., 2:177-224.

Friedman, I., and G.I. Smith, 1972. Deuterium content of snow as an index to winter climate in the Sierra Nevada area. Science, 176:790-793.

Hattersley-Smith, G., J.E. Keys, H. Serson, and J.E. Mielke, 1970. Density stratified lakes in northern Ellesmere Island. Nature, 225:55-56.

Hislop, R. and H.R. Krouse, 1973. Unpublished O^{18} data from Peyto Glacier and vicinity.

Holecek, G. and H.R. Krouse, 1973. Unpublished O^{18} data on the Spring Creek Basin, Alberta.

Koerner, M.R., S. Paterson, and H.R. Krouse. A δO^{18} profile in ice formed between the equilibrium and firn lines. Nature (in press, 1973).

Krouse, H.R., and J.L. Smith, 1973. O^{18}/O^{16} abundance variations in Sierra Nevada snowpacks and their use in hydrological research. Internat. Symp. on Role of Snow and Ice in Hydrology, Banff, Canada, 1972.

Mackay, J. R., 1972. Offshore permafrost and ground ice, southern Beaufort Sea, Canada. Can. J. Earth Sci., 9:1550-1561.

Macpherson, D. S., and H. R. Krouse, 1967. O^{18}/O^{16} ratios in snow and ice of the Hubbard and Kaskawulsh glaciers: Isotope techniques in the hydrologic cycle. AGU Monograph 11:180-194.

Meiman, J. R., I. Friedman, and H. Hardcastle, 1973. Deuterium as a tracer in snow hydrology. Internat. Symp. on Role of Snow and Ice in Hydrology, Banff, Canada, 1972.

Merlivat, L., R. Botter, and G. Nief, 1963. Fractionnement isotopique au cours de la distillation de l'eau. J. Chim. Phys., 60:56-61.

Moser, H., C. Silva, W. Stichler, and I. Stowhas, 1973. Measuring the isotopic content in precipitation in the Andes. Internat. Symp. on Role of Snow and Ice in Hydrology, Banff, Canada, 1972.

Paterson, W. S. B., 1973. Private communication.

Picciotto, E., R. Cameron, G. Crozaz, S. Deutsch, and S. Wilgain, 1968. Determination of the rate of snow accumulation at the Pole of Relative Inaccessibility, Eastern Antarctica: A comparison of glaciological and isotopic techniques. J. Glaciol., 1:273-287.

Reisenfeld, E. H. and L. T. Chang, 1936. Dampfdruck, Siedenpunkt, und Verdamfundswarme von HDO and H_2O^{18}. Z. Physik. Chem. B, 33:127-132.

Sharp, R. P., S. Epstein, and I. Vidziunas, 1960. Oxygen-isotope ratios in the Blue Glacier, Olympic Mountains, Washington, USA. J. Geophys. Res., 65:4043-4059.

Smith, J. L., D. W. Willen and M. S. Owens, 1965. Measurement of snowpack profile with radioactive isotopes. Weatherwise, 18:246-251, 287.

Smith, J. L., and D. W. Willen, 1969. Gage device for measurement of density profiles of snowpack. Patent No. 3,432,656. U. S. Patent Office, Washington, D. C.

Smith, J. L., H. G. Halverson, and R. A. Jones, 1972. Central Sierra Profiling snow gage: A guide to fabrication and operation. USAEC Report TID-25986. National Technical Information Service, U. S. Dept. of Commerce, Springfield, Va. 22151.

Thode, H. G., J. Monster, and H. B. Dunford, 1961. Sulphur isotope geochemistry. Geochim. et Cosmochim. Acta, 25:159-174.

Thurston, W. M., 1970. Automatic mass spectrometric analysis of the deuterium to hydrogen ratio in natural water. Rev. Scientific Instruments, 41:963-966.

West, K., 1972. H_2O^{18}/H_2O^{16} variations in ice and snow of mountainous regions of Canada. PhD Thesis, University of Alberta.

West, K. and H. R. Krouse, 1972. Abundances of isotopic species of water in the St Elias Mountains. Icefield Ranges Research Project Scientific Results, 3, Amer. Geog. Soc. and A. I. N. A., V. C. Bushnell and R. H. Ragle (Ed.), 117-130.

DEVELOPMENT AND FIELD TESTING OF A REMOTE RADIOISOTOPIC SNOW GAGE

D. C. Shreve
JRB Inc., La Jolla, California

A. J. Brown
Snow Surveys and Water Supply
Forecasting Section
California Division of Resources Development

ABSTRACT

In stream runoff forecasting and flood prediction it is necessary to have timely information about the water stored in the high mountain snowpacks. In addition to the total water equivalent, it may be useful to have information about the nature and conditions of the snowpack. Dry snow, for example, will absorb significant amounts of water before runoff begins, whereas wet snow will allow additional water to drain. Present methods of remotely measuring the water equivalent of the snowpack rely on pressure measuring devices located at or below ground level. These devices give a measurement of the total water equivalent in the snow above the device but not its condition. The development of a radioisotopic profiling gage by Dr. Smith has shown that detailed information above the location of the water throughout the snowpack could be obtained by measuring the attenuation of gamma radiation passing through the snow. Research scientists are now studying the physics of snow using such gages.

A simpler radioisotopic snow gage which provides information intermediate between the pressure device and the profiling gage has been developed by JRB Inc., and has undergone field tests conducted by the State of California Department of Water Resources at their developmental test station. The gage provides the water equivalent of the snowpack in each of ten vertical increments. Both a total water equivalent and a course vertical density distribution of the snowpack are obtained.

The gage consists of two 18-ft high vertical masts separated by 25 inches. One of the masts houses five radioisotopic sources and the other six Gieger Müller detectors spaced at equal intervals up its length. During a measurement period, ten collimated beams of gamma rays are directed diagonally from the sources to the detectors in a zig-zag pattern. The attenuation produced in each of the beams yields the water equivalent in that vertical increment.

Design considerations and constraints are discussed. Details of the installation, calibration, and data analysis are presented. Data accumulated for the past winter are compared with other measurements. The accuracy of the gage as compared to core sample measurements appears to be better than 10%. A discussion of the results of the field tests is presented.

6.9

INTRODUCTION

There is a need in the western states for managing the water stored temporarily as snow in the mountains. The rivers which supply most of the areas drinking water, irrigation water, and a sizable portion of its electrical power originates largely from the melting of these snowpacks. With the population of these states increasing, more land being irrigated, and energy demands growing, it is necessary that there be proper management of this vital natural resource.

In order to properly manage the runoff, one must be able to predict both the quantity and rate of runoff. One of the pieces of information the hydrologist needs for this forecasting is the quantity and condition of the snowpack. Presently, information of this sort is largely gathered by teams of men who sample the snow manually at various snowsources. Remote reasurements are presently made by pressure pillows which give only the total water equivalent of the pack but no information about its condition.

There have been several papers presented at this conference which describe new methods of increasing the information available to the hydrologist. Obviously, some of these methods will be used in the future as aids to the hydrologist.

This paper describes the design and field testing of a remote radioisotopic gage which should give information intermediate between the single number obtained by pressure pillows and the wealth of information obtained from a radioisotopic profiling gage.

BASIC DESIGN

The basic goal of JRB Inc. in designing this particular gage was to produce an instrument which would provide information about the conditions within the snowpack at a remote site. It was desired that the gage would provide both a measure of the total water equivalent and the snow densities of vertical increments throughout the pack. With this information the hydrologist could keep his finger on such things as pack ripening, new snow storms, pack draining, rain on snow conditions, and so forth.

Design criteria were based on the fact that the gage should operate unattended at a remote location. Reliability, accuracy, and low power consumption were therefore prime considerations. Another design constraint was radiation safety. Placing radioisotopic sources in unattended areas presented problems with tampering and accidental exposure.

The basic principle of operation of the gage is shown in Figure 1. A radioisotopic source of gamma rays is surrounded by a lead shield which directs a beam of gamma rays toward a detector. If there is snow between the source and detector, some of the gamma rays will be scattered or absorbed by the water molecules in their path. The detector measures the number of gamma rays that have passed

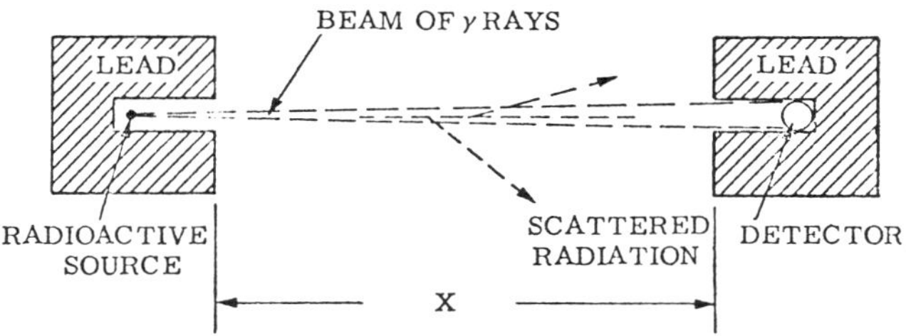

Figure 1. Basic principle of radioisotopic snow gage

through the snow in its direction. Each detected gamma ray produces an electrical pulse which can be counted electronically. The number of counts recorded in a fixed period of time can then be used to establish the water equivalent of the snow between the source and detector.

The design of the gage described here is shown in Figure 2. The gage consists of two vertical masts 18 ft high separated by 25 inches (27 inches pole-to-pole). There are five 10 mCi Cs 137 radioisotopic gamma ray sources (gamma ray energy of 0. 660 MeV) mounted at vertical intervals in one mast. The other mast contains six Gieger Müller (G. M.) detectors. All of the radioisotopic sources are normally stored in lead shields producing very little external radiation. When a measurement is to be made, the bottom most, middle, and top most sources (group 1 sources) are moved by individual solenoids from their stored positions to positions where each source emits two diagonal beams of gamma rays directed at the six detectors. These beams are shown schematically on Figure 2 as dot-dash lines. After a one-minute measurement period the group 1 sources are returned to their shielded positions and the remaining two sources (group 2 sources) are moved to their measurement positions and produce four gamma ray beams (shown as dashed lines in Figure 2). Another one-minute measurement is made and the group 2 source returns to their shielded positions. The number of counts recorded for the ten measurement intervals are then printed out by a digital printer. (The complications that might have been introduced by a radio data link did not seem to be justified for field testing the prototype instrument.)

The results of a complete measurement consisted of ten printed numbers -- the number of counts recorded for each interval. These numbers could then be used to obtain the average snow density in each of the ten vertical intervals. The water equivalents in each interval and the total water equivalent could then be obtained.

DESIGN DETAILS

The choice of radioisotopic source material, source strength pole spacing, and mast diameter were interrelated parameters which were

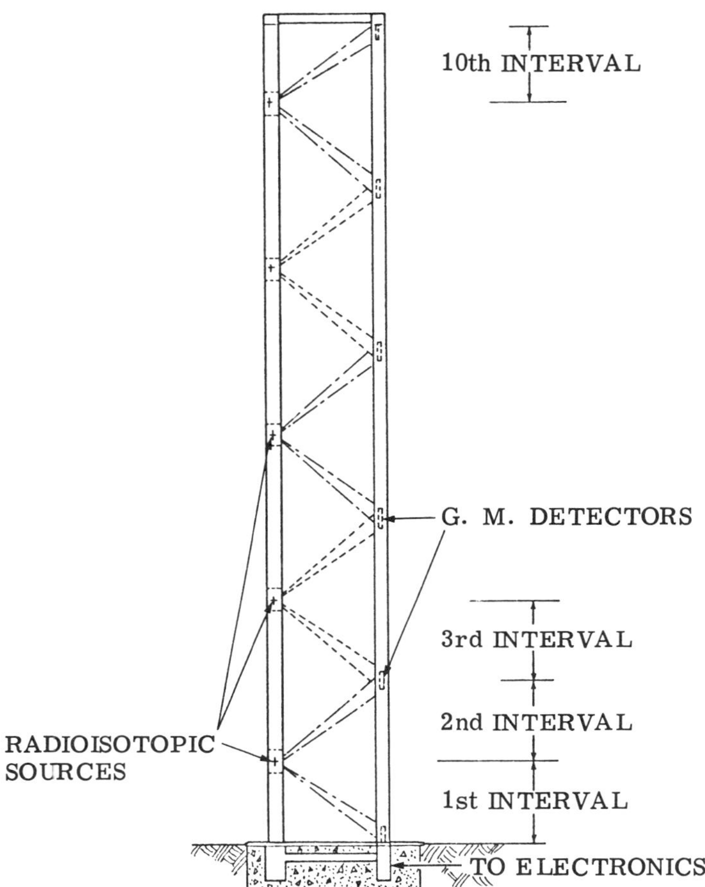

Figure 2. Sketch of radioisotopic snow gage

chosen first to provide sufficiently accurate measurements for maximum snowpack conditions, second to reduce the radiation fields around the sources to safe levels, and third to minimize the number of sources and detectors employed. The final compromise resulted in a choice of Cs 137 as the radioisotopic source, an individual source strength of 10 mCi, a pole separation of 27 inches, and a pole diameter of four inches.

Figure 3 shows a cut-away drawing of the lower portion of the snow gage. The masts are made of seamless steel coated with TFE teflon. Each mast is mounted in steel tubes embeded in a concrete footing. The sources are mounted in individual containers in one mast. Cables from the sources and detectors are fed in conduits to the electronics package and digital printer.

Since the radioisotopic source movement mechanisms were the only moving parts in the entire assembly, special care was taken in

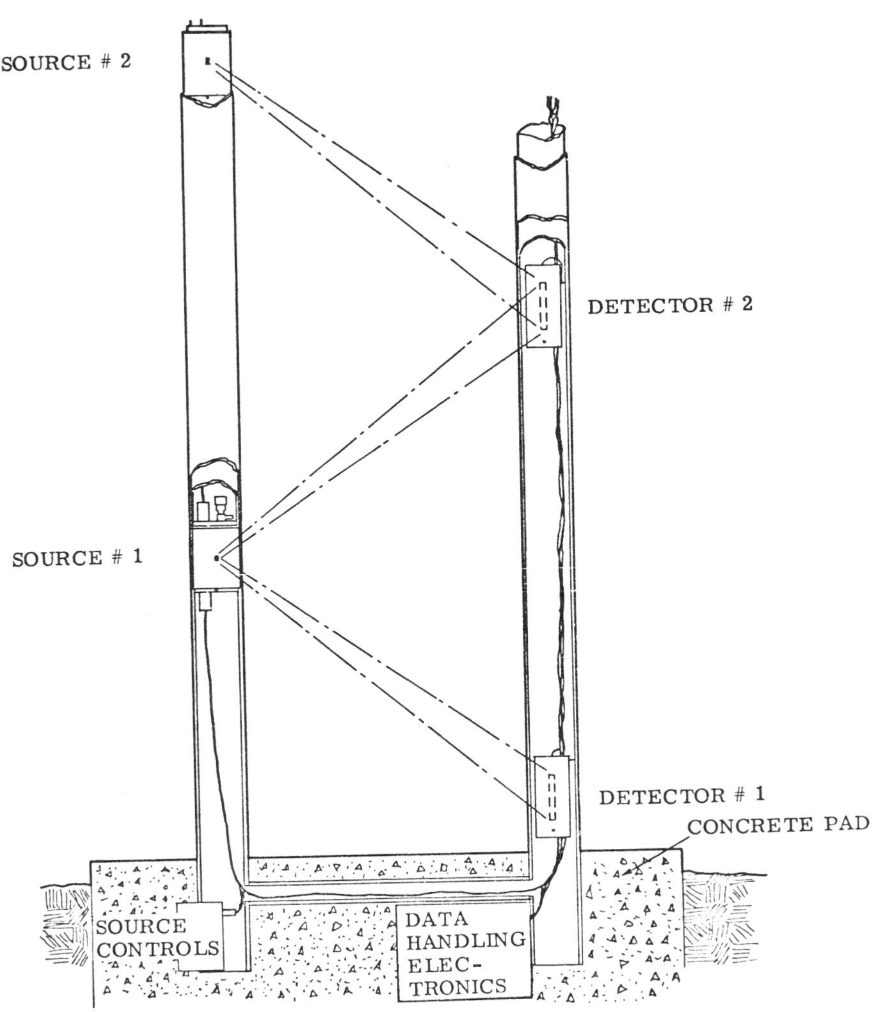

Figure 3. Sketch of lower portion of snow gage

their design and construction. Figure 4 shows a sketch of the source container. Normally the radioisotopic source was stored at the bottom of the shaft in the lead shield. During the measurement period the solenoid at the top of the container was activated raising the source to a position next to the two collimating holes. When power is removed from the solenoids the sources drop back into their stored positions to prevent condensation and possible freezing the entire movement mechanism, source, and lead shield were mounted in an air tight container which was filled with dry nitrogen before installation.

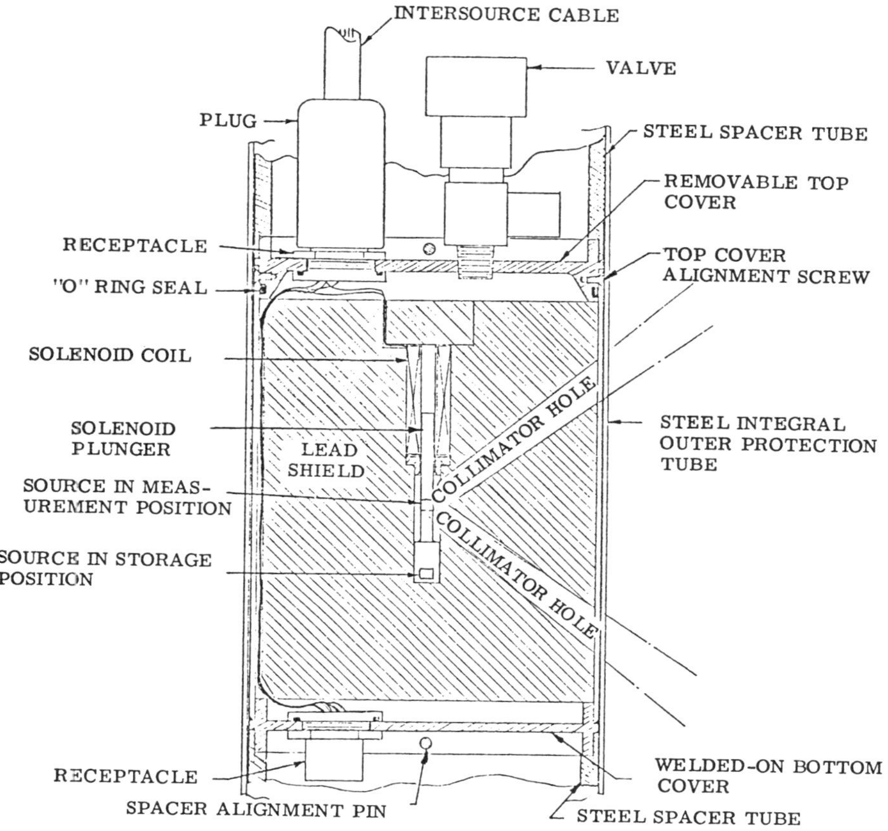

Figure 4. Details of source container

The electronics package consists of a control box, sequencer, gates and pulse shapers, scalers, and line drivers.

The control box allowed manual starting of the measurement sequence and readouts of the power supply voltage and the voltage for the G. M. tubes and electronics.

The sequencer was designed to control the automatic turnon, raising and lowering of the sources, routing of pulses from the G. M. tubes to the proper scalers, and the readout cycle of the digital printer. The gates and pulse shapers conditioned and routed the pulses from the detectors into scalers which stored the number of counts for each measurement.

An important aspect of the electronics is that the entire system is digital. The pulses from the G.M. tube are all the same size and treated as logic pulses. The rest of the processing is also done entirely with logic pulses. This fact largely eliminates temperature effects.

GAGE CALIBRATION

The gage was calibrated at the JRB facility in La Jolla and checked after installation. The calibration was accomplished in two steps. First, a replica of the first two intervals was immersed in water and the attenuation vs. depth measured. Second, the same set-up was used to measure the attenuation produced by plywood sheets. A comparison of these measurements yielded the water equivalent of each plywood sheet. These plywood sheets were then used to calibrate the gage and check its calibration after installation. Figure 5 shows a typical calibration curve of counts per minute vs. water equivalent. The deviation of the graph from a straight line (exponential decrease with water equivalent) is due to gamma rays from the nearby sources. Measurement with single sources showed only a very slight contribution from scattered photons for the maximum density snow.

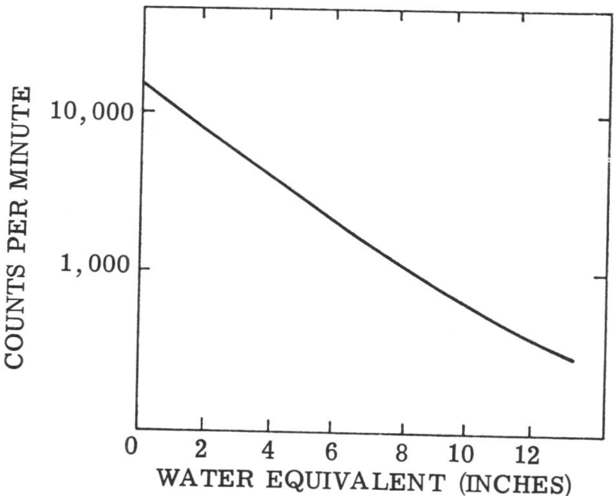

Figure 5. Typical calibration curve

FIELD TESTING

The prototype gage was installed at the State of California Department of Water Resources test site at the start of the 1972-73 snow season by JRB Inc. and DWR personnel. Figure 6 shows the final phases of this installation. Radiation signs are placed directly over the source positions. Figure 7 shows the gage during the middle of last December. The snow in this picture is just up to the top of the first measurement interval.

Figure 6. Snow gage after installation

Figure 7. Snow gage during early December

GAGE PERFORMANCE

The gage operated throughout most of last winter. Over 3,000 hourly measurements were made. Minor problems did arise; however, which caused some loss of data. Some data during the early portion of the winter were lost due to a slight malfunction of the thermoelectric generator which was used to charge the battery. This problem was not diagnosed correctly until the end of January, and once the problem was corrected almost continuous operation was obtained. The only loss of data after the middle of February, when the thermoelectric generator was replaced was one period when insufficient paper was left in the digital printer.

One problem did arise which necessitated removal of the sources this summer. During the latter part of the winter two of the source movement mechanisms sometimes failed to operate properly during the daylight hours. Inspection of the movement mechanism this summer showed that there had been a shift of the movement mechanism relative to the lead shield, presumably caused by mechanical vibration. The shift caused a binding in the left mechanism. Evidently the cooling at night either caused thermal contraction, which freed the mechanism or, lowered the resistance of the solenoid to the point where they had enough power to overcome the binding resistance. Fortunately, the operation during the evening, night, and morning hours allowed almost continuous records to be obtained.

DATA

As was stated earlier, snow between the source and detector caused the counting rate to decrease. Figure 8 shows the measured counts per minute for the fifth vertical interval during the period around the end of February. At the start of this period, the counts are what would be expected for no snow in this interval. The number of counts recorded is seen to decrease sharply during two periods indicating a build-up of snow. The first decrease, which lasts for 11 hours corresponds to an increase of snow in this interval corresponding to 1. 3 in. of water. The second decrease in the number of counts occurs 14 hours after the first count, lasts 12 hours, and increases the water equivalent of this interval by 1. 6 inch.

Data reduction was accomplished either by comparing the number of counts with the calibration curves, or by processing the numbers by a computer code which calculated the water equivalents.

Figure 9 shows a summary of the data obtained during last winter. Only one measurement a day is plotted. The bottom most points for any day represent the water equivalent in the lowest measurement interval which extends to 25. 6 inches. The water equivalent in this interval reaches a value of about 12 in. of water in the early portion of January and stays relatively constant until mid-April when it rises to about 13 in. of water. The second series of dots represent the water equivalent in the first 47. 2 inches. Similarly, the higher dots

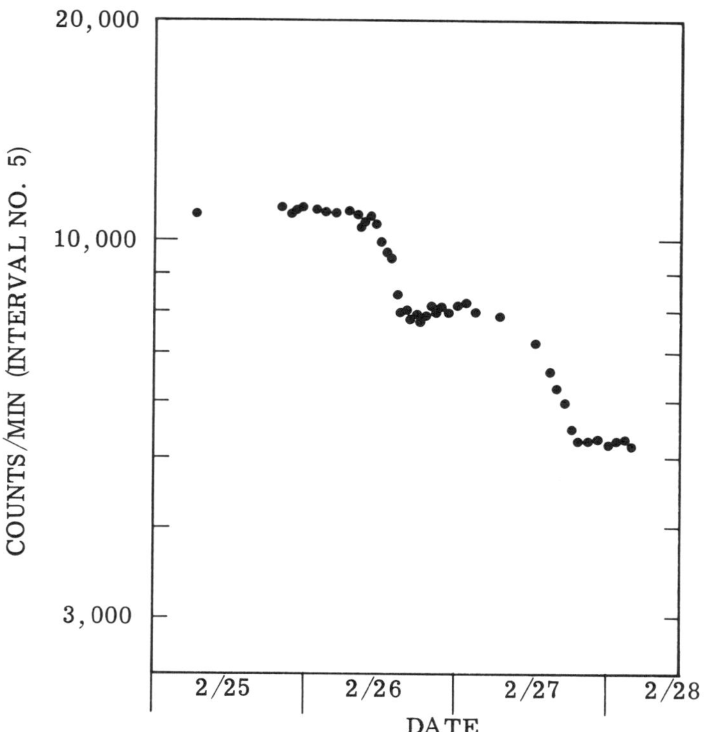

Figure 8. Observed counting rate in 5th interval

represent the water equivalents up to the top of the third interval, 68. 8 in.; the fourth interval, 90. 4 in.; the fifth interval 112 in.; and the sixth interval, 133. 6 inches. Only for a short period of time, around March 21st, did the snow pack exceed 112 in. in depth.

Several points should be made about the data. For example, in early December the snow first reaches into the second measurement interval (snow depth > 25. 6 in.). The pack reaches a water equivalent of about 11 in. of water then decreases slowely to a value of 10 inches. On December 16th and 17th, the total water equivalent increases to 12. 5 in. of water. The increase however, is entirely in the bottom interval as the water equivalent of the pack above 25-in. height rapidly decreases to zero. The occurrence corresponded to a known rain-on-snow condition which lowered the depth of the snow but increased its density.

The data around the end of February corresponds to the snow storms whose data were shown in Figure 8. The daily records do not show the discrete nature of the storms; however. During this period only slight increasea are seen in the lower intervals.

During the period of maximum water equivalent on April 5th (which occurs two wks after the maximum snow depth was reached on

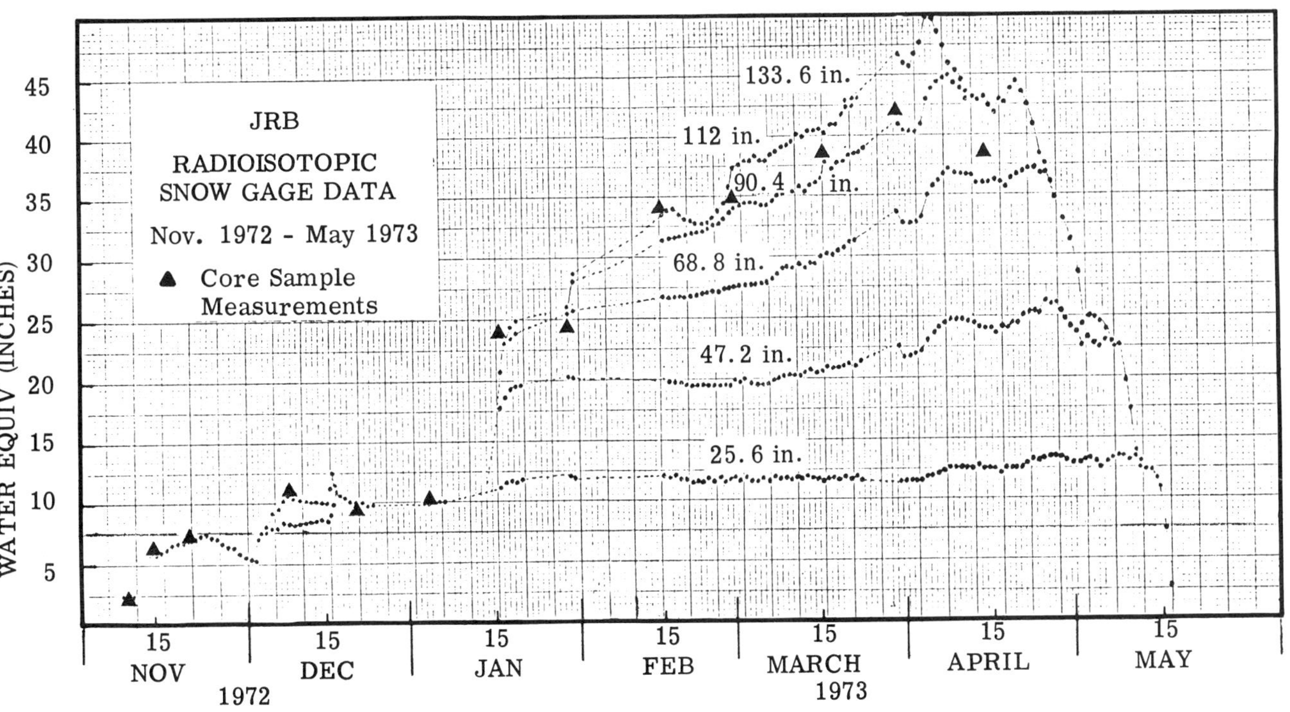

Figure 9. Data summary for 1972-1973 snow season

March 20) it is seen that the increase is mainly in the lower portions of the pack. Large effects are seen in the second interval which have to be caused by water draining from the top of the pack and being absorbed in the lower layers. After the pack has reached its maximum water equivalent, the lower levels only increase slightly showing that now the pack is starting to drain. The pack at this point is ripe and with the exception of snow storms on April 21st and May 2nd decreases rapidly.

Periodically, core sample measurements were taken in the pack near the gage. These measured values are shown as triangles in Figure 9. During the early part of the year the agreement is quite good; during the later part, the gage over estimates the water equivalent by about 10%. This discrepancy is probably due to calibration uncertainties for dense snow. The gage has been recalibrated and results are anticipated for this winter.

PRESENT STATUS

Presently the gage is again installed and taking data. The timing cycle which previously made measurements every hour has been modified to make measurements either hourly or every 12 hours.

Last year a radiation license had been obtained from the State of California based on an exemption to the regulations for the use of radioisotopic sources in uncontrolled areas. However, the U. S. Department of Agriculture was not in agreement with the exemption which had been based on the presumption that a person would have to lean against the radiation signs on the source mast for a period of four hours without moving, a situation the California Department of Public Health felt was unlikely. The Department of Agriculture radiation safety committee's criteria, however, is that one must keep the radiation levels in all uncontrolled areas such that if a person were to lean against the source mast continuously, without moving for an entire week, they would not receive more than the AEC allowed weekly dose. Only by completely changing the entire principle of the gage could this requirement be met, therefore the gage has been fenced.

SUMMARY OF DISCUSSION
(Papers 6.1 to 6.4)

During the discussion of these papers, the topics of interest included: A) The limits of airborne surveillance in terms of (1) the area that can be covered, (2) the thickness of snow that can be measured, and (3) the effects of uneven snow cover, or the capability of the system to integrate varying snow thicknesses seen by the sensor; B) Techniques for (1) measuring and correcting for atmospheric radon, (2) correcting for variations in background along flight lines, (3) measuring and correcting for variations in soil moisture, and (4) reducing data and the means of obtaining various factors including calibrations; C) The effects of (1) forestation, (2) noise to signal strength ratio, and (3) soil moisture contributions to runoff; and D) The physics of cosmic radiation as considered for possible use in making water equivalent measurements of thick snow and ice cover.

The authors' comments on these topics were:

A.

(1) Snow cover over areas as large as 2,500 km^2 has been measured and mapped in isothickness contours. (A day's operation is usually limited to 800 kilometers of flight line.)

(2) Practical limits of thickness are probably in the range of 30 to 40 cm of water equivalent.

(3) The detectors integrate the radiation from an area about 500 m in diameter; the effective water equivalent thickness is thus the average over the area seen by the detector.

B.

(1) Basically, two techniques are used to measure and correct for radon, although individuals differ somewhat in the details of their application. In one technique, radon is measured in the atmosphere by flying over large lakes where terrestrial radiation is absent. (Good atmospheric mixing is necessary to avoid errors.) In the other method, atmospheric particulates are collected on filters and counted; the count rates are subtracted as background.

(2) Variations in the strength of terrestrial radiation along the flight path affect the signals received by the detectors. Background - or baseline - information is obtained by flying the course without snowfall and storing the information for later use in calculating the attenuation of the signal by the snow thickness, or water equivalent.

(3) The authors all pointed out that it is important to obtain accurate soil moisture measurements along the flight lines by actual field sampling at the time the aerial survey is made. Corrections are then made by the computer. One speaker suggested that perhaps the soil moisture effect would not be a problem for hydrologists, since the uncorrected values would be more valuable for runoff predictions.

(4) Presuming that the cosmic contribution was flat for ^{204}Tl, high-altitude gamma spectra were used to obtain stripping coefficients. Cosmic Tl and Bi are scattered in the sensing crystal, attenuating their high energy. They are then counted in the lower energy part of the spectrum and, thus, contribute to the apparent 40K background. Stripping coefficients were considered as parameters to be optimized in a water content prediction equation, and the results obtained on this basis are quite in line with physical considerations.

C.

(1) The effects of forestation are treated as one of the variations in the background which is obtained during the pre-snow flight.

(2) Noise to signal strength ratios are recognized as limiting the method to 30 to 40 cm of water equivalent. The counting statistics (counts/unit time) deteriorate (scatter increases) with increasing snow depth. Thus, the percent of the count that can be attributed to terrestrial (as opposed to airborne) sources becomes so small that the real value due to terrestrial radiation is highly uncertain.

(3) Soil moisture contributions to run-off are complex and depend on variables related to terrain and groundwater hydrology.

D. The physics of the use of cosmic radiation for snow profiling has not been investigated and the validity of the concept has not been tested.

SUMMARY OF DISCUSSION
(Papers 6.5 to 6.9)

Papers 6.5, 6.7, and 6.9

Among the many points raised in discussion of these papers, were the following:

Profiling snow gauges cost up to $25,000 for one base station and one field station; probably this cost would be reduced if several were made on a commercial basis (Smith). No estimate is available for the radioisotope snow gauge (Brown).

There is essentially no limit on the length of time these gauges can operate in the field without being manned.

The vertical poles of the profiling gauge have very little effect on the snow pack until late in the melt season. At that time, the snow melts faster around the poles than elsewhere. The problem can be partially solved by the use of aluminum poles (Smith).

The condition of the snow pack at one location is probably fairly representative of an area if the point is well chosen. This relationship (a single point representing a given area) needs to be examined by many measurements, perhaps by portable gauges, preferably before the single point is chosen.

Understanding the timing of the melt is just as important a factor in good water resource management as knowing the total water inventory available for runoff.

Many comments were made concerning the use of thermoluminescent dosimeters (TLD's) for monitoring water equivalents. Some are summarized here (most comments are attributable to Grummitt and Prantl):

(1) In some instances, rather than use snow tubes, it may be desirable to place many TLD's in the field because of their relative simplicity of handling, because of weather conditions, or because of the desire to save time in the field.

(2) Rainout of radon daughters has only a small effect on the measurements; background measurements could be made over lakes; cosmic background could be obtained from TLD's in lead containers.

(3) Soil moisture conditions could possibly be measured by placing TLD's in, or on, the ground.

(4) Using a point source, such as ^{60}Co, could have advantages in some instances, since its radiation is not attenuated as rapidly as natural radiation. As low as 10 μCi could be used.

(5) It may be possible, with further development, to measure exposures for only a one-or-two-day period. TLD's are reusable and only cost about one dollar each.

(6) Time of snowfall (beginning or end of the measurement period) could be an important consideration in interpreting the integral readings.

Papers 6.6 and 6.8

One can follow the yearly moisture deposits in ice since the beginning of bomb testing (two decades ago) by examining the concentrations of fission products (such as ^{137}Cs) in the ice layers.

Melting tends to alter the stable isotope composition while it does not appear to influence as much the movement of fission products. In fact, in many cases, concentrations of fission products are found enhanced in the horizon after partial melting (Prantl).

The examination of stable isotope ratios is a useful method of dating snow layers for a variety of applications; however, many natural processes can cause isotope ratio variations. Careful, and often difficult and complex analyses are required if useful results are to be obtained (Krouse).

Session VII

Miscellaneous Techniques

Chairman:

M. F. Meier, Water Resources Division, U. S. Geological Survey

General Reporters:

R. W. Popham, National Environmental Satellite Service, NOAA

H. R. Krouse, Department of Physics, University of Calgary

ANALYSIS OF SNOW DISTRIBUTION USING TERRESTRIAL PHOTOGRAMMETRY

K. Blyth
R. B. Painter
Institute of Hydrology, United Kingdom

ABSTRACT

The feasibility of using terrestrial photogrammetry to measure the depth of shallow snowpacks is demonstrated. An operational system for a research catchment is described, and the potential of terrestrial photogrammetry in flood warning systems is discussed.

In the United Kingdom, snowfall is confined to the period October-April, and typically depths are less than 1 metre. Apart from relatively small mountainous areas above some 1200 m altitude, the snowpack will accumulate and completely ablate several times during the winter. Although the hydrological significance of snow is considerable, as a source of floodwater and in the recharge of aquifers, relatively little research effort is devoted to snow hydrology in the UK. Forecasting techniques for runoff from snowpacks are being investigated (Johnson and Archer, 1972), and limited work will shortly commence on the recharge of aquifers by snowmelt. The reliability of results from these studies will depend in part on the quality of the data available on the water equivalent of the snowpack. Conventional snow courses aimed at providing depth and density measurements often cannot cope with the high degree of variability of snow depth. Further, because of the frequent temporal variations of the snowpack, daily measurements are advisable; these are difficult to obtain by the use of snow courses, due to limitations of manpower.

These measurement problems have been encountered in the Institute's investigation of the hydrological consequences of a land use change from grass to coniferous forest in two small catchments in central Wales (Figure 1).

Any model relating precipitation and streamflow, requires accurate measurements of these variables, and one of the most likely sources of error is in the measurement of precipitation when this occurs as snow. Drifting precludes the use of the raingauge network to catch the snow as it falls, and even if access was always possible, snow courses would not provide the requisite accuracy of areal measurement. A requirement for frequent spatial measurement rules out the use of snow pillows and nuclear counting devices based on artificial sources of radioactivity. The choice of technique to be used, appeared to lie between stereo photography (Cooper, 1965,and Smith, Cooper & Chapman,1967) and monitoring the natural radioactivity of the soil. The latter offers substantial advantages (Bissell & Peck, 1973) but is not as yet sufficiently precise to deal with snow depths in the range 10-1000 mm. Stereo photography, using either a tethered balloon platform or a ground based system was discussed by Painter (1973) and terrestrial photogrammetry was chosen as the most likely solution to the problem.

This paper gives the results of an experiment set up to test the feasibility of terrestrial photogrammetry for measuring snow depth, and demonstrates the method's application in the analysis of snow distribution by comparison with conventional snow course measurements.

The operational system to be implemented in the research catchments is described, and finally the potential of terrestrial photogrammetry for monitoring snowpacks for flood warning purposes is discussed.

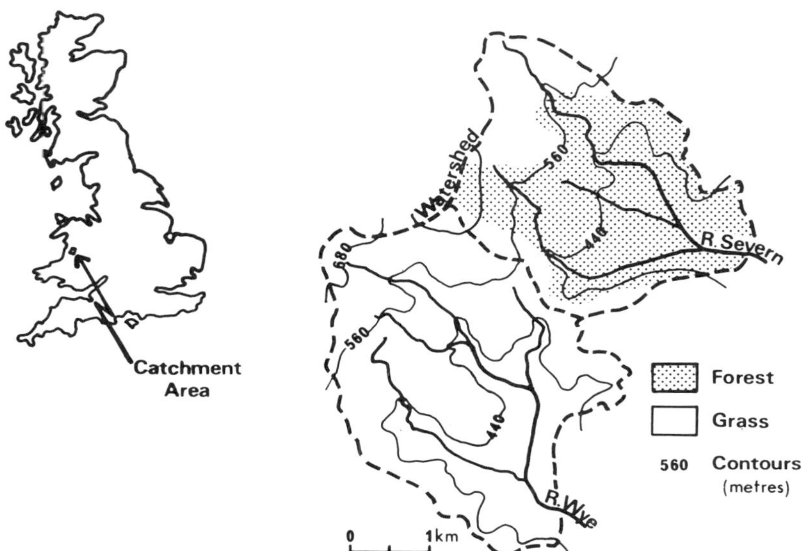

Figure 1. Location and nature of research catchments

FEASIBILITY OF TERRESTRIAL PHOTOGRAMMETRY

Terrestrial photogrammetry is a method by which the surface contours of an object are found by photographing it stereoscopically using ground based cameras, and analysing the photographs in a suitable stereo plotting machine. A feasibility study was carried out to establish whether terrestrial photogrammetry can be used to give areal estimates of snow depth, to find the errors involved, and to make recommendations on design practice for an operational network. Photogrammetric aspects of the tests are described by Blyth, Cooper, Lindsey and Painter (1973), and only a short summary is necessary here. The results are given, and are compared with those which could be obtained from measurements along a snow course.

Description of feasibility test

The tests were made on a N.E. facing slope, over an area 300 m x 100 m, ranging in altitude from 460 to 525 metres. Photographs of this area, first without snow and then with snow, were taken from each end of a 100 m base line, sited some 150 m from the near edge of the area, on the opposite side of the valley. Two cameras were used: the Zeiss UMK, format 130 x 180 mm, focal length 100 mm, and the Wild P32, format 60 x 80 mm, focal length 64 mm. Ground-truth measurements of snow depth were taken manually with metre rules at 1200 points on a grid within the test site, immediately after the photography.

To enable the stereo model to be scaled and levelled in the plotting machine, eight control points comprising circular targets were established on the test area. The position of each target was determined by theodolite and tellurometer to within ± 3 mm. When the stereo model had been restituted, a 5 metre square grid was set in a horizontal plane over the imagery and the heights of the ground and snow surface was found at each of the 861 intersections. Each height was assigned a standard error according to its distance from the camera base and to the nature of the ground surface at that point and subsequently its ease of heighting. A weighting for the snow depth at that point was computed as a function of the two standard errors. This weighting does not take account of random errors arising from ground control and from orientation of the model; nor do they account for systematic errors from lens and film distortions, or from instrument maladjustment. These errors are kept to a minimum however by choice of equipment and operational procedures. Figure 2 shows the format of the computer print-out of the results.

Results

Complete coverage of the test site was attained with the Zeiss UMK camera. Because of mist, however, only the lower half of the site was covered by the Wild P32 photography. In both sets of photographs, long grass in the foreground of the site prevented delineation of the ground surface by photogrammetry. Table 1 shows the mean snow depths attained by ground measurement, both over the entire area and over the parts of the area where photogrammetry was possible.

Camera and area covered	Mean snow depth (mm)		
		Ground measurement	
	Terrestrial Photogrammetry	Whole area	Where photogrammetry possible
UMK Whole test site	99.0 ± 6.7	108.6	107.6
UMK Lower half of test site	90.9 ± 8.2	97.2	102.7
P32 Lower half of test site	148.4 ± 15.8	97.2	102.7

Table 1. Snow depths obtained by terrestrial photogrammetry and ground measurement

Results with the Zeiss UMK are very satisfactory, but the Wild P32 gave a substantially higher value than was obtained by ground

PLYNLIMON TEST SITE

CAMERA: UMK

PHOTOGRAPHY: 16/2/73 PLATES A7 AND B7. 1/2/73 PLATES A3 AND B2

POINT	Z1	Z2	S1	S2	DEPTH	REMARKS	WT
34	64.400	64.375	0.053	0.053	0.025		178.1
35	65.138	65.088	0.053	0.053	0.050		178.1
36	65.075	65.05	0.053	0.053	0.025		178.1
37	65.075	64.963	0.025	0.053	0.112		289.5
38	65.362	65.375	0.025	0.025	-0.013	NEGATIVE DEPTH	773.1
39	66.213	66.162	0.025	0.025	0.050		773.1
40	66.875	66.862	0.053	0.053	0.013		178.1
41	66.038	68.162	0.053	0.053	-0.125	NEGATIVE DEPTH	178.1
42	69.737	69.650	0.056	0.056	0.088		159.1
43	61.587	61.312	0.027	0.056	0.275		258.7
44	61.587	61.350	0.027	0.056	0.238	IMPOSSIBLE	0.0
45	62.425	62.575	0.027	4.000	-0.150		0.0

Z1 is elevation of snow surface

Z2 is elevation of ground surface

S1 is standard error of snow surface measurement

S2 is standard error of ground surface measurement

WT is weighting given to point measurement

Figure 2. Format of computer print-out of results in feasibility study measurement. This difference is much greater than the relative standard errors of the two results suggest, and was due to mist obscuring all but the lower line of control targets thus preventing the stereo model from being correctly levelled. Figure 3 clearly shows the effect of this on snow depth measurement, and this discrepancy is attributable to a 3-4 second horizontal maladjustment of the model.

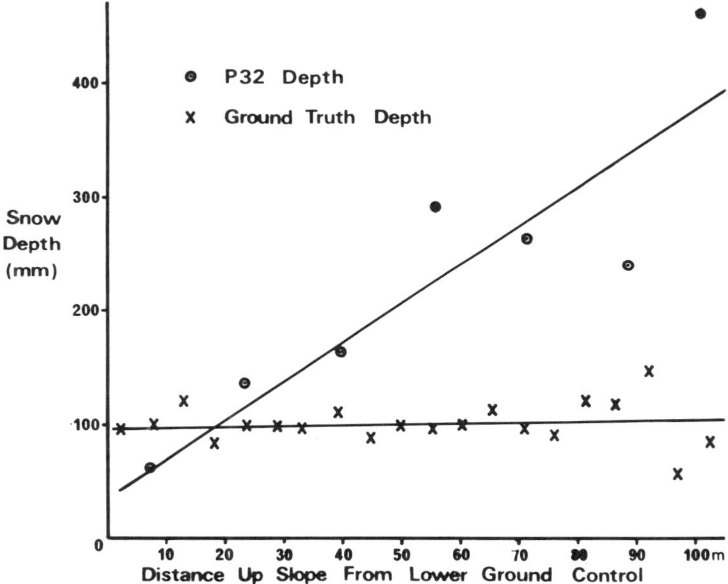

Figure 3. Effect of maladjustment of P32 on results

Comparison with snow course measurements

Over an area 300 x 100 m, a conventional snow course would at most include 8-10 depth measurements. From the ground measurements, a series of different snow courses, comprising 9 depth measurements, have been used to calculate the average snow depth over the test site. These are plotted in histogram form in Figure 4, and this exhibits a generally normal distribution. From a probability analysis, it was shown that only 25% of snow course readings lie within $\pm$ 10 mm of the true average depth, 45% within $\pm$ 20 mm and 66% within $\pm$ 30 mm. Thus, by comparison with Table 1, a conventional snow course would only offer a 1 in 4 chance of accuracy comparable with that obtained through terrestrial photogrammetry.

OPERATIONAL SYSTEM IN RESEARCH CATCHMENTS

The input of precipitation as snow into the catchments is required on a continual basis. In the grassed areas snow depth will be measured by terrestrial photogrammetry,and snow pillows will give a continuous point record of water equivalent. During photography, courses of snow cores will be taken in order to estimate areal variations in density. Runoff from the pillows will also be recorded to determine the timing of melt, and to distinguish between melt and further snow occurring between daily observations.

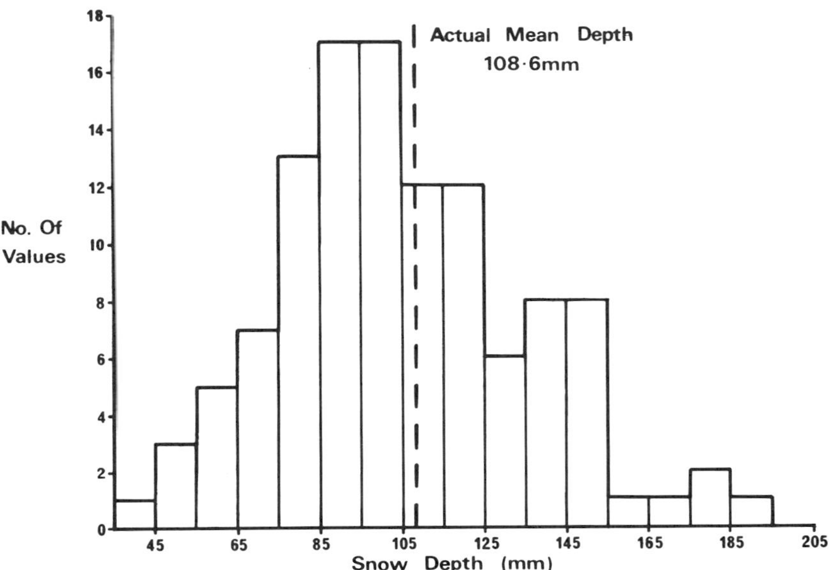

Figure 4. Distribution of snow depths obtained from different snow courses

Snow depth

Eight sites for photography are being set up on the grassed areas of the catchments, with aspects covering the eight cardinal compass directions, and a range of altitude, 370-630 m.

In addition to these ranges of aspect and altitude, sites were chosen to sample a variety of slopes, both convex and concave, and a number of gullies. Where possible, valley floors will be sampled, but often these are photogrammetrically unsuitable due to the presence of long grass; similarly hill tops are generally flat, and unsuitable for terrestrial photogrammetry. Where possible sites where chosen on valley sides, so that photography was possible from the opposite slope, to give a horizontal view and the maximum possible incident viewing angle. Apart from the problems caused by long grass and by high flat areas, the main photogrammetric limitations to ensure accuracy were dimensional, imposed both by the optimum type of stereoplotter used and for ease of working of the plotter operator. These limitations covered camera separation, both horizontal and vertical, and maximum and minimum viewing distances.

Both camera plinths and ground control markers must be robust and stable. Camera plinths comprise 230 mm cast iron pipe with flanges, set in concrete. The upper flange holds a brass plate onto which a Wild theodolite and P32 camera are fixed by a ball centering system; this enables quick and accurate centering of the camera to within ± 0.02 mm. Targets are sited towards each corner of the area to be viewed stereoscopically, with a further target in the centre. Each target stands some 2 metres above the ground, and comprises a rigid alloy plate bearing black and white markings designed to provide maximum contrast. Target sizes vary from 250-1000 mm diameter, according to their distance from the camera, taking into account the focal length of the camera and the size of the floating mark on the stereo plotter.

During photography, camera axes will be maintained parallel and horizontal by sighting the theodolite, bearing the camera, onto a target on the other camera base, then swinging through 90° to view the hillslope. After photography the theodolite and target are interchanged and the procedure is repeated.

Density of snow and runoff from snowpack

To convert the measured volume of snow into a water equivalent, snow cores will be taken adjacent to the test area, and weighed. Because snow density is spatially less variable than depth, a relatively small number of samples should be representative of the site. Cores will be taken along courses with a 100 mm diameter corer, at the same time as the photography.

These data will be supplemented by a continuous point record of snow water equivalent from a ground level catchment pressure pillow installed near each site; the runoff from each pillow is also to be measured. Due to the steep inclination of much of the catchment, the conventional liquid filled rubber pressure pillows are unsuitable. Instead, a 'pillow' system has been designed, which comprises a 2.5 metre square rigid wooden board resting on four interlinked oil filled hydraulic cylinders. Snow falling on the board exerts a pressure within the hydraulic system, and this pressure is measured by a pressure transducer. Pressure readings are taken every 15 minutes and are recorded on magnetic tape. The system can be easily set at ground level on slopes of large inclination. Effects of temperature on measurements are negligible, as compared to the conventional liquid filled pillow, where temperature compensation is necessary.

Meltwater from the pillow is collected in a trough set along its lower edge, and is measured via a tipping bucket device, to be recorded on magnetic tape. Snow depth on the pillow at the time of photography is measured by hand, to allow extrapolation of its record over the adjacent area.

System operation and data analysis

The system will be operated on a daily basis when snow lies, and two persons will be required. Camera stations have been sited next to or within short distances of roads to enable speed of operation. If access becomes a problem during deeper snow, a snow vehicle may be considered.

Photographs of the ground surface will be taken annually before the winter, to detect any significant changes in the land surfaces. Development will be carried out in accordance with the recommendations of the feasibility test, and photographs will be stored and analysed photogrammetrically at the end of each winter.

Extrapolation of the snow depths on the photographed areas to the rest of the catchment is likely to be achieved by a combination of approaches:

a. correlation with ground measurements taken above upper limit of photography

b. derivation of relationships between snow depth and physical characteristics such as aspect, altitude, and exposure

c. examination of hydrograph records and soil moisture readings during snowmelt, to establish the volume of melt.

POTENTIAL USE OF TERRESTRIAL PHOTOGRAMMETRY IN FLOOD WARNING SYSTEMS

Many of the most damaging floods in the United Kingdom result from heavy rain on snow on a frozen ground. Grinsted (1972) suggested an accuracy of 8% for forecasts of areal rainfall in 3 hourly periods over an area of 1000 km^2; updated estimates of lying snow are not available at anything like this accuracy.

Hydrological data for use in flood warning systems is required in real time. The potential and actual contribution of snowmelt to flood discharge will need to be measured and then telemetered to the control station. At present no reasonable technique is applied for this purpose in the United Kingdom, but it is considered that a combination of terrestrial photogrammetry and telemetering snow pillows would produce significant benefits. In the immediate future a system similar to that to be employed by the Institute on its research catchments could be used, provided development and interpretation of photographs could be achieved quickly. Looking to the future, the transmission of the photographic imagery by vidicon, and the development of automatic stereo plotters would further assist the forecasting process. Coupled with telemetering snow pillows, it is suggested that the water equivalent of a snowpack could be determined to better than 20%.

CONCLUSIONS

The feasibility of using terrestrial photogrammetry to determine the volume of shallow and frequently ablating snowpacks has been demonstrated. Its current application to the Institute's research programme is likely to be extended to examine the physical processes of snowmelt. As an operational tool in flood warning systems it clearly has immediate potential, and there are intriguing possible future developments which could increase further its value.

REFERENCES

Bissell, V.O. and Peck, E.L., 1973. An aerial measurement of snow water equivalent by terrestrial gamma radiation survey, Bull.Int. Ass.Hydrol.Sci. XVIII, 47-62.

Blyth, K., Cooper, M.A.R., Lindsey, N.E. and Painter, R.B., 1973. The measurement of snow depth by terrestrial photogrammetry (in press).

Cooper, C.F. 1965. Snow cover measurement, Photogr.Engrng., 31, 611-619.

Grinsted, W.A., 1972. The measurement of areal rainfall by the use of radar,Symp.Rem.Sens.Man's Envir. Bristol. England.

Johnson, P. and Archer, D.R., 1972. Current research in British snow-melt river flooding, Bull.Int.Ass.Hydrol.Sci, 4, 17, 443-451

Painter, R.B., 1972. Some present uses of remote sensing in monitoring hydrological variables. Symp.Rem.Sens.Man's Envir, Bristol, England.

Smith, F.M., Cooper, C.F. and Chapman, E.G., 1967. Measuring snow depth by aerial photogrammetry: evaluation and recommendations, Proc. 35th Western Snow Conf. Boise, U.S.A., 66-72.

CONCEPTUAL DESIGN OF A MULTIPURPOSE INSTRUMENT FOR WINTER STREAM METERING

Gee Tsang
Canada Centre for Inland Waters

ABSTRACT

The conceptual design of a multi-purpose instrument for winter stream metering is proposed. The instrument is for measuring:

1. the flow velocity at a point in a stream;
2. the concentration of frazil ice at the point; and,
3. the velocity of the frazil ice at the point.

The measurement of the flow velocity is by releasing a small volume of conductive solution into the stream. The passing of the conductive "puff" through two consecutive pairs of electrical poles enables the measurement of the velocity of the flow. The releasing velocity of the puff can be controlled so the velocity of the puff is indicative of the flow velocity. By releasing the conductive solution as a steady plume, the concentration of frazil ice is measured. The physical principle that the current through two electrical poles decreases as non-conductive frazil ice passes through the gap is used for frazil concentration measurement. When the instrument is switched to a sensitive branch, instantaneous point concentration of frazil ice is measured. By switching the instrument to a branch containing a low frequency pass, average concentration of frazil over a time period may be obtained. By recording the instantaneous frazil concentration at two points a distance apart and measuring the time displacement of the two recordings, the velocity of the suspended frazil can be calculated. Conceptual circuit and block diagrams for the composing sensing units and the final composite instrument are presented.

INTRODUCTION

Most streams at high latitudes freeze in winter. In some regions, an ice cover of several metres thick is not uncommon (Strilaeff 1972). The presence of an additional boundary at the top, which itself changes with the meteorological conditions, makes the flow of water under an ice cover a much more complicated problem than the simple open channel flow in summer. The problem is further complicated by the loading of frazil in water. Frazil crystals are produced in open water sections of a stream when the flow is turbulent and temporarily supercooled. Not long after their formation, the frazil crystals agglomerate and float to the top. On the other hand, the turbulence in the flow transports the frazil clusters back to the lower layers. Eventually an equilibrium between the turbulence and the frazil distribution is reached. The resultant density gradient due to different frazil concentration at different depths inhibits turbulent mixing in the stream. It was found by Tsang (1970) that when the frazil loading is high, the velocity distribution profile for the upper part of an ice covered flow changes from the characteristic uniform distribution of a turbulent flow to an approximately parabolic distribution for the same rate of discharge. Frazil sometimes accumulates to the underside of an ice cover. Depending on factors which are not yet known, water may or may not flow through a frazil pack (Bennett 1968). Because of the influences of ice, the rating curve for a guaging station obtained in summer serves little purpose for winter flow calculations. A rating curve obtained from winter metering over several winters may only serve as a statistical tool as the ice conditions of a river change from year to year (Chin, 1966). The above thus lead to the necessity of frequent winter flow metering for an accurate budgeting of winter water resources.

The current practice of winter stream metering is to drill holes on the ice cover and measure the flow velocity by conventional current meters. Although power augers have been used for ice drilling in many cases, the drilling of numerous relatively large holes (20-30 cm dia.) through an ice cover meters thick is still a time-consuming process, not to mention the physical hardship of working in subfreezing temperature using heavy tools. As a current practice in Canada, if a river section for flow metering is found to be frazil packed, the field crew will abandon the crossing and look for a frazil free one. By so doing, sometimes much time is wasted. For conventional current meters, usually an averaging time of about a minute is required for measuring the velocity at one point. The setting up of the instrument requires several additional minutes. Thus, winter stream metering is a tedious and time-consuming process. The situation is further aggravated by the short daylight time in winter. Thus, it is seen from the above, for a more effective winter stream metering, a better instrument and an improved procedure are needed.

To save time and to simplify the procedure, studies have been made in the past to obtain the average velocity of a vertical by two point sampling (May, 1966) and one point sampling (Strilaeff, 1972). The universal applicability of the one and two-point sampling methods has yet to be proven. In any case, the use of conventional instrument for flow velocity measurement does not reduce the time for ice drilling and the time for averaging.

The above discussions are mainly concerned with problems of field operations. For both field and laboratory measurements, although the flow velocity at a point can be accurately measured when the flow is frazil free, an accurate velocity measurement may not be obtained if the flow is frazil loaded as the effect of frazil on the accuracy of a conventional flow meter or a probe is not known. The knowledge of frazil concentration at a point is important for explaining and predicting the hydraulic behaviour of a stream in winter. In Iceland, some success has been obtained in measuring the average frazil concentration over a depth by measuring the conductivity of frazil laden water (Kristinsson, 1970). The Icelandic instrument should be advanced for measuring point frazil concentration and a theoretical basis should be provided to guide the design. For heat budgeting and ice balance calculations of a section of a channel, one also needs to know the velocity of the suspended frazil ice, which may not move at the same velocity as the ambient flow, especially when the concentration is high. An instrument for measuring the velocity of suspended frazil has not yet been developed.

This paper studies the physical principles and the construction of an instrument which may be used for fast measurement of flow velocity, frazil concentration and the velocity of the suspended frazil, both in the field and in laboratories. The investigation is limited to conceptual. More work and effort are required before a workable instrument is developed.

CONCEPTUAL DESIGN OF A MULTI-PURPOSE INSTRUMENT

As mentioned in the above section, for a proper winter flow monitoring, the velocity of flow, the concentration of frazil ice and the velocity of the suspended frazil should be measured. In this section, the sensors for measuring each of the above quantities will be discussed first. A composite instrument which combines all the sensors will be proposed at the conclusion of the section.

1. Conceptual design of the velocity measuring component

It is mentioned in Introduction that time is a critical factor in winter stream metering. Thus, a desirable winter metering instrument should have a fast response time. It should also be small in size so drilling large holes is not necessary. In addition, it should be simple to operate and sufficiently rugged for field operation.

Most instruments presently used for velocity measurement are based on mechanical principles. The flow either imparts momentum to a rotor, or forces a water column to rise by converting its velocity head into pressure head. Because of the inertia of the moving parts, be it solid or liquid, a length of time is required to reach the equilibrium state. For a fast response, it appears that the involvement of an inertial moving part for velocity measurement should be avoided. Electrical sensing thus offers a good alternative.

Pure water is electrically non-conductive. When a very small amount of salts is dissolved in it, such as the fresh water in natural streams, it becomes a weak conductor. If a pair of electrical poles of a small distance apart is placed in a natural stream, a weak electrical current will flow between the poles.

Now if a patch of conductive solution is introduced upstream of the electrical poles, then when the solution passes through the gap, an increase in electrical current in the circuit will show. By placing two pairs of poles streamline-wise in the stream at certain distance apart and measuring the time lapse between the two current increases, the velocity of the solution between the two pairs of poles can be calculated.

Based on the above principle, an instrument based on the theoretical concept of the diagram shown in Fugure 1a is proposed for measuring the velocity of flow at a point. The instrument consists of a solution injection circuit and an electrical sensing circuit. When the switch on the electrical circuit S_1 is pressed, the solenoid L_1 is activated, which pushes the piston downwards to inject a fixed volume of solution into the stream. After its release, the solution mixes with the ambient water and travels downstream. When the solution cloud passes the electrical poles, momentary current increases result. These electrical signals are picked up by feeding the potential drop across a constant resistance R_1 to the vertical input of an oscilloscope or other recording and displaying device. A triggering device may be used in the oscilloscope so the first signal is always located at the extreme left of the screen. A memory oscilloscope may be used so the display on the screen may stay for a long duration for an accurate measurement.

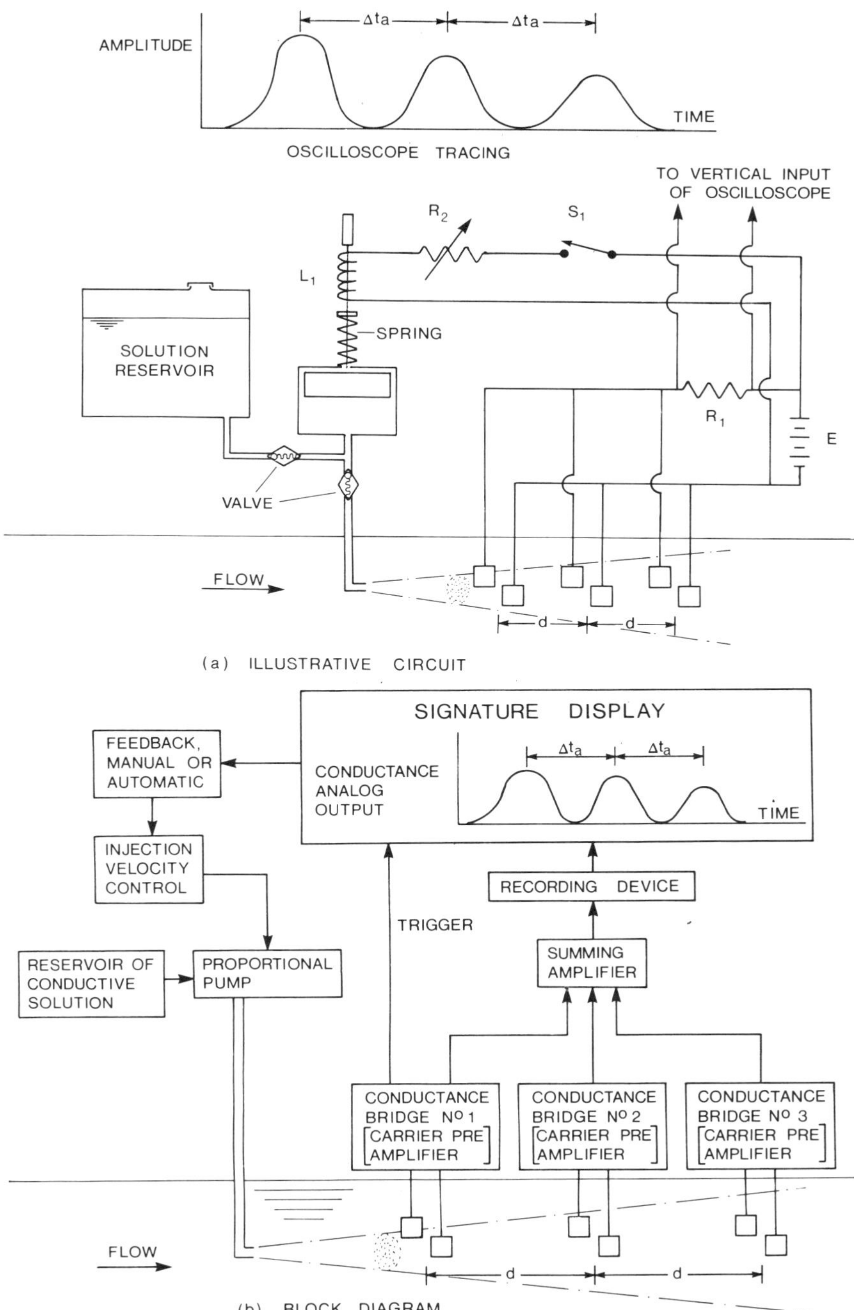

Figure 1. Component for velocity measurement

7.2

Unless the injecting velocity of the solution is exactly the same as the ambient flow, the solution cloud will be accelerated or decelerated, resulting in the time lapse between the first and the second peak Δt_a being either greater or less than the time lapse between the second and the third peak Δt_b. The issuing velocity of the solution can be controlled by adjusting the variable resistance R_2. By continuously adjusting R_2 and at the same time observing the display on the screen of the oscilloscope until $\Delta t_a = \Delta t_b$, the desired injecting velocity is obtained and the velocity of flow may be calculated from

$$V = \frac{d}{\Delta t_a} = \frac{d}{\Delta t_b} \quad , \qquad (1)$$

where d is the distance between two pairs of consecutive poles.

The circuit diagram shown in Figure 1a is for illustrative purposes only. The hardware design of the component will be much more complicated and a task of its own. A better presentation of the velocity measuring component of the instrument may be shown by the block diagram shown in Figure 1b.

The measurement of flow velocity using tracers is really nothing new. Its use is proposed here because the concentration of frazil in the flow and the velocity of the frazil may also be measured using the same tracer as shown in the following sections.

2. Conceptual design of the frazil concentration measuring component

To measure the concentration of frazil at a point, a pair of electrical poles may be used and the tracer solution may be introduced upstream of the poles as a steady plume. The solution from the nozzle, after mixing with the ambient water, passes through the electrical poles and causes an electrical current to flow between the poles. The principle of the tracer releasing system and the electrical sensing circuit is shown by the illustrative diagram Figure 2a.

In Figure 2a, the electrical current flowing between the poles can be shown to be given by

$$I_{(t)} = \frac{E}{R_1 + \dfrac{1}{\displaystyle\int_A \frac{dA}{\int_S \rho ds}}} \quad , \qquad (2)$$

where E is the potential difference across the terminals of the battery, R_1 is the on-line resistance, S is the separation between the two pole plates, A is the area of the pole plates and ρ is the resistivity of the space between the pole plates and is a function of both space and time. The variation of the resistivity in the gap between the two poles is due to different frazil ice concentration in the flow. When water freezes, the dissolved impurities are rejected. Ice, therefore, is practically non-conductive. The variation of the amount of frazil ice suspended in the conductive plume thus changes the conductivity of the gap between the two electrical poles.

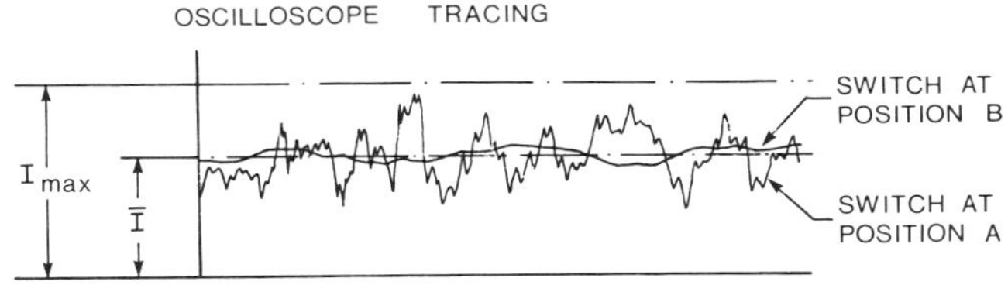

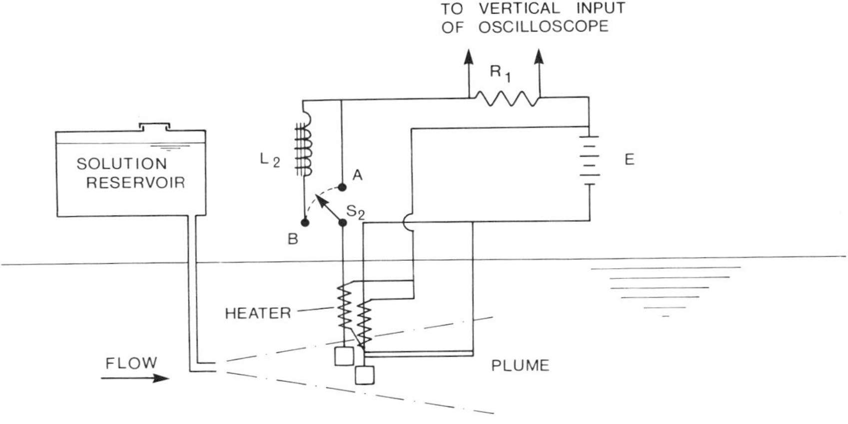

(a) ILLUSTRATIVE CIRCUIT

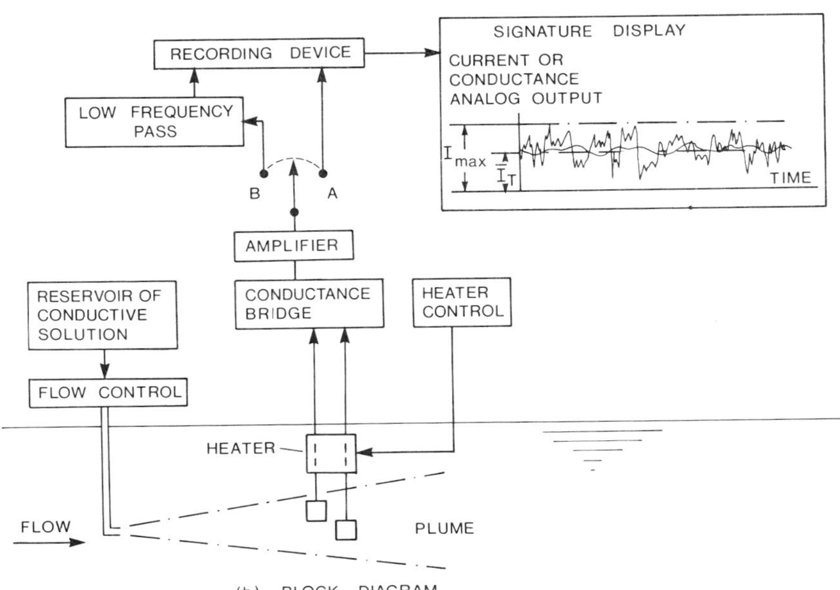

(b) BLOCK DIAGRAM

Figure 2. Component for frazil concentration measurement.

7.2

When the 3-way switch S_2 is switched to position A, the electrical circuit has little inductance and is sensitive to the variation of frazil concentration in the flow. As the distribution of frazil ice in flow is not homogeneous, a fluctuating tracing will show on the screen of the oscilloscope. If the time for sampling is sufficiently long and the size of the pole plates is of reasonable small size, it is very probably during the sampling period there is a short moment at which the flow passed through the gap between the two poles carries no frazil. In this situation, the resistivity of the gap will be minimum and uniform throughout the gap. Denoting the minimum resistivity by ρ_{min}, the maximum current flowed between the gap will be

$$I_{max} = \frac{E}{R_1 + \frac{\rho_{min} S}{A}} \quad . \tag{3}$$

When the 3-way valve S_2 is switched to position B, the electrical circuit is connected to a large inductance L_2 and becomes insensitive to the momentary variation of frazil loading. The current passing through the poles will show the average current in the sampling period. From equation (2), this average is given by

$$\bar{I}_T = \frac{E}{R_1 + \bar{\rho}_T S/A} \quad , \tag{4}$$

where the subscript T indicates the length of the sampling time.

If the sampling time is large compared to the time required for a fluid element to flow through the gap, in a statistical sense the frazil ice may be considered as uniformly distributed in the space. The suspended frazil thus will have the effect of reducing the conductive cross-sectional area. The resistance of the gap between the two poles, therefore, will be given by

$$\bar{R}_T = \frac{S\bar{\rho}_T}{A} = \frac{\rho_{min} S}{(1 - \bar{C}_T)A} \quad , \tag{5}$$

where $\bar{C}_T$ is the percentage of the cross-sectional area occupied by frazil and is equal to the average volumetric concentration of frazil ice in the flow in time period T.

The substitution of the above equation in equation (4) gives

$$\bar{I}_T = \frac{E}{R_1 + \frac{\rho_{min} S}{(1-\bar{C}_T)A}} \quad . \tag{6}$$

The divisions of the above equation by the equation (3) gives

$$\frac{\bar{I}_T}{I_{max}} = \frac{R_1 + \frac{\rho_{min} S}{A}}{R_1 + \frac{\rho_{min} S}{(1-\bar{C}_T)A}} \qquad (7)$$

If the on-line resistance R_1 is small compared to the resistance of the gap, it may be dropped from the above equation and one has

$$\bar{C}_T = 1 - \frac{\bar{I}_T}{I_{max}} \qquad (8)$$

Thus, from the amplitude of the tracing on the screen of the oscilloscope, one can measure the concentration of frazil ice in the flow.

The heaters shown in Figure 2a around the electrical poles are for preventing the suspended frazil ice from adhering to the poles.

The relationship shown by equation (8) was used by Kristinsson (1970) as a hypothesis supported by laboratory experiments. A probe was made by winding two parallel wires on an insulated rod heated internally. The instrument measured the electrical current conveyed by the river water only, no foreign solution was introduced. Because of the weak current conducted by the natural water, the probe was rather long (60 cm.). The instrument was used in Iceland to measure the rate of frazil discharge in natural rivers.

The introduction of a conductive solution increases the conductance between the poles and by which reduces the size of the probe for point measurement.

3. Conceptual design of the component for measuring the velocity of suspended frazil ice

To measure the velocity of the suspended frazil ice, another pair of poles may be added to and the high inductance may be eliminated from the illustrative circuit for frazil concentration measurement as shown in Figure 3a.

As mentioned in the last section, when the conductive plume flows through the gap of the first pair of poles, a fluctuating tracing will be shown on the screen of the oscilloscope. The amplitude of fluctuation depends on the variation of the concentration of suspended frazil ice. When the same train of water element passes through the second pair of poles, another fluctuating tracing will be shown on the screen. If the separation between the two pairs of poles is small so the measurements at the two points are correlated, two tracings similar in shape will be shown on the screen. Thus, by feeding the signals from the two pairs of poles to the vertical input of a two-beam oscilloscope which is subject to a common horizontal sweep and measuring the displacement of the two tracings on the oscilloscope, the velocity of the suspended frazil ice may be measured.

The component circuit, in a more realistic block diagram form, is shown in Figure 3b.

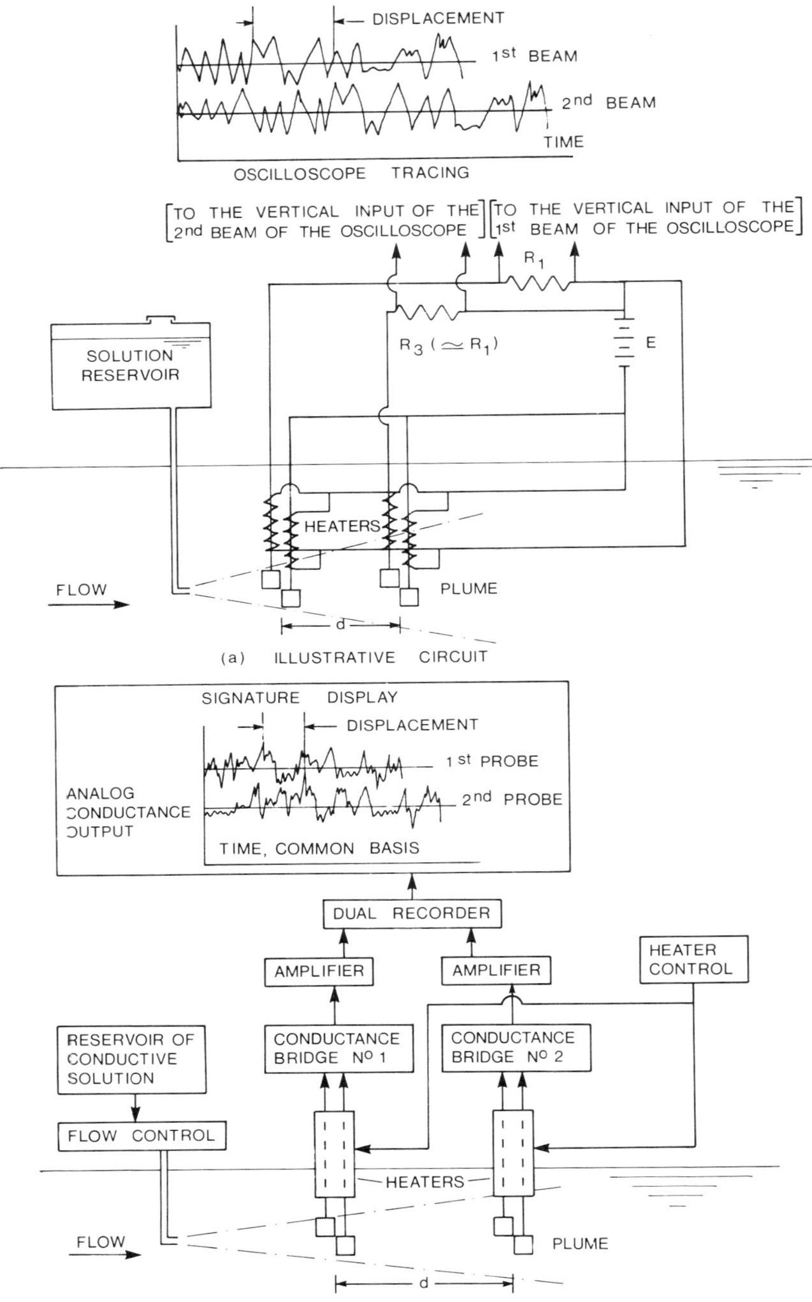

Figure 3. Component for measuring velocity of suspended frazil.

7.2

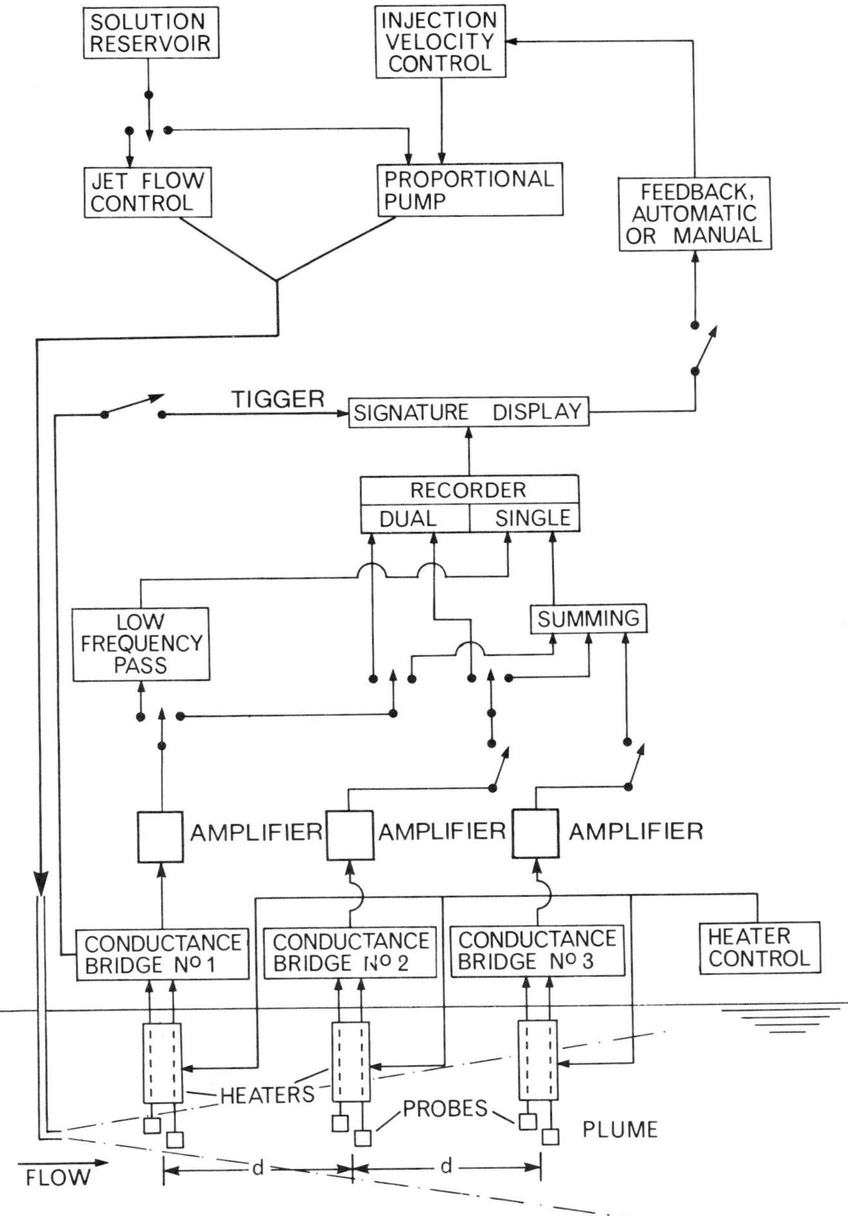

FIGURE 4 Composite block diagram of the multipurpose instrument.

7.2

4. Composite circuit of the multi-purpose instrument

By putting the component block diagrams together, a composite block diagram for an instrument for measuring flow velocity, frazil concentration and the velocity of the suspended frazil ice is obtained as shown in Figure 4. By connecting and opening appropriate linkages of the circuit, the three different functions of the instrument may be performed.

DISCUSSIONS

The concept of the multi-purpose instrument mentioned above appears to be theoretically workable. Of course, much work has yet to be done for the hardware design of the instrument. The ability of the instrument to measure the instantaneous velocity and frazil concentration at a point is very important for this will permit us to study the time correlation of frazil concentration $\overline{C_{(t)}\, C_{(t+\tau)}}$ at a point and the spatial correlation of frazil concentration $\overline{C_{(x_i)}\, C(x_i+\Delta x_j)}$ between two points. The study of these two correlations will open a frontier comparable to the study of turbulence in flow.

Although the proposed instrument is for measuring frazil ice, it may also be used to measure the concentration and the velocity of other non-conductive sediments. For the instrument proposed here, electrical conductance is utilized as the basis of sensing. The sensing of other physical or chemical properties, which are different between water and ice, may equally be used as the basis of the instrument.

REFERENCES

Bennett, R.M., "The stage-discharge relationship under ice cover for the Peace and Slave Rivers". M.Sc. Thesis, Utah State University 1968.

Chin, W.Q. "Hydrology of the Takhini River Basin, Y.T., with special reference to accuracy of Winter Stream records and factors affecting winter stream flow". Internal Report No. 2. Studies Section, Vancouver, B.C., Department of Energy, Mines and Resources, Oct. 1966.

Kristinsson, B., "Ice Monitoring Equipment", Proc., I.A.H.R., Symposium on ice and its action on hydraulic structures, Reykjavik. September 1970, p.p.1.1, 1 - 14.

May, R.D. "Report on activities in Calgary District during winter of 1965-66". Internal Report, Arctic rivers work group, Water Resources Branch, Department of Energy, Mines and Resources, 1966.

Strilaeff, P.W. "Measurement of discharge under ice cover". Proc. International Symposia on the role of snow and ice in hydrology. Banff. September 1972 (in press).

Tsang, G. "Change of velocity distribution in a cross-section of a freezing river and the effect of frazil ice loading on velocity distribution". Proc., I.A.H.R. Symposium on ice and its action on hydraulic structures, Reykjavik. September 1970 p.p. 3.2, 1-11.

AN ELECTRO-OPTICAL INSTRUMENT FOR MEASURING TOTAL PRECIPITATION AND SNOW PACK WATER CONTENT

John C. Bass
Bass Engineering, Inc.

ABSTRACT

A unique optical lever instrument originally developed for oceanographic studies is described. This instrument has been recently receiving wider use in the field of hydrology. An example of a current application to a total precipitation gauge is described as well as a possible future application to a dual installation for the measurement of total precipitation and snow pack water content.

The development of the optical lever sensing system began at Gulf General Atomic in 1967. The system was originally developed to provide a means for accurately measuring ocean tides in the offshore regions.

Several tide measuring instruments were available at the time, however they all suffered from similar problems. The major difficulty was that of obtaining an instrument with sufficient full scale capability while at the same time providing high resolution and accuracy.

The optical lever system which was developed and subsequently patented, was a new approach and provided a solution to these problems. Instead of either the single full scale or multiple scales obtained through variable gain, the optical lever design allowed the full range of the instrument to be divided into a multiplicity of small equal increments. With automatic range switching this technique allows the user to maintain the high accuracy and resolution desired over the entire dynamic range of the instrument.

An example of this is the STG/100 shallow water tide gauge shown in Figure 1. This instrument typically has a total dynamic range of sixty feet of water level change which is read out in thrity individual full scales, each of which represents approximately a thirty inch change in level. The resolution, even on a small chart recorder, is better than 0.3 inches. With a digital recording system, the resolution can be increased by at least an order of magnitude.

In addition to the STG/100, instruments developed using the optical lever principle were successfully designed and used in the field of oceanography as magnetometers, deep sea tide gauges, and shallow water wave gauges. The results obtained by these instruments and their operation have been reported in several trade journals and papers.

Of great importance for the oceanographic application was extreme ruggedness to withstand shipboard handling and excellent long term stability characteristics to be in keeping with the long term deployments in the ocean. The basic design criteria and method of construction is still used today through the entire product line including the hydrology instruments.

It was not until the period of late 1969 to early 1970 that an

Figure 1. STG/100 Tide Gauge.

optical lever instrument was used in the field of hydrology. The instruments first used were tide gauges, unmodified from their original configuration, applied to measure stream level as a function of time. One gauge was installed at Alpha Site to provide a demonstration of its application to the problem of determining snow pack water content.

In this case of the instrument at Alpha Site, a standard STG/100 tide gauge was connected to an "Oregon" type snow pillow. The tide gauge was burried beneath the ground and data was recorded on the small chart recprder inside the tide gauge. The instrument was installed near the end of the season and data was taken for only about a month. The data received was very encouraging.

The first optical lever instrument designed specifically for hydrological use was designated the L/200. The unit was a reservoir level gauge equipped with a digital punch tape recorder, with a full scale range of 200 feet. It was installed in Del Valle Reservoir.

The principles of operation for the optical lever system are essentailly the same for all water level measuring instruments. A multi-turn helical bourdon tube is the primary sensing element. Bourdon tubes of this type produce an angular rotation which is directly proportional to the pressure acting on the tube.

A simplified schematic of a typical optical lever sensor is shown in Figure 2. Hydrostatic pressure resulting from the height fluid column acts directly on the bourdon tube. For certain applications, such as tide and reservoir or stream sensors, a hydraulic wave filter is interposed between the fluid column and the bourdon tube to eliminate high frequency oscillation such as might be caused by wind waves or other surface phenomena.

A mirror bar magnet assembly, and a set of damping vanes are mounted on the Bourdon tube extension. Unrestrained rotation of the bourdon tube results in the rotation of the mirror, bar magnet, and damping vanes about the axis of the Bourdon tube helix. A multi-turn coil is positioned directly behind the mirror and bar magnet.

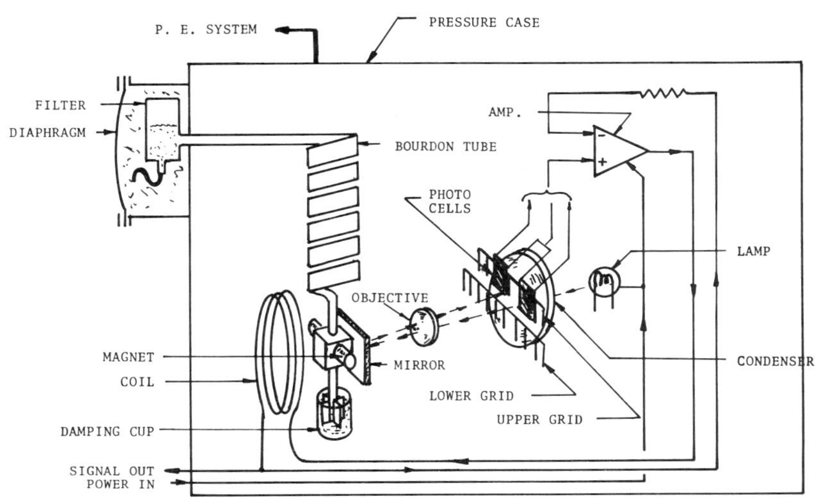

Figure 2. Schematic of Optical Lever Sensor

The damping vanes extend into a cup which contains a high vicosity silicon oil and is located directly below the bourdon tube.

A small incandescent light bulb is mounted at the end of the sensor opposite the bourdon tube and coil. Light emanating from the bulb passes through a condensing lens and uniformly illuminates the lower half of the specially prepared glass grating plate.

The image of the lower half grating is focused by the objective lens, reflects from the mirror attached to the end of the bourdon tube, passes back through the objective, so that a virtual image of the lower half grating is superimposed on the upper half grating. A pair of matched photo cells mounted behind the upper half grating receive the residual light which pases through the upper half grating.

As shown in Figure 3, both the upper and lower halves of the grating are constructed of equal width lines and spaces. The lower half grating being uniform over its entire width. The upper half, however, is divided into two parts; one half is exactly like the lower grid while in the other half the lines and spaces interchange. This configuration allows one photocell to be completely darkened and the other fully illuminated. In the "NULL" position, both photo cells are illuminated equally.

The output signal from the photocells is applied differentially to the input of a high gain operational amplifier. The output of the amplifier is then applied across the winding of the multi-turn coil to produce an electro magnetic field. The field of the coil interacts with the field of the bar magnet mounted on the bourdon tube.

The interacting magnetic fields provides a mechanical feedback which causes the bar magnet to rotate until the output signals of the two photocells are "NULLED". The voltage across the coil required to "NULL" the system is taken as the sensor output and is

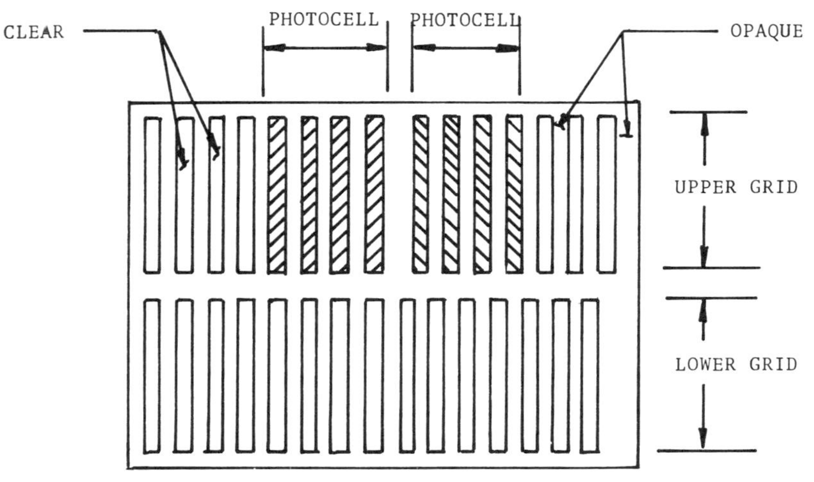

Figure 3. Optical Lever Grating.

linear with head change.

One of the unique features of the system is that it can provide multiple "NULLS". This allows one to design a sensor which can measure a large change in head with a much greater accuracy than can normally be expected from a single scale sensor by dividing the range of the sensor into many small equal increments. Each increment is then measured separately and a composite is taken for the total range.

The multiple "NULL" feature is obtained by having an excess of "active" grids in the lower half of the grid plate of those in the upper half. Extinguishing the light or shorting the feedback circuit between readings allows the bourdon tube to rotate freely with changes in head. When the circuit is restored to obtain a new reading, the optical system aligns and "NULLS" to the nearest set of grid lines.

In a situation in which the water level is increasing linearly with time and the sensor output is sampled at regular intervals, the output of the sensor would show a linear increase from minimum to maximum as shown in Figure 4, until the limit of the scale is reached. The output signal would then revert to a minimum and increase again to maximum over the next incremental scale. This process repeats until the maximum range of the sensor is reached.

The number of times the sensor shifts scales ("NULLS") depends on the design of the particular sensor. Sensors are currently manufactured with from one to as many as thirty-three equal scale shifts over the range of the sensor. The single scale ranges are available from four inches of water column to two hundred feet of water column, and can be readily changed by adjusting various system parameters such as grid spacing, bourdon tube stiffness, or lense focal length.

For most applications in the field of oceanography and hydrology, it is easy to keep track of the scale shifts of an instrument by simple inspection of the record. This is usually true where the change in fluid column is slow compared with the frequency of data acquisition. In cases where the data is taken at long or infrequent intervals, such as in a telemetry station, an additional channel

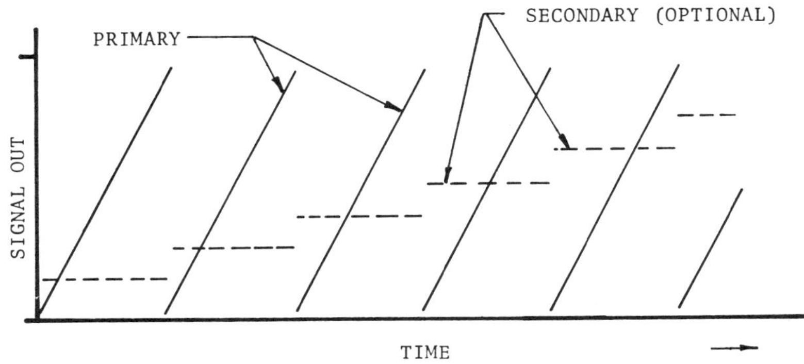

Figure 4. Output of Typical Optical Level Sensor.

of information which provides a secondary output, can be added. The secondary output is in the form of a discrete voltage output for each scale. This added information channel can then be used to determine if one or more scales have passed between interrogations.

Thermal drift of the sensor is miminized by utilizing inherently low drift electronics. High thermal mechanical stability is obtained by fabricating the bourdon tube from Ni Span C, a low thermal expansion nickle/iron alloy.

Mechanical hysteresis associated with many pressure sensors is eliminated in the optical lever sensor sinch coupling between the sensing element and the output is optical and magnetic in nature, not mechanical. The only hysteresis present in the optical lever sensor is the negligible amount associated with the flexure of the metal bourdon tube, and that is well below the level of error introduced external to the instrument. Late in 1972, the need was recognized for an instrument to accurately measure total precipitation and snow pack water content for both local recording and telemetry. There appeared to be a need for instruments capable of measuring total water columns in excess of 200 inches to a precision of 0.01 inches. Since most good transducers available at that time in the 0.5% F.S. accuracy range and could not possibly provide the accuracy required. It was at this time that the PG/100T was developed. This instrument was designed to have a total range in excess of 200 inches of water, using multiple individual full scales of about 10 inches of water. The precision is plus or minus .01 inches.

The PG/100T sensor can be installed on a 200 inch precipitation tower as shwon in Figure 5. This unit is now installed and under test in the southern Sierra. It contains its own all season battery power supply, timing units, and bubbler control to prevent ice formation. The system shown is self recording. It was designed, however, for conversion to a telemetry system at a later date.

The hydrostatic pressure created by the precipitation collected within the gauge tube is transmitted to the PG/100T sensor through a silicon oil filled diaphragm and line gauge tube inside the tube cleanout as are the fittings for the gauge tube drain, a bubbler tube and an outlet for a sight glass. A pressure equilization system is provided in the form of a large sealed flexible rubber

Figure 5. Installation of 200 Inch Total Precipitation Tower.

diaphragm. It can be seen mounted on top of the instrument enclosure. One side of the equilization diaphragm is connected to the interior of the sensor and the other side is connected to the inside of the gauge tube at a point above the overflow and below the inlet orifice. The pressure equilization unit is designed to prevent apparent changes in water levles due either to changes in atmospheric pressure or to wind blowing across the gauge tube orifice. The construction of the diaphragm eliminates uptake of moisture inside the sensor during operation.

A station for obtaining data from both a total precipitation gauge and one or more snow pillows to measure snow pack water content can be built using a single PG/100T sensor. The PG/100T sensor makes the system feasible since it has the range and resolution capability to cover the requirements of both systems. The advantage of such a system is that any errors introduced by using two different types of sensors can be eliminated.

A block schematic of this system is shown in Figure 6. The PG/100 sensor would be installed below the level of both the gauge tube and the snow pillow. A pressure line from the snow pillow is connected to the orifice area of the gauge tube as would the pressure equilization line from the sensor.

The snow pillow pressure line and the gauge tube line would be connected through diaphragms to an electrical driven selector valve. The common on the selector valve would be connected to the pressure port on the PG/100T sensor.

By alternating the position of the selector valve with an

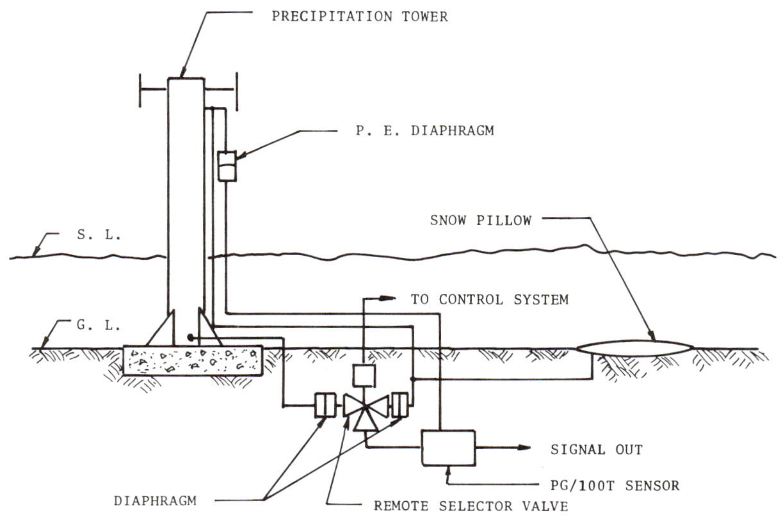

Figure 6. Schematic of Dual Total Precipitation and Snow Pack Water Content Station.

electronic timing circuit, the hydrostatic head on each of the lines can be read individually without transferring fluid between the units. The connection from the snow pillow to the gauge tube orifice would cancel errors due to changes.

This arrangement shown can be used with either a local recording system where the two readings would be multiplexed on the recorder or with a telemetry system. This system would require a delay of the order of 3 to 4 seconds between parameter readings.

In summary, we have shown the unique optical lever system which was originally developed for oceanographic studies and how it has evolved into what should prove to be a very useful tool in the field of hydrology. We are continuing to develop optical lever instruments as the needs and requirements grow, and as we gain more experience with the instrument in field trials.

REFERENCES

Filloux, Jean H., October 20, 1970. Optical Lever and Detecting System and Plate Assembly for Use Therewith. U.S. Patent 3,535,538.

Filloux, Jean H., November 10, 1970. Monitoring Apparatus. U.S. Patent 3,538,772.

Filloux, Jean H., 1969. Bourdon Tube Deep Sea Tide Gauges. Proceedings International Symposium on Tsunami and Tsunami Research, International Union of Geodesy and Geophysics.

Filloux, Jean H., 1970. Deep Sea Tide Gauge with Optical Readout of Bourdon Tube Roataions. Nature Vol. 226, No. 5249, pp. 935-937.

MEASURING SNOWFALL, A CRITICAL FACTOR FOR SNOW RESOURCE MANAGEMENT

Paul A. Rechard
Roy E. Brewer
Alan Sullivan
University of Wyoming Water Resources Research Institute

ABSTRACT

With the advent of detailed computer modeling techniques for determining the effects of watershed management practices, the ability to quantify the precipitation input is of critical importance. The measurement of precipitation in the form of snow is an observation task which to date has proven unsatisfactory. It has been shown quite conclusively that under windy conditions the catch of snowfall by usual precipitation gages may be eighty percent in error.

The findings of this study indicate that the catch of a gage placed in a manmade site shield utilizing the blow-fence design, termed the Wyoming Shield, can duplicate that of a gage placed within a forested opening. The advantages of the Wyoming Shield are that it is always uniform, i.e. not subject to the natural variations in timber protection, it provides freedom of choice for siting a gage, and it is relatively inexpensive.

As a general guide, it has been found that unshielded gages catch per season from one-third to one-half the amount of precipitation that a well-protected gage in a forest opening catches. A gage with only a free swinging Alter-type shield catches from two-thirds to three-fourths the amount a well-protected gage catches, while a gage shielded with the Wyoming Shield configuration will catch about the same amount.

With the advent of detailed computer modeling techniques for determining the effects of watershed management practices, quantification of the precipitation input to the watershed becomes of critical importance. One such computer model, mentioned for example only, is the snowmelt simulation model of Leaf and Brink (1973 a and b) in which the amount of precipitation is important to define the snow on the ground available for melt and is used to modify the pack water equivalent and radiation balance during the melting season. Schmidt (1972) has shown that considerable quantities of snow may sublimate during blowing conditions, and Tabler shows that some of the sublimation loss can be reduced by snow fences; thus fencing is a viable watershed management technique in windy areas. The amount of snowfall is a principal component in the model used to determine the fencing height and spacing, as well as the amount of sublimation loss.

In order to design a snow fence system or to determine the benefits from snow fencing, it is necessary to estimate how much snow might be relocated, and how much water equivalent might be lost in the transport process. Tabler's (1973) model indicates that average values for the total amount of blowing snow at a point can be calculated from:

$$\overline{Q} = 1/2\ \overline{\theta}\ \overline{P}\ \overline{R}_m \text{ when } \overline{R}_m \geqq \overline{R}_c$$

where $\overline{Q}$ = average water equivalent of the blowing snow,

θ = average snow transfer coefficient or the fraction of precipitation in the form of snow which blows,

$\overline{P}$ = average water equivalent of precipitation in the form of snow,

$\overline{R}_m$ = maximum transport distance or maximum distance a snow particle will travel before it sublimates,

$\overline{R}_c$ = contributing distance;

and the amount of water lost through sublimation is:

$$\overline{Q}_e = \overline{\theta}\ \overline{P}\ (R - \overline{R}_m/2) \text{ when } R \geqq R_m$$

where $\overline{Q}_e$ = average water equivalent lost through evaporation or sublimation, and

R = upwind distance without major barriers ($R \geqq R_m$).

In order to make use of these relationships, it is necessary to have approximate values for the terms in the equations, and as can be noted, precipitation has a direct influence in the equations.

The measurement of precipitation in the form of snow is an observation task which to date has proven unsatisfactory. Rodda (1971) observes that "the problems of measuring snowfall by the use of gauges are acknowledged to be so great that a considerable amount of attention has been devoted to the development of alternative methods of measurement, particularly measurement of snow lying on the ground." Yet, there are problems of measuring the snow lying on the ground (Work, et al., 1965), problems of selecting a representative site for attempting to measure the snow lying on the ground (Bartos and Rechard 1973), and even problems of relating the amount of snow on the ground to the amount of precipitation. As an example of these latter two types of problems, during the winter of 1972-73, a year of above normal precipitation and below normal temperatures for southeastern Wyoming, the snow course on Pole Mountain near Laramie gave readings which indicated the snowfall was 3-, 4-, or even 500 percent of normal. Yet, precipitation measurements nearby were indicating that snowfall was about 150 percent of normal. Why this discrepancy? While not scientifically documented, there appear to have been three primary reasons: 1) the snow course is at the edge of a clearing used extensively by snowmobilers, and visual observations showed that the machines compacted the snow on the snow course; 2) the winter was colder than normal and therefore, more of the snow stayed on the ground to be measured at subsequent intervals; and 3) there was less wind than in most years which resulted in the snow staying on the ground and not sublimating. In this instance at least, the gage measurements were more representative of snowfall than was the snow course measurement.

Among the possible sources of error in the measurement of precipitation, especially snowfall, wind is the factor which by far overshadows the combined effects of all other influences (Kurtyka, 1953). Because of this fact, over the years there have been many attempts to shield precipitation gages from the adverse effects of wind. The literature is repleat with such studies, and Kurtyka (1953), Israelsen (1967), and Larson (1971) have summarized the extensive writing on the subject. It is the purpose of this paper to report on the design and utilization of a unique type of precipitation gage wind shield constructed with snow fencing material.

The basic concept for the project developed during the first year of a study investigating the possibility of using snow fences as gage shields. Initially, the snow fences were installed in the conventional manner with the gage placed downwind at the point of maximum wind reduction (about 5 x height). (It has been shown that this configuration will work if enough snow storage space is provided upwind [Rechard,1972].) During the following summer wind tunnel studies were made of various configurations of the "blow-fence", or inclined fence design, using geometric scaling only. Because of the difficulty of a rigorous similitude analysis, the wind tunnel studies did not attempt to quantify the design considerations. Tufts of string were used to observe the action of the air currents, and the shielding configuration which resulted in the least turbulence over the orifice of the gage was chosen for prototype testing. In this configuration the gage is placed in the center of two concentric rings of snow fencing. The inside ring is 10 feet (3.1 m) in diameter inclined at 45° with the bottom of the fence level with the bottom of the gage. The outside ring is 20 feet (6.1 m) in diameter inclined at 60° from the horizontal with the bottom of the fence 2 feet (0.6 m) higher than the bottom of the gage. The original design had the bottom of the Universal weighing recording gage 2 feet (0.6 m) above the ground.

Field tests indicated that the prototype acted very much like the model except that in the wind tunnel the floor was smooth, whereas in the field the rough ground and vegetation apparently caused turbulence and verticle wind flow between the gage and the Alter shield. To overcome this, the whole structure, gage and shield, was raised so that the bottom of the gage was 5 feet (1.5 meters) above the ground.

Two study areas have been used for field testing the design configuration which has been termed a "blow fence" or preferably, "Wyoming Shield." One study area is on Pole Mountain in the Medicine Bow National Forest, 17 miles (27 km) east of Laramie, Wyoming at an elevation of 8,100 feet (2,470 meters). Annual snow-fall amounts(October - May) at this site vary from about 6 to 16 inches (160 to 400 mm) water equivalent. The area is comprised of grasslands with scattered rock outcrops and is subject to frequent winds of relatively high speeds and prolonged duration. The terrain slopes gently down to the east (downwind) and there is an effective barrier to blowing snow about 1,100 feet (335 meters) west (upwind) of the gage location.

The other study area is the Snowy Range of the Medicine Bow Mountains in the Medicine Bow National Forest, 35 miles (56 km) west of Laramie, Wyoming, with an average elevation of 10,000 feet (3,170 meters). There are four specific sites in the Snowy Range Observatory ranging in elevation from 9,900 feet (3,020 meters) to 10,600 feet (3,230 meters).

In selecting the sites, an important consideration was the presence of well-protected, small openings in forests near the site to be used for the installation of a gage defined as the "standard." The assumption must be made that the small forest opening represents the best available shielding and that the gage so located will catch as close to the "true" catch as is presently feasible. This is only an assumption as Rodda (1971) emphasizes, "there is still no method of measuring to a known degree of accuracy the quantity of precipitation falling at a particular point on the earth's surface."

Even though a small forest opening is probably the best shielding available, there are problems in selecting the proper one.

The original "standards" for two of the sites on Snowy Range were selected after detailed field inspection by Institute and U. S. Weather Bureau personnel. Even so, at one site there was snow drifting into the gage and at another, the gage caught significantly less snow than was on the ground in adjacent openings. The difficulty experienced in locating suitable standard sites reinforces the need for a controlled, consistent standard, in place of having to rely on nature. Also, it can be noted in the precipitation table that at Knight Science Camp the unshielded and Alter shielded gages in a well-protected forest opening did not catch the same amount of precipitation. This would seem to indicate that even in a forest opening there are wind currents which affect the catch of gages. Mass curves of gage catch at two research sites, Pole Mountain and Knight Science Camp, are shown in Figures 1 and 2.

The results of the field studies of the Wyoming Shield indicate that gages without any kind of shield catch much less than the standard (even down to 10 percent of standard, and on the average only 30 to 50 percent of the standard catch), and while gages with only an Alter shield do much better, they still catch only from 65 to 75 percent of the standard. The Wyoming Shielded gage consistently catches within 10 percent of gages placed in small forest openings.

The consistency of the catch is shown in Figures 3 and 4 which are plottings of standard catch vs. Wyoming Shielded catch at Pole Mountain and Snowy Range Knight Science Camp, respectively. From these curves for 47 storms during the 1972-73 season, it can be seen that when related to the standard catch, Wyoming Shielded gage 14 on Pole Mountain has a correlation coefficient, R, of 0.97 and Standard Error, S, of 0.08, while unshielded gage 6 has an R of 0.88 and an S of 0.16. The total seasonal catch of gage 14 is within 3 percent of the standard, while gage 6 caught only about one-half the standard.

On Snowy Range at the Knight Science Camp, the Alter-shielded gage, in an unprotected site, 103A-2, compared quite well with the standard ($R = .97$, $S = .14$). This gives rise to the possibility of calibrating the Alter-shielded gage with a standard and then assuming that the gage would consistently catch a certain percentage, say 75 percent, of the amount it should have caught. This percentage seems to hold quite well for both the Pole Mountain and Snowy Range areas for seasonal totals; however on a storm by storm basis the relationship is not consistent. Also, the calibration would need to be confirmed for each location, so that in many cases a Wyoming Shielded gage would still be needed as a standard.

Thus, it is believed that in windy areas,existing gages should be calibrated against a gage within a "Wyoming Shield." and that for research or modeling purposes the Wyoming Shield should be used to define the precipitation input.

The authors would like to acknowledge the financial support of the National Weather Service (NOAA) for the research reported herein.

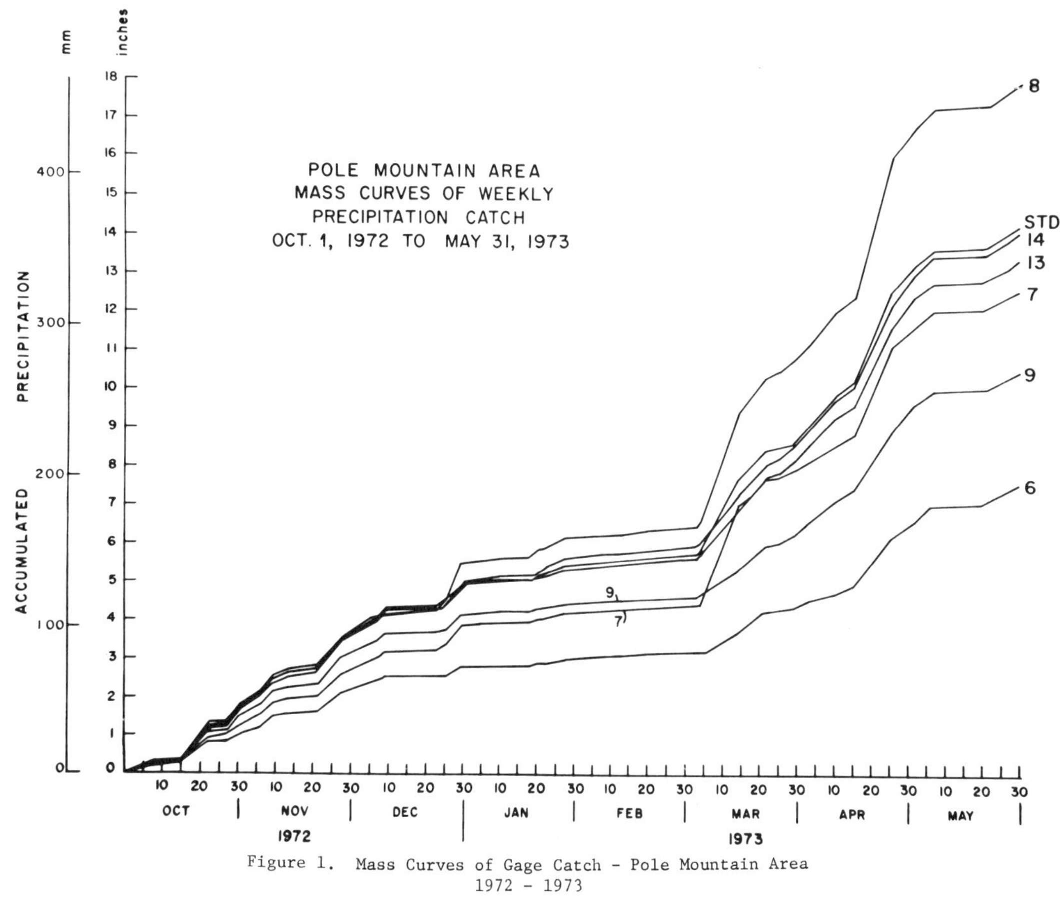

Figure 1. Mass Curves of Gage Catch - Pole Mountain Area
1972 - 1973

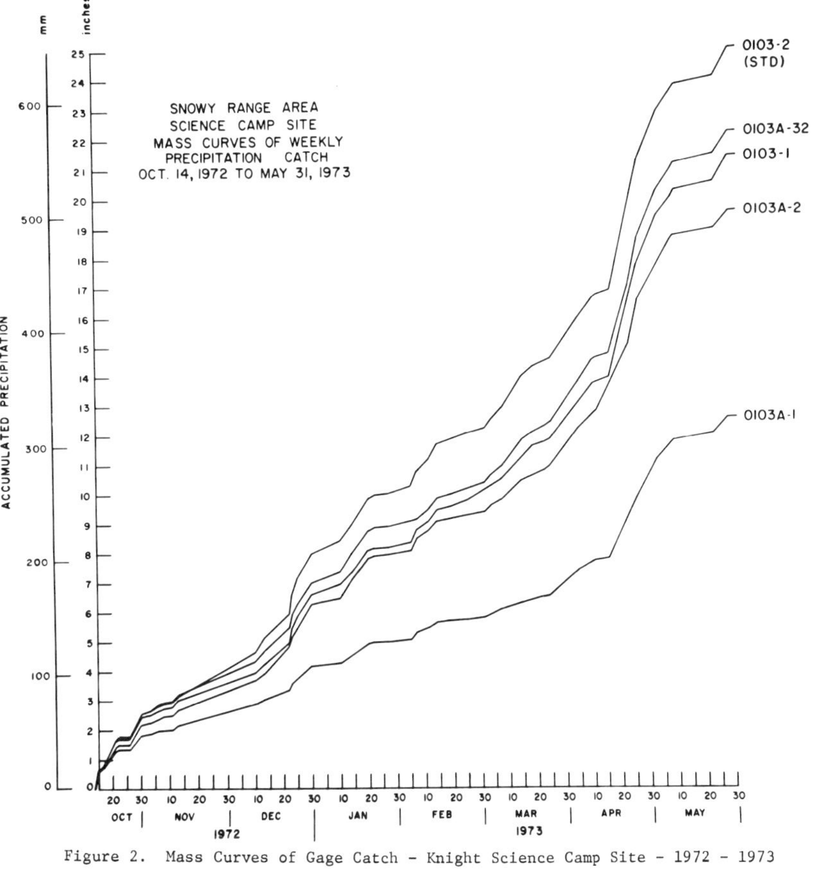

Figure 2. Mass Curves of Gage Catch - Knight Science Camp Site - 1972 - 1973

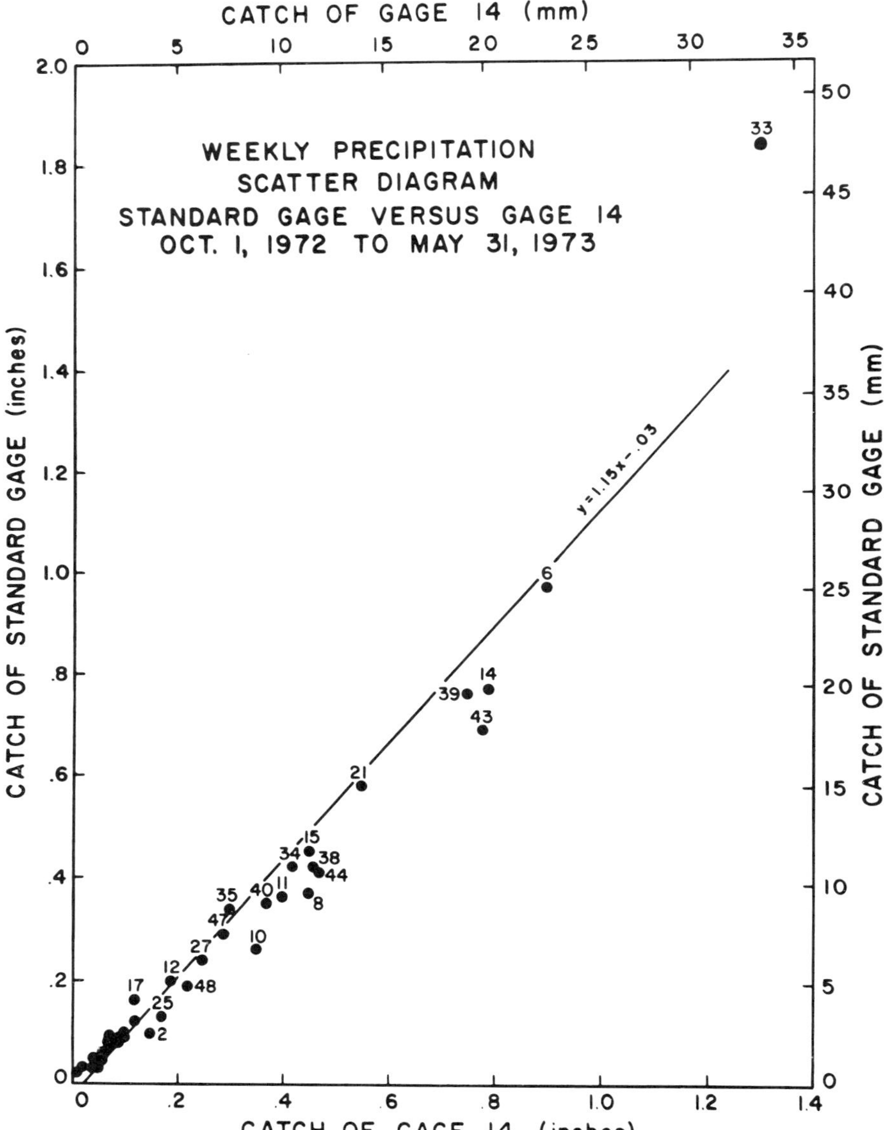

Figure 3. Plotting of Catch by Gage 14 vs. Standard - Pole Mountain Area - 1972 - 1973

7.4

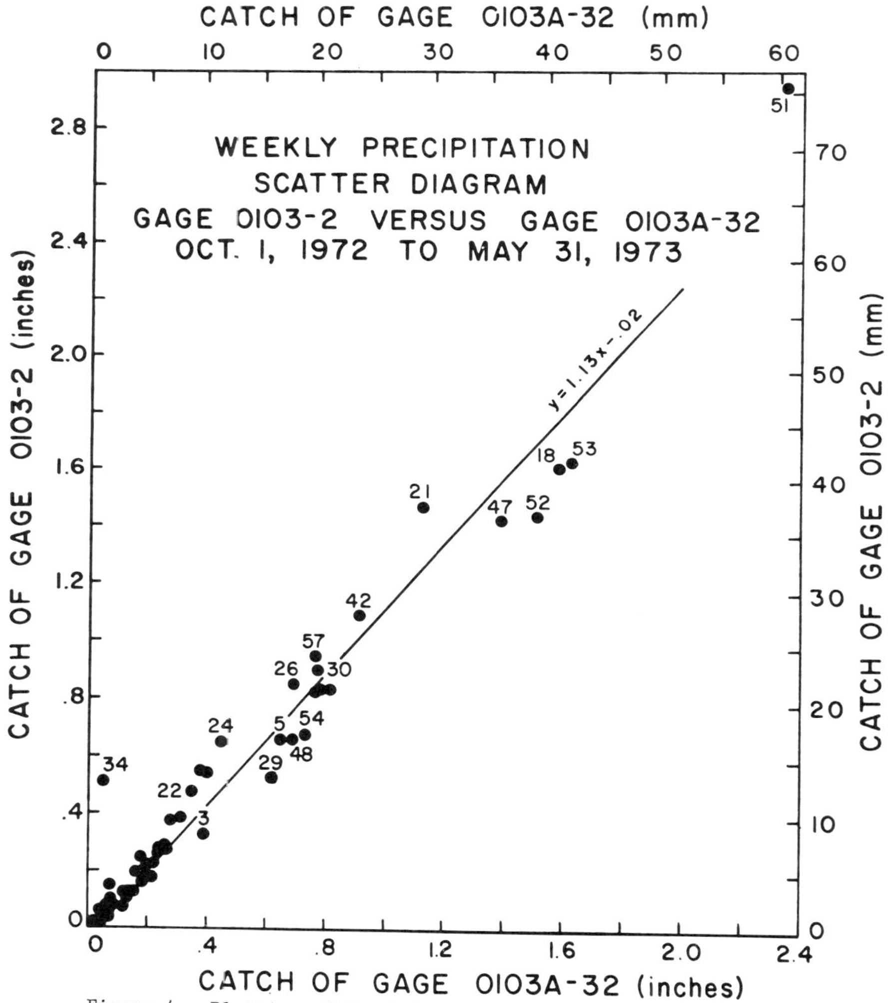

Figure 4. Plotting of Catch by Gage 103A-32 vs. Standard Knight Science Camp Site - 1972 - 1973

7.4

MONTHLY AND SEASONAL PRECIPITATION

Pole Mountain and Snowy Range Observatories, Wyoming

October 1972 - May 1973

Pole Mountain - Elevation 8,100 feet (2,470 m)

Mo.	Unshielded No.6		Alter Shield No.7		Wyo.Shield No.13		Wyo.Shield No.14		Standard	
	in.	mm.	in.	mm.	in.	mm.	in.	mm.	in.	mm.
Oct.	1.02	25.9	1.24	31.5	1.75	44.4	1.71	43.4	1.66	42.2
Nov.	1.09	27.7	1.34	34.0	1.78	45.2	1.90	48.3	1.78	45.2
Dec.	0.68	17.3	1.29	32.8	1.37	34.8	1.43	69.0	1.50	38.1
Jan.	0.22	5.6	0.32	8.1	0.46	11.7	0.63	16.0	0.54	13.7
Feb.	0.07	1.8	0.13	3.3	0.18	4.6	0.18	4.6	0.18	4.6
Mar.	1.31	33.3	3.61	91.7	2.64	67.1	2.69	68.3	3.26	82.8
Apr.	2.21	56.1	3.68	93.5	4.22	107.2	4.44	112.8	4.53	115.1
May	0.95	24.1	0.95	24.1	0.96	24.4	1.10	27.9	1.02	25.9
Total	7.55	191.8	12.56	319.0	13.36	339.3	14.08	357.6	14.47	367.5

Snowy Range - Knight Science Camp - Elevation 10,000 feet (3,048 m)

Mo.	Forest Opening				Windy - Open Area					
	Unshielded		Alter Shield		Unshielded		Alter Shield		Wyo.Shield	
	in.	mm.	in.	mm.	in.	mm.	in.	mm.	in.	mm.
Oct.P	2.48	63.0	2.58	65.5	1.85	47.0	2.20	55.9	2.60	66.0
Nov.P	0.54	13.7	0.59	15.0	0.31	7.9	0.49	12.4	0.61	15.5
Dec.P	3.57	90.7	4.82	122.4	2.00	50.8	3.58	90.9	3.78	96.0
Jan.	1.57	39.9	1.99	50.6	2.79	70.9	1.66	42.2	1.89	48.0
Feb.	1.56	39.6	2.21	56.1	0.80	20.3	1.41	35.8	1.89	48.0
Mar.	2.07	52.6	3.74	95.0	1.61	40.9	2.80	71.1	3.43	87.1
Apr.	7.54	191.5	6.90	175.3	3.71	94.2	4.34	110.2	6.42	163.1
May	2.01	51.0	2.11	53.6	1.39	35.3	2.97	75.4	1.96	49.8
Total	21.34	542.0	24.94	633.5	14.46	367.3	19.45	494.0	22.58	573.5

Snowy Range - Little Brooklyn Lake - Elevation 10,400 feet (3,170 m)

Mo.	Forest Open		Windy - Open Area							
	Alter Shield		Unshielded		Alter Shield		Blow Fence		Wyo.Shield	
	in.	mm.	in.	mm.	in.	mm.	in.	mm.	in.	mm.
Oct.P	2.96	75.2	1.44	36.6	1.80	45.7	2.76	70.1	2.78	70.6
Nov.	3.42	86.9	0.79	20.1	1.91	48.5	3.64	92.5	3.33	84.6
Dec.	3.62P	92.0	1.59	40.4	2.11	53.6	5.06	128.5	-	-
Jan.	2.53	64.3	1.22	31.0	1.01	25.6	2.44	62.0	2.61	66.3
Feb.	2.40	61.0	0.66	16.8	0.93	23.6	2.12	53.8	1.81	46.0
Mar.	4.62	117.4	1.28	32.5	1.89P	48.0	4.13	104.9	4.12	104.6
Apr.	7.07	179.6	2.85	72.4	3.46P	87.9	6.63	168.4	4.45P	113.0
May	2.67	67.8	1.13	28.7	1.39	35.3	2.27	57.7	3.84	97.5
Total	29.29	744.0	10.96	278.4	14.50	368.3	29.05	737.9	22.94P	582.7

P = Partial

REFERENCES

Bartos, L. R. and P. A. Rechard, 1973. An Evaluation of Areal Reconnaissance and Index Snow Sampling Techniques on a Small Subalpine Watershed. Paper presented at 41st Western Snow Conference, Grand Junction, Colorado, April 17-19, 1973.

Israelsen, C. E., 1967. Reliability of Can-type Precipitation Gage Measurements - A State-of-the-Science Study. Utah Water Research Lab. Report No. 2, Utah State University, College of Engineering, Logan, Utah. 73 p.

Kurtyka, J. C., 1953. Precipitation Measurements Study. Illinois Water Survey Report of Invest. No. 20, State Water Survey Division, Urbana, Illinois. 178 p.

Larson. L. W., 1971. Precipitation and Its Measurement: A State of the Art. Water Resources Research Institute Series No. 24, University of Wyoming, Laramie, Wyoming. 74 p.

Leaf, C. F. and G. E. Brink, 1973. Computer Simulation of Snowmelt within a Colorado Subalpine Watershed. USDA Forest Service Research Paper RM-99. Rocky Mountain Forest and Range Experiment Station, Fort Collins, Colorado. 22 p.

Leaf. C. F. and G. E. Brink, 1973. Hydrologic Simulation Model of Colorado Subalpine Forest. USDA Forest Service Research Paper RM-107. Rocky Mountain Forest and Range Experiment Station, Fort Collins, Colorado. 23 p.

Rechard, P. A., 1972. Are Existing Measurements of Precipitation Input to Watersheds Adequate for Management Decisions? National Symposium on Watersheds in Transition, Fort Collins, Colorado, June 19-22, 1972. 98-106.

Rodda, J. C., 1971. The Precipitation Measurement Paradox - The Instrument Accuracy Problem. Reports on WMO/IHD Projects, World Meteorological Organization, No. 316. 42 p.

Schmidt, R. A., Jr., 1972. Sublimation of Wind-Transported Snow: A Model. USDA Forest Service Research Paper RM-90. Rocky Mountain Forest and Range Experiment Station, Fort Collins, Colorado. 24 p.

Tabler, R. D., 1973. Evaporation Losses of Windblown Snow, and the Potential for Recovery. Paper presented at 41st Western Snow Conference, Grand Junction, Colorado, April 17-19, 1973.

Work, R. A., _et al_., 1965. Accuracy of Field Snow Surveys, Western United States, Including Alaska. U. S. Army Cold Regions Research and Engineering Laboratory Technical Report 163. 43 p.

THE SNOW MOISTURE INTEGRATOR

George R. Alger
Michigan Technological University

Henry S. Santeford
USNC/IHD, National Academy of Sciences

ABSTRACT

The Snow Moisture Integrator is a device developed at Michigan Technological University for automatically measuring and recording changes in moisture content of a snow sample exposed to ambient weather conditions. Changes in moisture content resulting from additional snowfall, condensation onto the snow sample, or evapo-sublimation from the sample may be measured to 0.001 inches (0.0254 mm) water equivalent.

The device has been field tested in the relatively humid Lake Superior region of Upper Michigan and the much dryer alpine region of central Colorado. The utility of the device for measuring light snowfalls, evapo-sublimation losses, and condensation increases was shown at both installations.

The concept of a device that could measure and record changes in moisture content of a snow sample was generated during a two year (1968-70) snow study in the Upper Peninsula of Michigan. Extremely large increases in snowpack water equivalent were noted on several occasions (in one instance 41% increase) without any precipitation. Also, unexpectedly large spring runoff amounts were measured relative to both the water equivalent of the snowpack at the beginning of the melt season and to the seasonal precipitation (Santeford, et. al, 1972). These observations stimulated the development of an inexpensive device that might accurately measure and record moisture exchanges between the snowpack and the atmosphere.

The Snow Moisture Integrator consists of two vertical conduits connected by a horizontal tube. The vertical conduits and the horizontal connecting tube contain a non-freezing liquid, such as ethylene glycol, which provides for a hydraulic response transfer from one vertical conduit to the other. A platform holding the snow sample is connected to a vertically positioned plunger located in one of the vertical conduits. The plunger is free-floating, yet its diameter is sufficiently close to that of the containing conduit that small vertical movements of the plunger (resulting from a change in weight of the snow sample) result in large vertical displacements of the fluid surface. The second vertical conduit serves as a stilling well for the float attached to a simple stage recorder. A sketch of a typical field installation is shown in Figure 1.

A wide range of sensitivity levels can be obtained by varying the size of the various components. One such combination consists of a 3 foot (0.914 m) by 3 foot (0.914 m) sample container mounted on top of a 5 1/2 inch (13.97 cm) diameter plunger. The vertical conduit containing the plunger has a 6.0 inch (15.24 cm) inside diameter while the stilling well housing the stage recorder float has an inside diameter of 1 1/2 inches (3.81 cm). With this arrangement a change of 0.01 feet (0.30 cm) on the stage recorder corresponds to an equivalent weight change of the sample of 0.0012

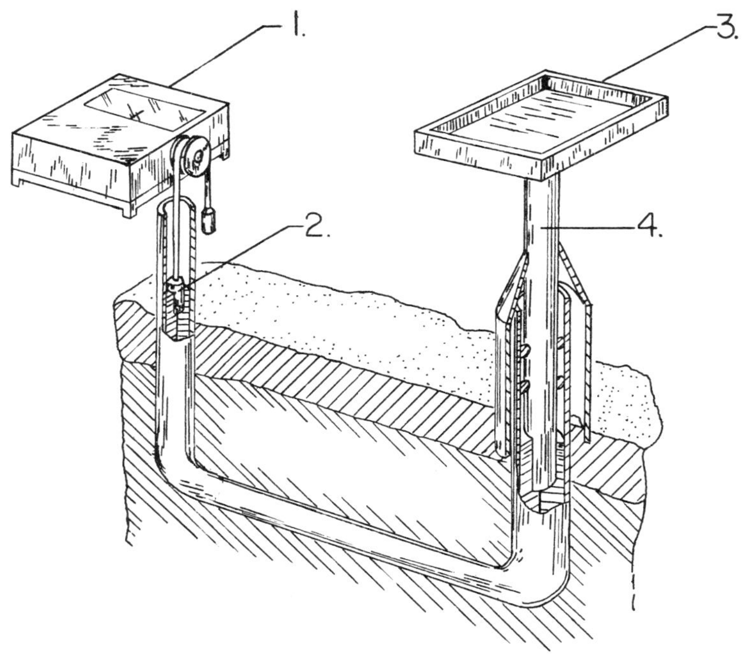

Figure 1. Sketch of the snow moisture integrator.

inches (0.0030 cm) of water. Thus even very small moisture changes resulting from light snowfalls, frost, condensation, or evapo-sublimation can be easily detected on the stage recorder chart.

When the device is constructed of a plastic material such as plexiglass or PVC, and the hydraulic fluid (such as ethylene glycol) is shielded from direct solar radiation, a correction for temperature changes is not necessary under most field conditions. The device is not truly self compensating, in that a device of the dimensions mentioned above will indicate a change on the recorder chart of 0.005 feet (0.15 cm) when exposed to a temperature change of 40°F (22.2°C). This error in the recorder readings -- one half of the smallest chart division -- is generally much less than the indicated changes on the recorder and can be neglected for most field installations.

In order for the snow moisture integrator to perform properly, care must be taken during the site selection and field installation. The site should have the same basic characteristics as those required for the installation of a precipitation gauge. During the installation, the lower portion of the vertical conduits is buried in the ground and provides the necessary support mechanism. Care must be taken to ensure that the conduits are precisely vertical, and that the sample container is at an appropriate elevation. The sample should be as close to the surface of the snowpack as possible. The elevation of the sample can be regulated to a moderate degree by regulating the volume of fluid contained in the system. When the volume of fluid is increased, the plunger supporting the sample assembly is displaced upward. By decreasing the volume of fluid, the sample assembly can be lowered. Such adjustments in the elevation of the sample assembly are limited to a range of approximately 2 feet (60 cm).

The snow moisture integrator has been field tested in the Upper Peninsula of Michigan (near Houghton, and also near Ishpeming) and in the Colorado Rocky Mountains (near Climax). Because of the different climatic conditions of the various sites, the usefulness of the device was not the same at all locations.

In the Michigan studies the device was placed in a small opening in the hardwood forest and operated during the late winter and early spring. When the maximum daily temperature was below 30°F (-1°C) excellent correlation existed among the readings on the snow moisture integrator, a recording precipitation gauge, a non-recording precipitation gauge, and snow surveys performed in the general vicinity of the gauges. (A special lysimeter study was being conducted at the same site.) The integrator produced an excellent record of the snowfalls as well as the evapo-sublimination loss from the sample to the atmosphere. In these studies transport by the wind was seldom a factor to be considered, the moisture exchange between the snow sample and the atmosphere was easily discernible on the integrator's recorder chart. On clear sunny days with low temperatures and low relative humidities, an average moisture exchange from the snowpack to the atmosphere of 0.03 inches/day (0.08 cm/day) was indicated. On several occasions moisture losses from the snow sample as large as 0.15 inches/day (0.38 cm/day) were recorded. Since these larger losses occured on very cold days when wind conditions at the gauge site were less than 5 mph (2.2 m/s), the indicated loss could only be attributed to a vapor exchange between the sample and the atmosphere and not a transport of solid particles by the wind.

During the melt season in the Michigan studies, large amounts of condensation onto the snow sample were recorded. In one instance an increase of 0.29 inches (0.74 cm) of water equivalent was recorded in a two day period. This increase occured without precipitation and thus was not recorded on the standard precipitation gauges. However, snow surveys performed in the immediate area of the integrator (and for the watershed as a whole) indicated an even larger increase than that shown on the integrator. The discrepancy between the integrator readings (0.29 inches) and the snow surveys adjacent to the integrator (0.57 inches) was attributed to the distance between the snow sample and the snow surface. The snow sample attached to the integrator was located approximately 2 feet (0.6 m) above the snowpack surface and as such was not exposed to the same environmental conditions. (Subsequent studies have shown that the location of the sample in regard to the snow surface is a prime factor in determining the correlation between the integrator response and the actual snowpack conditions.)

In the Colorado studies, wind -- and thus transport of snow particles -- was a much greater factor in interpreting the indicated response of the integrator. Similar evapo-sublimation losses were recorded yet could not be substantiated (from independent measurements) to the same degree as with the Michigan studies. However, the integrator did prove to be a very useful device for measuring the moisture content of the light snowfalls and frost which frequently occured during the early morning hours. No significant amounts of condensation were indicated.

REFERENCES

Santeford, H.S., G.R. Alger, and J.G. Meier, 1972. Snowmelt energy exchange in the Lake Superior region. Water Resources Research, Vol. 8, No. 2, p. 390-397.

Santeford, H.S., 1973. Management of windblown alpine snows, A Ph.D. Dissertation. Colorado State University, Fort Collins, 182 p.

ELECTRONIC MEASUREMENTS OF SNOW SAMPLE WETNESS

William I. Linlor
Ames Research Center, NASA

James L. Smith
Pacific Southwest Forest and Range Experiment Station, Forest Service, USDA

ABSTRACT

Several electronic methods described here can readily measure the wetness of snow. The first method is based on the change in capacitance of a snow sample before and after freezing, which is directly related to the amount of moisture initially present. The second is based on the losses in a resonating electrical circuit (or Q) and depends on the amount of liquid-phase water in the snow being measured. Field tests for both systems have been made to demonstrate the principles. Preliminary calibration curves are presented. A field unit is described that can obtain in situ values for snow wetness.

The amount of liquid-phase water in snow ("wetness") is an important factor in runoff prediction and water resource management. Various systems (calorimeters, centrifuges, etc.) are available for measuring wetness. Although each method has its proponents, none has been uniformly accepted. This paper describes two electrical methods for measuring the wetness of snow: one based on measuring the capacitance of a sample before and after freezing it and the other on in situ measurement of the dielectric loss of a sample in a high-frequency field.

The capacitance of a given test unit is proportional to the net dielectric constant of the medium between the electrodes. For dry snow – a mixture of air and ice – the factors that affect the net dielectric constant include the individual dielectric constants of the air and the ice, the relative proportions of each, the temperature, the oscillation frequency at which the measurements are made, and the "form factor" (also called "formzahl"), which depends on the shapes and orientations of the crystal. For wet snow, the situation is more complicated, even though the temperature is not a variable, being essentially 0°C because of the effect of melting. The dielectric constant of pure water at 0°C is 87.7, but when the water is in contact with ice crystals, intramolecular forces can affect the dipole moment of the water. Nevertheless, the dielectric constant of snow is greatly increased by the presence of liquid-phase water, and this fact has been the basis for previous methods of determining snow wetness by Gerdel (1954) and Ambach and Howorka (1966) from capacitance measurements. Gerdel's equation for the dielectric constant of a unit volume of wet snow is

$$k_s = k_i V_i + k_w V_w + k_a V_a \qquad (1)$$

The individual dielectric constants and volumes of snow, ice, water, and air are represented, respectively, by k and V with subscripts s, i, w, and a.

Gerdel states: "The capacitance meter readily detects the presence of water in the snow in amounts of a fraction of one per cent. It is not possible to produce a uniform artificial mixture of snow and water with a known low free water content to serve as a standard and since no satisfactory precision method for measuring the liquid water present in a natural snow cover is available it has not been possible to calibrate the meter for all snow types."

Gerdel used his capacitance meter to measure the relative amounts of water in a given snowpack under various wetness conditions.

Ambach and Howorka give equations for the dielectric constant as a function of snowdensity and wetness:

$$k_s(\text{dry}) = 1 + 2.2g \quad (g \text{ in gm/cm}^3) \tag{2}$$

$$k_s(\text{wet}) = 1 + 2.22g + 0.236W \tag{3}$$

The volume percent of liquid-phase water is represented by W and k_s represents the dielectric constant of snow. However, the effect of the "form factor" of the snow is not included in equations (2) and (3).

The method proposed here to obtain the wetness of snow involves measurements of capacitance before and after a sample is frozen. The unit can consist of two electrodes separated by dielectric spacers that also serve as side walls. All four edges that enter the snow are sharpened. Since the effect of the form factor on the capacitance is present in both measurements, in the subtraction process the net contribution vanishes. In other words, only the capacitance attributable to the original wetness is manifested by the value before freezing minus the value after freezing. After the unit has been completely inserted into the snow and is uniformly filled, it is carefully removed and its capacitance measured. The unit is then placed in a chamber maintained at a temperature below freezing (e.g., at −20°C) to freeze the snow by heat conduction through the electrodes. A convenient, portable, freezing apparatus can consist of an insulated box with dry ice (solid carbon dioxide). After it is completely frozen, the test unit is removed and its capacitance remeasured. Although knowledge of the density is not required, it can be obtained for the sample by weighing the snow since the volume is known.

The information thus obtained includes the density of the snow, the dielectric constant of the snow plus liquid-phase water, and the dielectric constant of the snow plus frozen-out water. The method was tested during the spring of 1973 at the Central Sierra Snow Laboratory (CSSL), Soda Springs, California. For example, a snow bank was sprayed with 0° water and allowed to drain for about an hour. The snow was then placed between electrodes separated by 1 inch (2.54 cm), with dimensions of 12 inches X 16 inches (30.5 cm X 40.7 cm), having a capacitance of 80 micro-micro-farads (mmf) when empty. With the wet snow present, the capacitance was 193 mmf. After freezing in a dry-ice box, the snow reached a temperature of about −20° C, and the corresponding capacitance was 136 mmf. From these data, the dielectric constant for the wet snow was calculated to be 2.41, for the dry snow it was 1.70, leaving a contribution of 0.71 for the dielectric constant of the liquid-phase water initially present in the snow. The density of the snow was 0.42 gm/cm³. To obtain the volume percent of liquid-phase water initially present in the snow, suitable calibration curves relating the wetness to the increase in dielectric constant are necessary. (This research is planned for the winter of 1973-74.)

Alternative methods of freezing the snow sample were also tested, such as circulating cold air through the snow sample with perforated electrodes so that the dielectric constant could be measured during the freezing process. An ordinary household vacuum cleaner provided sufficient air flow through a snow sample in the form of a cylinder having a diameter of 10 inches (25.4 cm) and a height of 1 inch (2.54 cm). The air was passed through a tube containing dry ice. However, the heat-conduction method of freezing appears to be more convenient and simpler.

Some of the problems encountered in producing dielectric constant versus wetness curves were explored at CSSL. A walk-in refrigerated room was maintained at 0°C, where all weighing and electrical measurements were made. A capacitor was employed having aluminum electrodes 13 inches X 13 inches (33.0 cm X 33.0 cm) separated by a Plexiglas frame 1 inch (2.54 cm) high and 0.5 inch (1.27 cm) thick, giving a working volume of 12 inches X 12 inches X 1 inch (30.5 cm X 30.5 cm X 2.54 cm or

2.36×10^3 cm^3). The empty weight of the unit was 1465 gm, and its capacitance was 75 mmf. Dry snow at −3.5°C was placed in the unit; it weighed 1267 gm, yielding a dry density of 0.536. The capacitance of the unit with the dry snow was 104 mmf, for which the dielectric constant was 1.39.

Next the snow was removed from the unit and placed in a mixing bowl, where 0°C water was sprayed on it. After it was thoroughly mixed by insulated gloved hands, the snow was replaced in the capacitance unit and reweighed. The increase in weight in grams, due to the water added, multiplied by 100 and divided by the volume of 2.36×10^3 cm^3 represents the volume percent of water added. The capacitance of the unit with wet snow present minus the capacitance of the unit with dry snow present divided by the capacitance of the empty unit represents the increase in dielectric constant produced by the liquid-phase water in the snow. Such data were obtained in successive steps (water added, mixed, weighed, and capacitance measured) and are shown in Figure 1 where the volume percent of water added is the abscissa and the increase in dielectric constant is the ordinate. The measurement frequency was 3.84×10^6 Hz. The measured capacitance values have an uncertainty of about ±1 mmf, which is equivalent to ±0.02 in the dielectric constant.

The experimental data include the effect of the initial temperature of the snow (−3.5°C). The specific heat of ice is 0.5 Cal/gm/°C. With the density of 0.54 gm/cm^3, the amount of heat to bring the snow to 0°C is 0.95 Cal/cm^3. The heat of fusion of pure water is about 80 Cal/gm. To heat the snow to 0°C requires about one volume percent of water freezing to ice. This implies that, for accurate results, the snow should be very close to 0°C initially.

Other effects to be considered include unmeasured heat input to the snow, such as the effect of heat conduction from the gloved hands, the work of mixing, human body radiation, breath exhaling, etc. These have an opposite effect to that discussed in the preceding paragraph. For future calibrations using the mixing method, a "control" sample of snow will be subjected to the same operations and environment as the test sample, except no water will be added.

If it is assumed that Figure 1 can be applied to the snowbank example previously described, for which a change in the dielectric constant of 0.81 was measured upon freezing the sample, the wetness value of 5.5 volume percent is obtained; for a density of 0.42 gm/cm^3, this corresponds to a wetness of 13 percent by weight (i.e., grams liquid water/grams dry snow multiplied by 100).

So far, the discussion of calibration has included only the mixing of dry snow with known amounts of water. Clearly, comparisons should be made with measurements using other systems, such as calorimeters, centrifuges, etc.

A more elegant method of calibration (not yet tested) could be based on the use of "heavy water" (D_2O) in a refrigerated room. With a large volume of snow at slightly below freezing (about −0.1°C), a known amount of heavy water would be sprayed so that the upper portion of the snow would become saturated. After suitable time intervals, samples from various layers of the snow would be selected, and capacitance values measured before and after freezing. These changes in dielectric constant would then be related to the heavy water present in each sample, using a mass spectrometer or other nuclear abundance instrument to obtain the relative amounts (i.e., grams D_2O/grams H_2O). The homogeneity of the D_2O distribution may be similarly evaluated by dividing the snow sample from the test unit into small portions and measuring them on the nuclear instrument.

Additional moisture measurements have been performed using foam rubber as the medium to which water is added to determine the reproducibility and precision of results. No freezing or melting considerations are involved. Results are shown in

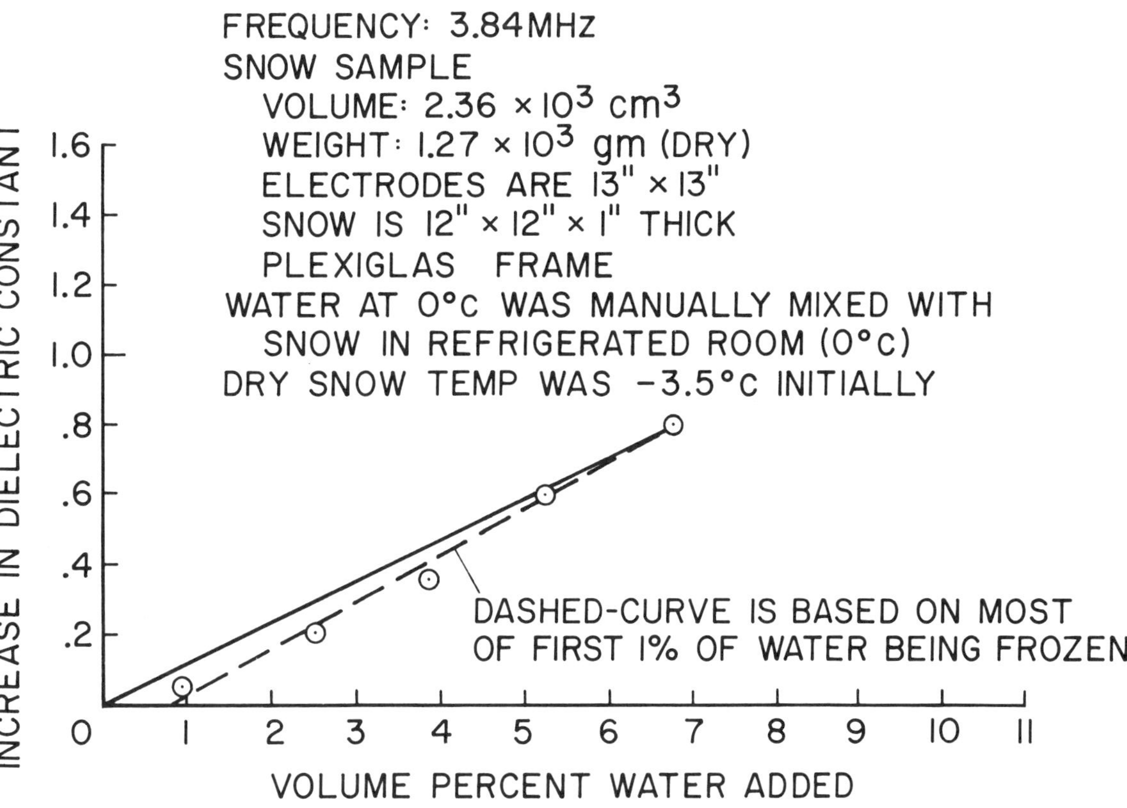

Figure 1. Increase in the dielectric constant of snow produced by water added.

Figure 2, where the volume percent of water added is the abscissa and the increase in dielectric constant is the ordinate. The foam rubber had a dry density of only 0.0180 gm/cm^3 and a dielectric constant of 1.082. The dimensions of the foam rubber and the aluminum electrodes were 13 inches × 13 inches (33.0 cm × 33.0 cm) with a height of 1 inch (2.54 cm), giving a volume of 2.73×10^3 cm^3. The frequency used was 3.68×10^5 Hz. Runs were made on successive days, as indicated by the X, O, and Δ points.

Figures 1 and 2 show that a given volume percent of water (e.g., 5%) added to foam rubber did not increase the dielectric constant much more than when the same volume percent of water was added to snow (0.72 and 0.60, respectively). Because the density of foam rubber (0.018) was much less than the density of snow (0.42) for the samples tested, one might have expected that the moisture in the foam rubber would exhibit more nearly its pure-state dielectric constant value of 87.7 instead of the effective value of only about 15.

It is concluded from the preliminary calibration curve for snow and the comparison curves for foam rubber that the wetness of snow can be measured conveniently and accurately with the system based on the capacitance change caused by freezing the snow sample.

IN SITU MEASUREMENTS BASED ON "QUALITY FACTOR"

A disadvantage of sampling methods for measuring snow wetness is that a pit must be dug to obtain the samples; this is difficult, time-consuming, and expensive for routine data acquisition. For this reason, an alternative method based on "quality factor" (hereafter termed Q) measurement is described below, which could be used in conjunction with the customary snow-tube measurements on established snow courses. The Q as well as the capacitance (hereafter termed C) of the snow can be measured with the same set of electrodes. The term "capacitance unit" will be used to describe the electrodes plus an intervening dielectric, even though the unit will be used to measure the Q.

Because the cylindrical hole left by the snow tube is "available at no additional cost," we have designed a cylindrical capacitance unit that can be attached to a snow tube and inserted into the snow. The snow tube itself acts as one electrode, having a diameter of 2 inches (5.08 cm), and is surrounded by a coaxial sleeve acting as the other electrode 4 inches (10.2 cm) in diameter and 4 inches (10.2 cm) long. As this cylindrical capacitor is lowered into the snow, measurements are made at each desired depth; the sampling is done in situ, without extracting the snow sample or freezing it.

Let us now consider the principle of operation for the Q measurement. For a series circuit containing resistance R, inductance L, and capacitance C, when oscillating at a frequency F, $Q = 2\pi FL/R = (2\pi FCR)^{-1}$. A commercially available instrument called a Q-Meter gives readings for Q and C when a capacitance unit is connected to its terminals.

The Q is proportional to the ratio of energy storage to energy dissipation per cycle in the resonating circuit. If the energy dissipation is increased, the Q decreases. When the dielectric between the plates of a capacitor is snow, most of the dissipation in a high-frequency field is produced by the presence of liquid-phase water, thus the Q is high for dry snow and low for wet snow. From a measurement of the Q for a given snow as the dielectric, and with a suitable calibration curve, it appears that the wetness may be determined immediately, without the need for a freezing cycle. (Experimental proof of this hypothesis will be sought during the 1973-74 winter season at CSSL.)

Preliminary measurements of the relationship of the Q to snow wetness have been made in connection with the calibration curve already described, for which known

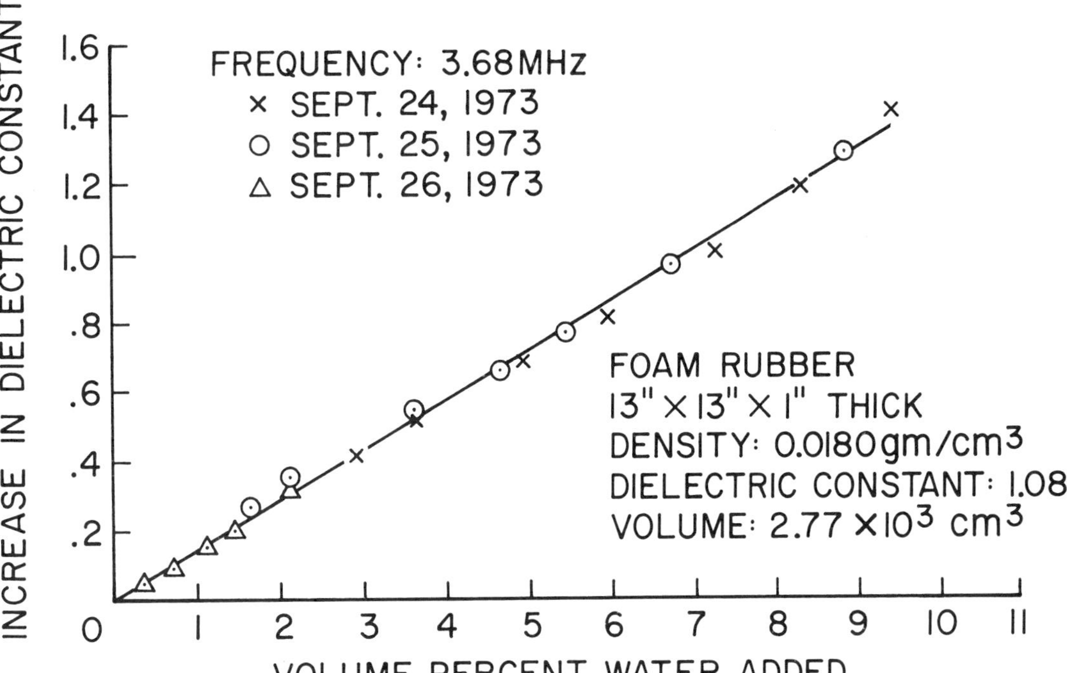

Figure 2. Increase in the dielectric constant of foam rubber produced by water added.

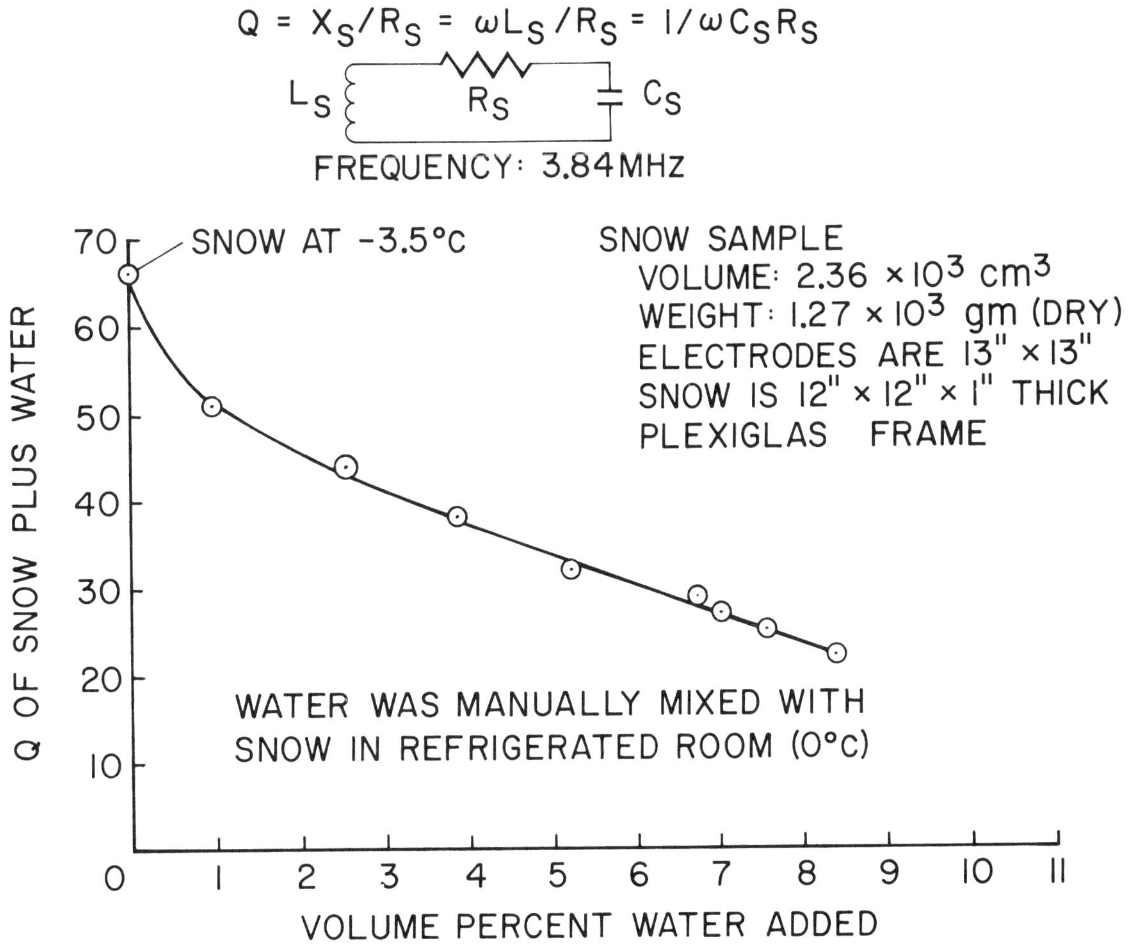

Figure 3. Dependence of Q factor of snow on water added.

amounts of 0°C water were manually mixed with initially dry snow. Readings for C and Q were taken for each moisture condition; the C dependence has previously been discussed.

In Figure 3, values for Q for dry snow and progressively wetter snow are plotted versus volume percent of water added. Q decreases rapidly as water is added, then it exhibits a more or less uniform rate of decrease in the range of 2 to 8 volume percent of water added.

Because Figure 3 is the only experimental curve to date for snow which shows the dependence of Q on known amounts of wetness, tests were run in the laboratory with foam rubber as the dielectric in a capacitor. As described before, the foam rubber had a dry density of 0.018 gm/cm^3, and a dielectric constant of 1.082. The electrodes and foam rubber were 13 inches × 13 inches (33.0 cm × 33.0 cm) with a height of 1 inch (2.54 cm), giving a volume of 2.77×10^3 cm^3. The frequency used was 3.68×10^6 Hz. The values of Q versus volume percent of water added are shown in Figure 4.

Although the results to date have been quite encouraging, this method of measuring snow wetness must be tested under the normal variety of field conditions before firm conclusions can be drawn. A small amount of conductive impurity in snow, for example, may affect the Q values significantly. However, the use of multiple frequencies for the measurements may provide additional information so that effects produced by liquid-phase water may be identified in a suitably precise and convenient way.

For both methods of wetness measurement described here, a *functional calibration* approach may be useful: the wetness measurements can be correlated with measured runoff data during the ripening and saturated melting episodes.

REFERENCES

Ambach, W., and F. Howorka, 1966. Avalanche activity and free water content of snow at Obergurgl. International Symposium on Scientific Aspects of Snow and Ice Avalanches, pp. 65-72.

Gerdel, R. W., 1954. The transmission of water through snow. Transactions of the American Geophysical Union, Vol. 35, No. 3, pp. 475-485.

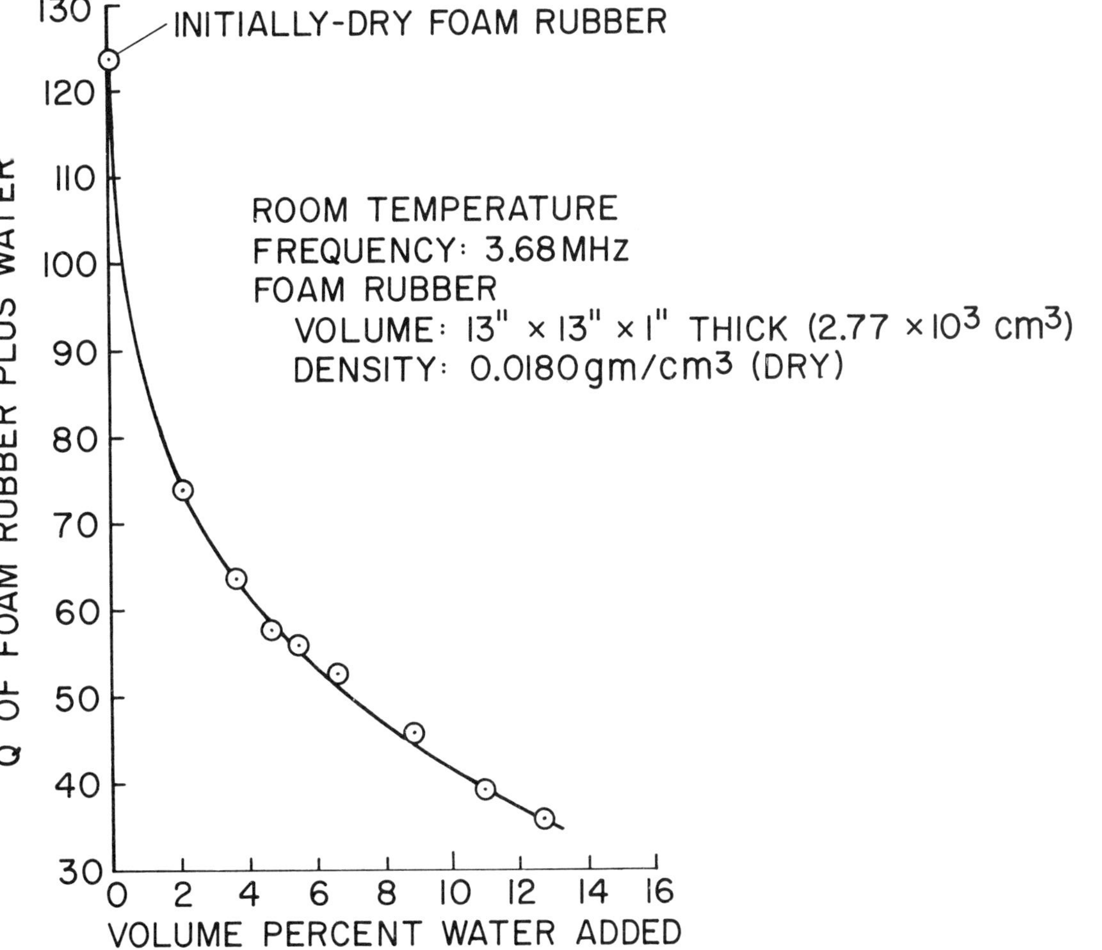

Figure 4. Dependence of Q factor of foam rubber on water added.

MICROWAVE PROFILING OF SNOWPACK FREE-WATER CONTENT

William I. Linlor
Ames Research Center, NASA

Mark F. Meier
U. S. Geological Survey, Tacoma, USDI

James L. Smith
Pacific Southwest Forest and Range Experiment Station, Forest Service, USDA

ABSTRACT

A microwave system is proposed to measure the amount of liquid-phase water in a snowpack operating in the range 1 to 10×10^9 Hz. Attenuation of the beam between source and receivers is produced by the water in the snow. The relationships of frequency, distance, and volume percent of water are calculated for an assumed detector sensitivity, together with the estimated cost of a representative system. A laboratory test is described that shows the attenuation for snow at maximum wetness at 9.35×10^9 Hz. A configuration is proposed that involves a vertical tube containing microwave and radioactive sources and another vertical tube containing microwave and γ-ray detectors, so that density and wetness profiles are obtained simultaneously over essentially the same path. With both density and free-water profiles available, the condition of the snowpack can be accurately determined and, combined with meterological information, useful predictions can be made regarding the melting and water discharge rates from snowpacks.

For snowpack water resource management to be effective, time-progessive measurements of water equivalent and the amount of liquid-phase water ("free-water content" or "wetness") are required. The amount of wetness, which can range from zero to a maximum of about 15 to 20% or more by weight, is an important quantity for predicting the melting rate and runoff. Present methods of measurement (such as calorimeters or centrifuges) require sampling and are difficult to automate. This paper shows that microwave techniques can be used in remotely operated, data-gathering systems to obtain in situ profiles of snowpack free-water content. The measurement is based on the attenuation produced by liquid-phase water when present in snow.

The proposed microwave system may use various frequencies from 10^9 to 10^{10} Hz. From both the theoretical and experimental points of view, electromagnetic (EM) measurements of free water in snow in this range can present very complicated problems because of the anomalous dielectric dispersion produced by the relaxation effects of the water dipoles, the molecular binding of water and snow molecules, scattering from snow particles or layers, and subtle instrumental errors such as mode mixing.

Any attempt to describe the proposed system in fine detail would be premature until field tests are completed for actual in situ measurements and the results compared to those obtained from sampling. However, the general characteristics of the system can be outlined based on present knowledge. The systems to be described are expected to have an accuracy better than obtainable from existing techniques, and thus would be useful for snowpack water resource management purposes.

The system proposed here is "free-wave" propagation between sources and receivers. In its simplest form, the system consists of a source that can be moved vertically in a dielectric tube, and a receiver about a meter away, also arranged so that it can be moved vertically in its dielectric tube. Thus a horizontal beam would scan the snowpack when the source and receiver are moved vertically in synchronism. The arrangement would resemble physically and in principle the snowpack density profile system of Smith *et al* (1972). It may be possible to combine the microwave source with a radioactive source (described by Smith *et al.*) in a common vertical tube, together with a microwave receiver and a γ-ray detector in the other tube, so that a density profile and a wetness profile are obtained simultaneously over essentially the same path. Such measurements would serve to determine the snowpack properties without physically affecting it in any way. With both density and free-water profiles available, the condition of the snowpack can be determined. Combined with meteorological information, predictions can be made regarding the melting and water discharge rates from snowpacks.

An EM system with a single source and receiver may encounter such difficulties as variable reflection at the source/snow interface (when time-progressive changes in density and wetness alter the effective dielectric constant), source power variations, receiver sensitivity drift, insufficient dynamic range, etc. A possible solution to some of these problems is to use a multiplicity of receivers, spaced at different distances from the common source. This would avoid the initial interface reflection difficulty because ratio measurements from two receivers at known separation permits the calculation of attenuation. Spacing the receivers at logarithmic intervals may reduce the dynamic range problem: e.g., receivers at 30, 90, 300, 900 cm, etc. The selection of actual receiver spacing would depend on the wavelength and the wetness.

The effect of scattering the beam, similar to absorption in regard to loss of intensity, must be determined. This may be done by taking measurements at night or during subfreezing weather, when the wetness is low. Under such conditions, scattering mechanisms of dry snow or ice layers would have their full effects, but the free-water beam loss would be greatly reduced. In this way, the two types of beam losses could be determined.

Now consider the quantitative aspects of the proposed system. Assume that the snowpack is melting, so the temperature is essentially uniform at 0°C. From the "relaxation spectrum" of river water (i.e., having impurities equivalent to 0.01 N aqueous NaCl) at 0°C (discussed by Von Hippel (1954)), the attenuation versus frequency is calculated (in dB/cm), as shown in Figure 1. If the water is dispersed in the snow, the attenuation would not necessarily be proportional to the volume percent of the wetness; however, making such an assumption would yield a reasonable estimate. Thus, at a frequency of 10^{10} Hz, each volume percent of free water in snow would have an attenuation of about 0.5 dB/cm; at other frequencies, corresponding values are shown in Figure 1.

To determine the actual attenuation properties of a snow sample with maximum wetness, the following measurement was made. As shown in Figures 2 and 3, snow with a density of about 0.5 gm/cm^3 and a temperature of -15°C was placed in a cell with inside dimensions of 12 inches X 12 inches X 1 inch thick (30.5 X 30.5 X 2.54 cm). This volume was enclosed in Plexiglas plates 1/4 inch (0.64 cm) thick, with sidewalls 1/2 inch (1.27 cm) thick. The cell was placed at a 45° angle to minimize the effects of surface reflection and to provide for water drainage from the snow when it became saturated. The snow was heated gradually in the room temperature environment (18°C) by conduction through the Plexiglas plates. The oscillator was held at constant power at 9.35×10^9 Hz. Transmission, reflection, and phase were measured.

While the snow sample was below 0°C, the transmission was essentially independent of time. The values for transmission and reflection were practically unchanged from those obtained for the empty unit, which indicates that scattering and absorption from the snow were both negligibly small at this temperature.

After about an hour, the snow began to melt, as evidenced by the readings for transmission. As the melting progressed, during a 15-min interval the transmission

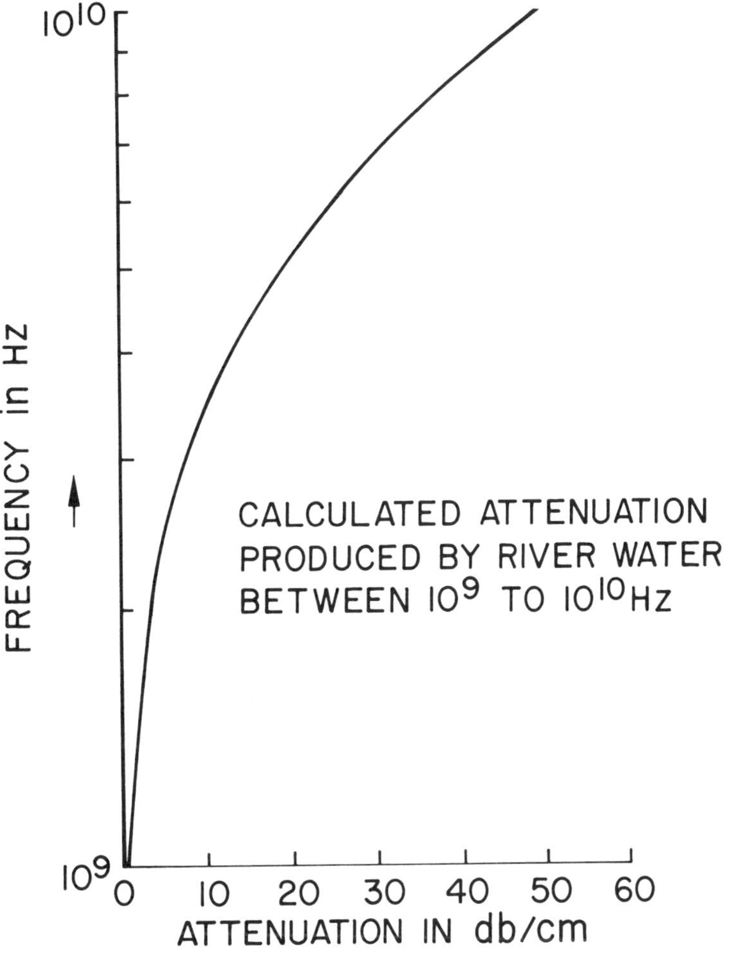

Figure 1. Attenuation of river water as a function of frequency.

Figure 2. Microwave system to obtain snow sample attenuation.

Figure 3. Snow sample in Plexiglas container.

decreased by 10 dB. When the snow sample was saturated (provision was made for water to drain from the unit), an attenuation value of 5 db/cm was obtained. If the wetness at this time is assumed to be 20% by weight, and the snow density is 0.5 gm/cm^3, the volume percent wetness obtained is 10%. In Figure 1, at 9.35×10^9 Hz, 45 dB/cm is obtained for 100% wetness, which, proportionally, becomes 4.5 dB/cm for 10% volume wetness. Thus, the measured value agrees with the theoretical value.

From the value obtained for the attenuation of a saturated snow sample at 9.35×10^9 Hz, the results can be extended to the case of a snowpack. In a saturated snowpack, a microwave beam operating at 9.35×10^9 Hz would be attenuated by 500 dB in a distance of 100 cm, or by a factor of 10^{-50}. Obviously, when such saturated snow conditions prevail, a lower frequency should be used, or a lesser distance, or both, compared to the situation when the snowpack is only slightly wet.

Consider next the attenuation value of 10 dB as the criterion, (even though a practical value could be larger or smaller) with a distance of 100 cm. The question to be asked is: at what frequency would the attenuation be 0.1 dB/cm for saturated snow? From the previous assumption that saturated snow has 10% volume wetness, multiply 0.1 by 10 to obtain 1 dB/cm for river water. In Figure 1, this occurs at 10^9 Hz.

Having established the frequency suitable for saturated snow, next consider the other extreme, namely, snow with very little moisture. The method of calculation is similar to the preceding one and will therefore be omitted. Instead, Table I presents the frequencies that correspond to various degrees of wetness, based on the attenuation value of 10 dB occurring over the distance 100 cm.

TABLE I. FREQUENCY AND SNOW WETNESS FOR 0.1 DB/CM SNOW ATTENUATION

Volume Percent	dB/cm	Frequency
10	1	1.0×10^9
1	10	3.4×10^9
1/2	20	5.1×10^9
1/4	40	8.5×10^9
1/5	50	10.0×10^9

These values are calculated on the basis of the stated assumptions and should be regarded as guidelines only. It is interesting to note that microwave ovens operate at the frequency of 2.45×10^9 Hz.

Let us next make an approximate calculation of the required radiated power of the microwave source. Until a complete system has been designed, built, and tested, preliminary calculations should be regarded as approximate. However, some idea of the order of magnitude of the quantities (and therefrom the instrument requirements) would be useful to indicate whether the cost of the proposed system might be prohibitive or encouraging.

The relations derived are available in standard references such as Skolnik (1970). Assume a simple half-wave dipole antenna for conservatism with a gain G of 2.15 dB (a microwave horn with a 15 dB gain or other beam-forming antenna could alternatively be used, with a corresponding gain factor), a separation distance D of 1 m between source and receiver, and a frequency of 3×10^9 Hz (midway in the frequency range of interest) for which the free-space wavelength λ_f is 0.1 m. In snow with a dielectric constant $\epsilon = 2$, the wavelength becomes $\lambda_f/\sqrt{2} = \lambda_s$. The ratio of received power P_r to radiated power P_t, (also known as path loss) is

$$\frac{P_r}{P_t} = \frac{1}{(4\pi)^2} \left(\frac{\lambda_s}{D}\right)^2 G^2 = 41 \text{ dB}.$$

Assume that an attenuation of 10 dB is caused by the snow absorption, and another 5 dB is caused by unspecified mechanisms such as scattering, ice layers, etc., yielding a total loss in transmission of -56 dB.

The signal/noise ratio of the detector depends on the level of sophistication of the equipment. At room temperature, the thermal-noise power spectral density is 4×10^{-9} μW/MHz, or -114 dBm in a 1-MHz bandwidth. For a 40 dB noise figure and a 1-MHz bandwidth, the noise level is $-114 + 40 = -74$ dBm. A comfortable signal/noise ratio of 20 dB with this noise level requires a signal of −74 + 20 = −54 dBm. From this, together with the transmission loss of −56 dB previously calculated, the required radiated power is obtained: −54 − (−56) = + 2 dBm, or approximately 2 mW.

From the preceding calculation, it appears that the power requirement for the microwave source is quite low if the stated assumptions apply. Verification must await system tests. To give some idea of the cost and performance capability of commercially available units,* the Varian YIG-tuned transistor oscillator model 6010S has the following characteristics: electrically tunable frequency range of 2.0 to 4.0 GHz, output power to a matched load of 10 mW, input power of a few watts, outside dimensions of 1.2 inch X 1.2 inch X 1.2 inch, (3.05 cm X 3.05 cm X 3.05 cm) and weight of 7 oz (200 gm). Its list price is about $600.

Detectors such as the Hewlett-Packard model 8470A have a sensitivity of 0.4 mV/μW, a frequency range of 10 MHz to 18 GHz, a length of 2-1/2 inches (6.35 cm), a diameter of about 1 inch (2.54 cm), and a weight 4 oz (114 gm). Its list price is about $200. Other components such as source modulation and signal–measuring equipment might require an additional $500. Together with assembly costs, a prototype system perhaps would cost from $2000 to $4000. After the first system has been tested, the cost may be reduced considerably.

Let us next analyze the performance of a system operating at 1.0×10^9, 2.45×10^9 (microwave oven frequency), and 1.0×10^{10} Hz. The following assumptions are made for these three illustrative cases:

(a) The attenuation produced by the wet snow is 20 dB or greater.
(b) Attenuations of unspecified origins total 5 dB.
(c) The signal needed is −54 dBm (as in the previous example).
(d) The snow dielectric constant is 2.0.
(e) The antenna gain of source and receiver is 2.15 dB each.
(f) Snow impurities are similar to those in river water (Figure 1).

Tables II, III, and IV list the required radiated power at the respective frequencies of 1.0×10^9, 2.45×10^9, and 1.0×10^{10} Hz. The snow wetness ranges in Table II are 2 to 20 volume percent; in Table III, 0.5 to 8.0 volume percent; and in Table IV, 0.1 to 1.0 volume percent.

TABLE II. RADIATED POWER DEPENDENCE ON DISTANCE AND SNOW WETNESS AT 1.00×10^9 HZ

Volume Percent Water	Path Length (cm)	Path Loss (dB)	Snow Attenuation (dB)	Unspecified Attenuation (dB)	Total Transmission Loss (dB)	Source Radiated Power (dBm)	Source Radiated Power (Watts)
2.0	1,000	-59	-20	-5	-84	30	1.00
4.0	500	-53	-20	-5	-78	24	0.25
5.0	400	-51	-20	-5	-76	22	0.16
10.0	200	-45	-20	-5	-70	16	0.04
15.0	133	-41	-20	-5	-66	12	0.02
20.0	100	-39	-20	-5	-64	10	0.01

*Trade names and commercial enterprises or products are mentioned solely for necessary information. No endorsement by NASA, USDI, or USDA is implied.

TABLE III. RADIATED POWER DEPENDENCE ON DISTANCE AND SNOW WETNESS AT 2.45×10^9 HZ

Volume Percent Water	Path Length (cm)	Path Loss (dB)	Snow Attenuation (dB)	Unspecified Attenuation (dB)	Total Transmission Loss (dB)	Source Radiated Power (dBm)	Source Radiated Power (Watts)
0.5	800	-57	-20	-5	-82	28	0.64
1.0	400	-51	-20	-5	-76	22	0.16
2.0	200	-45	-20	-5	-70	16	0.04
3.0	133	-42	-20	-5	-67	13	0.02
4.0	100	-39	-20	-5	-64	10	0.01
5.0	100	-39	-25	-5	-69	15	0.03
6.0	100	-39	-30	-5	-74	20	0.10
7.0	100	-39	-35	-5	-79	25	0.32
8.0	100	-39	-40	-5	-84	30	1.00

TABLE IV. RADIATED POWER DEPENDENCE ON DISTANCE AND SNOW WETNESS AT 1.00×10^{10} HZ

Volume Percent Water	Path Length (cm)	Path Loss (dB)	Snow Attenuation (dB)	Unspecified Attenuation (dB)	Total Transmission Loss (dB)	Source Radiated Power (dBm)	Source Radiated Power (Watts)
0.1	400	-51	-20	-5	-76	22	0.160
0.2	200	-45	-20	-5	-70	16	0.040
0.3	133	-40	-20	-5	-65	11	0.013
0.4	100	-39	-20	-5	-64	10	0.010
0.5	80	-38	-20	-5	-63	9	0.008
0.6	67	-37	-20	-5	-62	8	0.006
0.7	57	-37	-20	-5	-62	8	0.006
0.8	50	-36	-20	-5	-61	7	0.005
0.9	44	-36	-20	-5	-61	7	0.005
1.0	40	-35	-20	-5	-60	6	0.004
0.4	100	-39	-20	-5	-64	10	0.010
0.5	100	-39	-25	-5	-69	15	0.032
0.6	100	-39	-30	-5	-74	20	0.100
0.7	100	-39	-35	-5	-79	25	0.320
0.8	100	-39	-40	-5	-84	30	1.0
0.9	100	-39	-45	-5	-89	35	3.2
1.0	100	-39	-50	-5	-94	40	10.0

The calculated results of Tables II, III, and IV are intended to illustrate how the required source power depends on distance, degree of wetness, and frequency. Clearly, much additional theoretical and experimental information is needed before an "optimum" system is found. Component costs, sizes, reliability, etc., are important as well as the power requirement. Safety at high powers and compatibility with FCC regulations are of prime importance, although these are expected to represent problems no greater in the field than in the laboratory, where such matters are routinely encountered.

As one would expect from the attenuation curve for water as a function of frequency, if measurement of water content for 0.1 to 1.0 volume percent is desired, the frequency of 1.0×10^{10} Hz is indicated. If the ripening history of a snowpack is desired, a system with an adjustable frequency over about 1 to 3×10^9 Hz appears to be a good choice.

For some applications, knowledge of the wetness profile may not be needed, but rather the integrated effect through the snowpack e.g., vertically or diagonally. In such cases, the source can be in an enclosure above the snow, with detectors arrayed at various distances and angles. The advantage of such arrangements would be no moving parts. Depending on the circumstances, a wide variety of other source-and-receiver combinations is possible.

The microwave wetness-measuring system can be calibrated by taking a profile, then sampling the snowpack with other techniques (such as calorimeters, centrifuges, or the electronic systems described by Linlor and Smith,(1973)). Also, as described by Linlor and Smith (1973), the calibration could be accomplished with "heavy water" (D_2O) in a refrigerated room. With a large volume of snow at slightly below freezing (about −0.1°C), a known amount of heavy water (at 0°C) would be sprayed so that the upper portion of the snow would become saturated. After suitable time intervals, microwave profiles would be taken, and the readings at various levels of wetness would be compared to measurements of the D_2O/H_2O ratio obtained with nuclear abundance instruments.

So far, only the attenuation of the microwave beam by absorption has been discussed. It is also planned to investigate the phase increments produced by snow to determine whether the mass (i.e., the water equivalent) can be measured by this type of information.

In summary, microwave measurements of snow wetness appear capable of yielding reasonably precise profiles without affecting the physical properties of the snowpack. Simultaneous density profiles obtained from γ-ray instrumentation, together with meteorological information, provide a basis for predicting the melting and water discharge rates from snowpacks, which is important for water resources management.

REFERENCES

Linlor, William I., and James L. Smith, 1973. Electronic measurements of snow sample wetness. Proc. Interdisciplinary Symposium on Advanced Concepts and Techniques in the Study of Snow and Ice Resources, Monterey, California.

Skolnik, Merrill I. *ed,* 1970. Radar handbook. New York: McGraw-Hill Publishing Co. 700 p.

Smith, James L., Howard G. Halverson, and Ronald A. Jones, 1972. Central Sierra profiling snow gage: A guide to fabrication and operation. TID-25986, Division of Isotopes Development, U. S. Atomic Energy Commission. Available from National Technical Information Center, U. S. Dept. of Commerce, Springfield, Virginia, 2215-.

Von Hippel, Arthur R., 1954. Dielectrics and waves. Cambridge, Mass.: The M.I.T. Press. 284 p.

METEOR BURST COMMUNICATION

Raymond E. Leader
The Boeing Company

ABSTRACT

This paper will discuss the application of Meteor Burst Communication technology to remote data acquisition and control. This type of system is ideally suited for low data rate telemetry such as that encountered in hydrological, meteorological, and oceanographic data sensing.

Ionized trails in the upper atmosphere resulting from meteor bursts can be utilized to reflect or re-radiate short VHF radio messages between point-to-point stations located anywhere within 1200 miles. Reliable message exchange in excess of 100 per hour can be realized.

A typical meteor burst system would consist of a base station transceiver and a central processor with a number of remote radio stations designed for data collection or external control.

INTRODUCTION

A growing need exists for a new family of telemetry systems to gather the geophysical environmental data required by the scientific, governmental and industrial communities. Fields of interest include convservation, oceanography, meteorology, public health, hydroelectric power resource management, etc.

The geophysical data acquired by these services in many cases have several common features; usually the rate of change of the phenomena being measured is fairly low; i.e., the data rate associated with the measurement is low. For example, measurements of ocean current precipitation levels, river or lake water levels, smog content and radiation levels require only modest channel capacity when compared with telephony, television, computer-to-computer communication, etc. In addition to low entropy, the data usually does not require real time transmission. Other characteristics that apply in varying degrees also come to mind. These may be listed as follows:

1. Measurements frequently require placement of sensors in difficult, remote terrain, and beyond line-of-site ranges.
2. Primary power is at a premium.
3. Sensor stations are frequently unmanned, widely separated and sometimes difficult to service.

A frequently overlooked radio propagation medium that is attractive for these applications is that of VHF propagation via scattering from meteor trails. Most previous meteor burst systems have been used for teletype "quasi-continuous" data communications. In many respects, however, it is an ideal medium for low entropy, remote telemetry applications for long ranges. A complete description of forward meteor scatter communication is beyond the scope of this paper as a number of excellent reports exist in the literature.

A meteor burst communication system (MBCS) utilizes as a means of radio signal propagation ionized meteor trails that exist in altitudes of the 80 to 120 Km region of the earth's atmosphere. These trails reflect or actually re-radiate the RF energy (usually in the low VHF range of from 40 to 100 MHz) from a transmitting source to a recovery terminal. The height of the trails allows over-the-horizon communication at distances up to 1200 miles (see Figure 1). However, because the ionized trails exist for only short periods of time (usually from a few milliseconds to a few seconds) communication is intermittent and high speed digital pulse transmission techniques must be used to convey the information. The system is particularly suited to long range, low rate data acquisition applications but it can also support a low speed (100 words/minute) teletype link. Shorter operational ranges, over mountainous terrain, are also a capability due to the height of the reflective meteor trail.

At the present time there are several methods of data communication over long ranges. These methods include HF ionospheric scatter links, VHF repeater links, microwave links, telephone lines, and satellites.

HF scatter systems are plagued by ionospheric losses which are very frequency dependent. This requires that the operating frequency be changed to minimize the losses. Another problem with HF systems is that they suffer from large amounts of fading which can be due to several ionospheric phenomena; e.g., movements of the ionosphere causing interference fading, rotation of the axis of the polarization ellipses, time variations in ionospheric absorption, focusing, and skipping of the signal due to maximum usable frequency (MUF) failure. Periods of this fading are highly irregular and can vary from a fraction of a second to a few hours depending upon the cause. Fading and ionospheric losses can render an HF system useless for long periods of time.

VHF repeater networks and microwave links are line-of-site, thus requiring the use of a large amount of repeaters if any range is to be covered. These present an installation and maintenance problem especially in remote mountainous areas. Initial cost of the system is also high. Microwave systems are also geographically fixed, thus preventing system changes or expansion without great cost.

Telephone lines suffer from all the same problems as microwave links with respect to installation, maintenance, and flexibility.

Synchronous communication satellites are the only means of exceeding the long range data communication properties of MBCS. Synchronous satellites are those which maintain a fixed position in the sky relative to the earth. These satellites are more costly and complex than a typical orbiting satellite which is not effective for daily communication systems. The obvious high cost associated with synchronous satellites eliminate their use for most applications.

HISTORY

Communication by scattering radio signals off of ionized meteor trails has been investigated by a number of independent agencies over the past thirty years, and as a result of this investigation, a number of MBCSs have been built.

Early systems built almost entirely for the military have had poor reception because of the intermittent nature of the medium. Message waiting times for these early systems were longer than could be tolerated, and with the advent of communication satellites, MBCSs were pushed to the back burner. Another

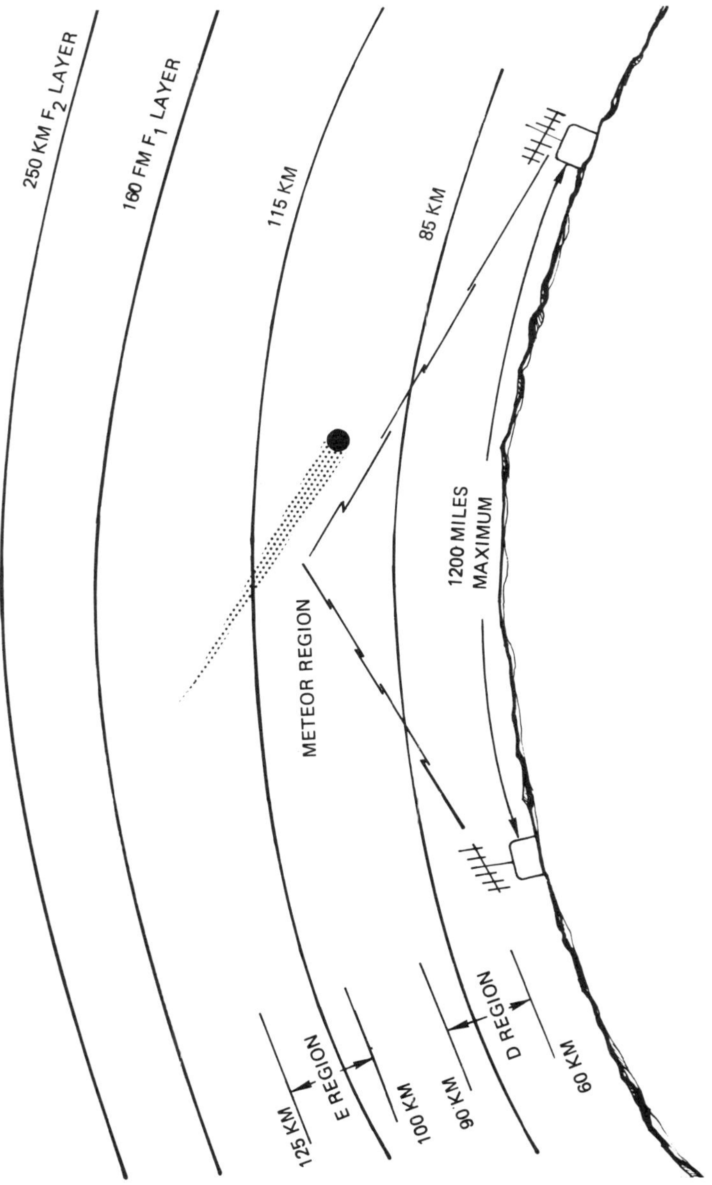

Figure 1: METEOR BURST COMMUNICATION SYSTEM

factor in this early rejection of MBCS was the high cost and complexity of the equipment, which resulted mainly because of the amount of control electronics required. With the advent of the minicomputer and integrated logic circuitry, this objection has been overcome.

Investigation into the phenomenon of ionization by meteors began in the thirty's and continued until a fairly sophisticated, well-defined theory was developed and experimentally verified. This theory is on the formation of ionized meteor trails, their radiant distribution, and the scattering of radio signals off these trails in both the backscatter (radar) mode and forward scatter (communication) mode. This article will deal only with the forward scatter mode, and a brief review of the meteor scatter theory follows. A more detailed review can be found in the reference listed at the end of this article.

METEOR TRAIL IONIZATION

Billions of ionized meteor trails are produced daily in the earth's atmosphere at heights of 80 to 120 Km. These trails diffuse rapidly and usually disappear within a few seconds. However, during their brief existence, they will reflect radio waves in the VHF frequency range.

The meteor, as it enters the upper atmosphere traveling at speeds of 10 to 75 Km/second, possesses a large amount of kinetic energy. As it begins colliding with air molecules, much of this kinetic energy is converted into heat which effectively vaporizes atoms from the surface of the parent meteor. These vaporized atoms, which are traveling at about the same speed as the meteor, are further restricted by air molecules expecially as they progress further into the atmosphere. This results in the transformation of kinetic energy into the energy of ionization which effectively strips electrons from the atoms leaving a trail of positive charged ions and free electrons. It is the electrons that reflect or re-radiate radio waves.

This ionization is distributed in the form of a long, thin paraboloid of revolution with the particle at the head. The electron line density (electrons/meter) is proportional to the mass of the meteor, and ranges from 10^{18} electrons/meter to about 10^{10} electrons/meter. Typical trails are 25 Km long and have an initial trail radius of about 1 meter.

PHYSICAL PROPERTIES OF METEORS

In this article, meteors are defined as extraterrestrial objects that travel in orbits (eliptical) around the sun, and at some point in these orbits enter the earth's atmosphere. These objects can be divided into two basic classes:

Shower Meteors - These are groups of particles all moving with the same velocity in fairly well-defined orbits around the sun. To an observer on the earth, they are the most spectacular and appear to radiate from a common point in the sky (called the radiant). Shower meteors account for only a small fraction of the total incidence of meteors.

Sporadic Meteors - These are particles that move in random orbits around the sun and account for the vast majority of meteors that are used in radio work. Their radiants and times of occurrences are random and cannot be cataloged as shower meteors can; however, it is known that they are not uniformly distributed in the sky but are mostly confined to within about 20° of the ecliptic plane (the plane of the earth's orbit about the sun). Also, the intersection of the meteor orbits with the earth's orbit are not uniformly distributed but are concentrated so as to produce a maximum of intersections in August and a minimum in February, with about a 4:1 variation. The rate of incidence of sporadic meteors is further dependent upon the time of day with the morning hours being more active. On the morning side of the earth, meteors are swept up by the forward motion of the earth in its orbit around the sun. On the evening side, the only meteors reaching the earth are those which overtake it. A daily variation of about 4:1 can be expected.

CHARACTERISTICS OF REFLECTED SIGNALS

Radio signals received on a meteor-burst link are basically of two forms. The most prevalent of which is the underdense trail reflection. In this case, the reflecting meteor is characterized by a relatively low electron line density ($q < 10^{14}$ electrons/meteor). An underdense trail does not actually reflect energy, instead the radio waves pass through the trail exciting individual electrons as it does. These excited electrons act as small dipoles re-radiating the signal at an angle equal but opposite to the incident angle of the trail and radio signal.

Signals from underdense trails rise to an initial peak value in a few hundred microseconds then decay exponentially in amplitude. Decay times from a few milliseconds to a few seconds are typical. The decay in signal strength is due to distructive phase interference caused by the radial expansion or diffusion of the trail's electrons.

Meteor trails with electron line densities greater than 10^{14} electrons/meteor actually reflect signals and are called overdense. Here the line density is so great that radio signals cannot penetrate them and are actually reflected instead of re-radiated. There are no distinctive patterns associated with overdense trails except they usually reach higher amplitudes than underdense and usually last longer. Signal fading often occurs due to reflections from two parts of the trail destructively interfering. Ionosphere winds blow these trails around causing some of the fading. The period of the fading is usually greater than a few hundred milliseconds and thus does not affect the typical intermittent MBCS. A fairly good illustration of this type of trail is the contrail left by a high altitude jet and the effects of wind on it. Due to the unpredictable nature of the signals, no useful theory has been developed to describe them.

Other means of communication over the typical MBCS are airplane reflection at distances of less than 100 miles. These usually last many seconds and are characterized by a great deal of signal fading. Sporadic E also contributes long periods of communication time.

As stated above the typical MBCS will utilize all methods of propagation but will mainly rely on underdense meteor reflection.

SYSTEM DESCRIPTION

A MBCS system is composed of two types of terminals, a base or master, and remote stations. Normally, one master station communicates with a number of remote stations. Figure 2 is a block diagram of the two stations. Boeing has recently developed, fabricated and field tested such a system. The specific hardware is described below.

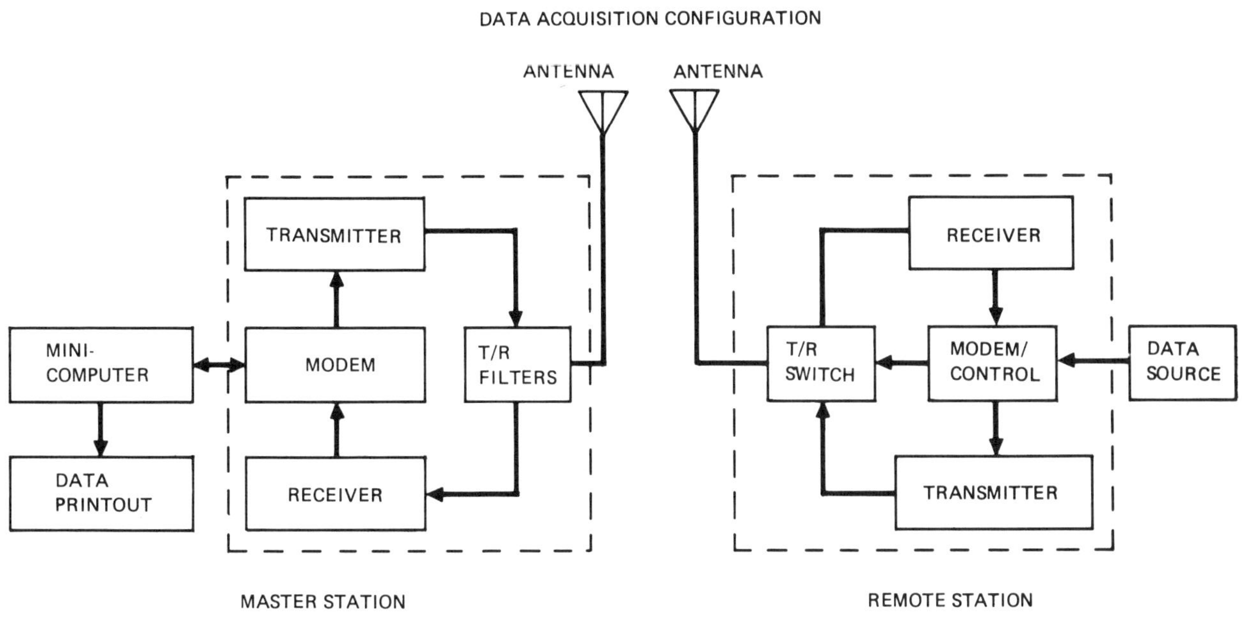

Figure 2: METEOR BURST COMMUNICATION SYSTEM

MASTER STATION

I. Communication Unit - The receiver/modulator assemblies are completely transistorized and are of modular construction. The receiver is solid state of modern design. The transmitter has a controlled output of 2000 watts of RF power. The communication unit modulates the carrier with information obtained from the control unit and will transmit the data via the antenna. Furthermore, the signal from the remote unit is received, demodulated and clocked to the control unit for processing.

The unit is of modular construction using an integrated packaging technique. Components are mounted on 9.25 x 4.5 edge connector cards that are housed in a card file that fits a standard 19-inch rack.

II. Control Unit - The control unit consists of a minicomputer that is interfaced with the communication unit of the master station. Through software, the control unit formats and processes the codes that are transmitted and received by the communication unit. A teletype printout is used to display the received data and the keyboard is used to input the different command sequences to the selected remote station. The computer is programmed to automatically control the entire sequence.

III. Power Unit - The master station transmitter is powered with 220 VAC single phase power. In the mobile configuration, the 110 VAC is obtained from one side of the 220 VAC line and is used to power the various power supplies and test equipment. The power supply produces the required voltage for the electronics of the various modules.

IV. Antenna Unit - The antenna unit is comprised of two (2) stacked, five (5) element yagi antennas and a duplexer designed to operate continuous duty. The antennas feed a high power duplexer section allowing full duplex operation at the base station.

REMOTE STATION

I. Communication Unit - The transmitter/receiver assembly is completely transistorized and is of modular construction. The electronics are designed for minimum power consumption. The receiver is a modern solid state design. The modular construction of the power amplifier section offers a selection of RF power outputs. The remote station can transmit from 20 watts to 300 watts of RF power.

The communication unit will receive, demodulate, and clock the data to the control unit. The communication unit also receives data from the control unit and modulates it on the remote unit carrier frequency, and forwards it to the antenna unit via the T-R switch.

II. Control Unit - The control unit formats and processes messages that are transmitted and received. The logic module is designed with COS-MOS logic for minimum power consumption when in the operational mode. Jumper wires are used to program the station address.

III. Power Unit - The remote power unit is powered by battery, and can either be continually trickle charged, or operated in isolation. The length of isolation time and interrogation rate will determine the battery size requirements.

IV. Antenna Unit - The antenna unit is comprised of one (1), five (5) element yagi (or a dipole) and a T-R switch. The antenna feeds the T-R switch section allowing only simplex operation at the remote station.

OPERATIONAL SEQUENCE

A typical operational sequence starts when the master station continually probes the sky for a suitable meteor trail. The probe consists of digital information that is modulated on the master station VHF carrier frequency and is continually being transmitted. The information includes a preamble, recognition code, and a remote station address code. The coded digital signal is received at the remote station when the meteor trail is properly oriented and is suitably ionized. The remote station is in the standby configuration for maximum battery life. When the signal is received, the remote station becomes operational and the digital information decoded. If during the decoding it is determined that the addressing is incorrect, the station will return to the standby mode. However, if correct, the remote station will transmit a block of data to the base station over the same meteor trail. When the master station receives the remote signal, the computer will examine the addressing and, if correct, will accept the data.

PERFORMANCE

Performance of a MBCS is measured in the number of error free data receptions at the master station. The following performance was obtained by field testing the system described in this paper over the past one and one half years.

Performance will vary with a number of parameters which include transmitter power, operating RF frequency, receiver sensitivity, and antenna gains. The following performance data is based on parameters which reflect a reasonable hardware configuration. That is, no high power transmitters or large antenna configurations.

Diurnal Characteristics - Operation varies over the daily cycle as shown in Figure 3. Nearly 400 receptions per hour were received in the early morning hours, and 100 per hour in the evening hours. This 4:1 daily variation has been demonstrated to be consistent throughout the year.

Yearly Characteristic - Performance over a yearly cycle also has a four to one variation, with a maximum in August and a minimum in February. If the curve of Figure 3 were plotted with data during February, the morning maximum would be approximately 100 receptions per hour and the evening minimum at 25 receptions per hour.

Range Characteristics - Performance over the possible operating range of 0 to 1200 miles varies as shown by the curve in Figure 4. Actual testing was not performed beyond 1000 miles. Data was sampled at the same time of day to provide comparable results for the ranges measured. The 160 messages per hour shown is not indicative of the maximum daily message rate. The basic reason for the reduced performance at short and long ranges was the reduction of common sky at the 80 to 100 Km altitude region between the two stations. Therefore, the number of usable meteor is reduced as compared to mid-range (400 to 800 mile) operation.

At some range, as the two stations become closer and closer together in range, a continuous master station signal can be detected at the remote station. The range at which this can occur is highly dependent on the nature of the horizon angle of each of the stations. This range was detected to be as far as 150 miles given a fairly low (less than two degrees) horizon profile between the stations.

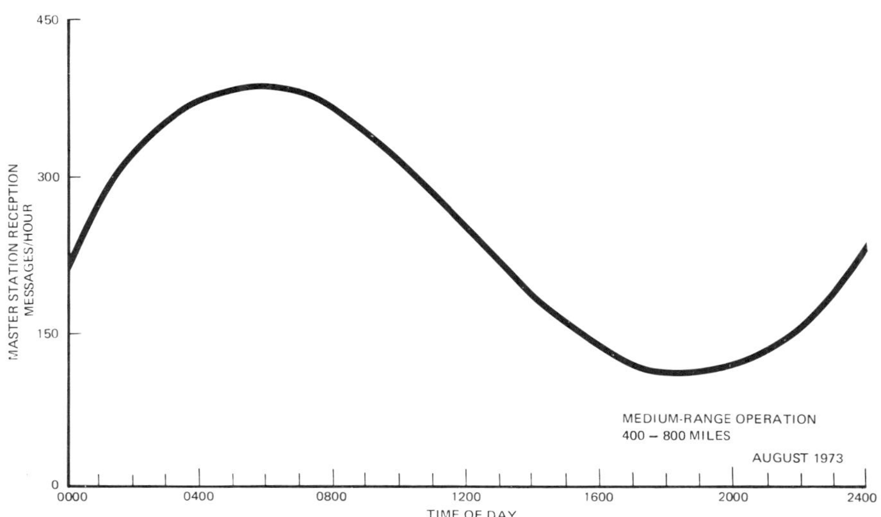

Figure 3: METEOR BURST COMMUNICATION SYSTEM DIURNAL DATA RATE VARIATION

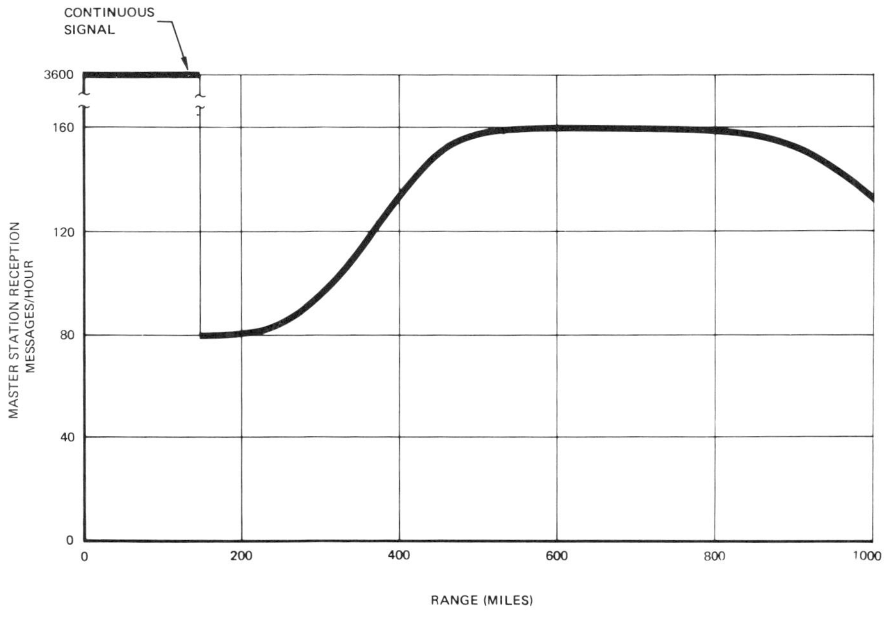

Figure 4: METEOR BURST COMMUNICATION SYSTEM DATA RATE VS RANGE

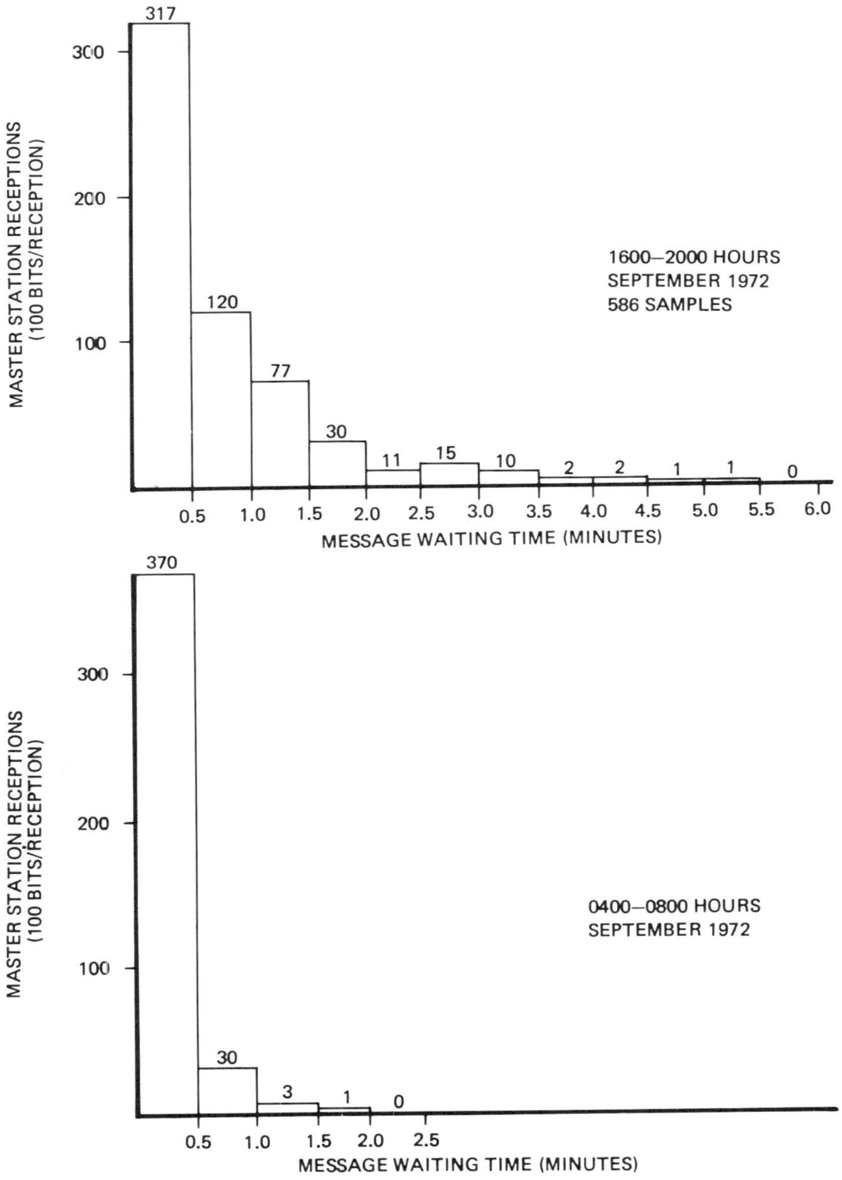

Figure 5: METEOR BURST COMMUNICATION SYSTEM MESSAGE WAITING TIME

Message Waiting Time - An important consideration in the application of Meteor Burst Communications is the expected waiting time between messages. Figure 5 plots this statistic during the daily minimum performance hours, 1600-2000, and the peak hours from 0400 to 0800. Ninety percent of the wait intervals are less than two minutes during the least active period of the day (1600-2000).

SUMMARY

Application of Meteor Burst Communication technique to remote data collection provides a very cost competitive approach where the requirements fit within the Meteor Burst statistical capabilities.

An example of an ideal application is the communication link to remote stations being installed in the Northwest mountain areas to gather hydrological and meteorological data. The Bonneville Power Administration (BPA) plans to install meteor burst communications to a number of their hydromet remote stations. BPA can thus monitor remotes throughout their mountainous domain (Washington, Oregon, Idaho, and Montana) from a single central control station. No relays or intermediate communication means will be required.

REFERENCE

McKinley, D. W. R., "Meteor Science and Engineering" McGraw-Hill, New York, 1961

SUMMARY OF DISCUSSION
(Papers 7.1 to 7.5)

The common denominator used to group the papers in this session is the instrument approach for in situ measurements of specific hydrologic phenomena.

The paper by Blyth and Painter (7.1) on the use of terrestrial photogrammetry for snow distribution analyses stimulated little discussion, perhaps because of the uniqueness of the approach. Painter noted that this was one of the first major efforts by the U. K. Institute of Hydrology to incorporate snow measurements into hydrologic research and operational activities in the U. K.

The University of Wyoming Water Resources Research Institute paper (7.4) on measuring snowfall in a man-made site shield (termed the Wyoming Shield) also stimulated little substantive discussion. It was generally acknowledged that this type of snow gage shielding increases the probable accuracy of precipitation measurements as compared to other devices, and permits increased freedom of choice in siting a gage rather than dependence upon the existence of a suitable natural site.

The snow moisture integrator described in the paper by Alger and Santeford (7.5) precipitated a rather lengthy discussion, particularly with regard to what was actually being measured, and the effects of varying the height of the sample container above ground, or varying the size of the container itself. As one of the authors pointed out, the device was designed primarily to measure condensation, although light snowfalls, frost, or evapo-sublimation could be measured as well. It was acknowledged that the height of the container could have an effect, although not a very significant one, on these measurements. The size of the container could not be increased without significant structural changes and, based on previous experiments with smaller containers, any increase in size would not appreciably increase measuring accuracies.

Bass's (7.3) electro-optical instrument for measuring total precipitation and snow pack water content, although difficult to describe, is evidently easily understood when seen in actual operation. There were no substantive comments on this paper.

Tsang's paper (7.2) offering a conceptual design for a multipurpose winter stream measuring instrument was not discussed at length, although the importance of measuring winter streamflow and the difficulties encountered - particularly with regard to frasil ice - were emphasized in the comments made.

SUMMARY OF DISCUSSION
(Papers 7.6 to 7.9)

Paper 7.8 by Theakston emphasized highway design. In the discussion, Theakston presented other slides showing simulations of snowfall distribution in forested and cleared areas, at airports, and on buildings and ski slopes. Concern was expressed about scaling problems and the extent to which theoretical considerations were invoked. In answering these concerns, Theakston emphasized that in construction, information is required relating to the steady state condition in which wind-scouring prevents further buildup of snow. The amount that can be built up, and its distribution, are the critical factors in determining roof design or highway contouring. Hence, theoretical concepts are essentially ignored in Theakston's approach.

In the discussion of papers 7.6 and 7.7, concern was expressed about the dielectric constant of water changing considerably with frequency. The authors made the point that water attached to snow has a behavior quite different from free water. In the former case, ϵ is fairly constant over a rather large frequency range, having a value of the order of 20 (compared to 87.7 for pure water). A paper relevant to the discussion, entitled "Determination of dielectric constant of some frost-prone soils," by A. S. Jumikis, was brought to the attention of the participants. This paper was not included in the symposium but is published in these Proceedings (Appendix A).

In the discussion, it was ascertained that in paper 7.6 the Q value was measured, rather than frequency shift or another parameter. The purpose of the experiments with foam rubber was established as twofold: (1) to assure that the change in the physical properties measured was indeed related to increasing water content (as opposed to some other property that may have existed in the snow-water system); and (2) to test the reproducibility and precision of a system in which the solid matrix remained constant.

In the discussion of paper 7.9, Leader showed that the components for the Meteor Burst Propagation (MBP) System are impressively small, with the master station easily fitted into an Econoline van, and the remote battery-operated electronics package having approximate dimensions of 3 cm x 20 cm x 30 cm. The effects of the aurora were discussed. Leader pointed out that the aurora could be used for reflecting signals with the MBP system. Interference between two stations trying to relay information at the same time had a very low probability of occurrence, he noted.

APPENDIX A

DIELECTRIC CONSTANTS IN THE MANAGEMENT OF FREEZING SYSTEMS,
Alfreds R. Jumikis

ELECTRICAL PARAMETERS OF SOME FROST-PRONE SOILS,
Alfreds R. Jumikis and W. A. Slusarchuk

NOTE: The first of these two papers is a summary and expansion of the discussion concerning the papers, Electronic Measurement of Snow Sample Wetness (Linlor and Smith), and Microwave Profiling of Snowpack Free-water Content (Linlor, Meier, and Smith). The second paper, which was submitted for the Symposium but was not included in the program, is presented here because of the great interest in it expressed during the discussion.

DIELECTRIC CONSTANTS IN THE MANAGEMENT OF FREEZING SYSTEMS

Alfreds R. Jumikis
Department of Civil and Environmental Engineering
Rutgers University, New Brunswick, New Jersey, U.S.A.

INTRODUCTION

It is a generally well-known fact that the evolution of any scientific and engineering discipline does not take place by merely following a rigid, preconceived plan and schedule, but that these disciplines grow more or less concurrently along different directions, depending upon the time, need, and the many essentially different points of view of the researchers. Also, the theories thus developed often turn out to be interrelated with other, overlapping scientific disciplines, and/or are similar, and that they may influence one another either beneficially or adversely, depending upon circumstances.

This article constitutes the author's discussions, comments and remarks made at the Interdisciplinary Symposium on Advanced Concepts and Techniques in the Study of Snow and Ice Resources held at the Asilomar Conference Grounds at Monterey, California, on December 2-6, 1973. It draws together the author's impressions and thoughts of the competently presented but fragmented information in a theoretical framework, based on the author's own experience. This applies in particular to the nature of the materials (soil, ice, snow) and some physical processes involved, rather than to the technology used.

There were several papers presented at this Symposium describing new methods for increasing information in snow and ice studies. The papers revealed that a complete set of parameters must be known throughout the snowpack and/or ice, and not merely on the surface, i. e., all pertinent calculations demonstrate the need for independent knowledge of the necessary parameters in the complex borderline disciplines such as thermal soil mechanics, snow mechanics, ice mechanics, and snow and ice hydrology.

Discussions during the Symposium revealed also the following:

a) the necessary parameters to know are, among other things, the dielectric constant of the material and its physical properties, such as porosity, moisture content, temperature and thermal properties. The selection of proper oscillating frequencies, measurement accuracy, is strongly affected by soil (snow) moisture changes;
b) for different applications, different frequencies must be used, and factors that determine the choice of frequency should be examined.

Also, parameters should be studied for selecting the maximum energy to impart to sheets of material of corresponding thickness.

All this leads to the requirement of standardizing a

method of testing procedure, the kind of equipment to use, calibration, frequency to use, and power density. In the interpretation of test results, moisture data concerning the state of the soil (snow) at the same depth from prior calibration are needed.

In practice, knowledge is required of the temperature of the surface in order to determine the emissivity from a radiometer output. Also, in general, as the dielectric constant becomes larger, the more effect surface roughness has on the emissivity.

In snow and ice studies, and in the case of calculating the temperature of the land surface by means of spaceborne remote sensors, there is a need to know prior to such tests the dielectric constants of the air, snow, ice and soil.

SOME FACTORS AFFECTING THE DIELECTRIC CONSTANT

The magnitude of the dielectric constant is known to depend upon many basic physical factors. Some of them are as follows:

1) Nature of soil, rock, snow and ice, including type of material, whether non-cohesive or cohesive, fine or coarse-particled, dense or loose. Also composition: sand, silt or clay; mineralogical components. Surface quality of the soil, snowpack or ice.
2) Moisture and ice content in soil, rock, snow and ice.
3) Physical properties of soil, snowpack and ice, among them permeability to water; permeability of snowpack to air; electrolyte in solution present in soil; and shape and orientation of mineral and ice crystals.
4) Specific surface of soil and of snow.
5) Amount of unfrozen water in a frozen soil, viz., dense, unfrozen film of water surrounding the soil particles.
6) Temperature of soil, snow and ice.
7) Cryogenic texture of soil and snow (finely or coarsely crystallized ice in soil or snow).
8) Dielectric constants of the component parts making up the composite of the soil or snowpack (relative proportions of individual dielectric constants of mineral matter, water, ice and air fractions in the soil, for example).
9) Regional geothermal heat (ground or soil heat).
10) Radiation.
11) Microclimate temperature just above the surfaces of the ground and snowpack.
12) Thermal properties of the layered system composed of the microclimate atmospheric air; the snowpack or ice layer; and frozen and unfrozen soil (thermal conductivity, diffusivity, heat capacity of these materials, for example); as well as the process of heat exchange between the layered soil-snow system to the microclimate atmosphere.
13) Contact-interface temperature between the bottom of a snowpack and the ground surface (snow layer as a

thermal insulator on top of the soil, viz., ground surface). Also, the contact interface in one kind of material of a certain thickness, consisting of one dense part on top of a loose part, or vice versa, adds to the difficulty in obtaining consistent measurements.

14) The humidity of the air and the ambient temperature also influence remotely sensed electrical measurements, and hence the dielectric constants of the materials detected.
15) The type of instrument used, and the method of electrical measurements applied, as well as oscillation measurement frequencies chosen for determination of dielectric constants also affect the research results obtained from soil, snow and ice studies.
16) Besides the above factors, an external field may bring about two or more possible influences: a) molecules having permanent dipole moments may become aligned along the field direction, and one speaks then about orientational polarization; and b) distances between ions or atoms may be influenced by an external field.

All these and possibly a host of other factors as yet unaccounted for influence the magnitude of the dielectric constants determined either from laboratory tests or *in situ*. Of course, in a complex heat-exchange process such as frost penetration into the soil, or the performance of a compound snow-soil-water-temperature system, it is practically necessary to ignore some of the contributing factors that numerically seem to have insignificant effects on the result, or when some of the contributing factors are difficult or even impossible to evaluate numerically or are not available at all as yet. For example, today the effect of radiation is a very complex factor to assess; hence there is hardly any basis for a numerical evaluation. Also, some authors contend that in studying freezing soil systems the soil heat (geothermal heat) can be ignored because its small quantity contributes little to the end result; only the latent heat of fusion (melting) is usually reckoned with. These and other simplifications generally aim at devising simple methods, rules of study and designs that are clear and simple enough to be commonly accepted and used in practice.

Of course, some assumptions made for the approximations used in soil, snow and ice studies have in some cases oversimplified the actual processes in the thermal systems, such as soil-water-temperature, snow-water-temperature, or ice-water-temperature. However, such assumptions have the advantage of being simple and explicit, and can be justified as a means of approach to a better understanding of the measurements made for determining the factors involved in the snow and ice forming and performing processes, as well as in determining the dielectric constants of soil, water, snow and ice encountered under various physical conditions.

Of course, the usefulness of such an analysis must

be verified by experiment and practice. Recalling the dictum of Sir Francis Bacon, ". . . that which is most useful in practice is most useful in theory."

GENERAL OBSERVATIONS CONCERNING SOME FACTORS IN RESEARCH IN FREEZING

Only a few general observations will be touched upon here, such as the ground heat, cryogenic texture of soil, specific surface of soil particles, unfrozen water in frozen soil, thickness of ice layers (and frost penetration depth), contact-interface temperature between soil and a snow cover, snowpack, and dielectric constants.

Ground Heat. Soil temperatures have been studied at Rutgers University, College of Engineering, in 32 highway soils, all under the same climatic conditions. If in some instances, for example in calculating theoretically frost penetration depth in soil, the ground (soil) heat may be ignored, there are observations indicating that the upward-flowing ground heat is a quite significant factor on paved roads during thawing periods. For example, Figure 1

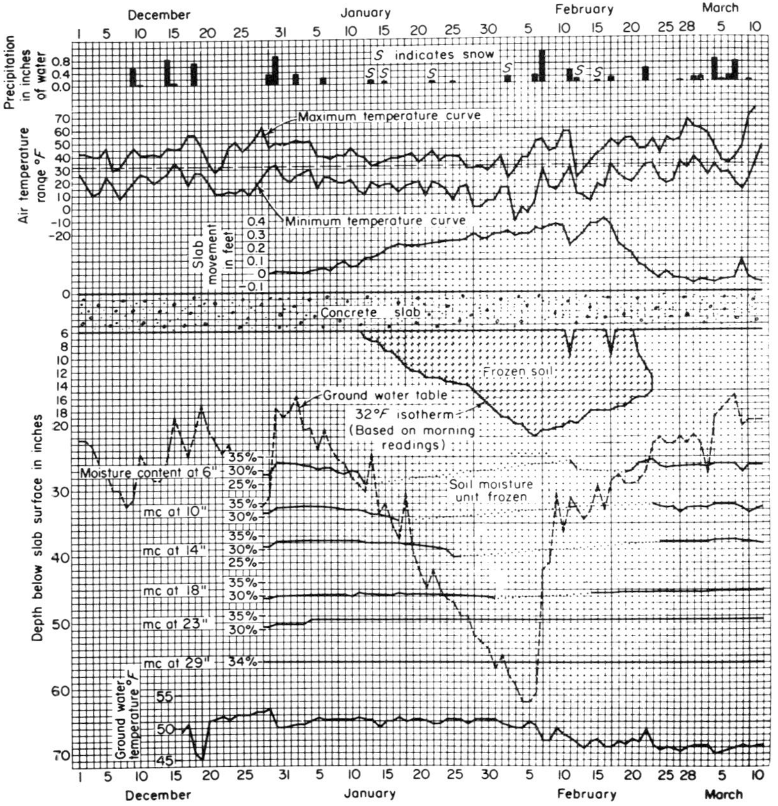

Figure 1. Changes in sandy, silty-clay soil on freezing.

illustrates results obtained in the author's outdoor experimental studies on frost action in a sandy, silt-clay soil classed as A-5 according to the Highway Research Board soil classification system. Notice from the figure that thawing of the frozen soil directly underneath a concrete pavement slab occurs from the top downward during thaw periods. However, some thawing takes place also from the bottom of the frozen soil. It thus appears that in such instances the upward-flowing soil heat cannot be ignored. As a frozen layer of soil thaws from its top and bottom, this fact is of particular importance; there may form "pockets" of thawed soil at high moisture content, saturating the soil, and thus causing it to lose its load-bearing capacity.

This observation implies a special drainage problem concerning base and subbase courses to insure traffic safety on roads. Similar thawing conditions of snow-soil and ice-water course systems (from above and from below) can be encountered in snow and ice hydrology. In such instances interpretation of terrestrial and remote-control measurements requires a good understanding of the physical conditions and the dielectric constant of the material prevailing at the particular time when the measurements were made.

One can cite another example demonstrating that soil heat from below cannot be ignored. In an attempt to reduce frost penetration depth in highways, some research agencies recently began experimenting with introducing anti-frost insulation courses in soil beneath the pavement. However, results obtained from such preliminary experiments so far have indicated that although reduction in frost penetration depths was observed, the road surface became covered with icing, meaning that the heat from the soil below the insulation was retarded by the insulation, and could not thaw or prevent the icing, whereas pavement surfaces in the same locality (adjacent to the experimental sections) and under the same climatic conditions but with no added insulation course in the soil were free from surface icing.

Cryogenic Texture of Frozen Soil. The cryogenic texture of soil is also a factor affecting the magnitude of the dielectric constants of rock, soil, snow and ice, and combinations thereof. Figure 2-I illustrates the three basic types of cryogenic textures in soil: A) homogeneous, B) layered, and C) mesh-like. A homogeneous texture is characterized by the presence of uniformly distributed, very fine ice crystals cementing the soil particles into a frozen mass of soil. The homogeneous texture forms upon quick, intense freezing of the soil, bringing a considerable amount of the soil mass, viz., soil pore water, below freezing temperature. The centers of crystallization of water into ice are formed simultaneously at various depths, and freezing propagates rapidly into the mass of the soil, thus forming a dense, homogeneous soil texture (dense ice on loose snow). In contradistinction, slow cooling of soil in the presence of water migrating

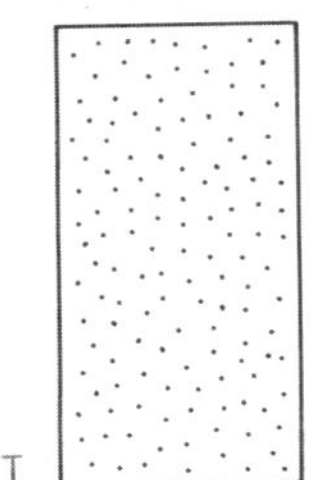

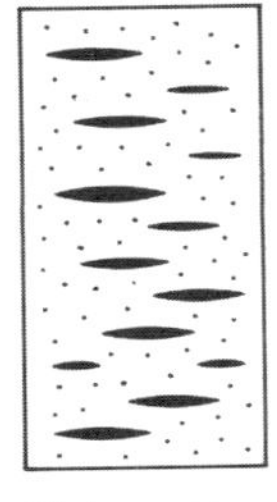

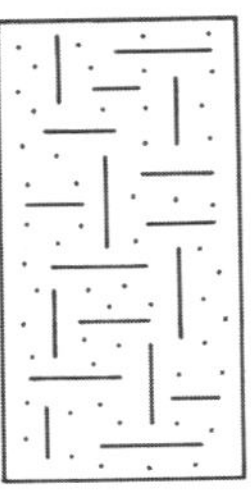

I

II

Figure 2. Cryogenic changes in soil. I) Schematic rendering. II) Variations in texture of frozen soil: top, homogeneous texture formed during a cold spell (quick freezing); bottom, layered, rhythmically banded texture--ice lenses (slow freezing) over unfrozen base. Author's study.

from groundwater to the cold front brings about relatively large segregated ice lenses--or layered ice--in the frozen zone. Figure 2-II shows a fine-particled soil sample first frozen rapidly (top, homogeneous part), and later slowly, bringing about neatly segregated, rhythmic banding of ice lenses (lower, layered part). Because of these two kinds of textures (densities, water and ice contents, and temperatures), there exists a contact interface between the two kinds of textures, an interface which affects the remote measurements made in soil, snow and ice studies. With the same reasoning, texture thus affects the dielectric constants of the materials studied.

The mesh-like cryogenic structure of soil is a very inhomogeneous one, thus affecting the dielectric constant magnitudes in a very uncertain manner. As a rule, the mesh-like texture forms upon freezing of oversaturated, fine-particled soils. Because of the large soil moisture content, there occurs simultaneously, even upon an insignificant cooling, a large quantity of irregularly oriented ice crystals. Their growth takes place unimpeded on account of the surrounding water, and forms a network of mesh-like inclusions of ice in the soil, within the cells of which moist soil is contained. The more rapid the cooling, the denser is the network of ice inclusions, and the finer are the moist cells in the texture thus formed.

Thus temperature and moisture regimens regulating ice formation in various types of soil have a great influence on the formation of texture of the freezing material.

Specific Surface of Soil Particles. Unfrozen Water in Frozen Soil. Fine-particled soils such as silts and clays with well-developed specific internal surface permit the presence of large quantities of unsaturated, bound (stressed), unfrozen water freezing at various temperatures below 0° C. In such soils, there exist all conditions for soil moisture migration along a thermal gradient, resulting in considerable soil moisture redistribution (Figure 3), plenty of ice segregation and formation of ice lenses. The immobile part of unfrozen film water in soil, because of the great density of such bound (stressed) water, brings about a large dielectric constant of the composite of such soil as compared with the dielectric

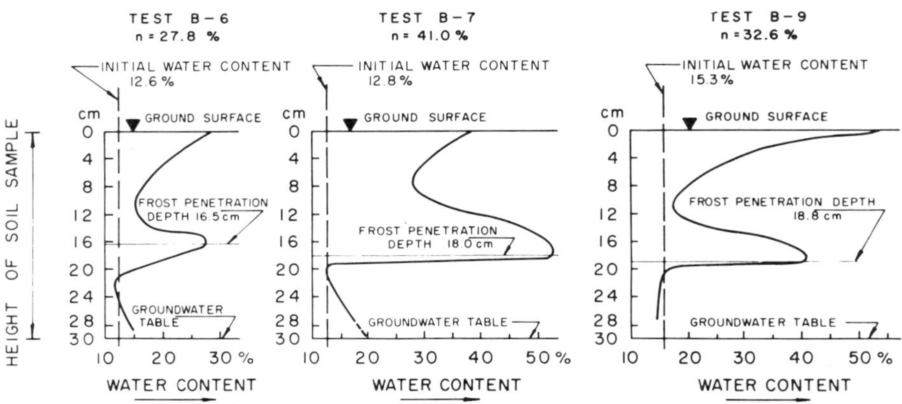

Figure 3. Soil moisture distribution in soil before and after freezing.

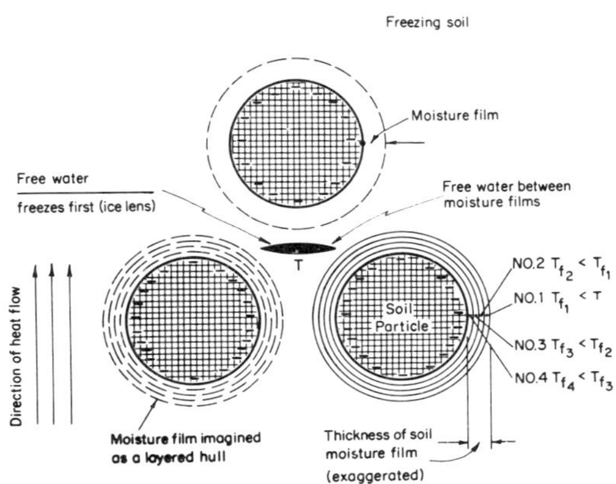

Figure 4. Diagrammatic illustration of freezing of soil moisture.

constants of the individual component parts composing the soil.

Under natural conditions, the various kinds of soil water, such as free water in bulk and film water, are seldom present singly. They usually exist simultaneously and have different freezing points. The free water in bulk in the voids of the soil is the first type of soil water to freeze (Figure 4), then the less-stressed water films (farther away from the surface of the soil solid particle), and so on. At the same time and temperature, imagine the layered parts of the water film, which are nearer to the surface of the soil particle, to be liquid, being subjected to a more intense attraction and having, therefore, a lower freezing point (see Figure 4). Thus layer No. 1, whose freezing point is $T_{f1} < T$, freezes first (T = freezing point temperature of free water). Then, as the temperature drops, layer No. 2 freezes at $T_{f2} < T_{f1}$, then layer No. 3, and so on. Thus, upon freezing, a fine-particled soil may contain a considerable amount of stressed, unfrozen water affecting the magnitude of the dielectric constant as compared with a dry soil.

Thickness of Ice Layer. In soil and foundation engineering, it is often necessary to know the maximum depth to which frost will penetrate a given type of soil and the rate of penetration. The depth depends upon the type of soil, its moisture content, thermal properties of the soil, and upon the cold temperatures of the microclimate atmosphere.

In ice and water management, hydrologists are interested to know the thickness of the ice sheet. The thickness ξ of the ice sheet, viz., the frost penetration depth

(ξ) in soil, can be calculated by Neumann's well-known formula devised for thickness ξ of formation of ice in calm polar waters:

$$\xi = m\sqrt{t} ,$$

where t is elapsed time of freezing and m is Neumann's coefficient of proportionality, to be determined from the following transcendental equation:

$$(0.5)\ L \rho_s w \pi m = \frac{K_1}{\sqrt{\alpha_1}} (T_f - T_s) \frac{e^{-(m^2/4\alpha_1)}}{G(m/2\sqrt{\alpha_1})} -$$

$$- \frac{K_2}{\sqrt{\alpha_2}} \frac{e^{-(m^2/4\alpha_2)}}{1 - G(m/2\sqrt{\alpha_2})} ,$$

wherein the following quantities must be known:

L = latent heat of fusion of water (ice);
ρ_s = density of ice;
w = soil (ice) moisture content by dry weight of soil, in decimal fractions;
$K_{1,2}$ = thermal conductivity of frozen and unfrozen soil, respectively (ice and water);
$\alpha_{1,2}$ = coefficients of thermal diffusivity of frozen and unfrozen parts of the layered system, respectively (= temperature conductivity);
T_f = 0° C = 32° F = freezing temperature;
T_s = surface temperature (= microclimate temperature);
e = 2.7182 . . . = base of natural logarithm system, and

$$G(m/2\sqrt{\alpha_{1,2}}) = G(\beta) = \frac{2}{\pi} \int_0^{\beta} e^{-\beta^2}\, d\beta$$

is Gauss' error function or the probability integral, all in consistent units.

Another simple formula for calculating thickness ξ of an ice sheet in calm water, considering one (ice) layer only, is due to Stefan:

$$\xi = \left[(2K_1/L\rho_i)\ (T_f - T_s)\ t \right]^{\frac{1}{2}} = m_s \sqrt{t} ,$$

where all symbols are as before, ρ_i = density of ice, and m_s = Stefan's coefficient of proportionality.

Thus, in general, the thickness of the ice sheet is proportional to the square root of the time t elapsed since the beginning of the application of freezing surface temperature (microclimate) T_s.

Contact Interface Temperature between a Snowpack and Soil. This is the problem of heat exchange between soil and snow, and between snow and the atmosphere--in other words, the problem of snow cover as an insulating material on the ground surface. The question of knowing the soil surface temperature underneath a snow cover is of very great importance, not merely in thermal soil mechanics and the geo-cryogenic sciences and geothermics but also in agroclimatology, meteorology, hydrology, forestry.

The contact-interface temperature also affects electrical measurements made in studies of soil, snow and ice, as well as the magnitude of the dielectric constants of these materials. The contact-interface temperature T_{cont} between the ground surface and a snowpack can be calculated approximately based on the principle that the linear temperature gradients in soil (i_{soil}) and in snow (i_{snow}) are inversely proportional to the respective coefficients of their thermal conductivities, K_{soil} and K_{snow}, i. e.,

$$\frac{i_{soil}}{i_{snow}} = \frac{K_{snow}}{K_{soil}} .$$

Also,

$$i_{snow} = \frac{T_{cont} - T_{air}}{x_{snow}} .$$

If T_{fr} = freezing isotherm (say 0° C) at frost penetration depth ξ (viz., freezing temperature of water in voids of soil);

T_{cont} = contact-interface temperature;

T_{air} = microclimate temperature immediately above the surface of the snowpack;

x_{snow} = thickness of snow cover (snowpack), and

ξ = frost penetration depth (thickness of frozen soil layer),

all in consistent units, then the contact-interface temperature calculates as

$$T_{cont} = T_{fr} - i_{soil}\,\xi = T_{fr} - i_{snow}\frac{K_{snow}}{K_{soil}}\,\xi =$$

$$= T_{fr} - (T_{cont} - T_{air})\frac{K_{snow}}{K_{soil}}\frac{\xi}{x_{snow}} , \quad \text{or}$$

$$T_{cont} = T_{fr}/[1 + (K_{snow}/K_{soil})(\xi/x_{snow})] +$$

$$+ T_{air}/[1 + (K_{soil}/K_{snow})(x_{snow}/\xi)] .$$

If it can be assumed that $T_{fr} \approx 0°$ C, then

$$T_{cont} = T_{air}/[1 + (K_{soil}/K_{snow})(x_{snow}/\xi)].$$

Here T_{air}, K_{soil}, K_{snow}, x_{snow} and ξ must be known.

The contact-interface temperature T_{cont} can also be calculated based on the principle that the contact interface snow/soil should receive as much heat from the soil below as it loses to the snow cover above the contact interface if there is an upward flow of heat. Equating the heat flow rates through soil and through snow, and observing the continuity of heat flow from soil through snow,

$$T_{cont} = \frac{(K_{soil} \cdot T_{soil})/\sqrt{\alpha_{soil}} + (K_{snow} \cdot T_{snow})/\sqrt{\alpha_{snow}}}{K_{soil}/\sqrt{\alpha_{soil}} + K_{snow}/\sqrt{\alpha_{snow}}}.$$

Here T_{soil} = initial temperature of soil;
T_{snow} = initial temperature of snow;
K_{soil}, α_{soil} = thermal conductivity and diffusivity of soil, respectively;
K_{snow}, α_{snow} = thermal conductivity and diffusivity, respectively, of snow cover.

Then $T_{soil} > T_{contact} > T_{snow}$. T_{soil}, T_{snow}, K_{soil}, K_{snow}, α_{soil} and α_{snow} must be known.

Dielectric Constant. Besides texture, the dielectric constant of a porous medium such as soil, snowpack and ice depends upon its structure. There is a need for systematic studies of dielectric constants of soils, snow and ice in the bulk condition (soil solids, water and gas), at various temperatures and frequencies.

The author's studies of electrical parameters of fine-particled, frost-prone soils have shown that the dielectric constants of bulk soils are sensitive to temperature, water content, porosity of the soil and to the electrical oscillating frequency applied to the soil sample. The soil samples were tested at six equilibrium temperatures, +30° C, +20° C, +10° C, 0° C, -10° C and -20° C, and at four frequencies, namely, 0.1, 1.0, 10 and 100 kc/sec. The relative dielectric constants (dimensionless) of the five bulk soils (solids + water + air) studied were found to vary from 1.91 to 10.2 for a water content of 0.0% (oven-dry soil), and from 2.47 to 57.2 for a water content of 1.0%.

The dielectric constants of the soil solids were found to vary from 2.52 to 5.84, and the average dielectric constants of the stressed water film in these tests varied from 22.0 to 2,819.0, all depending upon soil type, porosity, moisture content, temperature and frequency.

The dielectric constants of the bulk soil, of the solids and of the water film increase with an increase in temperature, with a decrease in soil porosity and with decrease in frequency. Also the dielectric constant of the bulk soil increases with an increase in water content.

CONCLUSIONS

Because water management research, hydrology and runoff studies in one way or another usually end up with some engineering projects such as constructing water-storage reservoir dams, flood control, regional irrigation and/or drainage facilities, and development of hydraulic power schemes, and for the purpose of good cooperation, mutual understanding and ease of communication among the engineering and scientific professions, the following would be desirable:

a) to qualify, define and agree on soil porosity and soil moisture classification systems;
b) to define terms such as dielectric constant (absolute, relative, total, net, effective, bulk, compound, component, composite, individual, for example).

Because each material has its own individual dielectric constant (a property of the material), an important agreement must be made as to what frequency to use for determining the dielectric constant, whether low, high, microwave frequency, or that of a molecular vibration frequency of the particular material.

It is envisioned that the dielectric constant data on soil, water, snow and ice would facilitate agreement between the observed and theoretically calculated quantities not merely in soil mechanics, but also in the various bordering disciplines such as snow and ice hydrology.

ELECTRICAL PARAMETERS OF SOME FROST-PRONE SOILS

Alfreds R. Jumikis
Department of Civil and Environmental Engineering
Rutgers University, New Brunswick, New Jersey, U.S.A.

W. A. Slusarchuk
National Research Council, Ottawa, Canada

INTRODUCTION

General Notes. Dielectric constant and electroconductance determinations on various organic and inorganic materials are a common practice in nearly all branches of technology for characterizing certain of their physical properties. Dielectric constant determination of porous materials, such as dry and moist, frozen and unfrozen soils, is now also receiving increasing attention from those interested in the earth sciences; cold regions research; snow, ice and permafrost studies; various phases of thermal soil mechanics, and by man's various engineering activities in the Arctic, Antarctic and on the moon. In addition, the various engineering problems peculiar to northern communities in Alaska and Canada, for example, are offering new challenges to soil and foundation engineers. In these regions, the environmental factors, such as freezing temperatures, for example, require completely new design and construction methods as compared with conventional ones under moderate climatic conditions.

It is an indisputable fact that research techniques hitherto used in soil mechanics and thermal soil mechanics have recently added considerably to the knowledge and better understanding of the behavior of soil under the action of load, water and temperature. There is, therefore, a very good reason to believe that independent soil parameters such as the dielectric constant and electrical conductance of soils, for example, could also assist the soil engineer in his thermal soil mechanics work with frozen and unfrozen soils, among them also the relatively dry, loose, dusty, frozen and unfrozen polar desert soils of the Arctic and the cold desert soils of the Antarctic (Tedrow, 1966a, b; 1971).

Dielectric Constant. The dielectric constant D of a soil material is an electrical property which is a measure of the ability of a dielectric to store electric energy in the presence of an electrostatic field. In general, the dielectric constant of a soil material is a basic property reflecting its molecular configuration, composition, texture and porosity.

Electrically, the (relative) dielectric constant D is the dimensionless ratio of the permittivity ε over the permittivity ε_o of free space (vacuum) (Gamow and Cleveland, 1960; Morgan, 1967):

$$D = \varepsilon/\varepsilon_o \, .$$

For vacuum, the relative dielectric constant is D = 1.0. It is nearly equal to 1 for air (Eshbach, 1952). In these studies, a cylindrical capacitor was used. By means of it, the dielectric constant D was determined as

$$D = \frac{C_{xp}}{C_o} = \frac{\text{Capacitance of capacitor containing a dielectric}}{\text{Capacitance of capacitor containing air}}$$

where C_{xp} and C_o are in pica-Farads (pF) (General Radio Co., 1965).

Electrical Conductance. The electrical conductance $\lambda = 1/R$ in mhos of a material is the inverse of its electrical resistance R (in ohms) to current flow (Freiberger, 1960). The conductance values for the soil samples tested were obtained from capacitance and dissipation factor measurements based on bridge balancing operations.

USE OF DIELECTRIC CONSTANT AND ELECTROCONDUCTANCE IN THERMAL SOIL MECHANICS

Some of the uses of electrical parameters in thermal soil mechanics are as follows:

1) Soils are classified as to their performance under various temperature conditions.
2) Some soil properties, such as porosity and moisture conditions, can be characterized.
3) In frost-prone soil systems, the dielectric constant and electroconductance are vitally necessary for a better understanding of the soil moisture migration phenomenon under the effect of a thermal gradient (Jumikis, 1966).
4) These parameters are needed in the theory of thermo-osmotic and electro-osmotic studies for dewatering silty soils to lay foundations in temperate and permafrost regions (Jumikis, 1962; 1971).
5) Such data are used in electro-osmotic pile-driving (Jumikis, 1971).
6) The electrical parameters are needed for calculating the electrokinetic (zeta) potential (Jumikis, 1969).
7) They are used in electro-osmotic stabilization of fine-particled soils.
8) Electrical parameters of dry soils may have a use in dry climatic regions, among them the dry polar desert soils of the Arctic and the cold desert soils of the Antarctic (Tedrow, 1966a, b; 1971), as well as for soils under dry, high and low temperature conditions, such as encountered in lunar and possibly other extra-terrestrial environments.
9) Electrical parameters are also necessary for grounding electronic and radio equipment, telecommunication devices, various electrical installations, electrical generating plants, and distribution networks.
10) In general, electrical parameters of soils are needed in research and further development of the engineering discipline of thermal soil mechanics, and snow and ice hydrology.

REVIEW OF PUBLICATIONS ON DIELECTRIC CONSTANTS OF SOILS

A review of the pertinent literature reveals that in the civilian sector very little significant data are available on dielectric constants for engineering soils. Also, a review of the six volumes on Progress in Dielectrics (Birks and Hart, eds., 1959-1961; 1962-1965) reveals absolutely no information on dielectric constants of soils. Nor does the Digest of Literature on Dielectrics (Nat. Acad. Sci., 1964) contain any such information. A very few dielectric constant data at constant temperatures were reported by von Hippel (1954). These data pertain to bulk clay-water mixtures. Work done by Smith-Rose (1933) pertains to bulk electrical properties of soil for alternating currents at radio frequencies.

A paper by Saxton (1952), and researches of a more recent date by Schwan et al. (1962), Schwarz (1962), and Hoekstra and O'Brien (1969) all deal with dielectric dispersions of colloidal soil particles in electrolyte solutions, and montmorillonite clay in suspensions at constant temperature. These works, too, pertain to bulk properties of such solutions and slurries. Recently the dielectric constants and conductances of some dry frost-prone soils were studied experimentally by Jumikis (1969). The electrical parameters of dry and moist soils at very low moisture contents were studied by Slusarchuk (1971).

The Problem. For the reasons described above the authors endeavored to study the electrical parameters of dry and moist soils experimentally as a function of soil type; its porosity; moisture content; frequency and temperature. Five New Jersey fine-particled, frost-prone soils were used in these studies. Electrical parameters were determined on bulk soil and soil solid particles. An attempt was also made to shed some light on the order of magnitude of the dielectric constant and electroconductance of the water film surrounding the soil solid particles. Scrutiny and adoption of associated instrumentation within the limits of available resources were undertaken, and a methodology for determining the electrical parameters was developed.

INSTRUMENTATION

Main Component Parts. The main component parts of the instrumentation used in these studies were: 1) a dielectric soil sample holder-capacitor, 2) a hot-cold temperature chamber, 3) a capacitance bridge for measuring the soil capacitance, dissipation factor and resistance, 4) an oscillator with varying frequencies, and 5) other pertinent assessories (General Radio Co., 1966; 1967). The General Radio Company's type 716-C Schering capacitance bridge was used (General Radio Co., 1965).

SOIL MATERIALS IN THESE STUDIES

Because polar desert soil samples from the Arctic

Table 1

DESCRIPTION OF SOILS STUDIED IN FROST ACTION RESEARCH

Name of Soil	HRB Class	Texture of Soil	Origin of Soil
Dunellen	A-2-4	A silty sand with traces of gravel.	Derived from stratified glacial outwash, mostly Triassic rock fragments
Lansdale	A-7-6	Well-graded sand-silt-clay mixture containing considerable gravel.	Derived from underlying Triassic shale, sandstone, argillite.
Merrimac	A-1-b	Mixture of coarse, medium and fine sand containing gravel, clay, some silt.	Derived from gneissic glacial materials reworked by water.
Montalto	A-2-4	Sandy gravel with considerable silt and clay.	Derived from basalt and diabase. Gravel is large angular fragments.
Penn	A-2-4	Well-graded mixture of friable shale fragments, gravel to clay sizes.	Derived from Triassic shales. Angular fragments may have been broken up by compacting in pits during placement of sample.

were unavailable, and to develop a testing procedure, the authors resorted to the study of five New Jersey fine-particled, frost-prone soil types. The description of these soils studied is given in Table 1. Some of their physical properties are compiled in Table 2. The specific surface area and some of the predominant minerals of the soil types tested are given in Table 3. The specific surface area, in m^2/g, was determined by the Sor-BET method (American Instrument Co., 1969).

Each soil sample was tested at six equilibrium temperatures, +30° C, +20° C, +10° C, 0° C, -10° C and -20° C; and at four frequencies, namely, 0.1, 1.0, 10 and 100 kc/sec. The soil electrical parameter studies were performed in the Thermal Soil Mechanics Laboratory, College of Engineering, Rutgers University, under National Science Foundation grants.

EXPERIMENTAL RESULTS

Treatment of Experimental Data. The soil type chosen to be reported here is the Penn soil. Some of these results are shown in Figures 1 through 6 as examples. No results were obtained for water contents more than 1.0% because the dissipation factors exceeded the range of the bridge.

Table 2

PHYSICAL PROPERTIES OF THE SOILS TESTED

Name of Soil	HRB Class.	Munsell Color Code Dry	Munsell Color Code Wet	Spec. Grav. G_s
1	2	3	4	5
Dunellen	A-2-4	5 Yr6/3	5 Yr3/3	2.66
Lansdale	A-7-6	10 Yr6/3	7.5 Yr4/4	2.78
Merrimac	A-1-b	10 Yr3/2	7.5 Yr3/2	2.66
Montalto	A-2-4	7.5 Yr3/4	5 Yr3/4	2.66
Penn	A-2-4	5 Yr5/4	2.5 Yr3/4	2.76

Name of Soil	Percent Silt	Percent Clay	Percent Silt & Clay	Consistency LL	Consistency PL	Max. Dry Density (lb/ft^3) (t/m^3)	Opt. Moist. Cont. (%)
1	6	7	8	9	10	11	12
Dunellen	10	10	20	16	0	120/1.92	12
Lansdale	23	32	55	41	15	95/1.52	26
Merrimac	11	--	11	NL	NP	129/2.07	9
Montalto	9	19	28	32	9	114/1.82	17
Penn	16	19	35	31	7	106/1.70	17

Table 3

SPECIFIC SURFACE AREA AND PREDOMINANT MINERALS OF SOIL TYPES

Name of Soil	Specific Surface Area (m^2/g)	Predominant Minerals
Dunellen	5.7	Quartz, feldspars, muscovite
Lansdale	17.5	Quartz, feldspars
Merrimac	2.1	Quartz, feldspars
Montalto	22.6	Quartz, feldspars, muscovite
Penn	22.4	Quartz, feldspars, muscovite

In the graphs in Figures 1 through 6 the curves are the analytical second-order, best-fit curves found by means of the statistical method of least squares (Spiegel, 1961). These least-squares parabolic curves were computed by the IBM 1130 computer and plotted by an IBM 1627 (X-Y) plotter connected to that computer (Slusarchuk, 1971).

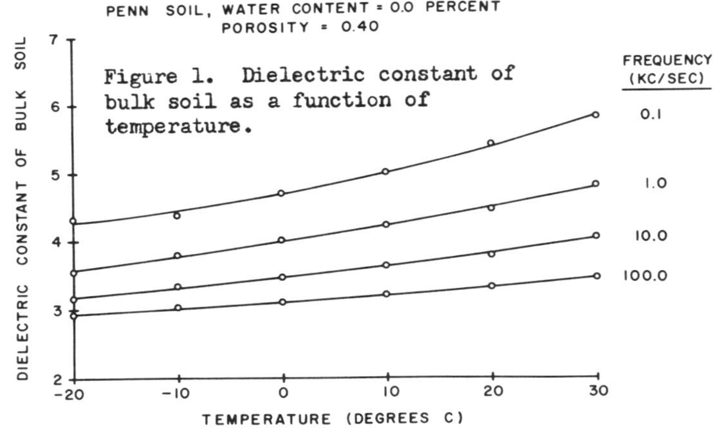

Figure 1. Dielectric constant of bulk soil as a function of temperature.

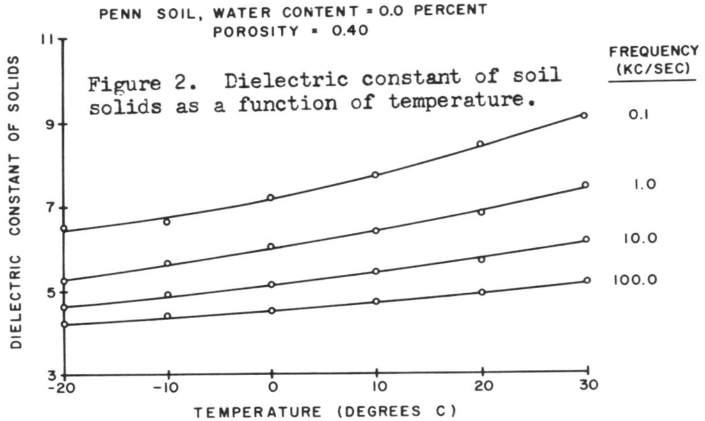

Figure 2. Dielectric constant of soil solids as a function of temperature.

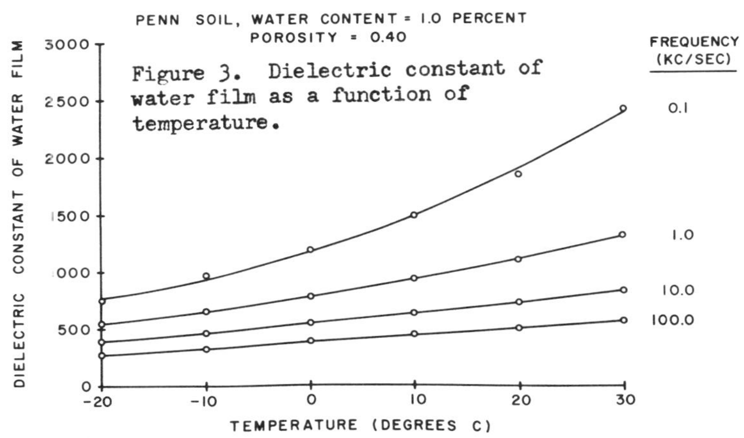

Figure 3. Dielectric constant of water film as a function of temperature.

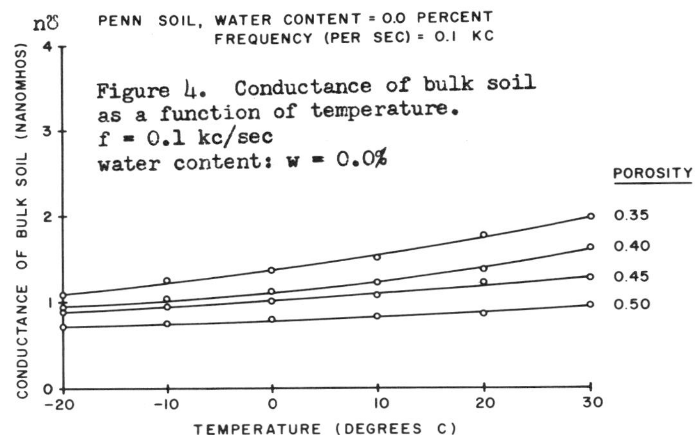

Figure 4. Conductance of bulk soil as a function of temperature. f = 0.1 kc/sec water content: w = 0.0%

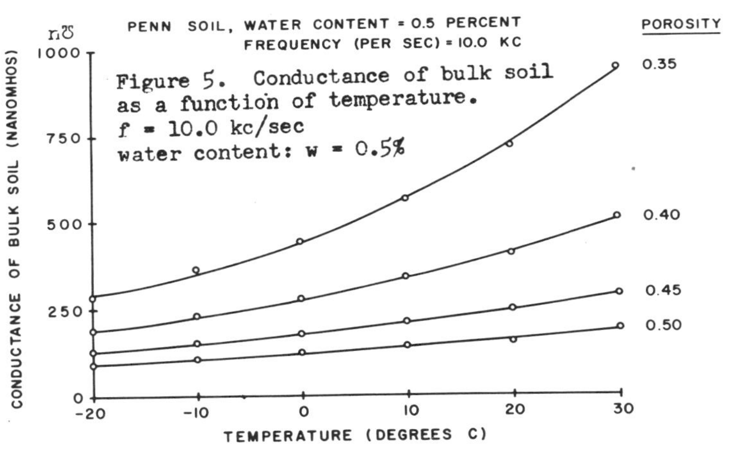

Figure 5. Conductance of bulk soil as a function of temperature. f = 10.0 kc/sec water content: w = 0.5%

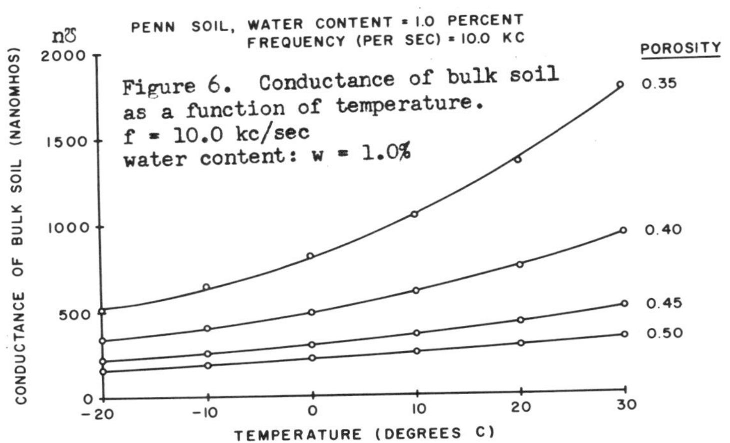

Figure 6. Conductance of bulk soil as a function of temperature. f = 10.0 kc/sec water content: w = 1.0%

Dielectric Constant of Bulk Soil. The data show that for a particular frequency the dielectric constant values increase with increasing soil temperature and increasing water content. This observation for a porous medium such as soil agrees with the opinion of Srivastava and Varshni (1956) that the dielectric constant of solids decreases as the temperature decreases. In these experiments, the dielectric constant also increases with decreasing porosity of the soil.

For dry and moist soils the effect of frequency is also corroborated by the few data tabulated by von Hippel (1954).

The general dielectric constant-temperature relationships of the soil-air and soil-air-water systems by the authors' tests agree also with the relationships of many materials other than soil. For example, they agree with those of aluminum oxide, talc, soapstone, ceramic materials and quartz (von Hippel, 1954; Sutton, 1960), as well as with the dry soils previously studied experimentally at Rutgers University (Jumikis, 1966).

Dielectric Constant of Soil Solids. The data given in Figures 1-6 and Table 4 show that the magnitudes of dielectric constant data of the soil solids are greater than those of the bulk soil. The dielectric constant of soil solids should be regarded as a component of the total dielectric constant of the bulk soil. Here the effect of the porosity of the bulk soil on the dielectric constant can be seen. In other words, this means that the dielectric constant of the solid particles of a dry soil depends upon the amount of solid matter surrounding them. The more solid matter surrounding a soil particle, the greater its dielectric constant.

Dielectric Constant of Water Film. The average water film thickness, calculated as the total volume of water added to the dry solids in the soil sample divided by the total surface area of the solids in the soil sample, did not exceed 20 Ångstrom units (see Table 3 for specific surface areas). As shown by Dorsey (1940), the tabulated diameters of a water molecule range from 2.34 to 6.4 Å, depending on the researcher and method of measurement. Thus, using these water-molecule diameters, it can be seen that the average water film was only a few water molecules thick.

From the data it is seen that the dielectric constant of the water film increases with increasing temperature, with decreasing porosity and with decreasing frequency. The slopes of the curves are gentle and the greatest slopes occur at the higher temperatures. These results are similar to the results obtained for dielectric constant vs. temperature for the solids, the bulk soil and other solid materials referred to, but are not similar to the results of water in bulk. As given by Dorsey (1940), the dielectric constant of the bulk water decreases with temperature, which is opposite to the effect of temperature on the dielectric constant of the water film found in this experimental research. Thus

Table 4

TEST RESULTS FOR PENN SOIL

Water Content: w = 0.00%; Porosity: n = 0.50%

Dielectric Constant of Bulk Soil

°C	Frequency (kc/sec)			
	0.1	1	10	100
-20	2.7505	2.4052	2.1911	2.0618
-10	2.8587	2.4919	2.2560	2.1050
0	2.9449	2.5782	2.2990	2.1482
10	3.0966	2.7077	2.3854	2.1913
20	3.2266	2.7722	2.4502	2.2127
30	3.3555	2.9014	2.5147	2.2557

Dielectric Constant of Soil Solids

°C	Frequency (kc/sec)			
	0.1	1	10	100
-20	4.5011	3.8105	3.3823	3.1237
-10	4.7175	3.9839	3.5120	3.2100
0	4.8898	4.1564	3.5980	3.2964
10	5.1933	4.4154	3.7709	3.3826
20	5.4532	4.5445	3.9004	3.4255
30	5.7110	4.8028	4.0295	3.5115

Conductance (in Nanomhos) of Bulk Soil

°C	Frequency (kc/sec)			
	0.1	1	10	100
-20	0.7059	5.3193	28.9672	203.8597
-10	0.7390	5.4951	31.6003	227.9704
0	0.7941	5.9786	35.1111	249.8889
10	0.8184	6.6159	38.6218	278.3824
20	0.8493	7.0993	42.1324	300.3002
30	0.9485	7.9563	46.9593	333.1762

Water Content: w = 0.00%; Porosity: n = 0.45%

Dielectric Constant of Bulk Soil

°C	Frequency (kc/sec)			
	0.1	1	10	100
-20	3.7259	3.2500	2.8844	2.6038
-10	3.9421	3.4009	2.9703	2.6682
0	4.0501	3.4872	3.0133	2.7114
10	4.0924	3.5518	3.0346	2.7328
20	4.5248	3.8109	3.2507	2.8622
30	4.8936	4.0483	3.4016	2.9483

Table 4 (Continued)

Dielectric Constant of Soil Solids

°C	Frequency (kc/sec) 0.1	1	10	100
-20	5.9563	5.0910	4.4262	3.9160
-10	6.3494	5.3653	4.5824	4.0332
0	6.5457	5.5223	4.6606	4.1117
10	6.6226	5.6398	4.6992	4.1506
20	7.4088	6.1108	5.0922	4.3859
30	8.0793	6.5424	5.3665	4.5424

Conductance (in Nanomhos) of Bulk Soil

°C	Frequency (kc/sec) 0.1	1	10	100
-20	0.8779	7.2751	45.2041	306.8755
-10	0.9529	8.1541	52.6636	361.6682
0	0.9992	8.5496	55.9545	385.7764
10	1.0742	8.9450	59.4647	409.8842
20	1.2175	10.0875	68.0204	482.2053
30	1.2572	11.1640	76.7949	545.7572

Water Content: w = 0.00%; Porosity: n = 0.40%

Dielectric Constant of Bulk Soil

°C	Frequency (kc/sec) 0.1	1	10	100
-20	4.3118	3.5540	3.1668	2.9297
-10	4.3756	3.7917	3.3395	3.0375
0	4.7002	4.0075	3.4686	3.1017
10	5.0029	4.2231	3.6413	3.2095
20	5.4134	4.4600	3.7920	3.3171
30	5.8005	4.8050	4.0511	3.4679

Dielectric Constant of Soil Solids

°C	Frequency (kc/sec) 0.1	1	10	100
-20	6.5197	5.2567	4.6113	4.2162
-10	6.6260	5.6529	4.8991	4.3958
0	7.1671	6.0126	5.1144	4.5029
10	7.6715	6.3718	5.4021	4.6825
20	8.3557	6.7666	5.6534	4.8618
30	9.0009	7.3418	6.0851	5.1132

Table 4 (Continued)

Conductance (in Nanomhos) of Bulk Soil

°C	Frequency (kc/sec) 0.1	1	10	100
-20	0.9331	7.4289	45.6429	304.6837
-10	1.0301	8.3518	54.4188	374.8182
0	1.1139	9.3845	63.4135	449.3325
10	1.2175	10.5489	71.9690	515.0773
20	1.3652	11.9548	81.6207	593.9672
30	1.6099	13.6022	93.2457	685.9999

Water Content: w = 0.00%; Porosity: n = 0.35%

Dielectric Constant of Bulk Soil

°C	Frequency (kc/sec) 0.1	1	10	100
-20	4.6574	3.9431	3.5133	3.2114
-10	4.9372	4.2017	3.6639	3.2974
0	5.3265	4.4606	3.8364	3.4050
10	5.7587	4.7412	4.0304	3.5124
20	6.2108	5.0645	4.2678	3.6851
30	6.7071	5.3873	4.4834	3.8142

Dielectric Constant of Soil Solids

°C	Frequency (kc/sec) 0.1	1	10	100
-20	6.6267	5.5279	4.8666	4.4022
-10	7.0572	5.9258	5.0984	4.5345
0	7.6563	6.3241	5.3637	4.7000
10	8.3211	6.7557	5.6621	4.8653
20	9.0166	7.2531	6.0275	5.1310
30	9.7802	7.7497	6.3591	5.3295

Conductance (in Nanomhos) of Bulk Soil

°C	Frequency (kc/sec) 0.1	1	10	100
-20	1.0764	8.8572	54.4188	368.2432
-10	1.2506	10.2852	65.1686	440.5663
0	1.3608	11.5154	75.0401	519.4602
10	1.5107	12.7675	86.2270	607.1151
20	1.7620	14.3928	96.7550	690.3823
30	1.9581	16.3034	108.3787	784.5998

Table 4 (Continued)

Water Content: w = 1.00%; Porosity: n = 0.50%

Dielectric Constant of Bulk Soil

°C	Frequency (kc/sec) 0.1	1	10	100
-20	8.3076	6.7703	5.4974	4.4795
-10	9.8477	7.6731	6.1225	4.9551
0	11.4464	8.6606	6.7905	5.4297
10	12.7215	9.5190	7.4144	5.8829
20	13.8936	10.5449	8.0819	6.3574
30	16.2676	11.6091	8.7462	6.8313

Dielectric Constant of Water Film

°C	Frequency (kc/sec) 0.1	1	10	100
-20	406.2693	319.3960	242.4235	177.9193
-10	510.0130	378.5738	283.0585	209.2734
0	619.5369	443.9088	328.3308	240.5479
10	701.0575	496.8620	367.3594	270.2812
20	776.6635	566.5013	411.0660	303.0796
30	939.2372	634.3616	454.5574	334.3185

Conductance (in Nanomhos) of Bulk Soil

°C	Frequency (kc/sec) 0.1	1	10	100
-20	2.3634	20.6943	154.2055	1334.4365
-10	3.5258	25.8515	183.7975	1483.3562
0	4.7790	31.6202	214.9160	1691.3758
10	5.7890	36.4432	247.7783	1873.0898
20	5.8329	43.4539	278.4407	2070.0973
30	9.3343	51.7717	322.2282	2291.1427

Water Content: w = 1.00%; Porosity: n = 0.45%

Dielectric Constant of Bulk Soil

°C	Frequency (kc/sec) 0.1	1	10	100
-20	11.2121	8.8188	7.0951	5.6664
-10	13.0532	9.9768	7.8486	6.2495
0	15.1929	11.3528	8.7736	6.8962
10	17.8756	12.9470	9.8041	7.6289
20	20.9274	14.6632	10.8531	8.3388
30	24.5905	16.5780	12.0053	9.0686

Table 4 (Continued)

Dielectric Constant of Water Film

°C	Frequency (kc/sec) 0.1	1	10	100
-20	499.1580	371.9714	281.8336	205.6887
-10	606.5998	438.5986	325.9720	239.9736
0	740.6501	523.7177	384.1572	279.8164
10	914.6776	624.6059	450.6869	326.7158
20	1088.0405	721.0779	505.9541	365.1921
30	1305.7465	832.0213	572.1949	407.7557

Conductance (in Nanomhos) of Bulk Soil

°C	Frequency (kc/sec) 0.1	1	10	100
-20	3.7238	28.7692	213.1631	1824.9274
-10	4.8230	35.7857	249.9687	2026.3208
0	6.6662	41.9208	295.9580	2304.2728
10	9.3343	55.0527	352.8680	2617.1582
20	12.8356	69.0352	418.4905	2967.1298
30	17.9815	86.2546	499.3569	3382.5629

Water Content: w = 1.00%; Porosity: n = 0.40%

Dielectric Constant of Bulk Soil

°C	Frequency (kc/sec) 0.1	1	10	100
-20	16.6101	12.4245	9.5695	7.3881
-10	20.1745	14.4903	10.9004	8.3143
0	24.1885	16.7542	12.3342	9.3248
10	29.5594	19.5169	13.9745	10.4194
20	35.8659	22.5424	15.6481	11.4684
30	45.6815	26.5058	17.7275	12.7054

Dielectric Constant of Water Film

°C	Frequency (kc/sec) 0.1	1	10	100
-20	746.6775	539.1175	389.9511	272.5406
-10	957.4660	649.5357	459.9436	321.9739
0	1180.0410	773.1554	538.6805	379.0356
10	1485.5747	926.8097	627.2944	438.6056
20	1841.1017	1095.0380	719.1987	495.4400
30	2409.1901	1313.3764	829.1816	561.0604

Table 4 (Continued)

Conductance (in Nanomhos) of Bulk Soil

°C	Frequency (kc/sec) 0.1	1	10	100
-20	7.1700	47.6139	333.1721	2704.6622
-10	10.4460	62.0474	403.1829	3098.3381
0	15.1263	79.9391	488.4336	3579.2847
10	22.8407	105.5929	601.9609	4147.3565
20	34.6242	138.8779	732.7328	4758.7142
30	57.8631	191.2717	921.8774	5587.6811

Water Content: w = 1.00%; Porosity: n = 0.35%

Dielectric Constant of Bulk Soil

°C	Frequency (kc/sec) 0.1	1	10	100
-20	24.5066	16.9039	12.3717	9.1443
-10	31.0133	20.3176	14.3953	10.4963
0	38.9272	24.1496	16.5549	11.9077
10	50.6616	29.1212	19.1341	13.4425
20	55.0477	35.1114	22.0491	15.1008
30	57.1695	43.4379	25.6986	16.9251

Dielectric Constant of Water Film

°C	Frequency (kc/sec) 0.1	1	10	100
-20	1111.6317	727.0539	498.0123	334.6881
-10	1458.7160	903.0939	602.5153	405.3004
0	1878.2058	1102.4068	713.4037	478.0485
10	2508.0648	1363.9928	846.4847	557.6734
20	2727.7685	1679.9648	995.9085	640.6415
30	2819.0201	2126.0254	1187.4063	735.2128

Conductance (in Nanomhos) of Bulk Soil

°C	Frequency (kc/sec) 0.1	1	10	100
-20	14.5224	80.5927	510.2787	3863.3651
-10	22.6091	111.0124	645.5791	4540.4229
0	34.6441	150.2745	808.8967	5369.6171
10	56.9952	210.1057	1036.7966	6350.4016
20	38.5121	295.8256	1336.9304	6807.6413
30	51.5500	442.2290	1762.9101	9695.8350

the water film (oriented and stressed water molecules) appears to behave more like a solid than a liquid with regard to its dielectric constant properties.

Another interesting observation is that the dielectric constant values of the water film at a particular porosity and frequency do not appear to depend upon the thickness of the water film used in these tests. Of course, only two water contents were used, other than 0.0%, so only two water film thicknesses were obtained. Thus concrete conclusions cannot be drawn on this point at this time. It appears, though, that the water contents experimented with were too low for bringing about the effect of the soil water on the magnitude of the dielectric constant of the water film. However, the equipment limitations did not permit greater water contents to be used in the experiments. As the water content increased the dissipation factor and electrical conductance increased, and at these higher values the capacitance bridge could not be balanced at the frequencies of 0.1, 1, 10 and 100 kc/sec. Greater water contents could have been used if the experimental frequency range could have been extended to higher values.

It was thus found in this research that the dielectric constant values of the water film are greater than the values of the dielectric constant of pure bulk water, which usually has a value of approximately 80 at these frequencies (Dorsey, 1940). As shown by Dorsey, the dielectric constant of water changes abruptly as it freezes. This fact could be used to determine the amount of unfrozen water present in soil at temperatures below freezing.

Electrical Conductance Data. For the Penn soil, the electrical conductance data are given in Table 4 for the bulk soil and soil solid-air and soil-air-moisture systems. They are shown as functions of soil porosity, water content, temperature and frequency. Their test ranges were the same as those used in dielectric constant determinations.

Electroconductance of Bulk Soil. The conductance increases with increasing temperature, with increasing frequency and with decreasing porosity of the soil (Table 4). For a given porosity and frequency, the conductance increases with water content.

At a frequency from 0.1 to 1.0 kc/sec, the electrical conductance values are small, and the slopes of the curves are very gentle. From 10 to 100 kc/sec the conductance values become large, and the slopes of their curves become steep as the 100 kc/sec frequency value is approached.

CONCLUSIONS

1) The electrical parameters, that is, the dielectric constant and conductance, as shown in this research are sensitive to temperature, water content, porosity of the soil and to the electrical frequency applied to the soil sample.

2) The dielectric constants of the five bulk soils studied were found to vary from 1.91 to 10.2 for a water content of 0.0% (dry soil) and from 2.47 to 57.2 for a water content of 1.0%.

3) The dielectric constants of the soil solids were found to vary from 2.52 to 5.84, and the dielectric constants of the water film varied from 22.0 to 2,819.0, all depending upon soil type, porosity, moisture content, temperature and frequency.

4) The dielectric constants of the bulk soil, of the solids and of the water film increase with an increase in temperature, with a decrease in soil porosity and with a decrease in frequency. Also the dielectric constant of the bulk soil increases with an increase in water content.

5) Just as the dielectric constant is sensitive to temperature, frequency, water content and porosity, so is the electrical conductance.

6) The electrical conductances of the bulk soil varied from 0.4 to 965.8 nanomhos. All of these values depended upon soil type, porosity, water content, frequency and temperature at which the soil was tested.

7) Experiments at higher frequencies and greater soil moisture contents should be pursued by means of instruments having proper capacity.

REFERENCES

American Instrument Co., 1969. Sor-BET, bulletin 2404, Silver Springs, Maryland.

Birks, J. B., and J. Hart, eds., 1959-1961. Progress in dielectrics, London, Heywood & Co., Ltd., vols. 1-3.

Birks, J. B., and J. Hart, eds., 1962-1965. Progress in dielectrics, New York, Academic Press, Inc., vols. 4-6.

Dorsey, N. E., 1940. Properties of ordinary water-substance, New York, Reinhold Publ. Corp., pp. 43, 357-372.

Eshbach, O. W., 1952. Handbook of engineering fundamentals, 2d ed., New York, John Wiley & Sons, Inc., pp. 9-36.

Freiberger, W. F., 1960. The international dictionary of applied mathematics, Princeton, N. J., D. Van Nostrand Co., p. 161.

Gamow, G., and J. M. Cleveland, 1960. Physics, Englewood Cliffs, N. J., Prentice-Hall, pp. 189-192.

General Radio Co., 1965. Capacitance bridge, type 716-C, operating instructions, West Concord, Mass.

General Radio Co., 1966. Null detector, type 1232-A, instruction manual, West Concord, Mass.

General Radio Co., 1967. Oscillator, type 1310-A, instruction manual, West Concord, Mass.

Hoekstra, P., and H. W. O'Brien, 1969. The dielectric properties of clay suspensions in the frequency range from 50 Hz to 20 kHz, Hanover, N. H., Corps of Engineers, U. S. Army, Cold Regions Research and Engineering Laboratory, CRREL Res. Rept. 266.

Jumikis, A. R., 1966. Thermal soil mechanics, New Brunswick, N. J., Rutgers University Press, pp. 85-92.

Jumikis, A. R., 1962. Soil mechanics, Princeton, N. J.,

D. Van Nostrand Co., Inc., pp. 222-226.
Jumikis, A. R., 1971. Foundation engineering, Scranton, Pa., Intext Educational Publishers, pp. 273-277; ibid., pp. 633-635.
Jumikis, A. R., 1969. Concerning an analytical verification of the streaming potential function, in A report to the National Science Foundation on frost action in frost-prone soils.
Jumikis, A. R., 1969. Determination of dielectric constant and electroconductance of some frost-prone soils, New Brunswick, N. J., Rutgers University, College of Engineering, Bureau of Engineering Research.
Morgan, B. H., 1967. Conductors and dielectrics (in Electricity and magnetismus), in Handbook of the engineering sciences, J. Potter, ed., Princeton, N. J., D. Van Nostrand Co., Inc., vol. I, pp. 300-302.
National Academy of Sciences-National Research Council, 1964. Digest of literature on dielectrics, Washington, D. C., Superintendent of Government Documents. Publication No. 1342, vol. 28.
Saxton, J. A., 1952. Dielectric dispersion in pure liquids at very high radio-frequencies, Proc. Roy. Soc., London, 213A:473-492.
Schwan, H. P., G. Schwarz, J. Maczuk and H. Pauly, 1962. On the low-frequency dielectric dispersion of colloidal particles in electrolyte solution, J. Phys. Chem., 66:2626-2635.
Schwarz, G., 1962. A theory of the low frequency dielectric dispersion of colloidal particles in electrolyte solution, J. Phys. Chem., 66:2636-2642.
Slusarchuk, W. A., 1971. An experimental study of dielectric constant and electrical conductance of some frost-prone soils, Ph. D. thesis, Rutgers University, pp. 19-43.
Smith-Rose, R. L., 1933. The electrical properties of soil for alternating currents at radio-frequencies, Proc. Roy. Soc., London, 140A:359-377.
Spiegel, M. R., 1961. Statistics, New York, Schaum Pub. Co., pp. 217-221.
Srivastava, V., and P. Varshni, 1956. Variation in dielectric constant with temperature, Physica, vol. 22, 7:584-586.
Sutton, P. M., 1960. The dielectric properties of glass, in Progress in dielectrics, London, Heywood & Co., vol. 2, pp. 140-145.
Tedrow, J. C. F., 1966a. Polar desert soils, Proc. Soil Science Soc. Amer., vol. 30, 3:381-387.
Tedrow, J. C. F., 1966b. Antarctic soils, in Antarctic soils and soil forming processes, Antarctic Res. Ser., vol. 8, Amer. Geophys. Union, pp. 161-177.
Tedrow, J. C. F., 1971. Soils of the high Arctic landscape, in Polar desert symposium, Amer. Assn. Advancement Sci., 138th meeting, Philadelphia.
von Hippel, A. R., 1954. Dielectric materials and applications, Cambridge, Mass., M. I. T. Press, pp. 300-425.

APPENDIX B

LIST OF PARTICIPANTS

AMOROCHO, Prof. J., Dept. of Water Science & Engineering, Univ. of Calif., Davis, California 95616. (916) 752-0685.

ARMSTRONG, Dr. Richard L., INSTAAR, Univ. of Colorado, Box 456, Silverton, Colorado 81433. (303) 387-5769.

ASSUR, Dr. A., USA CRREL, P.O. Box 282, Hanover, New Hampshire 03755. (603) 643-3200.

BARNES, George, Calif. Dept. of Water Resources, 1416 9th St., Sacramento, California 95814.

BARNES, James C., Environmental Research & Technology, Inc., 429 Marrett Rd., Lexington, Massachusetts 02173. (613) 996-5236.

BARTOS, Louis, U.S. Forest Service, 88 West First South, Provo, Utah 84601. (801) 374-7301.

BASS, John C., Bass Engineering, Inc., 3446 Kurtz St., San Diego, California 92110. (714) 299-4509.

BELL, David L., AIDJEX/ Univ. of Washington, 4059 Roosevelt Way N.E., Seattle, Washington 98105. (206) 543-6613.

BERGMAN, Jim, U.S. Forest Service, P.O. Box 245, Berkeley, California 94701. (415) 486-3456.

BILELLO, Michael, U.S. Army, CRREL, Hanover, New Hampshire 03755. (603) 643-4105.

BISSELL, Vernon C., NOAA, National Weather Service, 8060 13th St., Silver Spring, Maryland 20910. (301) 427-7640.

BOWERS, Thomas J., Regional Hydrologist, National Weather Service, 632 Sixth Ave., Anchorage, Alaska 99501. (907) 265-4718.

BOWLEY, Clinton J., Environmental Research & Technology, Inc., 429 Marrett Rd., Lexington, Massachusetts 02173. (613) 996-5236.

BROWMAN, Ludvig G., 664 S. 6th E., Missoula, Montana 59801. (406) 543-5234.

BROWN, A. Jean, Calif. Dept. of Water Resources, 1416 9th St., Sacramento, California 95814. (916) 445-2196.

BRYAN, M. Leonard, Environmental Research Institute of Michigan, P.O. Box 618, Ann Arbor, Michigan 48107. (313) 483-0500.

BURNASH, Robert J. C., National Weather Service, Rm. 1631, 1416 9th St., Sacramento, California 95814. (916) 449-2462.

BURSON, Zolin G., E G & G, P.O. Box 1912, Las Vegas, Nevada 89101. (702) 739-0587.

CARROLL, Tom, Inst. of Arctic & Alpine Research, Univ. of Colorado, Boulder, Colorado 80302. (303) 443-2211.

COULSON, C. H., B.C. Water Resources Service, Parliament Bldg., Victoria, B.C., Canada. (604) 387-3432.

CROOK, Art, Soil Conservation Service, USDA, Rm. 217, 204 E. 5th Ave., Anchorage, Alaska 99501. (907) 274-7626.

DROWDER, W. Kent, U.S. Army CRREL, Hanover, New Hampshire 03755. (603) 643-3200.

DUCRUC, Jean Pierre, Univ. Laval, Dept. d'Ecologie et Pedologie. Quebec, G1K 7P4, Canada. (418) 656-5405.

DUNBAR, Moira, DRB Canada, Defense Research Establishment, Ottawa, Ontario K1A 0Z4, Canada. (613) 996-7051.

DYCK, Eric, Div. of Hydrology, Univ. of Saskatchewan, Saskatoon, Saskatchewan S7N 0W0, Canada. (306) 343-3191.

EAGLESON, Dr. Peter S., Dept. of Civil Engineering, Rm. 1-290, M.I.T., Cambridge, Massachusetts 02139. (617) 253-7103.
ENGLAND, Dr. Tony, U.S. Geological Survey, Branch of Regional Geophysics, Federal Center, Denver, Colorado 80225.
FERGUSON, H. L., Atmospheric Environment Service, Environment Canada, 4905 Dufferin St., Downsview, Ontario, Canada. (416) 667-4610.
FRITZSCHE, Allen E., E G & G, P.O. Box 1912, Las Vegas, Nevada 89101. (702) 736-054.
GANONG, W. F., Dept. of the Environment, Ice Div., Atmospheric Environment Service, 4905 Dufferin St., Downsview, Ontario, Canada. (416) 667-4727.

GLOERSEN, Per, NASA/Goddard Space Flight Center, Code 652, Greenbelt, Maryland 20771. (301) 982-6362.
GOLDING, Dr. D. L., Canadian Forestry Service, 5320- 122 St., Edmonton, Alberta T6H 3S5, Canada. (403) 435-7242.
GRASTY, Bob, Geological Survey of Canada, 601 Booth St., Ottawa, Ontario, Canada. (613) 994-5309.
GRUMMITT, W. E., Atomic Energy of Canada, Chalk River, Ontario K0J 1P0, Canada. (613) 687-5581.
HALVERSON, Dr. Howard, U.S. Forest Service, P.O. Box 245, Berkeley, California 94701. (415) 486-3459.

HANSON, Arnold M., Office of Naval Research, Arctic Program (415), Arlington, Virginia 22217. (202) 692-4118.
HART, George, Forest Science Dept., Utah State Univ., Logan, Utah 84322. (801) 752-4100.
HEINDL, Dr. Leo, USNC/IHD, National Academy of Sciences, 2101 Constitution Ave., N.W., Washington, D.C. 20418. (202) 389-6248.
HENDRICK, Robert L., USDA, Agricultural Research Service, 150 Kennedy Dr., So. Burlington, Vermont 05401. (802) 864-7414.
HIBLER, Dr. W. D., III, USA, CRREL, Hanover, New Hampshire 03755. (603) 643-3200.

HISLOP, Randy, Dept. of Physics, Univ. of Calgary, Calgary, Alberta, Canada.
HORNBECK, James, Northeast Forest Experiment Station, P.O. Box 640, Durham, New Hampshire 03824. (603) 868-9697.
HUBBARD, Dr. John E., State Univ. College-Brockport, Brockport, New York 14420. (716) 395-2636.
HUNTER, H. I., B. C. Water Resources Service, Parliament Bldg., Victoria, B.C., Canada. (604) 387-3434.
ISAACSON, Raymond E., Atlantic Richfield Hanford Co., P.O. Box 250, Richland, Washington 99352. (509) 942-2827.

JIRBERG, Russell J., NASA - Lewis Research Center, 21000 Brookpark Rd., Cleveland, Ohio 44135. (216) 533-4000, x201.
JOLLY, Dr. J. P., Montreal Engr. Co., Ltd., P.O. Box 777, Montreal, 114, Canada. (514) 395-3636.
JONES, E. Bruce, M. W. Bittinger & Assoc., Inc., P.O. Box Q, Fort Collins, Colorado 80521. (303) 482-8471.
JONES, Ron, U.S. Forest Service, P.O. Box 245, Berkeley, California 94701. (415) 486-3459.
JUMIKIS, Dr. Alfreds R., Dept. of Civil Engineering, Rutgers Univ., New Brunswick, New Jersey 08903. (201) 932-1766.

KOERNER, Fritz, Energy Mines & Resources, 880 Wellington St., Ottawa, Canada. (613) 996-2114.

KOTRAS, Tom, Arctec, Inc., Columbia, Maryland 21044. (301) 730-1030.

KROUSE, Dr. H. Roy, Dept. of Physics, University of Calgary, Alberta T2N 1N4, Canada.

LAKE, Robert A., Frozen Sea Research Group, Dept. of the Environment, Marine Sciences Branch, 825 Devonshire Rd., Victoria, B.C., Canada. (604) 388-3676.

LANGHAM, Dr. E. J., Institut National de la Recherche Scientifique, Universite du Quebec, Case Postale 7500, Quebec 10, Quebec. (418) 657-2554.

LAUER, Donald T., School of Forestry & Conservation, Univ. of Calif., Berkeley, California 94720. (415) 642-6000.

LEADER, R. E., The Boeing Co., MS 84-43, P.O. Box 3999, Seattle, Washington 98124. (206) 655-2121.

LIMPERT, Fred A., Bonneville Power Administration, P.O. Box 3621, Portland, Oregon 97208. (503) 234-3361.

LINKLETTER, Dr. George O., Desert Research Institute, Lab. of Atmospheric Physics, Stead Campus, Reno, Nevada 89507. (702) 972-1676.

LINLOR, Dr. Bill, NASA, Ames Research Center, Moffett Field, California 94035. (415) 965-5538.

LOGAN, Lloyd A., Ministry of Environment, 135 St. Clair Ave., W., Toronto, Ontario M4V 1P5, Canada. (416) 965-6194.

LOIJENS, Harry S., Inland Waters Directorate, Dept. of the Environment, Ottawa, Ontario K1A 0E7, Canada. (819) 997-2369.

LOUGEAY, Ray, Dept. of Geography, S.U.N.Y. at Geneseo, Geneseo, New York 14454. (716) 243-2258.

LUTHER, Stephen G., Lab. for Applications of Remote Sensing, Purdue Univ., West Lafayette, Indiana 47906. (317) 749-2052.

LYON, Ron J. P., Dept. of Applied Earth Sciences, Stanford Univ., Stanford, California 94305. (415) 321-2300, x2747.

MALE, Dr. D. H., Div. of Hydrology, Univ. of Saskatchewan, Saskatoon, Saskatchewan, Canada. (306) 343-3726.

MARKHAM, W. E., Dept. of the Environment, Ice Forecasting Central, 473 Albert St., Trebla Bldg., Ottawa, Ontario K1A 0H3, Canada. (613) 996-5236.

McANDREW, Donald, Soil Conservation Service, 1155 Johnson Pl., Reno, Nevada 89502. (702) 784-5304.

McCLAIN, Dr. E. Paul, NOAA, Suite 300, 3737 Branch Ave., S.E. Washington, D.C. 20031. (301) 763-5981.

MEIER, Dr. Mark F., U.S. Geological Survey, 1305 Tacoma Ave., S., Tacoma, Washington 98402. (206) 593-6502.

MEIMAN, Dr. James, Colorado State Univ., Dept. of Earth Resources, Fort Collins, Colorado 80521. (303) 49106592.

MILLER, Len, Avalanche Control Consultants, 3537 Overhulse St., N.W., Olympia, Washington 98502. (206) 866-2235.

MOREY, Resford, Geophysical Survey Systems, 16 Republic Rd., N., Billerica, Massachusetts 01862. (612) 667-4341.

NEWHALL, George N. NRM Corp., P.O. Box 827, Davis, California 95616. (916) 756-4607.

NORUM, D. I., Univ. of Saskatchewan, Dept. of Agric. Engineering, Saskatoon, Saskatchewan S7N OWO, Canada. (306) 343-3191.

OMMANNEY, Simon, Glaciology Div., Environment Canada, Ottawa, Ontario K1A 0E7, Canada. (613) 997-2476.

O'NEILL, Desmond, Canada Dept. of the Environment, Atmospheric Environment Service, P.O. Box 42, Moncton, N.B., Canada. (506) 858-2278.

PAINTER, Dr. Ralph, Inst. of Hydrology, Maclean Bldg., Crowmarsh Gifford, Wallingford, Berks., England (Crowmarsh 4247).

PEARSON, Gregory L., Soil Conservation Service, 511 N.W. Broadway, Portland, Oregon 97209. (503) 246-7357.

PECK, Dr. Eugene, National Weather Service/NOAA, Gramax Bldg., 8060 13th St., Silver Spring, Maryland 20910. (301) 427-7619.

POPHAM, Robert W., NESC/NOAA, Federal Office Bldg. No. 4, Rm. 2049, Suitland, Maryland 20233. (301) 763-5237.

POULIN, Ambrose, USA ETL, 5613 Alta Vista Rd., Bethesda, Maryland 20034. (202) 227-2513.

PRANTL, Dr. Friederike A., Canada Centre for Inland Waters, Radiochemistry Labs., P.O. Box 5050, Burlington, Ontario, Canada. (416) 637-4258.

PRICE, Edgar P., U.S. Bureau of Reclamation, 2800 Cottage Way, Sacramento, California 95825. (916) 484-4939.

QUINN, Dr. Frank H., Lake Survey Center/NOAA, 630 Federal Bldg. & U.S. Courthouse, Detroit, Michigan 48226. (313) 226-7463.

QURESHI, A. Saleem, Water Resources Branch, Federal Dept. of the Environment, 2-1309 Henryfarm Dr., Ottawa, Ontario, Canada. (613) 997-1509.

RANGO, Dr. Albert, NASA/GSFC, Code 650.3, Greenbelt, Maryland 20771. (301) 982-6481.

RECHARD, Dr. Paul A., Wyoming Water Resources Research Inst., Box 3038, Univ. Sta., Laramie, Wyoming 82070. (307) 766-2143.

REINKING, Roger F., Earth & Planetary Sciences Div. Code 6022, Naval Weapons Center, China Lake, California 93555. (714) 939-2440.

RHEA, J. Owen, Dept. of Atmospheric Sciences, Colorado State Univ., Ft. Collins, Colorado 80521. (303) 49108675.

SANTEFORD, Henry S., USNC/IHD, National Academy of Sciences, 2101 Constitution Ave., N.W., Washington, D.C. 20418. (202) 389-6248.

SCHERTLER, Ronald J., NASA Lewis Research Center, 21000 Brookpark Rd., Cleveland, Ohio 44135. (216) 433-4000, x 201.

SCHMUGGE, Tom, NASA/Goddard Space Flight Center, Mail Code 652, Greenbelt, Maryland 20771. (301) 982-6360.

SLAUGHTER, Dr. Charles W., USA CRREL, Ft. Wainwright, Fairbanks, Alaska 99703. (907) 353-9102.

SMITH, James L., Snow Project, U.S. Forest Service, P.O. Box 245, Berkeley, California 94701. (415) 486-3456.

STODDART, R. B. L., Canadian IHD Secretariat, c/o Inland Waters Directorate, Environment Canada, Ottawa, Ontario K1A 0H3, Canada. (819) 997-1487.

THEAKSTON, Prof. Frank H., School of Engineering, Univ. of Guelph, Guelph, Ontario N1G 2W1, Canada. (519) 824-4120, x2431.

THOMAS, Billy J., U.S. Army, Corps of Engineers, 210 Custom House, Portland, Oregon 97209. (503) 221-3757.

TSANG, Dr. Gee, Canada Centre for Inland Waters, Hydraulics Div., P.O. Box 5050, Burlington, Ontario, Canada. (416) 637-4735.

WANKIEWICZ, Anthony, Univ. of British Columbia, Dept. of Geography, Vancouver, B.C., Canada. (604) 224-1493.

WEEKS, Dr. W. F., U.S. Army, CRREL, P.O. Box 282, Hanover, New Hampshire 03755. (603) 643-3200, x261.

WELSH, Dr. J. Patrick, U.S. Coast Guard, R & D Center, Avery Point, Groton, Connecticut 06340. (203) 445-8501.

WENDLER, Dr. Gerd, Geophysical Institute, Univ. of Alaska, Fairbanks, Alaska 99701. (909) 479-7378.

WHIPKEY, Ronald Z., USDA, Agricultural Research Service, 150 Kennedy Dr., S. Burlington, Vermont 05401. (802) 864-7414.

WHITING, Robert E., Calif. Dept. of Water Resources, 1416 9th St., Sacramento, California 95814. (916) 445-5140.

WIESNET, Donald R., NOAA/NESS, Suite 300, 3737 Branch Ave., S.E., Washington, D.C. 20031. (301) 763-5981.

WILLIAMS, Larry, Institute of Arctic & Alpine Research, Univ. of Colorado, Boulder, Colorado 80302.

WILSON, Jim, Center for the Environment & Man, 275 Windsor St., Hartford, Connecticut 06120. (203) 549-4400.

YOUNG, Gordon J., Environment Canada, Glaciology Div., Ottawa, Ontario, Canada. (819) 997-2469.